HANDBOOK OF SAFETY PRINCIPLES

Wiley Essentials in
Operations Research and Management Science

Founding Series Editor
James J. Cochran, Louisiana Tech University

Wiley Essentials in Operations Research and Management Science is a unique collection of international research written by recognized experts and features both state-of-the-art developments in the field and classical methods that are appropriate for researchers, practitioners, policy makers, academicians, and students alike. Inspired by the *Wiley Encyclopedia of Operations Research and Management Science* (EORMS), this authoritative series covers topics of current research and practical interest and has a global emphasis on specific and key areas of study within the diverse field of operations research and management science. This broad collection of books provides wide-ranging and complete coverage within individual books and unifies the most important and key advances in the theory, methods, and applications within a specific area of study. Each book in the series successfully upholds the goals of EORMS by combining established literature and newly developed contributions written by leading academics, researchers, and practitioners in a comprehensive and accessible format. The result is a succinct reference that unveils modern, cutting-edge approaches to acquiring, analyzing, and presenting information across various subject areas.

Published *Wiley Essentials in Operations Research and Management Science*

Cox • Breakthroughs in Decision Science and Risk Analysis
Möller, Hansson, Holmberg, and Rollenhagen • Handbook of Safety Principles

HANDBOOK OF SAFETY PRINCIPLES

Edited by

Niklas Möller
Sven Ove Hansson
Jan-Erik Holmberg
Carl Rollenhagen

WILEY

This edition first published 2018

Registered Office
John Wiley & Sons, Inc., 111 River Street, Hoboken, NJ 07030, USA

Editorial Office
111 River Street, Hoboken, NJ 07030, USA

For details of our global editorial offices, customer services, and more information about Wiley products visit us at www.wiley.com.

Wiley also publishes its books in a variety of electronic formats and by print-on-demand. Some content that appears in standard print versions of this book may not be available in other formats.

Library of Congress Cataloging-in-Publication Data
Names: Möller, Niklas, editor. | Hansson, Sven Ove, 1951- editor. | Holmberg, Jan-Erik, editor.
Title: Handbook of safety principles / edited by Niklas Möller, Sven Ove Hansson, Jan-Erik Holmberg, Carl Rollenhagen.
Description: Hoboken : Wiley, 2017. | Series: Wiley essentials in operations research and management science; 9 | Includes bibliographical references and index. |
Identifiers: LCCN 2017024321 (print) | LCCN 2017049290 (ebook) | ISBN 9781118950708 (pdf) | ISBN 9781118950715 (epub) | ISBN 9781118950692 (hardback)
Subjects: LCSH: Industrial safety–Management. | Accidents–Prevention. | Industrial hygiene. | BISAC: BUSINESS & ECONOMICS / Production & Operations Management. | TECHNOLOGY & ENGINEERING / Industrial Health & Safety. | TECHNOLOGY & ENGINEERING / Industrial Engineering.
Classification: LCC HD7261 (ebook) | LCC HD7261 .H36 2017 (print) | DDC 658.4/08–dc23
LC record available at https://lccn.loc.gov/2017024321

Cover image: "Work with care" Courtesy of Library of Congress
Cover design by Wiley

Set in 10/12pt TimesLTStd by Aptara Inc., New Delhi, India

Printed in the United States of America

10 9 8 7 6 5 4 3 2 1

CONTENTS

PREFACE

The history of this volume goes back several years to a discussion among the four editors about the lack in the safety literature of a comprehensive overview of safety principles. The discussion took place at the Academy for Nuclear Safety at KTH (The Royal Institute of Technology), where we have collected a group of researchers from as diverse fields as philosophy, psychology, and risk analysis. Our main aim is to promote communication between practitioners and researchers in all aspects of safety and security, and one of our activities is a yearly workshop on a central safety topic. In 2013, the theme of our symposium was *Safety Principles*. The discussions before, during, and after the symposium confirmed the need for a thorough and comprehensive volume about safety principles. We decided that a handbook with leading experts in the field would be the best way to sum up and discuss the vast and complex landscape of principles of safety. To our great satisfaction, this idea was strongly supported both by the colleagues whom we invited to contribute to this handbook and by Wiley-Blackwell, who generously offered to include the handbook in their book series *Wiley Essentials in Operations Research and Management Science* and in the web-based *Wiley Encyclopedia of Operations Research and Management Science (EORMS)*.

This book consists of 32 chapters in total. The first two are introductory: "Introduction," putting the handbook in a broader context, and "Preview," providing an overview of the contents of the handbook. The 30 main chapters of the handbook are then categorized into five parts: Safety Reserves, Information and Control, Demonstrability, Optimization, and Organizational Principles and Practices.

We would like to thank all the contributors for excellent cooperation, and not least for their many substantial comments on the overall theme of the handbook as well as each other's chapters. All the chapters were thoroughly discussed on our workshop in Stockholm, April 22–23, 2015, followed by several e-mail exchanges and personal meetings. We would like to thank Kathleen Pagliaro and Susanne Steitz-Filler for invaluable editorial help and for their support and belief in the project.

Stockholm and Esbo
September 2017
Niklas Möller, Sven Ove Hansson,
Jan-Erik Holmberg and Carl Rollenhagen

LIST OF CONTRIBUTORS

Håkan Alm is professor emeritus in engineering psychology at Luleå University of Technology in Luleå, Sweden, where he was a professor from 2002 until 2016. His research interests cover many areas such as cognitive psychology, traffic psychology, new technology and safety, human work conditions, risk perception, and safety in complex systems. He has publications in international journals, book chapters, and technical reports. He has been a supervisor for 10 PhD students. His teaching activities cover a broad spectrum in psychology and engineering psychology.

Bo Bergman is professor emeritus at Chalmers University of Technology, Gothenburg, Sweden and retired in 2015 from a chair in Quality Sciences. From 2012 to 2015, he was a guest professor at Meiji University, Tokyo, Japan. His career started with 15 years in the aerospace industry, during which time he also became a PhD in mathematical statistics from Lund University, Lund, Sweden in 1978, and was a part-time professor in reliability at the Royal Institute of Technology, Stockholm, Sweden, (1981–1983). In 1983, he became a professor of quality technology at Linköping University, Linköping, Sweden, and in 1999, he was appointed the SKF professor in quality management at Chalmers University of Technology. Bergman was a co-founder of the Centre for Healthcare Improvement (CHI) at Chalmers and its first director (2004–2009). As a professor, he has supervised a large number of PhD students, many of whom are now professors themselves. Professor Bergman is a member of the International Statistical Institute (ISI) and an academician of the International Academy for Quality (IAQ).

Jon Bokrantz is a PhD student in the area of production service systems and maintenance at the Industrial and Materials Science, Chalmers University of Technology. He has a background in production engineering, and his research focuses on maintenance in digital manufacturing.

Neelke Doorn is a full professor Ethics of Water Engineering at the Technical University Delft, the Netherlands, with a background in civil engineering (MSc), philosophy (MA, PhD), and law (LLM). She is editor-in-chief of *Techné: Research in Philosophy and Technology* (official journal of the Society for Philosophy and Technology). Her current research concentrates on moral issues in risk governance, with a special focus on water-related risks.

In 2013, she was awarded a personal Veni grant for outstanding researchers from the Netherlands Organization for Scientific Research (NWO). She was shortlisted for the Engineer of the Year Award 2014 from KIVI NIRIA, a Dutch professional engineering organization, for her work on the interface of ethics and engineering.

Ann Enander is a licensed psychologist and professor of leadership psychology at the Leadership Center of the Swedish Defence University in Karlstad, Sweden. Her research has primarily been concerned with issues of risk perception and communication, emergency preparedness and crisis management at the local, regional, and national level. Her empirical work encompasses studies of a number of crises and critical events including the Chernobyl disaster, the Kobe earthquake, the 2009 A(H1N1) pandemic, and other technological and natural disasters. She has published more than 100 articles, book chapters, and books. She is a fellow of the Royal Swedish Academy of War Sciences, and currently government-appointed member of the Advisory Monitoring Board of the Swedish Chemicals Inspectorate and of the Swedish Defense Recruitment Agency. She is also a past president of the Society for Risk Analysis Europe.

Roger Flage is an assistant professor of risk analysis at the University of Stavanger, Norway. He has a PhD in risk management and societal safety, a master's degree in offshore technology with specialization in offshore safety, and a bachelor's degree in health, safety, and environmental engineering. He has also worked as a consultant in the field of risk assessment and risk management. His research focuses on risk and uncertainty assessment, integrated risk management and risk-informed decision-making, and maintenance modeling and optimization. He is a member of the editorial board of the journal *Risk Analysis*.

Pieter van Gelder is a professor of safety science at the Faculty of Technology, Policy, and Management of Delft University of Technology and director of the TU Delft Safety and Security Institute. He is also the chairman of the ESRA Technical Committee on Safety from Natural Hazards. Van Gelder has been involved in research and education on safety and reliability since 1991. His research interests are in risk analysis and optimization of systems, processes, and structures. He teaches 4th and 5th year courses at TU Delft and conducts research on new methods and techniques in risk analysis. Van Gelder has authored and co-authored over 300 papers and several books in the field of risk and safety and has supervised over 50 MSc students and 15 PhD students.

E. Scott Geller, PhD, is an Alumni Distinguished Professor in the Department of Psychology at Virginia Tech, and senior partner at Safety Performance Solutions, Inc. He authored, edited, or co-authored 41 books, 82 book chapters, 39 training programs, 259 magazine articles, and more than 300 research articles addressing the development and evaluation of behavior-change interventions to improve quality of life on a large scale. His most recent 700-page book *Applied Psychology: Actively Caring for People*, published by Cambridge University Press, reflects the mission of his teaching, research, and scholarship throughout his 49-year career. He was awarded the statewide Virginia Outstanding Faculty Award by the State Council of Higher Education, and he has received lifetime achievement awards from the International Organizational Behavior Management Network (in 2008) and the American Psychological Foundation (in 2009). The College of Wooster awarded E. Scott Geller the honorary degree Doctor of Humane Letters.

Gudela Grote is a professor of work and organizational psychology at the Department of Management, Technology, and Economics at the ETH Zürich, Switzerland. She received her PhD in industrial/organizational psychology from the Georgia Institute of Technology, Atlanta, GA, USA. A special interest in her research is the increasing flexibility and virtuality of work and its consequences for the individual and organizational management of uncertainty. She has published widely on topics in organizational behavior, human factors, human resource management, and safety management. Professor Grote is associate editor of the journal *Safety Science* and the president of the European Association of Work and Organizational Psychology.

Jan M. Gutteling is an associate professor of crisis and risk communication at the University of Twente (UT). He received his training as a clinical psychologist with an extended minor in social psychology from Utrecht University, and his PhD from the UT. His research focus is on the understanding of risk perception and the application of this understanding in crisis and risk communication. His studies are primarily quantitative and empirical, and aim (i) to develop social psychological models of risk, or (ii) to establish experimentally how and under which circumstances communication and information influence risk perception and risk-related behavior. His recent research themes are environmental risks and physical safety issues, modern biotechnology and genomics, water safety management (flood risks), and new communication tools in disaster management. His teaching focuses on risk management (perception/communication) in the broader societal context, as well as in occupational safety and health, at the Master and PhD levels. He has published approximately 100 papers in reviewed journals, books, book chapters, and scientific reports, including *Exploring Risk Communication*, the first European book on risk communication (in 1996).

Ibrahim Habli is a Lecturer in Safety-Critical Systems at the University of York, England. His expertise lies in expertise lies in the design and assurance of safety-critical systems, primarily within the aviation, automotive, and healthcare domains. He currently holds an Industrial Fellowship Award from the Royal Academy of Engineering, funding a collaborative project with the English National Health Service on understanding the relationship between Health IT and patient safety. He teaches on York's postgraduate programs in safety-critical systems engineering. He is currently a member of the DS/1 Dependability

Committee at BSI, the committee on safety case development within the Motor Industry Software Reliability Association (MISRA) and the Goal Structuring Notation (GSN) Standardization group. He was a member of the Joint EUROCAE/RTCA committee responsible for developing the international aerospace guidance DO-178C.

Sven Ove Hansson is a professor in philosophy at the Royal Institute of Technology, Stockholm. He is editor-in-chief of *Theoria* and of the book series *Outstanding Contributions to Logic*. He is also member of the editorial boards of several journals, including *Philosophy and Technology*, *Techné*, and *Synthese*. His research on risk and safety focuses on ethical and epistemological aspects. His other philosophical research includes contributions to the philosophy of science and technology, decision theory, logic, and moral and political philosophy. He is a member of the Royal Swedish Academy of Engineering Sciences and past president of the Society for Philosophy and Technology. He has published well over 300 papers in refereed international journals and books. His books include *The Ethics of Risk* (Palgrave Macmillan, 2013), *Social and Ethical Aspects of Radiation Risk Management* (edited with Deborah Oughton, Elsevier Science, 2013), and *The Role of Technology in Science* (edited, Springer, 2015).

Lars Harms-Ringdahl has been engaged in safety management as a researcher, consultant, and teacher for many years. He works at the Institute for Risk Management and Safety Analysis in Stockholm, and has been professor at the Royal Institute of Technology, Stockholm, Sweden and at Karlstad University, Karlstad, Sweden. He has been involved in several fields such as industrial safety, fire prevention, patient safety, and societal risk management. He has a special interest in methodologies for safety analysis, event investigations, and safety management. Examples of publications are *Safety Analysis—Principles and Practice in Occupational Safety* (Taylor & Francis, 2001) and *Guide to Safety Analysis for Accident Prevention* (IRS Riskhantering AB, Stockholm, Sweden 2013).

Erik Hollnagel is a professor at the Institute of Regional Health Research, University of Southern Denmark and Senior professor of Patient Safety at Jönköping University, Sweden. He is also adjunct professor at Central Queensland University (Australia), visiting professor at the Centre for Healthcare Resilience and Implementation Science, Macquarie University (Australia), and professor emeritus at the Department of Computer Science, University of Linköping (Sweden). His professional interests include industrial safety, resilience engineering, patient safety, and complex socio-technical systems. He has published widely and is the author/editor of 24 books, including five books on resilience engineering, as well as a large number of papers and book chapters. Erik has been President of the European Association of Cognitive Ergonomics (1994–2000) as well as co-founder and past President of the Resilience Engineering Association.

Jan-Erik Holmberg is senior consultant and office manager at Risk Pilot AB in Espoo, Finland. He is also an adjunct professor at the Royal Institute of Technology, Stockholm, and he gives lectures in reliability and risk analysis at Aalto University in Espoo and Lappeenranta University of Technology. He has over 25 years' experience in nuclear power plant safety analyses and probabilistic safety assessment. His research on probabilistic safety assessment focuses on mathematical methods, risk-informed decision-making, human reliability analysis, and analysis of digital instrumentation and control systems. He has more than 70 scientific and technical papers in leading technical journals and conferences and over 10 reviewed articles in peer-reviewed journals.

Mohd Umair Iqbal is a PhD scholar in the Department of Chemical Engineering at Indian Institute of Technology Gandhinagar, India. He has obtained his master's degree from the same college. He has obtained his bachelor's degree from the National Institute of Technology, Srinagar, India. He is involved in various research activities encompassing the field of safety and risk assessment. His main area of interest is the study of human reliability. He is actively researching the performance and reliability of control room operators. His research on process safety focuses on dynamic error.

Tim Kelly is professor of high integrity systems at the Department of Computer Science in the University of York. He is best known for his work on system and software safety case development, particularly his work on developing the Goal Structuring Notation (GSN)—an argumentation notation and method for safety case arguments. His research interests include safety case management, software safety analysis and justification, modular certification, certification of adaptive and learning systems, and the dependability of "Systems of Systems." He has supervised many research projects in these areas with funding that spans industry, government, research councils, and the European Union. He has published over 150 papers on high integrity systems development and justification in international refereed journals and conferences. He has also been involved in supporting the development of a number of international standards in the area of system and software safety assurance (such as the automotive standard ISO 26262).

Urban Kjellén is an associate professor of safety management at the Norwegian University of Science and Technology. He has 30 years of industrial experience primarily in various HSE management positions in investment projects and at the corporate level in the oil and gas, light metal and hydropower industries. He has published 35 papers in refereed international journals and books on risk analysis and HSE management of design and construction. His books include *Prevention of Accidents and Unwanted Occurrences—Theory, Methods, and Tools in Safety Management* (CRC Press, 2017) and *Occupational Accident Research* (Elsevier, 1984). Urban Kjellén has been a member of various standardization committees, and has been instrumental in the development of the design standard Norsok S-002 for the working environment.

Jean-Christophe Le Coze is a safety scientist with an interdisciplinary background, including engineering and the social sciences. He works at INERIS, the French national institute for environmental safety, where he is in charge of a research program. His activities combine ethnographic studies and action research programs in various safety-critical systems, with an empirical, theoretical, historical, and epistemological orientation. Outcomes of his research have regularly been published in the past 10 years, including the book *Trente ans d'accidents. Le nouveau visage des risques sociotechnologiques* (Thirty Years of Accidents: The New Face of Sociotechnological Risks).

Yan-Fu Li is a professor at the Chair on Systems Science and the Energetic Challenge at Laboratoire Genie Industriel, CentraleSupélec, Université Paris-Saclay, France. Dr. Li completed his PhD research in 2009 at the National University of Singapore, and went to the University of Tennessee as a research associate. His current research interests include reliability modeling, risk assessment, and optimization. He is the author of over 70 publications, all in refereed international journals, conferences, and books. He is a senior member of IEEE.

Niklas Möller is an associate professor at the Royal Institute of Technology, Stockholm. His research interests lie in value questions in the philosophy of risk, moral philosophy and political philosophy. Möller received his PhD in philosophy at KTH in 2009, after which he worked 2 years at Cambridge University as a post-doctorate researcher. Thereafter, he worked as a research scholar at the Department of Philosophy at Stockholm University, before returning to KTH. Möller has published numerous articles in international peer review journals such as *Journal of Applied Philosophy*, *Philosophical Studies*, *Social Theory & Practice*, *Ethical Theory & Moral Practice*, *Ethics Policy & Environment*, *Journal of Philosophical Research*, and *Risk Analysis*.

Leena Norros, research professor (emerita), is an industrial psychologist working on human factors in complex industrial systems. She received a Dr. rer. nat. from the Technical University of Dresden, Germany, and a PhD in psychology from the University of Helsinki, Finland. She created a human factors research team at the VTT Technical Research Centre of Finland and lead the team 25 years. Her main interest is understanding work activity in real-life contexts, for which she has created new concepts and methods. She acts as docent at Helsinki University and lectures on human factors there and at Aalto University. She has published widely and participates actively in international forums of human factors/ergonomics.

Anna-Lisa Osvalder is a professor in Human Machine Systems and senior lecturer in Ergonomics at Division Design & Human Factors at Chalmers University of Technology in Gothenburg, Sweden. She is also guest professor at Department of Design Sciences at Lund University in Sweden Her research focuses on human factors engineering and specifically the interaction between users and technology in various complex socio-technical systems with safety and comfort aspects in focus (nuclear and process control, medical technology, vehicles and transportation, IT-systems). Methods for analytical evaluations of use errors, usability problems, and comfort aspects have been developed in her research to be used when designing proper ergonomic products and working environments with effectiveness, safety, and human well-being in focus.

Jinkyun Park has worked as a senior researcher since 2000 in the Integrated Safety Assessment Division of the Korea Atomic Energy Research Institute (KAERI), whose major role is to conduct R&D activities pertaining to the risk assessment and management of nuclear power plants based on PSA (probabilistic safety assessment) and HRA (human reliability analysis). His main research interest is to scrutinize the nature of human performance, for instance, by (i) collecting human performance data under simulated emergencies using the full-scope simulator of nuclear power plants, (ii) analyzing the characteristics of human performance data with respect to various kinds of performance shaping factors, and (iii) characterizing team dynamics under simulated emergencies in term of several aspects including their communication characteristics and cultural differences. He has published over 50 papers in major international journals and the book *The Complexity of Proceduralized Tasks* (2009).

Luca Podofillini is a senior scientist in the Risk and Human Reliability Group of the Paul Scherrer Institute (Switzerland). He has a nuclear engineering degree and a PhD in nuclear engineering from the Polytechnic of Milan (2004). His activities include human reliability analysis (HRA) research and regulatory support tasks for the Swiss Federal Nuclear Safety Inspectorate. His research addresses the development of quantitative models of human performance in industrial systems, with focus on errors in decision-making, dynamic safety assessment, and collection of data in simulated emergencies. He is a co-author of about 30 papers in international scientific journals. He is chair of the ESRA (European Safety and Reliability Association) Technical Committee for Human Factors and Human Reliability and a member of the board of the HRA Society. He has been the chair of the Technical Program Committee of the European Safety and Reliability (ESREL) 2015 international conference. Since 2016, he is associate editor of the *Journal of Risk and Reliability*.

Andrew Rae is a lecturer at Griffith University in Brisbane, Australia, and manages the Safety Science Innovation Lab. His research critically examines claims about the effectiveness of activities such as risk assessments and incident investigations, and looks for new ways to evaluate safety practices. He is also involved in improving the public understanding of safety through speaking, podcasting, and blogging, and has published academically on the topic of storytelling in safety education. He is a fellow of the Higher Education Academy, and a committee member of the Australian Safety Critical Systems Association.

Teemu Reiman is currently working as a safety culture manager at the new Finnish nuclear power company Fennovoima based in Helsinki. Reiman also has his own consultancy company. Reiman has a doctoral degree in psychology from the University of Helsinki, and a title of Docent (Adjunct Professor) from the Aalto University. He made his dissertation in 2007 on safety culture evaluations of nuclear power plant maintenance organizations. He has previously worked as a senior scientist at VTT Technical Research Centre of Finland, specializing in safety management and safety culture. At VTT, Reiman acted as a project manager and a researcher in several national and international research and consultancy projects covering a wide range of topics from safety culture and management to resilience and properties of complex adaptive systems. Reiman has experience from various safety-critical domains including nuclear power, conventional power, transportation, metal industry, oil industry, and healthcare.

Genserik Reniers is a professor (in a part-time capacity) at the University of Antwerp, Belgium, responsible for safety and risk teaching and research. He also is a professor at the Safety and Security Science Group of Delft University of Technology in the Netherlands. His expertise is situated within the field of engineering and management of safety and security problems within the chemical and process industry.

Steve Roberts is co-founder and senior partner at Safety Performance Solutions, Inc. (SPS) Blacksburg, VA 24060, USA. He earned an MS in industrial/organizational psychology from West Chester University and an MA and a PhD in applied/experimental psychology focusing on organizational behavior management from Virginia Tech under the guidance of Dr. Scott Geller. His specific areas of expertise include the design, implementation, and evaluation of people-based safety processes, the assessment of organizational culture, increasing employee involvement, organizational management systems design, organizational leadership development, and reducing human error. Before co-founding SPS in 1995, Dr. Roberts was a research associate with Management Systems Laboratories of the Industrial Engineering Department of Virginia Tech and a safety consultant with Make-A-Difference, Inc. He has taught research methods at Virginia Tech, and served as a project manager at the Center for Applied Behavior Systems of the Psychology Department at Virginia Tech. He is a regular consultant and speaker across a wide variety of organizations.

Carl Rollenhagen is an adjunct professor at the Royal Institute of Technology, Stockholm, and a human factors specialist. He is the founder of the Safety Management Institute (SMI) at the power company Vattenfall. Together with other Swedish colleagues, he has introduced the concept of Man-Technology-Organization (MTO) for the nuclear industry in Sweden. MTO is today applied in many branches in Scandinavia and represents a cross-domain systemic perspective on risk and safety. His research focus and associated publications (books, scientific articles, conference contributions) span over a wide area including safety culture, safety analysis, root-cause analysis, and safety management.

Paula Savioja, DSc (Tech), acts as a senior inspector at the Radiation and Nuclear Safety Authority Helsinki, Finland STUK. Previously, she was the research team leader and a research scientist in the Human Factors in Complex Systems team at the VTT Technical Research Centre of Espoo, Finland. Savioja received a masters' degree in engineering from the Department of Automation and Systems Technology of Helsinki University of Technology Espoo, Finland in 2003. In 2014, she completed her doctoral studies with distinction in Aalto University School of Science Espoo, Finland. In her dissertation, she developed a systemic usability concept and evaluation methods to help the deployment of user-centered design approaches within complex socio-technical systems. Savioja has worked as a research scientist in various research projects at VTT between 2001 and 2014. She has expertise from several industrial domains including nuclear power production, maritime industry, metals and machines, process industry, and fire and rescue services. Savioja is author of over 70 scientific and other papers and reports.

Anders Skoogh is an associate professor at the Industrial and Materials Science in Chalmers University of Technology. He is a specialist in the management and analysis of production data and a research group leader for Production Service and Maintenance Systems. Anders is also the director of Chalmers' Master's Program in Production Engineering and a board member of Sustainability Circle with responsibilities for research collaboration.

Rajagopalan Srinivasan is professor of chemical engineering and coordinator of the Safety Centre at the Indian Institute of Technology, Gandhinagar, India. His research program is targeted toward developing AI-inspired systems engineering approaches for design and operation of complex systems such as chemical plants and supply chains. His research in process safety seeks to develop a fundamental science-based understanding of human error. His research has resulted in over 385 peer-reviewed journal articles, book chapters, and conferences publications. He is a member of the editorial board of *Process Safety and Environmental Protection* and several other journals.

Luca Talarico graduated as an industrial engineer in utility management at the University of Calabria (Italy) in 2005. In 2007, he obtained his master's degree in management engineering with a focus on integrated logistics at the University of Calabria. In 2011, he achieved a postgraduate degree in business administration (MBA) at the Alma Graduate School of Bologna. In 2015, he completed his PhD in risk management for routing problems at the University of Antwerp (Belgium). In his former professional career, he worked as a project manager and logistics expert at Ceva logistics and the Auchan group.

Risto Tiusanen, DSc (Tech), works at VTT Technical research centre of Finland as a senior scientist in the research area of life cycle solutions in Tampere. He has over 25 years' experience in the fields of system safety and risk management. His research interests include risk assessment processes to support machinery life cycle management and development of risk analysis and risk evaluation methods for complex machinery and industrial automation systems. He is an internationally recognized specialist on machinery safety and system safety issues. His project experience includes participation as the VTT project manager or the principal analyst in a number of national and international research projects and direct contract research projects with industrial partners in several industrial sectors. He has published over 20 scientific articles or publications, about 30 conference papers and several other publications. He has recently (2014) published his doctoral thesis "An Approach for the Assessment of Safety Risks in Automated Mobile Work Machine Systems."

Björn Wahlström joined the Technical Research Centre of Finland (VTT) in 1971. In 2008, he retired from VTT as a research professor in systems engineering. During his career at VTT, he participated in several national and international projects connected to automation, simulation, systems engineering, and nuclear safety. From 1989 to 1991, he was on leave from VTT working with the International Institute for Applied Systems Analysis (IIASA) in Austria. He was the coordinator of two EU-projects: "Organizational Factors: Their Definition and Influence on Nuclear Safety" (ORFA, completed in 1999) and "Learning Organizations for Nuclear Safety" (LearnSafe, completed in 2004). After his retirement, he worked until 2014 as a consultant for Vattenfall in Sweden through his company Bewas OyAb. In 2015, he joined the Systems Analysis Laboratory of the Aalto University, where he is teaching safety management. He has written more than 300 papers, articles, and book chapters.

Torbjörn Ylipää is a lecturer at the Industrial and Materials Science in Chalmers University of Technology, Göteborg. He has a Licentiate of Engineering degree and has performed research and education within the maintenance and safety area for more than two decades. Social sustainability in maintenance and production service systems and maintenance are his two main areas of interest.

Enrico Zio received an MSc degree in nuclear engineering from Politecnico di Milano in 1991, an MSc in mechanical engineering from UCLA in 1995, and PhD degrees in nuclear engineering from Politecnico di Milano and MIT in 1996 and 1998, respectively. He is currently director of the Chair on Systems Science and the Energetic Challenge of the Foundation Electricite' de France (EDF) at CentraleSupélec, Paris, France, full professor and president of the Alumni Association at Politecnico di Milano, visiting professor at MIT, distinguished guest professor at Tsinghua University, Beijing, China, adjunct professor at the University of Stavanger, Norway, the City University of Hong Kong, Beihang University and Wuhan University, China, and co-director of the Center for Reliability and Safety of Critical Infrastructures (CRESCI), China. His research focuses on the modeling of the failure–repair–maintenance behavior of components and complex systems, for the analysis of their reliability, maintainability, prognostics, safety, vulnerability, resilience, and security characteristics, and on the development and use of Monte Carlo simulation methods, soft computing techniques, and optimization heuristics.

1

INTRODUCTION

Niklas Möller, Sven Ove Hansson, Jan-Erik Holmberg, and Carl Rollenhagen

Principles for action have a much more important role in safety engineering and safety management than in most other disciplines. In practical safety work, we refer to principles such as fail-safe, safety barriers, safety factors, system redundancy, resilience, inherent safety, and many others. Much of the academic literature on safety, safety standards, and regulations recommends the use of one or other such principle. Many of the best-known contributors to the safety literature owe their fame to their roles as originators or promoters of one or other safety principle.

1.1 COMPETITION, OVERLAP, AND CONFLICTS

But the field is not characterized by consensus on which safety principles we should use. To the contrary, the literature on these principles abounds with divergent and sometimes conflicting recommendations. The overall picture is a rather confused one, due to competition, overlap, and conflicts among the principles.

Handbook of Safety Principles, First Edition. Edited by Niklas Möller, Sven Ove Hansson, Jan-Erik Holmberg, and Carl Rollenhagen.

It is not uncommon to hear presentations in which one of the safety principles is expanded to include all aspects of risk and safety enhancement so that it becomes *the* overarching principle under which the others can be subsumed. Quite a few of the principles have been presented with such ambitions—general quality principles, integrated risk management, and safety culture to name just a few—but obviously at most one of the principles can be superordinate to all the others. There seems to be a certain element of modishness in the coming and going of safety principles, and the field may not be entirely devoid of factionalism.

More often than not, one and the same safety measure can be presented as based on various principles. Terminologies also differ between industry branches and engineering specialties. For instance, what is called "inherent safety" in the chemical industry is called "substitution principle" in many industries that use chemical products, "passive safety" in the nuclear industry, and "primary prevention" in health-related applications. These principles seem to be close in meaning, but how large is the overlap? Can they perhaps even be described as one and the same principle but under different names?

Conflicts between the principles are far from uncommon. The principle of cost–benefit optimization tells us not to reduce low radiation doses if the reduction is costly, but at least some interpretations of the ALARA ("as low as reasonably achievable") principle tell us to reduce them. The principle of multiple safety barriers sometimes recommends an extra layer of safety that the principle of simplicity would dissuade us from since it makes the system more complex and difficult to manage in a safety-critical situation. Sometimes, even two applications of the same safety principle can lead to a conflict. For instance, the substitution principle recommends that we replace flammable substances by less flammable ones and toxic substances by less toxic ones. In the choice between two substances, one of which is less flammable and the other less toxic, this will lead to a conflict.

1.2 A NEW LEVEL IN THE STUDY OF SAFETY PRINCIPLES

As we see it, the study of safety principles has to be taken to a new and more comprehensive level. It is not sufficient to study the principles one at a time, and promotion of single principles needs to be replaced by unbiased comparative investigations. There is no lack of topics for such studies. We need to find out the relationships between the different principles, not least how they overlap and how they may run into conflict with each other. We also need to learn how they are conceived and applied by those who are supposed to be helped by them in their daily work (not only how they are conceived by their most fervent champions). And most importantly, we need to know if they make a difference in practice. What effects, if any, does their application have on safety outcomes? In short, the academic literature on safety principles should become much less advocacy-based and much more evidence-based.

We see this book as a first step toward that new level in the study of safety principles. Most of the major safety principles are presented, and they are all dealt with on an equal basis. We have asked the authors to compare the principle(s) they present to other safety principles. We have also asked them to clarify the limitations and weaknesses of the various principles, and to inform the reader of whatever empirical evidence there may be of the effects of using the principles in practice. The book contains a significant amount of comparative material, and we hope that it will also serve as an inspiration for more comparative studies of safety principles in the near future.

1.3 METAPRINCIPLES OF SAFETY

Does it make any difference which safety principle(s) we appeal to, and which of them we choose as an overarching principle for safety management? We believe that it can indeed make a difference. The reason for this is that the different safety principles put emphasis on different components of safety management. There are many possible "metaprinciples" which may be used for bringing out differences in emphasis between safety principles. We have found the following simple list of basic tasks in safety management useful, and will in this section illustrate how it brings forward an interesting pattern for the principles of safety covered in this handbook:

1. *Inventorize*. Identify and assess specific safety problems in planned or existing systems.
2. *Capacitate*. Investigate what capacities the system has to deal with safety-related problems and how those capacities can be improved. Many of these principles are applied in the design phase but can also be implemented as a consequence of applying problem-finding principles in existing systems.
3. *Prioritize*. Set priorities among the potential improvements.
4. *Integrate*. Make safety management coherent and comprehensive, for example, by using general quality principles and integrated safety management principles.

Each of these tasks is an important component in safety management. Therefore, the safety principle(s) applied in safety management should sustain the performance of each of them. We will call them metaprinciples since they will be used to evaluate many of the common safety principles.

In Figure 1.1, we have placed three of the safety metaprinciples at the vertices of a triangle, and we have introduced some well-known safety principles at different places in the triangle. The diagram illustrates how these three principles give rise to different approaches to practical safety work. Notably, some safety principles are close to one of the vertices. Such a safety principle will in practice only support one of the metaprinciples, and it is therefore in obvious need of supplementation. One example is the principle of experience feedback. This is a principle with a strong focus on inventorizing. By studying previous accidents, incidents, and other events, we can learn much on how to avoid similar events in the future. But obviously,

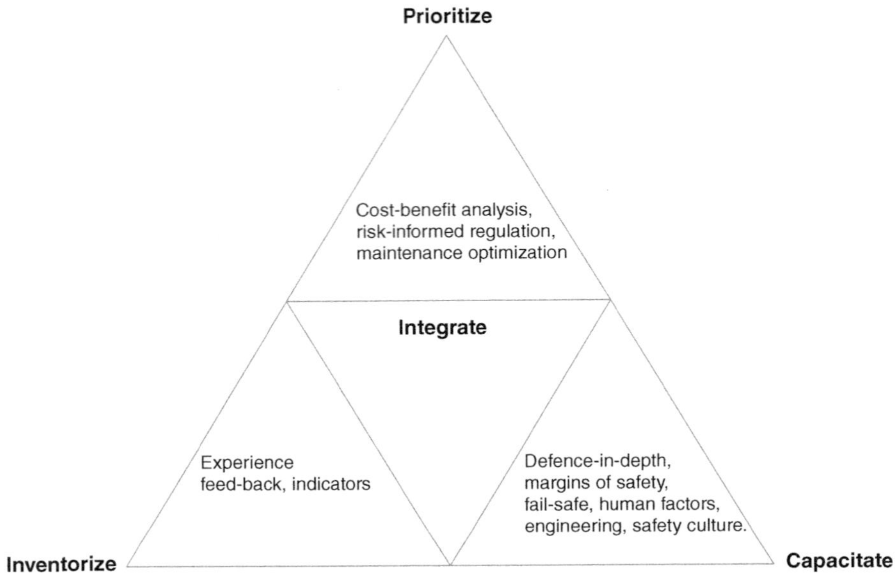

Figure 1.1. Four metaprinciples of safety. Specific safety principles can be characterized according to whether they have a dominant focus on finding real or potential safety problems (inventorize), providing capacity and resources to cope with real or potential problems (capacitate), or to make priorities (prioritize). At the center of the figure are principles that describe how to integrate components of the other principles.

this does not necessarily teach us about the capacity of the system to deal with the events (particularly unforeseen ones), and neither does it tell us much about how to prioritize among different problems in need of solution.[1] Therefore, the principle of experience feedback is not sufficient to guide safety management as a whole. We will have to add other principles that provide guidance for capacitation and prioritization.

Similarly, the principle of cost–benefit optimization is very close to the vertex representing prioritization. It is a rather complete (but of course not uncontestable) principle for priority-setting of safety measures, but it does not help us much in identifying safety problems or pinpointing general capacity improvements that can increase our preparedness for unidentified hazards. Therefore, cost–benefit optimization is not suitable as an overarching principle to cover all safety work.

In the middle triangle of Figure 1.1, we have positioned safety principles that are very general and usually contain parts of the other three metaprinciples. These principles often give general advice about how to structure safety management and the other metaprinciples. For example, general quality principles give advice that

[1] It should, however, be recognized that the principle of experience feedback is also used to identify good practices and solutions to previously identified safety problems.

can be applied to most other specific safety principles. Integrated safety management principles are also of this type.

Not surprisingly, principles that are close to each other in the diagram are more likely to overlap than principles at larger distance from each other. For instance, inherent safety and the substitution principle are close to each other in the diagram. This is because they both provide means to avoid both known and unknown dangers. These two principles tend to yield the same recommendations. For instance, both tell us to replace a flammable substance by a non-flammable one, which is an advantage both in known and unknown accident scenarios. The two principles are quite close to each other, and one might sensibly ask if they coincide or if one of them is a variant of the other.

In other cases, principles that are close to each other in the diagram run into conflict rather than overlapping. This applies for instance to cost–benefit optimization and best available technology. These two principles tell us how to prioritize, but they give us different advice on that topic. When a new, safer technology becomes available, best available technology will advise us to implement it, but cost–benefit optimization will often give contradictory advice. Other potential relations and potential conflicts between safety principles will be discussed further below.

1.4 OTHER WAYS TO CHARACTERIZE SAFETY PRINCIPLES

While our list of metaprinciples provides a way to see how safety principles overlap and what parts of safety management they cover, it is far from the only way to describe and categorize safety principles. As we see it, there is a need for a considerable methodological pluralism in the characterization of safety principles. For the purpose of dividing this book into main sections, we have adopted a more traditional approach, collecting the chapters *thematically* into five parts: Safety Reserves, Information and Control, Demonstrability, Optimization, and Organizational Principles and Practices. Some of these categories seem to correlate rather closely to our suggested metaprinciples. All of the principles treated in the first part, Safety Reserves, have a rather strong capacitating focus, for example, whereas most of the principles covered in the Optimization part focus on prioritizing. Other parts correlate more diversely and can be placed in the middle. The chapters in the Organizational Principles and Practices part, for example, have some emphasis on capacitation, but include many safety principles with a wholesale approach which cover all the three components positioned at the corners.

There are several alternative ways of categorizing safety principles. One common division is procedural. Different phases in the lifecycle of a system require different principles; typically, the three basic system phases selected are *design*, *operation*, and *decommissioning* (IAEA, 1986).

Another categorization of safety principles attempts at dividing hazard reduction into categories according to their priority. The basic idea here is that a hazard should if possible be *eliminated*. If that is not possible, it should be *isolated*, *controlled*, or *limited*. Haddon (1980a, 1980b; cf. also Runyan, 2003; Saleh et al., 2014) arguably provides the classic account of this strategy. On his model, safety is analyzed through the three basic concepts of *threat*, *barrier*, and *object of value,* and his strategy, in which he utilizes the so-called energy model of accidents (Saleh et al., 2010), includes:

1. Reducing the energy in the system
2. Controlling the energy in the system
3. Separating the energy source temporally and spatially from the object of value
4. Enhancing the damage resistance of the objects of value (Saleh et al., 2014)

Bahr (1997) provides a more recent example of a similar strategy (in turn based on NASA, 1993), suggesting the following taxonomy:

1. "Designing out" the hazard
2. Safety devices
3. Warning devices
4. Special procedures and training

First, Bahr writes, we should "design out" the hazard from the system. If that is not possible, we should control the hazard using various fail-safe devices; for example, pressure valves relieving the system of dangerous pressure build-up. When designing out or controlling is not an option, warning devices (e.g., smoke alarm) and procedures (e.g., emergency shutdown) and training should be used (Bahr, 1997).

Another suggested list of covering principles, similar in that it also focuses on substantially different strategies of risk reduction, is given by Möller and Hansson (2008). They divide a large number of engineering safety principles into four covering principles:

1. Inherently safe design
2. Safety reserves
3. Safe fail
4. Procedural safeguards

Inherently safe design is the design strategy to minimize the inherent dangers in the process as far as possible. The general idea here is that potential hazards are excluded rather than just enclosed or otherwise coped with (cf. Chapter 17). Safety reserves is the strategy of making constructions strong enough to resist loads and disturbances exceeding those that are intended; for example, by employing explicitly chosen, numerical safety factors. (Chapters 3–6 treat different aspects of this strategy.) The covering principle of safe fail entails that the system should fail "safely"; internal components may fail without the system as a whole failing, or the system fails without causing harm. "Fail-safe," "fail-silence," and "negative feedback" denote different variants of this principle. (Hammer, 1980). Procedural safeguards refer to control mechanisms for enhancing safety, ranging from general safety standards and quality assurance to training and behavior control of the staff. (Several chapters in this handbook, and Part V in particular, treat procedural safeguards.)

Another categorization focuses on the temporal dimension involved in decision-making. Here, a fundamental division is between principles focusing on *passed experience* (such as in Chapter 7), *current states of the system*, or *projections to the future*.

Safety principles may also be categorized in relation to the object of regulation. Here, a distinction can be drawn between principles directed at the *technical system*, the *human agent*, and the *organization*.

Many other ways to characterize safety principles are available in the literature. For further categorizations, see, for example, Saleh et al. (2014), Jackson and Ferris (2013), Jackson (2010), Khan and Amyotte (2003), and Kletz (1978, 1998).

1.5 CONFLICTS BETWEEN SAFETY PRINCIPLES

Safety principles are not conflict-free. However, we should distinguish between on the one hand conflicts between principles as such and on the other hand conflicts between applications of principles. To exemplify the former type of conflicts, we can suppose that a company has adopted the following two principles: All employees have the right to report any safety concern to the chief safety manager, and all communications on safety must be checked by the responsible foreman before they are disseminated outside of the department. These two principles are obviously in conflict. The example is contrived, and the reason why it is contrived is that this type of conflict appears to be unusual. Safety principles that are used in practice tend not to be in conflict in this way.[2]

The other type of conflict, between applications of principles, is much more common. Two principles may seem to be perfectly compatible, but there may still be practical cases when they cannot both be satisfied. For instance, in workplaces with risks of poisonous gas leakage, we may wish to implement the principle that it should be possible to evacuate the building in a very short time. In workplaces where a terrorist attack is comparatively likely and can have disastrous effects, we may wish to implement the principle that unauthorized access should be virtually impossible. It is not difficult (although sometimes expensive) to implement one of these two principles. However, implementing them both is often very difficult. Therefore, when they both

[2] Note though that safety and security principles often conflict, since they may have different goals. Security procedures may, for example, attempt to keep an intruder from getting away in case of an incident (locking down a site after a breach), while safety procedures should do the exact opposite, that is, help people to abandon a site in case of an incident.

need to be applied, we have a conflict, but strictly speaking, it is not a conflict between the principles but between their applications in a particular situation.

In practice, it is the latter type of conflicts (between applications of principles) that we have to deal with, rather than conflicts of the former type (between principles as such). We can further distinguish between three types of conflicts here.

First, there are conflicts between two applications of the same safety principle. We have already given an example of this: The principles of inherent safety requires both that we replace toxic substances by less toxic ones and that we replace flammable substances by less flammable ones. These two specifications run into conflict if the least toxic alternative is not also the least flammable one.

The second type concerns applications of different safety principles. For instance, we may have one safety principle requiring that all safety-critical procedures should follow pre-determined protocols, and another safety principle requiring that all employees should be encouraged to take initiatives and continuously improve the safety of work processes in which they take part. Although these two principles can be combined it is difficult to do so, and in most practical cases they will give rise to conflicts.

The third type of conflict concerns applications of a safety principle and some other principle that is supposed to be upheld on the workplace. Safety measures are sometimes costly (and the savings they induce tend to be difficult to demonstrate beforehand). Therefore, conflicts between safety principles and the ubiquitous principle of cost minimization are common.

Other conflicts that are rather typical in many sociotechnical systems are represented by the following examples:

- The principle of automation of controlling important safety functions may be in conflict with the safety principle that operators should have full control and situation awareness about what happens in a technical process.
- The principle of diversified safety systems may be in conflict with the principle of simplicity in designs.
- Principles of clear lines of accountability in a line organization may be in conflict with the principle that decisions shall be taken in those groups that hold the strongest expertise in a certain domain.
- Principles of documentation and details in descriptions of a safety management system may be in conflict with the principle of striving for oversight and simplicity.
- The principle of applying detailed step-by-step instructions may be in conflict with principles that strive for learning and competence in safety related work.

1.6 WHEN CAN SAFETY PRINCIPLES BE BROKEN?

Since conflicts involving safety principles are so common, it seems unavoidable that sometimes safety principles can legitimately be broken. But on the other hand, many if not most, severe accidents resulted from violations of safety principles. Therefore, we need guidelines (or metaprinciples) for legitimate violations of safety principles. We propose four such guidelines:

1. Barring exceptional circumstances, violations of safety principles can only be legitimate if the purpose is to *improve safety*. In a conflict between two safety principles, one of them will have to yield, at least in the short run. In a conflict between a safety principle and some other principle (such as cost minimization), the safety principle should be upheld.
2. Conflicts between safety principles should as far as possible be solved in a *risk-minimizing way*. For instance, in the above-mentioned conflict between the aims to reduce toxicity and risks of fire, an analysis showing that one of these two risks is larger than the other provides a weighty argument on how to proceed.
3. Conflicts between safety principles are often a sign that *more thorough changes are needed* that will make it possible to satisfy all of the conflicting principles. For instance, having to choose between a highly toxic and a highly flammable substance is an unsatisfactory situation. The short-term choice between conflicting safety principles should therefore be followed by development work aiming at finding ways to comply fully with both of them.
4. All violations of safety principles should be done *openly*, and discussed with everyone who is concerned. Such open discussions diminish the risk that safety principles are given up for no good reasons.

1.7 SAFETY IN CONTEXT

In this introductory chapter, our aim has been to put the notion of safety principles in context, suggesting a simple set of metaprinciples as well as pointing to the multitude of ways in which safety principles may be characterized. By covering a large number of safety principles this book actualizes how principles may conflict, and we have sketched a number of different types of conflicts between principles, as well as addressed the question when safety principles can rightly be broken. Our treatment has by necessity been on the abstract side of things, and before turning to the main content of this book, we would therefore like to remind the reader that while principles may provide action guidance and structure, the actual decision situation in which we find ourselves comes with a unique context (cf. Jackson and Ferris, 2013). In the individual case, we always have to look carefully at the context and take a stand on the salient factors. There will always be a judgment involved on which principles to apply, how, and when. This judgment can be fine-tuned by carefully studying our most informed accounts of the principles of safety. To this we now turn.

REFERENCES

Bahr, N. J. (1997). *System Safety Engineering and Risk Assessment: A Practical Approach.* Washington, DC: Taylor & Francis.

Haddon, W. (1980a). Advances in the epidemiology of injuries as a basis for public policy. *Public Health Reports*, 95(5), 411–421.

Haddon, W. (1980b). The basic strategies for preventing damage from hazards of all kinds. *Hazard Prevention*, 16, 8–11.

Hammer, W. (1980). *Product Safety Management and Engineering*. NJ: Prentice-Hall.

IAEA (1986). *General design safety principles for nuclear power plants: A safety guide*. International Atomic Energy Agency, Vienna.

Jackson, S. (2010). *Architecting Resilient Systems: Accident Avoidance and Survival and Recovery from Disruptions*. Wiley Series in Systems Engineering and Management, A.P. Sage (Series Ed.). Hoboken, NJ: John Wiley & Sons.

Jackson, S., and Ferris, T. (2013). Resilience principles for engineered systems. *Systems Engineering*, 16(2), 152–164.

Khan, F. I., and Amyotte, P. R. (2003). How to make inherent safety practice a reality. *The Canadian Journal of Chemical Engineering*, 8(1), 2–16.

Kletz, T. (1978). What you don't have, can't leak. *Chemistry and Industry*, 6, 287–292.

Kletz, T. (1998). *Process Plants: A Handbook for Inherently Safer Design*. Taylor & Francis.

Möller, N., and Hansson, S. O. (2008). Principles of engineering safety: Risk and uncertainty reduction. *Reliability Engineering & System Safety*, 93(6), 776–783.

NASA (1993). Safety policy and requirements document. NHB 1700.1 (V1-B). NASA, Washington, DC.

Runyan, C. W. (2003). Back to the future – revisiting Haddon's conceptualization of injury epidemiology and prevention. *Epidemiologic Reviews*, 15(1), 60–64.

Saleh, J. H., Marais, K. B., and Favaró, F. M. (2014). System safety principles: A multidisciplinary engineering perspective. *Journal of Loss Prevention in the Process Industries*, 29, 283–294.

Saleh, J. H., Marais, K. B., Bakolas, E., and Cowlagi, R. V. (2010). Highlights from the literature on accident causation and system safety: review of major ideas, recent contributions, and challenges. *Reliability Engineering and System Safety*, 95(11), 1105–1116.

2

PREVIEW

Niklas Möller, Sven Ove Hansson, Jan-Erik Holmberg, and Carl Rollenhagen

This handbook consists of 32 chapters divided thematically into one introductory part and five main parts: Safety Reserves, Information and Control, Demonstrability, Optimization, and Organizational Principles and Practices. The introductory part consists of this and the previous chapter, Chapter 1, in which we introduced the theme of the handbook. In Chapter 1, we noted that while there is strong consensus as to the usefulness of the main safety principles we treat in this handbook, there is no comparable consensus on how the safety principles are related to each other or whether or not some of them are more central than others. We present some accounts of overarching hierarchies in the literature, as well as suggest a simple list of basic tasks in safety management which we find useful. With these four "metaprinciples of safety," specific safety principles can be characterized according to whether they have a dominant focus on finding real or potential safety problems (inventorizing), providing capacity and resources to cope with real or potential problems (capacitating), making priorities (prioritizing), or making safety management coherent and comprehensive (integrating). Furthermore, we discuss conflicts between principles and the question when a safety principle may be broken.

Handbook of Safety Principles, First Edition. Edited by Niklas Möller, Sven Ove Hansson, Jan-Erik Holmberg, and Carl Rollenhagen.

2.1 PART I: SAFETY RESERVES

In Chapter 3, Erik Hollnagel discusses *resilience engineering*. Hollnagel traces the origins of the concept of resilience from material science to industrial safety. He then characterizes the modern concept of resilience through the four potentials that are the basis for resilient performance: responding, monitoring, learning, and anticipating. Hollnagel continues with an account of how resilience can be managed, and how various styles of safety management correspond to four different types of organizations. Hollnagel concludes that safety management should pay more attention to high-probability low-impact events since this is the very foundation of resilient performance.

Defence-in-depth is a concept widely applied in safety-critical technological areas. In Chapter 4, Jan-Erik Holmberg treats its history and contemporary usage. It was originally a military strategy in which the forces were distributed into several layers so that a single attack could not destroy the defensive forces. The contemporary usage of the concept of defence-in-depth is generalized and refers to having multiple protective measures for a given safety objective, such that the objective is achieved even if one of the protective measures fails. Holmberg stresses how defence-in-depth applies several fundamental safety principles, such as redundancy, diversity, physical separation, and successive barriers. The effectiveness of defence-in-depth, Holmberg points out, depends on two main features: the capability (1) to minimize the likelihood of holes and hazards in each barrier level and (2) to minimize the dependencies between barrier levels. Holmberg concludes that one weakness of defence-in-depth is that it can lead to complex design solutions, which is not necessarily optimal from an overall risk assessment point of view.

A related concept is *safety barriers*, which is the topic of Chapter 5. Lars Harms-Ringdahl and Carl Rollenhagen describe how the concept has developed from its original physical conceptualization to a much more generalized one in modern usage. A simple description of a safety barrier is that it is something that prevents an accident from occurring. However, the authors point out, there are different understandings of what this really means. The concept has evolved within different technical traditions and various theoretical considerations, and the authors argue that the width of the concept can to a large extent be explained by the multitude of existing assumptions of what constitutes a safety barrier. Harms-Ringdahl and Rollenhagen suggest a number of parameters for describing these varying notions, and stress the need for carefully defining the meaning of "safety barrier" in all applications, when any doubt of the interpretations could arise.

A safety reserve is "an extra distance or difference between actual conditions and conditions that are expected to cause an accident or some other type of damage," according to Neelke Doorn and Sven Ove Hansson in Chapter 6: Factors and Margins of Safety. Safety factors and safety margins are two of the oldest approaches in engineering to achieve safety under uncertain and variable conditions. A safety factor is a numerical value with which the risk indicator should be multiplied in order to reach a safety reserve, whereas a safety margin is an additive measure, that is, a numerical value added to the risk indicator. The authors discuss the theoretical assumptions underlying the safety factor and safety margin approaches and illustrate their use with examples from structural engineering, geotechnical engineering, and toxicology. The safety factor approach has increasingly been replaced with probabilistic risk assessment (PRA) as a tool for dimensioning safety measures. The authors compare two approaches and argue that they should be seen as complementary rather than as competitors.

2.2 PART II: INFORMATION AND CONTROL

In this section of the handbook, information and control concerning safety is the basic theme. In Chapter 7, Urban Kjellén discusses *experience feedback* as a fundamental principle in safety management. This is a process which involves the collection, analysis, and distribution of information in order to modify and improve activities. It is accomplished in safety management through various activities such as accident reporting and investigation, safety inspections, and risk assessments. A review of the history of safety management shows that this activity cannot be seen in isolation from the parallel developments in quality management; experience feedback occupies a central role in both areas. Experience feedback is also an important dimension in the concept of safety culture. The chapter discusses experience feedback from four complementary theoretical perspectives: the feedback cycle, models of an information system, the diagnostic process, and knowledge management. Four areas of application are reviewed, namely safety practices in plant operation, risk assessment, and transfer of safety-related experience to new construction projects and from users to design. Kjellén argues that there is a general lack of adequate scientific evidence on the effects on the risk of accident of various methods of experience feedback, and that this is mainly due to the methodological difficulties in controlling for the varying conditions in real-life processes.

In Chapter 8, Drew Rae reviews basic principles used to construct various forms of safety indicators, that is, metrics used to indicate safety and risk. Potential benefits and risks associated with using indicators are discussed. Many safety decisions rely on accurate knowledge about the current state of safety. Both the absolute amount of safety and the trend toward increasing or decreasing safety are used to judge the effectiveness of safety activities, to intervene in ongoing operations, and to target risk reduction initiatives. Measures of safety also have a political and regulatory role in demonstrating the effectiveness and value of the safety function. Since safety cannot be measured directly, a wide variety of metrics have been used to indicate the presence or absence of safety. These metrics are based on different theoretical models of what safety is and how it can be achieved. Organizational theories of accident causation highlight the value and risk of safety indicators. Rae argues that used appropriately, these indicators can challenge organizational beliefs about safety and provide intelligence about current and future threats. However, indicators can also reinforce incorrect assumptions and provide a false assurance of safety. As well as the role of indicators in providing information, the act of measurement itself is a driver of staff and management behavior. Where measurements are closely aligned with desired behaviors, these behaviors may be reinforced. Where measurement is indirect, particularly where it is focused on outcomes rather than actions, it can provide a "perverse incentive" for counterproductive behaviors such as concealment and bureaucratization.

The large field often referred to with the labels "human factors" and "human factors engineering (HFE)" is discussed in several chapters in this section. Lena Norros and Paula Savioja present *principles of human factors engineering* in Chapter 9. In this methodologically oriented chapter, a new interpretation of HFE is developed. The authors suggest that the standard interpretation of HFE as a process of systematic application of human factors knowledge needs a redefinition. HFE should be considered a constructive practice within which epistemic principles and ways of argumentation, known to characterize "design thinking," could be applied. It is argued that this new perspective would improve the realization of high-quality human factors standards, upon which the community agrees, but which have shown to be difficult to fulfill. Drawing on the recent international discussion and the authors' own experience of HFE projects, four generic principles are proposed that should guide the transformation of HFE from an applied science to a constructive design practice. These principles are: HFE is design thinking; HFE studies the human being as a manifold entity; HFE focuses on technology in use; and safety is achieved through continuous HFE. Each principle is described and its theoretical foundations are made explicit. The use and implementation of each principle is elaborated drawing on the authors' research. In this connection, novel concepts are introduced to HFE. With the above qualifications, HFE can participate in an integrated design culture that is needed to ensure the safety and efficiency of complex sociotechnical systems.

How different tasks are allocated between people and machines is a subject that has attracted attention both from an engineering and a human factors perspective. In Chapter 10, Björn Wahlström discusses *safety automation*. Safety automation is designed and implemented as a part of instrumentation and control (I&C). I&C systems serve as intermediates between human operators and the physical system. Safety automation has an important role in protecting physical systems from excursions into unsafe states. Automated controls on the one hand keep the physical system in safe regions of its state space and on the other bring it back to safety from occasional excursions into unsafe regions. The efficacy of safety automation builds on the quality of its design process. The focus of this chapter is on the design of safety automation and safety principles that are applied in this process. Safety automation is an integral part of the I&C and Wahlström suggests that it is therefore practical to broaden the discussion to the design of I&C and to the closely connected design of physical processes. Safety automation, the author explains, has its roots in the development of simple control loops, which were developed to protect engines against dangerous speeds. This technology worked as an enabler for the design and construction of increasingly complex processes, which in turn relied on the advancement of project management and systems engineering.

Risk communication is discussed in Chapter 11 by Jan Gutteling. Risk communication is part of risk management. Risk is a central issue for policy areas as diverse as health, safety, technology, environment, and finance—important issues for the average individual in day-to-day life. Stakeholders in the context of risk communication are very diverse. They could be members of the public, the media/press, governmental organizations, NGOs, or even workers in an organization. Gutteling emphasizes that risk perception is a central issue in risk communication. Over the years, communication processes in controversial contexts and in crises have gained importance. In this chapter, the complexity of risk communication is discussed, both from the perspective of a developing academic discipline and from a more practical perspective with a number of examples. Several risk communication models are presented, illustrated with empirical studies. Limitations, new focus points, and new approaches are described.

Sven Ove Hansson discusses the *precautionary principle* in Chapter 12. The precautionary principle tells us that we may have reason to take preventive action against a potential danger even if there is scientific uncertainty whether the danger is real or not. This approach is uncontroversial in most safety applications, but has turned out to be controversial in some areas, in particular chemical safety. This is also the area in relation to which the principle is most often discussed. The precautionary principle was formulated in the 1970s and 1980s. It has been incorporated into European law and is also endorsed in several international treaties on environmental protection. This chapter discusses how the principle is defined, how it is related to theories of scientific evidence, its limitations, and its relation to other safety principles.

Themes related to human factors are continued in the following two chapters. In Chapter 13, Jinkuyn Park discusses principles that can be used to develop *operating procedures*. One of the effective measures to enhance the operational safety of socio-technical systems is to provide a series of good procedures that are helpful for accomplishing required purposes or goals. However, a procedure seems to be a double-edged sword since it can be a major cause for the non-compliance behaviors of operating personnel, which can result in incidents and accidents in socio-technical systems. For this reason, based on the results of recent studies, additional principles are issued in this chapter with two kinds of promising rules, which could be considered for developing a good procedure.

In Chapter 14, Anna-Lisa Osvander and Håkan Alm take a broad view on the development of human machine system interfaces and how new technology has opened new possibilities (but also risks) in the interaction between people and technology. Looking back in time can give us a useful perspective on technological development and its impact on human machine systems. Human use of technology before the industrial revolution was characterized by systems that were limited in size, had uncomplicated technology with linear cause and effect relations, and because of that, were easy to understand. In many cases, it was possible to see how different parts of the systems used were connected and had impact on other parts. The information an operator needed to control and maintain the system was often possible to obtain with direct perception (using visual, auditory, and other input channels). The shift from analog to digital technology has radically changed this. It has resulted in possibilities to increase the information flow to the operator of a system, and increase the complexity of the system to be controlled. Since information can be distributed quickly between many subsystems, and what happens within the system is not directly visible for operators, the task to control a process or system has become increasingly complex and abstract. Since processes are only possible to understand indirectly, using different user interfaces, the demands on human cognitive resources (e.g., perception, attention, interpretation, decision-making) have increased.

2.3 PART III: DEMONSTRABILITY

In Chapter 15, Bo Bergman investigates how principles from quality management can be used in safety work. The chapter has a specific focus on the notion of quality improvement, and how it can be used in support of safety. The understanding of variation, as developed in the quality discourse, can usefully be applied to safety. Two approaches to variation, control charts, and robust design methodology are discussed in relation to their applicability in the safety context. The organizational learning cycle PDSA (plan-do-study-act) is discussed, and so is the role of motivation in quality improvement. Applications to healthcare and patient safety are used as examples.

By a safety case is meant "the argument and evidence supporting the claims about the acceptable safety of the system in operation in a defined context." In Chapter 16, Tim Kelly explains how the notion of a safety case was developed in response to a problem observed in several severe accidents, namely that the operator had dealt with safety in a reactive way, focusing on the regulator's detailed requirements rather than on a systematic and thorough consideration of safety in all its aspects. A safety case-based approach to safety assurance demands that the developers and operators of safety critical systems develop and present (typically for approval by a regulator) the argument and evidence to support the claim that their system is acceptably safe to operate.

With the ever-increasing complexity of manufacturing systems, safety cannot be ensured through add-on measures alone. In Chapter 17, Rajagopalan Srinivasan and Mohd Umair Iqbal argue that safety has to be built into the design. This principle is called inherently safer design. Inherently safer systems seek to eliminate hazards rather than control them. The chapter discusses the various sub-principles of inherent safety such as minimization, substitution, simplification, and moderation. Their implementation is illustrated with examples from diverse fields including the chemical and food industries, offshore oil and gas exploration, and electrical safety. Furthermore, the chapter highlights how inherently safe design relates to other approaches to safety.

Several principles for efficient maintenance have been developed, including total productive maintenance (TPM) and reliability centered maintenance (RCM). These and related concepts are discussed by Torbjörn Ylipää, Anders Skoogh, and Jon Bokrantz in Chapter 18: Maintenance, Maintainability, and Inspectabillity. It has a strong focus on the role of maintenance workers in manufacturing industries with a high level of automation including the "Internet of Things," internet connectivity for virtually any device or artifact. Maintenance workers have a key role in ensuring that the automated equipment is always available. However, their closeness to technical equipment, especially when it is broken and in need of repair, increases the safety risks that they are exposed to.

2.4 PART IV: OPTIMIZATION

This part discusses optimization of design, operation, and maintenance of safety-critical systems given that criteria for optimization can be agreed on. First, in Chapter 19, Pieter van Gelder gives an overview of the principles and issues of *risk-informed regulation*. A number of safety domains in which risk-informed regulation is adopted are discussed, such as safety against flooding and safety against environmental risks. Risk-informed regulation involves the estimation of probabilities and consequences (loss of life, economic damage) of undesired events and decision-making whether the calculated risk is acceptable for society. Experiences from the Netherlands are used to exemplify how risk-informed regulation has been implemented. The history of risk-informed regulation dates back to the early 1960s and an assessment of flooding risks. The regulation was originally based on water level exceedance probabilities, followed by a structural failure probability approach to flood defenses, and is currently being transformed to norms on drowning probabilities of the people being protected against flooding. In the safety domain against environmental or external industrial risks, regulation is based on an individual risk norm and a group risk norm. In the chapter, these norms are reviewed and some weaknesses are addressed. Recommendations for an improved, consistent, and homogeneous risk-informed regulatory framework are finally made.

In Chapter 20, Jan-Erik Holmberg presents methods and applications of *quantitative risk analysis* (QRA), which is a necessity for any optimization. Risk analysis is a broad area of methods and applications, including risk assessment (identification, evaluation, and quantification of risk) and risk management (decision-making). In the narrow sense, QRA includes a calculation of the two components of risk: the magnitude of the potential loss and the probability that the loss will occur. Engineering QRA covers risk assessment of hazardous socio-technological processes, where the losses can be described in terms of human, environmental, or economic losses. One of the objectives is to show compliance with safety objectives expressed with numerical probabilistic criteria, called acceptable risk. Further objectives are to identify weak points, search for design improvements, and even to optimize the system. In engineering risk analysis, there is usually a regulatory framework, including a regulator (safety authority) who stipulates the requirements for QRA. The choice of QRA methods is an application area-specific issue. In decision-making, risk acceptance criteria are commonly used together with the expected utility principle to choose between acceptable options, for example, cost–benefit analyses are performed.

In Chapter 21, Risto Tiusanen complements the discussion of risk analysis methods by introducing principles and approaches to *qualitative risk analysis*. The two primary functions of a qualitative risk analysis are risk identification and risk estimation. Qualitative risk analysis included methods for examining the probabilities and consequences of risks. In this chapter, the theoretical basis for qualitative risk analysis principles, typical qualitative methods, and experiences of their use are presented. In addition, overall systems engineering viewpoints on systematic risk assessment and the need for qualitative risk analysis methods in different safety engineering domains are discussed. Qualitative risk analysis methods have certain strengths, for example, they are relatively quick and easy to set up; they allow risks to be considered very early in the system lifecycle; they provide means to systematically and thoroughly examine a system, process, or procedure; and a simple risk matrix provides a rapid ranking of risks into different significance levels. Practical experiences have also shown that systematic analysis of system operations and system functions by using qualitative PHA, OHA, and HAZOP methods bring out a great deal of useful information that is not directly related to safety but has links to system availability, system usability, and system reliability.

Cost–benefit analysis, discussed by Genserik Reniers and Luca Talarico in Chapter 22, is a systematic way for decision-makers to ensure a proper trade-off by comparing the costs of additional risk reduction and the potential benefits that an investment could potentially bring. Any type of company or public organization might face operational risks, which are usually classified into three categories: very small risks where no further investment in risk reduction is necessary, very large risks with an outcome so unacceptable that these risks need to be reduced immediately, and intermediate risks that fall between the previous two risk categories. Focusing on the ALARA-area (between unacceptable and negligible risk), several models based on cost–benefit analysis (CBA) are described that can be used to support safety investment decision-making. These models are grouped into two categories depending on the type of accident and its severity. Each scenario might trigger major consequences with injured people or fatalities. One should bear in mind that depending on the type of risk, there always exist a maximum and a minimum safety level, which can both be seen as economic constraints varying over time within an organization or even within society. Safety levels and economic constraints differ between organizations.

Mathematical *optimization principles* reliability and safety are introduced by Enrico Zio and Yanfu Li in Chapter 23. The search for optimal system design solutions, operation procedures, and maintenance strategies that minimize cost and maximize reliability dates back to the 1960s. In recent years, safety has gained increased importance in highly hazardous activities and critical infrastructures, with a demand to be maximized. Multiple attributes must be considered in the search for optimal design solutions, operation procedures and maintenance strategies, rendering the optimization problem multi-objective. This chapter presents a holistic framework to multi-objective optimization, introduces the principles for problem solution, defines performance measures of the solution methods, and presents the methods for selecting the final preferred solution(s).

The application of optimization principles in maintenance is further discussed by Roger Flage in Chapter 24. *Maintenance optimization* refers to the development and use of theory, principles, methods, and models to find maintenance policies that are in some sense optimal. System safety is influenced by system reliability and hence by maintenance performed on the system. This influence may be more or less explicitly considered and accounted for in the optimization process. In this chapter, the author reviews and discusses maintenance optimization with emphasis on its relation to safety. The review ranges from pure cost benefit optimization, via quantitatively accounting for safety beyond the expected value calculations of cost–benefit optimization through constrained optimization, to a broader decision basis covering aspects beyond traditional cost–benefit models, such as considerations of the potential for major accidents, the type of uncertainties involved, and lack of knowledge of important risk influencing factors. The relation to other safety principles is also covered, including the ALARP principle and the use of risk acceptance criteria.

Human reliability analysis (HRA), discussed by Luca Podofillini in Chapter 25, is a set of analysis methods addressing human failures to perform safety-relevant tasks, generally when interacting with a technical system. The personnel tasks considered take place either during normal system operation, or shutdown, or in response to abnormal situations. Maintenance and test tasks are also in the scope of HRA. The most typical application areas are nuclear power, aerospace, oil and gas, and transportation industries. The chapter presents the underlying quantification models of the main HRA methods. A common assumption in HRA methods is that it is possible to develop a model that allows quantification of the probability of failure in a specific task, by specifying some parameters or elements of the model representing the relevant performance conditions. Since HRA results are used to inform decisions, with safety-relevant implications, there are continuous efforts by the HRA community to systematically assess their quality. A recurrent concern is the variability of the results, across analysts and across methods. The sources of this variability are diverse: different scopes and aims of the methods, imprecise guidance to apply the methods and limited empirical foundation of the methods, reliance on subjective assessment (which, to some extent, is always inevitable for many risk analysis applications).

In the final chapter of part IV (Chapter 26), Sven Ove Hansson compares three safety optimization principles: *ALARA, BAT, and the substitution principle*. ALARA (as low as reasonably achievable), also called ALARP (as low as reasonably practicable), was developed as a principle for radiation protection in the 1950s. It was largely a response to new information showing that even quite low doses of radiation might increase the risk of cancer. The ALARA principle is also used in workplace health and safety, in particular in the United Kingdom. BAT (best available technology) was introduced in emissions control in the 1960s, mainly as a means to avoid that legislation lagged behind technological improvements that would have made stricter requirements possible. The substitution principle has a background in safety work in the chemical industry in the 1970s but came to the fore as a regulatory principle for chemicals in the European Union in the early 2000s. It requires that dangerous substances be replaced by less dangerous alternatives. These three principles have much in common. They can be called improvement principles since they all require that safety be improved when that is possible, in contrast to optimization principles that emphasize the need to weigh safety against other values. This chapter discusses how each of the principles is defined, interpreted, and implemented, and compares them in terms of important issues of interpretation such as the scope that they are assumed to leave for economic considerations.

2.5 PART V: ORGANIZATIONAL PRINCIPLES AND PRACTICES

Some safety principles have a high level of abstraction, aimed for supporting the application of more specific principles. In this last section of the handbook, several such integrating principles are presented. The section begins with a chapter on *safety management* by Gudela Grote (Chapter 27). Safety management concerns all organizational activities that aim at controlling risk and keeping it at an acceptable level. Individual safety management practices should be aligned within a safety management system to promote an integral approach to managing safety. The core principle of any management system is the implementation of feedback loops that permit continuous monitoring and improvement of the components of the management system. In this chapter, the origins of safety management systems in quality control and the most frequently encountered components of such systems are described and evidence for their use and effectiveness is discussed. Some recommendations for designing and implementing safety management systems are proposed, especially regarding adequate consideration of contextual factors such as the kinds of safety to be managed, the general approach to managing uncertainty in the organization, and the regulatory regime within which safety is managed. By discussing safety management in connection with these factors, contingencies are outlined that can help decision-makers in companies to tailor safety management to their own situation and can support regulators in drawing up and evaluating safety management requirements for different industries.

Safety culture, a concept much discussed and used in safety related contexts is treated by Teemu Reiman and Carl Rollenhagen in Chapter 28. The concept of safety culture was first coined to explain the Chernobyl nuclear accident in the late 1980s. From the very beginning, the concept aimed at highlighting and illustrating that safety cannot ever be guaranteed by technical means alone, but rather depends heavily on management, leadership, and so-called human and organizational factors. Today, the concept of safety culture has become established in safety management applications in all major safety-critical domains. Nevertheless, there is a large variance in the conceptualizations of safety culture. Despite the variance in exact definitions, a consensus exists that safety culture is a multi-dimensional construct. Safety culture theories build on previous research on organizational culture, especially the model created by Edgar Schein. Also, the interdependent nature of individual behavior and cultural assumptions is often emphasized. This means that the deep core of culture, shared values, beliefs, and assumptions not only relate to each other but also causally influence practices, structures, personnel's behavior, and attitudes, as well as technology. Furthermore, these elements may influence each other as well as reflect back on the shared values, beliefs, and assumptions. Thus, safety culture is constantly in the making.

Steve Roberts and Scott Geller discuss *principles of behavior-based safety* in Chapter 29. Behavior-based safety (BBS) focuses on identifying and providing interpersonal feedback for the critical behaviors that may lead to or prevent injury, analyzing the factors that contribute to critical safety-related behaviors, and developing interventions to increase the frequency of safe behavior and decrease that of at-risk behavior. BBS has traditionally involved line-level employees, with support from the leadership. However, a more comprehensive view of BBS includes identifying, analyzing, and improving the critical safety-related behaviors of employees at all levels of the organization, from wage workers to supervision and executive employees. BBS also considers a full range of factors that contribute to safe and at-risk behavior, including leadership, physical environment/conditions, systems, behaviors, employee engagement, and internal person factors.

Planning for future emergencies and building up a capacity to manage unexpected events and crises is relevant for all types of organizations. In Chapter 30, Ann Enander discusses some of the challenges intrinsic to planning for the unknown and dealing with critical events. Examples from the national, community, and organizational levels are discussed. Although the context for implementation varies, there are some key generic features common to most emergency plans. Evaluations of actual crisis incidents also reveal some recurring themes which emerge as challenges when emergency plans have to be put into practice.

Safety standards provide an important basis on which deployment decisions of safety-critical systems are made. In Chapter 31, Ibrahim Habli discusses challenges and principles of *safety standards.* Generally, standards provide a set of objectives and mandates or recommend specific means for compliance. They also specify items of evidence that have the potential for satisfying the objectives. However, safety standards vary between regions, domains, and technologies in terms of factors such as the level of prescription, the qualitative and quantitative nature of the safety requirements, and the relation between safety criticality and engineering rigor. The chapter examines these factors and discusses how they relate to fundamental concepts and practices in safety, such as risk analysis and safety cases. This chapter concludes with examples of assurance principles for the safety approval of software systems and complex automotive applications.

In the last chapter of the handbook, Jean-Christophe Le Coze examines the topic of *managing the unexpected*. Le Coze takes as a departure an interpretation of the social sciences of the past 30–40 years, through which he discusses the definition and conceptualization of the unexpected in the context of high-risk systems. Following this strategy, four threads are identified which help characterize the unexpected, corresponding to four rationales established by three authors who pioneered the field, Barry Turner, Charles Perrow, and Jens Rasmussen. These rationales correspond to the ideas of (1) technology out of control, (2) executives (and regulatory) failure, (3) fallible (cognitive) constructs, and (4) emergent properties (due to self-organized adaptive agents). They are named using four major intellectual figures of the social sciences, Jacques Ellul, Karl Marx, Thomas Kuhn, and Ross Ashby who provided a conceptual background widely shared among authors in the field, beyond the three selected pioneers. Finally, this chapter introduces conceptual developments that have been made in the search for properties allowing actors in high-risk systems to manage the unexpected. The author groups these properties under three main categories, (1) building favorable power configurations, (2) confronting our cultural constructs, and (3) keeping sight of the relations between parts and whole.

Part I

SAFETY RESERVES

3

RESILIENCE ENGINEERING AND THE FUTURE OF SAFETY MANAGEMENT

Erik Hollnagel

3.1 ON THE ORIGINS OF RESILIENCE

The terms "resilience" and "resilience engineering" appeared in safety discussions around 2000 (Woods, 2000). The first major effort of clarifying what resilience engineering might mean for the traditional safety practices—accident prevention and safety management—was an invited symposium held in the Swedish town of Söderköping in October 2004. The discussions and afterthoughts from that symposium were documented in Hollnagel et al. (2006). Since then there have been several books, multiple meetings of various types, numerous papers, and even the creation of an international organization—the Resilience Engineering Association.

The term resilience has, however, previously been used in several different ways (McAslan, 2010). It can be found, possibly for the first time, in the beginning of the nineteenth century as a way of describing a property of timber and as an explanation of why some types of wood were able to accommodate sudden and severe loads without breaking. About four decades later, a report to the British Admiralty referred to a measure called the *modulus of resilience* as a means of assessing the ability of materials to withstand severe conditions.

Handbook of Safety Principles, First Edition. Edited by Niklas Möller, Sven Ove Hansson, Jan-Erik Holmberg, and Carl Rollenhagen.

In 1973, Holling introduced the term resilience as a measure of the ability of an ecosystem to absorb changes so that it could continue to exist despite a changing environment. Holling made a useful distinction between resilience and stability, where the latter was defined as the ability of a system to return to its equilibrium state after a temporary disturbance. In other words, a stable system would be able to withstand the "slings and arrows of outrageous fortune" and then continue as if nothing had happened. It would be robust or impervious, but not resilient. (A similar type of discussion took place in the early days of resilience engineering (e.g., Pavard et al., 2008) but that has now subsided.) Resilience and stability were seen as two important properties of ecological systems, rather than as synonyms for the same property.

The use of the term resilience in ecology later led to a distinction between engineering resilience and ecological resilience. According to this, engineering resilience considered how ecological systems could exist close to a stable steady state with resilience being the ability to return to that state following a perturbation. Ecological resilience was used to describe conditions far from any stable steady state, where instabilities could flip a system from one regime of behavior to another. The meaning of engineering resilience was thus close to the ideas of robustness and imperviousness, which corresponds reasonably well to what can be achieved by classical safety engineering. "True"—or ecological—resilience is concerned with how a system can "persist through continuous development in the face of change and … innovate and transform into new more desirable configurations" (Folke, 2006). Ironically, ecological systems cannot really be resilient in this sense since they are unable to think about the future and anticipate what may happen. (Although human ecological systems ought to be an exception to that, they have often proved otherwise.)

As a testimony to the versatility—or is it resilience?—of the term, "resilience" was also used in psychological studies of children in the early 1970s as a synonym for stress resistance. It soon occurred frequently in psychology, and was many years later, in 2007, defined as "the capacity to withstand traumatic situations and the ability to use a trauma as the start of something new" (Tisseron, 2007). The business community finally picked it up in the beginning of the twenty-first century and used resilience to describe the ability dynamically to reinvent business models and strategies as circumstances change (Hamel and Valikangas, 2003).

A common theme in the multiple uses of resilience was the ability to survive more or less intact and to resist stresses, harmful influences, and unfavorable conditions. The thinking about resilience typically referred to a juxtaposition of two situations: on the one hand, materials, individuals, organizations, or systems where resilience was absent and where adverse outcomes therefore might happen, and, on the other hand, materials, individuals, organizations, or systems where resilience was present and where adverse outcomes therefore could be avoided. This was also the case in the early 2000s when resilience engineering was proposed as an alternative to the error tabulation that was, and in many cases still is, the mainstay of conventional safety approaches (Woods and Wreathall, 2003). A decade later, the scene has changed and resilience engineering has come into its own as a new way of understanding how complex socio-technical systems work and not just as an alternative to traditional safety (Hollnagel, 2014).

3.2 THE RESILIENCE ENGINEERING UNDERSTANDING OF "RESILIENCE"

The full meaning of resilience, however, cannot be expressed by a dichotomy or by juxtaposing resilience with "safety," "robustness," "brittleness," etc. Discussions about safety matters have often been couched in terms that imply absolute differences or absolute states. The best example of that is the zero accident vision (Zwetsloot et al., 2013), the ideal that the goal of safety is to ensure that nothing goes wrong, whether counted as accidents, incidents, lost time injuries, etc. While the zero accident vision is a noble ideal, it is quite impractical and actually goes against the spirit of resilience engineering.

The zero accident vision follows logically from the *causality credo*, which is the tacit assumption that outcomes can be understood as effects that follow from prior causes. As an argument, the causality credo has three parts:

- Acceptable and unacceptable outcomes have different causes. The reason for unacceptable outcomes (accidents, incidents) is that something has gone wrong. Similarly, the reason for acceptable outcomes is that everything worked as it should.
- It is possible to find the causes of unacceptable outcomes through rigorous reasoning, provided enough evidence is collected. When the causes have been found, something can be done to either eliminate or neutralize them. (Strangely enough, we rarely bother to find out why things go well.)
- Since all unacceptable outcomes have causes and since all causes can be found, it follows that the zero accident vision is achievable.

Looked at more closely, the zero accident vision implies a "*hypothesis of different causes*," according to which the causes of unacceptable outcomes (failures) are different from the causes of acceptable outcomes (successes). This must be so since the elimination of causes of unacceptable outcomes otherwise would affect also the occurrence of acceptable outcomes, hence be counter-productive. In a zero accident vision, the purpose of safety management is to ensure that the organization is free from unacceptable losses and risks, which in turn favors a piecemeal, reactive approach. This is illustrated by a recent report from the Australian Transport Safety Bureau, which defines a safety management system "simply as a planned, documented and verifiable method of managing hazards and associated risks" (Thomas, 2012).

Resilience engineering disagrees with the assumption that we can describe the world in terms of dichotomies and that it makes sense to characterize systems in the same manner. Resilience is not a quality (or quantity) as such and therefore not something that is either absent or present in a system. Neither is resilience simply a synonym for successful outcomes as the following quotations show.

> "Outcome is a powerful signal, and we tend to associate resilience with successful outcomes. From this superficial view, a system with a positive outcome seems resilient, while one with a negative outcome does not. However, if resilience engineering is to progress from being a descriptive field to an applied one improving the overall performance of complex socio-technical systems, some way of recognizing cues or indicators of resilience will be necessary. Separating resilience from success—process from outcome—will assist in clarifying the concept of resilience regardless of the outcome of an event."
>
> (Fairbanks et al., 2013, p. 159)

> "Success and resilience should be treated as two different orthogonal dimensions of performance... Maintaining a clear separation of process and outcome, and restricting the notion of resilience to a system's processes (i.e., what the system does), should help ... avoid the sorts of narrow optimizations that can reduce small common problems at the price of increasing large, rare ones."
>
> (Ibid., p. 167)

Resilience is in practice used as a single term to characterize the potential of a system or an organization to perform in a certain way. However, this is merely a convenience and does not imply that resilience *is* a single quality. By separating resilience from the acceptability or unacceptability of outcomes, from things that go right or things that go wrong, it is recognized that resilience in principle is never completely absent. There may, of course, be types of performance that nearly always lead to unacceptable or undesirable outcomes and where it therefore is tempting to declare that resilience is too low. But this is only a truncated way of saying that the organization's potential to perform in a resilient manner was not realized in practice. The decision about whether outcomes are acceptable or unacceptable is furthermore relative rather than absolute (cf., Amalberti, 2013).

Resilience (or more accurately, the potential for resilient performance—although this is too long to write every time) is, however, not about avoiding failures and breakdowns, that is, it is not just the opposite of a lack of safety. The purpose of the argument that failures are the flip side of successes (Hollnagel, 2006) was not to propose yet another dichotomy but rather to point out that things that go wrong happen in, basically, the same way as things that go right. The main difference is in the value of the outcomes, not in how the outcomes come about. This was famously expressed by Ernst Mach almost a century ago, when he noted that "Erkenntnis und Irrtum fließen aus denselben psychischen Quellen; nur der Erfolg vermag beide zu scheiden" (Mach, 1917).

3.3 THE FOUR POTENTIALS FOR RESILIENCE PERFORMANCE

Resilient performance cannot be accounted for simply by referring to some mystical or mythical quality called resilience. Yet, it is still necessary to describe how resilient performance comes about, both so that it is possible to determine whether an organization has appropriate potentials for resilient performance, and to propose concrete ways in which these potentials can be improved or engineered. Resilience engineering recognizes these needs and proposes that they can be addressed by considering four fundamental potentials or abilities, sometimes referred to as the four cornerstones of resilience (Hollnagel, 2009). The four potentials are derived from the definition that a system is resilient if it can adjust its functioning prior to, during, or following events (changes, disturbances, and opportunities), and thereby sustain required operations under both expected and unexpected conditions.

- An organization must first of all know what to do when something happens. It must be able to *respond* to regular and irregular events, disturbances, and opportunities. This can happen either by implementing a prepared set of responses or by adjusting ongoing functioning to the situation.
- An organization must know what to look for, that is, how to monitor that which is or can become a threat in the near term. It must be able to *monitor* that which happens and recognize if something changes so much that this may affect the organization's ability to sustain its current operations. The monitoring must cover both that which happens in the environment and that which happens in the organization itself, that is, its own performance.
- An organization must learn from—and remember—what has happened. It must be able to *learn* the right lessons from the right experience—from acceptable outcomes (successes) as well as from unacceptable ones (failures). The essence of learning is that the organization can modify what it does—how it responds, monitors, anticipates, and even how it learns—so that it increases the likelihood that future actions go well.
- An organization must finally know what to expect. It must be able to *anticipate* developments—threats as well as opportunities—that lie further into the future, beyond the range of current operations. (Monitoring is about what happens now; anticipation is about what may happen in the future, either in the short term or in the long term.) The anticipation must, of course, include a realistic assessment of the long-term effects of planned changes.

The four potentials are obviously not independent of each other. For instance, the effectiveness of an organization's response to something depends on whether it is prepared (i.e., able to monitor) and whether it has been able to learn from past experience. Indeed, if there is no monitoring, every case where a response is needed will be a surprise, which hardly is a desirable condition for any organization. Each of

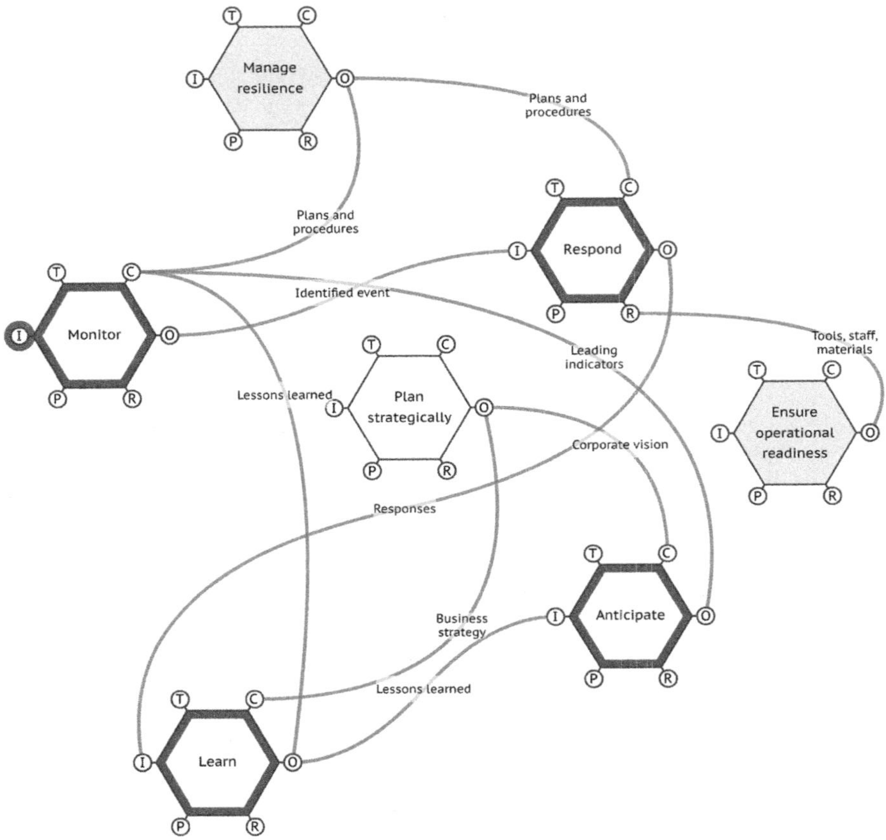

Figure 3.1. A FRAM model of the four basic potentials for resilient performance.

the four potentials also depends on other things. While it is beyond the scope of this chapter to develop these couplings or dependencies in detail, an illustration of what it might look like is shown in Figure 3.1, using the formalism of the FRAM (Hollnagel, 2013).

For any given organization, the proper balance or mix of the four potentials depends on the nature of its operations and the specific operating environment (business, regulatory, environmental, social, etc.). The four potentials must be developed to different degrees for different organizations, depending on the organization's overall ambition or goal, the current situation, etc. If, for example, the operating (or business) environment deteriorates rapidly, it obviously becomes more important to keep the operations going than to digest the lessons learned or to prepare for the future, although neither of these must be completely neglected.

Since the four potentials refer to tangible and well-known types of activity (at least for responding, monitoring, and learning), it is quite easy to describe them more

TABLE 3.1. The Resilience Analysis Grid: Questions Relating to the Ability to Monitor

(a)	What systems are in place to establish and maintain a set of indicators of safety and security performance? On what basis are these created?
(b)	What systems are in place to assess whether the indicators used are adequate and relevant?
(c)	How are indicators selected and utilized by relevant people and departments/ processes?
(d)	What systems are in place to create and revise monitoring indicators when necessary or relevant? On what basis are these indicators revised?
(e)	What systems are in place to analyze or extract relevant information from these indicators?
(f)	What systems are in place to ensure information from indicators are given or communicated to relevant people, departments, or processes?
(g)	What systems are in place to ensure communicated information leads to meaningful, useful, or necessary actions/measures being taken to maintain safety and security? What systems are in place to ensure such processes are properly resourced?

concretely. In this way, each ability can be specified in operational terms by means of a set of underlying or constituent functions (or sub-potentials), which then can be assessed in practice. An assessment of this kind can be used both to describe how well an organization currently *is* doing and to prescribe how well it *should* be doing. A concrete example of that is the resilience analysis grid (RAG) (cf., ARPANSA, 2012; Hollnagel, 2011a), which can be used to determine how well an organization performs, or is ready to perform, at a given moment. An example of a subset of specific issues from ARPANSA's Holistic Safety Guidelines is shown in Table 3.1. The same assessment can be used to propose concrete ways to develop specific sub-functions of a potential—without forgetting that the potentials are mutually dependent.

3.4 SAFETY MANAGEMENT SYSTEMS

Safety management systems should ideally be based on a small set of well-defined principles that make clear both what safety *is*, how it can be *measured*, and how it can be achieved and *managed*. Safety management systems have, however, on the whole evolved with little overall guidance from a haphazard collection of "best practices" found in a wide variety of industrial domains. Since the driving concern always has been to prevent or avoid accidents, with theories and models added as an afterthought, efforts to provide safety have been linked to the current understanding of why accidents happen and what the major causes are.

Hale and Hovden (1998) have described three ages of safety thinking where each age represents the common understanding of how things go wrong and what the major causes are. In the first age, nominally lasting from the beginning of the second industrial revolution in the late eighteenth century to the 1970s, thinking was in terms of linear cause–effect relations and the main causes were failures of technology (ruptures, breaks, explosions, and collapses). The focus on technology as the main—or even only—source of both safety problems and their solutions was maintained, although with diminishing success, until 1979, when the accident at the Three Mile Island nuclear power plant demonstrated that safeguarding technology was not enough. The TMI accident made it necessary to include a consideration of human factors—or even of *the* human factor—in safety thinking and safety management. In the second age of safety thinking, technology was rapidly replaced by a vaguely described idea of human factors as the main source of root causes. The thinking about how accidents could happen did, however, not change except that simple linear models were augmented by composite models where "active failures" and "latent conditions" could be combined. This gave some peace of mind but only for a short while. The combination of human ingenuity and technological hubris soon produced working environments that were only partly comprehensible and where safety management therefore became constrained by its own simplified assumptions. The third age of safety thinking made safety management itself a focus and introduced new causes, such as the influence of organizational failures and safety culture, without challenging the continued belief in linear causal thinking.

The practice of safety management was for many years guided by prescriptive regulation as set out by government bodies. Nicknamed the "boots, belts, and buckles" approach, the assumption was that safety could be managed effectively if organizations adhered to government regulation or recognized "best industrial practices."

In the late 1970s, safety management shifted from an approach based on following prescriptive legislation to an approach where organizations took responsibility for their own management of their unique risk profile. This was mainly due to a number of catastrophic accidents (e.g., the Seveso disaster in 1976 and the Piper Alpha accident in 1988) which made clear that prescriptive regulation was inadequate for complex socio-technical systems. The acknowledgement that the organization and its culture was at least as important as technical malfunctions and human factors, corresponding to a move from an engineering to a social science paradigm, also challenged regulations. Semantically, emphasis was now put on "self-regulation" where organizations were required to ensure that they took all reasonably practical steps to ensure the health and safety of their workforce.

The change from prescriptive regulation to organizational responsibility also meant that safety management no longer could be reactive only. This is reflected in the current distinction between three types of safety management called *reactive*, *proactive*, and *predictive*, respectively. Reactive safety management looks at the past and relies on the use of lagging indicators. Safety is basically taken care of by analyzing events that have happened and by trying to prevent that they will happen again, either through regulation or new "best" practices. Proactive safety management involves looking at the present or the current situation in an attempt to recognize known risks and hazardous conditions through the analysis of the organization's processes. Predictive safety management, finally, involves looking for possible future problems that may arise as a result either of external developments or of changes made by the organization itself. Predictive safety management differs from proactive safety management by looking for that which is partly unknown although still imaginable (cf., Westrum, 2006).

There have been important changes in the practice of managing safety during the last century or so, both in terms of how it is done and in terms of the main targets—the potential sources of lack of safety. The focus has, however, remained the same, namely to prevent a loss of safety, cf., the prototypical definition of safety as the "freedom from unacceptable risk." The preferred solution has been to strengthen specific safety systems and thereby bolster safety margins in the case of exceptional events. Resilience engineering turns this on its head, and focuses on situations where safety is present and where the organization therefore functions as it should, rather than situations where safety is absent. Since this changes the meaning of safety, it also changes the practice of safety management.

3.5 DEVELOPING DEFINITIONS OF RESILIENCE

The term resilience has a long history and has been used by different sciences, as described in the introduction to this chapter. The starting point for resilience engineering was to find a way to work with safety without being constrained by linear causal thinking and without immediately invoking the concepts of failures and errors. The first definition therefore indirectly referred to the established view of safety, and therefore unintentionally carried along some of the connotations of that.

> "The essence of resilience is … the intrinsic ability of an organization (system) to maintain or regain a dynamically stable state, which allows it to continue operations after a major mishap and/or in the presence of a continuous stress."
>
> (Hollnagel et al., 2006)

The definition retained the traditional juxtaposition of two states—one of stable functioning and continued operations and one where the system has broken down. The functioning state was either due to a response to a serious disturbance or an ability to resist continued stress (from the organization's environment), but not the "normal" functioning state.

While this definition served as a useful start for discussions, it soon became clear that it was too narrow and that it defined resilience in contrast to something else, rather than as a phenomenon in its own right. Through many discussions—written as well as verbal—a new definition was presented in Hollnagel (2011b) as:

> "The intrinsic ability of a system to adjust its functioning prior to, during, or following changes and disturbances, so that it can sustain required operations under both expected and unexpected conditions."

In this definition, the emphasis on risks and threats had been reduced, and replaced by a reference to "expected and unexpected conditions." The focus had also changed from "maintaining or regaining a dynamically stable state" to the ability to "sustain required operations." The emphasis was now on how an organization functioned rather than on how it recovered from harmful or stressful conditions. There was, however, still a bias toward the "dark side" in the use of "changes and disturbances." This has been eliminated in the following definition, which is the most current.

> "A system is resilient if it can adjust its functioning prior to, during, or following events (changes, disturbances, and opportunities), and thereby sustain required operations under both expected and unexpected conditions."

The important change in the definitions has been to broaden the scope of resilient performance. It is no longer the ability to recover from threats and stresses, but instead the ability to perform as required under a variety of conditions—advantageous as well as potentially harmful.

The emphasis on opportunities is important for the change from protective safety (the freedom from things that go wrong) to productive safety (the facilitation of things that go right)—and ultimately for the dissociation of resilience from safety that hopefully will leave the sterile discussions and the stereotypes of the past behind. Resilience is about how well systems perform, not just about how they remain safe, cf., the discussion of Safety-I and Safety-II (Hollnagel, 2014). And a system that is unable to make use of opportunities is not in a better position than a system that cannot respond to threats and disturbances—at least not in the long term.

The above definition is probably not the final one. Although resilience engineering started as a contrast to conventional (protective) safety thinking, it soon became something in its own right. Resilience engineering is about the potential for resilient performance *per se*, how we can recognize it, how we can assess (or measure) it, and how we can improve it. The discussions should therefore focus on what resilience (or rather, the potentials for resilient performance) *is*, rather than on what it *is not*.

3.6 MANAGING THE POTENTIALS FOR RESILIENT PERFORMANCE

Since resilience is entirely different from safety, a system to manage the potentials for resilient performance (for lack of a better or shorter term) will clearly be different from a safety management system.

Safety management systems typically emphasize or highlight a single item or aspect, not only in terms of indications or manifestations (failures and the lack of safety), but also in terms of approach. If the purpose is taken to be the reduction and possible elimination of adverse outcomes, then the approach taken must of course be consistent with the understanding of how things happen. Since the second principle of the *causality credo* assumes that there is a symmetry of values between causes and effects—in the sense that an adverse outcome must be due to a preceding adverse event, a failure or a malfunction—then the understanding of how failures and malfunctions (or even "human errors") happen becomes critical. The preferred explanations have gone from the technology, to the human factor, and then to the organization and safety management, as described by Hale and Hovden's account of the three ages or safety thinking. In each case, the change was required because the accepted explanations were unable to account for new types of accidents. This has in the beginning of the twenty-first century led to explanations couched in terms of complex adaptive systems.

Resilient performance does not characterize a state or a condition—what a system *is*—but how processes or performance are carried out—what an organization *does*. Resilient performance is more precisely about maintaining a balance among the four potentials, not only after an adverse event has taken place but continuously. If, for instance, an organization focuses primarily on responding, as in handling unforeseen or difficult situations, and therefore neglects monitoring, then it is not considered to have the potentials for resilient performance. The reason is simply that neglecting monitoring will increase the likelihood that performance is disturbed by unforeseen events, which will lead to reduced productivity and/or jeopardizing safety.

The management of the potentials for resilient performance must necessarily take a holistic (or systemic) view. Rather than beginning by specific factors—such as the workplace, the technology, the safety culture—the management of the potentials looks at how the organization functions as a whole. This, of course, does not render the typical OHS and HSE concerns invalid or unimportant. Managing the potentials for resilient performance is a complement to established safety management practices rather than a replacement of them.

The four basic potentials for resilient performance resilience are a natural starting point for understanding how an organization functions: how it responds, how it monitors, how it learns, and how it anticipates. But the four potentials must always be seen together rather than one by one (cf., Figure 3.1). The top-down perspective that the integration of the four potentials provides also suggests a useful way to distinguish between four types of organizations, which here are simply called organizations of the first, the second, the third, and the fourth kind.

3.6.1 Organizations of the First Kind

It is a fundamental requirement to any organization or system that it can react appropriately when something happens, particularly when something unexpected happens. Failure to that will sooner or later lead to the demise of the organization—even those that are "too large to fail." Organizations that can react appropriately and therefore sustain their existence are called organizations of the first kind.

Reacting when something happens requires the abilities to monitor and to respond. Monitoring is needed to determine that the situation requires some kind of reaction or intervention. And responding is necessary to implement the reaction or intervention. The two must necessarily go together. An organization that passively reacts whenever something happens—whenever a condition passes a certain threshold—will by definition always be surprised and therefore always reactive. This may work as long as the frequency of events is so low that one response can be completed before the next is required. Unfortunately, only few organizations in today's world can enjoy that luxury.

While organizations of the first kind may survive, at least temporarily, they do not really have the potentials for resilient performance. They may possibly appear so because they are able to avoid disabling accidents and disasters for a time, cf., the discussion about the separation of resilience from outcomes.

3.6.2 Organizations of the Second Kind

While the ability to respond is fundamental, it is also necessary to be able to modify the responses based on experience. Organizations of the second kind are therefore those that can manage something not only *when* it happens but also *after* it has happened. This means that the organization can learn from what has happened, and can use this learning to adjust both how it monitors—what it looks for—and how it responds—what it does.

Learning is necessary in order to be sure that the organization is monitoring the proper signs, signals, and symbols. Learning is also necessary to adjust the responses both in terms of what they are and in terms of when they are given (and for how long, etc.). Both types of learning are necessary, because the context or environment in which the organization works, and in which it must operate, keeps changing. The higher the stability of the context or working environment is, the less is the need to modify monitoring and responding, and the less therefore is the need to learn. But no environment is perfectly stable, even though organizations sometimes seem to find excuses for avoiding to learn (Baumard and Starbuck, 2005).

Management on the level of an organization of the second kind is necessary for management on the level of an organization of the first kind. Clearly, without learning, the response management will become less and less adequate, until it, in the end, no longer is effective. Learning also requires an effort and an investment that ties up resources that could have been used elsewhere. It may therefore be seen as something that is not strictly necessary. Yet without learning, an organization will perish.

3.6.3 Organizations of the Third Kind

One effect of monitoring is that an organization may detect a developing situation in time and therefore be able to respond before it has become too serious (or before it has become too late). On the level of the individual, this is known as management by exception (Dekker and Woods, 1999). The dilemma of management by exception is that an early intervention may seem appealing but be difficult to justify, while a late intervention is more difficult because the situation will have deteriorated further and become more complicated and possibly also more costly to control. But responding ahead of time may also be thought of in a wider context, where the response is based on expectations or anticipations of what may happen, even if future events cannot be monitored *per se*.

Organizations of the third kind are those that can manage something *before* it happens, by analyzing the developments in the world around and preparing itself as well as possible. A typical example of that is an organization that tries to anticipate changes in customer needs, or in regulations (carbon emissions, for instance), or a growing number of refugees, and try to be ready for that. Responding before something happens requires the ability to anticipate and/or predict. Anticipation in turn requires indicators that can be used to make the predictions, that is, indicators which somehow are correlated with future events. They are therefore called leading indicators—although the indicators themselves obviously must be of the present or the past.

Anticipation also requires taking a risk. Initiating an action (or a response) before there is something to respond to (either at all or of a large enough magnitude), introduces the risk that the prediction/anticipation may have been wrong and that the effort therefore is wasted. When you respond after something has happened, this uncertainty does not exist. On the other hand, waiting for too long may require a larger response, hence be cost-ineffective in its own right.

3.6.4 Organizations of the Fourth Kind

While organizations of the third kind have the potential for resilient performance, it is possible to become even better by considering not only what happens *between* the organization and its environment (business, operations, etc.), but also what happens *in* the organization itself. Organizations of the fourth kind are those where the anticipation includes the organization itself—not only in the sense of monitoring itself (which would be a stage in the first type of management) or learning about itself (which would be a stage in the third type of management), but considering how the world responds or changes when the organization makes changes, how these responses may affect the changes, and so on.

The need for this is obvious from the many examples of organizations that sanguinely consider only the good or useful consequences of changes it make, but overlook or are ignorant about other possible effects. This phenomenon has been recognized for many years and was delightfully named "imperious immediacy of interest" by Merton (1936). This was found in "instances where the actor's paramount concern with the foreseen immediate consequences excludes the consideration of further or other consequences of the same act" (Ibid., p. 901).

3.7 RESILIENCE MANAGEMENT: LP-HI OR HP-LI?

People who are responsible for safety are everywhere concerned about the so-called low-probability, high-impact (LP-HI) events. (These are sometimes also called low-probability, high-risk events.) If we, for argument's sake, think of outcomes as being normally distributed, the LP-HI events correspond to the left tail of the normal distribution curve. (They are also the ones that are found in the "red corner" of the conventional risk matrix.) Safety management and safety sciences focus almost exclusively on these events, and the reason is not hard to understand. The high impact, or high risk, means that the outcomes correspond to large losses. (Strangely enough, there is little concern for the LP-HI events on the right tail.)

As pointed out by the Safety-I, Safety-II debate, the traditional focus on LP-HI events means that the study of safety and safety management are concerned with situations where safety, by definition, is low or even missing. In other words, the study of safety and safety management deals with states that, by definition, are unsafe rather than with states that, by definition, are safe.

Resilience engineering and resilience management point out that it is not sufficient to look at the LP-HI events, regardless of whether the outcomes are negative or positive. It is also necessary to look at the HP-LI events—the high-probability, low-impact events. These are the outcomes that constitute the main area of the normal distribution curve, cf., Figure 3.2. This leads to the recommendation that the resources that are available for learning to become safer should focus on events based on their frequency rather than their severity. That which happens often is in the long run more important than that which happens rarely.

The recommendation to base learning on frequency of occurrence rather than severity of outcome does not mean that every activity that has an acceptable outcome must be studied. Instead, it means that learning should be based on the patterns that can be found in everyday performance. To do so does not require any kind of

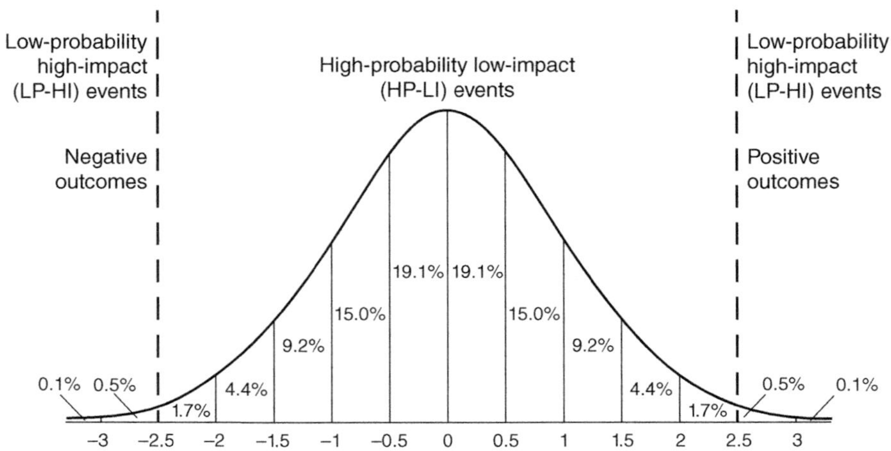

Figure 3.2. Event distribution versus event probability and impact.

extraordinary effort, but only that we learn to perceive the things that we usually disregard because they are prosaic and unglamorous.

Resilience engineering also points out that the meaning of *impact* must change. In the LP-HI framework that represents safety thinking in general, the impact is always negative: death, destruction, and losses. But everyday activities, work that is safe, also has an impact, namely the acceptable outcomes that we expect from everyday work and that indeed are the reason why work is done. Therefore, that an organization is safe does not only mean that few things go wrong, but also means that the organization delivers the desired and intended outputs or services. The impact is therefore that which happens, not only when something fails but also when it is successful. For high-probability events, for things that happen all the time, the negative impact will of course be low. No system and no organization could survive for long if serious adverse outcomes happened frequently. This is among other things what classical safety management is for. But for high-probability events, for everyday work, there is also a positive impact, namely the acceptable outcomes. The magnitude of the impact may be small for each event but it is positive. This means that the accumulated impact of the many everyday events can be quite large. The purpose of resilient management is precisely to manage that which works and which is safe because it works, in other words the HP-LI events. The benefit is therefore not (just) that serious losses are avoided, but also that productivity is ensured, hence that small—or perhaps not so small—gains accrue. Safety management and resilience management both contribute to a smaller number of adverse events, that is, to a higher level of safety in the traditional sense. But safety management does it by trying to prevent things from failing or going wrong, that is, a Safety-I approach. Investments in safety are therefore a cost, because they do not contribute directly to productivity. Resilience management does it by trying to make sure that things go right, that is, a Safety-II approach. This reduces the number of things that go wrong—because something cannot go right and wrong at the same time. But it also contributes directly to productivity, by the accumulation of small gains.

REFERENCES

Amalberti, R. (2013). *Navigating Safety. Necessary Compromises and Trade-offs – Theory and Practice*. Dordrecht, The Netherlands: Springer Verlag.

Australian Radiation Protection and Nuclear Safety Agency [ARPANSA] (2012). *Holistic Safety Sample Questions v1 (OS-LA-SUP-240V v1)*. Canberra, Australia: Australian Government.

Baumard, P., and Starbuck, W. H. (2005). Learning from failures: Why it may not happen. *Long Range Planning*, 38: 281–298.

Dekker, S. W. A., and Woods, D. D. (1999). To intervene or not to intervene: The dilemma of management by exception. *Cognition, Technology and Work*, 1(2), 86–96.

Fairbanks, R. J., Perry, S., Bond, W., and Wears, R. L. (2013). Separating resilience from success. In: E. Hollnagel, J. Braithwaite, and R. L. Wears (Eds.). *Resilient Health Care.* Franham, UK: Ashgate.

Folke, C. (2006). Resilience: The emergence of a perspective for social-ecological systems analyses. *Global Environmental Change*, 16, 253–267.

Hale, A. R., and Hovden, J. (1998). Management and culture: the third age of safety. A review of approaches to organizational aspects of safety, health and environment. In A. M. Feyer, and A. Williamson (Eds.), *Occupational Injury. Risk Prevention and Intervention.* London: Taylor & Francis.

Hamel, G., and Valikangas, L. (2003). The quest for resilience. *Harvard Business Review*, 81(9), 52–65.

Holling, C. S. (1973). Resilience and stability of ecological systems. *Annual Review of Ecology and Systematics*, 4, 1–23.

Hollnagel, E. (2006). Resilience: The challenge of the unstable. In: E. Hollnagel, D. D. Woods, and N. C. Leveson (Eds.), *Resilience Engineering: Concepts and Precepts* (p. 9–18). Aldershot, UK: Ashgate.

Hollnagel, E. (2009). The four cornerstones of resilience engineering. In: C. P. Nemeth, E. Hollnagel, and S. W. A. Dekker (Eds.), *Preparation and Restoration* (pp. 117–134). Aldershot, UK: Ashgate.

Hollnagel, E. (2011a). Epilogue: RAG—The resilience analysis grid. In E. Hollnagel, J. Pariés, D. D. Woods, and J. Wreathall (Eds.). *Resilience Engineering in Practice.* Farnham, UK: Ashgate.

Hollnagel, E. (2011b). Prologue: The scope of resilience engineering. In E. Hollnagel, J. Pariés, D. D. Woods and J. Wreathall (Eds.). *Resilience Engineering in Practice.* Farnham, UK: Ashgate.

Hollnagel, E. (2013). *FRAM: The Functional Resonance Analysis Method for Modelling Complex Socio-Technical Systems.* Farnham, UK: Ashgate.

Hollnagel, E. (2014). *Safety-I and Safety-II: The Past and Future of Safety Management.* Farnham, UK: Ashgate.

Hollnagel, E., Woods D. D., and Leveson N. C. (Eds.). (2006). *Resilience Engineering: Concepts and Precepts.* Aldershot, UK: Ashgate.

Mach, E. (1917). *Erkenntnis und Irrtum. Skizzen zur Psychologie der Forschung. Dritte Durchgesehene Auflage.* Leipzig, Germany: Johann Ambrosius Barth.

McAslan, A. (2010). *The Concept of Resilience. Understanding its Origins, Meaning and Utility.* Adelaide, Australia: The Torrens Resilience Institute.

Merton, R. K. (1936). The unanticipated consequences of purposive social action. *American Sociological Review*, 1(6), 894–904.

Pavard, B., Dugdale, J., Saoud, N. B. B., Darcy, S., and Salembier, P. (2008). Underlying concepts in robustness and resilience and their use in designing socio-technical systems. In: E. Hollnagel, C. P. Nemeth, and S. W. A. Dekker (Eds.) *Resilience Engineering Perspectives: Remaining Sensitive to the Possibility of Failure* (Vol. 1). Ashgate Publishing, Ltd.

Thomas, W. J. W. (2012). *A Systematic Review of the Effectiveness of Safety Management Systems.* Canberra, Australia: Australian Transport Safety Bureau.

Tisseron, S. (2007). *La Resilience*. Paris: PUF.

Westrum, R. (2006). A typology of resilience situations. In E. Hollnagel, D. D. Woods, and N. G. Leveson (Eds.), *Resilience Engineering: Concepts and Precepts*. Aldershot, UK: Ashgate.

Woods, D. D. (2000). Designing for resilience in the face of change and surprise: Creating safety under pressure. Plenary Talk, Design for Safety Workshop, NASA Ames Research Center, October 10.

Woods, D. D., and Wreathall, J. (2003). *Managing Risk Proactively: The Emergence of Resilience Engineering*. Columbus, OH: Institute for Ergonomics, The Ohio State University.

Zwetsloot, G. I. J. M, Aaltonen, M., Wybo, J.-L., Saari, J., Kines, P., and Op De Beeck, R. (2013). The case for research into the zero accident vision. *Safety Science*, 58, 41–48.

4

DEFENSE-IN-DEPTH

Jan-Erik Holmberg

4.1 INTRODUCTION

Defense-in-depth is historically a military strategy where instead of defending in a single strong line, the forces are distributed into several layers so that a single attack cannot destroy the defensive forces. In the second and third century, the Romans developed a mobile military system to meet invaders within the Roman territory (Luttwak, 1976).

In today's civil sector safety management applications, defense-in-depth is understood quite differently from a military strategy. Defense-in-depth is a widely applied safety principle in practically all safety-critical technological areas even if different terms are used in some contexts, for example, "lines" or "layers of protection."

In a safety management context, defense-in-depth means having more than one protective measure for a given safety objective, such that the objective is achieved even if one of the protective measures fails. The protective measures can be anything from inherent safety features; use of multiple barriers; engineered safety features; principles and procedures followed in design, construction, operation, maintenance

Handbook of Safety Principles, First Edition. Edited by Niklas Möller, Sven Ove Hansson, Jan-Erik Holmberg, and Carl Rollenhagen.

and decommission of the system. In this sense, defense-in-depth covers more or less everything which affects safety.

Fleming and Silady (2002) distinguish three usages of the term defense-in-depth:

1. *Design Defense-in-Depth*. Design feature to have multiple and physical lines of defense between the hazard and the public.
2. *Process Defense-in-Depth*. Incorporation of the defense-in-depth thinking into the licensing requirements. Although there is a relationship between these requirements and the detailed design features that are reflected in design defense-in-depth, they are not one and the same as they are controlled by different stakeholders in the process.
3. *Scenario Defense-in-Depth*. Strategies to prevent initiating events from occurring and from progressing to accidents, and strategies to mitigate the consequences of events and accidents.

In the following discussion, we focus on usages 1 and 3 and pay less attention to usage 2 since it is discussed in other chapters of the handbook (risk-informed regulation, standards, and safety management principles). Several application areas are discussed, providing a variety of quite different interpretations of defense-in-depth yet sharing the main principles, such as multiple barriers, accident prevention and mitigation and safety/security regulation, polices, and standards.

4.2 UNDERLYING THEORY AND THEORETICAL ASSUMPTIONS

4.2.1 Definitions and Terminology

Defense-in-depth is based on an accident model consisting of initiating events and the following event sequence where depending on the success or failure of barriers, the end state is an accident or a safe state.

Initiating events can be called threats or hazards. Types of initiating events can vary from technical system internal failures, external events to malicious actions. Defense-in-depth strategy relies on the identification of the set of relevant initiating events against which barriers are needed. Set of considered initiating events can be called the design basis (of defense-in-depth).

Barriers can be called lines of defenses, protective layers, safety functions or safety systems, and they should not be understood only as physical hinders. In this chapter, the general term barrier is used for any protective measure. Barriers are further discussed in Chapter 5 of the handbook.

Accident is a negative outcome from the point of view of human beings, environment, or economy of an enterprise. Accidents can be further categorized into several classes depending on the severity of the consequence, which is usually related to the number of failed barriers.

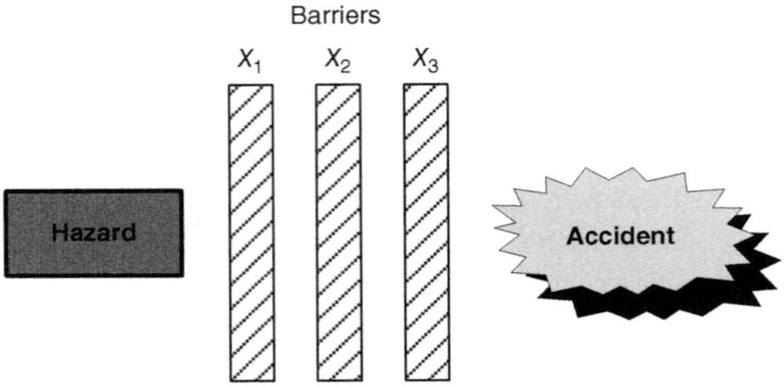

Figure 4.1. Defense-in-depth as a design of multiple barriers.

Figure 4.1 illustrates a simple defense-in-depth solution consisting of three barriers.

4.3 REDUNDANCY, DIVERSITY, AND SEPARATION PRINCIPLES

Redundancy is duplication or generally multiplication of components of a system to increase the reliability of the system. Redundancy is especially effective if redundant components do not have dependencies, such as common support system, common maintenance, or common environment. Dependencies can be avoided by introducing technological diversity between the components, by physical separation, and by functional isolation (no common support systems).

Probabilistically, the effect of redundancy can be represented as follows. Let a protection system consist of n barriers $X_i, i = 1, \ldots, n$, each of which has a failure probability p_i. Given that they form a serial system, the failure probability of the system is

$$P(S) = P(X_1 \cap \ldots \cap X_n), \tag{4.1}$$

where $P(S)$ is the system failure probability and $P(X_1 \cap \ldots \cap X_n)$ is the probability that all the barriers $X_1, \ldots, X_n$ fail.

If the barriers are independent, the system failure probability is a product of the failure probabilities of the barriers

$$P(S) = P(X_1) \cdot \ldots \cdot P(X_n). \tag{4.2}$$

Evidently, under the condition of independent barriers, the system failure probability decreases when barriers are added in the protection system. In reality, it can be hard to design fully independent barriers, and then the following relationship will hold.

$$P(S) = P(X_1 \cap \ldots \cap X_n) > P(X_1) \cdot \ldots \cdot P(X_n). \quad (4.3)$$

To have independent barriers is thus a desired property of defense-in-depth.

It should be noted that the redundancy principle is applied in defense-in-depth on two dimensions. To have several barriers is a kind of overall redundancy principle, and on that dimension, the diversity principle is followed as much as reasonably possible. On the other hand, redundancy principle can be applied for each barrier to increase its reliability. Within each barrier, diversity is not necessarily required.

Diversity or diverse redundancy is a central principle to reduce the influence of common cause failures. It uses different technology, design, manufacture, software, etc. Depending on the type of safety function, there are different possibilities to achieve diversity. Confinement type of purely passive physical structures can be diversified by building several different types of consecutive layers, which is a strategy, for example, spent nuclear fuel final repository. Electromechanical safety systems depend on power supply and I&C, which considerably limits the degree of diversity that can be achieved. One strategy could be to combine electromechanical safety system barrier with some type of passive system. Otherwise, we may not be able to claim full diversity.

Physical separation principle comes from the avoidance of spatial dependences between barriers, such as hazards that can destroy several redundancies at the same time. Fire, flooding due to leaking fluid systems, and missiles from breaking structures are examples of such hazards. Physical separation is an especially meaningful principle for barriers which are electromechanical systems.

There are several strategies for how to implement physical separation. Figure 4.2 shows three basic alternatives. In the first case, barrier systems at different levels of defense-in-depth are separated from each other. For instance, it is a common requirement that non-safety (operational systems) are physically separated from safety systems (to the extent reasonably possible). In the second case, redundancies within each barrier level are separated but different barriers are not separated. The logic is that even if a hazard can eliminate one redundancy at all levels of defense-in-depth, some defense-in-depth and diversity still remain. In the third case, both redundancies and barrier levels are separated from each other. This is the most effective way of separation, but it is also the most expensive and practically can be difficult to build.

Some remarks need to be made in relation to Figure 4.2. First, the "separation wall" should not be understood as a defense-in-depth barrier. It is a separation barrier that ensures independence between different parts of the defense-in-depth structure. Second, "separation wall" needs not to be an actual wall. Sufficient separation can be achieved, for example, by distance.

Third, the subsystems of the defense-in-depth structure (such as the A redundancy 1, A red. 2, B red. 1, B red. 2 of Figure 4.2) are often wide entities, which

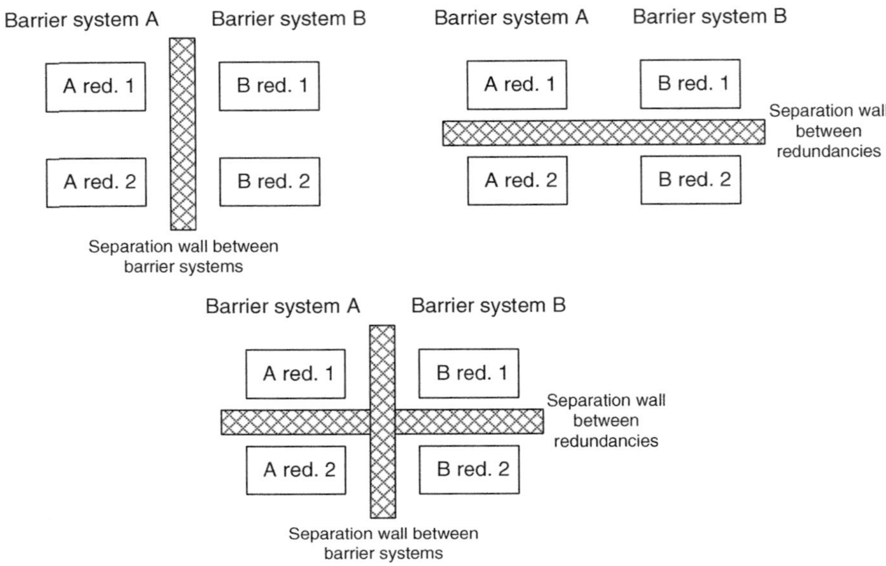

Figure 4.2. Implementation strategies for physical separation between barrier systems and their redundant subsystems.

makes the full physical separation difficult. A whole system function may consist of actuators (pumps, valves, etc.), I&C (processors, communication network), power supply (cables, power supply source), measurements for input signals, interface for operators, and so on. In practice, separation—like diversity—can only be partially achieved.

4.3.1 Principle of Successive Barriers and Reducing Consequences

In defense-in-depth, each barrier is effective in a certain order, that is, they are successively challenged by the threat. In nuclear power plant context, the orders are also called "levels" of defense-in-depth.

Figure 4.3 presents a simple event tree for such a design. The barrier X_1 is first challenged. If X_1 fails, there is some unwanted consequence, $C1$. If the second barrier X_2 fails, the consequence is $C2 > C1$, and so on.

The risk model of Figure 4.3 can be translated into a complementary cumulative probability function for the consequences, as illustrated in Figure 4.4. Frequency of an accident is used as the risk metric, which is a product of the initiating event frequency, F_I, and the failure probability of one or more barriers, $P(X_1 \cap \ldots \cap X_n)$. In this simple example, we assume that there is no uncertainty about the consequences, given the failure of certain numbers of barriers.

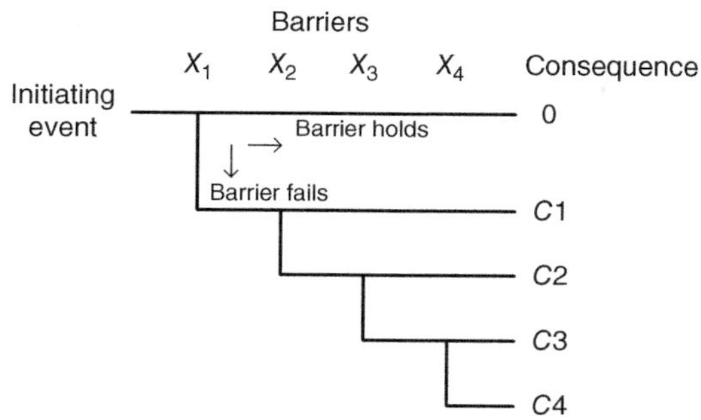

Figure 4.3. Event tree for a barrier system.

Principle of reducing consequences is associated with the usual property of defense-in-depth that each barrier has a different functional meaning. This principle is related to the requirement to have a barrier with respect to each consequence category. Probabilistically, numerical risk criteria are defined for various consequence categories (see Section 4.3.5).

4.3.2 Principle of Accident Prevention and Mitigation

Principle of accident prevention and mitigation is a developed form of the principle of reducing consequences. It is based on an accident model, where we distinguish between events and barriers before an accident and events and barriers after an accident. If the accident prevention succeeds, the event is only an incident (near-miss). Otherwise, an accident happens, and the effectiveness of accident mitigation barrier determines the level of consequences. Figure 4.5 illustrates an event tree for such a

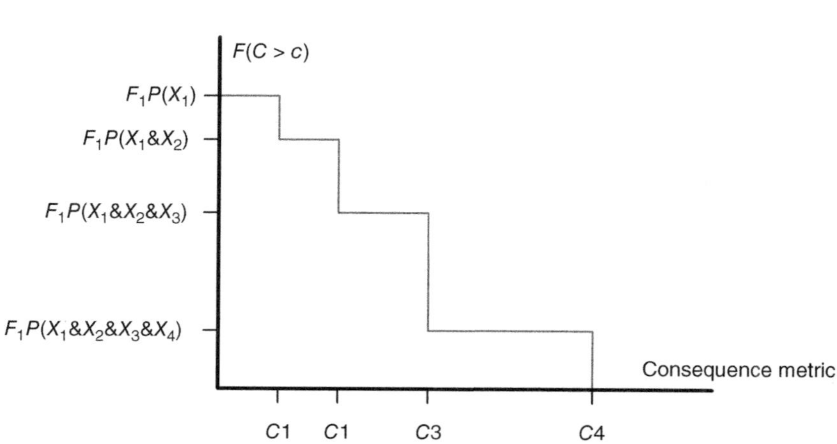

Figure 4.4. Complementary cumulative function of the consequences.

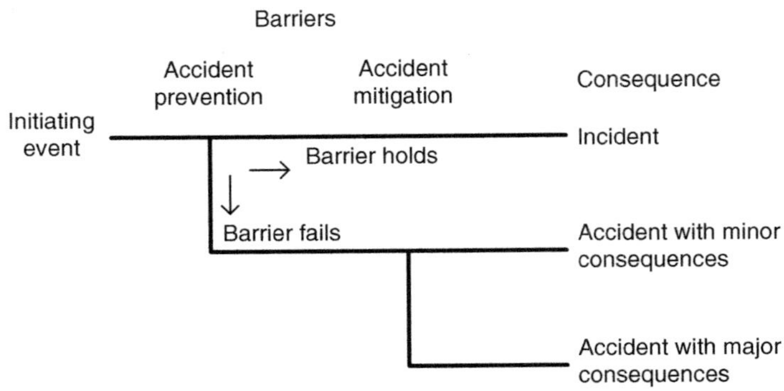

Figure 4.5. Event tree for an accident prevention and mitigation system.

defense-in-depth strategy. Bow-tie method is an alternative way of representing accident prevention barriers on the left-hand side and mitigation barriers on the right-hand side (Salvi and Debray, 2006); see Figure 4.6.

The meaning of accident respectively incident is application-specific and there can be ambiguity in the terminology. For instance, in the nuclear power plant safety nomenclature, an event is an "accident" if it involves a radioactive release (IAEA, 2009). Less severe events are called anomalies or "incidents."

The defense-in-depth principle requires that there are both accident prevention and mitigation systems, and that there is sufficient independence between them. Accident prevention and mitigation systems may have different technological and reliability requirements.

Figure 4.6 illustrates the scope of three industries: nuclear, railway, and offshore oil and gas. Nuclear power plant requirements cover both accident prevention and

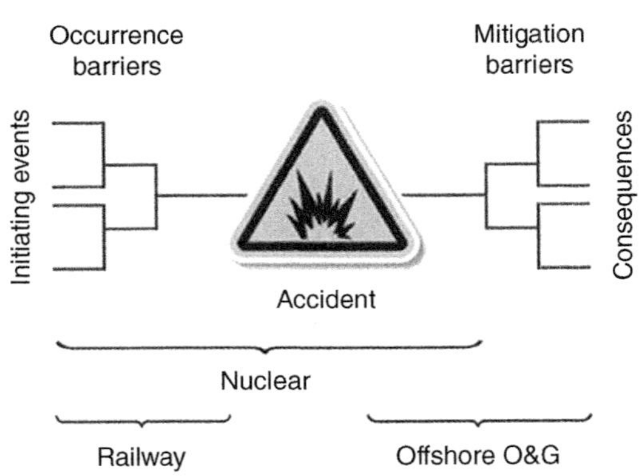

Figure 4.6. Scope of risk criteria and safety requirements in nuclear, offshore oil and gas, and railway industries.

mitigation barriers. In the offshore oil and gas sector, the number of accidents requiring mitigation is much higher than in the nuclear sector, resulting in high focus on consequence mitigation. In the railway sector, requirements are defined for the accident prevention systems (Bengtsson et al., 2011).

4.3.3 Classification of Barriers

Defense-in-depth is more than having multiple barriers even if a redundancy is one of the central principles. In defense-in-depth, the barriers usually have different functional roles, that is, we distinguish between accident prevention and accident mitigation barriers. In nuclear power plants, five levels of barriers are considered (IAEA, 1996).

Besides distinguishing between accident prevention and mitigation barriers, the barriers can be classified according to the implementation principle such as degree or levels of passivity or degree of automation. In many industrial applications, requirements are set for the implementation principle of different barriers. For some applications, a clearly physical, passive barrier is required. For others, automated protection system is required. There are also design rules when and when not manual intervention can be or must be counted on.

Degree of passivity depends on the need of a system or component to have any external input to operate (IAEA, 1991). If a component is not passive, it is necessarily an active one. The concept of passivity can be considered in terms of several categories. IAEA has defined the following four categories for passive systems:

1. Category A
   - No signal inputs, external power sources or forces
   - No moving mechanical parts
   - No moving working fluids

   For example, physical barriers against the release of fission products
2. Category B
   - No signal inputs, external power sources or forces
   - No moving mechanical parts
   - Moving working fluids, which are due to thermal-hydraulic conditions occurring when the safety function is activated

   For example, emergency cooling systems based on air or water natural circulation in heat exchangers immersed in water pools
3. Category C
   - No signal inputs, external power sources or forces
   - Moving mechanical parts, whether or not moving working fluids are also present

For example, check valves and spring-loaded relief valves

4. Category D: Intermediary zone between active and passive
   - Execution of the safety function is made through passive methods
   - External signal is permitted to trigger the passive process
   - "Passive execution/active initiation"
   - Energy only from stored sources
   - Active components limited to controls, instruments, and valves

   For example, emergency core cooling systems based on gravity-driven flow of water, activated by valves which break open on demand.

A system can have passive and active characteristics at different times. For example, the active opening of a valve initiates subsequent passive operation by natural convection.

Levels of automation have similarities with the levels of passivity. As an example, German Federal Highway Research Institute BASt-Expert-Group has defined five vehicle automation degrees (BASt, 2013):

- *Full Automation*. The system takes over longitudinal and lateral control completely and permanently. In case of a take-over request that is not followed, the system will return to the minimal risk condition by itself.
- *High Automation*. The system takes over longitudinal and lateral control; the driver is no longer required to permanently monitor the system. In case of a take-over request, the driver must take over control with a certain time.
- *Partial Automation*. The system takes over longitudinal and lateral control, the driver shall permanently monitor the system and shall be prepared to take over control at any time.
- *Driver Assistance*. The driver permanently controls either longitudinal or lateral control. The other task can be automated to a certain extent by the assistance system.
- *Driver Only*. Human driver executes manual driving task.

4.3.4 Safety Classification

Safety classification of system, structures, and components is a means to manage the quality requirements of items of a safety critical installation such as a nuclear power plant. Safety classification systems are dependent on the industrial field and national regulation. In many areas, the risk-assessment based classification of the standard IEC 61508 (IEC, 2010) is followed with industry-specific adaptions. The classification scheme is called safety integrity levels (SIL), which are derived from a risk assessment. Four safety integrity levels are defined and for each level are defined target failure probabilities (on demand systems) or target failure rates (continuous operation). SIL also implies requirement to apply redundancy and diversity in the design.

In the nuclear field, there is a national variation in the safety classification. For example, the International Electrotechnical Commission categorization (IEC, 2009) defines three safety categories A, B, and C, where the American standards of Institute of Electrical and Electronics Engineers uses classification process that only distinguishes between safety and non-safety systems (IEEE, 2003). The IAEA has generally adopted a distinction between safety systems, safety related systems, and non-safety systems (IAEA, 2002).

The relationship between defense-in-depth and safety classification is immediate in the sense that each barrier is assigned with a safety class, and all items belonging to one barrier have the same safety class (at least in the nuclear field applications). It follows that all items within one barrier have the same requirements for design, qualification, regulatory review, and QA procedures during all life cycle phases. Barriers can belong to different safety classes, and, in fact, use of different safety classes is considered beneficial both from the diversity point of view and from the optimal resource allocation point of view (graded QA).

4.3.5 Overall Safety Goals and Risk Acceptance Criteria vs. Defense-in-Depth

Safety goals and related risk acceptance criteria are overall qualitative and numerical targets given by the society or specified in the regulation for safety-critical installations. Qualitative target can be given, like "individual members of the public should be provided a level of protection from the consequences of nuclear power plant operation such that individuals bear no significant additional risk to life and health" (U.S.NRC, 1986). Numerical target can be, for example, that the individual fatality risk shall be less than 10^{-5} per year (Trbojevic, 2005)

Defense-in-depth is an implementation to achieve the goal. Numerical risk targets can be set for the whole defense-in-depth system, like in United Kingdom and the Netherlands (Bengtsson et al., 2011). In many cases, subsidiary or surrogate goals are used instead or in addition. A "subsidiary criterion" is a criterion on a lower technical level to assess in a simplified way the consequences on a higher level. The highest level is the consequences for the society or individual persons.

Figure 4.7 describes how risk criteria can be defined for nuclear power plants compared to the levels of defense-in-depth and probabilistic safety assessment (PSA). In many countries, societal or individual risk criteria are not applied, but numerical criteria are defined in terms of the core damage frequency and large release frequency/large early release frequency. These may be interpreted as subsidiary criteria for risk of offsite consequences in countries where level 3 PSA is not required. Subsidiary criteria are preferred due to the uncertainties in the risk assessment of offsite consequences (e.g., societal and individual risk) and that they

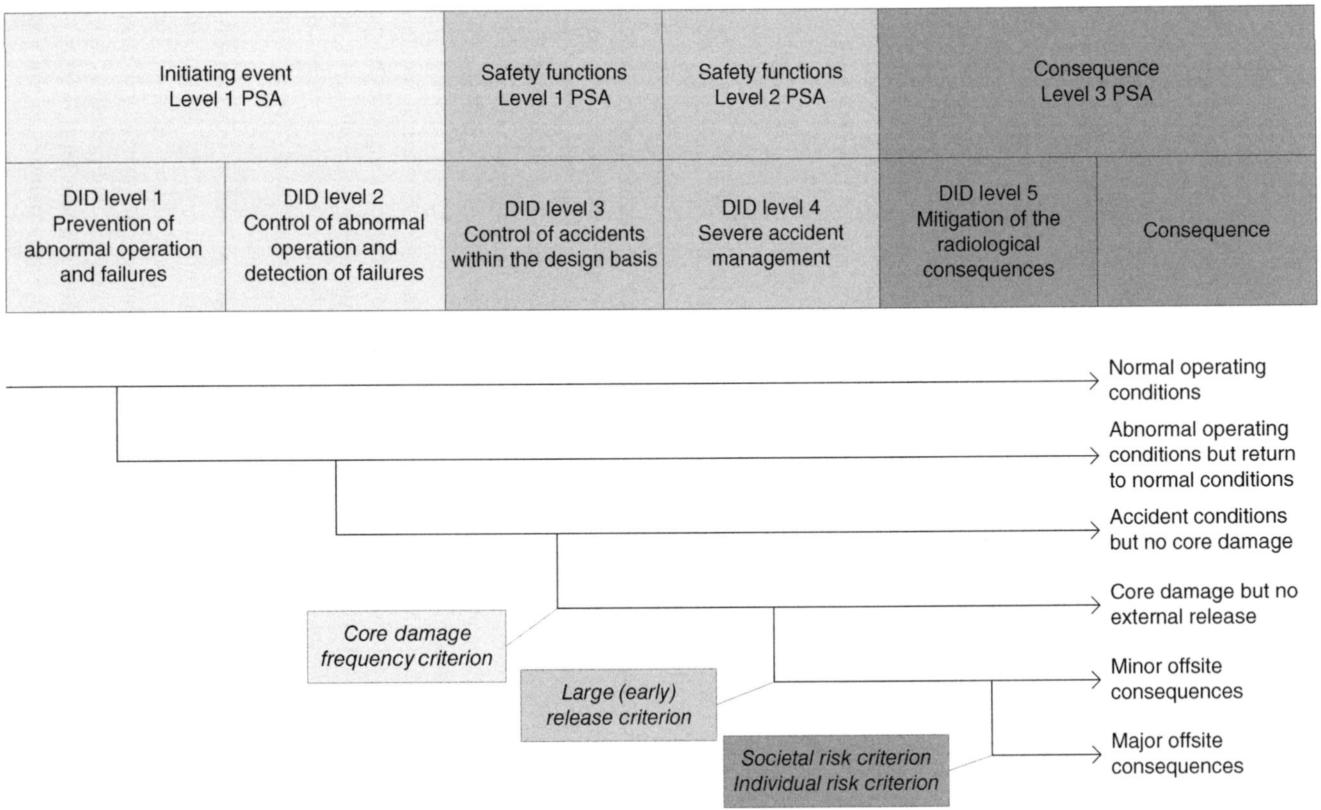

Figure 4.7. Levels of defense-in-depth for nuclear power plants, probabilistic safety assessment (PSA) and associated risk criteria.

explicitly put focus on defense-in-depth, in particular attention is paid to the accident prevention and mitigation.

4.4 USE AND IMPLEMENTATION

Principles of defense-in-depth strategy are followed everywhere in industrial safety management, that is, to focus on accident prevention and mitigation, to have several (different) barriers against accident, and to have a process of implementing requirements into regulation, design, and operation. Terminology is industry-dependent and also, the type of defensive measures depends on the type of threats. For nuclear safety, defense-in-depth is one of the most fundamental safety principles. For chemical plants, safety principles are similar. Information technology security has also adopted the same term, but barrier thinking is somewhat different due to different threats. In railway safety, focus is on accident prevention, where several barriers are required. Automotive industry applies both accident prevention and mitigation features calling them active versus passive safety (systems).

4.4.1 Nuclear Power Plant Safety

Defense-in-depth is perhaps most developed as a concept in the nuclear industry, where it is the cornerstone of the safety philosophy. The IAEA safety guide INSAG-10 (IAEA, 1996) defines defense-in-depth as follows: "A hierarchical deployment of different levels of diverse equipment and procedures to prevent the escalation of anticipated operational occurrences and to maintain the effectiveness of physical barriers placed between a radiation source or radioactive material and workers, members of the public or the environment, in operational states and, for some barriers, in accident conditions."

The objectives of defense-in-depth are

- To compensate for potential human and component failures
- To maintain the effectiveness of the barriers by averting damage to the facility and to the barriers themselves
- To protect workers, members of the public, and the environment from harm in accident conditions in the event that these barriers are not fully effective

Defense-in-depth is usually described in two ways for nuclear safety (IAEA, 1999):

- A system of successive physical barriers isolating the radioactive fuel from the environment (fuel matrix, fuel rod cladding, primary coolant boundary, reactor confinement)

- A system of successive levels of protection following the logic of the accident model
  - Level 1: Prevention of abnormal operation and failures
  - Level 2: Control of abnormal operation and detection of failures
  - Level 3: Control of accidents within the design basis
  - Level 4: Control of severe plant conditions, including prevention of accident progression and mitigation of the consequences of severe accidents
  - Level 5: Mitigation of radiological consequences of significant releases of radioactive material

In the safety analyses, it shall be demonstrated that for an identified set of initiating events:

- Deterministically, unwanted consequence is avoided even if one or more failures are postulated in the barriers. Failure criteria depend on the frequency of the (postulated) initiating event.
- Probabilistically, the frequency of the unwanted consequence is less than the numerical criterion (see Figure 4.7). Probabilistic criteria may be applied to individual levels, for example, for the reactor containment as a kind of conditional failure probability criterion.

The defense-in-depth concept is in practice incorporated everywhere in the safety management of nuclear power plant, like regulatory requirements and oversight procedures, QA activities at various system life cycle phases, safety classification of systems, structures and components, organizational structure and responsibilities and safety culture considerations. As Fleming and Silady (2002) point out, this all makes defense-in-depth a multi-faceted framework.

It should finally be noted that while the IAEA definitions can be considered generally accepted and followed in all countries, nuclear regulatory requirements differ between countries. Above all, the requirements for new reactors are much harder than they were for older reactors, see, for example, WENRA (2013). One new feature is that the defense-in-depth level 3 is divided into sublevels 3a and 3b, where 3b corresponds to additional safety features to cope with common cause failure in level 3a safety systems, which is a so-called design extension condition to be considered.

4.4.2 Chemical Industry

The safety philosophies of chemical plants and nuclear power are similar, due to similarity of threats; there are concentrated energy sources and process materials that under abnormal conditions can be harmful to the environment, people, and process equipment. Their analogous barriers (cf. levels of defense-in-depth at nuclear power

TABLE 4.1. Protection Layers in Chemical Process Safety

1	Process design
2	Basic process control system (BPCS), supplemented by operator supervision
3	Critical alarms and human intervention (operator-initiated corrective actions)
4	Safety instrumentation
5	Physical protection, for example, overpressure protection
6	Post-release physical protection, a dike for liquid spills
7	Emergency response at the plant and in the community

plant) can be found in the concept, called usually protection layers (CCPS, 2010), see Table 4.1.

4.4.3 Information Technology Security

In the area of information technology (IT), defense-in-depth is understood as an information assurance strategy (NSA, 2012). The purpose is to resist attacks against IT systems. Potential adversaries might include: nations, terrorists, criminals, hackers, or corporate competitors. Their motivations may include intelligence gathering, theft of intellectual property, denial of service, embarrassment, or just pride in exploiting a notable target. Classes of attack may include passive monitoring of communications, active network attacks, close-in attacks, exploitation of insiders, and attacks through the industry providers of one's IT resources.

In the simplest form, defense-in-depth means use of multiple complementary security products, so that a failure in one does not mean total insecurity. By using a multi-layered approach the likelihood of protecting confidential and highly sensitive information is increased. Breaches in such security measures—which do occur—can be more easily and efficiently detected in earlier stages. The organization then has ample time to effectively react to the threat, stopping it before it causes irreparable damages or losses.

The application of these services is based on the "Protect, Detect, and React" paradigm. This means that in addition to incorporating protection mechanisms, organizations need to expect attacks and include attack detection tools and procedures that allow them to react to and recover from these attacks (NSA, 2012). In this way, both preventive and mitigating barriers are considered.

An important aspect of the defense-in-depth principle is to achieve balance between people, technology, and operations (NSA, 2012). Responsible people must be aware of threats and follow the policies and procedures. Technologies provide the information assurance services. It is a "best practices" strategy in that it relies on the intelligent application of techniques and technologies that exist today. The strategy recommends a balance between the protection capability and cost, performance, and operational considerations. When more than one protective measure is applied for a given safety objective, the objective is achieved even if one of the protective measures fails. Operations focus on activities required to sustain an organization's security posture on a day-to-day basis.

The multi-layered approach includes features such as (US-CERT, 2009):

1. IT system architectural zoning
2. Firewalls
3. Intrusion detection systems
4. Security policies and procedures

4.4.4 Railway Safety

Railway safety is regulated by both national and international regulation (EU, 2014). In the European Union, there is EU Railway Safety Directive industry standards (e.g., EN 50129) as well as national rules, and the long-term intention is to harmonize them due to the fact to make it possible for the trains to cross national borders.

The Norwegian safety principles constitute a representative example of the implementation of defense-in-depth strategy. Here, there are six barriers against train collisions involving passenger trains on electric train lines (Elvik and Voll, 2013):

1. *Signals and Track Lock*. A lock is an automatic operation based on signalling whether a track is clear for operation (green vs. red light).
2. *Departure Routine*. This is a double check by train driver and conductor that the train can depart from the station.
3. *Traffic Management*. Violation of a stop signal is likely observed by traffic management.
4. *Automatic Train Control*. This is an automatic operation applied in electric lines and when activated will brake the train into a controlled stop.
5. *Direct Communication*. If the barrier four is failed, traffic managers can directly communicate with train drivers.
6. *Power-Cut*. If the train driver does not respond, traffic managers can turn off power.

All these barriers are accident prevention means, which is also the focus area of the safety regulation in railway safety. It could be also noted that there are clear dependences with the barriers (via signalling system, actions of train driver, and role of the traffic management). It illustrates the common feature that barriers are seldom fully independent of each other. In particular, railways is an interesting example of implementation of defense-in-depth when looking from the railway network perspective: a system of multiple trains, multiple actors, and geographically distributed. These features make the railway defense-in-depth system especially multidimensional and complex compared to industrial plant installations.

4.4.5 Automobile Safety

Automobile safety solutions can be divided into active and passive features. Active safety is used to refer to technology assisting in the prevention of a crash and passive safety to components of the vehicle (primarily airbags, seatbelts, and the physical structure of the vehicle) that help to protect occupants during a crash (https://en.wikipedia.org/wiki/Automobile_safety).

The division between active and passive safety is equivalent to the defense-in-depth principle of having preventive and mitigating barriers. It should be, however, noted that car driving safety is not only limited to the safety features of vehicles. The safety strategy is much more complex and includes also all road safety related aspects such as legislation, traffic control, safety campaigns, car inspections, accident investigations, industry standard work, etc. In this sense, automobile safety shares many similar features as railway safety.

4.5 EMPIRICAL RESEARCH ON USE AND EFFICIENCY

Since defense-in-depth is an overwhelming principle, it is hard to study the use and efficiency of it as a whole. On the other hand, all accident and incident statistics studies as well as individual event examinations provide insights on effectiveness of individual barriers as well as examples of cases where multiple barriers have failed. Also, trends in accident statistics can indicate improvements of implementation of safety barriers. Operating experience feedback is discussed in Chapter 7 of the handbook.

4.6 WEAKNESSES, LIMITATIONS, AND CRITICISM

Defense-in-depth is a powerful design against hazards as long as barriers are intact. In reality, the barriers have defects, which make the barrier system resemble a Swiss cheese having many holes. According to the Swiss cheese model, an accident happens when the holes of barriers in many layers momentarily line up to permit a trajectory of accident opportunity—enabling hazards to propagate into losses (Reason, 2000); see Figure 4.8.

Following the Swiss cheese model, we can see that there are two ways of minimizing the accident probability: (1) minimizing the likelihood of holes and hazards and (2) minimizing the dependencies between hazards and holes being in the same trajectory, that is, minimizing the dependencies between them.

The first target includes the usual design and operational challenge for socio-technological systems, that is, how to avoid active failures and latent conditions making the system unreliable (Reason, 2000). For both types of failures, any measure to minimize the failure probability or rate is relevant and there are a number of activities which need to be accounted during the life cycle of the system (specification, design, manufacturing, commissioning, operation, maintenance).

Figure 4.8. Swiss cheese model.

To avoid latent conditions, the question is how to detect them. Most problematic cases are design errors and erroneous or unintended system modifications. In those cases, the normal, planned methods to detect the availability of the barrier may be bypassed.

Regarding active failures, the system should be designed to tolerate single failures. This can often be achieved by introducing a transition to safe-state, building redundancy in the system (switch-over is made in case of failure), having an alarm system for operators including procedures to cope with the situation in the required time frame, and so on. Usually active failures are not considered as dangerous as latent conditions, but they may also include deceptive features, if the organization does not learn from repeating failures.

The target of having independent barriers implies a complexity challenge for the design since it is hard to avoid different types of dependences between the barriers. We may think if two barriers—one accident preventive and one mitigation system—are sufficient, we have solutions to design rather independent barriers. However, if we are required to have more barriers like in a nuclear power plant, we begin to be limited by space, physical separation means, material choices, technology providers, support system choices, maintenance and testing couplings, etc. The more barriers are required, it does not only lead to complex and expensive design, but there are practical limitations on how independent barriers can be.

Complexity of defense-in-depth strategy also leads to the question of balancing between safety/security and availability. Defense-in-depth implies thus a cost factor. Without a proper risk assessment, cost-ineffective barriers may be implemented.

Defense-in-depth may be also criticized for the fact that increasing protection does not always increase safety. Besnard and Hollnagel (2012) explain that increasing protection affects the perceived risk exposure, which affects the behavior so that performance efficiency is increased while keeping the perceived level of risk constant.

This phenomenon, which has been noticed, for example, in car-accident statistics, is called risk homeostasis hypothesis. Anti-lock braking system may tend to make people drive more aggressively. The lesson of the examples is that one should carefully consider both the intended and the unintended effects of implementing protection in sociotechnical systems.

A related feature of a multi-layer protection system is that it may be difficult to notice a local violation of a barrier since it does not necessarily have any immediate, visible effect. In systems designed according to the defense-in-depth strategy, the defenses are likely to degenerate systematically through time, when pressure toward cost-effectiveness is dominating (Rasmussen, 1997).

Effective preventive barriers may also lead to less experience feedback from management of incidents and accidents. From a learning and alertness point of view, it might be better that people get more experience on how to handle dangerous situations.

Defense-in-depth principle can be characterized as a deterministic and qualitative principle, which explicitly does not take into account probabilistic reasoning. It is a strict requirement to have each barrier. Weakness of one barrier cannot be compensated by making another barrier stronger. A nuclear power plant must have a reactor containment building regardless of the probability for a core damage accident. The reason for this is the underlying uncertainty that even a strongest barrier may fail in unexceptional conditions. It is hard to prove that one barrier is extremely reliable. However, from a risk decision-making point of view, it may be non-optimal to require a multi-barrier system rather than a single strong barrier.

4.7 RELATIONS TO OTHER SAFETY PRINCIPLES

Defense-in-depth obviously includes many other principles or vice versa other principles commonly refer to or rely on defense-in-depth, such as:

- *Safety Barriers*. Defense-in-depth relies on having more than one barrier preventing from unwanted consequences. Meaning of barrier is context- and application-dependent.
- *Redundancy*. Redundancy is a principle to duplicate critical system parts to increase the reliability of a system. Redundancy is applied in defense-in-depth both in the sense of having multiple barriers and of strengthening individual barriers.
- *Diversity*. Diversity between barriers is needed to avoid dependences between them, which is the main potential weakness of defense-in-depth. Whether or not sufficient diversity can be achieved may be hard to demonstrate.
- *Fail-Safe Design*. Defense-in-depth relies on barriers being intact. In case of detected defects in the barriers, the system should be directed to a safer state.

Therefore, fail-safe design is usually required for the safety systems, for example, loss of critical sensors actuates the system trip.

- *Robustness and Resilience*. Robustness and resilience are desired properties for barriers to cope with a large variety of threats, which can be difficult to fully anticipate.
- *Safety Assessment*. Defense-in-depth design is based on a safety assessment, where the potential threats are identified and where the sufficiency of the barriers is demonstrated against (regulatory) requirements. See nuclear power plant safety as an example.

REFERENCES

BASt. (2013). Legal consequences of an increase in vehicle automation. Consolidated final report of the project group. Part 1, Bundesanstalt für Straßenwesen, BASt-Report F83.

Bengtsson, L., Holmberg, J.-E., Knochenhauer, M., and Rossi, J. (2011). Probabilistic safety goals for nuclear power plants: Phases 2–4, Final Report. NKS-226. Nordic Nuclear Safety Research (NKS), Roskilde.

Besnard, D., and Hollnagel, E. (2012). Some myths about industrial safety. CRC Technical Report.

CCPS. (2010). *Guidelines for Safe Automation of Chemical Processes*. New York: American Institute of Chemical Engineers.

CENELEC. (2010). Railway applications—Communication, signalling and processing systems—Safety related electronic systems for signalling. EN 50129. CENELEC, Brussels.

Elvik, R., and Voll, N. G. (2013). Challenges of improving safety in very safe transport systems. *Safety Science*, 63, 115–123.

EU. (2014). Railway safety performance in the European Union. [Online]. European Railway Agency, Valenciennes. Available at: http://www.era.europa.eu/Document-Register/Documents/SPR2014.pdf

Fleming, K. N., and Silady, F. A. (2002). A risk informed defence-in-depth framework for existing and advanced reactors. *Reliability Engineering and System Safety*, 78, 205–225.

IAEA. (1991). Safety related terms for advanced nuclear plants. IAEA-TECDOC-626. International Atomic Energy Agency, Vienna.

IAEA. (1996). Defence-in-depth in nuclear safety. INSAG-10. International Atomic Energy Agency, Vienna.

IAEA. (1999). Basic safety principles for in nuclear power plants, 75-INSAG-3 Rev. 1. INSAG-12. International Atomic Energy Agency, Vienna.

IAEA. (2002). Instrumentation and control systems important to safety in nuclear power plants. IAEA Safety Standards Series No. NS-G-1.3. International Atomic Energy Agency, Vienna.

IAEA. (2009). *INES: The International Nuclear and Radiological Event Scale User's Manual* (2008 Edition). Vienna: International Nuclear Energy Agency.

IEC. (2009). Nuclear power plants — Instrumentation and control important to safety — Classification of instrumentation and control functions. IEC 61226, ed. 3.0. International Electrotechnical Commission, Geneva.

IEC. (2010). Functional safety of electricalelectronicprogrammable electronic safety-related systems (EEPES). IEC 61508, ed. 2.0. International Electrotechnical Commission, Geneva.

IEEE. (2003). IEEE standard for qualifying Class 1E equipment for nuclear power generating stations. IEEE Std. 323-2003. Institute of Electrical and Electronics Engineers.

Luttwak. E. N. (1976). *The Grand Strategy of the Roman Empire*. London: The John Hopkins University Press.

NSA. (2012). Defense in depth: A practical strategy for achieving Information Assurance in today's highly networked environments. Available at: https:www.nsa.govia_filessupportdefenseindepth.pdf

Rasmussen, J. (1997). Risk management in a dynamic society: A modelling problem. *Safety Science*, 27(23), 183–213.

Reason, J. (2000). Human error: models and management, *British Medical Association*, 320(7237), 768–770.

Salvi, O, and Debray, B. (2006). A global view on ARAMIS, a risk assessment methodology for industries in the framework of the SEVESO II directive. *Journal of Hazardous Materials*, 30(3), 187–199.

Trbojevic, V. M. (2005). Risk criteria in EU. In: *Advances in Safety and Reliability—Proceedings of the European Safety and Reliability Conference (ESREL 2005), Tri City (Gdynia-Sopot-Gdansk), Poland 27–30 June 2005* (Two Volume Set). CRC Press.

US-CERT. (2009). Recommended practice: Improving industrial control systems cybersecurity with defense-in-depth strategies. U.S. Department of Homeland Security.

U.S. NRC. (1986). Safety goals for the operation of nuclear power plants. 10 CFR Part 50. U.S. Nuclear Regulatory Commission, Washington D.C.

WENRA. (2013). Safety of new NPP designs. Study by Reactor Harmonization Working Group (RHWG). Western European Nuclear Regulators Association.

FURTHER READING

Franks, A. (2003). Lines of defence/layers of protection analysis in the COMAH context. HSE Report, Amey VECTRA, Warrington, UK.

Harms-Ringdahl, L. (2009). Analysis of safety functions and barriers in accidents. *Safety Science*, 47, 353–363.

Hellström, P. (2015). DiD-PSA: Development of a framework for evaluation of the defence-in-depth with PSA. SSM Report 2015:04. Strålsäkerhetsmyndigheten, Stockholm.

IAEA. (2005). Assessment of defence in depth for nuclear power plants. Safety Reports Series 46. International Nuclear Energy Agency, Vienna.

Jackson, S., and Ferris, T. L. J. (2013). Resilience principles for engineered systems. *Systems Engineering*, 16(2), 152–164.

Reason, J. (1997). *Managing the Risks of Organizational Accidents*. Aldershot, UK: Ashgate Publishing.

Saleh, J. H., Marais, K. B., Francesca, K. B., and Favaró, F. M. (2014). Safety in the mining industry and the unfinished legacy of mining accidents: Safety levers and defense-in-depth for addressing mining hazards. *Journal of Loss Prevention in the Process Industries*, 29, 283–294.

Sklet, S. (2006). Safety barriers: Definition, classification, and performance. *Journal of Loss Prevention in the Process Industries*, 19(5), 494–506.

5

SAFETY BARRIERS

Lars Harms-Ringdahl and Carl Rollenhagen

5.1 INTRODUCTION

There are many ways to name, describe, and model safety features of a system. The concept of barrier is found in both risk analysis and in many generic accident models. It is also an integral part of the principles of defense-in-depth where the general principle is to provide several levels of defense against system failures. In this chapter we shall review and discuss definitions and principles associated with the concept of barrier and its relation to similar concepts (e.g., safety function, defense, and layers of protection).

In everyday language, the term barrier is commonly used as something that involves a hindrance for reaching some positive state of affairs. However, in a safety context, a barrier is associated with something that blocks or in other ways hinders a *negative state* to occur or become worse. A simple description is that a safety barrier is something that prevents an accident to occur and/or reduces its consequences.

The research literature concerning barriers is often related to high-risk fields such as nuclear energy and major hazard sectors of process industry, such as chemicals, oil, and gas. Barriers are also commonly discussed in other areas such as medical

Handbook of Safety Principles, First Edition. Edited by Niklas Möller, Sven Ove Hansson, Jan-Erik Holmberg, and Carl Rollenhagen.

services and common workplaces. There is thus a good potential of learning across branches disciplines, and we will have a broad perspective in mind.

5.1.1 Classical and Radical Definitions of Barriers

The terminology related to the barrier concept varies between different branches (nuclear, transportation, chemical, etc.) as well as its use in accident models and methods for safety management. The variation of ideas and concepts related to safety barriers is sometimes confusing, which may lead to problems in communication both within and between different domains. We can distinguish two general views on barriers:

- The classical conceptualization of barriers.
- The radical conceptualization of barriers.

The classical conceptualization is understood as active or passive artefacts (e.g., physical structures and systems) which can constrain, block, or impede hazards from realizing their negative potentials. The radical interpretation of barrier extends the classical barrier concept so that almost any function that has safety significance can be classified as being a barrier. This practice has led to some confusion about how barriers can be defined and classified (Rollenhagen, 2011).

5.1.2 Examples

The two examples below illustrate what barrier could be at a common workplace.

5.1.2.1 Example 1: Building Site Consider a building site where a large construction is built up! Construction sites represent a complex activity with many types of hazards and a relatively high accident rate compared with other industries. Examples of hazards are mowing machinery, falling objects, working at height, toxic substances, and contact with electricity. Consequently, a large repertoire of safety barriers is relevant on building sites.

There are physical obstacles to prevent harmful contact with energy and toxic substances, such as doors, railings, and a variety of protective clothing such as hard hats and gloves. This represents a classical concept of physical barriers. Safety barriers in the form of interlocks and passwords also fit in this category.

There are also many other safety features such as warning signs, symbols, work permits, instructions, rules, and a risk management system with safety rounds. These are examples of barriers based on a more generalized radical concept of barrier.

5.1.2.2 Example 2: Installation with Chemicals Consider a technical system with a number of chemicals! One of these is lye (sodium hydroxide) which can cause considerable corrosion injuries and other damages. In order to reduce the risks associated with lye a number of barriers are in place.

The basic physical barrier is the containment which shall prevent contact with the surroundings and the operators. This containment consists of a stainless-steel tank, pipes in and out of the tank, and valves controlling the flow of lye in and out.

In order to operate the valves, there is an electronic control system which has important safety features. These can be seen as dynamic safety barriers which include supervising equipment, sensors, interlocks, an emergency alarm.

There are a number of operators involved, which are supposed to work safely with everyday routines. They also have a safety role by observing problems and correcting failures that occur. The management directs the work and has also an overall responsibility for safety of the installation. Operational and safety rules are essential in order keep up the safety for a longer period of time. And if something dangerous should happen, there is a need for emergency preparedness.

In the examples above, we see a mix of technical, organizational, and human resources, which are needed to obtain a safe workplace. In summary, depending on how we define the barrier concept, it can be restricted to certain physical arrangements or include almost everything that may prevent hazards and/or reduce their consequences.

5.2 ORIGIN AND THEORETICAL BACKGROUND

5.2.1 Energy and Sequence Models

The roots to the classical safety barrier concept are found in energy models, such as the ones suggested by Gibson (1961) and Haddon (1973). The core of these models is that accidents occur as a consequence of loss of control of the hazard (e.g., harmful energy). The control mechanism may include barriers, processes, procedures. In his 10 strategies for hazard control, Haddon (1980) mention physical barriers as one of the means to prevent accidents.

An early accident model (Heinrich, 1931) describes accident causation as a linear sequence of events and circumstances (called factors), where an accident is perceived as being a link in a causal chain. The underlying causal conceptualization in Heinrich model is thus a singular causal chain where event A leads to event B which leads to event C, etc. A closer reading of Heinrich reveals a rather broad scope of detailed circumstances that contributes to accidents, for example, management factors, work context, and unsafely designed machines. Among those is what we today would call physical barriers (called "guards" by Heinrich). The model of Heinrich is metaphorically often described as a limited set of five dominos consisting of five factors named:

- Social environment/ancestry
- Fault of person
- Unsafe act/mechanical and physical hazard
- Accident
- Injury

A further development of Heinrich model is represented by Bird and Germain (1985), who suggested a more complex sequence model. This describes a causal chain from management factors (inadequate safety programs and standards) over a set of basic causes (personal factors, job factors), which influence immediate causes (substandard acts and conditions), and which terminates in incidents/accidents and then loss. This is an example of a complex linear model thinking, which has been very influential in guiding various safety practices, such as root cause analysis of accidents.

A more recent example is the Accident Evolution and Barrier Function model (AEB) (Svenson, 1991). It supports a method for analysis of accidents that models the evolution toward an incident/accident as a series of interactions between human and technical systems.

Also techniques such as fault trees and event trees in risk analysis have an underlying binary logic, which is based on a sequence model with straight causal relationships.

The concept of barrier often has a place in these models as providing arrangements that may prevent or reduce potentially harmful events. Some research has focused explicitly on how hardware safety barriers are influenced by management and organizational factors (culture, competence, procedures, etc.) of various kinds. For example, a European research project ARAMIS (Salvi and Duijm, 2006) has explored how human and organizational factors can influence the reliability of different types of technical safety barriers (passive, automated, or involving human action).

5.2.2 Extended Models

As a consequence of criticism directed at previous accident models, various systemic control theoretical approaches have been suggested to model accident causation. These models draw on various sources found in system theory, system dynamics, control theory, and cognitive system engineering. They view accident causation as emergent phenomena due to complex non-linear interactions among system components and subsystems. These models usually depict a multitude of type of "factors" at different levels of granularity and involve feedback loops.

For example, Rasmussen (1997) describes accident causation as a control problem where different dynamic forces may cause the socio-technical system to migrate over safety boundaries. Rasmussen's model also takes a structural hierarchical approach, and it widens the system boundaries to include government, regulators, companies, management, staff and individual working groups, and the connections between these levels. Other models in line with these systemic ambitions can be found in Leveson (2004) and Hollnagel (2004).

Another example in this tradition is represented by Markert et al. (2013) who outline a methodology "to include the effects of deficiencies in management processes as common causes for the failure of safety barriers" (p. 287). These developments can be understood in the context of criticism toward safety approaches that have neglected the influence of human and organizational factors on risk. Other approaches that seek to combine more traditional technology oriented safety methods with a stronger focus on organizational and human factors is represented by, for example, Gambetti et al. (2012), Embrey (1992), Colombo and Demichela (2008), Davoudian et al. (1994), Øien (2001), Aven et al. (2006), Paté-Cornell and Murphy (1996).

In conclusion: In early linear accident models where the energy concept was central, the concept of barrier had a reasonable clear definition in terms of hardware/technological artefacts—the classical barrier concept. The development of multi-linear and systemic accident models has affected the view of what a barrier could be. The models have sometimes continued to apply the barrier concept, but in a much wider meaning than was originally intended.

5.3 DEFINITIONS AND TERMINOLOGY

A comprehensive review by Sklet (2006) found that there are no universal and commonly accepted definitions of terms like safety barrier, defense, defense in-depth, layer of protection, safety function, either in the literature or in regulations and standards. Actually, a wide scope of conceptualizations which varies among contexts and branches are found. This means that we will not establish a fixed definition, but rather give a number of examples.

5.3.1 Examples of Barrier Definitions

The energy model has a strong influence on the thinking. Johnson (1980) gave a general definition related to technical systems:

- *Barriers are physical and procedural measures to direct energy in wanted channels and control unwanted release.*

One type of definition has focused on the accident sequence, and how it can be interrupted. Svenson (1991) has given a definition based on that:

- *A barrier function represents a function that can arrest the accident evolution so that the next event in the chain will not be realized.*

It means that a *barrier function* is identified in relation to the system(s) it protects, has protected, or could have protected.

A similar but more general definition has been suggested by Hollnagel (1999):

- *A barrier is an obstruction or hindrance (or defense) that may prevent or lessen the impact of an unwanted consequence. This may include stopping, slowing down, restricting, limiting, or in some other way weakening an uncontrollable process.*

In the nuclear domain, a barrier is often restrictedly defined as physical passive obstructions (reactor tank, containment, etc.) distinguished from other safety devices (core cooling, reactivity control, etc.) called "safety systems." Barriers are seen as a part of the defense-in-depth concept which is described in Chapter 4 in this book. IAEA (2007) has defined:

- *A barrier is a physical obstruction that prevents or inhibits the movement of people, radionuclides, or some other phenomenon (e.g., fire), or provides shielding.*

5.3.2 Barriers and Barrier Systems

Sklet (2006) has considered safety barriers in carefully planned and designed systems such as large chemical installations and offshore platforms. He suggests that distinctions should be made between the terms *barrier* and *barrier function*, which easily can be confused. Sklet suggests three definitions related to safety barriers:

- *Safety barriers are physical and/or non-physical means planned to prevent, control, or mitigate undesired events or accidents.*
- *A barrier function is a function planned to prevent, control, or mitigate undesired events or accidents.*
- *A barrier system is a system that has been designed and implemented to perform one or more barrier functions.*

In these definitions, the barrier function describes the purpose of safety barriers, and it should have a direct and significant effect. A barrier function should preferably be defined by a verb and a noun, for example, "Close flow." These definitions emphasize that the barrier should be intended and planned to protect. According to this set of definitions, a function that has an indirect effect is not classified as a barrier function, but it can be described as a risk-influencing factor/function.

A similar definition has been proposed by Duijm (2009), which also considers major hazard installations:

- *A barrier function is a function to prevent, control, or mitigate the propagation of an event into and undesired condition or event.*
- *A safety barrier is a series of elements that implement a barrier function, each element consisting of a technical and human action.*

A third example is the Petroleum Safety Authority in Norway (2013), which provides similar definitions as those above and complement their definition of barriers with further important concepts associated with barriers:

- *Barrier*: Technical, operational, and organizational elements which are intended individually or collectively to reduce possibility for a specific error, hazard, or accident to occur or which limits its harm/disadvantages.
- *Barrier Element*: Technical, operational, or organizational measures or solutions which play a part in realizing a barrier function.

- *Barrier Function*: The task or role of a barrier. Examples include preventing leaks or ignition, reducing fire loads, ensuring acceptable evacuation, and preventing hearing damage.
- *Barrier Strategy*: Result of a process which, on the basis of the risk picture, describes and clarifies the barrier functions and elements to be implemented in order to reduce risk.
- *Barrier Management*: Coordinated activity to establish and maintain barrier so that they maintain their function at all times.
- *Performance Requirements*: Verifiable requirements related to barrier element properties to ensure that the barrier is effective. They can include such aspects as capacity, functionality, effectiveness, integrity, reliability, availability, ability to withstand loads, robustness, expertize, and mobilization time.
- *Performance-Influencing Factors*: Conditions which are significant for the ability of barrier functions and elements to perform as intended.
- *Context*: External and internal frame conditions which must be taken into account in barrier management.

5.3.3 Alternatives to the Barrier Concept

To further illustrate the width of definitions related to the barrier concept, below we have included some other definitions found in the literature.

5.3.3.1 Defense An alternative term is *defense*, which is defined by Reason (1997) as:

- *Various means by which the goals of ensuring the safety of people and assets can be achieved.*

Reason has made a distinction between *hard defenses*, such as physical barriers and alarms, and *soft defenses*, for example, regulations, procedures, and training. These defenses can be combined in various ways. A defense can be weakened by *active failures*, for example, unsafe acts, or by *latent conditions*, such as poor design. A combination of active failures, latent conditions, and *local circumstances* might cause an accident to occur.

This terminology comes from what is commonly known as "The Swiss Cheese Model" (Reason, 1997). This contains a defense-in-depth logic where accidents are seen as a consequence of combinations of active errors and latent conditions which breach system defenses (barriers and safeguards). This model has a strong connection to epidemiological accident models in the medical field (Gordon, 1949). In epidemiological accident models, the background causes to accident are perceived as analogous to spread of a disease where a combination of latent conditions (in persons and the environment) and harmful agents (such as a virus) creates the harmful condition in the host. Also in these models various barriers and defenses can be installed to protect the host.

Reason's work put emphasis on a complex set of causes and circumstances, such as decisions taken at the blunt end of a complex system and how the consequences of such decisions may create latent weaknesses in a system. However, the underlying dynamic perspective of this type of models has been criticized as being too static and linear (Hollnagel, 2004) to fully grasp the dynamics involved in event causation.

5.3.3.2 Protection Layers The chemical industry has a somewhat different approach. The Center for Chemical Process Safety (CCPS, 1993) has described general safety aspects in connection with chemical process control systems. A fundamental concept is *Protection layer*, although this is not explicitly defined. It "typically involves special process designs, process equipment, administrative procedures, the basic process control system and/or planned responses to imminent adverse process conditions; and these responses may be either automated or initiated by human actions." The description by CCPS shows how *Protection layers* are constituted in eight levels. These are arranged in the order of how they are activated in the case of an escalating accident.

5.3.4 Safety Functions

"Barriers" are often used in a physical and/or formal organizational sense. An alternative is *safety function* (SF), which is a rather common term used in different situations and with various meanings. One example of a technical definition comes from the standard on the functional safety of electronic safety-related systems (IEC, 2001). It covers the aspects that need to be addressed when electronic systems are used to carry out safety functions. It defines safety function as follows:

- *A function to be implemented by an E/E/PE safety-related system, other technology safety-related system or external risk reduction facilities, which is intended to achieve or maintain a safe state for the equipment under control, in respect of a specific hazardous event.*

Safety function is an important term in the nuclear field. IAEA (2007) describes it as:

- *A specific purpose that must be accomplished for safety.*

That is a highly abstract definition, but on a lower systems level they are further specified. Nineteen examples on safety functions are presented which are divided into three basic requirements (IAEA, 2000):

1. Control of reactivity.
2. Removal of heat from the core.
3. Confinement of radioactive materials and control of operational discharges, as well as limitation of accidental releases.

Safety function can also have a much wider meaning, and it has been suggested (e.g., Harms-Ringdahl, 1999) as a general concept to cover a broad spectrum of safety features. For that purpose a general definition has been suggested (Harms-Ringdahl, 2009):

- *A safety function is a technical or organizational function, a human action or a combination of these, which can reduce the probability and/or consequences of accidents and other unwanted events in a system.*

An alternative simpler and broader description has also been proposed (Harms-Ringdahl, 2013):

- *A safety function contributes to reducing risks in a system.*

5.3.5 Conclusion

A general conclusion from the above oversight is that various industries and branches use different definitions and perceptions of barriers. The attempts to find a clear conceptualization of barriers have led to a deconstruction into elements such as barrier function, barrier elements, barrier function systems, and also to put these concepts into context of safety management principles (barrier management). The reader should bear this in mind when we come to the next section which focuses on various classification systems of barriers.

5.4 CLASSIFICATION OF BARRIERS

5.4.1 General Considerations

This section will discuss different dimensions which are used in relation to barriers. Classification taxonomies and conceivable dimensions of barriers depend on the definition of the barrier concept. For example, a conservative definition of the barrier concept (such as often applied in the nuclear industry) leads to focusing on various dimensions of physical/technical barriers, whereas a more wide-ranging approach leads to other classifications and dimensions. This section will discuss different dimensions which are used in relation to barriers.

One variable that can be used as a basis for classification is the dimension from abstract to concrete.

The level of abstraction can range from a theory of functioning to a specific technical solution (see Table 5.1). Lind (2005) has addressed the importance of working on the right abstract level especially in specification of safety systems. This has been used explicitly in the general model of safety functions (Harms-Ringdahl, 2013).

TABLE 5.1. Level of Abstraction for Safety Functions

	Level of Abstraction	Example
1	Theory	Explodes if hot
2	General function, e.g., aim	Limit temperature
3	Principal function	Monitor temperature—one of several options
4	Functional solution	Thermometer—checked by operator or electronically
5	Concrete solution	Action by operator or by safety relay

Source: Harms-Ringdahl 2013. Reproduced with permission of Lars Harms-Ringdahl.

5.4.2 System Level Classification

An additional classification is based on the level of system. One example is Rasmussen's (1997) model of systems organizational hierarchy, which includes government, companies, management, staff.

A technical or functional system can also be described related to the systems hierarchy; from the whole system and down to the component level. Table 5.2 gives examples that include five levels. It is intended for safety functions, but can be applied also for other types of safety barriers.

5.4.3 Classification Related to Accident Sequence

Another classification is based on how the barrier intervenes with the accident sequence. In the defense-in-depth concept (see Chapter 4), an initiating event is assumed, which is followed by a sequence of events. A common division is *prevention, control, and mitigation* (e.g., Johnson, 1980). A variant on this is to *avoid, prevent, control, and protect,* as suggested by Duijm et al. (2004).

5.4.4 Physical and Non-physical Barriers

A basic distinction is between physical barriers and other barriers, which has been mentioned by several authors, like Johnson (1980). A similar division in hard and soft

TABLE 5.2. System Levels for Safety Functions

	System Level	Example
1	General system	Medical services in a country
2	Specific system or establishment	Hospital in city A
3	Department, or part of system	Maternity ward, or a children's ward
4	Activity or subsystem	Administration of medicine
5	Component, or sub-operation	Drug prescription by doctor

Source: Harms-Ringdahl 2013. Reproduced with permission of Lars Harms-Ringdahl.

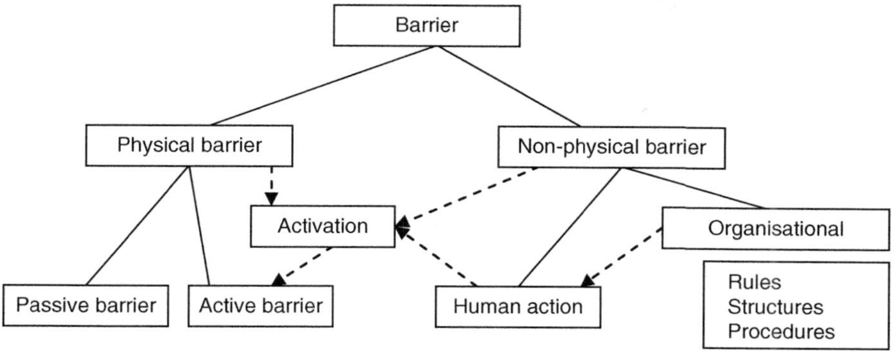

Figure 5.1. Classification of barriers in physical and non-physical.

defenses has been suggested by Reason (1997). A further example comes from Neogy et al. (1996), who make a division in physical barriers, procedural or administrative, and human actions. Figure 5.1 sketches some of these connections.

5.4.5 Administrative and Human Barriers

Non-physical barriers usually contain organizational elements as well as human actions. In technical installations and workplaces, there is usually a developed system of formal rules and defined responsibilities. These can be of varying quality, but in high-risk systems they are supposed to be well designed.

Another aspect is informal organizational factors which affect the quality and efficiency of the safety work. This has shown to be important and revealed by several accident investigations. The results are dependent on intentional conscious actors influenced by emotions, decision biases, values, social factors, culture, etc. Human variability and adaptability is here both a strength and a weakness depending on situation.

5.4.6 Passive and Active Barriers

One common classification is between passive and active barriers (Kjellén, 2000). A passive barrier can fulfill its safety function by its mere existence, such as a wall or a railing. Whereas an active barrier performs its function in response to a specific state or condition; for example, a fire sprinkling system can be triggered manually or automatically. Passive barrier may be permanently installed (such as containment at a nuclear station) or they may be mobile (such as a helmet). The distinction between passive and active barriers is however somewhat blurred depending on the interpretation of the meaning of "passive." For example, a mobile barrier needs human actions to be recognized and used properly, such as a pair of gloves.

The nature of the activation mechanism is another dimension that can be applied for further distinctions between different active barriers. Some active barriers are fully automated (detection, interpretation, action) such as a fully automatic shutdown system in nuclear plants. Whereas others are in need of human intervention to perform activities to fulfill the barrier function, such as a manual shutdown system.

5.4.7 Combined Models

One classification of barriers has been suggested by Hollnagel (2004). First, it first makes a distinction between barrier functions and barrier systems, that is, how the functions are implemented. The barrier systems are then classified depending on their nature, rather than on their purpose, location, or origin. Four categories are suggested:

1. Material barriers physically prevent an action or a harmful event to occur, such as walls, fences, railings, bars.
2. Functional (active or dynamic) barriers set up one or more preconditions that have to be met before actions are taken, such as locks, codes, passwords.
3. Symbolic barriers require a user that interprets symbols to understand their meaning, such as instructions, signs, permits.
4. Incorporeal barriers largely correspond to knowledge of the user but can also manifest as symbolic barriers, such as rules, guidelines.

However, as also recognized by Hollnagel, this is not a pure taxonomy, since a specific barrier system could fall in more than one category at the same time. For example, a fire door is a physical structure and thus a material barrier. For the fire door to be operative in case of a fire, it must meet the precondition of being closed (a functional barrier). The fire door should have a sign informing that it is a fire door (a symbolic barrier), and persons must know that it is important that the door remains closed during operation (an immaterial barrier).

Bellamy et al. (2007) have suggested a division in:

- Primary barriers
- Support safety barriers
- Safety management

The primary barriers are directly involved in a causal chain, the support barriers are used to support the primary barriers and safety management supports the integrity of the supportive barriers. Thinking along these lines can support a systemic perspective on barriers where concrete individual barriers are placed in a wider context of support, surveillance, and maintenance of barriers.

5.4.8 Purpose of Barriers

Another parameter is the purpose of the barrier. The intention is clear for barriers which are planned and designed to give some kind of protection, and this is especially obvious in a top-down perspective. However, in many cases the intention is unclear or has another goal than safety. For example, a tank with liquid is equipped with a pipe, which is intended to ventilate out evaporated gas. At the same time, it prevents overpressure in the tank during fill up. This other function might be very important for safety, but it is not intended.

For some authors, this intention is a mandatory condition in the definition of barriers. For example, Sklet (2006) defines a barrier as means planned to prevent, control, or mitigate undesired events or accidents. Other definitions do not state that the intention is mandatory, but it might be assumed but not spelled out.

5.5 METHODS FOR ANALYSIS OF SAFETY BARRIERS

There are several methods in which safety barriers are central or have an important role. Table 5.3 gives a few examples of methods that are oriented toward barriers and safety functions. They can be used for the analysis of systems and/or investigations of accidents. Short summaries of the methods are presented below, and more full descriptions of these methods can be found in a recent handbook for safety analysis (Harms-Ringdahl, 2013), and also in Chapters 4, 20, and 21 in this book.

TABLE 5.3. Examples of Methods Which Consider Safety Barriers

Method Name	Comments	Type
Energy analysis	Barriers are a fundamental part of the method	S
Event tree analysis	Analysis of barriers' role in how an event can cause damage	S
Fault tree analysis	Analysis of how barriers and safety features can prevent an accident	S
MORT	*Management Oversight and Risk Tree*	A & S
MTO event investigation	Summarizes the barriers in the accident sequence	A
Safety barrier diagrams	Also called Bowtie. A combination of fault tree analysis and event tree analysis	S (A)
Safety function analysis	Identifies and evaluates safety functions	A & S
Reliability techniques	A number of techniques to analyze and predict reliability of systems	S

Type—How the method is mainly used:
A, accident investigation; S, system analysis.

TABLE 5.4. Checklist for Safety Measures by Introducing Barriers

Barrier	Comments
Related to Energy	
1. Eliminate the energy	Work on the ground, instead of at a height
2. Reduce the magnitude	Lighter objects to be handled
3. Safer alternative solution	Less dangerous chemicals
4. Prevent the build-up of an extreme magnitude of energy	Control equipment
5. Prevent the release of energy	Container of sufficient strength
6. Controlled reduction of energy	Safety valve
Separation	
7. Separate object and energy	In space or in time
8. Safety protection on the energy source	Machine safeguards
Protection of the Object	
9. Personal protective equipment	Protective shoes, helmets
10. Limit the consequences when an accident occurs	Facilities for stopping the energy flow

5.5.1 Energy Analysis

In the method, energies which might cause injury or damage are identified. Barriers are a fundamental part of the method. They are considered in the evaluation if the particular energy is sufficiently under control, or if improvements are needed. The method includes a set of strategies which apply different types of barriers. These strategies contain both physical and non-physical barriers. Table 5.4 shows a checklist for finding safety measures which can be interpreted as introducing different types of barriers.

5.5.2 Event Tree Analysis

The method is used to study the alternative consequences of a defined event. It considers barriers and the course of events in a logical framework. The result is a logical tree, which starts with an initiating event showing the relations between alternative consequences. The method is binary, which means that an event is assumed to happen or not happen, and a barrier is assumed to work or fail. Chapter 4 explains the event trees and gives examples of and the relation to barriers. One common application (Rouhiainen, 1993) is to check a safety function to see whether or not an event gives rise to damage.

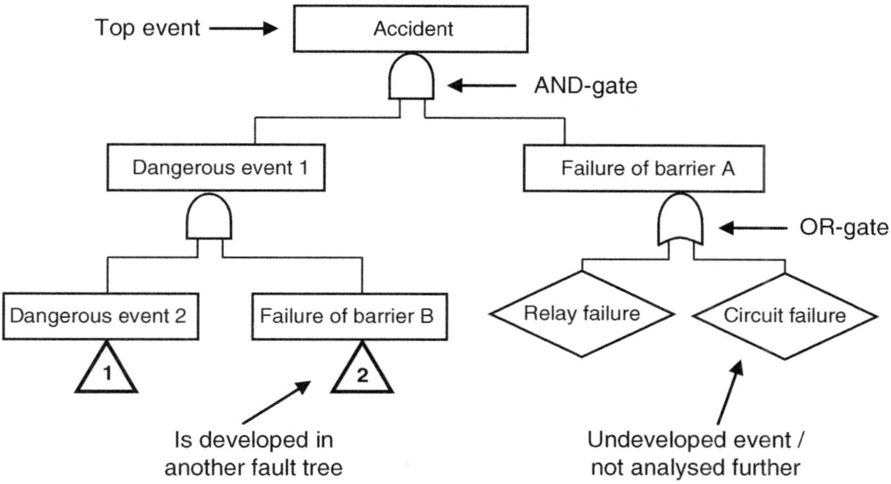

Figure 5.2. Sketch of a fault tree which includes two barrier failures.

5.5.3 Fault Tree Analysis

A fault tree is a diagram showing logical combinations of causes of a certain defined accident. It can be used to identify combinations of faults that can lead to the top event. It can also be used to estimate the probability of the defined accidents.

Fault Tree Analysis is based on binary approach. Either an event occurs or it does not; either a barriers works or not. An event statement can then be designated as *true* or *false*, and also expressed with logical values *1* and *0*. The principle of the tree is to combine events (and failure of barriers) through logical gates. Figure 5.2 presents a part of a fault tree, which shows how failure in barriers could lead to the top event.

The interpretation of the fault tree is that dangerous event 2 becomes dangerous event 1, if barrier B fails. Consequently, dangerous event 1 becomes accident, if barrier A fails.

5.5.4 Safety Barrier Diagrams

A group of methods focus on critical events in an accident chain and barriers related to that chain. The method is based on a combination of Fault Tree Analysis and Event Tree Analysis. The result is presented in a diagram, where the left part shows the causes of the critical event, and the right part shows its possible consequences. A common name is Bowtie diagram, since the diagram is like a bowtie in shape. The origin of this approach is *Safety Barrier Diagrams* (Taylor et al., 1989; Taylor, 1994; Duijm, 2009), which offers a way of presenting and analyzing barriers to accidents.

5.5.5 Management Oversight and Risk Tree

MORT is an acronym for *Management Oversight and Risk Tree,* which can be used for the analysis of systems and accidents (Johnson, 1980; NRI, 2009). A general idea is that "when an accident reveals errors, it is the system which fails." An important part of the method is to develop a logic diagram which provides a general problem description.

It is rather like a fault tree, and the same symbols are used. The top of the tree is an accident that has occurred. This can be due to an *assumed risk* or to an *oversight or omission*, represented by the two main branches of the tree, or both. For a risk to be assumed, it must have been analyzed and treated by the company management.

The other main branch examines organizational factors, which we here consider also as safety barriers. It has two subordinate branches; one is called *specific control factors* and focuses on what occurred during the accident. The second subsidiary branch treats *management system factors*.

5.5.6 MTO Event Investigation

MTO is an acronym for Man—Technique–Organization, and the MTO method for event investigation (Rollenhagen, 2011) is based on the HPES—*Human Performance Enhancement System* (INPO, 1990). These methods are representative of root cause analytical technique based on multi-linear accident models (Johnson, 1980). The methods MTO event analysis and HPES identify and analyze barriers, which exist but have failed. It also considers non-existing barriers that ideally could have prevented the event.

5.5.7 Safety Function Analysis

The concept of safety functions is a generalization of the idea of safety barriers. The method *Safety Function Analysis* (Harms-Ringdahl, 2003, 2013) is used for identification and evaluation of safety functions in a specific system. The identification can be based on an analytic study of a specific system or on results from an accident investigation. The evaluation is an important part of the method, and it concerns the suitability and the efficiency of the safety function.

5.5.8 Reliability Techniques

Reliability techniques are closely related to the thinking around safety barriers, especially in advanced technical systems with high demands on safety. It concerns the design of barriers and of barrier system, as well as their functionality and reliability. This is thoroughly discussed in Chapters 4 (Defense-in-Depth), 20 (Quantitative Risk Analysis), and 21 (Qualitative Risk Analysis). As an example, we could mention the method *Failure Mode and Effects Analysis* (FMEA), which can be used to analyze barriers and how they could fail.

5.6 QUALITY AND EFFICIENCY OF BARRIERS

In discussing the quality and efficiency of barriers, a separation is made between three main activity domains:

1. Design and installation of barriers
2. Operation of barriers
3. Maintenance of barriers

Each domain is associated with potential weaknesses in the management of barriers, which may result in degradation of the barriers. The last part of this section presents a summary of barrier management principles and quality dimensions.

5.6.1 Design and Installation of Barriers

Design is a complex task with a fundamental importance for the safety characteristics in the final system. In the design, a comprehensive view on the system is imperative so that the interacting parts and subsystems can function together. A technical component or subsystem (such as a technical barrier) may function according to specifications for that system in isolation. However, complex interactions between subsystems may nevertheless result in states that are difficult to predict in the design phase. Barriers generally make a system complex, and high complexity can itself impose risk to a system.

Design of safety barriers is concerned with safety, but many different aspects of safety may exist in parallel. This can result in conflicts between different safety goals. For example, barriers installed for security reasons may be in conflict with other safety goals. This was seen the March 24, 2015, when an Airbus A320 crashed in the French Alps killing 150 people. The accident investigation revealed that the co-pilot did not allow the captain to reenter into the cockpit. In this case, the barrier supporting a security goal (i.e., preventing unauthorized access into cockpit) was in conflict with other safety-related goals. Also, occupational safety goals may conflict with safety or security goals of other kinds.

Another aspect of barrier design concerns the possibility to detect latent weakness in barriers during operation. Some barriers have the capacity to automatically "inform" operators about the system's functional status during normal operations. For other barriers, it would be necessary to conduct regular tests of barrier functionality. Depending on the design, such tests are more or less easy to perform. Barrier tests are usually conducted at regular times or before startup, which expose technical barriers to potential degradation between tests.

A third class of potential weaknesses associated with barrier concerns the risk for *common cause failures* in barrier systems. This means that a single failure can degrade two or more barriers at the same occasion, which strongly reduces the safety performance. Such failures can be induced as:

- Consequences of weakness in design, for example, if redundant barriers use the same power supply
- During manufacturing
- Human errors during installation; the same installation mistake is made on several barriers

Human Factor issues related to design are a fourth class of design weaknesses that can compromise the efficiency and quality of barriers. Barriers that depend on human action in critical situations must be carefully designed from a human factor/ergonomics perspective. In general, the use of automation may detach operators from direct perception and knowledge concerning the state of production processes which, in some circumstances, can prevent situation awareness to emerge.

Weaknesses in barrier system can also arise as a consequence of redesign and upgrades of technology. For example, the reasons and assumptions behind implementation of a specific barrier system might have been weakly documented and/or forgotten during the life time of a technical system.

5.6.2 Management of Barrier Systems During Operation

An important aspect of barrier management concerns operation readiness before starting operations that involve barriers as protection. Examples can be found when production systems have been started without full readiness of the safety barriers. This might be revealed later during problematic events or later testing.

Human Factor issues are also involved when people do not use the barriers as intended, which can have many various reasons. This is also a design issue, when some barriers are designed so that a task becomes more time consuming and complicated when safety barriers are implemented in the system. Many systems can function without using installed safety barriers at all. Some barriers are add-on functions that do not influence normal production processes, because they are designed to intervene only in critical safety situations.

5.6.3 Maintenance of Barriers

Maintenance of barriers is critical for their effectiveness and quality. Several types of human errors can be made during maintenance and testing which can compromise later barrier functionality in events. The roots to such errors can stem from many different sources, such as weakness in the design which makes a barrier difficult to maintain, weaknesses in maintenance programs and other organizational issues, motivational factors, management factors.

5.6.4 Summary of Barrier Management Principles

There are several examples of problems and weaknesses related to safety barriers which were discussed above. Based on these the following principles for barrier management are suggested:

- Barriers should be made as simple as possible and it should be easy to understand the assumptions behind the design.
- Barriers should cover a wide range of initiating events whenever possible.
- Design of barriers should consider the possible conflicts that can arise from conflicting safety goals.
- Design of barriers should consider if the barriers are likely to be perceived as hindrances for normal production and if such perceptions make it likely that users easily can remove barrier functionality.
- Design of barriers should consider the sensitivity to human errors both in use and in maintenance, how this can be reduced and handled.
- Design of barriers should consider how the barriers can be tested and their efficiency evaluated.
- Administrative barriers can be expected to have relatively low efficiency, especially over long time. Therefore they need to be consciously planned and carefully evaluated, in order to get a sufficiently high efficiency over time.

Related to the management principles above, some quality dimensions can be tested addressed by the questions below:

- Does the barrier meet its functional requirements in terms of the situations it is designed for (efficiency, scope of situation the barrier can handle)?
- What is the barrier's ability to execute its specified functionality when needed (reliability, availability)?
- Given that the barrier execute its intended functionality when needed—to what extent can this functionality be maintained during progress of an accident scenario (capacity, effectiveness, robustness)?
- How much resources are needed and how easy is it to evaluate and maintain the barrier (barrier evaluation, maintainability, simplicity)?
- How much is the barrier dependent on human activity to release its functionality during operation?

5.7 DISCUSSION AND CONCLUSIONS

This chapter has shown a number of parameters that could characterize barriers. The definitions vary between various technical traditions, and the disparity might be something we have to live with. A conclusion is that all users need to be aware of the variation, and whenever needed clarify how the terms are used. The discussion is summarized in a few themes below:

- The classical and radical meaning
- Types of system
- Parameters in describing barriers

5.7.1 The Classical and Radical Meaning

Both the classical and the radical meaning of the barrier concept are associated with pros and cons. The classical meaning, based on a technological perspective, is easier to define in comparison with the radical interpretation.

An argument against the radical interpretation of barrier is that it can be difficult to draw boundaries between the different parts of a barrier system, that is, to maintain a separation between barriers on the one hand, and various functions that can support the barriers on the other hand. Sklet (2006) has in a discussion of safety barriers emphasized that they should be understood as directly related to the event sequence or accident scenario and not include risk-influencing factors which affect barrier performance (human and organizational influences, etc.). By a similar token, The Petroleum Safety Authority Norway (2013) state the regulators position of what is not considered a barrier: "There is little point in including or considering 'everything of importance' as barriers or barrier elements. Such interpretation will not contribute to more conscious barrier management and follow-up" (p. 26).

A radical interpretation of barriers has the benefit of considering safety-related human and organizational processes, which easily are neglected in an analysis with a technical focus. Whereas safety technology is relatively easy to specify and control departing from well-known principles, much less is understood regarding human nature and organizations. Socio-technical systems consist of people, technology, and organizational functions in interaction with numerous feedback loops among these factors, which make the modelling of such system highly complicated. The radical barrier concept may support a more holistic and systematic analysis of such organizations than the classical interpretation.

The classical view of barrier often adopts a view of linear causality where feedback loops are missing. Research traditions connected with a more radical meaning of barrier often reveals a more complex idea about causality since human and organizational factors are assumed to interact with technology in complex ways. However, it is sometimes tempting to try to model human and organizational factors departing from similar principles that are used for modelling technical systems. For example, a series of redundant physical barriers can be applied to strengthen the defense against release of toxic substances. However, the same principles may not necessarily function for a redundant set of human assessments aimed at detecting design faults. One reason is simply that several actors might falsely believe that some other actor in the chain will do the work.

5.7.2 Empirical Research on Use and Efficiency

There are a lot of research paper concerning barriers in well-defined hierarchical technical systems, for example, in nuclear power and major hazards installations, while common workplaces has got less attention (e.g., Hale, 2006; Sklet, 2006). This also means that most of the thinking about safety barriers is based on advanced technical systems.

Examples of barriers at more ordinary workplaces are presented in a study (Harms-Ringdahl, 2009, 2013), which is based on analysis of five incidents. The first incident occurred at an electrical power distribution station, the second at a railway station, and the last three occurred at a hospital, where three incidents with medication errors were investigated. These incidents were analyzed with the method Safety Function Analysis. It was observed that technical, organizational, and human safety features existed side by side. As an average, around 40 safety functions related to the incident were identified in the cases. A large fraction of these did not work properly—between 42% and 72% failed. Especially organizational safety functions on company and corporate level had a low efficiency.

5.7.3 General Conclusions

The differences between different definitions can be summarized in a set of parameters which characterize the actual safety barrier concept. This has been discussed in Section 5.4 and at other places. A general summary is given in Table 5.5.

The concept of safety barriers is an indispensable element in analyzing and designing systems where safety is important. However, there is a varying understanding of what it really means. This has evolved from different technical traditions and various theoretical considerations. Consequently, we see no point in trying to suggest a unifying definition for safety barriers.

The width of the concept can to a large extent be explained by various assumptions of what constitutes a "safety barrier." Such notions could partly be described by a set of parameters, which has been suggested above.

Therefore, it is essential to carefully define the meaning of "safety barrier" in all applications, when any doubt of the interpretations could arise.

TABLE 5.5. Summary of Parameters Related to Safety Barriers

Parameters	Comments
Level of abstraction	From theory and requirements to concrete solution (Table 5.1)
System level	From general system to component (Table 5.2)
Relation to accident sequence	A common division is *prevention, control, and mitigation*
Causality	Direct causality or influencing factor
Physical and non-physical barriers	This is a common basic distinction (Figure 5.1). Non-physical barriers usually contain organizational elements as well as human actions
Passive and active barriers	This is usually related to technical barriers, but it could also be relevant for administrative and human barriers
Purpose of barriers	Only barriers planned to protect are included, or alternatively also other functions with a safety function

5.7.4 Relations to the Other Chapters

The concept of safety barrier (this chapter) has relations to many of the other chapters in this handbook. An obvious relation exists between the present chapter and that of defense–in-depth (Chapter 4) since implementation of defense in depth make use of safety barriers. To detect weaknesses in barriers, it is also important to have an efficient feedback system (Chapter 7). Also the chapters describing principles of human factors engineering (Chapter 9), quality principles (Chapter 15), quantitative and qualitative risk analysis (Chapters 20 and 21), and human reliability analysis (Chapter 25), are some examples describing principles that are important for barriers.

REFERENCES

Aven, T., Sklet, S., and Vinnem, J. E. (2006). Barrier and operational risk analysis of hydrocarbon releases (BORA-Release): Part I. Method description. *Journal of Hazardous Materials*, 137, 681–691.

Bellamy, L. J., Ale, B. J. M, Geyer, T. A. W., Gossens, L. H. J., Hale, A. R., Oh, J., Mud, M., Bloemhof, L. A., Papazogulu, J. Y., and Whiston, J. Y. (2007). Storybuilder – a tool for the analysis of accident reports. *Reliability Engineering and System Safety*, 92, 735–744.

Bird, F. E. J., and Germain, G. L., (1985). *Practical Loss Control Leadership*. Loganville, GA: International Loss Control Institute, Inc.

CCPS. (1993). *Guidelines for Safe Automation of Chemical Industries*. New York: Center for Chemical Process Safety, American Institute of Chemical Engineers.

Colombo, S., and Demichela, M. (2008). The systematic integration of human factors into safety analyses: An integrated engineering approach. *Reliability Engineering and System Safety*, 93, 1911–1921.

Davoudian, K., Wu, J. S., and Apostolakis, G. (1994). The work process analysis model (WPAM). *Reliability Engineering and System Safety*, 45, 107–125.

Duijm, N. J., Madsen, M. D., Andersen, H. B., Goossens, L., Hale, A. R., Guldenmund, F. W., Hourtolou, D. (2004). ARAMIS project: Effect on safety management's structural and cultural factors on barrier performance. Chech Society of Chemical Engineering. In: 11th International Symposium on Loss Prevention and Safety Promotion in the Process Industries (pp. 1361–1368), Prague.

Duijm, N. J. (2009). Safety-barrier diagrams as a safety management tool. *Reliability Engineering and System Safety*, 94, 332–341.

Embrey, D. E. (1992) Incorporating management and organisational factors into probabilistic safety assessment. *Reliability Engineering and System Safety*, 38, 199–208.

Gambetti, F., Casalli, A., and Chisari, V. (2012). The human factor in process safety management. *Chemical Engineering Transactions*, 26, 279–302

Gibson, J. J. (1961). The contribution of experimental psychology to the formulation of the problem of safety—a brief for basic research. In: *Behavioural Approaches to Accident Research*. New York: Association for the aid of crippled children.

Gordon, J. E. (1949). The epidemiology of accidents. *American Journal of Public Health*, 39, 504–515.

Haddon, W., Jr. (1973). Energy damage and the ten counter-measure strategies. *Human Factors,* 15(4), 355–366.

Haddon, W., Jr. (1980). The basic strategies for reducing damage from hazards of all kinds. *Hazard Prevention*, 16, 8–12.

Hale, A. R. (2006). Method in your madness: System in your safety. Delft University, The Netherlands.

Harms-Ringdahl, L. (1999). On the modelling and characterisation of safety functions. In: G. I. Schueller, and P. Kafka (Eds.), *Safety and Reliability ESREL'99* (pp. 1459–1462). Rotterdam, The Netherlands: Balkema.

Harms-Ringdahl, L. (2003). Assessing safety functions—results from a case study at an industrial workplace. *Safety Science*, 41, 701–720.

Harms-Ringdahl, L. (2009). Analysis of safety functions and barriers in accidents. *Safety Science*, 47, 353–363.

Harms-Ringdahl, L. (2013). *Guide to Safety Analysis for Accident Prevention.* Stockholm, Sweden: IRS Riskhantering AB, Available at: www.irisk.se

Heinrich, H. W. (1931). *Industrial Accident Prevention: A Scientific Approach.* New York: McGraw-Hill.

Hollnagel, E. (1999). Accidents and barriers. *In: Proceedings of CSAPC '99.* Valenciennes, France: Presse Universitaires de Valenciennes.

Hollnagel, E. (2004). *Barriers and Accident Prevention.* Aldershot, UK: Ashgate publishers.

IAEA. (2000). Safety of nuclear power plants: Design—requirements. International Atomic Energy Agency, Vienna.

IAEA. (2007). Safety glossary: Terminology used in nuclear safety and radiation protection. International Atomic Energy Agency, Vienna.

IEC. (2001). Functional safety of electrical/electronic/ programmable electronic safety-related systems. IEC 61508. International Electrotechnical Commission, Geneva.

INPO. (1990). Human performance enhancement system. INPO 90-005. Institute of Nuclear Power Operations, Atlanta, GA.

Johnson, W. G. (1980). *MORT Safety Assurance System.* New York: Marcel Dekker.

Kjellén, U. (2000). *Prevention of Accidents Through Experience Feedback.* London: Taylor & Francis.

Leveson, N. (2004). A new accident model for engineering safer systems. *Safety Science*, 42(4), 237–270.

Lind, M. (2005). Modeling goals and functions of control and safety systems—Theoretical foundations and extensions of MFM. Technical University of Denmark.

Markert, F., Duijm, N. J., and Thommesen, J. (2013). Modelling of safety barriers including human and organisational factors to improve process safety, *Chemical Engineering Transactions*, 31, 283–288. DOI: 10.3303/CET1331048

Neogy, P., Hanson, A. L., Davis, P. R., and Fenstermacher, T. E. (1996). Hazard and Barrier analysis guidance document, Rev. 0. Report no. EH-33. Office of Operating Experience Analysis and Feedback, US Department of Energy.

NRI. (2009). *NRI MORT User's Manual (NRI-1)* (2nd ed.). Delft, The Netherlands: Noordwijk Risk Initiative Foundation, Available at: www.nri.eu.com

Øien, K. (2001). A framework for the establishment of organizational risk indicators. *Reliability Engineering and System Safety*, 74, 147–167.

Paté-Cornell, E. M., and Murphy, D. M. (1996). Human and management factors in probabilistic risk analysis: The SAM approach and observations from recent applications. *Reliability Engineering and System Safety*, 53, 115–126.

Petroleum Safety Authority Norway. (2013). *Principles for barrier management in the petroleum industry*. Petroleum Safety Authority, Norway. Available at: http://www.psa.no

Rasmussen, J. (1997). Risk management in a dynamic society: A modelling problem. *Safety Science*, 27, 183–213.

Reason, J. T. (1997). *Managing the Risk of Organisational Accidents*. Aldershot, UK: Ashgate.

Rollenhagen, C. (2011). Event investigation in nuclear power plants in Sweden: Reflections about a method and some associated practices. *Safety Science*, 49(1), 21–26.

Rouhiainen, V. (1993). Modelling of accident sequences. In: J. Suokas, and V. Rouhiainen, (Eds.), *Quality Management of Safety and Risk Analysis*. Amsterdam: Elsevier.

Salvi, O., and Duijm, N. J. (Eds.). (2006). Outcomes of the ARAMIS project [Special Issue]. *Journal of Hazardous Materials*, 130.

Sklet, S. (2006). Safety barriers: Definition, classification, and performance. *Journal of Loss Prevention in the Process Industries*, 19, 494–506.

Svenson, O. (1991). The accident evolution and barrier function (AEB) model applied to incident analysis in the processing industries. *Risk Analysis*, 11, 499–507.

Taylor, J. R. (1994). *Risk Analysis for Process Plan, Pipelines and Transport*. London: E and FN Spon.

Taylor, J. R., Becher, P., Pedersen, K. E., Kampmann, J., Schepper, L., Kragh, E., and Selig, R. (1989). *Quantitative and Qualitative Criteria for the Risk Analysis*, Copenhagen, Denmark: Danish Environmental Agency.

6

FACTORS AND MARGINS OF SAFETY

Neelke Doorn and Sven Ove Hansson

6.1 INTRODUCTION

A safety reserve is an extra distance or difference between actual conditions and conditions that are expected to cause an accident or some other type of damage. When applied to physical systems, it can be a physical distance to a dangerous place, an extra strength in a construction, or an extra difference between toxic doses and the exposures that are allowed to take place. Safety reserves may be quantitative or non-quantitative. In the former case they are often referred to as safety factors or safety margins.

It is important to keep in mind that the application of safety reserves is not limited to physical systems. They can equally apply to socio-technical systems, in which case they are often discussed under the label of a system's resilience (Chapter 3 in the current volume and Rasmussen, 1997, p. 192). The notion of resilience can be conceptualized in different ways (Doorn, 2017), ranging from a narrow interpretation that focuses on recovery and constancy to a broader interpretation that refers to a system's buffer capacity and its ability to withstand shocks and maintain its functions (Folke, 2006). The better the system is capable of self-(re)organization to accommodate external changes, the more resilient it is (Carpenter et al., 2001). Factors

Handbook of Safety Principles, First Edition. Edited by Niklas Möller, Sven Ove Hansson, Jan-Erik Holmberg, and Carl Rollenhagen.

influencing a socio-technical system's resilience may include the educational level of the people having to act in case of a threat or emergency, the availability of transport and shelter facilities, and the availability of information and communication facilities (Norris et al., 2008). A safety reserve in the context of a socio-technical system's resilience may be the availability of several instead of just one road to leave a dangerous area, backup facilities in case one communication service breaks down, etc. (Cutter et al., 2010).

In the remainder of this chapter, we focus on the use of quantified safety reserves for physical systems. The most common way to specify the size of a safety reserve is to introduce a numerical value ("safety factor") that some crucial parameter is divided or multiplied with.

Example 1

Consider the construction of an apartment building.[1] The building consists of reinforced concrete frameworks. Assume that we have to calculate the required thickness of one of the floors.

We first have to calculate the load, which consists of the weight of the floor itself (the so-called dead load), the variable load due to objects placed in the room and people moving around in it, and—if relevant—the environmental load due to weather conditions (say, wind and snow). This environmental load is only relevant for floors or facades facing the exterior of the building, so we can ignore it for the current floor. The building code prescribes both what load to use and what safety factor to apply. This factor is higher for a variable load, because of potential extremes, which are less likely for a permanent load. For the given example, the following values are prescribed:

Type of Load	Value of Load	Safety Factor
Weight reinforced concrete	25 kN/m^3	1.35
Variable load for dwelling	2 kN/m^2	1.50

With h referring to the thickness of the floor, the design load per square meter can be calculated as follows:

$$L_{\mathrm{DESIGN}} = 1.35 \times h \times 25\ \mathrm{kN/m^2} + 1.50 \times 2\ \mathrm{kN/m^2} = (33.75 \times h + 3)\ \mathrm{kN/m^2}.$$

The next step is to calculate the strength of the floor, which is of course very much dependent on the material properties. Depending on the expected lifetime of the structure, different material factors are prescribed. In order to obtain the design

[1] In this example we use a simplified version of the EN Eurocodes, a series of 10 European Standards, providing a common approach for the design of buildings and other civil engineering works and construction products. The example is a simplified version of the example used in Biasioli et al. (2014) to explain Eurocode 2, the code for concrete structures.

strength S_{DESIGN}, the given strength S_G should be divided by the relevant material factor. The required material factor for this floor is 1.2. Now assume that the strength S_G of the floor is related to the thickness of the floor h as follows (the thicker the floor, the stronger):

$$S_G = 65 \times h \text{ kN/m}^2,$$

which yields the following design strength S_{DESIGN}:

$$S_{DESIGN} = 65 \times h/1.2 \text{ kN/m}^2 = 54 \times h \text{ kN/m}^2.$$

Since the strength should obviously be equal to or higher than the load, we can now calculate the required (minimum) thickness of the floor as follows:

$$\begin{aligned} &S_{DESIGN} \geq L_{DESIGN} \\ &54 \times h \geq 33.75 \times h + 3 \\ &h \geq 0.15. \end{aligned}$$

Hence, the floor should have a thickness of 0.15 m at minimum.

This example shows how a safety factor is applied to both the design *load* of a structure and the design *strength* of a structure.

Example 2

In the geotechnical design of river dykes, one of the potential failure mechanisms of a slope is a slip circle; that is, a rotational slide along a generally curved surface (Figure 6.1).

The classic approach to determine the stability of a slope against sliding is to calculate for all possible sliding circles, the moment caused by the driving or destabilizing forces (i.e., the moment caused by the vertical arrows in Figure 6.2), and the moment caused by the resisting or stabilizing forces (i.e., the moment caused by the

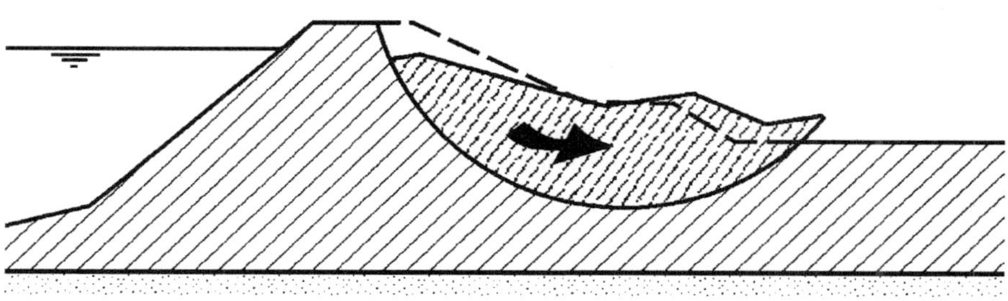

Figure 6.1. Slope instability of the inner slope. *Source:* TAW 2001. Reproduced with permission of Expertise Netwerk Water.

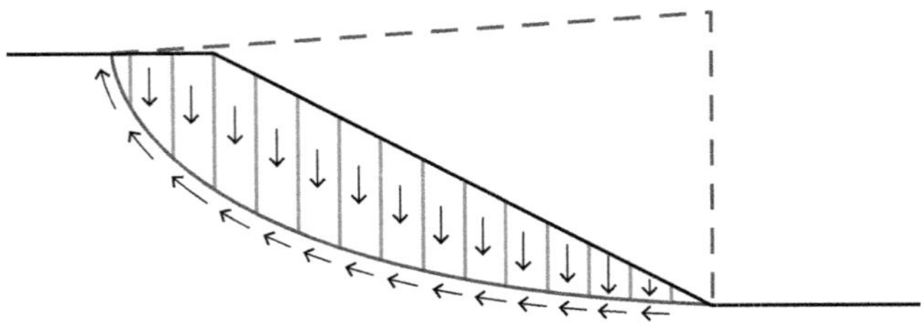

Figure 6.2. Potential sliding circle for geotechnical structure.

leftward-directed arrows in Figure 6.2). A slope is considered stable (or safe) if the ratio of the resisting momentum and the driving momentum is larger than a predefined safety factor. This safety factor is soil dependent. A safety factor of 1.25 is generally considered an absolute minimum; mostly a value higher than 1.5 is recommended (Loehr et al., 2006). If the ratio is lower than the required safety factor, then a flatter slope should be chosen and the calculation should be repeated. All engineers working with geotechnical materials are familiar with this iterative process of determining the maximum slope level (Terzaghi et al. 1996).

Example 3

Saccharin is a chemical substance that was discovered in 1878. It is much sweeter than sugar, and since it is not metabolized in the body it does not contribute to the dietary energy intake. It was used in World War I due to sugar shortage and became popular in the 1960s and 1970s as a calorie-free sweetener. It was believed to be non-toxic until studies in the early 1970s showed that it caused increased incidence of bladder cancer in male rats. In 1977, the US Food and Drug Administration proposed a ban of the substance, a proposal that met widespread opposition due to the uncertainty of the evidence of carcinogenicity in humans and to the fact that saccharin was at the time the only available sugar substitute. However, due to congressional action, the ban was never enforced (Klimko, 2011; Priebe and Kauffman, 1980). In Europe, regulations were based on laboratory data showing that daily doses not higher than 500 mg/kg bodyweight did not cause cancer in the laboratory animals. Such a dose is called a no observed adverse effect level (NOAEL). In 1977, the European Union's Scientific Committee for Food endorsed a temporary limit for human consumption in the form of an acceptable daily intake (ADI) of 2.5 mg/kg bodyweight, while also stating "this will have to be reassessed in the light of new information as it becomes available" (Scientific Committee for Food, 1977). The proposed value was 1/200 of the dose at which no cancer had been shown in animals, which means that a safety factor of 200 had been applied (Barlow and Schlatter, 2010; Larsen, 2012). This is twice the safety factor of 100 that is the established standard in food toxicology. The somewhat wider safety margin should be seen against the background of the serious nature of the disease it was intended to protect against.

Since 1977, evidence has accumulated that strongly indicates that the carcinogenic effects of saccharin are due to a mechanism of action involving crystal formation. This mechanism is specific to the male rat and requires high doses of the substance. It is not to be expected in humans, and the substance is therefore no longer considered a suspected human carcinogen (Dybing, 2002). In 1995, the Scientific Committee for Food decided to double the ADI to 5 mg/kg, stating that this value was "derived by applying a 100-fold safety factor" to the NOAEL of 500 mg/kg. However, they also said that the available studies "strongly indicate that saccharin is not related to bladder cancer in humans," in other words that the dose that they divided by the standard safety factor is not relevant for human risk assessment. The reason why they nevertheless used this value was that it "has not been possible to unequivocally demonstrate" the irrelevance for humans of the bladder tumors in male rats, and therefore the committee "wishes, as a matter of prudence, to take these lesions into account in setting an ADI" (Scientific Committee for Food, 1995). This decision takes a middle way that is quite interesting from a scientific point of view. If saccharin is not carcinogenic in humans (as most toxicologists believe), then the NOAEL level used to derive the ADI is wrong and should be replaced by a NOAEL representing the actual effects of the substance. If (contrary to the prevailing opinion among toxicologists) the substance is indeed carcinogenic, then the safety factor of 100 represents a much lower degree of caution than what is applied for other carcinogens. This illustrates how complex the use of safety factors can be and how much it can depend on issues at the science–policy interface that are neither purely scientific nor purely policy dependent.

In these and other examples, numbers representing a margin of safety are used as means to avoid an impending danger: physical harm and damage due to the collapse of a floor, flooding due to the collapse of a river dyke, a food-related cancer. These numbers are used as simple rules of thumb that are largely kept constant across contexts. Engineers use the same safety factors for different constructions of the same type, and toxicologists the same safety factors for different toxic substances.

6.2 ORIGIN AND HISTORY

The use of safety reserves goes back to prehistoric times. Ancient monuments bear witness to how builders have added extra strength to their constructions in order to be on the safe side. But we are concerned here with the use of explicit numerical values to design the safety reserves, a practice that can only arise in a context of design based on numerical calculations. The explicit use of safety factors in design calculations seems to have originated in the latter half of the nineteenth century. In the 1860s, the German railroad engineer A. Wohler recommended a factor of 2 for tension. In the early 1880s, the term "factor of safety" was in use; hence Rankine's *A Manual of Civil Engineering* defined it as the ratio of the breaking load to the working load, and recommended different factors of safety for different materials (Randall, 1976).

The use of safety factors has expanded to various branches of engineering design. Their use is partly non-codified engineering practice, but they have increasingly been codified in standard-setting documents such as building codes.

In toxicology, the use of explicit safety factors is even more recent (Dourson and Stara, 1983). The first proposal for a safety factor for toxicity seems to have been Lehman's and Fitzhugh's proposal in 1954 that an ADI be calculated for a food additive by dividing the chronic animal NEL (maximal no effect level) in mg/kg body weight by 100. (This term has now largely been replaced by NOAEL, see Example 3 above.) They thus defined a safety factor as the ratio of an experimentally determined dose to a dose to be accepted in humans in a particular regulatory context. If the NEL is 0.5 mg/kg body weight, the application of a safety factor of 100 will then result in a maximal allowed dose of 0.005 mg/kg body weight. This definition is still in use. Today, safety factors are widely used both in toxicology and ecotoxicology as a means to cope with uncertainty and variability in the response of organisms to toxic influence.

However, since the 1990s, the use of safety factors has often been criticized, and attempts have been made both in structural engineering and in toxicology to replace them with calculated failure probabilities. This gives us a good reason to reappraise the justification of safety factors and the ways in which they are used.

6.3 DEFINITIONS AND TERMINOLOGY

In order to achieve safety under uncertain and variable conditions we need to apply a *safety reserve*, a margin between the actual conditions that we will allow for and those conditions that would lead to failure. Safety reserves can be obtained with or without the use of numerical values that specify the size of the reserve. In both cases, it is important to note that safety reserves refer to specific types of failure. For instance, we can build a bridge that is strong enough for very heavy loads but it may still be unable to resist a relatively small sideways impact from a colliding ship.

In order to express a safety reserve numerically we need some *numerical risk indicator*, that is, a number that carries information about the magnitude of the risk we wish to avoid. To be useful, this variable has to satisfy two criteria. First, it has to be monotonically related to the probability or plausibility of failure. Second, it must be something that we can manipulate or incorporate into a design. The following are examples of such numerical risk indicators:

- The probability/plausibility of a collapse of a bridge increases as the *load* increases.
- The probability/plausibility of toxic effects from a food additive increases as the *dose* increases.
- The probability/plausibility of collision between cars, ships, or airplanes increases as the *distance* between them decreases.

Generally speaking, a *numerical safety reserve* is a function that we apply to that variable in order to determine the size of an adequate safety reserve (Clausen et al., 2006). In practice, two forms of numerical safety reserves are in use: safety factors and safety margins.

A *safety factor* is used multiplicatively, that is, we either multiply or divide the numerical risk indicator with it. For a safety factor to be useful, the risk we are trying to avoid should be at least roughly proportionate to the numerical risk indicator. For instance, consider two bridges, one of which is calculated to carry 10 tons and the other 100 tons. In the absence of more specific information, we would expect the effects of excess loads to be roughly the same for the same proportionate increase. For instance, we would expect a load of 20 tons on the first bridge to be comparable in risk with a load of 200 tons on the second (rather than with 110 tons on the second, which would indicate an additive relationship).

In structural engineering, a safety factor is most commonly applied to both the load (as a multiplication factor) and the strength or resistance of some specific component or structure (as a division factor). The value of a safety factor used in this way is always higher than 1. In some cases, the safety factor is defined as the ratio between the estimated design lifetime and the actual service lifetime. However, this is not common. Instead, the expected lifetime is usually taken into account in the material factor to be applied. The longer the design lifetime, the higher is the material factor.

It is important to note that a safety factor refers to a specific integrity-threatening mechanism. Most often, one safety factor will be required for resistance against shock or extreme load and another for fatigue resistance (i.e., resistance against relatively low but continuing load).

Furthermore, the definition of a safety factor is connected to a particular construction practice. The strength of steel, for example, can be expressed in terms of an ultimate tensile strength (i.e., the maximum stress that a material can withstand while being stretched or pulled before failing or breaking) but also in terms of a yield limit (i.e., the point at which extreme deformation starts without further increase in stress). In the American tradition, the safety factor has been applied to the ultimate tensile strength. However, in the European tradition it has been customary to apply the safety factor to the yield limit, which has in turn influenced the material development so that European steels in general have a higher yield limit to ultimate tensile strength ratio than the American ones. This shows that a safety factor can only be completely understood in the context of the construction practice in which it is used (Clausen et al., 2006).

In toxicology, safety factors usually refer to the dose as measured in relation to body weight, for instance mg of toxic substance/kg body weight. This is the standard measure used for interspecific (and interindividual) scaling. However, it has long been known that the use of body mass as a scaling factor is problematic (Pinkel. 1958), and some researchers have proposed that the surface area of the body should be used instead. Thus, the doses of toxic substances would be given as mg/m^3 rather than mg/kg. The reason for this is that the heat loss from warm-blooded animals is roughly proportionate to the surface areas of their bodies, and metabolic rates can be assumed to be roughly proportionate to the heat loss. The body surface is difficult to measure, but it is approximately proportional to the 2/3 power of the body weight ($weight^{2/3}$). However, empirical evidence indicates that metabolic rates are better correlated with the 3/4 power of the body mass ($weight^{3/4}$) (Travis and White, 1988). The choice of scaling factors for toxicological risk assessment is a scientifically complex and also controversial issue (Kang and Myung, 2011). In the absence of a consensus for a new praxis, the use of body weight still prevails.

A *safety margin* is used additively, that is, it is added to the numerical risk indicator. In order to avoid collisions between airplanes in the air, a safety margin in the form of a minimal distance is required. Surgeons removing a tumor also remove the tissue closest to the tumor. This "safety margin," or "surgical margin" is defined as the distance between the reactive zone of a tumor and the place of the surgical lesion. Typical values are 1–2 cm (Kawaguchi, 1995). The notion of a safety margin is also sometimes used in structural engineering, and is then defined as capacity minus load (O'Connor and Diamond, 1999). For some purposes, both the multiplicative (safety factor) and the additive (safety margin) approaches have been used. This applies to intestinal capacity to absorb nutrients and to the geotechnical issue of embankment reliability (Wolff, 1991).

The terminology for numerical safety reserves is far from unified. Sometimes the term "safety margin" is used for the general notion that we have called here "numerical safety reserve," that is, "in general an arithmetic relationship comparing resistance to load, whatever format it takes (safety factor, an expression of partial factors, or a difference in numbers)" (Knoll, 1976). What we have called "safety factors," that is, numerical values to be used multiplicatively, have also been called "uncertainty factors" (SCOEL, 1999), "margins of safety" (CSTEE, 2001), "factors of ignorance" (Mitroff, 1972), "contingency factors" (Capps and Thompson, 1993), and "assessment factors" (Chapman et al., 1998). Synonyms for (additive) "safety margin" include "reserve capacity" (O'Connor and Diamond, 1999) and "reserve strength" (Wolff, 1991). The terms "uncertainty factor" and "uncertainty margin" would arguably be preferable to "safety factor" and "safety margin" since these factors and margins are applied in order to cope with uncertainty but they do not necessarily give rise to safety. Currently, "assessment factor" and "uncertainty factor" are the most common terms used in toxicology, whereas "safety factor" dominates in safety engineering. We have chosen to use the latter term generically to cover both areas.

6.4 UNDERLYING THEORY AND THEORETICAL ASSUMPTIONS

In characterizing the sources of failure that safety reserves are intended to protect against, a number of distinctions from decision theory are useful. A decision is said to be made "under certainty" if the decision-maker knows, for each alternative under consideration, what will be the outcome if that alternative is chosen. Non-certainty is further divided into the categories of risk and uncertainty. The *locus classicus* for this subdivision is Knight, who pointed out that "the term 'risk', as loosely used in everyday speech and in economic discussion, really covers two things which, functionally at least, in their causal relations to the phenomena of economic organization, are categorically different" (Knight, 1935 [1921]). In some cases, "risk" means "a quantity susceptible of measurement," in other cases "something distinctly not of this character." He proposed to reserve the term "uncertainty" for cases of the non-quantifiable type, and the term "risk" for the quantifiable cases.

More recently, much focus has been put on even more extensive uncertainty than the probabilistic uncertainty emphasized by Knight. For the purposes of safety work we can distinguish between four levels of information concerning the negative effect(s) we wish to avoid:

1. *Certainty*: We are concerned with a known type of failure, and we know whether it will take place or not.
2. *Risk*: We are concerned with a known type of failure, and we know the probability that it will take place.
3. *Uncertainty*: We are concerned with a known type of failure, but we do not know the probability that it will take place.
4. *Great uncertainty*: We are concerned with some unknown or insufficiently characterized type(s) of failure, whose probability we do not know.

Safety factors and safety margins are best suited for the third of these four levels of uncertainty, that is, the level of probabilistic uncertainty. On the first level they are not needed, on the second they can be replaced by probabilities, and on the fourth it is usually difficult to define the numerical risk indicator to which the safety factor or margin can be applied.

6.4.1 Structural Engineering

It is generally agreed in the literature on structural engineering that safety factors are intended to compensate for five major types of sources of failure:

1. higher loads than those foreseen,
2. worse properties of the material than foreseen,
3. imperfect theory of the failure mechanism in question,
4. possibly unknown failure mechanisms, and
5. human error (e.g., in design) (Knoll, 1976; Moses, 1997).

These five types of sources of failure can be grouped under two distinctive types of uncertainty: those related to the stochastic nature of the world, in the risk assessment community often referred to as aleatory uncertainties, and those related to our state of knowledge, often referred to as epistemic uncertainties or state-of-knowledge uncertainties (Oberkampf et al., 2004).

The first two failure types account for aleatory variabilities, that is, they refer to the variability of empirical indicators of the propensity for failure. Aleatory uncertainty is unavoidable in structural engineering, or engineering more generally. Due to the non-deterministic nature of the world, we can never know with certainty that the load to which a construction is exposed will not exceed a certain limit. The weather conditions may, for example, be exceptionally strong. There is also always some variability in the material properties of a construction, which affects its strength. Aleatory probabilities are at least in principle accessible to probabilistic assessment, but these assessments may be difficult or impossible to perform on empirical grounds. Often the central parts of the relevant statistical distributions are fairly well known, but their tails can only be inferred under assumptions about the mathematical properties of the distribution that lack direct empirical evidence. This is the so-called *distribution arbitrariness* (Ditlevsen, 1994, Harris and Soms, 1983). Extreme value analysis often involves extrapolation to values beyond the largest or smallest observed value in order to assign probabilities to extreme events (Sanchez-Arcilla et al., 2008). Expert judgments (Slijkhuis et al., 1999) and bootstrapping techniques (Caers and Maes, 1998) are used to reduce the uncertainty of the tail-based estimates. However, bootstrapping techniques still require sufficiently long data records and a careful analysis of the influence of data sampling uncertainties (Van Noortwijk and Van Gelder, 1998).

The last three of the five failure types refer incomplete information or incomplete knowledge of some characteristic of the system or the environment. They are examples of epistemic uncertainty. Although an increase in knowledge or information can in principle lead to a reduction in such uncertainties, it is impossible to know for certain that this information will be sufficiently reliable and complete. There is always the possibility that some unknown failure mechanism has not been taken into account. It is difficult or impossible to represent these uncertainties in probabilistic terms, and therefore they belong to the category of (non-probabilizable) uncertainty. For instance, it would seem problematic to include in a calculation a probability that this calculation contains errors. It is recognized by many risk analysts that many types of uncertainties cannot be translated into probabilities (Ferson and Hajagos, 2004). In order to be used in practice, probabilistic analyses must include uncertainty representations of various types (Oberkampf et al., 2004). Different approaches have been developed for representing epistemic uncertainties in a probabilizable way.[2] However, comparisons of these different approaches with the same input data show that

[2] These include Monte Carlo simulations (Helton and Davis, 2002; Hofer et al., 2002), expert judgments and Bayesian belief systems (Stiber et al., 2004), iterative design processes (redesign) with different epistemic uncertainty treatments (Matsumura and Haftka, 2013; Villanueva et al., 2011), and a range of mathematical approaches that are collectively known as generalized information theory (GIT) (Klir and Wierman, 1998).

they can lead to very different representations in model predictions (Helton et al., 2004). This suggests that there is a limit to the representation of epistemic uncertainties in probabilistic terms. Safety factors are often the most practicable solution when probabilistic representations are misleading or too cumbersome to obtain.

6.4.2 Toxicology

In toxicology, safety factors are typically presented as compensations for various variabilities. The traditional 100-fold factor is often accounted for in terms of one factor of 10 for interspecies (animal to human) variability in response to the toxicity and another factor of 10 for intraspecies (human) variability in the same respect (Dourson and Stara, 1983). These two safety factors represent different types of relationships. Interspecies safety factors represent quotas between two stochastic variables, for example, the dose required for a certain effect in humans divided by the corresponding dose in an experimental species (the dose of a certain substance that will produce a certain effect in a randomly chosen individual being a stochastic variable), whereas intraspecies safety factors represent quotas between different percentiles, for example, in the density function for the dose required for a certain effect in humans.

In more recent approaches to toxicological safety factors, the number of subfactors is often larger, and may include factors compensating not only for variabilities but also for data deficiencies. Commonly used subfactors include factors for extrapolation from animals to humans, for intraspecies—normally human—variability, for different routes of exposure, for extrapolation from subchronic to chronic exposure, for extrapolation from LOAEL[3] to NOAEL,[4] and for inadequacy of databases (Gaylor and Kodell, 2000). The latter two subfactors will of course be different for different substances, depending on the data available for them. Therefore this approach has led to more sophisticated calculations in which the safety factor for a specific effect will be different for different substances.

Just as structural engineering, toxicological risk assessment is concerned with avoiding events with much lower probabilities than those that can be determined in actual experiments (or in epidemiological studies on exposed populations). In standard animal experiments (with typically ≤100 animals in each dose group) the doses that give rise to a 10% or 20% increase in cancer incidence can easily be determined, but it is in practice impossible to determine directly what doses give rise to, for example, an increase of 0.01% or 0.1% in that incidence. For effects that have no known threshold, such as the carcinogenic effects of radiation and many chemicals, this means that there is no safe dose that a safety factor can be applied to. Instead, protection against such effects is usually based on an assumed linear dose–response relationship, which means for instance that the excess number of cancer cases is assumed to be proportionate to the exposure. The accepted level is

[3] Lowest observed adverse effect level.

[4] No observed adverse effect level.

one with a calculated probability of fatality. Even if that probability is low, this is a different approach from that of a safety factor (that cannot be applied in this case).

6.5 USE AND IMPLEMENTATION

6.5.1 Three Types of Numerical Safety Reserves

From the viewpoint of their origins, numerical safety reserves, be it in the form of a safety factor or safety margin, can be divided into three categories:

1. *Explicitly Chosen Numerical Safety Reserves.* These are the numerical safety reserves most commonly referred to. They are used, for example, by the engineer who multiplies the foreseen load on a structure by a standard factor of, say, 3 and uses this larger value in the construction work. Similarly, the regulatory toxicologist applies an explicitly chosen safety factor when she divides the dose believed to be harmless in animals by a previously decided constant such as 100, and uses the obtained value as a regulatory limit. Explicitly chosen safety factors are also used, for example, in geotechnical engineering (Duncan, 2000; Wolff, 1991), ecotoxicology (Chapman et al., 1998), and fusion research (Wootton et al., 1997) (for plasma containment). Explicitly chosen safety margins are used in surgery (Elias et al., 1998; Kawaguchi, 1995) (the "surgical margin"), radiotherapy (Stroom, 2000) (to cope with set-up errors and internal organ motion), and air traffic safety (McKenna, 1999; Smith and Hancock, 1995).
2. *Implicit Numerical Safety Reserves.* We also use the term "safety reserve" to denote margins that have not been explicitly chosen but can, after the fact, be described as such. Implicit safety reserves have their origin in human choice, but in choices that are *not* made in terms of numerical safety reserves. As one example of this, occupational toxicology differs from food toxicology in that allowable doses are mostly determined in a case-by-case negotiation-like process that does not involve the use of generalized (fixed) safety factors. However, it is possible to infer implicit safety factors; in other words, the regulatory decision can be shown to be the same *as if* certain safety factors had been used (Hansson, 1998). Another example can be found in traffic safety research. The behavior of drivers can be described *as if* they applied a certain safety margin to the distance between their car and the car nearest ahead (Van der Hulst et al., 1999). (This margin is measured as the time headway, that is, the distance divided by the speed.)
3. *Naturally Occurring Safety Reserves.* These are the safety reserves of natural phenomena that can be calculated by comparing a structural or physiological capacity to the actually occurring loads. These safety reserves have not been chosen by human beings or by anybody else. They are our way of describing properties that have developed through evolution. Just as with

implicit safety reserves, naturally occurring safety reserves can often be described in terms of safety factors. Structural safety factors have been calculated for mammalian bones (Alexander, 1990; Rubin and Lanyon, 1982), crab claws (Palmer et al., 1999), shells of limpets (Lowell, 1985), and tree stems (Mattheck et al., 1993). Physiological uncertainty factors have been calculated, for example, for intestinal capacities such as glucose transport and lactose uptake (O'Connor and Diamond, 1999), for hypoxia tolerance in insects (Harrison et al., 1998), and for human speech recognition under conditions of speech distortion (Harris, 1959).

The reason why numerical safety reserves can be applied in descriptions of natural phenomena is that when we calculate loads—be it for natural or artificial artefacts—we do not consider more unusual loads. Resistance to unusual, unforeseen loads is as important for the survival of an organism as it is for the continued structural integrity of an anthropogenic artefact. If a limpet shell has extra strength, then it may resist predators even if its strength has been diminished due to infection by endolithic organisms. Similarly, the extra strength of tree stems makes it possible for them to withstand storms even if they have been damaged by insects. Hence, natural safety reserves—or more precisely: natural features that can be described in terms of safety reserves—are present although the physiological and mechanical capacities of animals and plants have been adapted to loads that will actually be encountered.

On the other hand, there is a limit to the evolutionary advantage of excessive safety reserves. Organisms with unnecessarily prudent safety reserves would be disadvantaged. Trees with large safety reserves are better able to resist storms, but in the competition for light reception, they may lose out to tender and high trees with smaller safety reserves (Mattheck et al., 1993). In general, the costs associated with excessive capacities result in their elimination by natural selection (O'Connor and Diamond, 1999).

There are at least two important lessons to learn from nature in this context. First, resistance to unusual loads that are sometimes difficult to foresee are essential for survival. Second, a balance will nevertheless always have to be struck between the dangers of having too little reserve capacity and the costs of having reserve capacity that is not used.

6.5.2 How Safety Factors are Determined

In the 1970s the unclear role of scientists taking part in risk policy decisions led to increasing awareness of the distinction between scientific assessments and policy decisions based on these assessments. This resulted in a standard view on the risk decision process that distinguishes strictly between its scientific and policy-making parts. This view was expressed in a 1983 report by the American National Academy of Sciences (National Research Council, 1983). The decision procedure is divided into two distinct parts to be performed consecutively. The first of these, commonly called *risk assessment*, is a scientific undertaking. It consists of collecting and assessing the relevant information and on this basis characterizing the nature and magnitude of the risk. The second procedure is called *risk management*. Contrary to risk assessment, this is not a scientific undertaking. Its starting point is the outcome of risk assessment, which it combines with economic and technological information pertaining to various ways of reducing or eliminating the risk in question, and also with political and social information. Based on this, a decision is made on what measures—if any—should be taken to reduce the risk. In order to protect risk assessments from being manipulated to meet predetermined policy objectives, it was proposed to separate risk assessment organizationally from risk management.

However, in spite of many official documents promoting this division between risk assessment and risk management, it is more often violated than obeyed. This applies to a high degree to safety factors and margins. Although these numerical values are based on policy and cannot be derived from science alone, they are usually not determined by policymakers. Instead, like other technical standards they are typically set by groups of experts who are entrusted both with collecting and interpreting the scientific data and with proposing regulation. In structural engineering, for example, the establishment of a European construction standard (Eurocodes) was characterized by organizational confluence of risk assessment and risk management (Clausen and Hansson, 2007). Similarly, in hydraulic engineering, Vrijling et al. (1998) developed a unified framework for the assessment of safety in terms of acceptable individual and societal risks levels, which they derived from accident statistics and an estimate of the value of human life. Although the authors admit that the final judgment is, in the end, a political one, the proposed approach tries to merge risk assessment and management into one decision procedure.

Experience shows that the standard distinction between scientific and policy-related issues is often difficult to uphold. Many of the technical issues involved in assessing risks are not properly speaking scientific but, in Alvin Weinberg's (1972) terminology, "transscientific," that is, they are "questions which can be asked of science and yet which cannot be answered by science." In many cases the science–policy border is deliberately blurred by stakeholders who try to transform value issues into issues of scientific fact, thus creating a "science charade" that can be difficult for the non-initiated to see through (Wagner, 1995). It may be useful to see the science–policy interface not as communication across a strictly delimited border but as a continuum of issues ranging from the purely scientific ones to policy issues with little or no scientific content, as shown in Figure 6.3 (Hansson and Aven, 2014). According to the view of the process shown in this figure, scientific experts do not only have to find and present scientific information, but also to make a broad risk evaluation that summarizes the scientific information in a way that fits into the decision-making process. Similarly, policymakers have to make their own evaluation of the scientific information as part of the process that leads up to a decision. The determination of safety factors has components in all four parts of this process. This applies not only to toxicological safety factors, whose value-ladenness has been much discussed, but also to safety factors in engineering which have not been the subject of much attention in discussions of the science–policy interface.

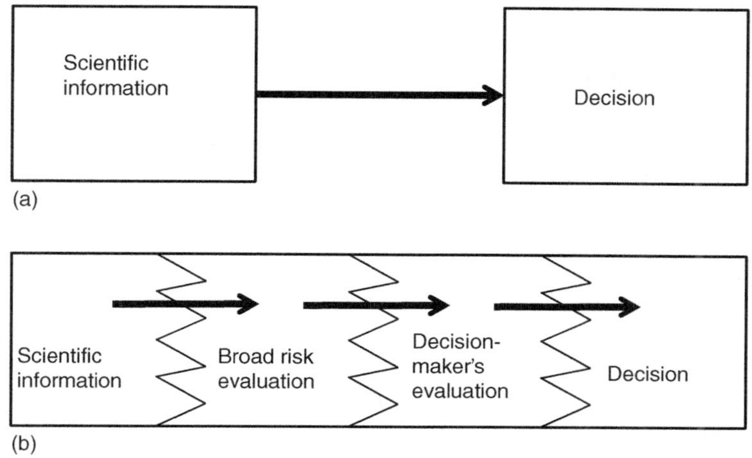

Figure 6.3. Diagram (a) shows the traditional view of the science–policy interaction in the determination of safety factors and diagram (b) an alternative, arguably more realistic picture based on Hansson and Aven (2014).

6.6 EMPIRICAL RESEARCH ON USE AND EFFICIENCY

6.6.1 Engineering

Not much empirical research has been performed on the use of safety factors in engineering. This can partly be explained by the fact that the need for some safety reserve is uncontroversial. Few engineers, if any, would deny the importance of applying at least some safety margin in the determination of the safety of a technological artefact. However, since the 1990s, the use of safety factors has often been criticized for being overly conservative. This has prompted some empirical research on establishing suitable context-specific safety factors, which are high enough to provide safety against the types of sources of failure mentioned in Section 6.4.1 but yet not overly protective.

Geotechnical engineering is one of the fields in which some empirical research has been performed to establish adequate safety factors, especially in relation to earth dams. As shown in the second example in Section 6.1, the safety of geotechnical structures is often expressed as the ratio between the resisting momentum and the driving momentum. The iterative method of establishing this factor does not easily translate into a design parameter that can be explicitly applied (cf. the numerical risk indicator mentioned in Section 6.3). Such a safety factor should rather be seen as the outcome of a safety assessment to be performed either before an earth dam will be utilized (Brady and Brown, 2004) or during its lifetime to determine the need for maintenance or improvements (Ishii et al., 2012). However, due to the complexity of the geological conditions and the material heterogeneity of the soil, the safety factors of real structures are usually difficult to obtain from laboratory tests or numerical analyses alone (Feng et al., 2011). Much research therefore goes into the development of methodologies for establishing safety factors for real structures (Potts and Zdravkovic, 2012; Sofianos et al., 2014; Xu and Niu, 2014). Many of these methods are hybrid, using both probabilistic and deterministic information, based on both laboratory tests and numerical models. Some research also goes into the establishment of a single overall safety factor against different failure mechanisms (Zhong et al., 2011). The work by Wang et al. (2013) is currently one of the few examples in which the safety factor of a geotechnical structure is derived analytically, that is, without the iterative process mentioned in Section 6.1. Wang et al. (2013) developed an analytical approach in which the outcome of a reliability analysis can be translated into a single safety factor.

Outside geotechnical engineering, the most common hybrid approach in which probabilistic information is combined with the safety factor approach is in the establishment of partial safety factors. Partial safety factors refer to the safety of specific components or elements. They are based on certain characteristic or nominal loading and strength variables for these components (Thoft-Christensen and Baker, 1982, p. 10). Currently, partial safety factors are increasingly established or calibrated on the basis of probabilistic information (so-called level-I probabilistic calculations), which allows for more specific partial safety factors. For the design of a wind turbine, for example, the effect of material fatigue on the strength of the rotor blade may be quite different from the effect of material fatigue on the strength of other moving or rotating objects (Veldkamp, 2008). Probabilistic models allow for predictions of the reliability of some components against failure which can then be translated into a suitable partial safety factor (Ronold and Christensen, 2001). This way of deriving partial safety factors is primarily meant to account for stochastic variability (higher loads than foreseen and worse material properties), not for epistemic uncertainty. It is therefore not surprising that most research into partial safety factors goes into areas where new materials are being used (Kala et al., 2009; Thieme et al., 2014; Toft et al., 2013) or where unique or new types of structures are being designed for which conventional load models do not suffice (Badalassi et al., 2014; Fischer and Schnell, 2010; MacKay and van Keulen, 2013; Moser et al., 2011).

6.6.2 Toxicology

Similar to the case in engineering, there is not much empirical research about the use of safety factors in toxicology. (Much more research has been devoted to the use of exposure limits, particularly in occupational safety.) However, in a couple of studies implicit safety factors have been calculated, with somewhat surprising results (Hansson, 1997; Schenk and Johanson, 2010).

In occupational toxicology just as in food toxicology, risk assessments often result in the determination of exposure limits intended to protect against excessive exposures. In occupational toxicology, these limits usually have the form of a maximal concentration in workplace air of the chemicals in question. Traditionally, safety factors are not used in occupational toxicology (but there is currently a movement toward their introduction). In Sweden, the process of setting occupational exposure limits (OELs) has been organized since 1978 in a way that attempts to distinguish between the scientific and the policy-related parts of the process. A Criteria Group provides scientific evidence on the basis of which the Regulations Group decides on the OELs. The Criteria Group determines the critical effect of each substance. This is the adverse effect that appears at the lowest exposure level. The Criteria Group also reports the information that is available on the exposure levels that give rise to the critical effect. If that information is precise, it is either the lowest concentration at which the effect is known to appear (EL, effect level) or the highest level at which it is reportedly known not to appear (NEL, no effect level). It can be assumed that the OEL is chosen to provide the desired safety margin to exposures that give rise to the critical effect. If an effect level has been used, we can assume that it has been divided by some factor in order to arrive at an exposure level, thus the ratio EL/OEL is the implicit safety factor. Similarly, if a no effect level has been used, then NEL/OEL is the implicit safety factor. If the safety margins are well chosen, then we would expect the following two hypotheses to be verified: (1) The safety factors are larger for serious, chronic effects than for less serious effects such as temporary eye irritation. (2) If for some specific effects there are both substances with an OEL based on effect levels and substances with an OEL based on no effect levels, then we should expect the safety factors to be higher in the former case, since we need a larger margin to an exposure level that is known to cause a noxious effect than to one that is known to be just below the levels that cause such an effect.

However, neither of these hypotheses was verified (Hansson, 1997). The implicit safety factors calculated as EL/OEL ratios were on average somewhat smaller than those calculated as NEL/OEL ratios. Even more surprisingly, the implicit safety factors for some diseases such as asthma and pulmonary fibrosis were smaller than for transient irritation effects on eyes and mucous membranes. A plausible conclusion from this study is that case-by-case determination of exposure limits has no guarantee of leading to adequate priorities. This can of course be used as an argument in favor of the consistent use of explicit safety factors. A more recent study of the European Union's OELs gave a very similar picture; in particular, safety margins did not increase with the severity of the effect as should have been expected (Schenk and Johanson, 2010).

6.7 WEAKNESSES, LIMITATIONS, AND CRITICISM

One weakness of safety factors is that *their scope is limited* to the failures that they are intended to protect against. A safety factor calculated to protect against one type of failure may be insufficient against another type of failure. The safety of levees provides an example of this. Traditionally, levees were designed to withstand a particular design water level (possibly in combination with a typical measure for the

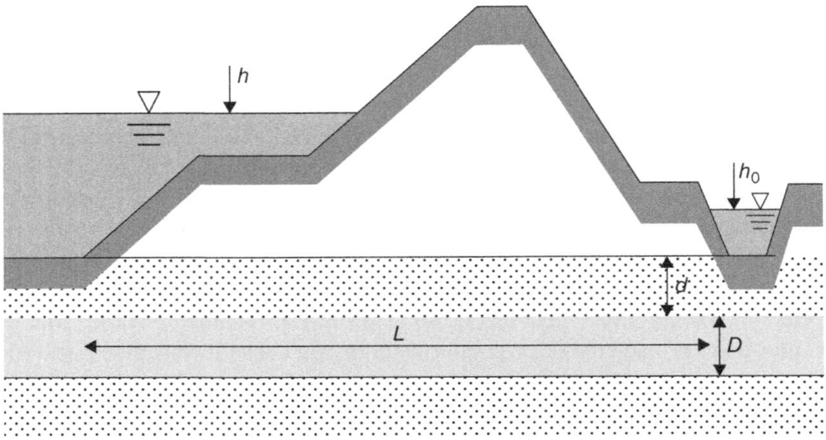

Figure 6.4. Seepage parameters. *Source*: Allsop et al. 2007. Reproduced with permission of HR Wallingford Limited.

design wave height). A safety margin was applied so that the total design water level would be lower than the crest level of the levee. This would prevent the occurrence of overtopping, that is, water flowing over the crest of the levee. During some severe flooding events in the late twentieth and the early twenty-first century, it turned out that many levees collapsed, not due to overtopping but due to piping, a failure mechanism that is caused by seepage through the structure, which in turn causes erosion at the landward slope of the levee. Piping occurs when the ratio of the water head (the difference in water level between the two sides of the levee; $h - h_0$ in Figure 6.4) and the seepage length (the length that the seeped water has to travel from one side of the levee to the other, L in the figure) is too high. Although this failure mechanism has been known for some time (Morris et al., 2008), until recently the design load conditions were still expressed in terms of the maximum water level. In this situation, applying a safety factor to heighten the crest of the levee would not protect against piping. More safety against piping could be provided by extending the seepage length L, for example, by broadening the width of the levee.

An obvious way to deal with this type of problem is to introduce additional safety factors, so that all safety-critical dimensions are covered by a safety factor. Alternatively, some dimensions can instead be covered by other safety criteria.

Another weakness of safety factors is that they *can be excessively protective* (Rackwitz, 2004). Unnecessary protection usually consumes resources. It can be costly, and arguably money spent on unneeded safety measures could be put to better use, for instance, to solve urgent safety issues elsewhere. It also consumes natural resources. More materials will be needed if we make bridges or dams excessively strong.

Several measures can be taken against this problem. Safety factors that are accustomed to particular constructions or types of constructions run a smaller risk of excessiveness than factors that cover widely different types of constructions. In some cases, probabilistic safety assessment can be used to calculate safety factors. In general, the more that is known about the risks and the damaging factors that may be involved, the better can we know what level of protection is really needed.

There are similar problems in toxicology. For instance, protection against possible negative health effects of pesticides or preservatives may have to be weighed against the positive effects of improved production and conservation of food. This can be used as an argument to adjust the safety factor to the positive effects of the substance. That is the method applied in the safety analysis of medical drugs; serious side effects will have to be accepted for potentially lifesaving drugs if no better alternative is available. In most other applications, however, such trade-offs are seldom made in an open and transparent way.

6.8 RELATIONS TO OTHER SAFETY PRINCIPLES

Safety factors are not the only method in safety engineering that aims at creating safety reserves. The same applies to safety principles such as inherent safety and multiple safety barriers. These principles have in common that they introduce some degree of redundancy in the system, which is often an efficient way to protect also against dangers for which meaningful probability estimates are not available. If one of the safety measures fails for some unknown reason or if an unforeseen failure mechanism is activated, then an additional defense can provide protection. Almost all other safety principles can be described as aiming at achieving safety reserves, but most of them do not do this numerically. In this section, we will compare safety factors and safety margins to two other numerical approaches, namely probabilistic safety analysis and cost–benefit (or risk–benefit) analysis.

6.8.1 Probabilistic Analysis

As was noted in Section 6.4, safety reserves are used to protect both against stochastic variability and against epistemic uncertainty. At first it may be surprising that safety factors are at all used for the former purpose. Why not use probabilistic risk analysis instead, since variabilities are probabilistic? There are at least four plausible answers to that question.

1. *Costs of Analysis*. Probabilistic models are usually based on complex calculations with many input variables. The costs for data acquisition and computation tend to increase rapidly with the number of input variables (Helton et al., 2004, p. 67). When the additional time also involves delays in the design and engineering process itself, the simplicity of the safety factor approach may

be an advantage, also from a cost–benefit point of view. For instance, in the building industry, the efficiency of the building process is often more important than the amount of material used.

2. *Risk of Mistakes*. The simplicity of the safety factor approach can reduce the likelihood of mistakes. The importance of simplicity in safety work is known from chemical plant design. Plants with inherently safer technologies tend to be simpler in design, easier to operate, and more tolerant of errors (Overton and King, 2006). Similarly, simpler calculation or design methods may be preferable to complex ones since they reduce the likelihood of mistakes in the calculations and hence, the likelihood of mistakes in the construction itself.
3. *Independence of Inaccessible Probabilities*. Many of the probabilities that we would need to know to make accurate probabilistic analyses are in practice inaccessible. This applies in particular to catastrophic accidents with very low probability. The safety factor approach has the advantage of not requiring access to these probabilities.
4. *Protection Against Uncertainties*. Safety factors are constructed to deal with both probabilistic and non-probabilistic uncertainty. Although stochastic variabilities are in principle probabilizable, for some of them it may in practice be impossible to obtain sufficiently reliable probability estimates.

In cases where both probabilistic and non-probabilistic uncertainties have to be taken into account, it is often advisable to factor out those for which probabilistic information is available, and perform a probabilistic analysis of them. In other words, whatever can reasonably be treated with probabilities should be treated so. To this can be added a component of residual uncertainty (both unknown probabilities and non-probabilistic uncertainty), or as Knoll (1976) called it, a "basic safety margin." The basic safety margin may either be introduced on the same level as the variabilities, as an "extra" variability, or it may be applied to the probability itself.

More generally, it should be emphasized that probabilistic analysis and safety factors are not two mutually exclusive approaches. Safety factors are indispensable for dealing with dangers that cannot be assigned meaningful probabilities (Doorn and Hansson, 2011). In many cases the best strategy is to combine the two, often as partial indicators for different types of hazards or different parts of a system (Doorn and Hansson, 2015). Safety factors can be based on probabilistic information, and safety reserves can be used as input values in probabilistic risk analysis (for instance by calculating the probability that a certain safety margin or safety factor will be upheld).

6.8.2 Cost–Benefit Analysis

Safety measures usually have disadvantages as well as advantages. The most obvious disadvantages of increased protection against health risks or mechanical failures are often increased costs. Therefore, the balance involved in the determination of a level of protection is often conceived as a competition between environment and safety on the one side and economy on the other. This is, however, a much too simplified picture. Countermeasures against a certain type of risk may at times bring about other risks, so-called *countervailing risks* (Graham and Wiener, 1995; Hansson, 2017). As already mentioned, overdesign usually has a price in terms of excess usage of energy and other natural resources.

Ruediger Rackwitz (2004) has pointed out that current safety factors and other acceptance criteria, as laid down in codes, standards, and regulations, have been set in a process that may have led to non-optimal results. As an alternative, he proposes direct cost–benefit analysis, based on the simple principle that a technical facility is optimal if and only if it maximizes the term $B(p) - C(p) - D(p)$, where p is the vector of all relevant parameters, $B(p)$ the benefit derived from the facility characterized by this vector, $C(p)$ the costs of design and construction, and $D(p)$ the (statistically expected) cost of failure.

Like other versions of risk–benefit analysis, this calculation requires that monetary values be assigned to all outcomes, including deaths, so that an overall value can be calculated for each alternative under consideration. Various methods to convert lives to monetary values have been devised, making use of expected earnings, actual sums paid to save lives, willingness to pay for reduced risks of death, etc. (The method recommended by Rackwitz is a life quality index that is based on the quality-adjusted life years used in some priority-setting practices in medical ethics (Nord, 1999).) The practice of converting lives into monetary units has obvious advantages in terms of computational convenience, but the normative validity of such procedures is far from clear (Hansson, 2007; Mishan, 1985). It is difficult to defend a "price" of human lives as anything but a technical necessity. Stuart Hampshire has warned that the explicit assignment of monetary values to lives may encourage "a coarseness and grossness of moral feeling, a blunting of sensibility, and a suppression of individual discrimination and gentleness" (Hampshire, 1972). This is a warning that should not be taken lightly. But it also has to be conceded that in a practice based on safety factors, risks of human lives are weighed against costs, albeit less transparently.

Some argue that economic optimization of risk management measures is in fact the main objective of probabilistic risk analysis (Guikema and Paté-Cornell, 2002). Traditional approaches in safety engineering, such as safety factors, provide regulatory bounds that may sometimes be overly conservative (Chapman et al., 1998). There is, for instance, no way to translate the difference between using the safety factor 2.0 and the safety factor 3.0 in the design of a bridge into a quantifiable effect on safety. Without a quantifiable effect (such as reduced expected number of fatalities) it is not possible to calculate the marginal cost of risk reduction, and therefore economic optimization of design is not possible. In contrast, a probabilistic safety analysis that provides accident probabilities as outcomes makes it possible to calculate the expected gains from a safer design. This is what is needed for an optimization in economic terms of the trade-off between risks and benefits (Moses, 1997; Paté-Cornell, 1996).

However, this approach puts very high demands on the probabilities that are outputs of the analysis. The availability of probability estimates answering to these demands differs between different contexts. Some relatively small and standardized infrastructure projects have effects that can be described fairly accurately in probabilistic terms. This applies, for instance, to some safety measures in road traffic such as central barriers on highways (Mak et al., 1998) or pedestrian crosswalks at intersections (Zegeer et al., 2006), for which the expected number of saved lives can be estimated with reasonable accuracy and weighed against the economic costs. On the other hand, in case of larger and more complex infrastructure projects, the probabilistic quantification of the effects of safety measures is often not considered accurate enough to be used as direct input into economic analysis. The safety of gravity dams, for example, is largely dependent on seismic activity and how the structure responds to it. Both can at most be quantified artificially and roughly, making it difficult to provide accurate accident probabilities (Abbas and Manohar, 2002). In cases like this it is therefore recommended to develop a robust structural design rather than an economically optimized one (Takewaki, 2005). Similar problems are faced in the design of other large infrastructure projects, such as flood defense structures and offshore facilities.

In summary, the use of numerical safety reserves can be overruled by cost–benefit calculations if these calculations are sufficiently accurate and reliable, but unfortunately there are many important areas of engineering where cost–benefit analysis of potential safety problems is too uncertain to be relied on for that purpose.

ACKNOWLEDGMENT

We would like to thank Carl Rollenhagen and Linda Schenk for valuable comments on an earlier version.

REFERENCES

Abbas, A. M., and Manohar, C. S. (2002). Investigations into critical earthquake load models within deterministic and probabilistic frameworks. *Earthquake Engineering and Structural Dynamics*, 31(4), 813–832.

Alexander, R. M. (1990). *Animals*. Cambridge: Cambridge University Press.

Allsop, W., Kortenhaus, A., Morris, M. W., Buijs,F., Hassan, R., Young, M., Doorn, N., Van der Meer, J., Van Gelder, P. H. A. J. M., Dyer, M. M., Redaelli, M., Utily, S., Visser, P., Bettess, R., Lesniewska, D., and Ter Horst, W. (2007). Failure mechanisms for flood defence structures. FLOODsite report number T04-06-01. Available at: www.floodsite.net.

Badalassi, M., Biolzi, L., Royer-Carfagni, G., and Salvatore, W. (2014). Safety factors for the structural design of glass. *Construction and Building Materials*, 55, 114–127.

Barlow, S., and Schlatter, J. (2010). Risk assessment of carcinogens in food. *Toxicology and Applied Pharmacology*, 243, 180–190.

Biasioli, F., Mancini, G., Just, M., Curbach, M., Walraven, J., Gmainer, S., Arrieta, J., Frank, R., Morin, C., and Robert, F. (2014). EUROCODE 2: Background and Applications: Design of

Concrete Buildings. Worked Examples. JRC Scientific and Policy Reports. Ispra European Commission.

Brady, B. H. G., and Brown, E. T. (2004). *Rock Mechanics for Underground Mining* (3rd ed.). Dordrecht, The Netherlands: Kluwer.

Caers, J., and Maes, M. A. (1998). Identifying tails, bounds and end-points of random variables. *Structural Safety*, 20(1), 1–23.

Capps, R. W., and Thompson, J. R. (1993). Statistical safety factors reduce overdesign. *Hydrocarbon Processing*, 72, 77–78.

Carpenter, S., Walker, B., Anderies, J. M., and Abel, N. (2001). From metaphor to measurement: Resilience of what to what. *Ecosystems*, 4(8), 765–781.

Chapman, P. M., Fairbrother, A., and Brown, D. (1998). A critical evaluation of safety (uncertainty) factors for ecological risk assessment. *Environmental Toxicology and Chemistry*, 17(1), 99–108.

Clausen, J., and Hansson, S. O. (2007). Eurocodes and REACH: Differences and similarities. *Risk Management*, 9(1), 19–35.

Clausen, J., Hansson, S. O., and Nilsson, F. (2006). Generalizing the safety factor approach. *Reliability Engineering and System Safety*, 91(8), 964–973.

CSTEE. (2001). Position paper on Margins of Safety (MOS) in human health risk assessment expressed at the 22nd CSTEE plenary meeting, Brussels, March 6–7, 2001. Available at: https://ec.europa.eu/health/node/42403_en

Cutter, S. L., Burton, C. G., and Emrich, C. T. (2010) Disaster resilience indicators for benchmarking baseline conditions. *Journal of Homeland Security and Emergency Management*, 7(1), 1–22.

Ditlevsen, O. (1994). Distribution arbitrariness in structural reliability. In: G. I. Schueller, M. Shinozuka, and J. T. P. Yao (Eds.), *Structural Safety and Reliability* (Vols. 1–3 - ICOSSAR '93, pp. 1241–1247). Rotterdam, The Netherlands: Balkema.

Doorn, N. (2017). Resilience indicators: Opportunities for including distributive justice concerns in disaster management. *Journal of Risk Research*, 20(6), 711–731. doi: 10.108013669877.2015.1100662

Doorn, N., and Hansson, S. O. (2011). Should probabilistic design replace safety factors? *Philosophy and Technology*, 24(2), 151–168.

Doorn, N., and Hansson, S. O. (2015). Design for the value of safety. In: M. J. Van den Hoven, P. Vermaas, and I. R. Van de Poel (Eds), *Handbook of Ethics and Values in Technological Design*. Dordrecht: Springer: doi: 10.10071978-1094-1007-6994-1006_1019-1001

Dourson, M. L., and Stara, J. F. (1983). Regulatory history and experimental support of uncertainty (safety) factors. *Regulatory Toxicology and Pharmacology*, 3(3), 224–238.

Duncan, J. M. (2000). Factors of safety and reliability in geotechnical engineering. *Journal of Geotechnical and Geoenvironmental Engineering*, 126, 307–316.

Dybing, E. (2002). Development and implementation of the IPCS conceptual framework for evaluating mode of action of chemical carcinogens. *Toxicology*, 181–182, 121–125.

Elias, D., Cavalcanti, A., Sabourin, J. C., Pignon, J. P., Ducreux, M., and Lasser, P. (1998). Results of 136 curative hepatectomies with a safety margin of less than 10 mm for colorectal metastasis. *Journal of Surgical Oncology*, 69(2), 88–93.

Feng, J., Wei, H., Jianwen, P., Jian, Y., Jinting, W., and Chuhan, Z. (2011). Comparative study procedure for the safety evaluation of high arch dams. *Computers and Geotechnics*, 38, 306–317.

Ferson, S., and Hajagos, J. G. (2004). Arithmetic with uncertain numbers: Rigorous and (often) best possible answers. *Reliability Engineering and System Safety*, 85(1–3), 135–152.

Fischer, A., and Schnell, J. (2010). Optimized partial safety factors for calculation of reinforced concrete components in existing structures. *Bauingenieur*, 85, 315–323.

Folke, C. (2006). Resilience: The emergence of a perspective for social-ecological systems analyses. *Global Environmental Change*, 16(3), 253–267.

Gaylor, D. W., and Kodell, R. L. (2000). Percentiles of the product of uncertainty factors for establishing probabilistic reference doses. *Risk Analysis*, 20(2), 245–250.

Graham, J., and Wiener, J. (1995). *Risk versus Risk*. Cambridge, MA: Harvard University Press.

Guikema, S. D., and Paté-Cornell, M. E. (2002). Component choice for managing risk in engineered systems with generalized riskcost functions. *Reliability Engineering and System Safety*, 78(3), 227–238.

Hampshire, S. (1972). *Morality and Pessimism*. Cambridge: Cambridge University Press.

Hansson, S. O. (1997). Critical effects and exposure limits. *Risk Analysis*, 17, 227–236.

Hansson, S. O. (1998). *Setting the Limit: Occupational Health Standards and the Limits of Science*. New York: Oxford University Press.

Hansson, S. O. (2007) Philosophical problems in cost-benefit analysis. *Economics and Philosophy*, 23, 163–183.

Hansson, S. O. (2017) Five caveats for risk-risk analysis. *Journal of Risk Research*, 20(8), 984–987.

Hansson, S. O., and Aven, T. (2014). Is risk analysis scientific? *Risk Analysis*, 34(7), 1173–1183.

Harris, J. D. (1959). Combinations of distortion in speech. *AMA Archives of Otolaryngology*, 72, 227–232.

Harris, B., and Soms, A. P. (1983). A note on a difficulty inherent in estimating reliability from stress strength relationships. *Naval Research Logistics*, 30(4), 659–663.

Harrison, J. F., Greenlee, K., and Lighton, J. R. B. (1998). Safety margins and the hypoxia sensitivity of insects. *FASEB Journal*, 12(4), S994.

Helton, J. C., and Davis, F. J. (2002). Illustration of sampling-based methods for uncertainty and sensitivity analysis. *Risk Analysis*, 22(3), 591–622.

Helton, J. C., Johnson, J. D., and Oberkampf, W. L. (2004). An exploration of alternative approaches to the representation of uncertainty in model predictions. *Reliability Engineering and System Safety*, 85(1–3), 39–71.

Hofer, E., Kloos, M., Krzykacz-Hausmann, B., Peschke, J., and Woltereck, M. (2002). An approximate epistemic uncertainty analysis approach in the presence of epistemic and aleatory uncertainties. *Reliability Engineering and System Safety*, 77(3), 229–238.

Ishii, Y., Ota, K., Kuraoka, S., and Tsunaki, R. (2012). Evaluation of slope stability by finite element method using observed displacement of landslide. *Landslides*, 9(3), 335–348.

Kala, Z., Melcher, J., and Puklicky, L. (2009). Material and geometrical characteristics of structural steels based on statistical analysis of metallurgical products. *Journal of Civil Engineering and Management*, 15(3), 299–307.

Kang, H. E., and Myung, G.L. (2011). Approaches for predicting human pharmacokinetics using interspecies pharmacokinetic scaling. *Archives of Pharmacal Research*, 34(11), 1779–1788.

Kawaguchi, N. (1995). New method of evaluating the surgical margin and safety margin for musculoskeletal sarcoma, analysed on the basis of 457 surgical cases. *Journal of Cancer Research and Clinical Oncology*, 121, 555–563.

Klimko, K. (2011). FDA's contradictory decisions related to the Delaney clause. Available at: http://nrs.harvard.eduurn-3:HUL.InstRepos:8963872

Klir, G. J., and Wierman, M. J. (1998). *Uncertainty-Based Information: Elements of Generalized Information Theory*. Heidelberg: Physica-Verlag.

Knight, F. H. (1935[1921]). *Risk, Uncertainty and Profit*. Boston, MA: Houghton Mifflin.

Knoll, F. (1976). Commentary on the basic philosophy and recent development of safety margins. *Canadian Journal of Civil Engineering*, 3(3), 409–416.

Larsen, J. C. (2012). Artificial sweeteners. A brief review of their safety issues. Nutrafoods, 11, 3–9.

Loehr, J. E., Finley, C. A., and Huaco, D. (2006). Procedures for design of earth slopes using LRFD. Report no. OR 06-10. University of Missouri, Columbia, MO.

Lowell, R. B. (1985). Selection for increased safety factors of biological structures as environmental unpredictability increases. *Science*, 228(4702), 1009–1011.

MacKay, J. R., and van Keulen, F. (2013). Partial safety factor approach to the design of submarine pressure hulls using nonlinear finite element analysis. *Finite Elements in Analysis and Design*, 65, 1–16.

Mak, K. K., Sicking, D. L., and Zimmerman, K. (1998). Roadside safety analysis program – a cost-effectiveness analysis procedure. *General Design and Roadside Safety Features*, 1647, 67–74.

Matsumura, T., and Haftka, R. T. (2013). Reliability based design optimization modeling future redesign with different epistemic uncertainty treatments. *Journal of Mechanical Design*, 135(9), 9.

Mattheck, C., Bethge, K., and Schafer, J. (1993). Safety factors in trees. *Journal of Theoretical Biology*, 165(2), 185–189.

McKenna, J. T. (1999). ATC, political pressures squeeze safety margins. *Aviation Week and Space Technology*, 151(5), 44–45.

Mishan, E. J. (1985). Consistency in the valuation of life: A wild goose chase? In: E. Frankel Paul, F. D. J. Miller, and J. Paul (Eds.) *Ethics and Economics*. Oxford: Basil Blackwell.

Mitroff, I. I. (1972). On the social psychology of the safety factor: A case study in the sociology of engineering science. *Management Science*, 18, 454–469.

Morris, M. W., Allsop, W., Buijs, F. A., Kortenhaus, A., Doorn, N., and Lesniewska, D. (2008). Failure modes and mechanisms for flood defence structures. In: P. Samuels, S. Huntington, W. Allsop, and J. Harrop (Eds.), *Flood Risk Management: Research and Practice* (pp. 693–701). London: Taylor & Francis Group.

Moser, T., Strauss, A., and Bergmeister, K. (2011). Partial safety factors for reinforced concrete structures verification of the shear capacity. *Beton- Und Stahlbetonbau*, 106(12), 814–825.

Moses, F. (1997). Problems and prospects of reliability-based optimization. *Engineering Structures*, 19(4), 293–301.

National Research Council. (1983). *Risk Assessment in the Federal Government: Managing the Process*. Washington, DC: National Academy Press.

Nord, E. (1999). *Cost-Value Analysis in Health Care: Making Sense out of QALYs*. Cambridge: Cambridge University Press.

Norris, F. H., Stevens, S. P., Pfefferbauam, B., Wyche, K. F., and Pfefferbaum, R. L. (2008). Community resilience as a metaphor, theory, set of capacities, and strategy for disaster readiness. *American Journal of Community Psychology*, 41(1–2), 127–150.

Oberkampf, W. L., Helton, J. C., Joslyn, C. A., Wojtkiewicz, S. F., and Ferson, S. (2004). Challenge problems: Uncertainty in system response given uncertain parameters. *Reliability Engineering and System Safety*, 85(1–3), 11–19.

O'Connor, T. P., and Diamond, J. (1999). Ontogeny of intestinal safety factors: Lactase capacities and lactose loads. *American Journal of Physiology – Regulatory Integrative and Comparative Physiology*, 45, R753–R765.

Overton, T., and King, G. M. (2006). Inherently safer technology: An evolutionary approach. *Process Safety Progress*, 25(2), 116–119.

Palmer, A. R., G. M. Taylor, and Barton, A. (1999). Cuticle strength and the size-dependence of safety factors in cancer crab claws. *Biological Bulletin*, 196(3), 281–294.

Paté-Cornell, M. E. (1996). Uncertainties in risk analysis: Six levels of treatment. *Reliability Engineering and System Safety*, 54(2–3), 95–111.

Pinkel, D. (1958). The use of body surface area as a criterion of drug dosage in cancer chemotherapy. *Cancer Research*, 18, 853–856.

Potts, D. M., and Zdravkovic, L. (2012). Accounting for partial material factors in numerical analysis. *Geotechnique*, 62(12), 1053–1065.

Priebe, P. M., and Kauffman, G. B. (1980). Making governmental policy under conditions of scientific uncertainty: A century of controversy about saccharin in congress and the laboratory. *Minerva*, 18(4), 556–574.

Rackwitz, R. (2004). Optimal and acceptable technical facilities involving risks. *Risk Analysis*, 24(3), 675–695.

Randall, F. A. (1976). The safety factor of structures in history. *Professional Safety*, January: 12–28.

Rasmussen, J. (1997). Risk management in a dynamic society: A modelling problem. *Safety Science*, 27(2): 183–213.

Ronold, K. O., and Christensen, C. J. (2001). Optimization of a design code for wind-turbine rotor blades in fatigue. *Engineering Structures*, 23(8), 993–1004.

Rubin, C., and Lanyon, L. (1982). Limb mechanics as a function of speed and gait. *Journal of Experimental Biology*, 101(1), 187–211.

Sanchez-Arcilla, A., Gonzalez-Marco, D., Doorn, N., and Kortenhaus, A. (2008). Extreme values for coastal, estuarine, and riverine environments. *Journal of Hydraulic Research*, 46, 183–190.

Schenk, L., and Johanson, G. (2010). Use of uncertainty factors by the SCOEL in their derivation of health-based occupational exposure limits. *Critical Reviews in Toxicology*, 40(9), 791–798.

Scientific Committee for Food. (1977). Saccharin (Opinion expressed on 24 June 1977). Reports of the Scientific Committee for Food, Fourth Series, Commission of the European Communities. Available at: http://ec.europa.eufoodfsscscfreportsscf_reports_04.pdf

Scientific Committee for Food. (1995). Opinion on saccharin and its sodium, potassium and calcium salts (Expressed on 2 June 1995). Annex III to document III515797, CSADDEDUL148-FINAL, European Commission, Directorate-General III. Available at: http://ec.europa.eufoodfsscoldcomm7out26_en.pdf

SCOEL. (1999). Methodology for the derivation of occupational exposure limits: Key documentation. EUR 19253 EN, European Commission.

Slijkhuis, K. A. H., Van Gelder, P. H. A. J. M., Vrijling, J. K., and Vrouwenvelder, A. C. W. M. (1999). On the lack of information in hydraulic engineering models. In: *Safety and Reliability: Proceedings of the ESREL '99*. Rotterdam: Balkema.

Smith, K., and Hancock, P. A. (1995). Situation awareness is adaptive, externally directed consciousness. *Human Factors*, 37(1), 137–148.

Sofianos, A. I., Nomikos, P. P., and Papantonopoulos, G. (2014). Distribution of the factor of safety, in geotechnical engineering, for independent piecewise linear capacity and demand density functions. *Computers and Geotechnics*, 55, 440–447.

Stiber, N. A., Small, M. J., and Pantazidou, M. (2004). Site-specific updating and aggregation of Bayesian belief network models for multiple experts. *Risk Analysis*, 24(6), 1529–1538.

Stroom, J. (2000). Safety margins for geometrical uncertainties in radiotherapy. *Medical Physics*, 27(9), 2194.

Takewaki, I. (2005). A comprehensive review of seismic critical excitation methods for robust design. *Advances in Structural Engineering*, 8(4), 349–363.

TAW. (2001). Technisch Rapport Waterkerende grondconstructies: Geotechnische aspecten van dijken, dammen en boezemkaden. Delft, Technische Adviescommissie voor de Waterkeringen (TAW)Expertise Netwerk Water (ENW).

Terzaghi, K., Peck, R. B., and Mesri, G. (1996). *Soil Mechanics in Engineering Practice* (3rd ed.). London: John Wiley & Sons.

Thieme, M., Boehm, R., Gude, M., and Hufenbach, W. (2014). Probabilistic failure simulation of glass fibre reinforced weft-knitted thermoplastics. *Composites Science and Technology*, 90, 25–31.

Thoft-Christensen, P., and Baker, M. J. (1982). *Structural Reliability Theory and Its Applications*. Berlin: Springer-Verlag.

Toft, H. S., Branner, K., Mishnaevsky, L., and Sorensen, J. D. (2013). Uncertainty modelling and code calibration for composite materials. *Journal of Composite Materials*, 47(14), 1729–1747.

Travis, C. C., and White, R. K. (1988). Interspecific scaling of toxicity data. *Risk Analysis*, 8, 119–125.

Van der Hulst, M., Meijman, T., and Rothengatter, T. (1999). Anticipation and the adaptive control of safety margins in driving. *Ergonomics*, 42(2), 336–345.

Van Noortwijk, J. M., and Van Gelder, P. H. A. J. M. (1998). Bayesian estimation of quantiles for the purpose of flood prevention. In: 26th International Conference on Coastal Engineering, Copenhagen, ASCE.

Veldkamp, D. (2008). A probabilistic evaluation of wind turbine fatigue design rules. *Wind Energy*, 11(6), 655–672.

Villanueva, D., Haftka, R. T., and Sankar, B. V. (2011). Including the effect of a future test and redesign in reliability calculations. *AIAA Journal*, 49(12), 2760–2769.

Vrijling, J. K., van Hengel, W., and Houben, R. J. (1998). Acceptable risk as a basis for design. *Reliability Engineering and System Safety*, 59(1), 141–150.

Wagner, W. E. (1995). The science charade in toxic risk regulation. *Columbia Law Review*, 95, 1613–1723.

Wang, P., Liu, D. Y., Huang, H. B., and Liu, D. S. (2013). Spoke model for calculating reliability index and safety factor of slopes. Mathematical Problems in Engineering, 2013, 7.

Weinberg, A. M. (1972). Science and trans-science. *Minerva*, 10, 209–222.

Wolff, T. F. (1991). Embankment reliability versus factor of safety: Before and after slide repair. *International Journal for Numerical and Analytical Methods in Geomechanics*, 15, 41–50.

Wootton, A. J., Wiley, J. C., Edmonds, P. H., and Ross, D. W. (1997). Compact tokamak reactors. *Nuclear Fusion*, 37(7), 927–937.

Xu, J. C., and Niu, F. S. (2014). Safety factor calculation of soil slope reinforced with piles based on Hill's model theory. *Environmental Earth Sciences*, 71(8), 3423–3428.

Zegeer, C. V., Carter, D. L., Hunter, W. W., Stewart, J. R., Huang, H., Do, A., and Sandt, L. (2006). Index for assessing pedestrian safety at intersections. In: *Transportation Research Record, No. 1982: Pedestrians and Bicycles* (76–83). Washington: Transportation Research Board, National Academy of Sciences.

Zhong, D. H., Sun, Y. F., and Li, M. C. (2011). Dam break threshold value and risk probability assessment for an earth dam. *Natural Hazards*, 59(1), 129–147.

FURTHER READING

Burr, A., and Cheatham, J. (1995). *Mechanical Design and Analysis* (2nd ed., Section 5.2). Prentice-Hall.

ECHA. (2012). Guidance on information requirements and chemical safety assessment. Chapter R.8: Characterisation of dose [concentration]-response for human health. Version: 2.1. Available at: http://echa.europa.eudocuments1016213632information_requirements_r8_en.pdf

Elishakoff, I. (2004). *Safety Factors and Reliability: Friends or Foes?* Dordrecht: Kluwer Academic Publishers.

Schenk, L., and Johanson, G. (2011). A quantitative comparison of the safety margins in the European indicative occupational exposure limits and the derived no-effect levels for workers under REACH. *Toxicological Sciences*, 121(2): 408–416.

Part II

INFORMATION AND CONTROL

7

EXPERIENCE FEEDBACK

Urban Kjellén

7.1 INTRODUCTION

Experience feedback is a basic principle in any system for the management of safety. The most obvious example is the use of experience from accident investigations in the design of measures to avoid similar accidents in the future. Experience feedback is closely linked to the principles of continual improvement in the area of quality management, where feedback from a process is evaluated against organizational goals and used to implement step-wise improvements in an ongoing process (Juran, 1989).

The focus in this article is on experience feedback as a process managed at the organizational level in order to prevent accidents. Experience feedback always involves learning and development of competence at the individual level in an organization. This aspect is an important step for an organization to learn, but the management of individual knowledge and competence lies outside the scope of this chapter.

7.1.1 Example

Figure 7.1 shows an illustrative example of the effects of improved experience feedback as part of a management strategy to improve the control climate at a plant. An

Handbook of Safety Principles, First Edition. Edited by Niklas Möller, Sven Ove Hansson, Jan-Erik Holmberg, and Carl Rollenhagen.

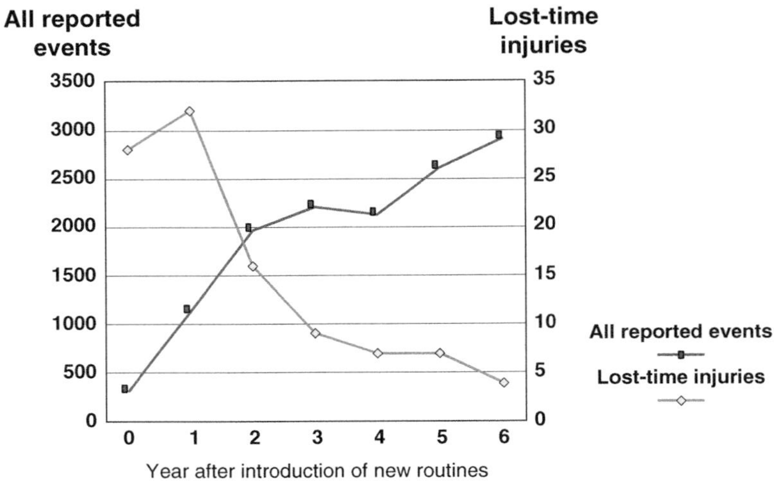

Figure 7.1. Development in the reporting of lost-time injuries and incidents at an aluminum smelter after the introduction of new routines. *Source*: Kjellén 2000. Reproduced with permission of Taylor & Francis.

aluminum plant went through a major restructuring process, including the introduction of a comprehensive system for the management of safety, based on the principles of Internal Control (see further Section 7.2). The safety management system included elements to improve experience feedback from the workplaces through the reporting and follow-up of incidents. The figure shows how the reporting of incidents increased because of this initiative while there was a simultaneous 90% reduction in the number of lost-time injuries.

7.2 ORIGIN AND HISTORY

Experience feedback from observations of effects of actions and learning followed by adaptation of behavior to avoid reproducing past mistakes is part of the human traits both at the individual and group/organizational levels. Learning to avoid accidents in the future is simply a survival strategy of the human species.

The health-care sector accounts for an early example of systematic experience feedback for use in improvements (Bergman, 2012). Early in the twentieth century, E.A. Codman founded "The End Result hospital," where all problems related to medical treatment were documented and published in order to make improvements in the treatment of patients possible.

A review of the origin and history of experience feedback in the prevention of accidents cannot be done without putting the concept into a context of developments in the fields of quality and safety management and safety culture.

In the advent of Industrial Safety Management, Heinrich devised an approach to accident prevention through 10 axioms that was based on the principles of Scientific Management (Heinrich, 1931; Taylor, 1911). Although experience feedback is not explicitly mentioned, his approach advocates learning from non-injury accidents as well as unsafe acts to devise accident prevention techniques. Unsafe acts were regarded by Heinrich as the dominant causes of accidents. Even if some of Heinrich's axioms today are regarded as misleading, his rational and systematic approach has set the standard for subsequent developments in safety management.

Similarly, the registration of data on accidents and feedback to decision-makers of statistics on safety performance has a long history. This subject was placed on the agenda in 1923 at the First International Conference of Labour Statisticians. The resolution from the conference covered the classification of accidents and the formulas for calculation of frequency and severity rates. In the United States, the (ANSI) Z16 Committee was formed in 1926 to revise existing standards on methods of recording and measurement of work injury experience.

The parallel developments of quality management have had significant impact on the management of safety as we know it today. Product quality control according to the principles of scientific management became a prerequisite for mass production. During World War II, the US militaries introduced statistical quality control techniques to speed up production without compromising quality and safety. In the quality revolution in Japan, following World War II, American advisors Joseph Juran and Edwards Deming lifted the quality management concept, which at that time focused on product inspections and testing. They introduced methods to implement refinements at the organizational and management levels to improve product and service quality in response to continuous feedback.

The statistical quality control techniques were adapted to the safety field by pioneers like Thomas Rockwell. In response to the so-called "problem of safety performance measurement," he devised a behavioral sampling technique that applied statistical sampling techniques in the collection and analysis of data on deviations from accepted safe work practices and conditions (Rockwell, 1959). Important principles in these so-called Behavioral-Based Safety (BBS) programs are to reinforce safe behavior through target setting and experience feedback to the workers at the sharp end (Komaki et al., 1978).

Another important milestone is the introduction by Gibson of the concept of unwanted energy transfer as an injury agent in most accidents (Gibson, 1961). Haddon followed up by defining 10 generic principles for the prevention of harm (injury) from transfer of energy (Haddon, 1970). Researchers working for the US Atomic Energy Commission (AEC) merged the energy model by Gibson and Haddon with the US military standard for quality control. This new concept of safety management was documented in a number of publications from the System Safety Development Centre (SSDC) working for AEC and later the US Department of Energy (DOE). The so-called MORT (Management Oversight and Risk Tree) handbook represented a milestone (Johnson, 1980). Although experience feedback was not an explicit concept in MORT, a management system model similar to the so-called Plan-Do-Check-Act (PDCA) cycle formed one of the two cornerstones of MORT (the other was the energy model).

In the late 1970s after the discovery of oil on the Norwegian continental shelf a decade earlier, the Norwegian Petroleum Directorate (NPD) took a strategic decision on Self-Regulation and Internal Control. The oil companies were required to develop safety management systems to ensure compliance with the safety regulations. These should primarily be risk based and of a goal-setting type as opposed to prescriptive regulations. NPD stood in close contact with consultants from SSDC in developing requirements to such systems. The approach selected by NPD was also inherent in the goal-setting regime selected by the UK authorities in the Health and Safety at Work Act of 1974, where society sets the overall goals on safety and the company can find its own ways on meeting them.

Disasters in Europe like at Flixborough in 1974, Seveso in1976, and Piper Alpha in 1987 resulted in an overall revamp of European regulations in line with the risk-based and goal-setting regime of Internal Control. An example is the so-called "Seveso Directive" (European Directive 82/501/EEC).

Central to the regulations at the European level and the national regulations in the United Kingdom and Norway is that safety shall be managed in a systematic way in accordance with the principles of the international ISO 9000 series of quality management standards or similar national standards. Typically, such systems consist of a number of elements based on the PDCA cycle (Deming, 1993). Figure 7.2 shows an example of a safety management system including elements according to OHSAS 18001 (OHSAS Project Group, 2007). The model demonstrates the central role of experience feedback in a safety management system (Hudson, 2001).

The international standard on the management of occupational health and safety, OHSAS 18001, first published in 1999, is compatible with the principles of ISO 9000 series of quality management standards (OHSAS Project Group, 2007). Again, the PDCA cycle and the principles of continual improvement are central elements in this standard.

Safety culture came into focus in the wake of the Chernobyl disaster that demonstrated the significance of management and human factors on safety performance (IAEA, 1991). It is seen as a concept that describes the shared values within an organization that define the attitudes and behaviors of its members of significance to health and safety performance (Cooper, 2000). A central attribute of a good safety culture is an organization's ability to learn and improve through feedback systems for monitoring and analysis of accidents and incidents (Health and Safety Laboratory, 2002). This way of thinking is closely linked to knowledge management (Nonaka and Takeuchi, 1995; Senge, 1990).

Resilience engineering has developed out of the need to better understand how a system adjusts to changes and disturbances (Hollnagel et al., 2006). The perspective of experience feedback in safety management is widened from why things go wrong

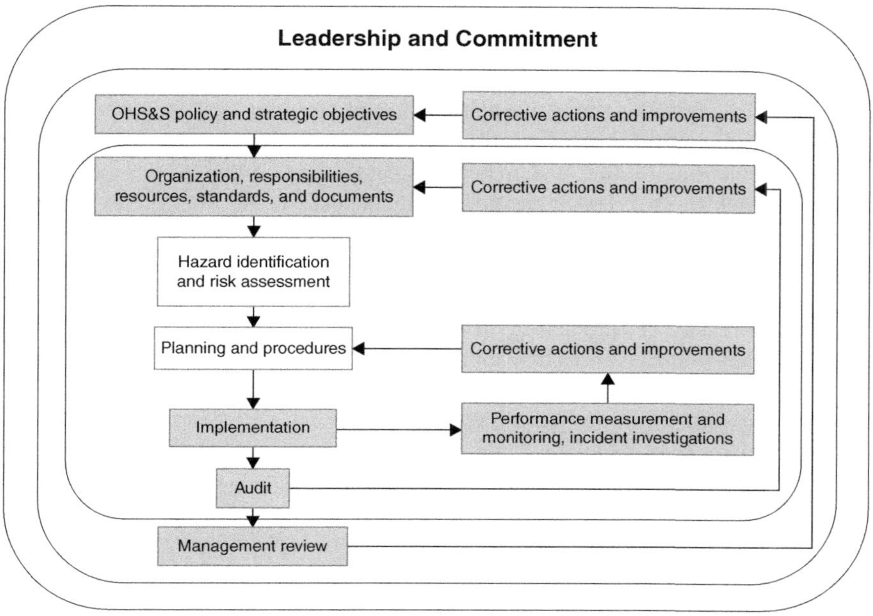

Figure 7.2. The elements of a typical safety management system.

to include experiences from successful adaptations. This will increase the scope of experience feedback extensively.

7.3 DEFINITIONS

There is no generally accepted definition of experience feedback in the safety literature. The concept consists of two words, each of which has different meanings depending on the context (Oxford University Press, 2012). The meaning of experience according to the dictionary coming closest to its use in this article is "the actual observation of facts or events, considered as a source of knowledge." Similarly, feedback is understood as "the modification, adjustment, or control of a process by a result or effect of the process."

A distinction is also made between negative and positive feedback. Negative feedback takes place when a process is controlled through input that is a function of the difference between the real and the desired output. The aim is to reduce this difference, and it is essential in maintaining stability in a system. In positive feedback, a system changes through amplification of divergences in output. It may lead to destabilization of the system but is also a condition for change (evolution, growth) and the ability of a system to reach a new point of equilibrium. Safety management is primarily based on negative feedback, that is, learning from "failures." Learning from "successes," involving positive feedback, is not so common but has achieved increasing attention (Hollnagel, 2014).

Feedforward control is the active anticipation of problems and their timely prevention, rather than after-the-fact reaction (Kreitner, 2009). Anticipatory regulation is based on prediction of future system behavior and requires more insight into a system than feedback control. This is exemplified by the use of risk assessments as input to the management of risks due to anticipated changes of a system.

We will here use the definition of experience feedback in Wiley's *Encyclopedia of Quantitative Risk Analysis and Assessment* (Melnick and Everitt, 2008). Experience feedback is defined as "the process by which information on the results of an activity is fed back to decision-makers as new input to modify and improve subsequent activities." Kamsu Foguem et al. (2008) use a similar definition from the perspective of knowledge management. Experience feedback is from their perspective seen as "an initiative to convey lessons learnt applicable to an operational, tactical or strategic level such that, when reused, this knowledge positively impacts on the results of the organisation."

It is important to note that both definitions involve learning and improvements, as opposed to the simple feedback loop aimed at bringing a system back to the predefined target.

7.4 UNDERLYING THEORIES AND ASSUMPTIONS

7.4.1 Feedback Cycle for the Control of Anything

A simple form of experience feedback is illustrated by the negative feedback cycle in Figure 7.3. Sensors provide measurements of the system (control subject), and the results are compared to norms. In case the results deviate from the norm, the gap is fed to the actuators that initiate corrective actions to move the system in a direction that provides results closer to the norm. A simple thermostat for temperature regulation is designed according to these principles, but the feedback cycle models basic principles in the management of quality (and safety) as well.

Juran's feedback cycle also illustrates that a certain degree of learning may take place in addition to the simple feedback cycle through "feedback to publishers of norms." This is when experiences are used to change the norms that govern the wanted behavior of the system. This characteristic of feedback systems has been elaborated by Argyris (1992), who distinguishes between single-loop and double-loop learning. Single-loop learning results in actions to eliminate or compensate for deviations but no learning is involved that changes the system to avoid deviations from occurring in the future. Double-loop learning affects the governing variables of the organization and its ability for prevention of recurrence through analysis and eliminating the causes of deviations.

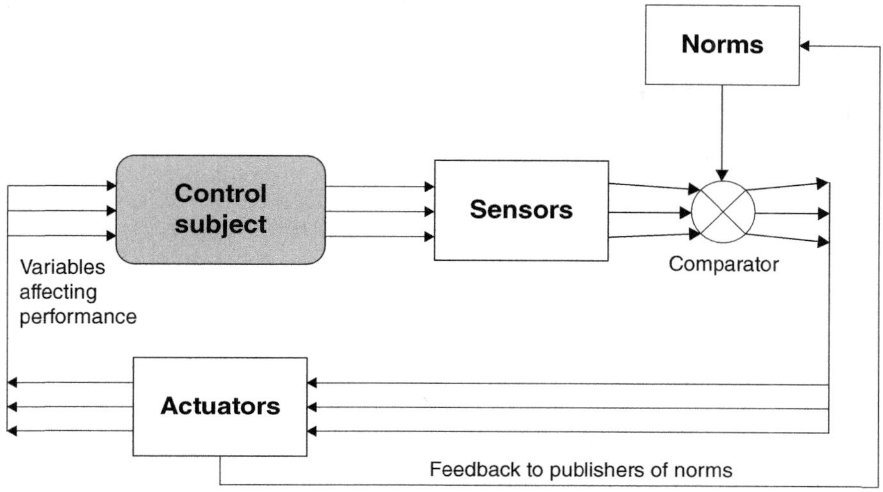

Figure 7.3. Feedback cycle for the control of anything. Adapted from Juran 1989.

There are many parallels in the management of safety. Safety inspections normally provide control of the technical standard of a workplace by identifying and correcting deviations from regulatory requirements or company standard (Kjellén, 2000). If only single-loop learning is accomplished, the same deviations will repeat themselves and show up in subsequent inspections.

W.R. Ashby (1956) has contributed to our understanding of how a control system must match the complexity of the control subject to exercise control. His two laws of requisite variety may be summarized as follows (Kjellén, 2000). To exercise full control, an analyst must be able to:

- Take at least as great a variety of actions as the observed system can show;
- Take precisely the correct set of action alternatives to counter those changes generated by the system; and
- Collect and process information and decide on and implement measures at a rate at least equal to the rate of change of the system.

Again, there are clear implications of Ashby's laws to the control of accidents through experience feedback. Data collection in safety management such as accident investigations and safety inspections need to be adequately comprehensive to cover the different factors affecting the risk of accidents. Safety inspections also need to show adequate variety and need to be frequent enough to match the rate of change at the workplace, etc. Delays in experience feedback and the implementation of countermeasures must not be larger than the rate of change of the system.

The simple feedback cycle is used here to analyze safety practices in a company. It is also applicable in very complex settings, where different authorities and corporate organizations interact to manage risk in hazardous industrial processes (Rasmussen, 1997; Wahlström et al., 1989).

7.4.2 Safety Information Systems

The safety information system model of Figure 7.4 illustrates the necessary subsystems to accomplish experience feedback and decision support in a safety management system (Kjellén, 2000).

The feedback according to the model goes from the workplace to decision-makers at different levels of the organization. Much of this information is informal and not necessarily documented. The model focuses on the formal processes, whereby explicit safety-related information is collected through various tools such as accident investigations, safety inspections, and reporting of unwanted occurrences. The information is processed and distributed for decision-making. With the support of a human information processing analogue, the model helps in analyzing filters in data collection, processing, and distribution of information. The designers of the information system have to balance the decision-makers' needs of information and

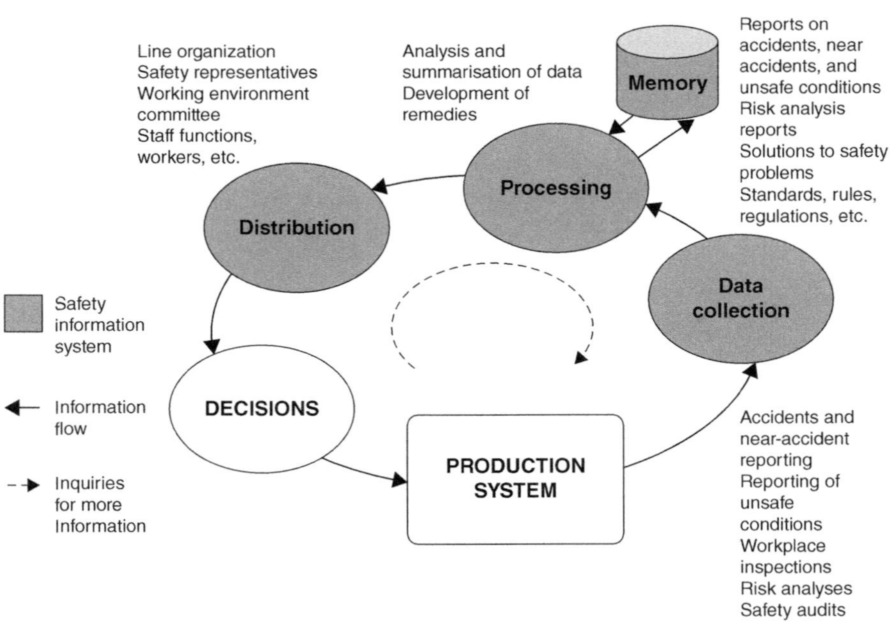

Figure 7.4. Safety information system. *Source*: Kjellén 2000. Reproduced with permission of Taylor & Francis.

prevention of information overload, at the same time avoiding the filtering of essential information in decision-making.

In an evaluation of existing experience feedback mechanisms in various companies by application of the safety information systems model, the following criteria for efficient experience feedback were identified (Kjellén, 2000):

- *Data Collection*: reliable, accurate, and comprehensive
- *Distribution and Presentation*: relevant to decision-making, easy to survey, timely, available when needed
- *The Total System*: involving easily understood and acceptable methods, promotion of involvement, cost-efficient

The promotion of involvement is an important aspect. In order to assess the effects of a safety information system, it has to be analyzed in relation to the social context where it is used (Maslen and Hayes, 2015). The issue of a community of practice of specialists that maximize knowledge sharing and learning is discussed in Section 7.4.4.

7.4.3 The Diagnostic Process

Another relevant theoretical concept in analyzing experience feedback is the diagnostic process. It is defined as a decision cycle including three steps: identification of symptoms, determining causes, and prescription of remedy (Kjellén, 2000). A symptom is a deviation of a system's behavior from what is considered normal.

Figure 7.5 illustrates a well-recognized approach to quality management based on the diagnostic processes, the PDCA wheel. As mentioned above, both the ISO

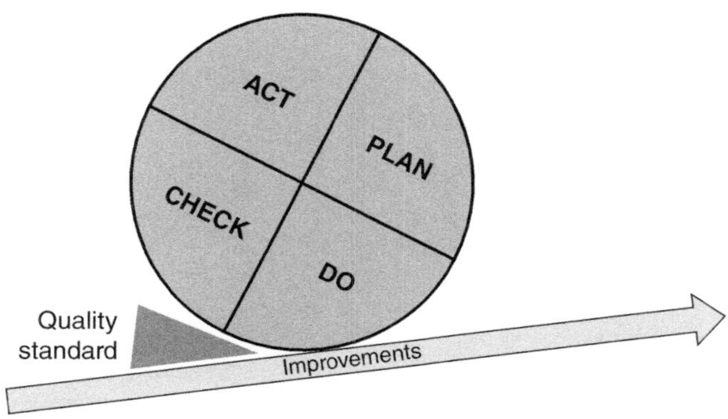

Figure 7.5. The PDCA cycle and the quality standard required to prevent fall back. *Source*: Deming 2000. Public domain. Kjellén 2000. Reproduced with permission of Taylor & Francis.

9000 series of quality management and the OHSAS 18001 health and safety management system standard are based on this quality management concept.

The application of the PDCA cycle involved successive steps to:

- *Plan*: where are we? where do we want to go? how do we get there?
- *Do*: communicate and train; secure resources; execute
- *Check*: do we follow the plan? do we meet our goal?
- *Act*: implement corrective actions; analyze experience and sum-up lessons learnt; standardize

The PDCA cycle helps in describing systematic safety management processes such as accident investigation and risk analysis. It is one of the several so-called "Problem Solving Processes" stemming from the diagnostic process (Kamsu Foguem et al., 2008). Hale's problem-solving cycle is an example of a safety application (Van der Schaaf et al., 1991).

7.4.4 Knowledge Management

The areas of organizational learning and knowledge management have contributed during the two latest decades to our understanding of the necessary conditions to accomplish efficient systems for experience feedback, both negative and positive (Sandven and Vik, 2013). Figure 7.6 illustrates the processes of experience feedback and knowledge creation in an industrial organization. The functions for transforming information gained through experience into explicit knowledge are distributed within the organization, which is typically the case in networking organizations. Experience is gained by a community of practice of members organized in projects, and

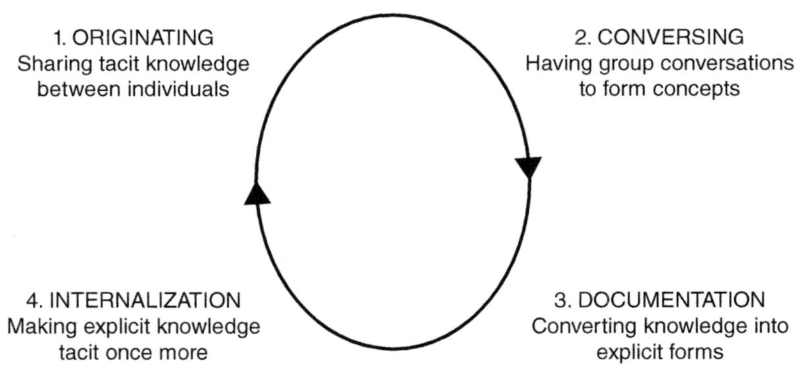

Figure 7.6. Organizational knowledge creation through a learning spiral. *Source*: Von Krogh et al. 2000. Reproduced with permission of Oxford University Press, USA.

transferred to a functional organization responsible for the knowledge base of the company (Kjellén, 2004).

The individual members gain experience through participation in project work. This experience is personal, subjective, and context related. It represents so-called tacit knowledge, that is, knowledge that has not been articulated. In the process of transforming tacit into explicit knowledge and for best practice to develop, the experience needs to be communicated and discussed between individuals and externalized (Nonaka, 1994). This process of externalization needs to take place between colleagues in a community of practice. This involves codification, articulation, and decontextualization. The functional organization will be responsible for documentation and justification of the experience in manuals, procedures, guidelines, specifications, etc. and make this available to the community of practice.

Nonaka identifies pre-conditions for experience feedback and learning to take place in a community of practice:

- Committed members need to experience benefits of participation in experience sharing through joy, competence growth, and improved work processes;
- Trustful and caring organizational environment;
- Not too uniform and not too disparate community members in terms of personal experience and qualifications;
- Moderate degree of changes (fluctuations and chaos) in the community of practice to provide for new input and unexpected opportunities and allow for continuity and time for dialogue; and
- Overlapping experience and responsibilities in order to amplify and crystallize experience.

7.5 USE AND IMPLEMENTATION

7.5.1 Safety Practice in an Operational Setting

In the context of safety management as practiced in any operation with potential for accidents, experience feedback involves practices for performance measurement and monitoring, and evaluation of compliance (OHSAS Project Group, 2007). Examples are:

- Accident reporting and investigation
- Reporting of unwanted occurrences (incidents, unsafe acts, and conditions)
- Safety inspections
- Audits and management reviews

Figure 7.7 shows the ideal scope of experience of these different practices seen in relation to a generalized accident analysis framework.

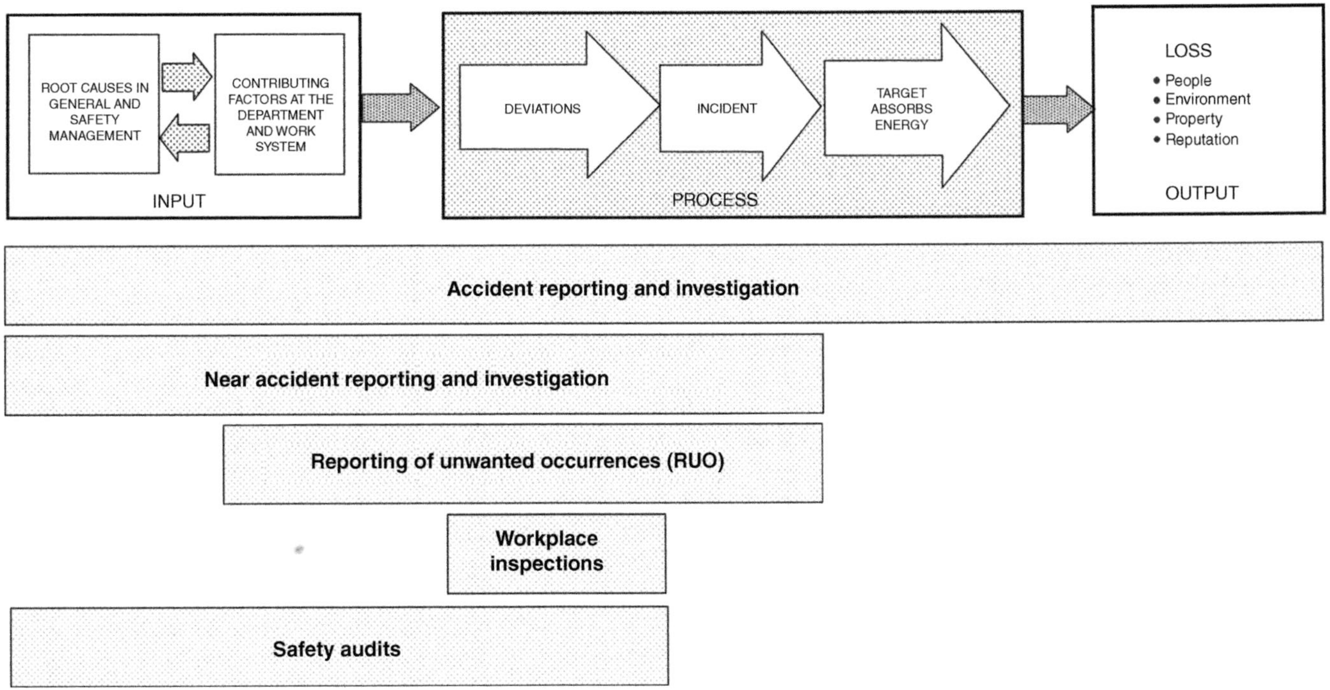

Figure 7.7. Ideal scope of different safety practices for experience feedback. *Source*: Kjellén 2000. Reproduced with permission of Taylor & Francis.

Data from the various safety practices are summarized in so-called safety performance indicators for use by management to monitor and control the risk of accidents. Examples of such indicators are (Kjellén, 2000):

- The Lost-Time Injury Frequency Rate (LTI-rate), defined as the number of injuries from accidents resulting in lost-time beyond the day of the accident per one million hours of work
- The Total Recordable Injury Frequency Rate (TRI-rate), which differs from the LTI-rate by including injuries resulting in medical treatment (beyond first aid) and transfer to alternative work assignment.
- Rate of Reporting of Unwanted Occurrences (RUO-rate), defined as the average number of reports from the employees on near accidents and unsafe acts and conditions per employee and year.

There are various factors limiting the efficiency and effectiveness of the experience feedback-based control and prevention of accident risks (Kjellén, 2000):

- The reliability in the reporting and follow-up of experience. This applies, for example, to under-reporting of recordable accidents, employees' unwillingness to report near accidents and unsafe acts and conditions (unwanted occurrences), failure to perform planned safety inspections and safety audits, and unsystematic mapping of deviations in these activities.
- Inadequate accuracy in reporting of facts about the experiences.
- Inadequate coverage of experiences through filters and biases. Examples are:
    - Accident investigations only covering the phase from loss-of-control and development of loss, and failing to identify deviations, and contributing factors and the workplace and root causes in the organization and management systems;
    - Failure to report incidents or unwanted occurrences where human errors have played an central role due to ignorance about the significance of the reports, shame, and blame and ultimately disciplinary action (Sanne, 2008);
    - Concentrating workplace inspections on a few obvious deviations like poor housekeeping and unsatisfactory scaffolding.
- Dilution of the system with data on a variety of issues related to personnel and process safety, making it difficult to identify those critical incidents that represent warning signals of potential major loss (Maslen and Hayes, 2015).

The safety performance indicators such as the LTI-rate and TRI-rate suffer from problems of inadequate validity as measures of the accident risk, delays in detecting changes in the risk level, and lack of robustness against manipulation (Kjellén, 2009). The LTI-rate, for example, gives the same consequence rating of a minor accident resulting in a few days of absence and an accident resulting in permanent disability. Delays in detecting a change in the risk level is due to the statistical uncertainties of small data sets from accident reporting, since accidents are relatively rare events. The LTI-rate is especially vulnerable to manipulation, simple because it can be defined as non-recordable if the injured person is given an alternative and less demanding job assignment in order to return to work on the day after the accident.

These different limitations are well known, and various means of improving experience feedback have been developed. An example is the comprehensive guidelines on experience feedback from unwanted occurrences on nuclear installations developed by IAEA (2006). ISO 19011 defines best practice on system audits that is applicable to the area of safety (ISO, 2011). On a more general level, there are three areas of intervention to improve the quality of experience feedback:

- Personnel, that is, selection, education and training, incentive schemes
- Organization and procedures
- Instruments and tools, such as checklists, forms, web-based solutions for data recording and retrieval.

The external framework conditions also play an important role (Kjellén, 2000). Insurance scheme and compensation to the victim has a significant influence on reporting reliability. Signals from external actors on the importance given to safety and feedback on reporting performance have shown to affect reporting behavior in contracted work (Kongsvik et al., 2012).

There is an extensive literature that describes different solutions like training programs, organization in multi-disciplinary problem-solving groups, use of auditing techniques, incentive programs to promote reporting of unwanted occurrences, etc. There are no solutions that fit any company. Companies select different solutions, depending on size, type of production, tradition, management philosophy, regulatory and insurance framework, etc.

Feedback of experience related to failures in the general and safety management systems of the organization having a determining influence on the conditions that have resulted in accidental loss requires special considerations. These so-called root causes are by their nature difficult to investigate, both because of their remoteness to the sequence of events and the sensitivities involved in investigating decision failure at the managerial level. Auditing techniques like ISO 19011 are designed to manage such obstacles by promoting a structured and transparent investigation technique. The composition of the audit team is essential. Independence is critical to induce trust, meaning that nobody in the investigation team shall have a stake in the outcome of the investigation. The reputation and competence of the investigation team members are also critical factors (Kjellén, 2000; Rollenhagen, 2003).

It is argued from a resilience engineering perspective that there is a need to move from "Safety-I to Safety-II" in accident investigations (Hollnagel, 2014).

According to the traditional "Safety-I" perspective, safety is understood as the absence of accidental losses due to failures and malfunctions. In the "Safety-II" perspective, on the other hand, safety is defined based on the organization's ability to adapt successfully under varying conditions. "Safety-II" investigations will provide experience feedback, not only on why things occasionally go wrong but also on why losses are avoided in similar situations involving successful adaptations.

7.5.2 Risk Assessment

Experience feedback in the area of safety is generally reactive, that is, building on experience made in the past. Experience of safety performance may also be used in risk assessments in assessing and managing future behavior of a system through processes called feedforward or anticipatory control (see Section 7.3). Risk assessments are applicable to systems in any phase of its life cycle.

Figure 7.8 shows the model of the risk assessment process according to ISO 31000. Examples of needs of data based on experience feedback are shown in the figure. A distinction is made between experience databases and experience carriers, where (Kjellén, 2008):

- Experience databases are compilations of data from different sources of experience such as accident and incident reports, failure rates of different components, data on exposure such as number of hours of operation.

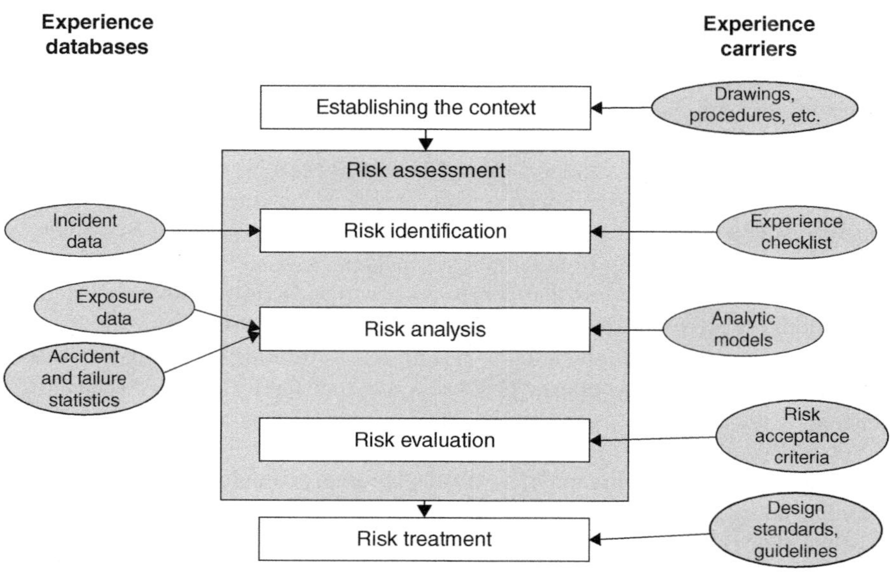

Figure 7.8. Examples of explicit experience used in risk management.

- Experience carriers represent explicit knowledge at a higher level of synthesis, where the raw experience data has been aggregated and decontextualized. Examples are drawings, procedures, standards or specifications, experience checklists on hazards, experience-based analytic models, and criteria for accepted risk. The latter are often experience based and reflect the level of tolerance of stakeholders.

Tacit experienced-based knowledge play an important role in risk assessment, and utilization of this knowledge is accomplished in the risk assessment team serving as an arena of experience exchange. The model of organizational knowledge creation in Figure 7.6 may explain this process. Individuals with operational experience from situations similar to those of the analysis object share tacit knowledge about hazards, causes, expected frequencies, etc. This shared knowledge is converted into concepts suitable for the context of the risk assessment through group discussions and documented in the risk assessment. The group processes allow for participation in the risk assessment process of stakeholders with an adequate variety of knowledge and values to promote validity, reliability, comprehensiveness, and accuracy of the results. A typical composition of a risk assessment team involved in the design of a new system is a risk analyst, systems designers from relevant disciplines, and operations and maintenance personnel at the sharp end of a similar system in operation.

7.5.3 Transfer of Experience to New Construction Projects

Construction is associated with a high risk of accidents compared to many other branches of industry. In Norway there were 4–6 fatalities per 100,000 employed in construction in 2011–2013 compared to 1–7 for all business. The corresponding figure for construction-related accidents in the EU was 13 (2001).

This may not only be explained by the fact that construction work requires management of significant energy sources such as movement of heavy machinery, working at height, electricity from temporary installations, and explosives. Another plausible explanation is the fact that construction is managed by temporary project organizations, being mobilized at start up and de-mobilized when construction is completed. Experience in cooperation as a team has to be built up during a critical part of the project, and experience feedback and transfer to the next project is hampered by the temporary assignment of the construction team members (Sandven and Vik, 2013).

The learning spiral of Figure 7.6 is applicable to construction projects as well, if learning is organized by the construction company to takes place in an interaction between a "permanent" functional or basis organization of the company and the community of practice of the construction project organization. This is illustrated by Figure 7.9 below.

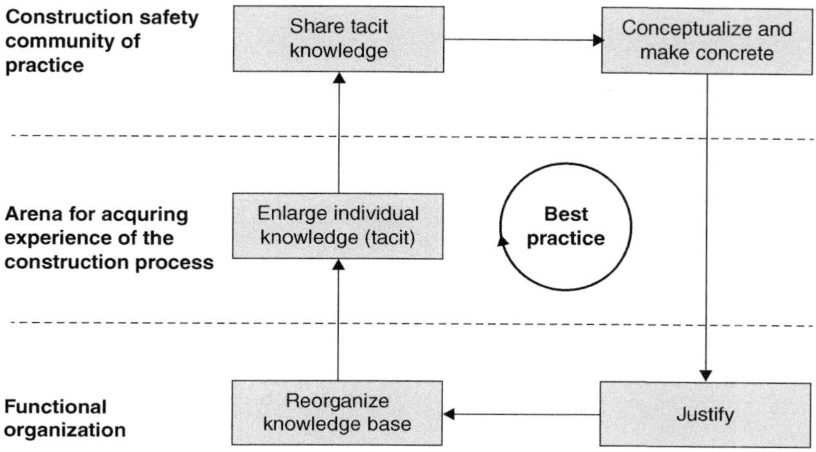

Figure 7.9. Development of best practice in construction through a learning spiral.

The procurement process in construction project seen from the owner's perspective is shown in Figure 7.10. This process is essential in understanding how a systematic management of safety is accomplished. As in the case of risk assessment, experience databases are distinguished from experience carriers; both being based on experience feedback from earlier construction projects to the Owner's functional organization for project management.

The Specification on Contractors' management of safety is a core document from the Owner's perspective. It defines the Contractual requirements to safety in construction work, including requirements to safety practices such as accident investigations and safety inspections, and requirements to the safety standard of construction work such as crane handling, work at height, and tunnel work. The constructability risk assessment summarizes the Owner's identification and assessment of hazards in construction based on the design, the conditions at the construction site, and experiences from similar projects in the past.

A study of two construction projects reports positive safety and general performance results of an approach as shown in Figure 7.10 (Kjellén, 2012). The first project produced poor safety and overall performance results. Because of these experiences, a proactive approach was selected in the second project, involving use of the contracting process to ensure adequate conditions. The second project delivered satisfactory safety and overall performance results (cost, progress) at moderate costs for follow-up.

7.5.4 Transfer of Experience from the Users to Design

Issues in organization of experience feedback to designers of new industrial systems have many similarities to those of a construction project. Design is often organized

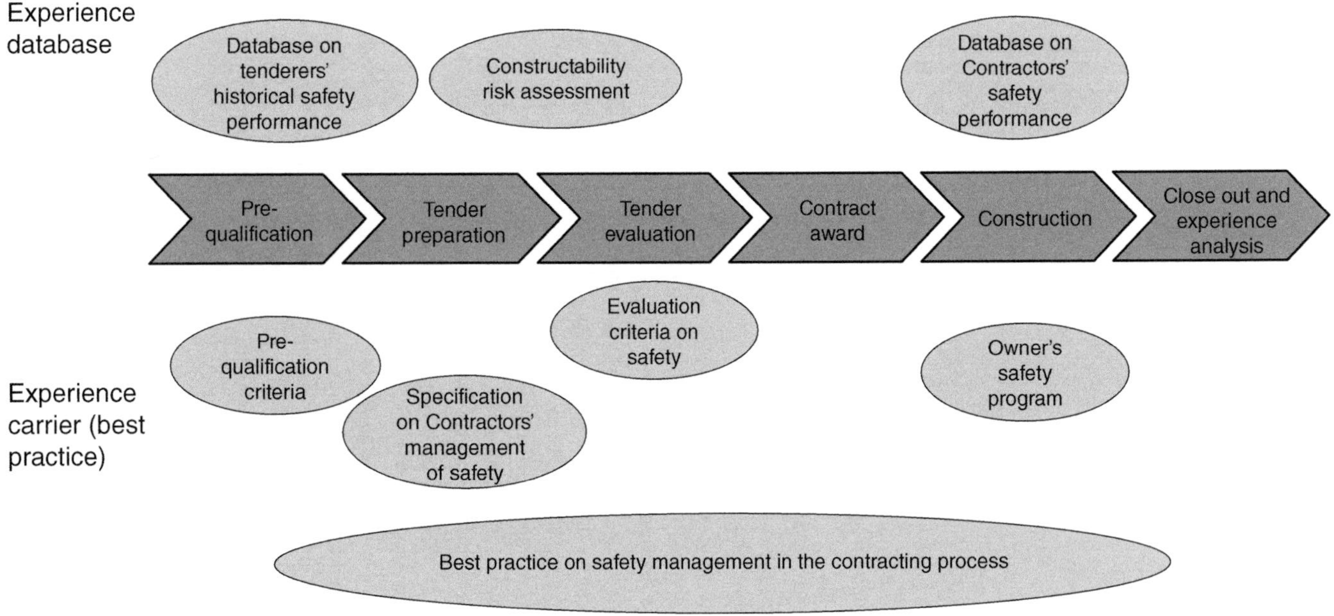

Figure 7.10. Areas of experience feedback in procurement process (examples).

in a temporary project organization by an engineering company, being separate from the owner's organization for operation and maintenance of existing plants. Experience feedback to the design organization is dependent on the cooperation between the engineering company and the owner, and the owner's ability to summarize operational experience in a language that is possible to communicate to the design team.

Figure 7.11 shows an example on how systematic experience feedback to the designers may be accomplished within the framework of a traditional design organization of an engineering company (Kjellén, 2002). Experience feedback to the design organization takes place in two different ways, through experience carriers such as technical specifications and checklists, and through person-to-person contact between designers and operations personnel in risk assessments, design reviews, verifications, and common design teams.

The operations organization maintains experience databases such as databases on incidents and accidents and downtime, but this detailed experience is not communicated directly to the design team.

The technical specification on safety represent the operations organization's "memory" on how to get design right and avoid design-related accidents in the new or modified plant. The arenas of experience exchange allow the operations personnel to share their tacit knowledge of relevance to safety in design with the designers. They, on the other hand, will have to translate this knowledge to explicit design criteria for implementation in the new plant, as illustrated by the learning spiral in Figure 7.6. This interaction is critical in implementing goal-oriented requirements to design.

Husmoen (1997) has identified enabling factors in bridging knowledge between design and operations organizations in offshore projects:

- Physical integration of operations personnel in the project organization;
- Involvement of design and operations personnel with in-depth knowledge of each other's area of expertise; and
- Mutual sympathy and respect.

7.6 EMPIRICAL RESEARCH ON USE AND EFFICIENCY

There is a general lack of evaluation studies of adequate scientific quality to provide evidence on the effects of various methods for experience feedback in safety management (Lindberg et al., 2010).

Experience feedback is a basic element in the prevailing guidelines and standards for occupational health and safety management systems (OHSMS). It is reasonable to assume that indications of the effectiveness of experience feedback may be derived from evaluation research into such systems. Robson et al., (2007) concluded from a systematic review of the research literature into the effects of OHSMS on employee

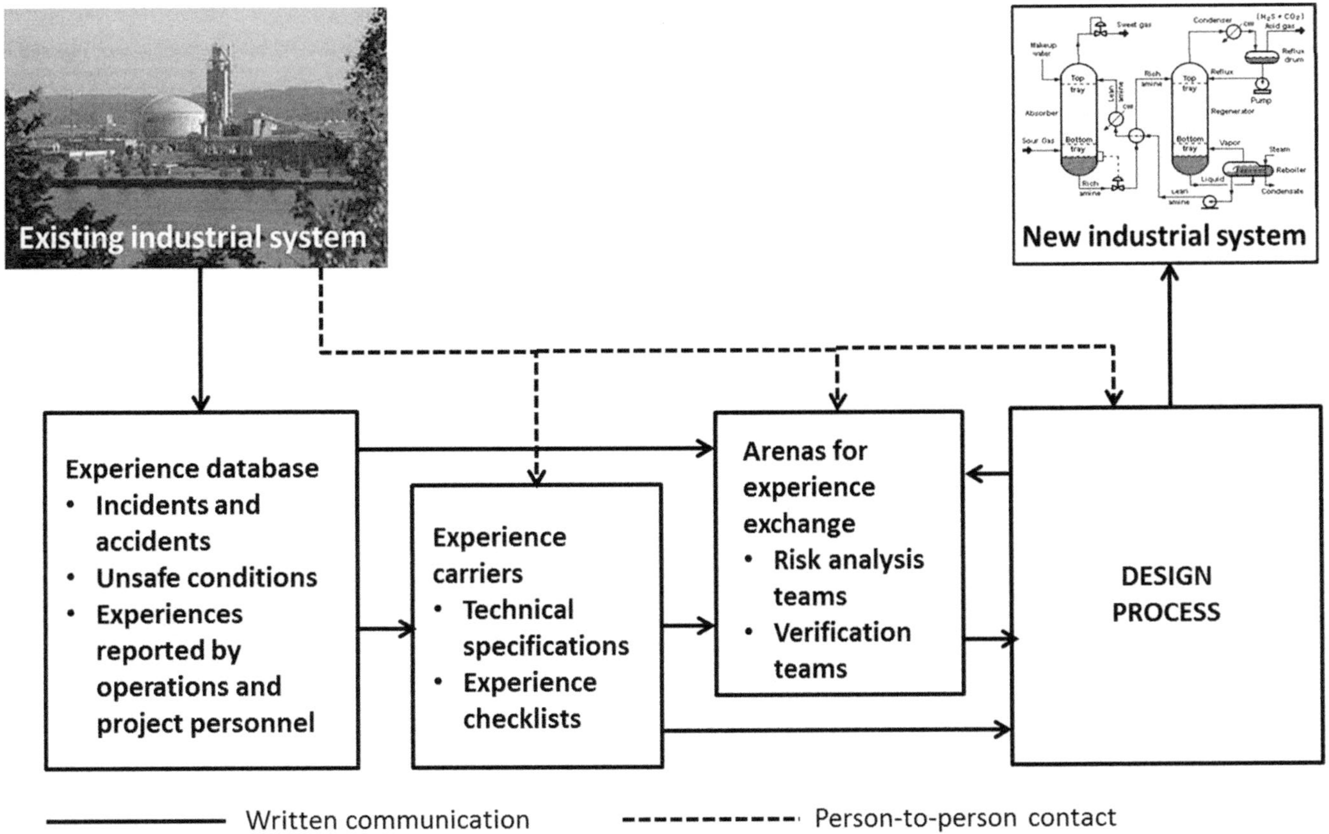

Figure 7.11. Experience feedback from operations to design. *Source*: Kjellén 2002. Reproduced with permission of Taylor & Francis.

health and safety that the reviewed studies generally reported a positive impact. Most of the studies showed methodological limitations, and the body of evidence was insufficient to draw any generalizable conclusions. Gallagher et al. (2001) came up with similar conclusions in a study of the effectiveness of OHSMS in Australia and barriers to their implementation. It was concluded that OHSMS can deliver more healthy and safe workplaces under the right circumstances. Such success is conditional on kind of system used, senior management commitment, integration into general management system, and employee participation.

This conclusion is further supported by the findings from a retrospective case study of an aluminum plant (Kjellén et al., 1997). The plant had undergone a major restructuring process during the study period of 10 years. This included the implementation of a safety management system based on the Norwegian regulatory requirements for internal control. The plant experienced improved economic results due to a combined effect of better control of expenditures for personnel, materials, and energy and reduced losses due to accidents, sick leave, and reclaims. The Company's investments in the management of safety were profitable as part of a comprehensive strategy to improve the "control climate" in the plant, but not seen in isolation through reduced costs of accidents.

Fernández-Muñiz et al. (2009) have through an extensive literature review identified "communication and transfer of information from the workplace" as one of the six critical aspects of a good OHS management system. The six OHSMS factors were evaluated in a questionnaire study of 455 Spanish firms, showing a positive relation between an adequate safety management system and safety performance as well as economic performance and competitiveness.

7.7 RELATIONS TO OTHER SAFETY PRINCIPLES

The management and control of safety of industrial organizations is implemented as a control system involving experience feedback (Juran, 1989; Wahlström et al., 1989). This basic principle of feedback control plays an important role in most safety principles. The significance of experience feedback to risk assessment and safety in design are already discussed in Sections 5.1.2 and 5.4, respectively. Here we will focus on a few others.

7.7.1 Safety Management

As we have seen in the historical review, experience feedback was a basic principle in the management of safety through the different phases of its development. A standard such as OHSAS 18001 identifies different experience feedback loops including incident reporting and feedback systems, risk and hazard assessment and management systems, and audits and management reviews (OHSAS Project Group, 2007).

7.7.2 Resilience Engineering

Resilience engineering represents a further development and complement to traditional safety management (Chapter 3). Resilience is a concept that differs from safety through its focus, not on freedom from accidents, but on a system's ability to sustain required operation under both expected and unexpected conditions. Experience feedback is characterized by an ability of an organization, not only to react when something happens and implement corrective actions, but also to modify response based on experience. This corresponds to double-loop learning according to Argyris (1992). A resilient system also needs to show predictive behavior by being able to anticipate future events and prepare accordingly.

7.7.3 Safety Indicators

Juran's (1989) feedback cycle for the control of anything represents the fundament for the use of safety indicators in the management of safety. Safety indicators (Chapter 8) use aggregate data from an ongoing industrial process to regulate it. Safety indicators based on data from incidents and other unwanted events are by their nature lagging, since a sufficient period in time is required to produce reliable statistics. This causes problems from a control point of view, due to this time lag between the occurrence of events and corrective actions. Safety indicators based on, for example, safety sampling, measure the current level of risk, whereas safety indicators integrating the use of predictive models may allow for the anticipation of future developments in the level of risk, so-called leading indicators.

7.7.4 Safety Culture

Safety culture (Chapter 28) is a multi-dimensional construct. There is no consensus among researchers and practitioners on the exact dimensions that characterize a good safety culture, but the ability to learn from past experience is a widely used dimension. This involves a "no blame culture" and the promotion of reporting of near accidents and unwanted occurrences such as human errors. There is also a culture for promotion of involvement through experience sharing and learning from incidents in order to improve performance, issues that are important aspects in explaining why safety information systems succeed or fail.

REFERENCES

Argyris, C. (1992). *On Organizational Learning*. Cambridge, MA: Blackwell Publishers.

Ashby, W. R. (1956). *An Introduction to Cybernetics*. London: Chapman & Hall.

Bergman, B. (2012). *Ännu bättre vård—Vad kan vi lära från variationer i öppna jämförelser?* Stockholm, Sweden: Sveriges Kommuner och Landsting.

Cooper, M. D. (2000). Towards a model of safety culture. *Safety Science*, 36, 111–136.

Deming, W. E. (1993). *The New Economics for Industry, Government, and Education*. Boston, MA: MIT Press.

Deming, W. E. (2000). *Out of the Crisis*. Cambridge, MA: MIT Press.

Fernández-Muñiz, B., Montes-Peón, J. M., and Váquez-Ordás, C. M. (2009). Relation between safety management and firm performance. *Safety Science*, 47, 980–991.

Gallagher, C., Underhill, E., and Rimmer, M. (2001). *Occupational Health and Safety Management Systems—A Review of their Effectiveness in Securing Healthy and Safe Workplaces*. Sydney: National Occupational Health and Safety Commission.

Gibson, J. (1961). The contribution of experimental psychology to the formulation of the problem of safety. In: *Behavioral Approaches to Accident Research*. New York: Association for the Aid of Crippled Children.

Haddon, W. (1970). On the escape of tigers: An ecological note. *Technology Review*, 72(7), 44–50.

Health and Safety Laboratory. (2002). Safety culture: A review of the literature. Report HSL/2002/25. Health and Safety Laboratory, Sheffield, UK.

Heinrich, H. W. (1931). *Industrial Accident Prevention—A Scientific Approach*. New York: McGraw-Hill.

Hollnagel, E. (2014). *Safety-I and Safety-II: The Past and Future of Safety Management*. Aldershot, UK: Ashgate.

Hollnagel, E., Woods, D. D., and Leveson, N. G. (2006). *Resilience Engineering: Concepts and Precepts*. Aldershot, UK: Ashgate.

Hudson, P. T. W. (2001). Safety management and safety culture: The long, hard and winding road. In: W. Pearse, C. Gallagher, and L. Bluff (Eds.), *Occupational Health and Safety Management Systems*. Melbourne: Crown Content.

Husmoen, M. S. (1997). Bridging operation and design—The encounter between practical and discipline-based knowledge in offshore platform design. Doctoral thesis 1997:123, Norwegian University of Science and Technology, Trondheim, Norway.

IAEA. (1991). Safety culture. Safety Series No. 75-INSAG-4. International Atomic Energy Agency, Vienna.

IAEA. (2006). A system for the feedback of experience from events in nuclear installations. Safety Guide No. NS-G-2.11, International Atomic Energy Agency, Vienna.

ISO. (2011). Guidelines for auditing management systems. ISO 19011:2011(E). International Standards Organisation, Geneva.Johnson, W. G. (1980). *MORT Safety Assurance System*. New York: Marcel Dekker.

Juran, J. M. (1989). *Juran on Leadership for Quality—An Executive Handbook*. New York: The Free Press.

Kamsu Foguem, B., Coudert, T., Béler, C., and Geneste, L. (2008). Knowledge formalization in experience feedback processes: An ontology-based approach. *Computers in Industry*, 59, 694–710.

Kjellén, U. (2000). *Prevention of Accidents Through Experience Feedback*. London/New York: Taylor & Francis.

Kjellén, U. (2002). Transfer of experience from the users to design to improve safety in offshore oil and gas production. In: B. Wilpert and B. Fahlbruch (Eds.), *System Safety—Challenges and Pitfalls of Intervention*. Bingley, UK: Emerald Group Publishing.

Kjellén, U. (2004). Improving knowledge sharing and learning in an organisation of safety, health and environmental project engineers. In: J. H. Andriessen and B. Fahlbruch (Eds.), *How to Manage Experience Sharing: From Organisational Surprises to Organisational Knowledge*. Amsterdam: Elsevier.

Kjellén, U. (2008). Experience feedback. In: E. L. Melnick and B. S. Everitt (Eds.), *Encyclopedia of Quantitative Risk Analysis and Assessment*. John Wiley & Sons. ISBN 978-0-470-03549-8.

Kjellén, U. (2009). The safety measurement problem revisited. *Safety Science*, 47, 486–489.

Kjellén, U. (2012). Managing safety in hydropower projects in emerging markets—Experiences in developing from a reactive to a proactive approach. *Safety Science*, 50, 1941–1951.

Kjellén, U., Boe, K., and Hagen, H. L. (1997). Economic effects of implementing internal control of health, safety and the environment: A retrospective case study of an aluminium plant. *Safety Science*, 27, 99–114.

Komaki, J., Barwick, K. D., and Scott, L. R. (1978). A behavioral approach to occupational safety: Pinpointing and reinforcing safe performance in a food manufacturing plant. *Journal of Applied Psychology*, 63, 434–445.

Kongsvik, T., Fenstad, J., and Wendelborg, C. (2012). Between a rock and a hard place: Accident and near-miss reporting on offshore service vessels. *Safety Science*, 50, 1839–1846.

Kreitner, R. (2009). *Principles of Management*. Boston, MA: Houghton Mifflin.

Lindberg, A.-K., Hansson, S. O., and Rollenhagen, C. (2010). Learning from accidents—What more do we need to know? *Safety Science*, 48, 714–721.

Maslen, S., and Hayes, J. (2015). Preventing black swans: Incident reporting systems as collective knowledge management. *Journal of Risk Research*. doi: 10.1080/13669877.2015.1057204

Melnick, E. L., and Everitt, B. S. (2008). *Encyclopedia of Quantitative Risk Analysis and Assessment*. John Wiley & Sons. ISBN: 978-0-470-03549-8

Nonaka, I. (1994). A dynamic theory of organizational knowledge creation. *Organization Science*, 5, 14–37.

Nonaka, I., and Takeuchi, H. (1995). *The Knowledge-Creating Company*. New York: Oxford University Press.

OHSAS Project Group. (2007). Occupational health and safety management systems—Requirements. OHSAS 18001:2007.

Oxford University Press. (2012). *Oxford English Dictionary*. Oxford.

Rasmussen, J. (1997). Risk management in a dynamic society: A modelling problem. *Safety Science*, 27, 183–213.

Robson, L. S., Clarke, J. A., Cullen, K., Bielecky, A., Severin, C., Bigelow, P. L., Irvin, E., Culyer, A., and Mahood, Q. (2007). The effectiveness of occupational health and safety management system interventions: A systematic review. *Safety Science*, 45, 329–353.

Rockwell, T. H. (1959). Safety performance measurement. *Journal of Industrial Engineering*, 10(1), 12–16.

Rollenhagen, C. (2003). *Att Utreda Olycksfall—Teori och Praktik*. Lund, Sweden: Studentlitteratur.

Sandven, F. and Vik, L. E. (2013). *Erfaringslæring og kunnskapsoverføring* (in Norwegian) (Tromsø, Master Thesis, University of Tromsø).

Sanne, J. M. (2008). Incident reporting or story telling? Competing schemes in a safety-critical and hazardous work setting. *Safety Science*, 46, 1205–1222.

Senge, P. M. (1990). *The Fifth Discipline: The Art and Practice of the Learning Organization*. New York: Doubleday.

Taylor, F. W. (1911). *The Principles of Scientific Management*. New York/London: Harper & Brothers.

Van der Schaaf, T. W., Lucas, D. A., and Hale, A. R. (1991). *Near-Miss Reporting as a Safety Tool*. Oxford: Butterworth-Heinemann.

Von Krogh, G., Ichijo, K., and Nonaka, I. (2000). *Enabling Knowledge Creation*. Oxford: Oxford University Press.

Wahlström, B., Laakso, K., and Lehtinen, E (1989). Feedback of experience for avoiding low probability disaster. In: *Proceedings of an International Symposium on Feedback of Operational Safety Experience from Nuclear Power plants*. Vienna: IAEA.

FURTHER READING

Kjellén, U., and Albrechtsen, E. (2017). *Prevention of Accidents and Unwanted Occurrences—Theory, Methods and Tools in Safety Management*. Boca Raton, FL: CRC Press.

8

RISK AND SAFETY INDICATORS

Drew Rae

8.1 INTRODUCTION

A safety indicator is a metric that is used to estimate the past, current, or future level of safety risk. An indicator may be based on the recorded frequency of injury events, or on other circumstances that correlate with or predict injury events. While even a single observation can reveal danger, and thus be indicative of safety risk, this chapter will focus on quantitative indicators that aggregate multiple countable events to provide a risk estimate.

Safety indicators serve a variety of purposes. When collected using standardized definitions and processes, they allow insurers and regulators to make comparisons between different firms. They allow organizations to assess their own safety performance, either in comparison to other organizations or by identifying trends over time. Reliable indicators are also important for assessing the performance of safety improvement initiatives, and for validating hypotheses about the nature and causes of safety performance.

The most widely used safety indicator is the injury rate—the number of injuries per unit of time worked. Injury rates are one of a number of adverse event counts

Handbook of Safety Principles, First Edition. Edited by Niklas Möller, Sven Ove Hansson, Jan-Erik Holmberg, and Carl Rollenhagen.

collectively referred to as "lagging" indicators, because they provide information after the events they are intended to measure. Lagging indicators are considered useful to the extent that recent past risk may be similar to current and future risks. Injury rates, in particular, are controversial because they are not necessarily predictive of fatality and multiple fatality events, and because they can be easily distorted through variation in reporting and recording practice.

The alternatives to lagging indicators are the so-called "leading" indicators. In economics, the term "leading indicator" refers to a variable that predicts the future state of some other variable. In safety, "leading indicator" is used to refer to any metric that does not directly count harmful events.

It is not usually possible to validate leading indicators by finding correlations between leading and lagging indicators. This is particularly the case for high consequence events, which are too rare to allow their statistical distribution to be accurately modelled. As a consequence, leading indicators are selected based on theoretical models for how safety is achieved. There is no academic or industry consensus on which leading indicators are most appropriate, and it is a topic of ongoing research and debate.

8.2 ORIGIN AND HISTORY

The collection of statistics for the purpose of assessing safety risk dates back to the life insurance industry of the eighteenth century (Bühlmann, 1997). Actuarial tables were used to set premiums based on the annual risk of mortality for people of different ages.

In the late nineteenth century, economic and insurance models began incorporating the idea of a stochastic process—a set of random variables that develops over time. Stochastic processes are not predictable, but can be represented through probability distributions. The parameters of these distributions may be estimated by observing the outputs of the process or by understanding the inputs to the process. Of particular importance to safety measurement is the Poisson distribution, rediscovered by Ladislaus von Bortkiewicz in 1898 (Good, 1986). Von Bortkiewicz used a dataset of Prussian cavalrymen killed by the kick of a horse. This is a classic Poisson process: rare events that arrive at random intervals but according to steady probability, counted over many discrete units of time.

In 1923, the first International Conference of Labour Statisticians published a report on industrial accident statistics, which, in addition to appropriately setting insurance premiums and compensation, had an explicit objective to "provide information as to nature and causes of industrial accidents, so that steps may be taken to prevent their occurrence" (International Labour Office, 1923). The report considered it important to consider not just the number of injuries but the exposure—the number of hours worked—so that accident rates and risk could be determined and compared between countries and types of injury. The International Labor Office rejected the validity of assigning and measuring "moral cause" (i.e., blame) for accidents, and suggested recording only "immediate objective" causes.

Herbert Heinrich (1930) argued that even the more fundamental causes of accidents followed statistical patterns. He put forward the hypothesis, popularized through "Heinrich's Pyramid," that there was a statistical relationship between the number of accidents experienced by an organization and the number of "unsafe acts" committed by its workforce. If correct, this hypothesis would allow safety to be measured and controlled according to non-accident statistical data.

During World War II, the hitherto ad hoc practice of collecting and analyzing statistical data to make better decisions grew into the formal field of "operational research" (Thomas, 2015). Operational research was credited with improvements in radar operation, training, logistics, and anti-submarine warfare, brought about by identifying appropriate indications of performance and comparing trends in those indications under varying conditions. Operational research was viewed as a "scientific" contribution to decision-making.

After the war, operations research became increasingly embedded within management science, and was integrated with the emergent fields of "cybernetics" (Clemson, 1991) and Total Quality Management (Feigenbaum, 1956). In contrast to early operations research, which sought to identify the key independent variables that influenced outcomes, cybernetics recognized organizations as complex systems with interacting feedback loops. Total Quality Management sought to improve product quality by treating the production process as a cybernetic system. The idea of a "safety management system" (SMS) was derived from Total Quality Management, and SMS literature retains many of the same concepts of measurement and feedback (Hale et al., 1997).

In parallel to the development of cybernetics and complex systems theories of safety, there was increasing interest in quantitative risk assessment (QRA). QRA seeks to estimate the total risk of systems or situations based on the frequency of smaller, more frequent events. In the United Kingdom in 1939, reliability targets were set for aircraft to meet during development (Rechard, 1999). However, safety in general was not considered to be quantifiable, with authors such as Stieglitz (1966) emphasizing throughout the 1940s, 1950s, and 1960s that it was not practicable to either predict or demonstrate the achievement of quantitative targets.

This changed with the advent of inter-continental ballistic missiles (ICBMs). It was recognized that a single ICBM accident could present risk for the continuance of human civilization, and so estimation of that risk was necessary before the system could be deployed. Fault Tree Analysis was the first quantitative safety technique, and was applied on both the Minuteman Missile and Boeing 747 development projects (Ericson, 2005). The chemical and nuclear industries also developed QRA practices, which led in turn to the collection of statistics related to "precursor events." Precursors are events that, according to a quantitative risk model, are necessary but not sufficient to cause an accident. Assuming that the risk model is correct and complete, the rate of precursor events is an indicator of the likelihood of an accident.

Throughout the 1970s and 1980s there was growth in safety legislation and standardization, encouraged and reinforced by a series of notable accidents such as Flixborough (United Kingdom Department of Employment, 1975), Seveso (Homberger et al., 1979), and Piper Alpha (Cullen, 1990). Each of these accidents occurred in an industry that already had extensive safety regulation in place, requiring more sophisticated explanations for why the accidents happened than a simple lack of safety management (Hale and Hovden, 1998). The new regulation paid significant attention to the measurement and evaluation of safety management performance. With the exception of QRA models, most approaches to measurement were based on audits or on lagging indicators.

In 2003, the United Kingdom Health and Safety Executive (HSE) issued a report into three incidents at BP Grangemouth Refinery, Scotland, during 2000 (Health and Safety Executive, 2003). The findings emphasized the inadequacy of lagging indicators for predicting major accidents, and the importance of process safety indicators. Similar concerns were raised by the Baker Panel Report following the Texas City Refinery disaster (Baker et al., 2007). These accidents and reports sparked renewed interest in indicators for process safety, including HSE guidance (Health and Safety Executive, 2006) and a special issue of Safety Science devoted to leading indicators (Hopkins, 2009).

8.3 DEFINITIONS AND TERMINOLOGY

Not everything that has meaning can be directly observed. Many useful concepts exist primarily in the mind or minds that are noticing them, rather than as objective or physical objects. A *construct* or *construction* is something that relies on the human mind for its existence (Russell, 1914). A *social construct* is a category or concept that forms as a shared understanding amongst a group of people. Safety is such a social construct. It "does not exist 'out there', independent of our minds and culture, ready to be measured, but is a constructed human concept, more easily judged than defined" (Rochlin, 1999).

The process of taking an abstract concept and working out how to measure it is called *operationalization*. An *operational definition* specifies a concept based on how it is to be measured (Ribes-Iñesta, 2003). *Measurement* is the quantification of a property in accordance with standardized units. A *metric* is a formally defined system of measurement.

The commonly used definition of safety "A thing is safe if its risks are judged to be acceptable" provided by Lowrance (1976) is not operational, because it is based on social consensus ("judged to be acceptable") rather than on a metric.

A common way to operationalize social and psychological concepts is through *indicators*. An indicator is a metric that is believed, based on theoretical or social understanding, to *correlate* with a construct. Correlation is when two objects reliably appear at the same time or at different times, depending on whether the correlation is positive or negative. Thus, the appearance of the measurable object indicates the presence or absence of the construct. For example, the presence of injuries is negatively correlated with safety.

Most attempts to operationalize safety do so by defining safety in terms of the likelihood of undesirable events. This approach faces two major challenges:

1. The undesirable events are typically not specified, but instead are themselves indirectly defined in terms of the severity of their consequences.
2. Likelihood is itself a social construct, particularly when applied to events with imperfect knowledge of their *etiology* (how they are caused), or the current state of the systems in which they arise.

Unlike likelihood, frequency can be directly measured. Safety can therefore be operationalized as the *recordable injury rate (RIR)*. RIR measures the frequency of injuries of specific magnitudes over a specific time period. One criterion for recording an injury is if it results in *lost time*—a worker unable to work—leading to the metric *lost-time injury frequency rate (LTIFR)*.

RIR and LTIFR are *lagging indicators*. Lagging indicator are counts of adverse events. The term "lagging" refers to the fact that the indicators are caused by safety or its absence, in contrast to *leading indicators,* which are based on the causes of safety.

Usage of the terms *leading* and *lagging* is inconsistent across safety literature.

Events that indicate safety failings without an actual accident are sometimes considered *leading* (with respect to the accident) and sometimes considered to be *lagging* (with respect to safety activity).

Lagging indicators are usually only measures of *personal safety*. Personal safety is concerned with low energy events that typically cause minor injuries, and at most will result in a single fatality. *Process safety* is concerned with high energy events capable of causing multiple fatalities. Multiple fatality accidents are so rare that they can only be measured as lagging indicators across entire states and industries.

A further distinction can be made between *risk indicators* and *safety indicators*. Both are metrics that purport to measure safety, but they do so through different types of models. Risk indicators are based on accident causation models, which link risk-influencing factors to accident outcomes. Safety indicators are based on organizational models that link culture, management systems, and activities to the presence of safety (Øien et al., 2011). Both of these models will be discussed further in Section 8.4 of this chapter.

8.4 UNDERLYING THEORY AND THEORETICAL ASSUMPTIONS

8.4.1 Past, Present, and Future Safety

The frequency of injuries is a measure of past safety. Except in the case of very rare events, it is not controversial to link past frequency to past likelihood, and thence to past safety. If one hundred injuries happened in one hundred thousand hours of work, it is reasonable to calculate the likelihood of an injury as one in a thousand hours. We can then decide if this likelihood matches our idea of "safe."

However, for most purposes we are interested in estimates of present or future safety. The problem here is construct validity. How do we know that what we are measuring reveals what we think of as "safety"? There are two strategies available.

The first strategy defines safety using lagging indicators as an intrinsic part of the definition. All other indicators are justified by correlation with these lagging indicators, as shown empirically or by argument. Empirical justification uses historical datasets to show that the indicator is able to "predict" the lagging indicators. This is accomplished by measuring the correlations between the leading indicator and the time-shifted lagging indicators. Argument justification requires a claim that there is a causal link between the indicator and the likelihood of injury. The thing being measured may be a cause of injury, or may be believed to improve safety.

The second strategy is to define safety based on norms of design or practice. A situation is more or less safe not according to how much risk it presents, but according to how well it conforms to accepted standards of safety. For example, one workplace may be judged to be safer than another because of better safety practices rather than because of fewer accidents.

The state of evidence regarding safety practices (Rae et al., 2010) forces us to rely primarily on the second method. This is especially the case for major accidents, which are too rare to establish a frequency against which indicators can be validated. Indicators, therefore, are based on theoretical models for how safety is achieved. Scholars and practitioners of particular safety paradigms will select metrics consistent with that model of safety.

8.4.2 Outcome Indicators

Where there is a steady underlying likelihood of an event, and where the intervals between events are independent, the applicable statistical representation is a Poisson distribution. There may be considerable stochastic variation between time periods with no external cause for the differences. For example, an organization may experience 2 injuries one month, 10 injuries the next month, and 0 injuries the next, despite the risk of an injury being constant across all three months.

Since "injury" can refer to a range of events of different severities, strict definitions are required for consistent counting of events. Typically either "recordable" or "lost-time" injuries are used, with regulation or legislation specifying the circumstances under which an injury should be counted. total recordable incident rate (TRIR) and "lost-time injury frequency rate" (LTIFR) are common standards for injury counting.

Whenever a metric is formed by counting instances of something, it creates an equivalence class for those instances. By capturing all "lost-time injuries" in a single count, LTIFR is treating all lost-time injuries as the same type of thing. If, for example, some injuries resulted in a few hours away from work while other injuries left a worker unemployed for months, those differences would not be captured by recording LTIFRs (O'Neill et al., 2015).

This problem can be partially alleviated by creating subcategories, for example, by separately recording injuries according to their impact on the worker. McDonald (1995) proposes three classes of damage:

Class 1: The person's life is permanently altered (including fatality)

Class 2: The person's life is temporarily altered

Class 3: The person is inconvenienced

Fatalities and disabling injuries (Class 1) are not drawn from the same statistical distribution as minor injuries, because they often have different immediate causes. For example, in the construction injury most fatalities involve falls, while injuries are mainly cuts and blunt trauma to limbs and fingers (Fredericks et al., 2005; Jeong, 1998). Even the use of subcategories may be insufficient to reveal important differences between injuries. Within each category or class there are different types of injury with different causes, frequency, and severity.

The alternative to subcategories is to use a single metric that is automatically weighted. For example, the total cost of compensation or insurance takes into account the fact that different injuries have different impacts. However, using such a metric disguises the weights that are used, which may be unsuitable for the purposes of the metric. The proportion of the cost of an injury that falls upon a company, the individual, and society may vary considerably according to the regulatory and insurance environment, the availability of civil litigation for workplace injury, and the exact nature and cause of the injury. Cost to the company or the insurer may be a misleading indicator of the total cost of injury.

Despite the concerns about inappropriate categorization, it is in theory possible for injuries of one type and severity to provide an indication of the risk of injuries of another type and severity. This would be the case if there were a common systematic cause, such as attitude toward safety or the quality of safety management activity. This is a topic of considerable controversy, with studies showing both positive (Wright and van der Schaaf, 2004) and negative (Hunting and Weeks, 1993) correlations between minor injuries and fatalities.

Injuries and fatalities are not predictive of multiple fatality accidents, which do not occur frequently enough to provide a measure of the safety of a particular design or organization (Elliott et al., 2008; Hopkins, 2000). However, aggregation of such accidents across countries or industries can provide an indication of broad social trends (Evans, 1994).

8.4.3 Risk Models and Precursor Events

Risk models represent the sequences and combinations of events that can lead to an accident. An example of such a model is the safety risk model (SRM) for United

Kingdom Railways (Muttram, 2002). The SRM consists of a set of linked fault trees and event trees, monitoring and aggregating the risk from 110 "hazardous events." The US Nuclear Regulatory Commission uses standardized plant analysis risk (SPAR) models, based on fault trees, to assess the frequency of nuclear reactor accidents (Siu and Collins, 2008).

Risk models are not in themselves safety indicators. QRA is a tool to support design and has little to no validity as a measure of current safety (Rae et al., 2012). However, risk models do provide a rationale for drawing a causal link between less significant, more frequent events and serious accidents. If the model is a complete representation of the ways in which an accident can occur, then changes in the frequency of events within the model represent changes in the likelihood of the accident.

In practice, all risk models are incomplete (Rae et al., 2014). For a risk model to provide an effective source of safety indicators, it must be assumed that risks not captured by the model are comparatively stable, and that most variation in risk is associated with events within the model. Consider a model for automobile safety based upon random hardware failures and excluding software faults. Such a model may suggest that measuring the condition of the brakes, tires, suspension, etc. can indicate the safety of a car. This is correct so long as the software risk is constant. However, if the software risk varies over time, then indications based on the physical components will be misleading as to the changing safety of the car.

Each risk model, and therefore each set of precursor events, is bespoke to a particular industry or even to a particular plant or system design. Examples of precursors include:

- In railways: Signals passed at danger (SPADs)
- In air traffic control: Runway incursions and separation violations
- In process plants: Activation of automated protection systems
- In nuclear power: Unplanned emergency shutdowns

A "near miss" is any unplanned event that could have led to an accident (Jones et al., 1999). Unlike predefined precursors, near misses are recognized in hindsight, and are based on an implied risk model. Since there is no explicit model, for near misses to be an indication of safety it must be assumed that they are drawn evenly from the population of event combinations that could lead to an accident.

This "common cause" hypothesis was forward by Heinrich (1930) and popularized in the pyramid diagram of Bird and Germain (1966). In his original formulation, Heinrich suggested that there was a fixed ratio between the number of unsafe acts, the number of near misses, and the number of accidents. The empirical basis of the fixed ratio has been widely empirically discredited (Wright and van der Schaaf, 2004). While there is some evidence that near misses and accidents can arise from the same tasks and errors (Alamgir et al., 2009), this commonality is insufficient to make near misses a reliable indicator. Not just the types of causes, but also the distribution of causes must be the same to allow the number of near misses to indicate the likelihood of accidents.

8.4.4 Status of Physical and Procedural Controls

Process control models of safety view accidents as arising from dangerous exchange of energy (Haddon, 1968). Safety is achieved through protective features, collectively known as "controls" or "barriers." Two common representations of this model are the "Swiss cheese" metaphor by Reason (2000) and "layers of protection analysis" developed within the chemical process industry (Bridges and Clark, 2010).

The model suggests that accidents require the coincidental failure of multiple barriers. Each barrier is considered independently for the purpose of measuring control effectiveness, with overall safety measured as the combined effectiveness of all controls. Whereas precursors are unanticipated events during live operation, barrier status is measured through planned observations, inspections, and tests.

Hollnagel (2008) describes four main types of barriers: physical, functional, symbolic, and incorporeal. Physical barriers directly block the transport of energy. They include walls, containers, valves, seals, and insulation. The condition of a physical barrier can usually be measured by inspection, either direct observation or examination assisted by specialized equipment for detecting flaws.

Functional barriers enforce preconditions for potentially dangerous states and actions. They include physical interlocks, passwords, and automatic sprinklers. The operation of a functional barrier may not be visible, and so their condition is measured through testing.

Symbolic barriers operate by communicating information or warning to human operators. They include signage, warning lights, instructions, and alarms. Successful operation of a symbolic barrier requires an act of transmission (for example, the shining of an indicator light) and an act of interpretation (a human understanding the purpose of the light and taking appropriate action). Both aspects need to be tested to establish the correct functioning of a symbolic barrier.

Incorporeal barriers are rules or norms that guide behavior. They are not physically, functionally, or symbolically present in a system, and so can only be measured by observing the behaviors they encourage.

8.4.5 Safe Behaviors

Behaviorism considers safety as arising from specific "safe" and "unsafe" behaviors of employees. Safe behaviors are specific to particular industries, and cover issues such as conducting pre-task checks, storing equipment not in use, and wearing personal protective equipment. An "observation" is an instance of an observer filling out a checklist of safe and unsafe behaviors.

The most direct measure of safety under a behavior-based model is the "percent-safe"—the proportion of observations that find safe behaviors (Krause and Hidley,

1989). Percent-safe is itself an output from "upstream" factors that influence safety. These factors include the act of observation itself. Observation does not just measure safety, it is considered to positively influence safety. Frequency and quality of observations are therefore key behavioral safety indicators. The other main upstream factor is employee involvement, as measured by attendance at safety meetings, participation in observer training, and the number of safety meetings (Krause et al., 1997)

8.4.6 Amount and Quality of Safety Activity

Activity or safety management models view safety as a consequence of the performance of specific safety functions. Since the overall effectiveness of these activities is not measurable, it is perhaps more accurate to say that safety management models define safety in terms of conformance with norms of safety practice. The "appropriate" performance of the activities is used to judge the presence or absence of safety.

Numerous models of safety are available, each with their own normative activities and suite of associated metrics. In each case it is recognized that volume of activity is insufficient as a metric, since activities may be tokenistic. Volume metrics are always combined with metrics relating to appropriate performance (Hopkins, 2009).

As with the models themselves, some individual safety activities can be very difficult to measure. For example, the key attributes of QRA are repeatability and accuracy. Since most QRAs are only performed once, and there is no independent means of verifying accuracy, the quality of QRA can only be established through peer review (Rae et al., 2014). Review against a normative activity standard is a common measure for safety activity quality.

Other activities seek to improve safety through influencing an intervening measurable variable. For example, cultural interventions seek to improve safety by changing employee satisfaction and engagement with safety programs. These can be measured through turnover, absenteeism, survey responses, and participation rates.

An example of an activity-based metric suite is provided by Reiman and Pietikäinen (2012). As well as standard outcome metrics, they propose "drive indicators" and "monitor indicators" based on a model in which safety management activity controls and directs the organization as a sociotechnical system. Drive indicators are "measures of the fulfillment of the selected safety management activities." They are principally concerned with the presence and currency of documented procedures for safety management. Monitor indicators are concerned with the current condition of people and equipment, and include measures of satisfaction, competency, and correct function.

8.4.7 Organizational Drivers and Attributes

While behavioral safety measures safety according to "sharp end" behaviors, and safety management is concerned with measuring safety function activity, organizational models view safety as arising from non-activity attributes of the organization.

These include competency, the presence of a "just culture," pre-occupation with failure, and adequate resources. While these attributes may be causally related to the way in which safety activities are carried out, it makes more sense to measure them directly.

Schulman (2004) describes the two main suites of organizational attributes as "precluded events" and "resilience." Precluded events models suggest that safety arises from the identification and prevention of known hazards, and are driven by standardization, analysis, and control. Resilience suggests that safety arises from the positive capacity to adapt to both routine and unexpected variation in the conditions of work (Hollnagel et al., 2006), and is driven by improvisation, flexibility, and tacit knowledge.

Safety climate and safety culture metrics, not discussed further in this chapter, are based on a third model where safety behavior arises from and is best measured by collective attitudes and understanding of safety (Pidgeon, 1991). Climate and culture are typically measured by surveying staff about their own attitude, or about their perceptions of the beliefs of others.

8.4.8 Variability

Safety-II models suggest that accidents have no "special cause," but represent extreme outcomes from variation in normal work. Safe operation, therefore, can be maintained by identifying and monitoring the main sources of variation in work performance (Hollnagel, 2014).

Variation may be driven by any number of factors including supply chains, schedules, employee experience, and equipment condition.

8.5 USE AND IMPLEMENTATION

8.5.1 Metrics Collection

A program of measurement will typically combine indicators from multiple theoretical models. For example, an organization employing a safety-critical sensor system might measure:

- Maintenance funding (an organizational driver)
- Safety attitudes and perceptions (a cultural driver)
- The number of completed sensor inspections (a safety activity)
- The number of sensor faults not found on first inspection (safety activity quality)
- The number of sensor faults found during inspection (status of the barrier or control)
- The number of instances of undiscovered sensor faults leading to incidents or near misses (precursors)

Most metrics can be expressed as rates, where the numerator is the number of items counted, and the denominator is the number of opportunities to count an item. Frequently used denominators include time periods or work hours, number of inspections, number of projects, or number of staff. Rates are more useful than raw counts where the denominator may change due to seasonal or economical variation in work intensity. A construction company does not become "more safe" during winter if any decrease in injuries can be explained by decrease in work hours. Similarly, a maintenance organization that performs more inspections may find more faults—an increase in detected faults does not necessarily represent a decrease in safety.

Most applications of metrics are for the purpose of creating feedback loops. Each application involves collecting indicators, analyzing the indicators, and then applying some form of control action.

8.5.2 Incentives and Accountability

Safety incentives are a control action where feedback, in the form of performance indicators, is translated into rewards or punishment as drivers for future performance. Incentives include individual financial rewards, collective rewards, recognition programs, or inclusion of safety indicators within "balanced scorecard" performance evaluations.

Individual and collective incentives are closely linked to behavior-based models of safety (Geller, 1996; Krause et al., 1997). Balanced scorecards are linked to safety management and organizational models, which consider management performance to be more important that front-line behavior in creating safety (Mearns and Ivar Håvold, 2003). Whether applied to workers or management, the intent is to shift trade-offs between productivity and safety to be more in favor of safety. All organizations, it is argued, have explicit or implicit rewards for meeting deadlines and saving costs. It is necessary, therefore, to provide equal reward for safe behavior.

8.5.3 Benchmarking and Comparison

All safety indicators are vulnerable to deliberate or inadvertent distortion (discussed further in Section 8.7 of this chapter). Accordingly, comparison of metrics is useful only when they are collected under a uniform policy, ideally where there is no incentive to bias the reported statistics. Even where a single agency collects statistics for a single organization, there may be subtle pressure to demonstrate the success of safety initiatives through continuous improvement in safety performance indicators. No stochastic process exhibits a monotonic improving trend, yet it may be politically unacceptable to report "worse" results on a safety indicator.

To the extent that it is possible to uniformly collect safety indicators, they may be used to reveal differences in safety between subdivisions of an organization. For example, particular sites or projects may be over- or underperforming compared to the organization as a whole.

For event-driven measurements such as injuries, the number of events in each period is drawn from a Poisson distribution, and will vary even in the absence of any change in risk. There may be further sources of variation that do have causes. These are local or short-term shifts in the probability distribution due to work patterns. Examples include changes in project phase, different risk on different days of the week, and seasonal variations in the amount and type of work conducted.

To compare metrics, sampling must be performed over a sufficiently long period to smooth stochastic variation and short-term shifts in probability. A Poisson Test is then performed to determine if there is a statistically significant difference between the two sample sets.

The process for analyzing non-event metrics is similar. Metrics that aggregate Boolean states (e.g., behavioral observations or barrier reliability) follow binomial distributions. Survey instruments have multiple dimensions, where each dimension can typically be modeled as a normal distribution (Soares, 2002).

8.5.4 Safety Management System Performance Monitoring

For a single organizational entity, safety indicators can be used to detect trends and step changes in safety. A trend is a gradual shift over time, whereas a step change is an abrupt shift in statistical distribution brought about by a deliberate intervention or an unintended disturbance.

Detection of changes in signals is a well-investigated statistical problem (Basseville, 1988). The basic process is to make repeated comparisons of pairs of time periods. Each comparison tests whether the parameters of the statistical distribution are significantly different between the time periods.

"Online" change detection continuously adds data until a significant change is detected. There is a direct trade-off between the speed of detecting change and the likelihood of a false positive. For safety, both speed of detection and reduction in false positives are important, so there is no truly optimum test for significant trends or changes.

"Offline" change detection uses a full dataset and employs clustering algorithms to detect the most likely times that changes have occurred. Offline change-point detection methods are useful for establishing cause–effect relationships (for example, measuring the origin of a problem or the success of a safety intervention), where the time of change may be as important as the fact of change.

8.6 EMPIRICAL RESEARCH ON USE AND EFFICACY

8.6.1 Usage of Indicators

There is no direct evidence for which indicators private companies use, or how companies interpret indicators. However, a number of studies have investigated the public disclosure of safety metrics.

Brown and Butcher (2005) examined the public reports of 100 large New Zealand employers. Under the topic of "monitoring," 22 reports provided injury counts or rates. Fifteen mentioned the amount of sick leave. The only other potentially safety-related metric mentioned was staff turnover.

Peebles (2003) found that 47% of United Kingdom companies and 43% of public bodies made safety performance information publicly available, mostly in the form of outcome metrics such as lost-time injuries and fatalities.

These statistics are unsurprising, given the regulatory role of outcome indicators. The United States Occupational Safety and Health Administration (OSHA) requires recordkeeping for injuries and fatalities (Occupational Safety and Health Administration, 2014). While the United Kingdom Health and Safety Executive provides guidance on leading indicators (Health and Safety Executive, 2006) it requires reporting only on lagging indicators.

8.6.2 Efficacy of Indicators

While the validity of safety indicators appears to be an empirical question, it is in fact an issue of construct validity. The question is not "Does this indicator actually measure safety?" but "Does this indicator measure what I/we think of as 'safety'?" Construct validity of indicators rests in turn on the construct validity of models of safety, which are not themselves amenable to empirical investigation.

Some issues surrounding indicators have been studied using aggregated injury data. These investigations are primarily concerned with the relationships between indicators rather than the absolute validity of indicators. For example, it has been shown that there is no correlation of minor injuries with compensation claims (O'Neill et al., 2015). The possibility of correlation of minor injuries with fatalities has been the subject of considerable empirical investigation, with inconclusive results (Hunting and Weeks, 1993; O'Neill et al., 2015; Salminen et al., 1992; Wright and van der Schaaf, 2004).

Despite strong consensus that the rate of personal injuries does not represent the likelihood of process accidents (National Occupational Health and Safety Commission, 1994; Hopkins, 2009; Reiman and Pietikäinen, 2012) this is not an issue that can be investigated statistically, due to the low rate of process accidents.

8.7 WEAKNESSES, LIMITATIONS, AND CRITICISM

8.7.1 Underreporting and Distortion

There are numerous situations in which the appearance of good safety performance can be desirable, irrespective of the actual level of safety. Safety indicators can be applied as measures of individual and collective performance for determining bonuses and pay increments. Companies may publicly report indicators, or be required to provide them to customers and regulators. Staff who have spent time and effort on safety improvement may be predisposed to observe success. In all of these situations there is subtle or overt pressure to manipulate safety indicators (Azaroff et al., 2002; Oleinick et al., 1995).

Injury statistics are particularly vulnerable to distortion and manipulation, because of ambiguity regarding what counts as a "recordable" injury. Injury rates can be artificially decreased by:

- Worker concealment of injuries, to avoid blame and punishment
- Failure to record injuries reported by workers
- Non-recording of serious injuries that do not prevent work, such as disfigurement or loss of hearing
- Avoiding classification as a "lost-time" injury by assigning the worker to alternate duties
- Using rosters and shift work to reduce or eliminate work time lost due to an injury
- Using other forms of leave to avoid recording absence due to injury
- Changing the employment status of injured workers (e.g., firing the worker or shifting them from full time to casual labor)
- Excluding contract staff or customer staff
- Excluding injuries in particular categories, such as transport to and from worksites
- Excluding injuries that occur on sites managed by other organizations
- Under-recording time lost to chronic injuries
- Classifying injury as non-work-related illness
- Inflating the denominator—the official number of hours worked—to reduce the rate determined from a given number of injuries

While other safety metrics can be more difficult to manipulate, the same incentives and biases encourage distortion.

8.7.2 The Regulator Paradox and Estimation of Rare Events

Outcome indicators, precursors, and status of control indicators measure safety by counting events that reveal the absence of safety (Hollnagel, 2014). As systems become increasingly safe, the number of such events decreases, reducing knowledge about the size and nature of the residual risk. This is known as the regulator paradox—the more successfully a system is controlled, the less feedback is available to further improve control. In very safe systems, many safety indicators are ineffective for improvement, comparison, or trend detection (Amalberti, 2001).

Even for high risk injuries, the number of recordable events can be too low and too variable for statistically significant changes to be observed. The choice is between a highly sensitive test with many false positives or a long delay before a change in risk can be detected.

Recording high frequency events as a proxy for low frequency events compounds the construct validity problem. The low frequency events themselves only indirectly measure safety. Higher frequency events only indirectly measure the low frequency events. Empirical validity is out of the question for correlating events that are already too rare to exhibit statistically significant trends.

8.7.3 Confusion Between Process Safety and Personal Safety Indicators

While the hypothesis that minor accidents and fatalities have common causes is still defended in some quarters, there is no question that single person injuries and multiple fatality accidents have distinct etiologies—they arise from different hazards and energy sources. An indicator that is valid for personal safety is unlikely to be valid for process or system safety and vice versa, since they measure different events, controls, or organizational attributes. Improvement in personal safety indicators can in fact be a sign of increased focus on personal safety at the expense of preventing major accidents—one of the "decoy phenomena" discussed by Turner (1976) in his groundbreaking study of disaster incubation.

The summary findings of the Baker Panel review into the Texas City refinery explosion of 2005 include the following comment about personal safety indicators:

> "BP has emphasized personal safety in recent years and has achieved significant improvement in personal safety performance, but BP did not emphasize process safety. BP mistakenly interpreted improving personal injury rates as an indication of acceptable process safety performance at its U.S. refineries. BP's reliance on this data, combined with an inadequate process safety understanding, created a false sense of confidence that BP was properly addressing process safety risks."
>
> (Baker et al., 2007)

8.7.4 Unintended Consequences of Indirect Measurement

Whenever performance is judged on an indirect indicator rather than the property of interest, there is a risk of "managing the measurement" rather than managing the property of interest. This results not just in underreporting, but also in perverse incentives for management and staff behavior (Geller, 1996).

For example, a metric based on safety activity can increase the volume of activity but decrease the quality and integrity of the tasks performed. Metrics based on survey instruments can increase pressure on staff to provide the "correct" answers, decreasing open and honest communication about safety issues at the same time as they report increased "engagement."

Grote (2009) suggests that indicators need not be primarily descriptive, but can be designed for the express purpose of driving changes in behavior. Even for this purpose, there is a need for "sound knowledge about cause-and-effect relationships" if indicators are to have the desired effect.

8.8 RELATIONS TO OTHER SAFETY PRINCIPLES

Risk and safety indicators are intrinsically linked to all aspects of safety. Principles that operate through ensurance—that is, they directly try to improve safety—suggest safety indicators. Principles that operate through assessment and assurance—they seek to measure, understand, or demonstrate safety as an indirect path to improvement—are consumers and users of safety indicators.

8.8.1 Ensurance Principles

Any principle that can be applied to make a system or situation safer can, in theory, provide a dimension against which safety can be measured. Defense-in-depth, resilience engineering, safety barriers, margins of safety, behavior-based safety, and safety culture are examples of principles that suggest particular safety models, leading to safety indicators.

For example, a margin of safety is a distance between the intended state of a system and an unsafe state of the system. This provides a measure of safety—the "safety reserve," or current distance to the unsafe state. A behavior-based safety program, on the other hand, suggests that worker behavior is the key to ensuring safety. The first step in such a program involves defining specific behaviors of interest. The prevalence of these behaviors becomes a metric for monitoring and evaluating each intervention.

As discussed in Section 8.6.2, safety indicators are incapable of comparing the relative efficacy of different ensurance principles, because each principle encapsulates a different understanding of what safety is, not just how it is achieved. For example, safety culture theory suggests that the intent and meaning of actions can be more important than the actions themselves. Safety culture metrics therefore measure intent and meaning. Behavioral safety theory suggests that culture can influence behavior, but it is the behaviors themselves that represent safety. Safety culture and behavioral safety are incommensurate—the theories suggest different interpretations of the same reality, and so it is unreasonable to measure the efficacy of one approach using the instruments of the other.

It is for this reason that proponents of behavior-based safety can legitimately claim that behavioral observations provide a highly valid indicator of safety, with sufficient data to draw statistically valid conclusions about the efficacy of behavioral safety practices, while at the same time critics can legitimately question status of this evidence. Unless the foundational premise that "safe behavior" is an indicator of "safety" is accepted, behavioral observations have no construct validity, and the evidence for the efficacy of behavior-based safety practices is weak.

The same is true of all empirical claims about the efficacy of safety ensurance principles.

8.8.2 Assessment and Assurance Principles

Some safety principles operate indirectly, by providing information that encourages safety. Risk analysis (both quantitative and qualitative), human reliability analysis and safety cases are examples of such principles.

There is a symbiotic relationship between these principles and risk and safety indicators. Risk analysis and human reliability analysis use as inputs the historical frequency of precursor events. Risk indicator programs use risk analysis and human reliability analysis models to determine which precursor events should be measured.

Ideally, risk indicators are used to test risk models during operation, providing real-world information about risk to inform future risk assessments. In practice, data typically invalidates the structure of the risk models, calling the validity of precursors as a safety indicator into question (Downer, 2013; Rae et al., 2012).

The end result is information which can be fed back to the system or operation for which safety is a concern. The difference between "safety management" and a "safety management system" is this feedback. Risk and safety indicators provide information on the effectiveness of management activities. This feedback allows control and optimization of safety processes, as well as targeting and evaluation of safety improvement initiatives.

Safety cases and risk communication facilitate feedback on a larger scale. They provide information to a regulator or the broader community about safety, allowing judgments to be made about the acceptability of designs and activities. As with risk analysis, these tools have a symbiotic relationship with safety indicators. Metrics can provide evidence that specific safety goals have been met in the past, supporting an argument that they can be met in the future. Safety cases and risk communication programs also provide an indication of what evidence is required to demonstrate safety (i.e., what cmetrics need to be collected).

REFERENCES

Alamgir H. Yu S. GormanE. Ngan K., and Guzman, J. (2009). Near miss and minor occupational injury: Does it share a common causal pathway with major injury?*American Journal of Industrial Medicine*, 52(1), 69–75.

Amalberti, R. (2001). The paradoxes of almost totally safe transportation systems. *Safety Science*, 37(2–3), 109–126.

Azaroff, L. S., Levenstein, C., and Wegman, D. H. (2002). Occupational injury and illness surveillance: Conceptual filters explain underreporting. *American Journal of Public Health*, 92(9), 1421–1429.

Baker, J. A., Bowman, F. L., Erwin, G., Gorton, S., Hendershot, D., Leveson, N., Priest, S., Rosenthal, I., Tebo, P., Wiegmann, D., and Wilson, L. D. (2007). The report of the BP US Refineries Independent Safety Review Panel. Available at http://www.bp.com/liveassets/bp_internet/globalbp/STAGING/global_assets/downloads/Baker_panel_report.pdf

Basseville, M. (1988). Detecting changes in signals and systems—A survey. *Automatica*, 24(3), 309–326.

Bird, F. E., and Germain, G. L. (1966). *Damage Control*. New York: American Management Association.

Bridges, W. B., and Clark, T. (2010). Key issues with implementing LOPA. *Process Safety Progress*, 29(2), 103–107.

Brown, J., and Butcher, F. (2005). Reporting on occupational health and safety in annual reports: A look at disclosure policies in New Zealand. *New Zealand Journal of Employment Relations*, 30(1), 1.

Bühlmann, H. (1997). The actuary: The role and limitations of the profession since the mid-19th century. *Astin Bulletin*, 27(02), 165–171.

Clemson, B. (1991). *Cybernetics: A New Management Tool*. Tunbridge Wells, UK: CRC Press.

Cullen, T. H. L. (1990). *The Public Inquiry into the Piper Alpha Disaster*. London: Her Majesty's Stationary Office.

Downer, J. (2013). Disowning Fukushima: Managing the credibility of nuclear reliability assessment in the wake of disaster. *Regulation and Governance*, 8(3), 287–309.

Elliott, M. R., Kleindorfer, P. R., DuBois, J. J., and Wang, Y. (2008). Linking OII and RMP data: Does everyday safety prevent catastrophic loss? *International Journal of Risk Assessment and Management*, 10(1), 130–146.

Ericson, C. A., II. (2005). *Hazard Analysis Techniques for System Safety*. Wiley-Blackwell.

Evans, A. W. (1994). Multiple-fatality transport accidents: 1946–1992. *Transport Statistics Great Britain 1994* Available at http://trid.trb.org/view.aspx?id=458604

Feigenbaum, A. V. (1956). Total quality control. *Harvard Business Review*, 34(6), 93–101.

Fredericks, T., Abudayyeh, O., Choi, S., Wiersma, M., and Charles, M. (2005). Occupational injuries and fatalities in the roofing contracting industry. *Journal of Construction Engineering and Management*, 131(11), 1233–1240.

Geller, E. S. (1996). The truth about safety incentives. *Professional Safety*, 41(10), 34–39.

Good, I. J. (1986). Some statistical applications of Poisson's work. *Statistical Science*, 1(2), 157–170.

Grote, G. (2009). Response to Andrew Hopkins. *Safety Science*, 47(4), 478.

Haddon, W. (1968). The changing approach to the epidemiology, prevention, and amelioration of trauma: The transition to approaches etiologically rather than descriptively based. *American Journal of Public Health and the Nation's Health*, 58(8), 1431–1438.

Hale, A. R., Heming, B. H. J., Carthey, J., and Kirwan, B. (1997). Modelling of safety management systems. *Safety Science*, 26(1–2), 121–140.

Hale, A. R., and Hovden, J. (1998). Management and culture: The third age of safety. A review of approaches to organizational aspects of safety, health and environment. In: A.-M. Feyer and A. Williamson (Eds.), *Occupational Injury*. London: Taylor & Francis.

Health and Safety Executive. (2003). *BP Grangemouth Scotland 29th May—10th June 2000*. Health and Safety Executive. Available at http://www.hse.gov.uk/comah/bpgrange/images/bprgrangemouth.pdf

Health and Safety Executive. (2006). Developing process safety indicators: A step-by-step guide for chemical and major hazard industries, Great Britain

Heinrich, H. W. (1930). Relation of accident statistics to industrial accident prevention. *Proceedings of the Casualty Actuarial Society*, 16(33–34), 170–174.

Hollnagel, E. (2008). Risk + barriers = safety? *Safety Science*, 46(2), 221–229.

Hollnagel, E. (2014). *Safety-I and Safety-II*. Burlington, UK: Ashgate.

Hollnagel, E., Woods, D. D., and Leveson, N. (Eds.). (2006). *Resilience Engineering: Concepts and Precepts* (New edition). Burlington, UK: Ashgate.

Homberger, E., Reggiani, G., Sambeth, J., and Wipf, H. K. (1979). The Seveso accident: Its nature, extent and consequences. *Annals of Occupational Hygiene*, 22, 327–370.

Hopkins, A. (2000). *Lessons from Longford: The ESSO Gas Plant Explosion*. CCH Australia.

Hopkins, A. (2009). Thinking about process safety indicators. *Safety Science*, 47(4), 460–465.

Hunting, K. L., and Weeks, J. L. (1993). Transport injuries in small coal mines: An exploratory analysis. *American Journal of Industrial Medicine*, 23(3), 391–406.

International Labour Office. (1923). Methods of statistics of industrial accidents. In: International Conference of Labour Statisticians. Geneva.

Jeong, B. Y. (1998). Occupational deaths and injuries in the construction industry. *Applied Ergonomics*, 29(5), 355–360.

Jones, S., Kirchsteiger, C., and Bjerke, W. (1999). The importance of near miss reporting to further improve safety performance. *Journal of Loss Prevention in the Process Industries*, 12(1), 59–67.

Krause, T. R., and Hidley, J. H. (1989). Behaviorally based safety management: Parallels with the quality improvement process. *Professional Safety*, 34(10), 20.

Krause, T. R., Hidley, J. H., and Hodson, S. J. (1997). *The Behavior-Based Safety Process: Managing Involvement for an Injury-Free Culture*. New York: Van Nostrand Reinhold.

Lowrance, W. W. (1976). *Of Acceptable Risk: Science and the Determination of Safety*. Los Altos, CA: William Kaufmann.

McDonald, G. (1995). Focus—don't fiddle (the obscenity of the LTFR). In: Queensland mining industry safety conference, Yeppoon, Australia.

Mearns, K., and Ivar Håvold, J. (2003). Occupational health and safety and the balanced scorecard. *The TQM Magazine*, 15(6), 408–423.

Muttram, R. I. (2002). Railway safety's safety risk model. *Proceedings of the Institution of Mechanical Engineers, Part F: Journal of Rail and Rapid Transit*, 216(2), 71–79.

National Occupational Health and Safety Commission. (Eds.). (1994). *Positive Performance Indicators for OHS: Beyond Lost Time Injuries*. Australian Government Public Service, Canberra.

Occupational Safety and Health Administration. (2014). Code of federal regulations 29 part 1904 recording and reporting occupational injuries and illness. United States Department of Labor.

Øien, K., Utne, I. B., and Herrera, I. A. (2011). Building safety indicators: Part 1—theoretical foundation. *Safety Science*, 49(2), 148–161.

Oleinick, A., Gluck, J. V., and Guire, K. E. (1995). Establishment size and risk of occupational injury. *American Journal of Industrial Medicine*, 28(1), 1–21.

O'Neill, S., McDonald, G., and Deegan, C. M. (2015). Lost in translation: Institutionalised logic and the problematisation of accounting for injury. *Accounting, Auditing and Accountability Journal*, 28(2), 180–209.

Peebles, L. (2003). The provision of health and safety information in the annual reports, websites and other publicly available documents produced by the UK's top companies and a sample of government departments, agencies, local authorities and NHS trusts..: Health and Safety Executive, Great Britain.

Pidgeon, N. F. (1991). Safety culture and risk management in organizations. *Journal of Cross-Cultural Psychology*, 22(1), 129–140.

Rae, A. J., Alexander, R., and McDermid, J. (2014). Fixing the cracks in the crystal ball: A maturity model for quantitative risk assessment. *Reliability Engineering and System Safety*, 125, 67–81.

Rae, A. J., McDermid, J., and Alexander, R. D. (2012). The science and superstition of quantitative risk assessment. In: Annual European safety and reliability conference, Helsinki, Finland.

Rae, A. J., Nicholson, M., and Alexander, R. D. (2010). The state of practice in system safety research evaluation. In: IET system safety conference, Manchester, UK.

Reason, J. (2000). Human error: Models and management. *British Medical Journal*, 320, 768–770.

Rechard, R. P. (1999). Historical relationship between performance assessment for radioactive waste disposal and other types of risk assessment. *Risk Analysis*, 19(5), 763–807.

Reiman, T., and Pietikäinen, E. (2012). Leading indicators of system safety—monitoring and driving the organizational safety potential. *Safety Science*, 50(10), 1993–2000.

Ribes-Iñesta, E. (2003). What is defined in operational definitions? The case of operant psychology. *Behavior and Philosophy*, 31, 111–126.

Rochlin, G. I. (1999). Safe operation as a social construct. *Ergonomics*, 42(11), 1549–1560.

Russell, B. (1914). *The Relation of Sense-Data to Physics*. Nicola Zanichelli.

Salminen, S., Saari, J., Saarela, K. L., and Räsänen, T. (1992). Fatal and non-fatal occupational accidents: Identical versus differential causation. *Safety Science*, 15(2), 109–118.

Schulman, P. R. (2004). General attributes of safe organisations. *Quality and Safety in Health Care*, 13(suppl 2), ii39–ii44.

Siu, N., and Collins, D. (2008). PRA research and the development of risk informed regulation at the U.S. Nuclear Regulatory Commission. *Nuclear Energy and Technology*, 40(5), 349–364.

Soares, P. M. (2002). Statistical analysis applied to safety culture self-assessment. A survey example. International Conference on Safety Culture in Nuclear Installations, Rio de Janeiro.

Stieglitz, W. (1966). Numerical safety goals—Are they practicable? In: American Institute of Aeronautics and Astronautics Symposium on Deep Submergence Propulsion and Marine Systems, Forest Park, IL, USA, February 28–March 1, 1966.

Thomas, W. (2015). *Rational Action: The Sciences of Policy in Britain and America, 1940–1960*. MIT Press.

Turner, B. A. (1976). The organizational and interorganizational development of disasters. *Administrative Science Quarterly*, 21(3), 378–397.

United Kingdom Department of Employment. (1975). *The Flixborough Disaster—Report of the Court of Inquiry*. London: Her Majesty's Stationery Office.

Wright, L., and van der Schaaf, T. (2004). Accident versus near miss causation: A critical review of the literature, an empirical test in the UK railway domain, and their implications for other sectors. *Journal of Hazardous Materials*, 111(1–3), 105–110.

9

PRINCIPLES OF HUMAN FACTORS ENGINEERING

Leena Norros and Paula Savioja

9.1 INTRODUCTION

The knowledge area human factors/ergonomics is defined by the Human Factors and Ergonomics Society as follows: "Ergonomics (or human factors) is the scientific discipline concerned with the understanding of interactions among humans and other elements of a system, and the profession that applies theory, principles, data, and other methods to design in order to optimize human well-being and overall system performance." (HFES, 2016). Human factors engineering (HFE) is considered as synonym for ergonomics (HumanFactors, 2016), but typically the practical side of the human factors notion is emphasized, that is, HFE refers to a systematic application of the human factors knowledge for engineering purposes (Karwowski, 2006; NUREG0711, 2004). In some definitions of HFE important methodological issues are also mentioned. For example, the Encyclopedia Britannica names two general premises of HFE: The first is that the engineer must solve the problems of integrating humans into machine systems by rigorous scientific methods and not rely on logic, intuition, or common sense. The second premise is that, typically, design decisions cannot be made without a great deal of trial and error. It is concluded that while

Handbook of Safety Principles, First Edition. Edited by Niklas Möller, Sven Ove Hansson, Jan-Erik Holmberg, and Carl Rollenhagen.

human-factors engineering aims at substituting scientific method for guesswork, its specific techniques are usually empirical rather than theoretical. (Holstein, 2016).

In light of the current definitions, HFE refers to the practical side of the human factors concept, that is, to the application of human factors knowledge to design. In order to be effective in its practical aim three so-called high-quality criteria have been assigned to human factors: It should apply a systems approach, it should be design driven, and it should focus on the closely related outcomes, performance, well-being, and safety (Dul et al., 2012). These criteria are very relevant but reaching them might not be so easy. Indeed, observations have been made concerning the inefficiency of the human factors practice, leading to an underutilization of available human factors knowledge in solving of problems arising from the ongoing transformations in work and daily life (Dul et al., 2012; Wilson and Carayon, 2014).

In the safety-critical industry, HFE is interpreted to be closely connected with ensuring the safety of the complex systems. Strict regulation has been established by the authorities concerning the entire life cycle of these systems, that is, design, construction, operation, and decommissioning. HFE has a role included in the regulations, and certain HFE review guidelines and standards are applied worldwide. Most important one is the NUREG-0711 guideline labeled the Human Factors Engineering Program Review Model (NUREG0711, 2004). Other widely used standards and guidelines for HFE are those established by the International Electrotechnical Commission (IEC) and International Standards Organization (ISO) (IEC-1771, 1995; IEC-61839, 2000; ISO, 2006). These guides and standards are applied in most design projects in the nuclear power domain. They have also shaped the licensees' HFE practices concerning operation. The NUREG-0711 has also influenced the HFE practices in other safety critical domains, like medicine, defense, or aviation. In these cases, the guidance complements the strong human factors practices in these domains and their domain-specific regulations.

The NUREG-0711 (2004) was constructed for the HFE staff of the Nuclear Regulatory Commission (NRC) to evaluate HFE programs of applicants for construction permits, operating licenses, standard design certifications, combined operating licenses, and for license amendments. In all these different connections, the purpose is to verify that accepted HFE practices and guidelines are incorporated into the applicant's HFE program. The HFE review covers the entire design process, from planning and analysis to performance monitoring during operation, and is defined to have 12 review elements (see Figure 9.1).

The HFE review elements in Figure 9.1 have been interpreted as the generic content of HFE. Even today, when the NUREG-0711 document is under revision, these elements will form the core structure for HFE (O'Hara et al., 2012). The revision is caused by the digitalization and other technical development of the plant technology and will focus on the more detailed content of the reviews.

A comprehensive research effort commissioned by the US NRC is going on to support the revision of the HFE review guide (Fleger and O'Hara, 2015; O'Hara, 2014). In parallel to this research, also the international nuclear human factors

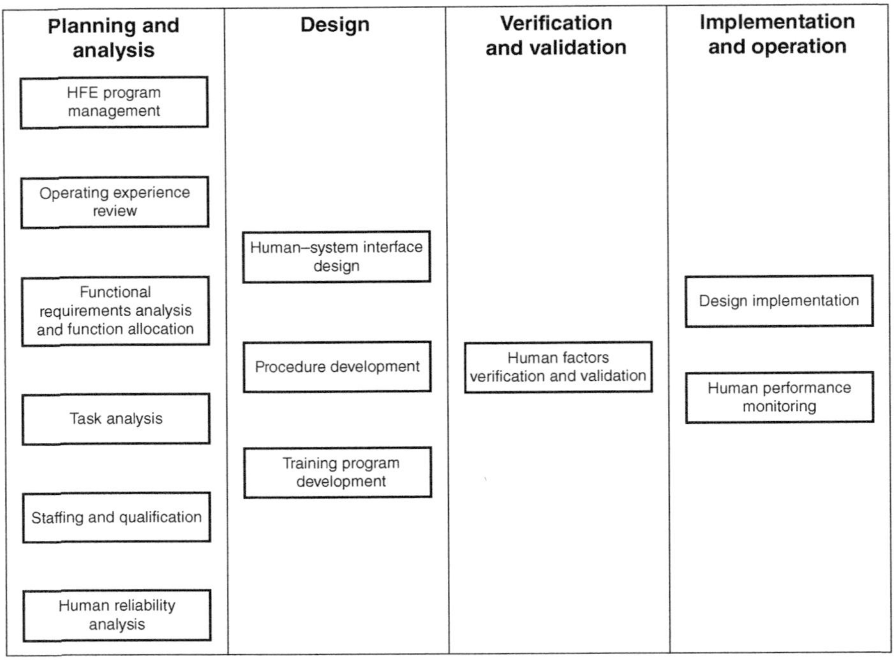

Figure 9.1. HFE program review model review elements. *Source*: NUREG-0711, 2004, Public domain.

community has made its own attempts to define the current HFE research needs (Heimdal et al., 2004; OECD/NEA, 2012; WGHOF, 2009). One of the topics that is unanimously defined to be a high priority research area is the NUREG-0711 element "human factors verification and validation," and, in particular, the integrated system validation concept (Fleger and O'Hara, 2015; O'Hara, 2014). Several issues of the integrated validation methodology appear to require revision, among the most important ones are the measures used in defining successful performance in tests, reference used in evaluations, and the type and coverage of scenarios for the tests. All these issues are deeply related to theories concerning human behavior. A further aspect concerning the methodology is that the integrated validation has traditionally been considered a singular comprehensive test of the safety of the system to be tested. Validation experience from recent years has shown that ensuring the safety of a complex and long-lasting design process would gain more of a stepwise HFE validation process.

In concert with the results by O'Hara and the international community, also our own experience of HFE demonstrates a need to improve the HFE methods and practices (Norros et al., 2015, p. 5–12). Also our findings show that HFE reviews must be a stepwise procedure, and that the review should intervene in the design. This conclusion would require major revisions into the prevailing review guidelines NEA (2016).

We also found that the traditional behavioral measures provide insufficient guidance for engineering of artifacts that do not exist yet. Therefore, a comprehensive characterization of good professional practice is needed, as well as context-sensitive measures of what is a good tool or technology. We found, further, that while the role of usability of technologies has been acknowledged as a performance and business factor by the modern product and service industry, it is still considered a secondary target by the safety engineers of high-reliability organizations. It appears also that many safety-culture specialists tend to neglect the significant influence of the technological artifacts on the users' behavior and culture. We conclude that the close connections between usability and safety should be understood better, and the too sharp contrast between the performance and safety goals alleviated.

In this article, we will claim that in order to fulfill the criteria of high-quality human factors, and to increase its effectiveness, HFE needs to enrich its perception of itself, that is, to reconsider what kind of intelligent activity it is. We see that HFE should adopt a methodology that changes HFE to something more than an application of existing scientific knowledge, that is, to a knowledge constructing design practice. The following methodological principles could lead HFE to this direction:

- HFE is design thinking
- HFE studies human as a manifold entity
- HFE focuses on technology in use
- Safety is achieved through continuous HFE

Presentation of these principles, elaborating their theoretical foundations, giving examples of how they have guided our own HFE work, and their practical effectiveness is the main content of this chapter. We will also discuss the weaknesses, limitations, and criticism of the principles and discuss their relations to other safety principles. We will finish the contribution by conclusions, including next steps of development of HFE.

9.2 PRINCIPLE 1: HFE IS DESIGN THINKING

9.2.1 Description

The widely accepted notion that human factors is both a scientific discipline and a design practice should not be taken to mean that human factors is *either* an academic *or* a practical activity, depending on the case, but a *unity of both*. Such a symbiosis could be realizable in HFE practice that is aware of its design identity, and expresses epistemic characteristic labeled as "design thinking" (Buchanan, 1992). Design thinking is capable of producing artifacts and solutions in conditions that are not comprehended completely and have contradicting demands. Design thinking creates new generic knowledge through a purposeful practice that changes the environment.

9.2.2 Theoretical Foundation

The impetus to formulate this principle lies in the recent discussion about the inefficiency of exploiting human factors in the development of modern work systems (Dul et al., 2012). A number of reasons were found for this situation, such as insufficient awareness of the value of human factors knowledge and exhibition of a weak demand for human factors practice among the stakeholders in leading positions in the design, management and operations. Where there was a demand for human factors knowledge, for example, in safety-critical organizations, its application was too limited in scope to bring optimal solutions. A further generic reason found was that the identity of human factors was unclear and its content appeared vague.

It was later proposed that it should also be reasonable to seek human factors *inherent* reason for its observed insufficient exploitation (Norros, 2014). Norros proposed that clinging to the validity ideals of knowledge typical in science, and a lacking awareness of design thinking as a possible model for acceptable reasoning in human factors studies, could be a major obstacle for effective utilization of the possibilities of human factors. This idea gained support from Kant, who identified a need to emphasize the epistemic advantages of engineering/design type of knowledge in comparison to science in HFE. Kant (2015, pp. 7–8, p. 90) bases his claim in an extensive account of the historical development of engineering craft in the USA that only in a later phase became enriched by the advances of science. He assumes that the advantages of design-oriented HFE would become evident when facing the issues of the future cyber-physical systems (CPS) based on nanotechnologies. Equal conceptions of the nature of HFE are advocated within the French-speaking ergonomics that is known to draw on philosophical sources that acknowledge the role of praxis in knowledge formation. For example, the prominent ergonomist Pierre Falzon advocates a design-oriented epistemology for human factors (Falzon, 2008). Corresponding epistemic analyses of the benefits of design thinking have been expressed within the human–computer interaction (HCI) tradition. For example, Kari Kuutti and Liam Bannon have analyzed the development of HCI and the nature of design activity. They came to the conclusions that design thinking could be an appropriate model for HCI practice (Kuutti, 2009; Kuutti and Bannon, 2014).

But what do we mean by design and what are the characteristics of design thinking? From the beginning of the 2000s, design has become a major phenomenon in the technologically developed post-modern society. Design refers to artistic shaping of various objects or artifacts. Today it also denotes creation of larger living environments, services, and work. Design is distinguished by its particular form of arguing, that became labeled design thinking. Design thinking appears in three spheres of life in the modern technological world, that is, in engineering, in marketing, and in the forms of graphical and industrial design (Margolin and Buchanan, 1998, Introduction, p. xiv).

One way to characterize design thinking is to compare it with science. The often cited scientist who articulated clearly the epistemic difference between design and science was Herbert Simon through his ground-braking book *The Sciences of the Artificial* (Simon, 1969). Indeed, today it is often taken more or less granted that design and science are very different intelligent activities. But, still, the relationship between the two forms of activity is also held as an open issue. For example, Farrel and Hooker forcefully argue for the identity of these two forms of activity (Farrel and Hooker, 2012, 2015). As a response to these ideas, Galle and Kroes (2014) provided a detailed analysis of the communalities and differences between the two activities, and ended up with emphasizing the distinctness of the two, here converging with the ideas of Simon. By our claim that HFE is design thinking, we do not want to enter into a philosophical discussion about the distinctness or identity of the two forms of activity, science and design. Instead, we draw the conclusion that features labeled as design thinking may basically be found within both science and design activity. We consider these features to bring advantages for HFE to meet its aims in design. So, let us look closer to the features of design thinking!

The characterizations of design thinking expressed by one of the most famous design theorists Richard Buchanan have proved very insightful. Already, in the beginning of 1990s, he proposed that in the thoroughly technologically mediated world, the discipline of design exemplifies a new form of rhetoric. "Design is a liberal art of technological culture, concerned with conception and planning of all the instances of the artificial or human-made world: signs and images, physical objects, activities and services, and systems or environments." (Margolin and Buchanan, 1998, Introduction, p. xiii). Galle and Kroes support Buchanan but they emphasize even more the *purpose-driven* and *artifact-oriented* nature of design (Galle and Kroes, 2014). Another well-known design theorist Klaus Krippendorf echoes these notions of design, but in his definition, the *meaning* of artifacts for people, or also design as an activity to *make sense*, is emphasized as a characteristic of design (Krippendorff, 1989, 1998, p. 156).

Central in all expressions of design thinking is that in the planning of new products, design *integrates knowledge* from different fields and disciplines while attacking problems that are in many ways *indeterminate,* that is, they are ill-defined, contain uncertainties and contradicting demands. These problems are defined in two ways in design, first, in an initial *generic* form, and, second, in the form of a *particular* proposal of a product that does *not yet* exist. (Buchanan, 1992, 1998).

In connection with the aim of bringing different views together, design thinking can be described a *visual* form of thinking in which connections between elements are grasped in a parallel way. The visual characteristic of design thinking is emphasized by Findeli who labels design visual intelligence (Findeli, 2001). *Models* and other design artifacts play a central role in design thinking. In the words of Buchanan, "Every designer's sketch, blueprint, flow chart, graph, three-dimensional model, or other product proposal is an example of such argumentation" (Buchanan, 1992, p. 19). In design thinking, the argument is a continuous interplay of signs, things in the world, thought and action; that is, a strict *distinction between thinking and acting is irrelevant.*

Drawing on Findeli (2001), Kuutti has indicated that in design thinking knowledge is created in all connections in which the artifact is involved, that is, *in design, production, and use* (Kuutti, 2009). Moreover, while design thinking materializes in the models and proposals, it is also true for design thinking that reasoning and knowledge in design are *only partially explicit* (Kuutti, 2009).

What regards the evaluation of what is good design, we may refer to Buchanan who states that three lines of reasoning must form a synthesis in design activity and in the product so that *"useful"*in human experience can be reached. These three lines are the *ideas of designers and manufacturers* about their products, the internal *operational logic of products*, and the *desire and ability of human beings to use products* in everyday life in ways that reflect personal and social value. (Buchanan, 1992, p. 19).

A further aspect concerning the nature of evaluation of design is the comparison between the typical quality criteria of science and those of design. The evaluation criteria of science relate to *truth of a theory or knowledge*. This reference does not make so much sense with regard to design outcomes, the evaluation criteria of which are rather qualifications relevant to *the purpose of the artifact proposal*. Hence, the evaluation criteria used in design may include categories like efficacy, efficiency, feasibility, safety, and beauty. (Galle and Kroes, 2014; Kuutti, 2009). Galle and Kroes continue that in the design of artifact proposals there are typically several relevant evaluation criteria between which *trade-offs* will have to be made. This implies that several quality dimensions are relevant at the same time and that the evaluation metrics have to be a complex of criteria.

9.2.3 Use and Implementation

Scientifically proven knowledge of well-defined human factors phenomena has its place in solving design problems. But due to its nature, scientific knowledge in its traditional form brings only limited value for HFE. Our claim is that HFE would gain in effectiveness by accepting the value and validity of design thinking as intellectual activity. The claimed effect of design thinking lies in its orientation to contextual constructive solving of complex problems with the purpose of developing the usefulness of the artifactual environment, and in the taking advantage of concepts and models. Objectivity, control, and independence are not the foremost qualifications for the design-oriented HFE, instead, purpose- and value-oriented evaluation, intervention, and dialogue should characterize the way HFE operates.

In the following Table 9.1, we demonstrate what kind of methodological characteristics HFE would adopt if taking advantage of those epistemic features of design thinking that were discussed above.

9.2.4 Empirical Research on Use and Efficiency

The adoption of the epistemology of design thinking in HFE is particularly suitable for the work in research institutions outside the academia. These institutions are

TABLE 9.1. Methodical Conclusions Concerning HFE Drawn from the Epistemic Features of Design Thinking

Epistemic Feature of Design Thinking	Methodological Conclusion with Regard to HFE
Understands and integrates different views.	Cross-disciplinary collaboration and dialogical thinking for increasing creativity of solutions.
Deals with indeterminate problems.	Experimenting as means to reduce uncertainty in the problem, and to construct knowledge.
Deals with particular objects but enables creation of generic concepts.	Qualitative data collection methods and conceptual generalization, organized in formative (developmental) projects.
Exploits models, schemes, and prototypes as means of creating knowledge.	Conscious use of modeling as an instrument for formative intervention with end user participation.
Knowledge is only partially explicit.	Tacit knowledge of designers and end users can be articulated and reflected on in modeling activity.
Demonstrates continuity between signs, things in the world, thinking, and acting.	Keeping doors open to naturalistic or ecological approaches to human activity that reject the information processing view to activity.
	Exploiting knowledge of artifacts through their design, production and use.
Orients to the purpose and usefulness of knowledge. This orientation is a synthesis of the designers' intentions, the material constraints of the product, and the users' capability and their experience of value for use.	Ethnography- and science-based modeling of the domain, tools, and the activity; accepting users' experience as valid information in defining the good outcome of design; involving users in design.
Is capable of trade-offs between different goals, and different quality criteria.	HFE should make the connections between safety, effectiveness and usability goals transparent, and enable trade-offs between them. Considering safety from a capability-creating (resilience) rather than failure-reducing perspective would emphasize the synergy between these targets.

organized according to societally meaningful application domains, for example, technology development, health, agriculture, environment, etc., and consequently require multi-disciplinary research. These institutions often have a role of technical support organizations (TSO), and, hence, they provide research-based expertise according to the needs of various national authorities. They also have close connections with the industry supporting the research needs of partners with difficult practical problems.

Having found ourselves in a research situation described above, we arrived at the conclusion that the problem-setting of the traditional human factors was too focused and its approach not sufficiently context-oriented. Hence, the development of a complementary research approach appeared necessary. The leading principle was that, in the HFE work, we should "have respect for the scientific value of practical problems." Our HFE domain is complex safety-critical industry and work, and we have worked on many safety-critical domains. As a thorough understanding of the problems of the specific domain under study has been our conviction, it has, of course, been a challenge for us to get acquainted with each domain. It has necessitated interdisciplinary work and collaboration with substance matter experts. We have accomplished studies mainly in the conventional and nuclear power production, in maritime, rail, and air traffic, in chemical, pulp and paper and manufacturing industry, in military and rescue services, and in health care. A first summary of the empirical work and the developing research approach was published in the book *Acting under Uncertainty—The Core-Task Analysis in Ecological Study of Work* (Norros, 2004). At this stage, the research approach was oriented toward analysis of work, whereas the second, recently published overview *Core-Task Design: A Practice-Theory Approach to Human Factors* adopted a stronger design orientation (Norros et al., 2015).

9.3 PRINCIPLE 2: HFE STUDIES HUMAN AS A MANIFOLD ENTITY

9.3.1 Description

The subject matter of HFE is human activity. In dealing with it, one needs to recognize that a human being is a different kind of entity compared to technology or the physical environment. Comprehending human being and his/her activity the physiological and neural *causes* of behavior need to be studied, but in connection to this, an adequate study also clarifies what the resulting behavior is a sign of, what it *means*. This assumes that human *intentionality* is observed when explaining behavior.

9.3.2 Theoretical Foundations

This principle focuses on the particular demands that human being as a research object puts on the methodology. The particularity originates in the fact that from an ontological perspective, that is, from the perspective of the nature of his/her being, the human is a manifold entity, that is, physiological and neural, behavioral, and mental phenomena all belong to the human being. Different scientific disciplines have specialized in the analysis of these phenomena and they elaborate different aspects of activity. Interaction between the disciplines is clearly advantageous for understanding particular expressions of activity. Even more fundamental, and less evident than the above, is the conclusion that the Finnish philosopher G. H. von Wright made regarding the ontological diversity of the human being (von Wright, 1998a, 1998b, pp. 43–47). He saw that the different perspectives are tightly tied with each other because mental phenomena, for example, a reported sensory experience, is needed to identify the existence of neural events—demonstrating *epistemic primacy* of mental to neural. Neural phenomena are needed as causes for accepting somebody's movements as behavior—demonstrating *causal primacy* of neural to behavioral. Finally, behavioral (often verbal) criteria are used as truth criteria of mental states; that is, they explain what happens when, for example, one is frightened (I tremble for your safety)—demonstrating *semantic primacy* of behavioral to mental. When these relationships between mental, neural, and behavioral are acknowledged in a psychological analysis it is possible to understand what behavior is a sign of, what it means. HFE needs to have this capability of grasping the meaning of causal phenomena for people in order to add value in design.

When seeking theoretical foundations for HFE in the psychological literature it would be wise to orient to those theories that acknowledge the ontological diversity of human activity and do not reduce activity to any of the three aspects alone, for example, to the causal explanations of behavior based on more and more detailed information of the neural phenomena. Those theories that grasp human as always embedded in his/her physical and social environment, and study human activity as a continuity of perceiving, acting, and interpreting, typically observe this ontological diversity. Examples of such theories of human conduct are pragmatist approaches (Bennet and Flach, 2011; Dewey, 2002; Peirce, 1998b), ecological approaches (Gibson, 1977; Neisser, 1976; Rasmussen, 1986), phenomenological approaches (Dourish, 2001; Merleau-Ponty, 1986), or also practice theory approaches (Nicolini, 2013; provides a review of 6 streams in practice theory).

The advantage for HFE of drawing on the above described theoretical foundations is connecting human activity to its environmental possibilities. This provides the possibility to express, in concrete terms, what good activity is: Good activity is appropriate taking into account the possibilities and constraints of the environment, which enables sustainability and development of the human–environment interaction. HFE aims at supporting good activity.

In HFE, it is usual to analyze and evaluate human activity from a behavioral perspective with emphasis on the *performance outcome*. In such an analysis, either a certain ideal path of acting, or the expected outcome of the performance is used as reference for evaluating the successfulness of activity. Examples of such measures are performance time and adequacy, errors, or deviations from an ideal course of actions. Also physiological or mental measures of work load, or some measures of situation awareness, are examples of performance outcome measures. In the safety context, outcome measures may also include reference to norms or rules, or statistical output measures like lack of accidents or occupational diseases, etc. Typical of these measures is that the good of performance is thereby defined from outside the studied practice in formal terms, and, hence, that the measures do not necessarily make sense for the actors themselves as signs of good practice. The logic underlying such externally defined measures is a causal relationship between bodily and situational factors on the observable behavior (including verbal).

A complementary view to the performance outcome perspective is to describe behavioral phenomena through the *meaning* people attach to them, and inform the analysis of the *intentions* people express through their behavior. According to this view, the whole human-environment system orients, adapts and develops according to the changing external and internal conditions and accordance with the intentions and the meaning that become available through them. Internal good behavior is to maintain appropriate acting in the changing world. Because intentions and meaning are experienced and expressed in behavior, it is possible to define behavioral markers that speak of the meaning and intentions they carry.

9.3.3 Use and Implementation

In safety-critical environments, it is, of course, important to take notice of the immediate causes on, and the eventual threat to safety (expressed as e.g., imposed time lags, or errors in behavior) that certain design proposals may have. Because negative outcomes are unwanted events, and they often are more salient than a fluent flow of optimal functioning of a system, it is understandable that orienting to failures has become a dominant way of thinking about safety. The term Safety 1 has been used by Hollnagel et al. to indicate this traditional conception of safety (2006). Due to the insufficiency of this conception for developing the safety of complex systems, attention has been drawn to those features of the system behavior that have positive, even though sometimes latent, or long-term effects on safety. This optional conceptualization of safety, labeled as Safety 2, has been described through the concept of resilience (Hollnagel et al., 2006). According to this concept, organizations have more or less developed capabilities to maintain and support system safety. Capabilities may be suitable organizational structures and rules, but also daily working practices of the personnel that create resilience into the system. Due to its focus on features that support the development of an organization, the resilience perspective would correspond with one of the criteria of the high-quality human factors, that is, with that of balancing between performance, well-being, and safety.

Yet, it is a challenging future task of HFE to develop a theory and methods to define resilience-informed features and measures of a human-environment system behavior. Some current research has focused on defining characteristics of resilience-driven practices (Back et al., 2008; Furniss et al., 2011). One possible attempt toward this direction is to develop indicators for good activity based on people's experiences and habitual ways of acting, as for example, we, and some other human factors experts have done (Norros et al., 2015; Savioja et al., 2014; Wright and McCarthy, 2010). These indicators express what is defined as internally good behavior by professional actors themselves. The logic underlying these measures is an attempt to understand what behavior means, and to use this knowledge as explanation of action. In order to carry out such analysis of work activity assumes that HFE experts take the accounts of people concerning their own behavior seriously, and as valid information.

9.3.4 Empirical Research on Use and Efficiency

The research approach that we use in analyzing work in complex safety-critical environments, and in evaluating design proposals aimed at improving safety follows the second HFE principle (Norros et al., 2015). This approach may be summarized by an analysis scheme that could be characterized as a contextual analysis of activity. According to this scheme, the analysis has three elements, that is, modeling, analysis and assessment (see Figure 9.2).

The *possibilities* of activity are *modeled* by defining the domain and situations with the aid of so-called core-task, activity system, and abstraction hierarchy models. By these models, the analyzed activity is embedded in its real environment. The modeling aims at understanding the general purposes of the work in certain domains, and at revealing the natural and social constraints and possibilities that the environment in this domain, and in the specific situations under scrutiny, provide for actual action. We use the domain models to create different content-related references that describe the possible field of acting. The field of the possible tends to extend when experience of real-life activity is gained. (See more details on modeling in the next section.)

In the *analysis of activity*, rich ethnographic data are collected by observations and video recordings, stimulated recalls, interviews, and questionnaires both in real work environments and in simulated work sessions. The analysis is a multi-perspective one and also the use of artifacts is studied. As a first view, the courses of behavior of each actor or team are described, and the variability of *performance* and its successfulness are identified. By taking a second view, the analysis also aims to understand the meaning of the observable behavior that habitual patterns of

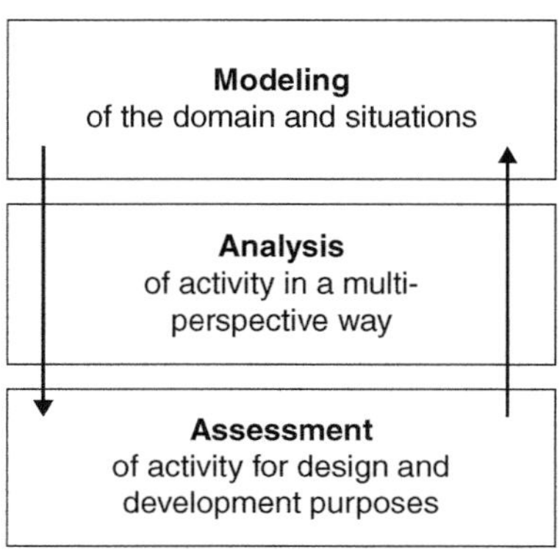

Figure 9.2. Simplified model of a contextual analysis of activity in HFE. Adapted from Savioja and Norros, 2008.

behavior, labeled *ways of acting*, reveal. These can be found by analyzing the variance in behavior in the critical points and situations of the work anticipated on the basis of the modeling. The ways of acting reveal differences in understanding the object of activity in the critical points and of situations. This analysis is enabled by exploiting a semiotic analysis of behavior. We identify triadic sign-object-interpreting operation structures, that is, habits that reveal the meaning of the observed performance (see this section and the next section). As a third view, we study the actors' personal relationship to the tools or artifact they use in work. This aspect is labeled *user experience*. The analysis should reflect how the users feel about the tools' contribution in the joint functioning of himself/herself and the tool. Also this analysis is contextualized by the models created in the modeling phase. The analysis of user experience draws on the belief that users as professional actors have a synthetic comprehension of the capabilities of the existing tools for their work, and also expectations about future tools. Therefore, even in situations where the analysis deals with a concept of a tool, expert users are able to anticipate the potential of the future tool for the development of work.

Assessment of activity for design and development purposes is the third part of the contextual analysis of activity. A concept was developed that serves as generic bases for the assessment. This is the *interpretive work,* and it serves the target of developing of expertise. It is identified through the analysis of ways of acting, during which we clarify what actually is the personal object of an actor's behavior in certain contexts and what epistemic attitude, or orientation, toward acting it reveals. Drawing on the ideas of Charles Sanders Peirce, we consider that an immediate and straightforward response to singular features of the object in a situation demonstrates a *reactive* way of working. In this case, perception already comprises action. If the response includes classificatory reasoning and matching the situation with known possibilities for action we have labeled the way of working as *confirmative*. In this case perception–action links are typical forms of behavior. In contrast to the former ways of working, in the third one, the response also includes the environment against which the perception and actions linked to it are tested. This way of acting is labeled *interpretive*. It draws on the possibility to test the validity of an action with regard to the actual state of the environment. In this connection, new knowledge is generated in the particular instance and action is potentially modified. With reference to Peirce's description of abductive reasoning as a knowledge creating feature of acting, we consider that abductive features are stronger in the interpretive way that in the other two ways of acting.

The well-known distinction of skill-based, rule-based, and knowledge-based activity by Jens Rasmussen (Rasmussen, 1986) has a connection to our distinction. Rasmussen was informed of the Peircean semiotic conception of activity, but in his definition of the three types of behavior, he leaned to the cognitive paradigm and characterized the differences in behavior as expressions of levels of internal information processing, and leaving out the differences in the relationship to the environment that the three forms of reacting portray. Thereby, he distanced himself, and even more did his followers, from semiotic view to the structure of activity. Therefore, the use of the distinctions of skill-, rule-, and knowledge-based behavior tends to reduce to an abstract classification of behavior without the actual content and context of the activity; hence the meaning of observed behavior remains unrevealed.

Our intention is to understand the three forms of acting from a semiotic perspective and to focus on the domain-related meaning of action. Hence, we consider that in the reactive and confirmative ways of acting there is a shallower and narrower relationship with the environment and less new knowledge is created, whereas in the interpretive way of acting a more comprehensive connection is created. The entire cycle in the human–environment interaction takes place, as a result of which the response is adapted in a situationally appropriate way, and, generic knowledge emerges in action.

Way of acting analysis may be used as a means to reflecting on practices, and as a starting point for development of practices. We have gained experience of this kind of use of the concept in characterizing differences in expert anesthetists' professional practices. We analyzed both the anesthetists' conceptual orientations to the patient, and the anesthetists' actual ways of conducting anesthetic treatment during surgical operations (Klemola and Norros, 1997, 2001, 2002; see summary in Norros, 2004). We have also used this type of analysis in characterizing nuclear power plant operators' conceptual orientations to their work (Norros et al., 2014), and, in characterizing these operators' actual behavior in difficult simulated disturbance situations (Savioja et al., 2014). In both cases, that is, anesthesia and nuclear power plant (NPP) production, different concrete indicators and assessment criteria were developed on the basis of the theoretical notion of interpretive work, and using collected empirical data. As a result, we were able to identify differences in the orientations and ways of acting among experts. The results gave bases to argue that the identified differences in practice could have safety- and quality-related consequences in critical situations.

9.4 PRINCIPLE 3: HFE FOCUSES ON TECHNOLOGY IN USE

9.4.1 Description

HFE aims at better teamwork between technology and human actors. This means that HFE is not focused on technology as such but considers technology as tools in activity in certain context and for certain purposes. Hence, HFE needs concepts that in new ways will capture the joint functioning of the human-technology-environment system. This is a real challenge for HFE but also a possibility to make a novel contribution to the existing human factors theory.

9.4.2 Theoretical Foundations

Developing the capability of HFE to solve the challenges that the technologically highly mediated work and daily life puts, has raised claims for change of paradigm in HFE (Hollnagel and Woods, 2005). The traditional "natural" distinction between human and technology as two separate elements draws the interest to identifying the internal processes of human brain, technical devices, and the computational connection with the elements in terms of their interaction. This perspective is a possible one but it neglects the wider connection of activity, and the possibilities and constraints in which it is situated. This causes difficulties to understand the meaning of interactions and misses the possibility of developing a realistic reference for the success of this interaction. One solution offered by different authors is that HFE could conceive functional units that parts of human, technology and environment form together. These units would be the new unit of analysis, and the task would be to discover the collaborative functions that they deliver for an appropriate and developing existence of the system (Bennet and Flach, 2011; Norros, 2014; Woods and Hollnagel, 2006). The need for revisions will become particularly pressing in the realization of embedded cyber-physical systems (CPS), in which the interface between human and technical elements disappears (Kant, 2015).

One way to concretize the joint functions is to consider technology from the point of view of its use. From this perspective, the technological and human elements become automatically inseparable, and their mutual determination becomes evident. This view is shared in the diversity of theories labeled collectively as the practice-theory approach by Nicolini (2013). One of the practice theories is the sociological approach by Wanda J. Orlikowski (2000). Her technology-in-practice approach analyzes how practice enacts emergent structures through recurrent interaction with the technology at hand, and forms structural properties of social systems. It is important that Orlikowski draws attention to the material features of technology and shows how technology-in-practice is both shaped by and shaping the use of material artifacts.

Our own practice-theory perspective provides a synthesis of ideas drawn from human-environment functional systems analysis (Gibson, 1977; Rasmussen, 1986), the cultural-historical activity theory (CHAT) (Engeström, 1987; Kaptelinin and Nardi, 2012), and, the semiotic perspective to the artifact mediation of activity (Peirce, 1998a). We labeled the approach as Core-Task Design (CTD). It was proposed as a possible answer to meet the high-quality challenges on human factors (Norros et al., 2015). Like Orlikowski, also we observe the material features of technology as determinants of practice, which makes both approaches more useful for HFE purposes than some other practice-oriented approaches.

In order to create concrete basis for the idea of a functional unity of human-technology-environment CTD devotes an intensive interest in defining the characteristics of the environment in concrete, as explained in the previous section on modeling. It is necessary to understand what parts of the environment are capable of offering possibilities for human action, and become meaningful as constraints, tools and objects of work. Drawing on these premises we developed the core-task concept (Norros, 2004, p. 17; Norros et al., 2015, p. 24). It defines the object of activity through consideration of the purpose the environment makes available and the functions that the human-environment system needs to fulfill for tackling the environment in an appropriate way. In the analysis of the ecology of human activity, several functional modeling techniques are used. The abstraction hierarchy modeling of work domains proposed by Jens Rasmussen has a central role here (Rasmussen, 1986). After an important contribution to this modeling approach by Morten Lind (1999) and Kim Vicente (1999), it has in recent years gained growing interest among young HFE specialists (Andersson, 2014; Kant, 2015; Naikar, 2013; Savioja, 2014).

Beyond the ecological and functional perspectives, the human–environment interaction is in the CTD also conceptualized and modeled as an activity system (Engeström, 1987). This is important for understanding the intentional and historical structuring of acting and the emergence of the hierarchical structure in real activity (activity, action, operation) (Leont'ev, 1978). The concept of activity also provided the important possibility for identifying how parts of the environment take mediating roles in acting, and function as tools. Continuing this line of thinking, it became possible in the CTD to articulate in generic terms what different mediating roles tools and technologies take in acting. Exploiting the notion of mediated action originally introduced by L. S. Vygotsky (1978), we proposed that a tool serves, first, an *instrumental function* while having an effect on the environment to bring about the anticipated outcome. The *psychological function* of a tool refers to the fundamental observation of Vygotsky that an external thing may act as a means of control of internal psychological phenomena, and thus shape the functioning of the mind. Further, a tool does not only mediate relations between the subject and the environment (serving the instrumental and psychological functions), but also acts as a sign that gets interpreted via the subject's actions aiming at establishing a shared belief of the situation. In this case, the tool serves a *communicative function.* It carries a meaning beyond the immediate practical purpose and rather speaks of enlarged human capabilities and culture. This tool function is very evident in the practices of using Internet-based information technical tools. All three tool functions are connected to the aim of fulfilling the purpose of work in an appropriate way, that is, so that the core task can be fulfilled and developed. Therefore, all tool functions are relevant for the goodness of any tool.

We consider that HFE should develop understanding of the ways in which the functional unity of human environment may manifest itself in concrete. The idea of using the concept of mediated action, and distinguishing three functions of tools in delivering the human-environment joint action, is an attempt toward this direction. Optional other ones within the current HFE are the so-called Joint Cognitive System (JCS) approach proposed by Hollnagel and Woods (Hollnagel and Woods, 2005; Woods and Hollnagel, 2006), the ecological interface design (EID) that systematically exploits Rasmussen's domain modeling (Jamieson et al., 2007), or also the interface design concept that likewise draws on Rasmussen but additionally explicitly exploits semiotic notions to elaborate the human–environment interaction (Bennet and Flach, 2011).

9.4.3 Use and Implementation

Usability is the concept typically attributed to technical devices to indicate the immediate experience of their use. But the value in use of a particular device may also be evidenced in how well the solution serves the whole socio-technical system in meeting its objectives. The concept of *Systems Usability* (SU) was developed for this purpose, that is, for connecting the features of technology to the overall performance of the socio-technical system, (see Figure 9.3) (Savioja, 2014; Savioja and Norros, 2008, 2013). SU is composed of two dimensions: The first is the functions of tools in fulfilling the core-task demands of the activity (Figure 9.3, vertical axis). How well the functions are realized can be evidenced by referring to the results of the usage of the tool. The activity of using the tool is the second dimension with regard to which the systems usability concept is defined (Figure 9.3, horizontal axis). As was indicated in the previous section, different perspectives need to be used to analyze human activity in a comprehensive way. Consequently, the three perspectives of activity, that is, performance, way of acting, and user experience, are utilized. Performance denotes to characteristics of behavior which can be rather objectively measured such as number of errors, task completion time, etc. Way of acting reveals how the tools are embedded in the professionals' operating practices developed over time to fulfil the objectives of work. Way of acting also informs of the tools' capability to endorse and nurture the development of the professional practice. User experience refers to the voice of the actual professionals. As actors and experts in the domain, their subjective

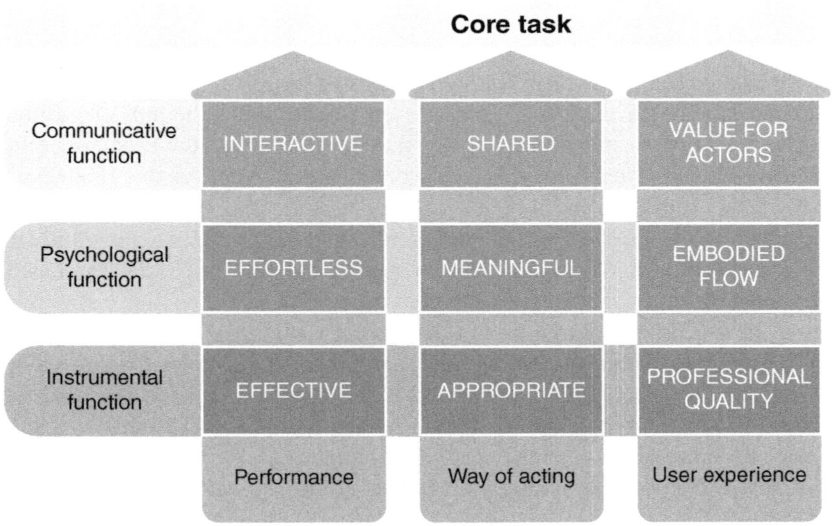

Figure 9.3. Generic systems usability indicators. *Source*: (Norros et al. 2015). Reproduced with permission of Morgan & Claypool Publishers.

experience matters: A professional tool must also feel like that to the user. As a result, a framework of systems usability indicators can be formulated (Figure 9.3).

9.4.4 Empirical Research on Use and Efficiency

In a design process, several phases gain from the conceptualization of SU. Designers themselves are inspired by the different roles technologies may have in an activity and thus they can avoid the pitfall of dealing with too narrow a design space which may lead to sub-optimizing, and to inconsistency between systems. Different kinds of evaluations are conducted for different purposes during the design, and it is important to set appropriate evaluation dimensions and criteria for these evaluations.

The SU indicators shown in Figure 9.3 are generic characterizations of the good quality of a tool in the different perspectives of activity. These characterizations have been broken down into more concrete SU measures that we have applied in a number of design and evaluation connections. Two comprehensive empirical references demonstrate the use of the systems usability approach in HFE:

The first reference is a validation project, in which an assessment-oriented use of the concept was dominant. The validation task was accomplished in collaboration with the power company Fortum in connection with a comprehensive modernization of the information and control systems and the main control room of the Fortum Loviisa nuclear power plant. The HFE team was composed of VTT and OECD/NEA Halden Reactor Project human factors experts. The HFE team developed a global validation plan, including detailed reference to the most important international guidelines and standards, that is, NUREG0711, IEC 61771, ISO/IEC 25062. On the basis of this, global plant specific plans were constructed for each validation phase (Laarni et al., 2013, 2014). Three specific plans were written, and two phases of validation accomplished. (The final validation phase was canceled due to a major change in the plant's overall renewal strategy.) The analysis approach described in this article was applied in the comprehensive empirical validation study. The systems usability concept was exploited as the quality criterion for the control room design, and as the theoretical reference for the evaluation. Because the validation comprised a series of validations there was a need to develop a way to systematically accumulate evaluation results. We borrowed the idea of a safety case and invented a conception of "usability case" for this purpose. The usability case was structured with regard to the tool functions as high-level systems usability claims under which more specific design claims were ordered. Confirmative or negative arguments were constructed to connect evaluation data to the claims (see overview in Laarni et al., 2014).

The systems usability concept has also been exploited for design purposes. In a European Commission project we, together with Électricité de France, embedded the SU concept in the development of a so-called tool-in-use model (Norros et al., 2015, Chapter 4) for an advanced nuclear power plant control center (Norros et al., 2011). The concept supported the HFE experts in formulating high-level design requirements on the basis of operator interviews and experimental testing of a high-fidelity prototype. This example dealt with design tasks that emerged early in the design, that is, during idea testing and concept design. In these design phases, the typical evaluation measures that draw on the experimental subjects' performance do not make sense because no actual use cannot be established. Instead, indicators developed on user experience (UX) may have an important role. Therefore, such indicators are embedded in the systems usability concept. A further design example in which the systems usability and user experience concepts were used was the development of a totally new concept for remote crane control (Koskinen et al., 2013).

9.5 PRINCIPLE 4: SAFETY IS ACHIEVED THROUGH CONTINUOUS HFE

9.5.1 Description

HFE is a continuous explicitly organized activity within an operating organization with the help of which preconditions for safety and efficiency are continuously created, including proactive reacting to early signs of deterioration of safety and efficiency. HFE facilitates gathering of experience of technology in use. Gained experience is effectively exploited for further engineering of the sociotechnical system during its entire lifecycle. Involvement of the users improves the quality and efficiency of engineering.

9.5.2 Theoretical Foundation

Human factors engineering should be an integral part of a modern enterprise interested in maintaining its safety and efficiency at a high level. HFE would gain of an advanced safety management activity in the organization, and of the adoption of the Safety 2 concept. This and the related concept of resilience are discussed extensively by Hollnagel in Chapter 3 of this volume. In this connection, we will argue that a resilience-oriented safety concept supports HFE activity better than the traditional failure-oriented concept.

The basic idea of the traditional safety concept is that safety is non-existence of failure. When the HFE function is organized according to this traditional notion, the HFE practice tends toward a reactive, discontinuous, and expert-driven mode of operation. HFE measures focus directly on identified, or anticipated, problems or failures. The measures are taken mainly by experts because it is difficult to involve users in the analysis and corrective work without blaming them for the failures. A failure-oriented safety concept facilitates the use of traditional human factors measures that orient to observable performance outcomes, often failures. Due to idealizing the scientific quality criteria, the traditional human factors also tends to focus on well-defined phenomena. Thus it takes distance from the messy contexts in which these phenomena are embedded and avoids the end users' views concerning the phenomena as subjective.

Turning to the Safety 2 concept shifts the analyst's focus to the *capabilities that support the development* of the activity of the personnel and of the organization. HFE that draws on this approach does not close the eyes for failures, but it shifts the attention to the practices of the people and organization that actually maintain appropriate output and increase the viability of the organization. Consequently, HFE measures need to be continuous and embedded in the daily productive functions of the organization. In this frame, the users are the main resource for the development because they, as members of the community of practice, have insight of the good toward which HFE should be oriented, and of the conditions its fulfillment requires. Outsiders have a strong tendency to focus on the failures because these are much more salient than those features of work that bring fluency and success.

The concepts of core-task design, interpretive work, and systems usability, introduced in the previous sections of this chapter, are working for Safety 2 because they are targeted to define the characteristics of human activity and technology-in-use that facilitate an appropriate functioning of the sociotechnical system. Looking from the perspective of resilience the objectives of efficiency and safety are most likely achieved together.

How then should the core-task design be implemented to create a continuous and co-creative HFE process that we assume to promote resilience-oriented safety of the system? We see that continuity in HFE is not simple repetition of certain HFE procedures at defined intervals. Instead, continuity must be a result of a flywheel mechanism that boosts improvement and development in the sociotechnical system.

9.5.3 Use and Implementation

We developed a Core-Task Design model including three interacting design functions (see Figure 9.4) to create the flywheel effect of HFE into design. The HFE practice defined by the model enables shifts of attention between three aspects of activity with the aid of methods designed for realizing the transitions, thus making the design process continuous.

According to the Core-Task Design model, HFE focuses on technology embedded in an activity. This activity is always considered from three different perspectives, as real-life *actual activity*, as *core activity*, and as *potential activity* (Figure 9.4, squares). In an ongoing design process continuous shifts between these perspectives take place. These are materialized by three design functions (Figure 9.4, arrows).

Actual activity refers to real work (or activity) in a defined environment of which operators or end users gain experience by participating in it, and analysts by observing it using different human factors methods. The first design function that the CTD model identifies is the *Understand-to-generalize* -function (Figure 9.4). It is a conceptualization and abstracting the experience of the existing actual activity into the core content of that activity. It summarizes generic features of this activity, and gives explanations of the internal regularities that are producing these features. Different kinds of models are needed to accomplish the understand-to-generalize function. As explained

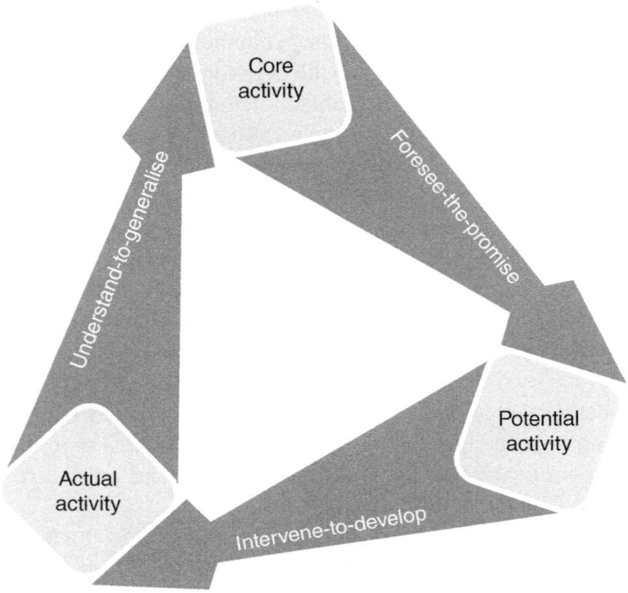

Figure 9.4. The Core-Task Design model that functions as a flywheel of HFE. The model is composed of three perspectives to activity (squares) and three design functions (arrows) that materialize the transition from one perspective to another. *Source*: (Norros et al. 2015). Reproduced with permission of Morgan & Claypool Publishers.

earlier, we utilize traditional modeling approaches, for example, the activity system model by CHAT (Engeström, 1987) and the functional abstraction hierarchy of Jens Rasmussen (1986), but we have also developed further models, for example, the so-called core-task function model, and the functional situation model (Norros et al., 2015). Raising the level of abstraction of the analysis by these models enables defining the core activity and understanding better the key features of the actual activity.

Identifying the core activity also opens a space for the identification of new possibilities and for the emergence of new features in a future activity. Because the core-task modeling focuses on the generic core functions of the work that can be assumed to be relevant also in the new forms of the work, it helps to avoid the design trap called the "task-artifact cycle" (Carroll et al., 1991) saying that the task changes along with new technology. The HFE specialists easily fall in this trap if they use regular descriptive task analysis in the definition of the requirements for future tools.

The design function that transforms the abstracted core activity toward a potential new activity was labeled as *Foresee-the-promise* (Figure 9.4). The focus of this function is to propose and test optional concrete features of future tools that enable the core activity. Completely new core features, or even new activities may emerge, during the design. This is possible if the design team succeeds to maintain a functional view as part of the design activity, in parallel with focusing on the particular aimed design proposals. The co-existence of generic concepts and particular solutions during design was expressed by the industrial designer Turkka Keinonen when he used "remote design" as the term for the development of conceptual solutions in the connection of producing particular design proposals, together with end users in a process of "immediate design" (Keinonen, 2007).

In order to create a new actual activity, the promises of new design proposals must be put into test in the real environment. This is what the design function labeled the *Intervene-to-develop* (Figure 9.4) is for. A real activity is a complex phenomenon. A change in some of its elements reveals the internal connectedness of various other elements and induces instability into the entire system. Ruptures and resistance for change are therefore inevitable. These also offer possibilities to experiment and test options, and to find solutions for building a new integrated whole, a new work activity. Intervene-to-develop exploits tools and results of the two other functions.

HFE meets great challenges to transform its traditional, and rather narrow and technically oriented practices into continuous and co-creative projects. HFE would need to consider the three aspects of activity, and to implement the three design functions. A strong involvement of the users and user organizations is particularly critical: Understanding-to-generalize function will be the more successful the better the human factors experts acknowledge the end users as experts of their own activity and encourage the end users to exploit the conceptual models as tools to reflect their own work. Equally, the end users are needed in anticipating the future potential in work on the whole. They could also express their experience of the developmental potential of different design concepts during the foresee-the-future function. In the intervene-to-develop function, the end users and user organizations are the agents of change whose creative powers the HFE practices need to facilitate. In all functions further methods are needed to support the participation of the users.

9.5.4 Empirical Research on Use and Efficiency

The principle of continuous development of safety through HFE can be demonstrated by some development processes accomplished together with the Finnish energy company Fortum. These projects focused on three topics: The HFE validation of human-technology systems; development of human competences in the operating organizations; and the improvement of the HFE process itself. (See details in Norros et al., 2015). The development projects were all initiated by the actual needs of the power company, and their final aim was to increase safety.

The generic lessons learned from the development projects can here be summarized by considering how well these projects could exploit the design ideas expressed in the foregoing section. In other words, we shall ask whether the projects actually showed signs of a flywheel effect into the HFE practices of the design and operating organizations of Fortum. To answer this question, we borrowed the concept of *formative intervention* from Engeström (2011). The author claimed that the mainstream design experiments aiming at developing new practices in organizations typically implement a linear process. This means that the aimed end product, and the means to reach it during the process, are pre-defined by the designers. Thus, while development projects involve many actors in the organization their role is often reduced to a verification of the development plan through accomplishing it. Such a process is not capable of exploiting the productive powers of the end users, nor to create something that goes beyond the specific design case.

We found Engeström's critique against organizational design experiments to the point also with regard to the existing conceptions of the role of HFE. The international guidelines for validations, the guidelines for development of training, or the conceptions concerning HFE reviews, all suffer more or less of those problems listed by Engeström. Yet, actual practice might portray progressive pressure for revisions of the conceptions and guidelines. A formative intervention that could reach more creative outcomes and create a fly-wheel effect should manifest following capabilities: It should comprehend the design problem holistically by developing and exploiting *conceptual tools* to master the actual problem, it should exploit the practical knowledge and *creative agency of the end users* and substance matter experts, it should aim at *generalizable outcomes* useful also in further circumstances, and it should *empower the end users* and substance matter experts to control the knowledge creation during the development. All this would also transfer the design experience of the end users and of the operative organization to become *part of normal productive practice.*

According to our experience, signs of the fulfilment of the above-listed features of a formative intervention were actually present in the projects we accomplished with Fortum. This became evident, for example, in the several development studies concerning operator and instructor training at the Loviisa power plant. Existing guidelines were used as starting point and conceptual means to involve a group of operators in the design of a generic NPP operator initial training concept, and a further concept for the simulator instructor training. As a further effort, a completely new training program for the operators of the digitalized control room and for the development of a new concept of operations was designed by a group of instructors (Laarni et al., 2011). The end users also continued their design-oriented activities while returning to their regular training practice.

We could also observe that a lack of formative features in the design and review guidelines were sometimes a reason for Fortum to make adaptations to the guidelines to make them more suitable to their needs. For example, Fortum expanded the existing guideline NUREG-0711 (NUREG0711, 2004). As indicated in the introduction, the HFE process described in the NUREG-0711 includes a validation function. This function is described as a comprehensive experimental test event during which the entire design of a control room is challenged at a full-scope simulator. The simulator needs to demonstrate high fidelity with the new design. The validation aims at a decision about the safety of the entire design and it should take place after the design of the technology and training of the operators have been finished. During the Fortum

Loviisa nuclear power plant automation renewal process this conception was found insufficient. It could not serve the needs of the stepwise, complex and comprehensive renewal of the automation and control room systems. Therefore, the Fortum HFE personnel together with external human factors specialists invented a multi-stage validation concept including a necessary number of sub-system validations before a final integrated validation. The challenge faced in the re-conceptualizing of the validation process was to find relevant reference to the stepwise evaluations, and to select a set of valid measures of operator performance that could deliver information of the usability of the designs and their impacts on the safety of operator performance (Laarni et al., 2014).

Design and development processes are extremely complex and contain many uncertainties. In all industrial domains, and in the high-reliability organizations in particular, well documented and tested procedures are a necessity and a great help in managing the design process and reaching a functional and safe outcome. Therefore design must be tied with a comprehensive quality assurance activity. In the information and control system (I&C) and control room renewal of the Fortum Loviisa plant the validation of the design followed carefully the procedures of the established quality process of Fortum.

Because design has a great influence not only on the inherent safety features of the technology but also on the forthcoming operations and use of the system, it is also very important to establish a close connection with the management and organizational developers of the operative organization. Fortum HFE accomplished an extensive interview study to make the operating and design organization reflect upon the HFE-awareness within the organization, and used this information as a starting point for a proposal for a continuous HFE process. (Laarni et al., 2014; Norros et al., 2015; Salo and Savioja, 2014).

9.6 RELATION TO OTHER SAFETY PRINCIPLES

HFE has a mediating role between design and operations. Having this role, HFE also offers a life cycle perspective to application of human factors knowledge and practices in the organizations. Therefore, HFE process has a potential to be combined with other measures aiming at increasing safety. HFE should establish living connections to the safety management functions of the organization. Within this global function, HFE would have to participate in the practices of operational experience review (OER). Due to its special expertise in human factors knowledge and methods, HFE could offer help in developing the tools of collecting and analysis of the OER data. The second safety management function that HFE should collaborate with is organizational development (OD) responsible of facilitating safety culture. Organizational developers are also responsible for the development of the competences of the personnel. It would be important that the developers would have better insights of the technical contents of the competences of the personnel, and an insight of the difference it makes for safety depending on the personnel's daily work practices. Organizational developers and HFE experts should achieve a shared understanding of the role of technology in shaping work practices, and organizational and safety culture. These ideas would support safety principles expressed in those chapters of the present book that discuss safety management (Grote, Chapter 27) and safety culture (Rollenhagen and Reiman, Chapter 28) in complex high-reliability organizations.

HFE plays an important role in the design of safety critical organizations. This statement is hardly questioned by anybody, but as the international discussion and our experience has shown, it is not that simple to fulfil this role effectively. The approach to HFE presented in this chapter draws on the Safety 2 concept (Hollnagel, Chapter 3). It is assumed that an orientation to resilience in safety work would be beneficial for strengthening the impact of HFE in design. The argument for this claim is that resilience draws attention to the capabilities of delivering the functions that the organization is aiming at, which are the natural targets of design, too. Hence, a resilience-orientation would facilitate a shared understanding of the aims of the design and, support participation in an integrated design process.

We have boosted the idea of a life cycle HFE. From this point of view, operations and design are in a constant interaction with each other. Adopting such a view to HFE in an organization is a strategic choice of the top-level management. It includes that the management perceives technology from a technology-in-use perspective, and appreciates resilience-orientation as an extension of the safety concept and a new means to improve safety and efficiency. Development of the management strategy toward these lines of thinking would require development of an even more comprehensive safety philosophy that also takes into account the economic aims and constraints of the organization in the global markets.

9.7 LIMITATIONS

A practice that exploits the Core-Task Design model according to which a continuous HFE process can be implemented faces two major challenges. The first deals with development of the empirical bases of the HFE on a continuous bases. The organization applying CTD-oriented HFE has to establish practices to accomplish the gathering and analysis of empirical data of the operations of the various organizational functions and tasks. It is also necessary to recruit personnel who plan and organize the data gathering, analyzes and draws conclusions of the acquired material, and involves the end users to participate in these measures. Beyond human factors, the HFE personnel have to master well the work domain and understand the principles of the information and control systems. Gaining this expertise on top of one's own background in human factors is time and effort consuming.

When using, for example, the systems usability framework, the personnel may face the situation that the used set of indicators, which are more or less independent from each other, may in some cases produce contradictory results. Therefore, the result of systems usability assessment always requires interpretation and contextual sense making. Systems usability evaluation does not produce one clear numerical result but rather a set of issues, some of which speak for the benefit of the new design, some are issues to be followed in use, and some which are to be changed before implementation. Therefore, the utilization of the results in design also calls for understanding of the activity and the role of technology in it.

Beyond the empirical work focusing on the operating activity, the HFE personnel also have to improve existing research and development tools and to ensure their methodical qualifications. Further efforts need to be devoted to the invention of new methods according to the developing needs of the HFE. The needed expertise in method development and the conceptual abstraction of the observed phenomena of the actual activity, create the second challenge for the successful exploitation of the CTD. It appears that HFE activity, when accomplished according to the lines of the CTD, assumes interest and competences for theoretical work, including continuous reflection of own ways of working. An inclination to research-like work would be a necessary characteristic of the HFE personnel, but this kind of work may appear useless or unprofitable in the industrial context, and therefore difficult to nurture.

All in all, it seems that the personnel involved in CTD-oriented HFE should not dislike being in a contradictory situations. The core demand of the HFE work is to cope with a diversity of requirements: There is a need to be content-oriented and practical on the one hand, and capable of abstracting generic features and theorizing on the other. The personnel should be interested in developing their special knowledge, but, at the same time, they should be open for dialogue with people from other backgrounds in order to create new knowledge together. They should also be prepared to critique on their work, and capable of defending their practices with good arguments, but also willing to learn from the critique.

The above-mentioned features of the CTD-oriented HFE can be counted as limitations when perceived from a perspective of the other option, that is, the more controlled exploitation of well standardized procedures within an objectivistic evaluation approach accomplished under a well-established division of responsibilities. But the same features can also be considered as characteristics of a realistic approach to technology assessment during its design and use, which is accomplished by personnel who are used to participating in a multidisciplinary adaptive knowledge creation process.

9.8 CONCLUSIONS

In this chapter, we have created a new interpretation of HFE. We claim that the standard interpretation of HFE as a process of application of human factors knowledge for engineering purposes does not resemble the real circumstances in which human factors experts work. We also see that maintaining within this traditional conception might be one cause for the observed underutilization of the potentials of human factors. We came up with the idea that it is rather within the constructive design-like practice where it would be possible to realize the ideal of high-quality human factors, that is, to adopt systems approach, to be design-driven, and to focus on a complex goal of performance safety and well-being. This practice could develop the argumentation of design thinking that seems adequate for reaching the high quality HF targets.

It was this constructive design-like practice we had in mind when we identified the HFE principles. This means that we do not propose that all human factors research should find these principles relevant and take them as guidance. There is certainly place for different kinds of scientific analyses of human conduct. Our principles, however, are aimed to support human factors experts in innovative design and development processes in practice. In that situation, it is necessary that HF experts find ways to orient themselves adequately according to the constraints, possibilities, and aims of the real design and development processes. We see that at least the listed four principles of understanding HFE could be helpful, that is, HFE is design thinking; HFE studies human as a manifold entity; HFE focuses on technology in use; and safety is achieved through continuous HFE. It is important to stress, once more, that human factors experts do not only deliver practical solutions to singular problems but they also create new knowledge of human conduct as it appears in the modern life and work environments.

While introducing the four HFE principles, we referred to some methods that were invented for the accomplishment of HFE according to the principles. These tools and their empirical use have been described in more detail elsewhere. It is clear that the proposed tools need methodical refinement and development. Moreover, as any research tools, also these will require continuous attention from the human factors specialists in HFE, due to the developing contents of the real-life activity and the new collaborative demands on design when new technological possibilities emerge.

We believe that HFE, as it has been described in this chapter, could have a positive contribution in a new engineering culture that the future development of the smart cyber-physical systems is foreseen to require. Such a new culture should be focused to create resilience into the systems during their entire life cycle and, consequently, it would also support the sustainability of the systems. The new culture should also be organized so as to form an integrated design activity, in which a diversity of design branches work together. When included in such a shared design activity and culture, HFE could make itself useful and develop into to a full-fledged design practice.

REFERENCES

Andersson, J. (2014). Representing human automation challenges. A model based approach for human factors engineering in industrial practice (Doctoral thesis, Chalmers University of Technology).

Back, J., Furniss, J., Hildebrandt, M., and Blandford, A. (2008). Resilience markers for safer systems and organisations. Proceedings of 27th International Conference, SAFECOMP 2008 Newcastle upon Tyne, UK, September 22–25, 2008, 99–112.

Bennet, K. B., and Flach, J. M. (2011). *Display and Interface Design, Subtle Science, Exact Design.* Boca Raton, FL: Taylor & Francis.

Buchanan, R. (1992). Wicked problems in design thinking. *Design Issues*, 8, 5–21.

Buchanan, R. (1998). Wicked problems in design thinking. In: V. Margolin, and R. Buchanan (Eds.), *The Idea of Design*. Cambridge, MA: The MIT Press.

Carroll, J. M., Kellogg, W. A., and Rosson, M. B. (1991). The task-artifact cycle. In: J. M. Carrol, (Ed.), *Designing Interaction: Psychology at the Human–Computer Interface*. Cambridge, UK: Cambridge University Press.

Dewey, J. (2002). *Human Nature and Conduct*. Toronto, Canada: Dover Publications, Inc.

Dourish, P. (2001). *Where the Action Is. The Foundations of Embodied Interaction*. Cambridge, MA: MIT Press.

Dul, J., Bruder, R., Buckle, P., Carayon, P., Falzon, P., Marras, W. S., Wilson, J. R., and Van Der Doelen, B. (2012). A strategy for human factror/sergonomics: developing the discipline and profession. *Ergonomics*, 55, 377–395.

Engeström, Y. (1987). *Expansive Learning*. Jyväskylä, Finland: Orienta.

Engeström, Y. (2011). From design experiments to formative interventions. *Theory & Psychology*, 21, 598–628.

Falzon, P. (2008). Enabling safety: Issues in design and continuous design. *Cognition Technology and Work*, 10, 7–14.

Farrel, R., and Hooker, C. (2012). The Simon-Kroes model of technical artifacts and the distinction between science and design. *Design Studies*, 33, 480–495.

Farrel, R. and Hooker, C. (2015). Designing and sciencing: Response to galle and kroes. *Design Studies*, 37, 1–11.

Findeli, A. (2001). Rethinking design education for the 21st century: Theoretical, methodological and ethical discussion. *Design Issues*, 17, 5–17.

Fleger, S., and O'Hara, J. (2015). Revision of human factors engineering guidance for conducting safety reviews of nuclear power plant control rooms: A status update and approach. In: 9th International Topical Meeting on Nuclear Plant Instrumentation, Control & Human–Machine Interface Technologies (NPIC & HMIT 2015), Charlotte, NC.

Furniss, D., Back, J., Blandford, A., Hildebrandt, M., and Broberg, H. (2011). A resilience markers framework for small teams. *Reliability Engineering and System Safety*, 96(1), 2–10.

Galle, P., and Kroes, P. (2014). Science and design: Identical twins?. *Design Studies*, 35, 201–231.

Gibson, J. (1977). The Theory of Affordances. In: R. Shaw, and J. Brandsford (Eds.), *Perceiving, Acting and Knowing. Toward and Ecological Psychology*. New York: Lawrence Erlbaum Associates.

Heimdal, J., Skraaning, G. Jr., and Braarud, P. O. (2004). *Integrataed System Validation: Status and Research Needs*. Halden, Norway: Institutt for energiteknikk, OECD Halden Reactor Project.

HFES. (2016). What is ergonomics? [Online]. Human Factors and Ergonomics Society. Available at: https://www.hfes.org/web/AboutHFES/about.html (Accessed April 1, 2016].

Hollnagel, E., and Woods, D. (2005). Joint Cognitive Systems. *Foundations of Cognitive Systems Engineering*. Boca Raton, FL: Taylor & Francis.

Hollnagel, E., Woods, D. D., and Leveson, N. (Eds.). (2006). *Resilience Engineering. Concepts and Precepts*. Aldershot, UK: Ashgate.

Holstein, W. K. (2016). Human factors engineering [Online]. Encyclopaedia Britannica. Available at: http://www.britannica.com/topic/human-factors-engineering (Accessed April 1, 2016).

Wikipedia contributors. (2016). Human factors and ergonomics [Online]. Available at: https://en.wikipedia.org/wiki/Human_factors_and_ergonomics (Accessed April 1, 2016].

IEC-1771. (1995). Nuclear power plants—main control room—verification and validation of design. International Electrotechnical Commission, Geneva.

IEC-61839. (2000). Nuclear power plants—design of control rooms—functional analysis and assignment. International Electrotechnical Commission, Geneva.

ISO. (2006). Software product quality requirements and evaluation—common industry format for usability test reports. International Standards Organization, Geneva.

Jamieson, G. A., Miller, C. A., Ho, W. H., and Vicente, K. J. (2007). Integrating task-and work domain-based work analysis in ecological interface design: A process control case. *IEEE Transactions on Systems, Man And Cybernetics*, 37, 887–905.

Kant, V. (2015). Extending cognitive work analysis for Cyber-physical systems (Doctoral dissertation, University of Waterloo).

Kaptelinin, V., and Nardi, B. (2012). Activity theory in HCI. Morgan & Claypool. Available at: www.morganclaypool.com

Karwowski, W. (Ed.). (2006). *Handbook of Standards and Guidelines in Ergonomics and Human Factors*. Mahwah, NJ: Lawrence Erlbaum Associates.

Keinonen, T. (2007). Immediate, product and remote design. In: 2007 IASDR Conference. Hong Kong.

Klemola, U.-M., and Norros, L. (1997). Analysis of the clinical behaviour of anaesthetists: recognition of uncertainty as basis for practice. *Medical Education*, 31, 449–456.

Klemola, U.-M., and Norros, L. (2001). Practice-based criteria for assessment the anaesthetists' habits of action. Outline for a reflexive turn in practice. *Medical Education*, 35, 455–464.

Klemola, U.-M., and Norros, L. (2002). Activity-based analysis of information characteristics of monitors and their use in anaesthetic practice. In: 11th European Congress on Cognitive Ergonomics, Catania, Italy.

Koskinen, H., Karvonen, H., and Tokkonen, H. (2013). User experience targets as design drivers: A case study on the development of a remote crane operator station. In: Proceedings of the 31st European Conference on Cognitive Ergonomics (ECCE '13).

Krippendorff, K. (1989). On the essential contexts of artifacts or on the proposition that 'Design is Making Sense of (Things)'. *Design Issues*, 5, 9–38.

Krippendorff, K. (1998). On the essential context of artefacts or on the proposition that "Design is Making Sense (of things)". In: V. Margolin, and R. Buchanan, (eds.), *The Idea of Design: A Design Issues Reader*. Cambridge, MA: The MIT Press.

Kuutti, K. (2009). Artifacts, activities, and design knowledge. In: S. Poggenpohl, and K. Sato, (Eds.), *Design Integrations: Research and Collaboration*. Chicago: University of Chicago Press; Bristol, UK: Intellect Ltd.

Kuutti, K., and Bannon, L. J. (2014). The turn to practice in HCI: Towards a research agenda. In: CHI 2014, Toronto, Canada. ACM Digital Library.

Laarni, J., Karvonen, H., Koskinen, H., Liinasuo, M., Norros, L., Savioja, P., Salo, L., Laakso, A.-M., and Lehtonen, M. (2013). A stepwise validation process for the main control room of Fortum Loviisa Nuclear Power Plant. In: *Enlarged Halden Project Group Meeting*, Storefjell, Norway.

Laarni, J., Norros, L., Savioja, P., and Rantanummmi, P. (2011). Development of training programs for nuclear power plant personnel – Lessons learned from Finnish development projects. In: *Enlarged Halden Project Group Meeting*, Sandefjord, Norway.

Laarni, J., Savioja, P., Norros, L., Liinasuo, M., and Wahlström, M. (2014). Conducting multistage HFE validations—Constructing systems usability case. In: International Symposium on Future I&C for Nuclear Power Plants/International Symposium on Symbiotic Nuclear Power, Jeju, Korea.

Leont'ev, A. N. (1978). *Activity, Consciousness, and Personality*. Englewood Cliffs, NJ: Prentice Hall.

Lind, M. (1999). Making sense of the abstraction hierarchy. In: Proceedings of the Cognitive Science Approaches to Process Control Conference (CSAPC99), Villeneuve d'Ascq, France.

Margolin, V., and Buchanan, R. (Eds.), (1998). *The idea of desisgn. A Design Issues Reasder.* Cambridge, MA: The MIT Press.

Merleau-Ponty, M. (1986). *The Phenomenology of Perception*. London: Routledge & Kegan Paul.

Naikar, N. (2013). *Work Domain Analysis. Concepts, Guidelines, and Cases*. Boca Raton, FL: Taylor & Francis Group.

NEA (2016). *Experts' Workshop on Human Factors Validation of Nuclear Power Plant Control Room Designs and Modifications*, February 19–21, 2015 (Draft Report), Paris, France, Nuclear Energy Agency, Working Group on Human and Organizational Factors.

Neisser, U. (1976). *Cognition and Reality*. San Francisco, CA: W.H. Freeman.

Nicolini, D. (2013). *Practice Theory, Work, and Organization. An Introduction*. Oxford: Oxford University Press.

Norros, L. (2004). *Acting under Uncertainty. The Core-Task Analysis in Ecological Study of Work*. Espoo, Finland: VTT. Available at: http://www.vtt.fi/inf/pdf/publications/2004/P546.pdf

Norros, L. (2014). Developing human factors/ergonomics as a design discipline. *International Journal of Applied Ergonomics*, 45, 61–71.

Norros, L., Koskinen, H., Hildebrandt, M., Paulus, V., and Gauthier, B. (2011). Methodology for evaluation of innovative concepts: Case FITNESS control station concept. Enlarged Halden Project Group Meeting, Sandefjord, Norway, OECD Halden Reactor Project.

Norros, L., Liinasuo, M., and Savioja, P. (2014). Operators' orientations to procedure guidance in NPP process control. *Cognition, Technology & Work*, 16, 487–499.

Norros, L., Savioja, P., and Koskinen, H. (2015). *Core-Task Design. A practice-theory Approach to Human Factors*. San Rafael, CA: Morgan & Claypool Publishers.

NUREG0711. (2004). *Human Factors Engineering Program Review Model*. Washington, DC: United States Regulatory Commission.

O'Hara, J. (2014). *Integrated System Validation: Models, Methods, and Issues*. Upton, NY: Brookhaven National Laboratory.

O'Hara, J., Higgins, J., and Fleger, S. (2012). Human factors engineering program review model (NUREG-0711) Revision 3: Update methodology and key revisions. *Eighth American Nuclear Society International Topical Meeting on Nuclear Plant Instrumentation, Control and Human-Machine Interface Technologies NPIC&HMIT*. San Diego, CA: American Nuclear Society.

OECD/NEA. (2012). Summary of survey and workshop results on areas of research in human factors for the design and operation of new nuclear plant technology. *Issy-les-Moulineaux*, France: OECD Nuclear Energy Agency.

Orlikowski, W. J. (2000). Using technology and constructing structures: A practice lens for studying technology in organizations. *Organizational Science*, 11, 404–428.

Peirce, C. S. (1998a). The Harvard lectures on pragmatism. In: *The Essential Peirce. Selected Philosophical Writings* (Peirce Edition Project). Bloomington, IN: Indiana University Press.

Peirce, C. S. (1998b). Pragmatism. In: *The Essential Peirce. Selected Philosophical Writings*. Bloomington, IN: Indiana University Press.

Rasmussen, J. (1986). *Information Processing and Human-Machine Interaction*. Amsterdam: North-Holland.

Salo, L., and Savioja, P. (2014). Development of a HFE program for an operating NPP: Balancing between existing design practices and human factors standards. *International Symposium on Future I&C for Nuclear Power Plants/International Symposium on Symbiotic Nuclear Power*. Jeju: Korea

Savioja, P. (2014). Evaluating systems usability in complex work: Development of a systemic usability concept to benefit control room design (Doctoral dissertation, Aalto University, School of Science). Available at: http://www.vtt.fi/inf/pdf/science/2014/S57.pdf

Savioja, P., and Norros, L. (2008). Systems usability—promoting core-task oriented work practices. In: E. Law, E. T. Hvannaberg, G. Cockton, and J. Vanderdonckt (Eds.), *Maturing Usability: Quality in Software, Interaction and Value*. London: Springer.

Savioja, P., and Norros, L. (2013). Systems usability framework for evaluating tools in safety-critical work. *Cognition Technology and Work*, 15, 255–275.

Savioja, P., Norros, L., Salo, L. and Aaltonen, I. (2014). Identifying resilience in Proceduralised Accident Management Activity of NPP operating crews. *Safety Science*, 68, 258–274.

Simon, H. (1969). *The Sciences of the Artificial*, Cambridge, MA: The MIT Press.

Vicente, K. J. (1999). *Cognitive Work Analysis. Toward a Safe, Productive, and Healthy Computer-Based Work*. Mahwah, NJ: Lawrence Erlbaum Publishers.

Von Wright, G. H. (1998a). *In the Shadow of Descart. Essays in the Philosophy of Mind*. Dordrecht, The Netherlands: Kluwer Academic Publisher.

Von Wright, G. H. (1998b). Soul, brain and behaviour (in Finnish). *Ajatus*, 55, 3–17.

Vygotsky, L. S. (1978). *Mind in Society. The Development of Higher Psychological Processes*. Cambridge, MA: Harvard University Press.

WGHOF. (2009). Research program topics on human performance in new nuclear plant technology OECD NEA working group for human and organizational factors. NEA No 6844, OECD.

Wilson, J., and Carayon, P. (2014). Special Issue: Systems Ergonomics/Human Factors. *Applied Ergonomics*, 45, 1–132.

Woods, D., and Hollnagel, E. (2006). *Joint Cognitive Systems - Patterns in Cognitive Systems Engineering*. Boca Raton, FL: Taylor & Francis.

Wright, P., and Mccarthy, J. (2010). *Experience-centred Design. Designers, users, and Communities in Dialogue*. Lexington, KY: Morgan & Claypool publishers.

FURTHER READING

Hollnagel, E. (Ed.). (2003). *Handbook of Cognitive Task Design*. Mahwah, NJ: Lawrence Erlbaum Associates.

Karwowski, W. (Ed.). (2006). *Handbook of Standards and Guidelines in Ergonomics and Human Factors*. Mahwah, NJ: Lawrence Erlbaum Associates.

NUREG0711 (2004). Human factors engineering program review model. United States Regulatory Commission, Washington, DC.

O'Hara, J. (2014). Integrated system validation: Models, methods, and issues. (BNL Technical Report No. 6859-1-2010, Rev 2). Brookhaven National Laboratory, Upton, NY.

Wilson, J., and Carayon, P. (2014). Special issue: Systems ergonomics/human factors applied. *Ergonomics*, 45, 1–132.

10

SAFETY AUTOMATION

Björn Wahlström

10.1 INTRODUCTION

This chapter deals with safety automation in large physical systems.[1] Safety automation is a part of instrumentation and control (I&C) of the physical system. The dividing line between safety and non-safety parts of the I&C is more a matter of convention than function. Therefore, a firm division between automation and I&C is not made in this chapter, notwithstanding that the word automation, at least in principle, should be restricted to applications where I&C has active control tasks. Passive control tasks are information presentations to human process operators, who react on indications and displays to use the I&C for manual control actions. Safety principles to be used in the design and operation of control rooms are described in other chapters (cf. Chapters 9 and 14) of this volume.

This chapter focuses on the design of safety automation and safety principles that are applied in this process. This is not intended to belittle the importance of operating

[1] In this chapter, the term "physical system" is used, depending on application domain, to denote production plants, industrial facilities, transportation systems, etc., which include but are not restricted to aeroplanes, power plants including nuclear, chemical factories, off-shore installations, trains, ships as well as various manufacturing factories.

Handbook of Safety Principles, First Edition. Edited by Niklas Möller, Sven Ove Hansson, Jan-Erik Holmberg, and Carl Rollenhagen.

the physical system and its I&C safely, but is merely an acceptance of the fact that design errors often have larger safety impacts than operational errors. This emphasis on design is motivated by a common understanding that half or more of incidents and accidents have a basis in design errors (Kinnersley and Roelen, 2007; Taylor, 2007a). This number is probably higher because errors in operation, maintenance, and modifications often have a background in the management system, which also is the product of a design process.

10.1.1 Purpose of Safety Automation

Safety automation, which is a more or less well defined part of the I&C, can be said to have two purposes: (1) to prevent the physical system to enter unsafe states and (2) to bring it back to a safe state if it, by some event or control action, has entered an unsafe state. The availability of such a division of the state space of the physical system implies that a thorough risk analysis (cf. Chapters 21 and 20) has been performed and that excursions from safe states can be detected (observability) and reacted on (controllability) within the I&C.

Considering the purpose of I&C on a general level, one may separate between the following:

- functioning as an *intermediate* between human operators and the physical system, that is, providing inputs to and outputs from manual control actions;
- automated functions to *maintain* the physical system and its subsystems in defined regions of their state space, for example, controls ensuring that positions, levels, temperatures, pressures, and flows are kept within prescribed regions;
- automated functions to *transfer* the physical system and its subsystems from one region of their state spaces to another region, for example, to bring subsystems and components from start up to normal operation.

I&C design has a close interrelation with the design of the physical system, because it is supposed to control state transitions of the physical system. To fulfill this purpose, the I&C will contain an implicit model of the physical system (Conant and Ashby, 1970). The I&C system is based on collecting and acting on information. Sensors convert physical variables such as levels, temperatures, pressures, and flows to internal signals, which can be analog or digital. The signals are forwarded to calculation units and further to control elements that act on the physical process (Figure 10.1).

The requirements on the I&C are set by the requirements coming from the design process of the physical system, where subsystems and components have state spaces defined by their major physical variables. On a general level, one may separate between normal (shut down, start-up, operational) states and unsafe states with a

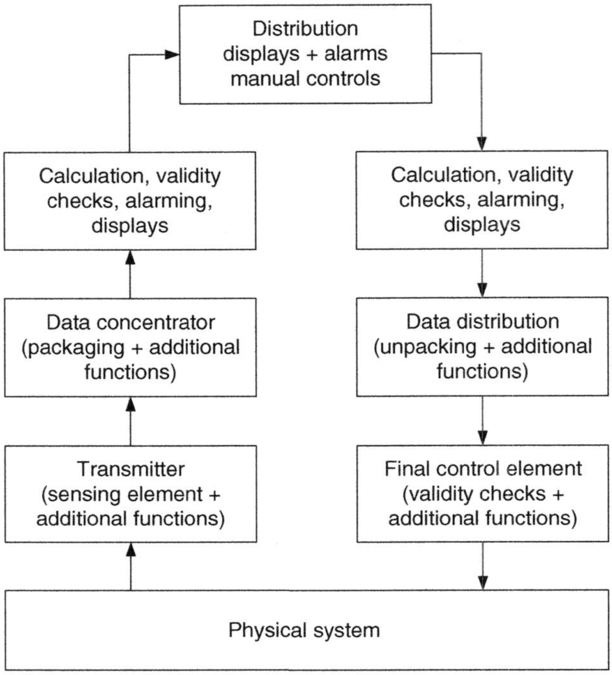

Figure 10.1. Functional components of an I&C system.

region of undecidable states between them. The safety automation should react with correcting controls if the physical system, its subsystems, or components enter unsafe regions of their state spaces (Wahlström and Rollenhagen, 2014).

A division of the state spaces of the physical process, its subsystems, and components relies on a comprehensive risk analysis. The general principles of safety engineering (eliminate, isolate, control, mitigate) are applied in the design of the physical system and its I&C. Thus states of system, subsystems, and components are allowed to move within their safe regions and are protected against transfers to unsafe states. A successful design of the physical system should lead to solutions that fulfill this high-level requirement. If design errors occur in the design of the physical system, its I&C, or safety automation, this very basic requirement may be challenged.

The design of a physical system as well as its I&C builds on practices of systems engineering, which applies the following general principles:

- A system of requirements is produced, which the physical system should fulfill (*requirements*).
- The design process is broken down into increasingly concrete design tasks or modules that are simple enough for single designers to handle (*modularization*).

- There is an integrative process where modules are combined into larger entities to make it possible to demonstrate that specific requirements are fulfilled. This phase ends in a demonstration that the entire system itself fulfills stated requirements (*integration*).
- Pre-existing products may be used in the physical system and its I&C provided that their quality and fitness for this purpose have been checked and confirmed by some competent body (*certification*).
- The involvement of a third party (regulator), who makes an assessment of available evidence (safety case) to be convinced that the system fulfills stated requirements and therefore can be considered safe enough (*regulatory acceptance*).

10.1.2 Functions of I&C Systems

The purpose of the physical system is to produce some commodity (product, service). This activity may, however, at the same time introduce risks due to component failures, human errors, and/or external events (flooding, extreme weather, earthquakes). The I&C systems can be considered as an enabler or support system, which makes it possible for the physical system to fulfill its intended function without undue risks for the society. The I&C includes automatic controls as well as functions for monitoring and manual control that give operators a possibility to follow and interact with what is going on in the physical system and its I&C. Control functions rely on the collection of data from the physical system that are assessed and acted on with algorithms implemented either manually or automatically to form control actions feed to the physical system through control elements (cf. Figure 10.2).

These manual and automatic control loops implement functionality to

- ensure that important state variables are kept within targeted values;
- initiate alarms when dangerous events or trends are detected;
- detect and automatically act on unwanted system excursions;

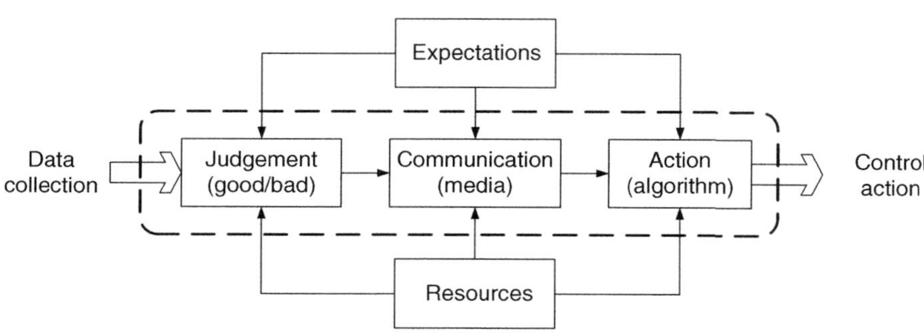

Figure 10.2. A generalized control loop connection within an I&C system.

- support operators through automatic control of subsystems and components;
- give possibilities for operator to intervene with corrective actions in dangerous situations;
- prevent operator errors;
- initiate automatic or semi-automatic sequence controls, which perform start-ups or shutdowns of the physical system, its subsystems, or components.

This functionality builds on a large range of physical components within the I&C system, which implement the following functions:

- sensors providing measurements of process variables, for example, temperatures, pressures, flows, velocities;
- transmitters and data concentrators forwarding measurements to computational units;
- communication units and cables connecting data concentrators and control elements to control rooms;
- control elements influencing components of the physical system, such as positioners, motors, pumps, and valves;
- computational units implementing various control algorithms;
- control rooms, where process variables are displayed as physical measurements and where operators can influence control elements of the physical system;
- display of binary signals to indicate states (on/off, available/not available, functioning/failure, high/normal, normal/low, etc.) of selected components or variables.

10.1.3 Allocation of Functions between Humans and Automation

The allocation of functions between human operators and automation is an important boundary to draw in the I&C design process. The so-called Fitts list from 1951 is still relevant when this allocation is made (Winter and Dodou, 2014). Basically, it suggests that automation should be used when rapid actions are needed, power is to be used, computations are needed, and a large capacity of short-term memory should be available together with a high assurance of dependability and reliability of the control. Control room operators use different strategies in their interactions with the physical system, which should be supported by appropriate means (Rasmussen, 1983). Development over the years has indicated that I&C designers sometimes have stretched the use of automation unnecessarily far and consequently left the human operators with tasks shattered into unrelated parts (Bainbridge, 1983). A remedy to this situation is to make the function allocation as a conscious decision after pondering alternatives between full manual and full automatic control (Hanes et al., 2015; Sheridan, 2001).

Task allocation between human operators and automation should be reflected in the requirements of the physical system and its I&C. Incidents and accidents caused by an unsuccessful task allocation have initiated the development of guidance by which a human-oriented design would be assured (Nachreiner et al., 2006). Simulator experiments with early design can be used to remove problems in the interfaces between human operators and the physical system (Carvalho et al., 2008). A bearing principle is that both manual and automatic controls are integrated into a hierarchy to make it possible for operators to switch between views ranging from an entirety to more detailed views (Rasmussen, 1985).

One specific difficulty in the task allocation between human operator and automation is that situations may emerge, where there is a conflict between human operators and safety automation. There are examples of such situations in the aviation, where perceptions of pilots have been different as compared to signals provided by the safety automation (cf. BEA, 2012).

10.2 ORIGIN AND HISTORY

Safety automation has its roots in the beginning of the industrialization in the eighteenth century. The industrialization brought external power to machines of all kinds. The economics of scale in industrial production facilitated larger and more complex physical systems to be used for an increasing number of products. Early design methods were mostly intuitive and resulted occasionally in sensational accidents. The need for both predictability and quality brought systematic methods into engineering design and project management. This development has facilitated present design processes used both for the physical system and its I&C. A complication in I&C design was introduced by a major technological revolution that was triggered by the developments in electronics, computers, and software. This change of technology necessitated major changes in thinking and design methodologies used for I&C and safety automation (Wahlström, 2015a).

10.2.1 Roots of Safety Automation

The roots of safety automation can be found in the advent of the steam engine and early industrialization that took place in England. Accidents with steam engines necessitated the introduction of automatic devices for controlling the rotational speed of the engine. The solution was a governor, which stabilized rotational speed by decreasing the steam pressure when rotational speed increased (Maxwell, 1867). The governor used rotational power of the engine as a source for the stabilizing control action. This was the first application of feedback control to ensure safety of an engine. It was based on two observations: (1) physical systems can, due to their nature, enter dangerous states and (2) rapid state transitions cannot be controlled manually.

The principle of feedback control has since then been applied wherever movements toward dangerous limits automatically should generate counterforces to maintain safety. Control engineering was developed from early controllers to become a discipline of its own. Today, control engineering is used in all technical domains and it has been integrated in the more general concept of systems engineering. Safety automation relies, in addition, on risk analysis and safety engineering to act on various threats with means of automatic control. Risk analysis and safety engineering use models of the physical system to suggest controls to eliminate, isolate, control, and/or mitigate identified threats.

I&C developed in parallel with the development of increasingly sophisticated industrial processes and equipment. The first solutions of stabilizing feedback were based on mechanical constructions. With the introduction of pneumatic controls, applications using PI- and PID-controllers took off and automatic control became an enabler of increasingly complex physical systems. Electronics offered an alternative path of development, which gained momentum with operational amplifiers and the advent of the transistor in the late 1940s. The transistor and later the integrated circuits made the development of digital computers possible. Analog I&C systems, both pneumatic and electronic, reached a large sophistication in the beginning of 1960s and they were used in large quantities for the control of industrial installations.

10.2.2 Systems Design

The science of design has been associated with the creation of artificial systems and components (Simon, 1996). Design, in general, contains large portions of creativity, but tedious experimentation may also be necessary before acceptable solutions are found. The design process can have good or bad impacts on the safety of physical systems (Hale et al., 2007). The possibility for one domain to learn from another has also been debated (Drogoul et al., 2007). The design of physical systems is usually associated with the term engineering design, a domain where relatively little academic research has been done (Horváth, 2004). The design of large physical systems, in the sense of this chapter, contains a range of different systems to be designed, that is, the physical system itself, its subsystems including its I&C system, and its management system as well as the management systems for the design process itself (cf. Figure 10.3).

A physical system that can be considered safety-critical should be designed with explicit safety goals (Pohjola, 2003) and there are many different views designers should be able to take (Hasan et al., 2003). The design of physical systems contains components of engineering design, software design, and service design (Kannengiesser and Gero, 2015). The design and implementation of information systems have also acknowledged the need for research in design science (Peffers et al., 2007).

Physical systems are characterized by a large number of subsystems that interact for the purpose of producing selected goods or services. The subsystems contain

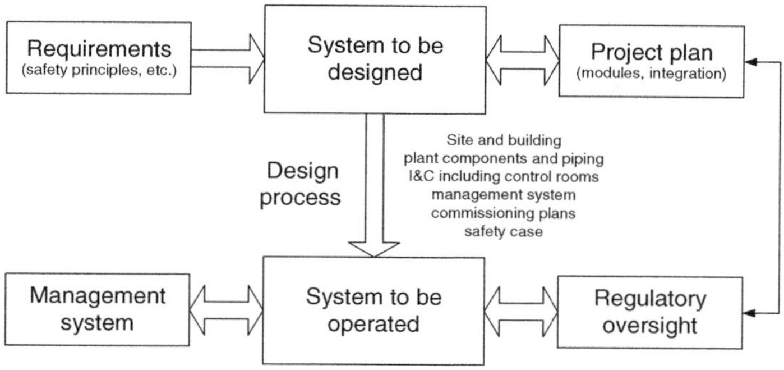

Figure 10.3. Design and operation of a physical system are two distinct phases in its lifetime.

a large number of components that form functional entities. The design of physical systems is typically carried out in projects restricted in scope and time. Project management can be considered as a special domain of management, where concerns for costs and timetables govern decisions made (Floricel and Miller, 2001). The challenge in design projects is to move from a set of requirements that are largely underdetermined and therefore includes a vast amount of alternative designs to be considered. Designers move through that space by making stepwise assumptions of good designs on which later concretizations can be built. Design is seldom initiated from scratch but from earlier successful designs that are modified, combined, and scaled up. The search for good designs is, to a large extent, based on heuristics that designers build with increase in experience (Chong et al., 2009). Usually, it is not possible to see full implications of initial assumptions and consequent design decisions. Therefore, there is often a need for modifications and iterations later in the design process (Zhang, 2013).

10.2.3 Typical Design Projects

Design and construction projects for large physical systems may easily cost hundreds of millions of Euros, involve thousands of people, and take several years to complete. Early building blocks for later success were military projects in which successful interactions between three new domains, that is, operations research, systems engineering, and project management, were achieved (Johnson, 1997). Because design projects are always unique in some sense, their organization and management are usually built for this purpose only (Georges and Romme, 2003). The challenge in design projects is to ensure a successful cooperation of engineers from several different technical domains. A design project should, in order to be successful, adhere to estimated cost and timetable and at the same time, produce good design and avoid design errors (Taylor, 2007b). Pressures for staying within cost and time frames unfortunately can make design errors increasingly likely. Uncertainties from various sources make it, however, difficult to obtain realistic estimates for project costs and duration (Saunders et al., 2015).

Industrial design and construction companies have over the years collected experience from both successful and less successful projects to arrive at present practices. Analysis and reanalysis of errors made in design and project management have over the years introduced a better understanding of demands on large design and construction projects (Kranakis, 2004). Poor feedback from earlier design projects has shown to hamper learning from experience (Busby, 1998). The practices used by design and construction companies are usually considered to be commercial secrets and they vary between companies. Taylor (2007b) and Kjellén (2007) describe typical design and construction projects in the chemical and offshore domains. Kirwan (2007) describes another example from the domain of air traffic management. A common view is that the design for safety should be integrated in the design process instead of being considered as something to be added later (Kidam et al., 2014; Kjellén, 2007). The design and construction projects of large physical systems can be shown to be far more complicated than expected (Ruuska et al., 2011).

10.2.4 Analog and Digital I&C

Digital control was introduced in the 1960s with the first process computers. At that time, I&C was analog and based on relays for binary control and on pneumatic and electronic systems for continuous control. The microprocessor introduced during the 1970s initiated a technical development leading to digital I&C and the new technology was rapidly adopted in most industrial processes (Wahlström et al., 1983).

The digital I&C systems made a division between application design and the design of hardware and software, which served as a platform for implementing a large range of applications. Application languages were designed to be structured around commonly used I&C entities, that is, functional blocks. The functional blocks were interconnected through software supplied by the platform. The platform contained hardware and software modules for connections of inputs from and output to the physical system, together with software for handling the functional blocks and their interconnections. The operating system, by which the execution of inputs, functional blocks, and outputs is connected to real time, is one important part of the platform software.

The change in I&C from analog to digital technology implied a move from continuous to discretized variables, serial instead of parallel execution, emphasis on time instead of frequency, and a move from hardware to software design. Digital technology also facilitates the use of floating point variables as compared with analog I&C, which used scaled variables.

The digital technology has several benefits as compared with analog, but it was proved more difficult to provide adequate evidence that the I&C would fulfill all required functions and never initiate spurious unwanted functions. This difficulty is aggravated by integrating large entities into the same functional units, which, at least in the beginning, was used in response to cost-cutting pressures. Such practices are in direct conflict with the safety principles of simplicity and independence. The change in technology enabled a far larger complexity in the I&C, which also was used by I&C designers. The complexity, however, also introduced the need to consider the adequacy of the design process in addition to product quality (Wahlström, 2015a).

The development of I&C systems has, during the last 40 years, been tremendous. The price of hardware has gone down and the functionality has gone up by several orders of magnitude. A simple projection is that this development will continue. Incentives for integrating several functions in the same hardware have disappeared, but on the other hand, there are pressures for additional interdependence in the I&C through benefits of increasingly complex control algorithms. The need for preparing evidence that the safety automation fulfills its requirements relies, however, on simplicity and independence. Today, I&C platforms are often supported by design and simulation software, which gives possibilities to build and test virtual control loops before their implementation.

10.3 DEFINITIONS AND TERMINOLOGY

Definitions and terminology connected to design of safety automation rely largely on concepts used in I&C design and, more generally, in the management of design projects. The design of a physical system has its starting point in the consideration of its whole life cycle, because overlooking distant requirements may produce considerable problems and costs in the operational phase. In assessing I&C design quality, it is important to separate process from product, because confidence that a system is safe relies both on confidence that a good design process has been used and that the product has gone through a comprehensive testing program. The operational phase carries its own requirements on maintenance, modifications, configuration management, documentation, and the safety case that should be reflected in the design process.

10.3.1 System Life Cycles

In design projects, an initial consideration is to create an understanding of the lifecycle of the physical system. This may start with a consideration of the applicability of the proposed site for the physical system. This phase will already require a risk analysis (Hassija et al., 2014). A typical division of the lifecycle of a physical system is to consider the phases of design, construction, manufacturing, commissioning, operation, maintenance, modifications, and decommissioning. It is important to consider the whole lifecycle because relative minor savings in the design phase may create considerably larger costs in later phases (Woodward, 1997). Work processes in each major lifecycle are typically controlled by its own management system. Safety should be a major concern during the whole lifecycle of the physical system (Kidam et al., 2014).

10.3.2 Process and Product

The separation between process and product is important in I&C design. The reason is that the final product is too complex to be tested in detail. It may be possible to build test programs that can achieve some restricted completeness in providing evidence that the final I&C system fulfills required functionality, but even that may prove impossible (cf. Section 7.1). To provide evidence that the I&C system does not in any situation show any unwanted spurious behavior is, however, impossible. Evidence that the I&C fulfills required functionality can at least partly be achieved with test programs that systematically go through different failure modes to verify correct behavior. This implies, however, that modules and larger entities are tested as part of the design process, because the testing of only the final product is not likely to produce the required confidence. Therefore, to provide evidence that the I&C system fulfills the two general requirements: (1) desired functionality is achieved and (2) no unwanted functionality will occur, it is necessary to consider both the process and the product.

Confidence in the process and the product can be supported by collecting evidence of their quality. This is possible even in the design of the first-of-a-kind physical system and its I&C, provided that the design process, to a large extent, can be characterized as an assembly of pre-existing modules, components, and subsystems. This has led to the use of certifications, where an accredited body issues a certificate that a person, a product, or a process has certain well-defined abilities or characteristics, which qualifies for certain tasks or functions. Consequently modules, components, and subsystems may be certified to be used as parts of the design. Similarly, designers may be certified to participate in the design and the used design processes certified to be used for the design of safety automation. Systems of accreditation and certification build on the existence of norms for well-defined processes and products.

10.3.3 Phases of Design

The design of a physical system is typically entered from general considerations of what it should be and do. Subsystems are defined and described with respect to their interactions in fulfilling the purpose of the physical system. The design process typically proceeds according to two dimensions, one of abstraction and the other of aggregation (Rasmussen, 1985; Wahlström et al., 1985). Three distinct levels of abstraction may be characterized by the terms of purpose, function, and form, and thus answering the questions why, what, and how (Cedergren and Theler, 2014). Similarly, three distinct levels of aggregation may be called system, subsystem, and component. With this characterization, the design process proceeds from abstract concepts to concrete designs and from the entire physical system through subsystems down to components (Figure 10.4).

A common approach in the design of physical system is to start from earlier designs, which are modified to meet applicable requirements. This may not be

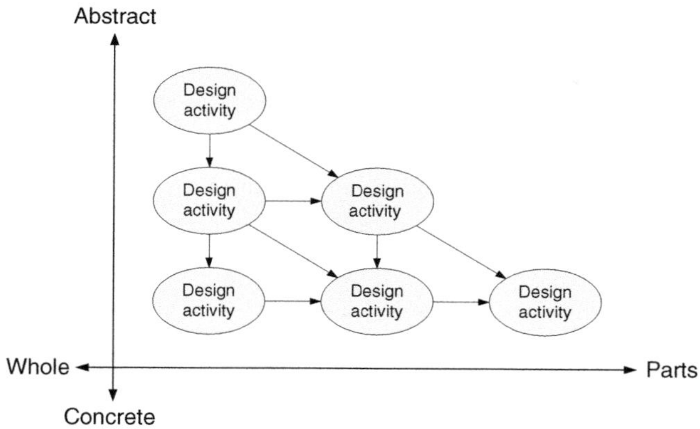

Figure 10.4. Design proceeds according to two dimensions—one from whole to parts and the other from abstract to concrete.

possible in the design of completely new physical systems, but even then it is often possible to reuse designs of earlier components and subsystems. The reuse of earlier designs has the advantage of being proven, but it also contains the danger that modifications are made without a proper understanding of earlier designs. Solutions that are reused should therefore be documented and understood. This is facilitated by databases and configuration management systems that major vendors use (Whyte et al., 2015).

10.3.3.1 Requirements Specifications A design process is typically entered by collecting and explicitly stating requirements that the finalized system should fulfill. The requirements may be functional (specification of inputs, behavior, and outputs) or non-functional (criteria that can be used to judge the suitability of a system for a specific purpose) and they may be targeted to any of the lifecycle phases. This means, for example, to require that certain conditions to be fulfilled for the site where the physical system is to be placed, to require access paths in buildings, to require environmental conditions for components, to avoid the use of dangerous materials, to require long operational times, or to require a manageable decommissioning of the physical system at the end of its life.

Requirements are created in top-down paths of increasing details, starting with requirements that apply to the physical system itself and continuing with requirements on subsystems and components (Hansen et al., 1998). Requirements may also be generated in a bottom-up path for prefabricated components and subsystems to ensure that they can be integrated in the final design. The requirements are used in the design in iterations of analysis, synthesis, and evaluations (cf. Figure 10.5), where alternative design solutions are generated and compared.

In most industrial domains, there are guides and norms that support the design process. There are also examples of user collaboration to create requirements to be

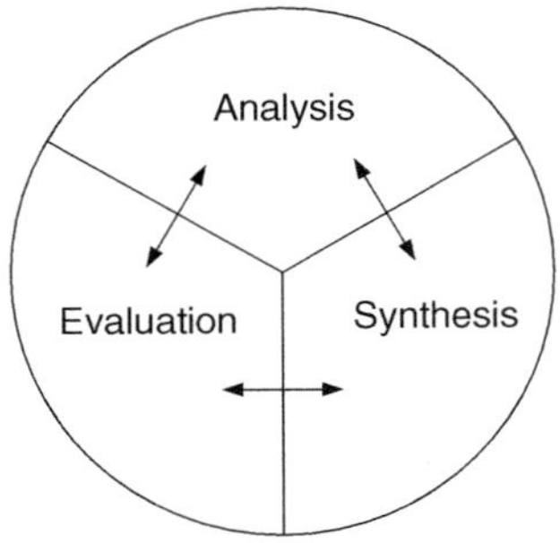

Figure 10.5. Design consists of three interacting activities.

placed on the physical system to be designed.[2] Such systems of requirements can be of large help when a procurement process is initiated.

10.3.3.2 Conceptual Design Conceptual design is initiated from a general concept (architecture) of the physical system and collected requirements. In this phase, there is an interaction between activities of analysis, synthesis, and evaluations in a search for good designs (Figure 10.5).

In the beginning of this phase, a detailed project plan is developed. This typically contains four different breakdowns of the design project as follows (Bachy and Hameri, 1997):

- *Product Breakdown.* A breakdown of suitable entities, which may be contracted out to subcontractors,
- *Work Breakdown.* A breakdown of tasks given to different designers participating in the project,
- *Organizational Breakdown.* A breakdown according to the management structure of the project,
- *Assembly Breakdown.* A breakdown according to assemblies and subassemblies that are manufactured and installed in due course of the project.

In the conceptual design phase functions, expected behavior, interconnections of functional entities, and resulting behavior are evaluated and brought to descriptions and representations to be used in next steps of the design process. In this phase, the modularization takes place. The division of the design into modules is important to make it possible to share tasks between several designers. Modules make it also possible to do iterations within a suitable entity in a search for good design solutions (Cho and Eppinger, 2005). Modularization will typically follow a division into

[2] http://www.europeanutilityrequirements.org/

systems, subsystems, and components, but this is seldom enough and further breakdowns are usually necessary. The aspiration in the modularization is to make the modules as independent as possible. When there are dependencies between modules, some restructuring of the modularization may be achieved (Steward, 1981). On the other hand, some interdependencies, at least on a macro level, will always remain because they are preconditions for the system to function as a whole (Geipel, 2012).

Interfaces between modules that are designed by different designers are vulnerable to misunderstanding and, consequently, to design errors. It is therefore important to keep interfaces between modules simple and clean. It is also beneficial to make consistency checks for information on both sides of interfaces between modules. The collaboration between designers can in this phase be supported by computerized tools (Wang et al., 2002). The functionality of modules and their interconnections are verified with inspections and tests to ensure that associated requirements are fulfilled.

In the conceptual design phase, it is possible to design for safety by considering possible failures within modules and their interfaces. Safety systems should, for example, be protected against failures of power supplies. Systems, subsystems, and components may require special I&C to make it possible to detect different kinds of failures, etc. Outstanding designers typically show both creativity in and passion for their task (Cross and Clayburn Cross, 1998). Creativity and passion for safe designs may in this connection be interpreted as the essence of safety culture in a design project (Macchi et al., 2013).

10.3.3.3 Detailed Design Detailed design starts when the conceptual design has been brought to a point where it is reasonably stable. The detailed design produces drawings, materials specifications, and instructions that are needed for coding, construction, installation, and testing of modules. In this stage, it may be necessary to modify parts of the design to correct earlier design errors or to cope with changes in requirements. Modifications in the middle of design project are, however, error-prone and should always be carefully scrutinized.

10.3.3.4 Integration Design projects rely on a large number of designers, which implies that their efforts should be integrated into a well-functioning whole. The integration starts when the detailed design has reached a stage, where single modules can be considered reasonably stable. The integration proceeds in steps where modules are combined into larger entities, which are subjected to thorough testing to verify proper functionality. This integration proceeds from components through subsystems to a system level, where the intended purpose of the system can be validated.

The integration process will proceed from single modules to larger entities until the functionality of all subsystems and components could be verified and finally the intended purpose of the physical system could be validated. The final steps in the verification and validation (V&V) of the design can often be carried out by simulation before the system has been built. More detailed testing can be initiated when parts of it exist as physical systems. Final tests are done after manufacturing in factory acceptance tests (FAT), and after installation in site acceptance tests (SAT), when the whole physical system has been finalized at its site.

10.3.4 Operations

The operational phase of the physical system can stretch over decades, which means that this phase should be properly reflected in the design process. In setting up a design and construction project, the usual practice is to plan for some reasonable period of operations, which gives the economic motivation for the project. However, in spite of the fact that this period is restricted in some sense to ensure the economic viability of the project, it does not imply that the technical life of the system has the same restriction. Instead, it is quite possible that modernizations and modifications give additional benefits of increased throughput or extended life of the physical system. A fair principle for such modifications is that their result should be comparable to new installations in respect to safety. Preconditions for such modifications to be viable are that systems have been maintained properly and that modifications are carried out with due carefulness. With regard to safety modification, it implies that a suitable configuration management system is in place and that documentation including the safety case has been kept up to date.

10.3.4.1 Maintenance It is usual to separate between corrective and predictive maintenance. The allocation of resources between corrective and predictive maintenance is a question of optimization (cf. Chapter 18). An emphasis on corrective maintenance may increase outage costs and too frequent predictive maintenance interventions may exchange fully functional components without due cause. The results of such an optimization depend on the accuracy of models used to predict wear and tear of subsystems and components.

10.3.4.2 Modifications During the operation of a physical system, additional operational experience is obtained. Some of that experience may reveal hidden design deficiencies, which would be advantageous to build away. Additional experience may also be available to indicate possibilities for improvements in system safety or economy. Vendor companies may improve their components and subsystems with various innovations. These possibilities can lead to interesting possibilities for system improvements.

Any modification of existing systems should be handled with the same rigor as the original design and construction project. A risk analysis should be carried out, alternative designs should be compared, and a detailed project plan should be created. In the decision to implement the proposed modification or not, it is important to consider optimistic projections of benefits with due skepticism.

Especially in modifications of the I&C system, it is important that proper and accurate documentation of the present installation is available. The technical development of digital I&C has been very rapid with rapid successions of product generations, which has resulted in early obsolescence of installed I&C. This fact has suggested the need to make major modernizations without large redesigns of the I&C.

10.3.4.3 Configuration Management Configuration management has to do with keeping track of design information, which is embedded in systems, subsystems, and components. Whenever modifications are made, small or large, they should be reflected in available documentation. For I&C, this means not only the configuration of control loops, but also their set points and information on measurements, cabling, hardware, control room indications, and control elements. To keep track of this information, it is necessary to have some sort of configuration management system that supports cross checks, listings, and documentation.

10.3.4.4 Documentation Documentation is an important part of each design process. A good practice is to ensure that major design decisions are documented and also the reasons for why they were made. The documentation should cover the entire physical system, its subsystems, and components. The documentation should also describe how the physical system should be operated and maintained. The documentation is one important part of the management system, which guides the operation of the physical system.

10.3.4.5 The Safety Case In physical systems where safety is important, the preparation of a safety case (cf. Chapter 16) is an essential part of the documentation. The safety case should include descriptions of the risk analysis connected to safety precautions as part of the design base, required operational limiting conditions, and arguments for fulfilment of regulatory requirements. When system modifications are made, it is important that the original safety case is consulted to ensure that proposed modifications do not violate safety principles that are built into the design. In the modification process, the safety case should be updated.

10.4 UNDERLYING THEORIES AND ASSUMPTIONS

The ability to predict system behavior in different situations provides the basis of risk analysis, which provides requirements on the physical systems, its I&C, and safety automation. The complexity of the present physical systems is a major obstacle in the design for safety. The concept of systems of systems (SoS) has, in academic literature, been proposed as an approach to remedy this difficulty. The systems are sociotechnical systems, which imply that in addition to models of the physical system, one has to create an understanding of its interaction with people, organizations, and information. This brings additional challenges in ensuring reliability, because not only the physical systems may fail, but people may err and organizations may not provide support to detect and correct such errors. Safety builds on principles of reliability engineering, where redundancy and diversity are used to achieve high reliability with unreliable components. Reliability engineering builds on operational experience and experience bases that can be extended through standardization of components and services. A specialization of companies provides an availability of world class performers for large design and construction projects, but it also makes it necessary to assess vendor capabilities. Successful design and construction of large physical systems assume an availability of refined methods for project management. Regulatory oversight provides a final check that the physical system can be considered safe enough.

10.4.1 Systems of Systems

An increasing amount of academic papers discuss complex systems and SoS (Amaral and Uzzi, 2007; Calvano and John, 2004; Harvey and Stanton, 2014; Maimon and Braha, 1996; Madni and Sievers, 2014). SoS may actually imply new approaches to risk analysis (Lock, 2012). The physical systems considered in this chapter have all the characteristics of these two concepts. At the same time, they are sociotechnical systems, where the behavior of people and organizations should be reflected in design decisions. Using the MTOI-model (Wahlström and Rollenhagen, 2014), there are some observations to be made:

- The M-system is formed by people at the sharp end (designers, operators, maintainers) and by people at the blunt end (managers, regulators, stakeholders, media, public), all have their own contributions to safety.
- The T-system is comprised of the physical system including its I&C and its successful operation depends on the success of applied design processes (physical system, I&C system, power supplies, management system) as well as successful operation and maintenance.
- The O-system defines the division of labor given in roles, authorities, and responsibilities and it is governed by a management system consisting of policies, system descriptions, instructions, and plans, whereas practices defined in the management system should be reasonably well reflected in actual practices.
- The I-system is embodied in system descriptions, data collection, and storage as well as communication routines. The functioning of the management system and the I&C are relying on information, which is generated, coded, stored, accessed, and scrapped in appropriate ways.

The MTOI-systems are artefacts in the sense they build on design activities for their structure and content. In a design for safety, it is important that threats are identified in a comprehensive risk analysis and acted upon using methods of safety engineering (elimination, isolation, control, mitigation). This implies a concurrent use of several safety principles such as defense-in-depth, a graded approach to safety, the single failure criterion, etc. In the control of the design processes, the safety principles should be explicitly stated in the management systems and their use should be thoroughly documented during all phases of the design process.

10.4.2 Building Reliability with Unreliable Parts

An important insight in early design and construction projects was that components could fail irrespective of efforts to make them reliable. This led to the use of redundancy and backup components to ensure that important functions can be maintained also in the case of failures. Redundancy, however, cannot protect against common cause failures (CCF), which means that it may also be necessary to introduce diversity in solutions to ensure a high reliability. Reliability engineering is today one of the standard methodologies to ensure system safety.

10.4.3 Reusability of Designs

An important principle in design is to reuse successful design solutions. Reusability can be achieved by standardization and adaptability in the design of modules, components, and subsystems. This can be achieved by

- agreeing on industry-wide norms and standards to be used in the design of certain products and services that can be used for selected applications;
- building modules, components, and subsystems with well-defined functionality and interfaces to ensure their use in a large range of applications;
- building configurability into modules, components, and subsystems through well-defined sets of parameters, which determine their functionality.

The design for reusability has the benefit of making it possible to collect a large experience base of modules, components, and subsystems. On the other hand, this implies a thorough understanding of differences between applications, where the modules, components, and subsystems are used. This question becomes important when operational experience of earlier designs is assessed. How large is the similarity of a new design as compared to earlier designs from which operational experience was obtained?

10.4.4 Vendor Capability

Globalization of industry has over the years brought a specialization of companies, which compete on international markets with specialized products and services. Some companies have specialized in the design and production of components to be used in physical systems and other in services connected to projects where physical systems are designed and constructed. The development has had the benefit of making it possible to create world-class products and services, but it has also had the drawback that it sometimes is difficult to establish a systems understanding in design and construction projects. To minimize economic risks with design and construction projects, it is important for financiers to assess vendor capability before signing delivery contracts. This difficulty is increased many times for design and construction projects in which a very large number of different vendors participate with large spans of contractual obligations.

10.4.5 Project Management

The principles used in the design of physical systems are connected to practical considerations of completing design and construction on time and within budget (Locatelli et al., 2014). Design of safety automation reflects safety concerns in requirements, specifications, and conceptual design. Risk identification and analysis is used as part of the design process, where candidate designs are compared and modified until acceptable solutions are found. Experience from earlier design projects is collected as a knowledge base of increasing detail to support libraries of design patterns, that is, good design solutions to common control problems. Design patterns usually contain implicit applications of certain safety principles.

The requirements specification phase is crucial for the success of a design project. It is necessary to specify important requirements up front; otherwise, important functionality may not be included. If new requirements are brought into the design process late in the project, it is usually difficult and costly to take them into account. A usual practice is to write requirements on a general level because a large number of detailed requirements can make it difficult to find viable designs. On the other hand, very general requirements can be difficult to translate into specific design solutions.

Modularization builds on the need to split the design into manageable parts. A successful modularization makes it possible to design different parts in parallel as long as interdependencies are taken care of. If interdependencies are minimized and interfaces are simple and clean, there are fewer possibilities for misunderstandings between interacting designers. A common practice is to use computerized tools to illustrate interactions between modules in terms of inputs, outputs, and timing sequences.

Successful design processes rely on early detection and correction of design errors. A good principle during design is that errors are corrected before the next step in the design is taken. Modifications or correction of errors may make it necessary to revise large parts of already completed design tasks.

Design as well as operation and maintenance of physical systems rely on the application of management systems (cf. Chapter 27) by which processes, activities, and tasks are controlled by more or less explicit principles and rules of conduct. The design process of the physical system is also supposed to build the basis of the management system, which is used in its operational phase. The management system can, with an analogy, be called the software of an organization.

10.4.6 Regulatory Oversight

Design, construction, and operation of safety critical system are usually regulated by national or international bodies. Regulation sets requirements on both process and product and specifies documentation that should be provided to support applications of building and operational licenses. Regulatory oversight, therefore, serves society by ensuring a sufficient safety by providing an independent assessment that the physical system fulfills appropriate requirements (Wahlström, 2007). For designers and operators of the physical system, regulatory oversight has the benefit of forcing a second thought on the sufficiency of collected evidence for safety. Depending on the industrial domain, the regulatory guidance has a varying degree of detail in the design of safety automation.

When the physical system is sited in a specific country, building and operational licenses are awarded by a national regulator, who specifies requirements and oversees that they are followed. If the physical system is used in several countries, in air or in sea, the requirements are usually defined in international agreements and norms, which designers and operators have to follow. In such cases, there are certification bodies that undertake inspections and reviews to verify compliance with specified requirements.

10.5 USE AND IMPLEMENTATION

Safety automation is designed as part of the I&C, which in turn is designed as an integrated part of the design of the physical system. The purpose of the safety automation within the I&C system is to initiate protective actions in abnormal situations and emergencies. Protection against internal failures of the safety automation is obtained using the principles of redundancy, separation, and diversity. Safety requirements for systems, subsystems, and components of the physical system define corresponding control requirements for I&C entities. The physical realization of the I&C depends on the selected vendor of the physical system. Many vendors have preferences regarding the selected I&C platform. This often proves helpful in the transfer of I&C requirements from the design of the physical process. The design of the I&C and the safety automation can be initiated after some initial considerations. Vendors as well as design practices differ between industrial domains, in spite of large similarities on a general level.

10.5.1 From Systems Design to I&C Design

The structure of the physical system in terms of subsystems and components provides the basis for defining entities of the I&C. According to the principle of a graded approach to safety, an important step in the design of the physical system is to create a classification document, which identifies the safety importance of various structures, systems, and components. This document also governs I&C design, where details of the design process depend on the safety class assigned to specific entities. I&C entities can largely be designed independently of each other, but parts of the safety automation may introduce need for interactions between subprocesses and components of the physical system.

I&C relies on availability of electric power. With regard to safety automation, this implies specific requirements to be taken into account for the safety automation. In the I&C design, requirements regarding isolation between redundancies, fire protection, and ventilation should be taken into account when locations of equipment, control rooms, and cabling ducts are determined.

10.5.2 Physical Realizations of I&C

I&C architecture and application are implemented with suitable platforms, which are designed and manufactured by I&C vendors. The I&C platforms consist of hardware and software and they offer standardized functions and modules for typical tasks such as data collection and storage, calculations, communication, interfaces to sensors and control elements, control room modules. Some vendors offer platforms targeted at safety automation. Such platforms have special modules for fault detection and management and they have gone through comprehensive V&V efforts.

The functionality of I&C is built on one or several platforms, which includes hardware and software for the implementation of required functions. Because I&C collects and acts on information, the physical variables and signals are coded using suitable internal representations of the platform. Earlier methods of coding were analog and used scaled representations of signals. Today, digital representations are used almost exclusively. One of the benefits is that it is possible to avoid scaled representations and instead use engineering values in a floating point representation.

10.5.3 Initial Considerations

There are many details that have to be settled before the design of the I&C architecture can be initiated. The selection of vendor is a decision with far reaching consequences, because the platforms may differ considerably in assumptions and structure. An important question is, for example, the availability of computerized systems for application design and testing. Also, the availability of a training simulator for the physical process may tip the benefit of selecting one vendor before another.

10.5.3.1 Vendor Selection In many domains, there are often only few I&C vendors that offer a systems approach together with safety certified platforms. Especially for safety automation, it may be necessary to compare solutions from two or more vendors. In such cases, a common practice is to enter small evaluation projects to assess the applicability of the solutions offered. I&C vendors are often familiar with safety regulation only in one or mostly in a few countries, which may introduce the need for additional evaluations. Vendors may use subcontractors for parts of the I&C as well as for the application programming. The subcontractors may also in turn use their own subcontractors, but too long chains of subcontractors are due to create problems in information transfer across contractual borders.

10.5.3.2 Requirements The first version of requirements for a physical system is developed by the customer and reflects applicable national and international regulations. The requirements are one of the most important documents in a call for tenders that are sent to a selected set of potential vendors. To support potential buyers, there has been cooperation between domain organizations to create common requirements that apply to the physical systems.

10.5.3.3 Tenders The next step of the design process is built on tenders from interested vendors. The tender would contain descriptions of the plant with its design and construction processes, details of subprocesses, and major components and principles for operating the physical system. The requirements and descriptions in the tenders are used by the customer to prepare a preliminary safety case for the physical system.

10.5.3.4 Project Initiation When the vendor has been selected, the construction project can be started. The detailed design of the plant systems is usually a cooperation between engineers from both the vendor and the customer. The participation from the customer aims at transferring local requirements to the vendor and to ensure that a comprehensive understanding of the physical system is available at start-up. This phase will initiate the design of the I&C system and its architecture. For the design of I&C and safety automation, there are national and international norms and guides available, which set requirements on both the design process and the final design. A common approach is to include a simulator in the design and construction contract. The simulator can be used for V&V activities and for operator training.

10.5.4 I&C Design

The design of modern I&C systems is governed by considerations distributed between two independent design processes, which are carried out by different parties at different times. The first one, that is, the design of the platform and its hardware and software, takes place before it is applied to concrete applications. The second one, the design of the application, may take place several years after the design of the platform has been completed. The application design can be divided into architecture and application design.

The change of technology from analog to digital implied a change in thinking, design methods, and V&V (Wahlström, 2015a). This caused a burden in many domains of application to implement the first-of-a-kind engineering project (FOAKE) (Patterson and Clarkson, 2015). Presently, there are many documents in the nuclear domain, which give guidance for design and implementation of digital I&C (IAEA, 2015; Wahlström and Duchac, 2015).

10.5.4.1 Architecture A safety philosophy of the I&C is established in the architecture, where control functions of subsystems and components are built using functional entities of the selected platform. The internal structure of the entities is determined by functional blocks and their interconnections. The I&C architecture is established by applying interconnections of hardware and software components that are available in the platform. Simplicity and independence are overriding principles in this process and builds on the separation between functions together with protection against internal failures. This is accomplished using redundancies in computational units, data buses, signaling, and power supplies. For important trip systems, it is common practice to introduce diversity between functions as an approach to protect against CCF. A recent document on principles and approaches for I&C architectures provides guidance for understanding systems of requirements in relation to the design of the physical system (EPRI, 2014). Using such guidance makes it easier to build credible safety cases.

Typical I&C architectures are hierarchical starting from field instruments and moving through a controller level to human machine interfaces and control rooms (Sisbot, 2011). Safety automation follows a similar architecture, but should be independent of non-safety I&C. Viel et al. (2015) propose that the architecture should be developed in three steps: (1) gathering safety requirements to be considered, (2) defining design rules that satisfy the requirements, and (3) designing an architecture that is compliant with applicable design rules in which due considerations are taken from other design domains of the physical system.

10.5.4.2 Application The application is built using software connections between functional blocks provided by the platform. Functional blocks are used to build displays for control room operators and to define limits where alarms should be initiated. One important part of the application design is to assign unique names to subsystems and components. These names are used to identify subsystems and components in the field, in the control rooms, and in a configuration management system. For application built directly with software, the selection of a suitable programming language should be considered (Halang and Zalewski, 2003; Zalewski et al., 2003).

Safety automation is designed in very much the same way as non-safety I&C, with the only exception that a more extensive V&V is required. For the signals used in the safety automation, a typical practice is to build online checks to ensure that signal values are mutually consistent.

10.5.4.3 Platform The platform is designed before it is utilized in concrete I&C projects. The platforms include common I&C functions such as controllers, calculation units, logic gates, signal conditioning, and so on. Functional blocks are interconnected with software to give specified functionality to selected controls. The platform design proceeds in similar ways as system design through requirements, conceptual design, detailed design, and V&V. Depending on the intended field of application, requirements may be more extensive if the platform should be able to support implementation of safety automation.

Because calculations should be carried out in real time, a common approach is to rely on cyclic execution of the application software. This approach makes it possible to maintain upper bounds for execution times. For safety automation, some vendors have taken the approach to have a restricted set of functional blocks that have been certified to be used for safety automation. Platforms for safety automation include many standard functions for fault detection and management including watch-dog timers, automatic switch over to standby units, possibilities to insert and remove hardware during run time, etc. In the development of the platform, it is important that testing takes place as the design is advancing (Tsai et al., 1997).

The platform integrates hardware and software. Due to rapid turnover of hardware generations, there have been cases where vendors have ported their software to new generations of hardware. Such solutions would require extensive V&V efforts, but appears to be economically motivated as compared with the alternative of designing a new platform.

10.5.4.4 Software The design of platform software is a typical software project in which good programming techniques should be used (Brooks, 1975). One important design principle is to ensure clean interfaces between software modules as well as a continuous V&V of both module design and module integration. In this stage, it is important to consider different types of software faults, because some may be more important than others (Hamill and Goseva-Popstojanova, 2015).

The design should start with requirements specifications using a suitable computerized environment, which provides possibilities for completeness and consistency checks. Such system may even provide run time environments, where early versions of integrated modules could be tested. Present software design practices rely on so-called object oriented programming (OOP) and special requirements modeling languages (Larman, 2005). There are arguments that the OOP should be amended with methods to ensure a proper trade-off between safety and development performance (Motet, 2009).

One important principle in software design is to ensure authority and authenticity of I&C users (operators, maintainers, system engineers) that interact with the software. Different roles should, for example, be defined for operators and for system engineers, who are authorized to make changes in I&C configuration, parameters, and set points. There should also be strict procedures for version management of the software and perhaps requirement that two persons work together when new software versionsare taken into operation. When the software is finalized, there should be assurance that patches and trapdoors from the design phase have been removed.

10.5.5 Practices in Different Domains

Safety automation is used in all industrial domains. Practices for I&C design are taught in control engineering, where courses are oriented toward specific control problems of selected domains. In addition, there are courses directed toward general control problems given in departments of control theory, systems engineering, and operations analysis. With the introduction of digital I&C, many practices from software engineering have been integrated into control engineering.

Design of safety automation relies on a combination of knowledge from the domains of safety and control engineering. This knowledge is further combined with the knowledge of the physical process for which the safety automation is designed. In I&C design projects, it is necessary to combine a thorough understanding of the physical system with the knowledge of the applied platform. This is usually achieved by involving design engineers from both the customer and the vendor of the I&C.

Design practices in general have benefitted from the development of computers and software. Today, computer-aided design (CAD) and computer-aided manufacturing (CAM) have yielded support for design and construction projects in most industrial domains. Development in the software domain has approached the possibility to build systems based on formal models and axiomatic design (Heo and Lee, 2007; Suh and Do, 2000).

10.6 RESEARCH ON USE AND EFFICIENCY

The design of I&C systems relies on a base of engineering experience and knowledge. There has not been much academic research on the efficiency of used approaches. Most of this information is actually considered as confidential with a commercial value and owned by vendors in the field. With reference to results from software design, an increasing complexity of the physical systems can be assumed to increase the probability of design errors (Basili and Perricone, 1984).

10.6.1 Estimates of Project Cost and Duration

Large projects of early software design were hampered with problems in meeting targets with regard to project costs and time. Some of the early software projects finished using more than 100% excess resources (time, man-power) as compared with initial estimates. These problems were apparently connected to several deficiencies in project planning (Boehm, 1976). Efforts were initiated to create better models for the economics of software development (Boehm, 1984). Early experience also indicated benefits of early prototyping as compared with the specifying approach (Boehm et al., 1984). Experienced software vendors are today able to give reasonable estimates of project cost and duration. The development of standards has apparently contributed to this development (Fenton and Neil, 1998). Early principles for reaching good designs involved detailed guidance in carrying out coding of software modules, which depended on the use of coding language, used operating system, and types of software. One well-known principle from that time, "keep it simple stupid (KISS)," is still applicable today. In spite of a largely positive development, especially large software projects in the health-care sector have still proved to be problematic.

10.6.2 Support Systems for Design and Construction

A common theme in research has been to create computerized support systems for tasks in design and construction that have proved tedious, difficult, and error-prone. There are many specialized systems intended to support requirements engineering, documentation, and configuration management. The requirements specification phase is crucial for I&C design and it is therefore natural that research activities have been directed toward this phase. In the software domain, there are support systems built on a database of requirements, which provide support in analyzing the requirements and their relationships. There are also systems that make it possible to animate requirements and thereby evaluate the functionality of systems to be built.

10.6.3 Benefits of Using Safety Principles

The benefit of using safety principles in I&C design is that both structural and empirical arguments can be used to provide arguments that design errors have been avoided and that test programs have been efficient. A well-structured design process, together with a comprehensive testing of intermediate products, can provide evidence for claims that requirements have been fulfilled. The use of safety principles can also provide a structure within which systems of requirements can be understood and analyzed.

Preconditions for these benefits to be realized is that the design process is governed with a safety management system, which is documented, understood, and used. It would, for example, be important that the management system contains specific guidance on how safety principles are used in different phases of design. It is also important that the path of the design is described in enough detail for its duration, to make it possible to collect evidence to be used for the safety case.

It is advantageous that a plan for preparing the safety case is developed in parallel with the preparation of the plan for the design process. Depending on the complexity of the system to be designed, the plan should give practical guidance in different phases of the design process, that is, requirements specification, levels of abstraction and aggregation, modularizations, design of major subsystems, path of integration with tests to be carried out, and plans for approval of the final design. For I&C

design, it would be important to cover interfaces between I&C design and the design processes of the physical system, human–machine interfaces, and buildings.

More specifically, plans for I&C design should describe the safety philosophy to be applied and I&C architecture to be implemented. Specifics of the selected I&C platform, which may include safety arguments from the platform vendor, should be described. The application software could implement algorithms for failure detection and handling, together with displays and automatics to support control room operators.

The design process should, at different levels of detail, implement relevant safety principles, that is, defense-in-depth (DiD), a graded approach to safety, redundancy, diversity, separation, the single failure criterion, the grace rule, etc. Especially, the DiD principle would assume that a certain degree of independence has been built into the I&C at different levels to ensure that CCF have been avoided.

10.7 WEAKNESSES, LIMITATIONS, AND CRITICISM

Present design practices do create safety automation, which is fit for its purpose. Still there are challenges to be addressed both in the design and the operation of the safety automation. The most important question is perhaps what can be considered to be safe enough. Based on the results in mathematical logic, there are limitations that cannot be overcome, which means that it is only possible to provide limited evidence that the safety automation is good enough. To some extent, the principle of a graded approach to safety may provide some help, but the possibility of CCF still generates questions that have to be evaluated thoroughly. Additional research to approach the question of how to ensure quality of design may provide some help. Considering regulatory oversight, the present situation of a large diversity in available systems of requirements introduces a global need for harmonization.

10.7.1 What is Safe Enough?

The main difficulty in design is to agree on what can be considered safe enough (Starr, 1969). There is a common belief that completeness, consistency, and correctness (C^3) in risk analysis and in corresponding safety precautions would be a realistic goal to strive for. This is not possible in the general case, which means that it is instead necessary to agree on what can be considered sufficient for a design to be safe enough (Wahlström, 2015a). Such agreements may include arguments on C^3, which are restricted to well-defined parts of the systems.

One of the underlying assumptions is that an objective view exists on what is required for a safety automation to be acceptable. This is not the case. The first difficulty is that regulation in different countries differ (Raetzeke and Micklinghoff, 2006). The second difficulty is that both assumptions and requirements are described in natural languages, which means that they always are due to interpretations. The integration of businesses in the world brings in a third difficulty in the use of several languages in a single design project.

10.7.1.1 Limitations of Reasoning In the preparation of evidence for safety of any design, there are three important limitations necessary to consider (Wahlström, 2015b), namely:

- Any set of requirements is incomplete.
- It is not possible to make qualitative predictions of software functionality.
- Engineered controls increase the complexity of a physical system.

The argument for the first limitation stems from a theorem of Gödel that states that theorems possible to formulate based on a set of axioms either contain contradictions or theorems that are not provable with the axioms. A set of requirements can therefore not cover all possible situations that may emerge; that is, completeness and consistency cannot be assured at the same time in a system of requirements.

The argument for the second limitation is connected to a theorem stating that it is impossible to say if a Turing machine will stop or not. Assuming that an arbitrary software module can be modelled as a Turing machine, the theorem can be interpreted to say that it is not possible to predict states that the software will attend without running the program.

The third limitation is based on Ashby's principle of requisite variety according to which a successful controller has to have the same complexity as the system it is set to control (Ashby, 1956). This can be interpreted that added control loops will bring additional complexity, which makes it increasingly difficult to verify that the system behaves correctly. A well-functioning controller can hide some system complexity, but it may in the case of failures put operators in situations, which they lack means to cope with.

10.7.1.2 A Graded Approach to Safety A graded approach to safety can be understood as a principle of its own (IAEA, 2014b). It can be interpreted to mean that subsystems and components that are important for safety should be given more design efforts as compared to non-safety subsystems or components. This principle may not be used to allocate design resources to subsystems or components, but it still gives general guidance in the planning of a design process. This principle is in practice implemented by safety classification of structures, systems, and components (IAEA, 2014a).

10.7.1.3 Independent Safety Barriers The defense-in-depth principle builds on independence between safety barriers (cf. Chapters 4 and 5). To be able to argue that a certain design is safe, it is necessary to argue that two or more safety functions cannot fail due to a single precursor (cf. the single failure criterion). Many incidents and accidents have shown that unexpected dependencies can introduce CCF. An argumentation that CCFs have been avoided would need structural evidence that interdependence of following types do not exist:

- physical connections through power supplies, cables, and locations
- operational connections through maintenance schedules, training, and education
- connections through the design process
- design errors in standard components, which are used in several functions
- common materials used in manufacturing, storage, and maintenance
- connections through common parts of the I&C software
- connections through equipment used for calibration of sensors

10.7.2 Quality of Design

The processes of modularization and integration in the design of complex system is sensitive to details, because a complex system can be broken down in several ways of which some may be more prone to errors than others. Especially when the three parties, customer, vendor, and regulator, have different views on how this breakdown should be done in the design, difficulties may occur.

There are many practical examples that the detection and correction of design errors late in the process has a potential to introduce new errors. The need for additional efforts will also have their influence on project costs and timetable, which is due to induce additional stress in the design and consequently make further design efforts more error-prone.

10.7.3 Field Programmable Gate Arrays

Field programmable gate arrays (FPGA) is a new hardware technology proposed to be used especially for safety automation (IAEA, 2016). An application of this technology brings in benefits as compared with microprocessor based I&C, but it also poses some new challenges. Without going into details, it is evident that the more generic safety principles in the design of safety automation are applicable also for this technology. On the other hand, it is equally clear that the design of FPGA-based systems will have its own set of principles to be applied to ensure successful results.

10.7.4 Cyber Security

A recent trend in the I&C systems is to connect them to the Internet and thereby rely on the additional functionality of distributed data processing and cloud based resources. Unfortunately this trend has the drawback of making systems vulnerable to attacks by adversaries operating through the internet. This has introduced the need for protecting I&C against cyber threats and consequently to new requirements to be taken into account in the requirements specifications. Protection against intelligent adversaries requires different measures as compared to protection against hardware failures and human errors. Relying on the assumption that potential hackers act rationally, protection can be built based on a cost benefit analysis (cf. Chapter 22). The probability of hacker success should be small enough in comparison with investments needed to succeed in mounted attacks.

The concerns connected to cyber security of software in computers rely on ensuring confidentiality, integrity and availability of systems (IAEA, 2011). This also assumes that the accessibility of systems can be restricted and that the authenticity of system users can be assured. Considering threats of cyber security it is important to detect attacks, to identify the damage they have caused and to restore compromised systems as soon as possible. Cyber security may require new safety principles to be applied. In practical terms a straightforward safety principle to protect the I&C against attacks is to isolate it completely from external connections.

10.7.5 Regulatory Acceptance

There have been controversies on what should be considered as adequate evidence for safety between on one hand regulators and on the other vendors and operators of physical systems. These controversies seem to be based both on less successful I&C projects and on a disagreement on the need for breadth and depth in safety argumentations (Wahlström, 2015a). The preparation of an I&C safety case presents several challenges (Jung, 2015; Morton, 2015). One specific challenge is connected to lack of harmonization in national regulatory systems (IAEA, 2002). Among the arguments for safety especially the possibility of CCF has to be thoroughly addressed (IAEA, 2009; Torok and Geddes, 2015). Present uneasiness with the possibility that the I&C contains design errors, which contribute to CCFs may, however, be somewhat exaggerated (EPRI, 2008). The human factors engineering (HFE) is also an important chapter in the safety case (O'Hara and Fleger, 2015).

It is possible to give believable evidence that completeness, consistency, and correctness (C^3) only in restricted cases (Wahlström, 2015b). Such cases could be established based on structural arguments regarding I&C architecture, on the interaction of modules and the internal structure of the platform software. A well-structured V&V program can also give empirical evidence that functions and subfunctions behave according to specifications by using data sets that provide a good coverage of possible situations in the applied test programs.

10.8 RELATIONS TO OTHER SAFETY PRINCIPLES

Safety automation relies on a systematic application of many safety principles. Design of complex system occurs on two hierarchical dimensions, one of abstraction stretches from general functions to details of their implementation and the other of aggregation stretches from system, through subsystems down to components. It can therefore be beneficial to speak of super principles and subprinciples of certain safety principles. In that way it is possible to create hierarchical trees of safety precautions that are applied in design processes. If such trees of safety principles are used, they would provide opportunities to argue that a specific design can be considered good enough. The extent safety principles are used in I&C design depends on the ingenuity of the designers involved. Some caution should, however, be used in building safety precautions, because building additional checks into I&C design has a tendency to increase system complexity. There may also be cases where applications of some lower level safety principle contradict higher level principles. Relationship between different safety principles can be ordered in different ways (Saleh et al., 2014). The division below follows the major groups as identified in the introductory chapter of this book.

10.8.1 Safety Reserves

Safety reserves imply that there is room for movement and actions before dangerous limits are reached. In the I&C this is typically ensured using alarms, where two limits may be used, one to issue a warning and the other to indicate danger. Robustness and resilience can be applied in the I&C architecture and application as well as in the platform design (cf. Chapter 3). The defense-in-depth principle is a very central design principle that is used both in the design of physical systems and in I&C design (cf. Chapter 4). It is used in the phase of conceptual design and in the design of I&C architecture through safety automation that reacts on combinations of signals using redundant and diverse control elements. Safety barriers and margins of safety can be built into subsystem and component by introducing trips and interlocks (Chapters 5 and 6). These principles are targeted to detailed design, the I&C application and design of platform software. Fail-safe designs can be built into the I&C application by defining target states for components in case of I&C failures.

10.8.2 Information and Control

The systems approach to safety implies the use of information and control. The control paradigm for safety implies the use of an objective function and a system model, together with criteria of observability and controllability (Wahlström and Rollenhagen, 2014). This means that there should be qualitative or quantitative measurements that serve as indicators of available safety margins (cf. Chapters 6 and 8), a model of how safety is constructed and means to initiate actions that drives the physical system in safe directions. Among the safety principles in this group one may speak about experience feedback (cf. Chapter 7), especially within design projects to ensure that good design principles are used. The feedback has also been crucial for creating an understanding of which design phases and tasks are error prone. HFE is an important field that applies to control room design as well as to design of operating procedures and the management system (Chapters 9, 13, and 27).

10.8.3 Demonstrability

Demonstrability has to do with the collection of evidence that the design process has the capacity to generate safe designs and that safety principles have been applied accordingly. The use of proven design and proven technology is a usual requirement in most design projects. A simple illustration of this principle is that most design projects start from earlier designs that are modified. This principle have in retrospect, at least in some industrial domains, caused delays in a transfer from analog to digital I&C. Simplicity is an important principle that should guide any design project from its beginning. Empirical demonstrations of safety involve making experiments with intermediate and final products to demonstrate that required behavior has been obtained. Experiments may be carried out using predefined tests or tests using stochastic inputs. I&C and safety automation will get its own chapter in the safety case (Chapter 16). In the argumentation it is important to use both structural and empirical evidence that the design is good enough. The safety case is the most important document with which designers and system vendors could argue that a physical system is safe enough. Digital I&C has unfortunately brought additional complexity to systems, which have led to disagreements on the sufficiency of collected evidence (Wahlström, 2015b). V&V is an important principle, which intends to ensure quality control of each step of the design process. The V&V activity should be planned and continuously in place during design projects to enable the collection of structural and empirical evidence to be included in the safety case. I&C has its own requirements for inspectability and maintainability. Especially modifications in earlier designs have proven to be error prone.

10.8.4 Optimization

Optimization has to do with situations, where two or more feasible options are available from which the best design should be chosen. Optimization implies the existence of an order relation "better than" on the set of options. Some caution should be exercised because optimization relies on models, which means that models with a poor ability to predict may bring results of optimizations into forbidden regions. Qualitative and quantitative risk analysis (cf. Chapters 21 and 20) are both important both in the design of the physical system and its I&C. In the conceptual design phase a qualitative analysis is often enough, but quantitative analysis is usually necessary at least to support argumentations of the safety case. Cost benefit analysis (Chapter 22) is typically used when modifications either of the physical system or the I&C are contemplated. Chapter 27 on regulation is relevant in adapting to national and international design requirements. Risk and safety indicators can be applied to inform control room operators on the state of the physical system or its I&C.

10.8.5 Organizational Principles and Practices

Many safety principles have to do with the organization of the I&C design process. The challenge is to get people to cooperate in complex tasks to achieve organizational goals, because there is a danger of misunderstandings in the interfaces. The management system is used to define goals, authorities, responsibilities, processes, activities and tasks. Organizational safety principles are for example the use of standards and design patterns, the establishment of emergency plans and procedures, safety management and safety culture. Standards (cf. Chapter 31) have an important position in design of both the physical system and its I&C. The physical system in the nuclear domain is governed by standards developed by the International Atomic Energy Agency, where fundamental safety principles are defined in the document (IAEA, 2006) and design requirements in the document (IAEA, 2012). Corresponding requirements for the I&C are given in a document (IAEA, 2014c), which is in a process of being published. Additional standards in the I&C field is published by the International Electrotechnical Commission (IEC) and Institute of Electrical and Electronics Engineers (IEEE). Safety management principles (cf. Chapter 27) are also highly relevant for the design of physical systems and their I&C, because design projects should be governed by a management system. Safety culture (cf. Chapter 28) in a design organizations has characteristics that have been identified in operational organizations, but also some own distinct characteristics (Macchi et al., 2013). There seems also to be difference in views between project managers and expert designers that can be reflected in characteristics of safety culture (Hayes, 2015). Finally managing the unexpected (cf. Chapter 32) is certainly important to pay due attention to, because any design project is most likely to be confronted with many unexpected issues to be resolved.

10.9 SUMMARY AND CONCLUSIONS

Safety automation is not a principle, but it is a physical system with a behavior that is defined in software. The use of safety automation can be seen as a safety principle, which makes it possible to implement other safety principles such as for instance defense-in-depth. Functionality of safety automation relies in the first place on the design process with which it has been designed and in the second place on how it is maintained during its operation. The design of safety automation builds on the application of many different safety principles. In the design of safety automation a crucial component is to avoid design errors and in this task the use of safety principles has a central role. The design process in itself relies heavily on experience feedback, where earlier mistakes and errors are removed gradually. Technological development has led to new design errors in digital I&C, which are removed when the technology is maturing. Similar evolutions have been seen in many other domains, where new technologies have been introduced. This observation may in turn suggest the use of the precautionary principle (cf. Chapter 12) in applying new technologies, to ensure that possible negative consequences are known before they are deployed.

Safety automation is an integrated part of the I&C. It relies on models, methods and tools used in I&C design. I&C has been characterized with a very rapid technological development during the last 50 years. The move from hardware to software design has been the major step, which still carries confusion in its need to cope with an increasing complexity of the systems. Research in SoS and complexity may provide components in resolving some of this confusion. Especially a consideration of safety principles in these domains may provide a fruitful route to investigate. One part of a solution lies in the use of computer support to investigate implications of early design decisions. In an assessment of findings from I&C and safety automation it seems clear that applied models, methods, and tools can be used also in the design of socio-technical systems in general. Especially the modeling languages used in software design may be applied to describe interactions between users and abstract functions. Similar modeling approaches can be used to specify, set requirements for and test interactions within and between organizations. In this context, it may also be possible to handle cyber security, which is a new threat introduced by an increased societal reliance on computers and networking.

REFERENCES

Amaral, L.A.N., and Uzzi, B. (2007). Complex systems—A new paradigm for the integrative study of management, physical, and technological systems, *Management Science*, 53(7), 1033–1035.

Ashby, W. R. (1956). *An Introduction to Cybernetics*. Chapman & Hall.

Bachy, G., and Hameri, A.-P. (1997). What to be implemented at the early stage of a large-scale project. *International Journal of Project Management*, 15(4), 211–218.

Bainbridge, L. (1983). Ironies of automation. *Automatica*, 19, 775–779.

Basili, V. R., and Perricone, B. T. (1984). Software errors and complexity: An empirical investigation. *Communications of the ACM*, 27(1), 42–52.

BEA. (2012). Final report on the accident on 1st June 2009 to the Airbus A330-203 registered F-GZCP operated by Air France flight AF 447 Rio de Janeiro - Paris. Available at: http://www.bea.aero/docspa/2009/f-cp090601.en/pdf/f-cp090601.en.pdf

Boehm, B. W. (1976). Software engineering. *IEEE Transactions on Computers*, 25(12), 1226–1241.

Boehm, B. W. (1984). Software engineering economics. *IEEE Transactions on Software Engineering*, 10(1), 4–21.

Boehm, B. W., Gray, T. E., and Seewaldt, T. (1984). Prototyping versus specifying: A multi-project experiment. *IEEE Transactions on Software Engineering*, 10(3), 290–303.

Brooks, F. P., Jr. (1975). *The Mythical Man-Month*. Addison Wesley Publishing Company.

Busby, J. S. (1998). The neglect of feedback in engineering design organisations. *Design Studies*, 19, 103–117.

Calvano, C. N., and John, P. (2004). Systems engineering in an age of complexity. *Systems Engineering*, 7(1), 25–34.

Carvalho, P. V. R., dos Santos, I. L., Gomes, J. O., Borges, M. R. S., and Guerlain, S. (2008). Human factors approach for evaluation and redesign of human–system interfaces of a nuclear power plant simulator. *Displays*, 29, 273–284.

Cedergren, A., and Theler, H. (2014). Studying risk governance using a design perspective. *Safety Science*, 68, 89–98.

Cho, S.-H., and Eppinger, S. D. (2005). A simulation-based process model for managing complex design projects. *IEEE Transactions on Engineering Management*, 52(3), 316–328.

Chong, Y. T., Chen, C.-H., and Leong, K. F. (2009). A heuristic-based approach to conceptual design. *Research in Engineering Design*, 20, 97–116.

Conant, R. C., and Ashby, W. R. (1970). Every good regulator of a system must be a model of that system. *International Journal of Systems Science*, 1, 89–97.

Cross, N., and Cross, A. C. (1998). Expertise in engineering design. *Research in Engineering Design*, 10, 141–149.

Drogoul, F., Kinnersly, S., Roelen, A., and Kirwan, B. (2007). Safety in design – can one industry learn from another?. *Safety Science*, 45, 129–153.

EPRI. (2008). Operating experience insights on common-cause failures in digital instrumentation and control systems. Report 1016731. Electric Power Research Institute, CA

EPRI. (2014). Principles and approaches for developing overall instrumentation and control architectures that support acceptance in multiple international regulatory environments. Report 3002002953. Electric Power Research Institute, CA

Fenton, N. E., and Neil, M. (1998). A strategy for improving safety related software engineering standards. *IEEE Transactions on Software Engineering*, 24(11), 1002–1013.

Floricel, S., and Miller, R. (2001). Strategizing for anticipated risks and turbulence in large-scale engineering projects. *International Journal of Project Management*, 19, 445–455.

Geipel, M. M. (2012). Modularity, dependence and change. *Advances in Complex Systems*, 15(6), 1–22.

Georges, A., and Romme, L. (2003). Making a Difference: Organization as Design. *Organization Science*, 14(5), 558–573.

Halang, W. A., and Zalewski, J. (2003). Programming languages for use in safety-related applications. *Annual Reviews in Control*, 27, 39–45.

Hale, A., Kirwan, B., and Kjellén, U. (2007). Safe by design: where are I now?. *Safety Science*, 45, 305–327.

Hamill, M., and Goseva-Popstojanova, K. (2015). Exploring fault types, detection activities, and failure severity in an evolving safety-critical software system. *Software Quality Journal*, 23, 229–265.

Hanes, L. F., Fink, R., and Naser, J. (2015). Human-automation function allocation. NPIC & HMIT 2015, Charlotte, NC.

Hansen, K. M., Ravn, A. P., and Stavridou, V. (1998). From safety analysis to software requirements. *IEEE Transactions on Software Engineering*, 24(7), 573–584.

Harvey, C., and Stanton, N. A. (2014). Safety in systems-of-systems: Ten challenges. *Safety Science*, 70, 358–366.

Hasan, R., Bernard, A., Ciccotelli, J., and Martin, P. (2003). Integrating safety into the design process: Elements and concepts relative to the working situation. *Safety Science*, 41, 155–179.

Hassija, V., Kumar, C. S., and Velusamy, K. (2014). Probabilistic safety assessment of multi-unit nuclear power plant sites – an integrated approach. *Journal of Loss Prevention in the Process Industries*, 32, 52–62.

Hayes, J. (2015). Taking responsibility for public safety: How engineers seek to minimise disaster incubation in design of hazardous facilities. *Safety Science*, 77, 48–56.

Heo, G., and Lee, S. K. (2007). Design evaluation of emergency core cooling systems using axiomatic design. *Nuclear Engineering and Design*, 237, 38–46.

Horváth, I. (2004). A treatise on order in engineering design research. *Research in Engineering Design*, 15, 155–181.

IAEA. (2002). Harmonization of the licensing process for digital instrumentation and control systems in nuclear power plants. TE-1327. International Atomic Energy Agency, Vienna.

IAEA. (2006). Fundamental safety principles. SF-1. International Atomic Energy Agency, Vienna.

IAEA. (2009). Protecting against common cause failures in digital I&C systems of nuclear power plants. NP-T-1.5. International Atomic Energy Agency, Vienna.

IAEA. (2011). Computer security at nuclear facilities. *Nuclear Security Series*, 17. International Atomic Energy Agency, Vienna.

IAEA. (2012), Safety of nuclear power plants: Design. SSR-2/1. International Atomic Energy Agency, Vienna.

IAEA. (2014a), Safety classification of structures, systems and components in nuclear power plants. SSG-30. International Atomic Energy Agency, Vienna.

IAEA. (2014b), Leadership and management for safety, draft general safety requirements.DS456. International Atomic Energy Agency, Vienna.

IAEA (2015), Technical challenges in the application and licensing of digital instrumentation and control systems in nuclear power plants, IAEA Nuclear Energy Series, No. NP-T-1.13, International Atomic Energy Agency, Vienna.

IAEA. (2016), Design of instrumentation and control systems for nuclear power plants, draft safety guide. DS431. International Atomic Energy Agency, Vienna.

IAEA. (2014c), Application of field programmable gate arrays in instrumentation and control systems of nuclear power plants. NP-T-3.17. International Atomic Energy Agency, Vienna.

Johnson, S. B. (1997). Three approaches to big technology: Operations research, systems engineering, and project management. *Technology and Culture*, 38, 891–919.

Jung, I. C. (2015). A vision for the safety and licensing of instrumentation and control for the future. NPIC & HMIT 2015, Charlotte NC.

Kannengiesser, U., and Gero, J. S. (2015). Is designing independent of domain? Comparing models of engineering, software and service design. *Research in Engineering Design*, 26, 253–275.

Kidam, K., Hussin, N. E., Hassan, O., Ahmad, A., Johari, A., and Hurme, M. (2014). Accident prevention approach throughout process design life cycle. *Process Safety and Environmental Protection*. http://dx.doi.org/10.1016/j.psep.2014.05.006

Kinnersley, S., and Roelen, A. (2007). The contribution of design to accidents. *Safety Science*, 45, 31–60.

Kirwan, B. (2007). Safety informing design. *Safety Science*, 45, 155–197.

Kjellén, U. (2007). Safety in the design of offshore platforms: Integrated safety versus safety as an add-on characteristics. *Safety Science*, 45, 107–127.

Kranakis, E. (2004), Fixing the blame: Organizational culture and the Quebec Bridge collapse. *Technology and Culture*, 45, 487–518.

Larman, C. (2005). *Applying UML and Patterns; An Introduction to Object-Oriented Analysis and Design and Iterative Development*. Prentice Hall.

Locatelli, G., Mancini, M., and Romano, E. (2014). Systems engineering to improve the governance in complex project environments. *International Journal of Project Management*, 32, 1395–1410

Lock, R. (2012). Developing a methodology to support the evolution of system of systems using risk analysis. *Systems Engineering*, 15(1), 62–73.

Macchi, L., Pietikäinen, E., Liinasuo, M., Savioja, P., Reiman, T., Wahlström, M., Kahlbom, U., and Rollenhagen, C. (2013). Safety culture in design. NKS-278.

Maimon, O., and Braha, D. (1996). On the complexity of the design synthesis problem. *IEEE Transactions on Systems, Man, and Cybernetics - Part A: Systems and Humans*, 26, 142–151.

Madni, A. M., and Sievers, M. (2014). System of systems integration: Key considerations and challenges. *Systems Engineering*, 17(3), 330–347.

Maxwell, J. C. (1867). On governors. *Proceedings Royal Society of London*, 16, 270–283.

Morton, W., Jr. (2015). Evaluating digital control systems in new reactors: Challenges and experiences. NPIC and HMIT 2015, Charlotte NC.

Motet, G. (2009). Risks of faults intrinsic to software languages: Trade-off between design performance and application safety. *Safety Science*, 47, 873–883.

Nachreiner, F., Nickel, P., and Meyer, I. (2006). Human factors in process control systems: The design of human–machine interfaces. *Safety Science*, 44, 5–26.

O'Hara, J., and Fleger, S. (2015). Integrated system validation: Models, methods, and issues. NPIC and HMIT 2015, Charlotte NC.

Patterson, S., and Clarkson, G. (2015). First of a kind engineering in digital I&C projects. NPIC and HMIT 2015, Charlotte NC.

Peffers, K., Tuunanen, T., Rothenberger, M. A., and Chatterjee, S. (2007). A design science research methodology for information systems research. *Journal of Management Information Systems*, 24(3), 45–77.

Pohjola, V. J. (2003). Fundamentals of safety conscious process design. *Safety Science*, 41, 181–218.

Raetzeke, C., and Micklinghoff, M. (2006). Existing nuclear power plants and new safety requirements—an international survey. *INIS*, 39(9)

Rasmussen, J. (1983). Skills, rules, and knowledge; signals, signs, and symbols, and other distinctions in human performance models. *IEEE Transactions On Systems, Man, and Cybernetics*, 13(3), 257–266.

Rasmussen, J. (1985). The role of hierarchical knowledge representation in decisionmaking and system management. *IEEE Transactions on Systems, Man, and Cybernetics*, 15(2), 234–243.

Ruuska, I., Aholaa, T., Artto, K., Locatellic, G., and Mancini, M. (2011). A new governance approach for multi-firm projects: Lessons from Olkiluoto 3 and Flamanville 3 nuclear power plant projects. *International Journal of Project Management*, 29, 647–660.

Saleh, J. H., Marais, K. B., and Favaró, F. M. (2014). System safety principles: A multidisciplinary engineering perspective. *Journal of Loss Prevention in the Process Industries*, 29, 283–294.

Saunders, F. C., Gale, A. W., and Sherry, A. H. (2015). Conceptualising uncertainty in safety-critical projects: A practitioner perspective. *International Journal of Project Management*, 33, 467–478.

Sheridan, T. B. (2001). Rumination on automation. *Annual Reviews in Control*, 25, 89–97.

Simon, H. A. (1996). *The Sciences of the Artificial* (3rd ed.). The MIT Press.

Sisbot, S. (2011). Execution and evaluation of complex industrial automation and control projects using the systems engineering approach. *Systems Engineering*, 14(2), 193–207.

Starr, C. (1969). Social benefit versus technological risk. *Science*, 165, 1232–1238.

Steward, D. V. (1981). The design structure system: A method for managing the design of complex systems. *IEEE Transactions on Engineering Management*, 28(3), 71–74.

Suh, N. P., and Do, S.-H. (2000). Axiomatic design of software systems. *Annals of the CIRP*, 49(1), 95–100.

Taylor, J. R. (2007a). Statistics of design errors in the process industries. *Safety Science*, 45, 61–73.

Taylor, J. R. (2007b). Understanding and combating design error in process plant design. *Safety Science*, 45, 75–105.

Torok, R., and Geddes, B. (2015). Managing common-cause failure issues in digital instrumentation and control systems. NPIC & HMIT 2015, Charlotte NC.

Tsai, B.-Y., Stobart, S., Parrington, N., and Thompson, B. (1997). Iterative design and testing within the software development life cycle. *Software Quality Journal*, 6, 295–309.

Viel, M., Salaun, P., Pierron, J.-Y., Gallois, J.-P., and Klein-Jaby, M. (2015). Design of I&C architectures for new nuclear power plants. NPIC and HMIT 2015, Charlotte NC.

Wahlström, B. (2007). Reflections on regulatory oversight of nuclear power plants. *International Journal of Nuclear Law*, 1(4), 344–377.

Wahlström, B. (2015a). Differences between analogue and digital I&C. NPIC and HMIT 2015, Charlotte NC.

Wahlström, B. (2015b). Safety principles in I&C design. NPIC and HMIT 2015, Charlotte NC.

Wahlström, B., and Duchac, A. (2015). The IAEA safety principles applied to NPP instrumentation and control. NPIC and HMIT 2015, Charlotte NC.

Wahlström, B., and Rollenhagen, C. (2014). Safety management – a multi-level control problem. *Safety Science*, 69, 3–17.

Wahlström, B., Heinonen, R., Ranta, J., and Haarla, J. (1985). The design process and the use of computerized tools in control room design. NKA/LIT(85)4. Nordic Liaison Committee for Atomic Energy, Stockholm, Sweden.

Wahlström, B., Juusela, A., Ollus, M., Närväinen, P., Lehmus, I., and Lönnqvist, P. (1983). A distributed control system and its application to a board mill. *Automatica*, 19, 1–14.

Wang, L., Shen, W., Xie, H., Neelamkavil, J., and Pardasani, A. (2002). Collaborative conceptual design - state of the art and future trends. *Computer Aided Design*, 34, 981–996.

Whyte, J., Stasis, A., and Lindkvist, C. (2015). Managing change in the delivery of complex projects: Configuration management, asset information and 'big data'. *International Journal of Project Management*. http://dx.doi.org/10.1016/j.ijproman.2015.02.006

de Winter, J. C. F., and Dodou, D. (2014). Why the Fitts list has persisted throughout the history of function allocation. *Cognition, Technology & Work*, 16, 1–11.

Woodward, D. G. (1997). Life cycle costing–theory, information acquisition and application. *International Journal of Project Management*, 15(6), 335–344.

Zalewski, J., Ehrenberger, W., Saglietti, F., Górski, J., and Kornecki, A. (2003). Safety of computer control systems: Challenges and results in software development. *Annual Reviews in Control*, 27, 23–37.

Zhang, L. (2013). Managing project changes: Case studies on stage iteration and functional interaction. *International Journal of Project Management*, 31, 958–970.

11

RISK COMMUNICATION

Jan M. Gutteling

11.1 INTRODUCTION

Risk communication is part of risk management (McComas, 2006). Risk is a central issue for policy areas as diverse as health, safety, technology, environment, and finance—important issues for the average individual in day-to-day life. An early risk communication definition is "*any purposeful exchange of scientific information between interested parties regarding health or environmental risks*" or "*the act of conveying or transmitting information between interested parties about levels of health or environmental risks; or decisions, actions, or policies aimed at managing or controlling such risks*" (Covello et al., 1988, p. 112). Based on these definitions, Covello (co)designed the "*Seven Cardinal Rules of Risk Communication*" guiding the risk communication policy of the US Environmental Protection Agency. The definition was perceived to focus on risk communication as "technical communication," which was identified as a problem (we will discuss why later in this chapter). The use of the word "scientific" does seem to imply a quality standard of the information in terms of its objectivity, reliability, and correctness. It is to be questioned, however, whether all risk stakeholders, the general public for one, have access to scientific

Handbook of Safety Principles, First Edition. Edited by Niklas Möller, Sven Ove Hansson, Jan-Erik Holmberg, and Carl Rollenhagen.

information or are able to use and interpret that information adequately, endangering the balanced exchange of information needed in the risk communication process. Later, Covello (1992) changed his definition to "*process of exchanging information among interested parties about the nature, magnitude, significance, or control of a risk*" (p. 359). Palenchar (2005) added to this the need for dialogue between communicators and stakeholders. Stakeholders in the context of risk communication are not a single group, but very diverse (Example 1 is an illustration of this). They could be members of the public, the media/press, governmental organizations, NGOs, or even workers in an organization when we are talking about occupational safety and health. When the communication is with workers in an organization, this is called internal communication.

11.1.1 Example 1

The Dutch summer of 2012 was characterized by finding asbestos in apartment buildings on July 22 and the immediate mandatory evacuation of 60 apartments in the city of Utrecht. A few days later, another 120 apartments in similar apartment buildings were evacuated voluntarily. The asbestos and the immediate mandatory evacuation led to a lot of concern among the inhabitants. A number of them were concerned about the health risks they might have been running for many years. Some inhabitants refused to return home even when measurements indicated that no asbestos was present in their apartment. On a daily basis, the municipality provided extensive information about the developments and on possible health risks; however, the concern remained. The press provided additional information on asbestos finds and evacuations in other cities in the Netherlands. Later that year, an independent committee reported that the large-scale evacuation of 180 apartments was unnecessary from the perspective of public health. The mayor of the city of Utrecht says in a reaction to this committee report that the municipality was acting out of a precautionary principle, because public health of the inhabitants was its most important concern. The National Ombudsman, however, concludes that the (mandatory) evacuations created uncertainty among those involved (Nationale Ombudsman, 2012). Furthermore, he concludes that authorities, when risks are involved, often take rather paternalistic decisions, more in particular when the authorities will be held accountable politically. Their—unsupported—assumption would be that citizens expect them to take those decisions. Another widely used argument by the (Dutch) authorities is that they do not want to invoke unduly worry or even panic among the public, which is supposed to happen when the public would know all the details. The Ombudsman calls this "the myth of panic," and says that the vast majority of people will not panic, but take relatively reasonable risk-related decisions.

11.1.2 Risk Perception, Awareness, and Communication

Risk management comprises the whole process of risk identification, estimation, evaluation, reduction, and control (Petts, 1992). According to Vlek and Cvetkovich

(1989), adequate risk management is a matter of organizing and maintaining a sufficient degree of (dynamic) control over a technological activity, rather than of continually, or just once, measuring accident probabilities and distributing the message that the risks are, and will be, negligibly low (see for more information on risk management, e.g., Chapter 7 on experience feedback in this book). Risk communication is seen as one of the several policy instruments to achieve risk management goals, even with various target groups and a different scope of the communication for each target group.

Risk communication as a field of interest is closely related to risk perception or awareness. Risk perception or risk awareness is usually described as an individual's subjective assessment of certain characteristics of a risk such as the severity of the risk in terms of negative consequences, the probability of occurrence of these consequences, and the individual's personal vulnerability (e.g., Slovic, 2000). Example 1 describes risk perception and perceived vulnerability issues in the asbestos case. Since the 1970s, psychological, sociological, and interdisciplinary risk perception studies focused on a broad domain of risk issues, among which are the application of nuclear energy to generate electricity; the production, use, and transport of chemical substances; human nutrition; health; innovative technological developments; and natural phenomena like extreme weather, climate change, or flooding.

Many early risk perception studies were motivated by the big gap between expert and non-expert risk assessment (see, e.g., Gurabardhi et al., 2004; Gutteling and Kuttschreuter, 2002; Slovic et al., 1982). In the early years of risk communication, it was observed (Plough and Krimsky, 1987, p. 4) that "*The emergence of risk communication as a research theme cannot be fully appreciated or accounted for without understanding its link to a set of issues that symbolize the discord between scientific experts and the public around the issue of risk. These tensions are played out in disputes between different research traditions on fundamental questions regarding the perception of risk and the essential nature of human rationality.*" It seems unlikely that everybody benefits equally from the stream of risk information developed over the last decades. Partly, this inequality coincides with the opinions opposing parties have of each other. Governmental officials, private sector officials, and some scientists seem to take for a fact that the general public is unable to understand technical risk information, and particularly the news media would contribute to a biased and inaccurate public risk perception or even panic (Mazur, 1984; Peltu, 1985). On the other hand, citizens and interest groups have been complaining that the government and the private sector are acting very passively in taking the right decisions to solve apparently simple problems (Covello et al., 1987). These situations are hindering a sound public risk discussion. It may result in experts being distrusted and their information ignored, or it can decrease the experts' motivation to provide risk information comprehensible to the public.

Due to all these issues, early studies in risk perception were aimed at finding explanations for individual differences in risk perception and the determinants of those differences. Among the determinants that were studied were knowledge of the subject matter and personal control over the risk (Slovic et al., 1982), individual information processing in terms of the use of heuristics (see, e.g., Tversky and Kahneman, 1974), social elements like trust in others as adequate risk manager (e.g., government, experts, or companies, see e.g., Renn and Levine, 1991), the availability of information in the public domain (either the journalistic media or the social media; see, e.g., Kuttschreuter et al., 2011; Terpstra et al., 2012), and cultural determinants (as in the Social Amplification model, see, e.g., Kasperson and Kasperson, 2005). More recently, the influence of affect (emotional reactions) on risk perception and the processing of risk-related information is studied (see, e.g., Visschers, Wiedemann et al., 2011).

11.1.3 This Chapter

Over the years, risk communication, more specifically the communication process in a controversial context or in a crisis, seems to have gained importance. In this chapter, the complexity of risk communication is discussed, both from the perspective of an academic discipline as from a more practical perspective. Several risk communication models are presented, illustrated with empirical studies in this area. New focus points and approaches are described. The focus of the chapter will be primarily on the safety issues, but some basic understanding of more general notions will be provided when needed.

11.2 THE ORIGIN AND HISTORY OF RISK COMMUNICATION AS ACADEMIC FIELD

In this paragraph, we describe the different perspectives on risk communication (in the safety domain). We will briefly describe early developments and focus on the changes in perspective over the years.

Risk communication has for long been based on top-down approaches. Governmental organizations (at local, regional, national levels) were providing information (sometimes even proactively) about risks, provided risk information during the aftermath of events, or were reacting to public concerns about situations (see Example 1, and Scolobig et al., 2015). These organizations had a substantial benefit due to their information position, which sometimes was described as a hegemony. This pattern has been observed in disaster risk management (Kievik and Gutteling, 2011; Scolobig et al., 2015); technology acceptance (e.g., Horst et al., 2007), workplace safety (e.g., Boyle, 2012), and many other domains. These organizations were fed with technical capacities and expertise, and with a centralized and hierarchical model of management, and were at a large "psychological or communicative" distance from the "target group." Depending on the urgency of the issues at hand, risk management and communication activities are nowadays being scrutinized by media and social media or those responsible are held accountable in the public domain (see Example 2).

11.2.1 Example 2

On Wednesday August 12, 2015, a series of very heavy explosions occurred in the Port of Tianjin (China), causing a massive loss of life (among which were many young fire fighters) and unprecedented material damage. Over one hundred people were killed and hundreds of others were injured (see Wikipedia, 2015). Fires caused by the initial explosions continued to burn uncontrolled throughout the weekend, repeatedly causing secondary explosions. After a few days, reports came out about the chemicals that were being stored at the site. An official spokesperson mentioned over 40 kinds of hazardous chemicals (approx. 3000 tonnes) among which were ammonium nitrate and potassium nitrate, which are known to have played a role in previous industrial catastrophes. A fire department spokesman confirmed that firefighters had used water in combating the initial fire, which may have led to additional detonations. The surrounding area was severely polluted, and several thousands of inhabitants of the affected area were mandatorily evacuated.

In terms of risk communication, the following is worth mentioning. Tianjin authorities banned editors and reporters from sharing information about the disaster on Weibo (Twitter) and WeChat, and websites were ordered to follow state media to use only official casualty numbers. The local television reported the initial explosions but was blamed for not reporting live, nor updating on the event. A great deal of specific information on the event, including the majority of early stage video was first released over social media sites. Major (international) media had drawn heavily from social media sources, greatly widening the audience. The Economist (cited on Wikipedia) noted, "*Social media fills in the blanks left by official narratives of the Tianjin disaster. The most remarkable feature of the aftermath of the explosions in Tianjin, in northern China, has been the extraordinary contrast between the official reaction to the crisis, which has been profoundly flawed, and the online reaction, which has entirely dominated the agenda.*" To deal with the demand for information, authorities started organizing press conferences, and (international) media showed images reflecting how the local population dealt with their hardship and their emotions of anger, fear, and despair.

11.2.2 Changing Notions about Communication

Many early risk perception studies were motivated to find a solution for the discrepancy in perception between experts and non-experts. The idea was, strictly top-down, to learn to synchronize public perceptions with expert opinions, with higher levels of technology or risk acceptance as primary spin off (Gurabardhi et al., 2004). However, studies indicated that the effectiveness of these top-down approaches was relatively low (with effectiveness being defined as meeting some preset goal in terms of attitude change, behavior change, increase in knowledge, etc.) (see, e.g., Freudenburg and Rursch, 1994; Johnson and Luken, 1987; Visschers et al., 2012).

Due to the lack of empirical success, scholars and practitioners have been studying new ways to get the job done. Rowan (1994) expanded the scope of this discussion by distinguishing between the *technical view* and the *democratic view* on risk communication. The technical view comprises a one-way, expert-to-lay public information flow, based on the premise that "*if people are given the facts, their subjective perceptions will begin to align with scientific judgments.*" From this perspective, the ideal risk communication applies the Factual Information Model in which the discrepancy between actual and subjectively perceived risk is assumed to be reduced by presenting technical, quantitative, or statistical (numerical) information (Cvetkovich et al., 1989). However, quite a few studies have shown major issues with this approach (Freudenburg and Rursch, 1994; Gurabardhi et al., 2004; Roth et al., 1990).

The democratic model of risk communication is governed by rules that guarantee a just and fair process, in which all parties have maximum participation and decision-making power. In the democratic view, persuasion is out of the question, particularly for governmental or industrial bodies, because the aim of communication should be mutual understanding and not the exertion of power. This is supposed to increase public trust, enable decision-making supported by all stakeholders, and strengthen democratic processes. This notion has led to a considerable body of empirical work focusing on the role of trust in risk communication (more on trust later in this chapter) in many studies related to the acceptance of technologies with uncertainties regarding risks and benefits.

As a consequence of these lines of thinking, more "people-centered" approaches were developed, which include, among others, increased stakeholder participation (individuals and also their representatives, and NGOs), responsibility shifts from the authorities to the public (increasing resilience), greater transparency in risk/uncertainty communication, and social/institutional capacity building (Scolobig et al., 2015). Resilience is described as: "*The ability of a system, community or society exposed to hazards to resist, absorb, accommodate to and recover from the effects of a hazard in a timely and efficient manner, including through the preservation and restoration of its essential basic structures and functions*" (UNISDR, 2009). Resilience depends on the availability of essential resources and the level of organization before and during an emergency. The relatively new focus on resilient citizens is part of the shift of responsibilities from governments to citizens. When citizens will prepare themselves, they will be able to respond better and faster to an emergency situation. Risk communication is seen as one of the tools to increase the resilience level of citizens (for more information on resilience, see also Chapter 9 on Human factors engineering, and Chapter 30 on Emergencies and Disasters).

The people-oriented approach is applied in several domains; not only with the traditional major risk controversies as nuclear energy (see USNRC, n.d.), but also with risk communication in the area of natural hazards (see, e.g., Basher, 2006), or the rail transport of hazardous substances through highly populated areas (Kievik et al., 2015).

In research, we have seen a shift in attention to information-seeking approaches, with the focus on end-users of risk information being the ones that take the initiative to search for that information, as opposed of being relatively passive consumers of that information when (or if) it is delivered by the authorities. Also, the arrival of social media in the area of risk communication helped to make a final end to the hegemony and sole information control of governmental institutions (Examples 2 and 3 illustrate this phenomenon). We come back to social media developments later in this chapter. Of course, the major, but as yet unanswered question is whether this approach will solve the risk communication issues that previous approaches did not?

11.2.3 Example 3

On Monday August 3, 2015 an industrial incident happened in the small Dutch city of Alphen a/d Rijn. Two enormous cranes were lifting a heavy part of a new bridge as part of its reconstruction, and the spectacular event had drawn many interested people to watch. The cranes were on two floating vessels, moved to quickly, lost their balance and collapsed with their load on the houses, shops, and restaurants on the bank on one side of the river. Immediately after the event, the social media were loaded with numerous pictures and videos of the event, as well as lay estimates of the number of casualties. At first, it was estimated that at least 20 people were expected to have died or got wounded, and the national media reported this number. Later that evening, it became clear that the only casualty was a dog (which was sad enough, but called a miracle looking at the devastation). More than 20 buildings were completely destroyed. That same night the national television interviewed an expert who indicated that the crane operators most certainly were to blame because of the mistakes they made. Journalists immediately started questioning local authorities about their risk management decisions even before the official investigation had been completed.

11.2.4 Conclusion

We conclude that risk communication takes different forms, has different goals, and various perspectives depending on its function: either as a mechanism to influence people's perceptions in value-laden discussions (e.g., the acceptance of Nanotech or Gene tech), or as a mechanism to proactively or actively inform people to be helpful to make safety decisions (either in society as a whole or as part of the safety culture in the workplace). There is a shift from top-down to bottom-up approaches in both practice and research.

11.3 UNDERLYING ASSUMPTIONS, CONCEPTS AND EMPIRICAL DATA ON RISK COMMUNICATION MODELS

11.3.1 Information versus Communication

Everything an individual receives through his or her senses is information. The receivers of information are free, considering their needs, knowledge, and experiences, to act or not to act on information as pleased. The term communication is used when the information has a symbolic value; it refers to something with some relevance and it has a particular meaning to the individual. The presence of a large petrochemical complex can actually diffuse different types of information, for example, the view or the smell. But the sight of such a complex also has a communicative meaning; for some people, it may refer to a steady job or economic prosperity, for others, it may remind them of unfortunate catastrophic events from the past. The concept of meaning is two-sided. On the one hand, the symbolic meaning is implemented (or "framed") by the sender of the information, but the receiver of the information also attributes an own meaning to the information, which could be quite different from the sender's intention. This implies communication can have both intended and unintended effects on receivers.

A popular distinction between communicative interventions is in (public) education, advertising, propaganda, and public relations. An important difference between these types of communication is in the pursued goal. Education is focused on improving general skills to make decisions. Advertising is aimed at the stimulation of sales, propaganda is aimed at disseminating cultural or political ideas, and public relations' goal is to promote the reputation of an organization. Generally, this is true for risk communication too. It is applied in advertising, propaganda, public relations, and public education.

Communication scientists have produced a large number of conceptual models that are needed to understand communication processes and to be able to optimize the effectiveness of the communication. Suppose a campaign is started to motivate people to take preparations for a natural hazard, for example, wildfires. People are best served with communication that will help them to make the right decisions, which is, as we will see in this chapter, not an arbitrary process.

We have already seen that communication involves the attribution of meaning. Communication usually is a dynamic process, in which persons successively act as source and receiver, so attention should be paid to the role of feedback and interaction. This adds a social aspect to the process. Other aspects that will be relevant for the attribution of meaning and thus for the effectiveness of the communication are existing knowledge and prior experiences, belief and value systems, emotions, and the opinions of significant others.

Social psychologists have been particularly interested in selection processes by receivers. Well known are the studies into the tendency to seek out information supporting one's decisions (this process is known as selective exposure), and the tendency to avoid information questioning one's prior decisions (selective avoidance). Both processes seem relevant for risk communication. In summing up, the empirical evidence on the selective exposure or avoidance hypotheses, Frey (1986) concluded:

> *"people are relatively open to new information (supporting and opposing their beliefs, decisions, or hypothesis) as long as they see a possibility for revision. When for varying reasons (e.g. external constraints, internal resistance to change), this possibility is*

not seen to exist, subjects tend to ignore non-supportive information and increase their preferences for information that lends them support"

(pp. 73–74).

11.3.2 Risk Communication Aims

Risk communication can have many different aims. Covello et al. (1987) made an early attempt to distinguish four areas in which risk communication is applied, namely:

- Informing and education.
- Stimulating behavioral change and taking protective measures.
- Issuing of disaster warnings and emergency information.
- Exchange of information and a common approach to risk issues.

With their first area of application, informing, and education, Covello et al. (1987) aim at providing risk information to the general public or parts of the public. In their view, this is a nondirective, but intentional activity in which the lay public receives explanatory information. The second area of application, stimulating behavioral change and taking protective measures, is used when studies indicate that particular behaviors, activities, or situations are hazardous to people's health and authorities want to advise the public to mitigate the risk (e.g., wear seat belts in cars, or do not drink and drive). The third area of application, the issuing of disaster warnings and emergency information, is aimed at emergencies due to man-made or natural disasters, such as explosions (see Example 2), wildfires, floods, hurricanes, and perhaps earthquakes. In Covello et al.'s view, the second and third areas are to be understood as directive communicative activities, motivating the public to adopt certain behavior. All three areas of application mentioned before are explicitly different from the fourth, which aims at an exchange of information and a common approach to risk issues by citizens, government, and other relevant actors.

From a communication point-of-view, several issues with Covello et al.'s approach can be identified. The first is the lack of specificity about the effects or consequences the risk communication could or should have. To establish the effects of risk communication or to improve risk communication efforts, communicators should be able to identify target groups and intended effects. Without that, communicators will never be able to evaluate the effectiveness of their work adequately, and will never be able to systematically improve their efforts. Evaluation procedures help to decide whether the observed effects of risk communication are in agreement with its pre-formulated goals (as in, e.g., the diagnostic approach a little bit further in this text). Second, it is also possible that risk communication has effects which are contrary to its goals; these unintended consequences of communication are referred to as side effects. In the worst case, these side effects are detrimental to the communicator's goals (like we have seen in Example 1, where local authorities wanted to avoid asbestos panic, but created it). Third, without some conception of what works and why, with whom, and under what circumstances, we are basically running a wheel of fortune. Sometimes, we get lucky and the communication works, but more often (much more often) we fail and we do not understand why we fail. That is why we need models and ideas.

In the rest of this chapter, we will introduce some models and ideas about risk communication. We will describe the mental models approach, the social amplification of risk framework (SARF), notions regarding trust in risk communication, socio-cognitive models, and information seeking models. We end this part with the role of the social media. Necessarily we will have to make choices here, which means that this overview is not complete. Models like CERC (Seeger et al., 2010), the risk attitude approach (Rimal and Real, 2003), and others are not described.

11.3.3 Diagnostic Risk Communication Studies

Several authors advocated a diagnostic or problem-solving approach to risk communication research. Ideally, such an approach would:

- Identify communication goals.
- Determine principal obstacles to those goals.
- Select research-based methods for overcoming or minimizing these difficulties and to achieve communication objectives.

A well-known diagnostic risk communication model is the four-step mental models approach (Morgan et al., 1992). This approach focuses on transferring cognitive elements (knowledge) as the major tenet of this approach is that risk information should be an addition to the receiver's mental model and should challenge misconceptions. The individual is in need of a basic level of knowledge about exposure, effects, and mitigation processes relevant to making informed decisions about the hazardous process. Furthermore, it is assumed that existing beliefs are used to interpret and process new information. To be effective, this should be presented adequately, with an appropriate text structure and reinforced with textual aids such as section headings. The mental model is also a persuasive tool aimed at developing effective risk messages. The four steps distinguished by Morgan et al. include:

- Open-ended elicitation of people's beliefs about a hazard, allowing expression of both accurate and inaccurate concepts.
- Structured questionnaires designed to determine the prevalence of these beliefs.

- Development of communications based on what people need to know to make informed decisions (as determined by decision analysis) and a psychological assessment of their current beliefs.
- Iterative testing of successive versions of those communications using open-ended, closed-form, and problem-solving instruments, administered before, during, and after the receipt of messages.

Studies revealed that risk communication designed according to the mental models approach did better on a number of indicators than risk communication that was not. The indicators were: fewer misconceptions, additional knowledge about the hazard, and better problem-solving skills (see, e.g., Atman et al., 1994; Bostrom et al., 1992; MacGregor et al., 1994; Read et al., 1994).

11.3.4 Social Amplification of Risk

The particular relevance of communication processes for risk management is underlined in the social amplification of risk framework (SARF; Kasperson et al., 1988; Pidgeon et al., 2003), which integrates technical assessment and the social characteristics of risk perception. SARF focuses on the interaction of hazard-driven events with psychological, social, institutional, and cultural processes, resulting in the amplification or attenuation of individual and social perceptions of risk, and the shaping of risk behavior (Kasperson et al., 1988). Behavioral patterns, in turn, generate secondary social or economic consequences extending far beyond direct harm to humans or to the environment, including significant indirect impacts such as liability, insurance costs, loss of trust in institutions, or alienation from community affairs. These secondary consequences may, according to SARF, develop into long-term anti-technology attitudes, alienation of the physical surrounding, apathy, or a loss of trust in risk managers. Examples of economic consequences are decreases of sales, property values, and tourism. Other possible secondary consequences are political unrest, changes in the nature of technology due to the risks, and an expansion of the risk controversy to other technologies or industrialization as a whole. SARF assumes that the process starts with either a physical event or the recognition of an adverse effect, which is communicated. Individuals, groups, and institutions collect and respond to this risk information, and act as amplification stations through behavioral responses or communication. Amplification differs among individuals in their roles as private citizens, and in their roles as employees or members of social groups and public institutions. Individuals not only serve as amplification stations according to their own personal values and risk perceptions, but are also part of social organizations which can serve as amplification stations too, according to the rules and values of that specific social group.

SARF's importance lies in the signaled relevancy of risk communication and the identification of possible secondary consequences, which presents a wider scope on the risk issue than previous models provided. At face value, the model offers a credible description of the way industrialized societies seem to respond to the risk issue. The major concern with the model is its complex nature and the difficulties this brings when we are searching for empirical support of the model. Probably for that reason, the model has been empirically tested in only some studies. For several of the model's elementary assumptions, empirical studies were conducted though (Burns et al., 1993; Renn et al., 1992; and more recently Mase et al., 2015).

11.3.5 Trust in Risk Communication

In many situations, we expect that the risk information receiver does not possess the knowledge or skills to assess its validity. In those circumstances, the perceived trustworthiness of the information is of eminent importance, and the way to get an idea about that is to assess (subjectively) the credibility of the person or organization providing the information (see, e.g., Kenrick et al., 2005). In general terms, we use a subjective perception of the source's expertise, honesty, and biases. However, not only the receiver assesses this, but also their social environment does, and the context in which the assessment is made (e.g., normal situation vs. a risk controversy) is important (see, e.g., Earle, 2010; Earle and Siegrist, 2006; Flynn et al., 1992). In risk literature, these issues are covered under the heading of "trust." Renn and Levine stated: "*trust in communication refers to the generalized expectancy that a message received is true and reliable and that the communicator demonstrates competence and honesty by conveying accurate, objective, and complete information*" (1991, p. 179). In Renn and Levine's definition, trust seems to incorporate an assessment of message content (it must be true and reliable) as well as a subjective assessment of the source of the information.

Several studies have found relations between trust and risk communication. Inevitably, to a certain degree, risk communication will convey technical or otherwise complex information. Therefore, one might assume source expertise to be an important determinant of the perceived credibility of risk communication sources. Traditionally, expertise is seen as an important credit to the information source. The assessment of source expertise seems to depend mainly on their level of formal education, intelligence, social status, familiarity with the issue, and professional abilities (McGuire, 1985). In controversial risk communication situations, the general rule that expert sources are more effective than non-expert sources might be wrong, in particular when the source's trustworthiness is being questioned, for example, when the receiver is convinced that the expert source is acting out of self-interest (McGuire, 1985). An expert source acting out of self-interest probably causes more reactance than a source without self-interest (see, e.g., Schwarz et al., 1986). This last phenomenon is also known as the boomerang effect (see, e.g., Snyder and Wicklund, 1976). The public's suspicion of self-interest is one of the source-related risk communication problems private-sector or governmental sources have to deal with. The important lesson to be learned here is that sources should be very careful in maintaining and improving a trusting relationship with their audience. Sources should rethink every action they plan with the possible consequences for this relationship in mind.

11.3.6 Socio-Cognitive Models

Socio-cognitive models like protection motivation theory (PMT) and extended parallel process model (EPPM) offer a completely different perspective on risk communication compared to the previously described approaches. Both find their roots in health communication (see, e.g., Gore and Bracken, 2005; Rogers, 1983; Witte and Allen, 2000). Both models have been studied extensively and meta-analyses of empirical studies (in the health domain) have found supporting evidence (Floyd et al., 2000; Witte and Allen, 2000). PMT has inspired researchers in disaster studies to develop Protective Action Decision Model (PADM; see, e.g., Lindell and Perry, 1992, 2012; Terpstra and Lindell, 2012). This line of research is characterized by the use of persuasive communication, which has more resemblance to technocratic approaches than the democratic approaches of risk communication.

Central notion in these theories is the concept of threat appraisal (or risk perception as it is called in the risk domain). A certain (increased) level of threat appraisal is essential to motivate the population "at risk" to take preventive measures to increase their resilience. This level may be reached by direct stimulation of ideas and notions (through the delivery of an advice how to act) or by stimulating the active searching of (risk or prevention) information. Acceptance of the advice is based on the individual's threat appraisal and coping appraisal. An irrelevant threat will be ignored, a relevant threat will successively be appraised for potential individual coping (on indicators as perceived *self-efficacy* "I believe I can cope adequately with the risk" and *response efficacy* "executing the advice will actually reduce or minimize the risk"). When coping is seen as potentially successful, the individual will adopt the risk-mitigating activity, otherwise fear control may lead to denial of the fear ("it is not going to happen to me"). Be aware that this last outcome resembles selective avoidance that was mentioned before.

From this perspective, several studies in the domain of risk and safety have been reported. Grothmann and Reusswig (2006) studied citizen preparation for flooding in Germany, as did Kievik and Gutteling (2011) and Zaalberg et al. (2009) in the Netherlands. Martin et al. (2007) applied these notions in the domain of uncontrollable wild fires in the United States, and Johnston and Warkentin (2010) focused on cyber security. The studies by Kievik and Gutteling (2011) and Zaalberg et al. (2009) indicate that coping appraisal—the combination of self-efficacy and outcome efficacy beliefs—is a solid predictor of individual preventive behavior.

11.3.7 Risk Information Seeking Models

Information seeking models are relatively new to the risk communication domain. Scholars have identified individual risk information seeking behavior as an important predictor of risk acceptance or risk-mitigation actions in times of crisis. These models may have an influence on the previous discussion of technocratic versus democratic types of risk communication. When stakeholders (e.g., citizens) inform themselves by finding and processing risk information (e.g., on the web) or share that information with others (e.g., through social media), information providers will have to make sure that their information is timely available otherwise it will not be considered for decision making. And "timely" in this context means when the individual search process starts. Technological developments in the last decade have enormously increased the individual risk information seeking potential; these include the 24/7 availability of risk information on the Internet, interactive tools of the web 2.0 and the social media, and increased mobility (smart phones, tablets), to name a few. This approach is based on the notion that risk communication is actually a bottom-up approach, with the initiative being taken by the information-seeking individual and not by the provider. Nevertheless, the provider's role is extremely important.

Motives for individual risk information seeking are acquiring knowledge, reducing risk or uncertainty, or to share or compare information with important others (ter Huurne and Gutteling, 2008). The information search process model assumes that individuals develop the intention to seek information in case of a perceived shortage in knowledge or understanding. This reasoning is also assumed in the risk information seeking and processing model (RISP; Griffin et al., 1999; Yang et al., 2014) that identifies the gap between present knowledge and needed knowledge as the psychological motive "information insufficiency." RISP has been studied several times, and has recently been meta-analyzed (see Yang et al., 2014). The underlying notion in RISP is that the individual needs a level of certainty on relevant issues. When uncertainty increases, a need for information develops and information-seeking behavior will occur. Uncertainty and trust in other people and the information they provide on risk seem very much interrelated. The individual's social network may influence the need felt for information and information seeking. When important others are assumed to have specific knowledge on important issues, the individual is stimulated to seek information to be able to compare his or her skills and knowledge with that of others. The reason: to acquire social support, approval, and respect which seems relevant in all sorts of situations—labor, family, friends, and dealing with risk (Griffin et al., 1999). Kahlor et al. (2006) confirm that perceived peer pressure to be well informed is a powerful predictor of information seeking. Also, affective reactions (emotions) may be a stimulant to information (Griffin et al., 2008).

The framework for risk information seeking (FRIS; ter Huurne, 2008) assumes that three cognitive factors (awareness factors) are predictive of information need in a risk setting. These factors are risk perception ("does a threat exist?"), personal involvement ("is the existing threat relevant to me?"), and self-efficacy ("am I capable of dealing with the threat adequately?"). Adequate behavior might be seeking additional information to take most effective measures to increase personal resilience. The combination of these awareness factors influences affective reactions, information sufficiency, and subjective norms that directly impact information-seeking behavior (see, e.g., Kievik and Gutteling, 2011; Kievik et al., 2012).

11.3.8 Risk Communication and Social Media

With ongoing developments in the information and communication technology, social media gained importance as a news/information source for large groups of the public (Palen and Liu, 2007). The social media are relevant not only for our understanding of individual information-seeking processes, but also for socio-cognitive concepts, as we will see further on. Using social media has several advantages for members of the public. First, social media are an easy accessible way to stay informed not only in normal times, but also in emergency situations (Palen et al., 2009; Terpstra et al., 2012; Vieweg et al., 2010; Yates and Paquette, 2011). Second, social media allow citizens to obtain the feedback and opinions of other users who are independent of organizations involved in the crisis, are presumed to be judged as peers and more similar to the average citizen, and hence may have more impact than the involved but distant official organizations. Such quick responses from peers with eyes-on-the-ground may be judged as very valid by those members of the public who need to make a decision on whether and how to engage in self-protective actions (Shklovski et al., 2008). Moreover, some media (e.g., Twitter) allow users to obtain information about the actual circumstances of the information source in relation to one's own location, thereby enabling better assessment of information relevance (Vieweg et al., 2010). Theoretically, these characteristics fit very well in the people-oriented focus on risk communication. Finally, the information is often distributed very quickly, allowing almost real-time information provision about (changes in) the current situation (Palen et al., 2009; Terpstra et al., 2012).

An interesting question related to the use of social media as a source of risk information is what impact they may have on the likelihood of the public to engage in self-protective actions. It is likely that these social media messages contain some peer feedback on the situation or on the advice provided through the official channels. Verroen et al. (2013) studied what the impact of this peer feedback (either supportive or opposing) might be on the willingness of recipients to engage in self-protective behavior. The results indicate that peer feedback can be important in helping to make decisions on preventive measures. When subjects were confronted with an official message that was appreciated as highly efficacious, the type of peer feedback did not have differential impact on the subjects' decisions. However, when the original advice was assessed low on efficacy, those who received supportive peer feedback were significantly more inclined to take preventive measures than those that received opposing feedback. In the current reality where social media are important sources for many people, authorities have to make sure that their advice is being regarded as adequate in terms of self-protective behavior. Because receivers with low self confidence in solving the problem may be very much influenced by opposing (or negative) feedback on the provided advice, and may eventually do nothing to reduce the risk.

11.3.9 Conclusion

In this chapter, we described several developments in risk communication, which all have their merits and which all have been successfully empirically studied. The presented theoretical notions do not provide a simple solution in the ongoing debate between the older top-down or technocratic approach versus the more bottom-up or democratic approaches. The mental models approach and the new studies on information seeking seem to support the bottom-up approach, while the socio-cognitive theories, focusing on persuasion, appear to be more in line with the top-down approach. Social media can provide many benefits for users, and theoretically they may be linked to information-seeking theories and socio-cognitive theories.

11.4 WEAKNESSES, LIMITATIONS, AND CRITICISM

In spite of several decades of theory developing, empirical (fact-finding) research, and practical experience with risk communication, the final word is still not said, and there are no easy recipes for any risk issue in which communication with stakeholders is seen as essential.

An important issue regarding risk communication is its complexity due to the specific risk domain that is being addressed and the type of risk information that is being delivered. In many early studies, the focus lay on presenting risk-related information in formats that were familiar to risk assessors, mainly quantitative information delivered in numbers, percentages, or whatever. In terms of communication effectiveness, this (to the non-expert) rather dull information appeared to be a "hot" issue. Put simply, it did not bring the expected clarity to the receivers of that information, and thus no solution to the health and safety problems they were experiencing.

Later the discussion evolved in the direction of risk uncertainty, which for many people not professionally active in the risk world, may be difficult to understand, and perhaps difficult to accept. In risk management, a distinction is made between absolute risk, that is the level of risk that exists without any risk control measure being operational, and residual risk. Risk control measures and the extent in which they are applied reduce the absolute risk level. The result is called residual risk, which is the risk level with all risk control measures operational. This last might be difficult to understand for non-experts because, in spite of risk control measures being applied, some risk still remains. Suppose we build a nuclear power plant along the coast in an area prone to earthquakes and the consequent tsunamis and during the construction, we apply risk control measures in such a way that the power plant can withstand earthquakes with a certain magnitude that will result in tsunami waves of 6 meters. The residual risk has the (relatively) small probability of having an earthquake with a higher magnitude, which may lead to higher tsunami waves that in turn might result in substantial damage to the nuclear plant. Although the residual risk is small (and earthquakes that will lead to tsunami waves higher than 6 meters may perhaps never have been observed in the area and therefore are deemed very unlikely), it is not zero. The fundamental issue here is to assess what is acceptable or tolerable as risk for all stakeholders (not only for experts or governmental decision-makers), and risk communicators have a difficult task here.

Of course, the exemplary theoretical and practical developments described in this chapter are not chosen at random. They represent the lines of thought that feed the discussion on risk communication for several decades now, about the fundamental choice for either a bottom-up or a top-down approach to risk communication. Perhaps these latest developments make it clear that an either or choice is not very practical, because studies indicate that both bottom-up and top-down types of communication have their merits, and might be effective in specific, unrelated types of situations or issues. The bottom-up approach seems very suitable for types of risk communication in which urgency is not at stake, but citizens want to be informed about risks, or want to inform themselves to be able to fulfill their information or other needs. The top-down persuasive form of risk communication is perhaps more suited for early warning and emergency communication. Of course, it is also clear that in these circumstances, the influence of citizens is becoming more and more important (e.g., because of their use of social media). And citizen input can surely help to improve the quality of emergency and preparative risk communication (see, e.g., Kievik et al., 2015; Scolobig et al., 2015).

Empirical work on risk communication is increasing in numbers over the years, as can be seen from a simple search in online databases like Scopus or Web of Science. A previous analysis (Gurabardhi et al., 2004) revealed that until the year 2000, risk communication was dominated by scholars (e.g., Paul Slovic, Baruch Fischhoff, and Lynn Frewer) from the Western world (United States, United Kingdom) and one very influential journal (Risk Analysis). In the present day, the same authors are very productive and cited frequently, and still rank highly on the list of the most productive individuals, and Risk Analysis is the most popular outlet. But many other scholars and new outlets joined the field, identifying risk communication as an area of increasing academic and practical importance. The risk issues that this empirical work is focusing on is also developing in line with developments in society, so there will be a continuous growth in domains of interest. In the early years, man-made hazards were important focus points; more recently, food and nutrition, cyber-crime, and international terrorism are topics of academic risk communication interest. We might expect climate change, critical infrastructure, tourism, and other issues as new fields of interest.

In the previously mentioned overview study (Gurabardhi et al., 2004), it was found that many published articles did not contain empirical work, but narratives. Fortunately, in the academic literature, one can observe now that scholars increasingly are working on testing hypotheses and adding to the overall notion of "what works, why and when." This will probably go on well into the next decade. As an example, the recent interest in risk information-seeking processes has led to the genesis of several hypothetical models and ongoing empirical work of which we described some in this chapter (e.g., risk information seeking and processing model; framework for risk information seeking; but there are others as well). Another recent development is focusing on the application of concepts like efficacy beliefs as precursors of individual and social resilience in the domains of natural hazards (e.g., flooding, wildland fires, and extreme weather), food-related issues, and risk concerns related to cyberspace and overall safety. This approach entails informing the public about how to avoid risks and providing functional information for increasing the level of resilience. This communication can, for instance, increase the risk awareness of the public in situations where an overoptimistic bias exists, or where the information provides clear pre-emergency instructions how to behave adequately and resiliently in disastrous situations. Next to studies on information seeking, efficacy beliefs, and resilience, it is to be expected that more studies will be published on the role and impact of social media.

11.5 FINAL WORD

The term "risk communication" has been used since 1984. In the context of everything that has been discussed in this chapter (e.g., the early attempts to present rather technical risk information, the subsequent discussion about more people-oriented approaches, studies making clear that risk communication is actually communication about a lot of other psychological and communicative aspects too, not just the risk aspects of the issue at hand), it might be worthwhile considering a new name for this activity that does justice to this complex human activity.

REFERENCES

Atman, C. J., Bostrom, A., Fischhoff, B., and Morgan, M. G. (1994). Designing risk communications: Completing and correcting mental models of hazardous processes, Part I. *Risk Analysis*, 14, 779–788.

Basher, R. (2006). Global early warning systems for natural hazards: Systematic and people-centred. *Philosophical Transactions of the Royal Society A*, 364, 2167–2182.

Bostrom, A., Fischhoff, B., and Morgan, M. G. (1992). Characterizing mental models of hazardous processes: A methodology and an application to radon. *Journal of Social Issues*, 48, 85–100.

Boyle, T. (2012). *Health and Safety: Risk Management*. Wigston, UK: IOSH Services Limited.

Burns, W. J., Slovic, P., Kasperson, R. E., Kasperson, J. X., Renn, O., and Emani, S. (1993). Incorporating structural models into research on the social amplification of risk: Implications for theory construction and decision making. *Risk Analysis*, 13, 611–623.

Covello, V. T. (1992). Risk communication: An emerging area of health communication research. In: S. A. Deetz (Ed.), *Communication Yearbook 15* (pp. 359–373). Newbury Park, CA: Sage.

Covello, V. T., Von Winterfeldt, D., and Slovic, P. (1987). Communicating scientific information about health and environmental risks: Problems and opportunities from a social and behavioral perspective. In: V. T. Covello, L. B. Lave, A. A. Moghissi, and V. R. R. Uppuluri (Eds.), *Uncertainty in Risk Assessment, Risk Management, and Decision Making* (pp. 39–61). New York: Plenum Publishing Corporation.

Covello, V. T., Sandman, P. M., and Slovic, P. (1988). *Risk Communication, Risk Statistics, and Risk Comparisons: A Manual for Plant Managers*. Washington DC.: Chemical Manufacturers Association.

Cvetkovich, G., Vlek, C. A. J., and Earle, T. C. (1989). Designing technological hazard information programs: Towards a model of risk-adaptive decision making. In: C. A. J. Vlek and G. Cvetkovich (Eds.), *Social Decision Methodology for Technological Projects* (pp. 253–276). Dordrecht, The Netherlands: Kluwer Academic Publishers.

Earle, T. C. (2010). Trust in risk management: A model-based review of empirical research. *Risk Analysis*, 30(4), 541–574.

Earle, T. C., and Siegrist, M. (2006). Morality information, performance information, and the distinction between trust and confidence. *Journal of Applied Social Psychology*, 36(2), 383–416.

Floyd, D. L., Prentice-Dunn, S., and Rogers, R. W. (2000). A meta-analysis of research on protection motivation theory. *Journal of Applied Social Psychology*, 30, 407–429.

Flynn, J., Burns, W. J., Mertz, C. K., and Slovic, P. (1992). Trust as a determinant of opposition to a high-level radioactive waste repository: Analysis of a structural model. *Risk Analysis*, 12, 417–429.

Freudenburg, W. R., and Rursch, J. A. (1994). The risks of "putting the numbers in context": A cautionary tale. *Risk Analysis*, 14, 949–958.

Frey, D. (1986). Recent research on selective exposure to information. In: L. Berkowitz (Ed.), *Advances in Experimental Social Psychology* (Vol. 19, pp. 41–80). Orlando, FL: Academic Press.

Gore, T. D., and Bracken, C. C. (2005). Testing the theoretical design of a health risk message: Reexamining the major tenets of the extended parallel process model. *Health Education and Behavior*, 32(1), 27–41.

Griffin, R. J., Dunwoody, S., and Neuwirth, K. (1999). Proposed model of the relationship of risk information seeking and processing to the development of preventive behaviors. *Environmental Research*, 80(2 II), S230–S245.

Griffin, R. J., Yang, Z., ter Huurne, E., Boerner, F., Ortiz, S., and Dunwoody, S. (2008). After the flood: Anger, attribution, and the seeking of information. *Science Communication*, 29(3), 285–315.

Grothmann, T., and Reusswig, F. (2006). People at risk of flooding: Why some residents take precautionary action while others do not. *Natural Hazards*, 38, 101–120.

Gurabardhi, Z., Gutteling, J. M., and Kuttschreuter, M. (2004). The development of risk communication. *Science Communication*, 25(4), 323–349.

Gutteling, J. M., and Kuttschreuter, M. (2002). The role of expertise in risk communication: Laypeople's and expert's perception of the millennium bug risk in the Netherlands. *Journal of Risk Research*, 5(1), 35–47.

Horst, M., Kuttschreuter, M., and Gutteling, J. M. (2007). Perceived usefulness, personal experiences, risk perception and trust as determinants of adoption of e-government services in The Netherlands. *Computers in Human Behavior*, 23(4), 1838–1852.

Johnson, F. R., and Luken, R. A. (1987). Radon risk information and voluntary protection: Evidence from a natural experiment. *Risk Analysis*, 7, 97–107.

Johnston, A. C., and Warkentin, M. (2010). Fear appeals and information security behaviors: An empirical study. *MIS Quarterly*, 34(3), 549–566.

Kahlor, L., Dunwoody, S., Griffin, R. J., and Neuwirth, K. (2006). Seeking and processing information about impersonal risk. *Science Communication*, 28(2), 163–194.

Kasperson, J. X., and Kasperson, R. E. (2005). *The Social Contours of Risk* (Vol. 1: Publics, Risk Communication and the Social Amplification of Risk). Virginia: Earthscan.

Kasperson, R. E., Renn, O., Slovic, P., Brown, H. S., Emel, J., Goble, R., Kasperson, J. X., and Ratick, S. (1988). The social amplification of risk: A conceptual framework. *Risk Analysis*, 8, 177–187.

Kenrick, D. T., Neuberg, S. L., and Cialdini, R. B. (2005). *Social Psychology–Unravelling the Mystery*. New Zealand: Pearson Education.

Kievik, M., and Gutteling, J. M. (2011). Yes, we can: Motivate Dutch citizens to engage in self-protective behavior with regard to flood risks. *Natural Hazards*, 59(3), 1475–1490.

Kievik, M., ter Huurne, E. F. J., and Gutteling, J. M. (2012). The action suited to the word? Use of the framework of risk information seeking to understand risk-related behaviors. *Journal of Risk Research*, 15(2), 131–147.

Kievik, M., Misana-ter Huurne, E., and Gutteling, J. M. (2015). Let's get together—the effect of personal experience with risk mitigating behavior on self-protectiveness of citizens. Paper presented at SRA Europe, Maastricht, The Netherlands, June 15, 2015.

Kuttschreuter, M., Gutteling, J. M., and de Hond, M. (2011). Framing and tone-of voice of disaster media coverage: The aftermath of the Enschede fireworks disaster in the Netherlands. *Health, Risk and Society*, 13(3), 201–220.

Lindell, M. K., and Perry, R. W. (1992). *Behavioural Foundations of Community Emergency Planning*. Washington, DC: Hemisphere Press.

Lindell, M. K., and Perry, R. W. (2012). The protective action decision model: Theoretical modifications and additional evidence. *Risk Analysis*, 32(4), 616–632.

MacGregor, D., Slovic, P., and Morgan, M. G. (1994). Perception of risks from electromagnetic fields: A psychometric evaluation of a risk-communication approach. *Risk Analysis*, 14, 815–828.

Martin, I. M., Bender, H., and Raish, C. (2007). What motivates individuals to protect themselves from risks: The case of wildland fires. *Risk Analysis*, 27(4), 887–900.

Mase, A. S., Cho, H., and Prokopy, L. S. (2015). Enhancing the social amplification of risk framework (SARF) by exploring trust, the availability heuristic, and agricultural advisors' belief in climate change. *Journal of Environmental Psychology*, 41, 166–176.

Mazur, A. (1984). The journalists and technology: Reporting about love canal and three mile island. *Minerva*, 22, 45–66.

McComas, K. A. (2006). Defining moments in risk communication research: 1996–2005. *Journal of Health Communication*, 11(1), 75–91.

McGuire, W. J. (1985). Attitudes and attitude change. In: G. Lindzey and E. Aronson (Eds.), *Handbook of Social Psychology* (3rd ed., Vol. 2, pp. 233–346). New York: Random House.

Morgan, M. G., Fischhoff, B., Bostrom, A., Lave, L., and Atman, C. J. (1992). Communicating risk to the public. *Environmental Science and Technology*, 26, 2048–2056.

Nationale Ombudsman. (2012). De asbestzomer van 2012 Over hysterie, bestuurlijk risico en horizontaal besturen. Den Haag, Rapport 2012/198.

Palen, L., and Liu, S. B. (2007). Citizen communications in crisis: Anticipating a future of ICT-supported participation. In: *Proceedings from the CHI Conference* (pp. 717–736). San Jose, CA/New York: ACM Press.

Palen, L., Vieweg, S., Liu, S. B., and Hughes, A. L. (2009). Crisis in a networked world: Features of computer-mediated communication in the April 16, 2007, Virginia Tech event. *Social Science Computer Review*, 27(4), 467–480.

Palenchar, M. J. (2005). Risk communication. In: R. L. Heath (Ed.), *Encyclopedia of Public Relations* (pp. 752–755). Thousand Oaks, CA: Sage.

Peltu, M. (1985). The role of communications media. In: H. J. Otway and M. Peltu (Eds.), *Regulating Industrial Risks* (pp. 128–148). London: Butterworths.

Petts, J. (1992). Incineration risk perceptions and public concern: Experience in the UK improving risk communication. *Waste Management and Research*, 10, 169–182.

Pidgeon, N., Kasperson, R. E., and Slovic, P. (2003). *The Social Amplification of Risk*. Cambridge University Press.

Plough, A., and Krimsky, S. (1987). The emergence of risk communication studies: Social and political context. *Science, Technology, and Human Values*, 12(3/4), 4–10.

Read, D., Bostrom, A., Morgan, M. G., Fischhoff, B., and Smuts, T. (1994). What do people know about global climate change? 2. Survey studies of educated laypeople. *Risk Analysis*, 14, 971–982.

Renn, O., and Levine, D. (1991). Credibility and trust in risk communication. *Technology, Risk, and Society*, 4, 175–217.

Renn, O., Burns, W. J., Kasperson, J. X., Kasperson, R. E., and Slovic, P. (1992). The social amplification of risk: Theoretical foundations and empirical applications. *Journal of Social Issues*, 48, 137–160.

Rimal, R. N., and Real, K. (2003). Perceived risk and efficacy beliefs as motivators of change: Use of the risk perception attitude (RPA) framework to understand health behaviors. *Human Communication Research*, 29, 370–399.

Rogers, R. W. (1983). Cognitive and physiological processes in fear appeals and attitude change: A revised theory of protection motivation. In: J. Cacioppo and R. Petty (Eds.), *Social Psychophysiology* (pp. 153–176). New York: Guilford Press.

Roth, E., Morgan, M. G., Fischhoff, B., Lave, L., and Bostrom, A. (1990). What do we know about making risk comparisons? *Risk Analysis*, 10, 375–387.

Rowan, K. E. (1994). The technical and democratic approaches to risk situations: Their appeal, limitations, and rhetorical alternative. *Argumentation*, 8, 391–409.

Schwarz, N., Kumpf, M., and Bussman, W. (1986). Resistance to persuasion as a consequence of influence attempts in advertising and non-advertising communications. *Psychology, a Quarterly Journal of Human Behavior*, 23, 72–76.

Scolobig, A., Prior, T., Schröter, D., Jörin, J., and Patt, A. (2015). Towards people-centred approaches for effective disaster risk management: Balancing rhetoric with reality. *International Journal of Disaster Risk Reduction*, 12, 202–212.

Seeger, M. W., Reynolds, B., and Sellnow, T. L. (2010). Crisis and emergency risk communication in health contexts: Applying the CDC model to pandemic influenza. In: R. L. Heath, and D. O'Hair (Eds.), *Handbook of Risk and Crisis Communication* (pp. 493–506). New York: Routledge.

Shklovski, I., Palen, L., and Sutton, J. (2008). Finding community through information and communication technology in disaster events. In: Proceedings of the ACM Conference on Computer Supported Cooperative Work, CSCW (pp. 127–136), San Diego, CA.

Slovic, P. (2000). *The Perception of Risk*. London: Earthscan Publications, Ltd.

Slovic, P., Fischhoff, B., and Lichtenstein, S. (1982). Why study risk perception?. *Risk Analysis*, 2(2), 83–93.

Snyder, M. L., and Wicklund, R. A. (1976). Prior exercise of freedom and reactance. *Journal of Experimental Social Psychology*, 12, 120–130.

ter Huurne, E. F. J. (2008). *Information Seeking in a Risky World. The Theoretical and Empirical Development of FRIS: A Framework of Risk Information Seeking*. Enschede, The Netherlands: University of Twente.

ter Huurne, E. F. J., and Gutteling, J. M. (2008). Information needs and risk perception as predictors of risk information seeking. *Journal of Risk Research*, 11(7), 847–862.

Terpstra, T., and Lindell, M. K. (2012). Citizens' perceptions of flood hazard adjustments: An application of the protective action decision model. *Environment and Behavior*, 45(8), 993–1018.

Terpstra, T., de Vries, A., Stronkman, R., and Paradies, G. L. (2012). Toward a realtime Twitter analysis during crisis for operational crisis management. In: Proceedings of the 9th International ISCRAM Conference, Vancouver, Canada.

Tversky, A., and Kahneman, D. (1974). Judgement under uncertainty: Heuristics and biases. In: D. Wendt, and C. A. J. Vlek (Eds.), *Utility, Probability and Human Decision Making* (pp. 141–162). American Association for the Advancement of Science.

UNISDR (United Nations International Strategy for Disaster Reduction). (2009). UNISDR terminology on disaster risk reduction, Geneva.

USNRC (United States Nuclear Regulatory Commission) (n.d.). Effective risk communication, Gaithersburg, MD.

Verroen, S., Gutteling, J. M., and de Vries, P. W. (2013). Enhancing self-protective behavior: Efficacy beliefs and peer feedback in risk communication. *Risk Analysis*, 33(7), 1252–1264.

Vieweg, S., Hughes, A. N., Starbird, K., and Palen, L. (2010). Microblogging during two natural hazards events: What Twitter may contribute to situational awareness. In: Conference on Human Factors in Computing Systems, Proceedings 2 (pp. 1079–1988), Atlanta, GA.

Visschers, V. H. M., Wiedemann, P. M., Gutscher, H., Kurzenhäuser, S., Seidl, R., Jardine, C. G., and Timmermans, D. R. M. (2012). Affect-inducing risk communication: Current knowledge and future directions. *Journal of Risk Research*, 15(3), 257–271.

Vlek, C. A. J., and Cvetkovich, G. (1989). Social decision making on technological projects: Review of key issues and a recommended procedure. In: C. A. J. Vlek and G. Cvetkovich

(Eds.), *Social Decision Methodology for Technological Projects* (pp. 297–322). Dordrecht, The Netherlands: Kluwer Academic Publishers.

Wikipedia (2015). 2015 Tianjin explosions [Online]. Available at: https://en.wikipedia.org/wiki/2015_Tianjin_explosions (Accessed January 17, 2016).

Witte, K., and Allen, M. (2000). A meta-analysis of fear appeals: Implications for effective public health campaigns. *Health Education and Behaviour*, 27(5), 591–615.

Yang, Z. J., Aloe, A. M., and Feeley, T. H. (2014). Risk information seeking and processing model: A meta-analysis. *Journal of Communication*, 64, 20–41.

Yates, D., and Paquette, S. (2011). Emergency knowledge management and social media technologies: A case study of the 2010 Haitian earthquake. *International Journal of Information Management*, 31(1), 6–13.

Zaalberg, R., Midden, C., Meijnders, A., and McCalley, T. (2009). Prevention, adaptation, and threat denial: Flooding experiences in The Netherlands. *Risk Analysis*, 29(12), 1759–1778.

FURTHER READING

Coombs, W. T., and Holladay, S. J. (Eds.). (2011). *The Handbook of Crisis Communication* (Vol. 22). John Wiley & Sons.

Heath, R. (1998). Working under pressure: Crisis management, pressure groups and the media. *Safety Science*, 30(1), 209–221.

Lundgren, R. E., and McMakin, A. H. (2013). *Risk Communication: A Handbook for Communicating Environmental, Safety, and Health Risks*. John Wiley & Sons.

McComas, K. A. (2003). Citizen satisfaction with public meetings used for risk communication. *Journal of Applied Communication Research*, 31(2), 164–184.

Slovic, P. (2010). *The Feeling of Risk: New Perspectives on Risk Perception*. Routledge.

Slovic, P., Flynn, J., and Kunreuther, H. (2013). *Risk, Media and Stigma: Understanding Public Challenges to Modern Science and Technology*. Routledge.

12

THE PRECAUTIONARY PRINCIPLE

Sven Ove Hansson

12.1 INTRODUCTION

Bisphenol A is a substance that is used in the production of plastics and epoxy resins. Human exposure occurs as a consequence of leakage from plastic products and thermal papers (e.g., receipts). The substance mimics the hormone estrogen, and in many animal studies it has caused negative health effects in the offspring of exposed females. Unfortunately, the results of these animal studies have been difficult to interpret. In the so-called standard tests, that is, animal tests standardized for regulatory purposes, the substance has only produced negative effects at levels much higher than what humans are currently exposed to. However, in research studies that measure other effects than the standard tests, bisphenol A has produced negative health effects at levels corresponding to actual human exposures. A controversy has emerged on the relevance of the two types of studies. On the one hand, the standard tests have not been constructed to detect effects on the hormone system, and they are therefore expected to be less sensitive to such effects than research tests tailored to measure them. On the other hand, the standard tests have been calibrated in studies of a large number of substances, which the research studies have not. The epidemiological studies that

Handbook of Safety Principles, First Edition. Edited by Niklas Möller, Sven Ove Hansson, Jan-Erik Holmberg, and Carl Rollenhagen.

have been performed on exposed humans are also difficult to interpret. Some of these studies have shown negative effects on exposed children whereas others have not.

Expert reviews performed by government agencies have reached different conclusions on the potential risks of bisphenol A. The European food agency EFSA has repeatedly concluded that it finds no evidence of a health risk at current levels of exposure. However, the agency notes that their judgment is preliminary due to scientific uncertainties. On the other hand, the French food authority Anses has concluded that in children of exposed mothers the substance causes an increased susceptibility of the mammary gland to tumor formation. Like EFSA, Anses emphasizes that its conclusions are far from certain.

In this situation of scientific uncertainty, several governments have taken action against the potential risks of bisphenol A exposure. For instance, the European Union has prohibited its use in baby bottles, and some member countries have forbidden other uses of the substance. The Risk Assessment Committee (RAC) of the European Chemicals Agency (ECHA) has proposed a prohibition of bisphenol A in thermal paper since the risk for workers is not adequately controlled. Sweden has prohibited bisphenol A in cans containing food for children under 3 years of age, and plans to prohibit it also in tubes for drinking water. France has decided to prohibit the substance in all articles intended to come in contact with food. These and other measures to reduce exposure to the substance have been undertaken in spite of the prevailing scientific uncertainty. They have been justified by the precautionary principle (Ågerstrand et al., 2014; Anses, 2013; Beronius et al., 2010).

This is how the precautionary principle is commonly used: as an argument for acting against a potential danger in spite of scientific uncertainty on whether the danger is real or not. No other safety principle has been so vehemently contested as the precautionary principle. It has been accused of being both irrational and unscientific. In this chapter, we are going to have a close look at what it means, how it can be justified, and its relations to other safety principles.

12.2 HISTORY AND CURRENT USE

Precautionary measures against potential but not fully proven dangers have a long history. We can easily imagine one of our ancestors reasoning like this: "I'm not sure, but there might be a saber-toothed tiger in that cave. Therefore I stay outside." The dictum *primum non nocere* (first do no harm) that stems from the Hippocratic Oath in the late fifth century BCE has been interpreted as a plea for caution in medical treatment (Ozonoff, 1999). Harremoës et al. (2001, pp. 14–15) mention a preventive health measure in London in 1854 as an example of a precautionary action. In early September that year, London was struck by cholera, and 500 people died in 10 days. The physician John Snow noted that a large number of those affected by the disease had drunk water from a pump on Broad Street. On his advice the authorities removed the handle of that pump, based on the reasonable, but far from confirmed, hypothesis that the disease might be caused by contaminated water from that source. (At that time most physicians believed cholera to be transmitted through air rather than water.) This precautionary action turned out successful, and the cholera epidemic was curbed. Another early example is the British Alkali Act of 1874, which "required that emissions of noxious gases from certain plants should be prevented, using the best practicable means, without any need to demonstrate that the gases were actually causing harm in any particular case" (Haigh, 1994, p. 241).

An early statement of what is now called the precautionary principle in environmental policy can be found in the Governmental Bill proposing the Swedish Environmental Protection Act that was adopted in 1969:

> "[T]he uncertainty of the hazardous potential of a substance must not affect the public, but rather the operator emitting the substance into water or air. This means that action should not wait until adverse effects become manifest. This also implies that anyone wanting to emit an unknown substance, which on reasonable grounds is considered hazardous, must prove that harm is not to be expected."
>
> (Proposition 1969:28, p. 210; Karlsson, 2006)

In much the same vein, the World Charter for Nature that was adopted by the UN General Assembly in 1982 stated:

> "Activities which are likely to pose a significant risk to nature shall be preceded by an exhaustive examination; their proponents shall demonstrate that expected benefits outweigh potential damage to nature, and where potential adverse effects are not fully understood, the activities should not proceed."
>
> (Wood, 1985, p. 994)

The phrase "precautionary principle" is commonly traced to West German environmental law in the 1970s, where it appeared under the name of *Vorsorgeprinzip* (Boehmer-Christiansen, 1994; Cameron and Abouchar, 1996; Freestone, 1991). According to some authors, the "precautionary concept found its way into international law and policy as a result of German proposals made to the International North Sea Ministerial Conferences" (Freestone and Ellen, 1996, p. 4). At any rate, the declaration from the Second International Conference on Protection of the North Sea in 1987 appears to have been the first major international document to make a clear reference to a "principle" of precaution. It was referred to in the document as "the principle of precautionary action," and explicated as follows:

> "The participants at the Second International Conference on the Protection of the North Sea, held in London on 24 and 25 November 1987, being the Ministers responsible for the protection of the North Sea of the Governments of Belgium, Denmark, France, Federal Republic of Germany, Netherlands, Norway, Sweden, United Kingdom, and the Member of the Commission of the European Communities responsible for environmental protection… [accept] that, in order to protect the North Sea from possibly damaging

> effects of the most dangerous substances, a precautionary approach is necessary which may require action to control inputs of such substances even before a causal link has been established by absolutely clear scientific evidence."
>
> (Ministerial Declaration, 1987)

On the next North Sea Conference in The Hague three years later, the ministers announced that they "will continue to apply the precautionary principle" (Ministerial Declaration, 1990). This document also contained a remarkably widened definition of the principle, namely "to take action to avoid potentially damaging impacts of substances that are persistent, toxic, and liable to bioaccumulate even where there is no scientific evidence to prove a causal link between emissions and effects."

Several other international conferences on the environment have made statements endorsing the precautionary principle, perhaps most importantly the Rio Declaration on Environment and Development that was a major outcome of the 1992 so-called earth summit in Rio de Janeiro.

> "Principle 15. Precautionary principle
>
> In order to protect the environment, the precautionary approach shall be widely applied by States according to their capabilities. Where there are threats of serious or irreversible damage, lack of full scientific certainty shall not be used as a reason for postponing cost-effective measures to prevent environmental degradation."
>
> (Report of the United Nations Conference, 1992)

In January 1998, 32 experts from Europe and Northern America met for three days at Wingspread, Wisconsin, to define the precautionary principle and identify its role in public health and environmental policies (Hileman, 1998; Raffensperger and Tickner, 1999). They agreed on the following definition that has continued to be influential, in particular, among policy makers and NGOs:

> "When an activity raises threats to the environment or human health, precautionary measures should be taken, even if some cause-and-effect relationships are not fully established scientifically.
>
> In this context, the proponent of an activity, rather than the public, should bear the burden of proof.
>
> The process of applying the Precautionary Principle must be open, informed and democratic and must include potentially affected parties. It must also involve an examination of the full range of alternatives, including no action."
>
> (Wingspread Statement, 1998)

The European Union and several of its member states have incorporated the precautionary principle into their legislations. Through the 1992 Maastricht Amendments to the European Treaty (Treaty of Rome, now known as the Treaty on the Functioning of the European Union), the precautionary principle was written into European legislation (Pyhälä et al., 2010, p. 206):

> "Union policy on the environment shall aim at a high level of protection taking into account the diversity of situations in the various regions of the Union. It shall be based on the precautionary principle and on the principles that preventive action should be taken, that environmental damage should as a priority be rectified at source and that the polluter should pay."
>
> (Consolidated version, 2012)

In February 2000, the Commission issued a Communication on the Precautionary Principle. According to that Communication, although the principle is mentioned in the treaty only in the context of environmental protection, "in practice, its scope is much wider, and specifically where preliminary objective scientific evaluation, indicates that there are reasonable grounds for concern that the potentially dangerous effects on the environment, human, animal, or plant health may be inconsistent with the high level of protection chosen for the Community." The principle is described as a decision-maker's principle for risk management, which "should not be confused with the element of caution that scientists apply in their assessment of scientific data." Its implementation should "start with a scientific evaluation, as complete as possible, and where possible, identifying at each stage the degree of scientific uncertainty." The measures taken may range from banning a substance or procedure to initiating research or issuing a recommendation. Six requirements on applications of the principle were laid down:

> "Where action is deemed necessary, measures based on the precautionary principle should be, *inter alia*:
>
> - *proportional* to the chosen level of protection,
> - *non-discriminatory* in their application,
> - *consistent* with similar measures already taken,
> - *based on an examination of the potential benefits and costs* of action or lack of action (including, where appropriate and feasible, an economic cost/benefit analysis),
> - *subject to review*, in the light of new scientific data, and
> - *capable of assigning responsibility for producing the scientific evidence* necessary for a more comprehensive risk assessment."
>
> (Communication from the Commission, 2000)

Each of these six principles is further explained in the document. For instance, the requirement that applications of the principle should be subject to review is said to imply that the principle will only be maintained as long as the scientific information is incomplete or inconclusive. Precautionary measures should be "periodically reviewed in the light of scientific progress, and amended as necessary." (cf. Hansson 2016.) In general, the document puts a strong emphasis on the use of scientific information in risk management, and restricts the application of the precautionary principle to cases of decision-relevant scientific uncertainty.

The precautionary principle has been particularly important in the Union's chemicals policy. In February 2001, the European Commission issued a White Paper outlining the strategy for a future chemicals policy. The objectives of the coming legislation would be "in line with the overriding goal of sustainable development and seek to make the chemical industry accept more responsibility by respecting the precautionary principle and safeguarding the Single Market and the competitiveness of European industry." The principle was said to stipulate that "action must be taken even if there is still scientific uncertainty as to the precise nature of the risks" (Commission White Paper, 2001). The resulting legislation contains legal instruments for precautionary measures in cases with insufficient but non-negligible scientific evidence that a substance poses a danger to human health and the environment. It also requires industry to produce more data in order to reduce the large scientific uncertainty that prevails concerning the potential adverse effects of most chemicals currently in use (Rudén and Hansson, 2010).

In the United States, the precautionary principle has been much less referred to by policy makers than in Europe. However, the difference seems to be larger in terminology than in actual practice. The American government has often taken protective measures based on incomplete scientific evidence. In some cases it has even done so at an earlier point in time and thus with less scientific evidence than what the European Union did (Wiener and Rogers, 2002). For instance, the American Environmental Protection Agency banned the use of chlorofluorocarbons (CFC) in 1978, several years before most European countries (Smith, 1998). The use of the growth hormone diethylstilbestrol (DES) in cattle was banned in the United States in 1979, but in the EU it was not banned until 1987 (Bridges and Bridges, 2001; Friedberger, 1994). However, in the American discussion, the precautionary principle is commonly described as a principle that allows regulators to ban a product in the absence of respectable scientific evidence that it might cause environmental problems. Environmental and public health measures that would be described in Europe as applications of the precautionary principle are often described in the United States as representing the idea "better safe than sorry." In this chapter, the terminological choice has been made to use the term "the precautionary principle" in the way it is used by those who claim that they apply and/or promote it. Readers used to another terminology may wish to substitute the term "better safe than sorry" for "the precautionary principle." The terminology is not important. How scientific uncertainty is dealt with in the protection of human health and the environment is important.

12.3 DEFINITIONS

There are two major forms of the principle, argumentative and prescriptive versions.[1] An argumentative version tells us what kinds of arguments we should use in our

[1] The distinction between prescriptive and argumentative versions of the precautionary principle was noted by Morris (2000, p. 1), but he used a rather misleading terminology for the two versions, calling them "weak" and "strong."

decision-guiding deliberations. The Rio Declaration is an example of this. As noted above, it requires that "lack of full scientific certainty shall not be used as a *reason* for postponing cost-effective measures to prevent environmental degradation" (italics added). This is not a directly action-guiding decision rule, but an argumentation-guiding rule for our decision-guiding deliberations. What it says is essentially that lack of full scientific certainty that a danger exists is not a valid reason against preventive action. In contrast, prescriptive versions of the principle enjoin us to take preventive action in various concrete situations. The Wingspread Statement is an example, saying that "precautionary measures should be taken" in certain situations. Similarly, the European Commission's White Paper stipulates that "action must be taken" under the conditions when the principle applies. In what follows, we will focus on prescriptive versions of the principle.

As was shown by Per Sandin (1999), most prescriptive versions of the precautionary principle share four common elements or dimensions. These formulations can be recast into the following if-clause, containing the four dimensions:

> "*If* there is (1) a threat, which is (2) uncertain, *then* (3) some kind of action (4) is mandatory."

The first two of these dimensions belong to the realm of scientific risk assessment. They jointly indicate *when* the precautionary principle can be applied. The last two belong to the realm of risk management. They tell us *how* to apply the principle. Following Ahteensuu (2008) we can summarize the first two dimensions as the *trigger* of the precautionary principle and the last two as the *precautionary response*.

By further specifying these four dimensions, the precautionary principle can be made more precise. This means that in order to clarify what a particular version or application of the principle requires, we have to answer the following four questions:

1. To what types of hazards does the principle apply? (the threat)
2. Which level of evidence (lower than that of full scientific certainty) should be required? (the level of evidence)
3. What types of measures against potential hazards does the principle recommend? (the type of action)
4. With what force are these measures recommended (mandatory, merely permitted, etc.)? (normative status)

Beginning with the *threat* dimension, the precautionary principle has been put forward as a principle for decision-making in environmental and health-related issues. However, the same pattern of thought is prevalent also in other areas of decision-making. For instance, governments, central banks, and business leaders take action against a potential financial crisis without being certain that such a crisis would otherwise occur. Similarly, military commanders take countermeasures against an expected enemy attack without being sure that the attack will actually take place.

(Examples of such countermeasures are mobilization, a heightened alert level, and a pre-emptive attack.) In safety engineering, actions against potential dangers are routinely taken in the absence of full evidence that the danger is real. For instance, if you have uncertain indications that an elevator cable might be damaged, you stop the elevator for maintenance without waiting for full evidence that there is a damage. In these and many other cases, a thought pattern similar to that of the precautionary principle is applied in areas where the term "precautionary principle" is not commonly used. The difference is that in these cases, the rationality of taking action without full evidence is seldom contested. Economists commonly agree that action should be taken against a potential financial crisis even in the absence of full evidence that it will otherwise take place.[2] A military commander who waits for full evidence of an enemy attack would be regarded as incompetent, and the same applies to a safety engineer who takes all machinery to be safe unless there is fully convincing evidence to the contrary. In contrast, measures against potentially harmful chemicals—the most common application of the precautionary principle—are often controversial. As will be discussed in Section 12.8.3, in this area proposed measures against uncertain dangers are often met with claims that such measures should wait until full evidence of the danger can be presented.

The justification for precautionary action against a threat is of course stronger, the more serious that threat is. This is indicated in the Rio Declaration that restricts the application of the principle to "threats of serious or irreversible damage." The European Commission's communication on the principle from 2000 specifies that the principle should be applied when there are "potentially dangerous effects" that "may be inconsistent with the high level of protection chosen for the Community." The latter wording implies a lower threshold in terms of the severity of the threat than that of the Rio Declaration. (However, the Communication puts much emphasis on proportionality, which means that precautionary actions against less severe threats will not be as far-reaching as those against more severe ones.) The severity of the threat required to trigger precautionary action may be different for different legislations or applications, and is certainly one of the factors to be included in a specification of the principle.

The second dimension is the *level of evidence* required for precautionary action. This dimension is also relevant for the preventive actions without full evidence that are taken in areas where the precautionary principle is usually not explicitly referred to. You would evacuate a public building when there is strong but yet insufficient evidence of an uncontrollable fire (for instance that the fire alarm has gone off), but not if the evidence is very weak (for instance that someone has seen a person walking around with a lighter in his hand). Similarly, public authorities endorsing the precautionary principle usually emphasize that it can only be triggered when there is fairly strong evidence of danger. For instance, the Swedish Chemicals Legislation from 1985 required a "reasonable scientific foundation" in order to trigger precautionary measures (JoU 1984/85:30, p.23). In its Communication from 2000 the European

[2] But see Schefczyk (2016).

Commission stated that the principle should be applied when there are "reasonable grounds for concern." The same document puts much emphasis on the use of scientific information, and it can therefore be inferred that the judgments of what grounds are reasonable will have to be based on science. Rumors or pseudoscientific statements would not constitute grounds for applying the precautionary principle. On the other hand, insufficient but worrisome scientific information may trigger its use. One example can be a situation in which there is scientifically reliable evidence that a substance causes severe liver disease in some animal species by some unknown mechanism. Since the mechanism is unknown it is not known whether it will act in humans as well.

In order to apply the precautionary principle it is necessary to distinguish between different levels of scientific evidence, and in particular between different levels below what is called "full" scientific evidence. One way to do this is to ask scientists to estimate the probability that a danger exists. Another, more common approach is to have them categorize the strength of evidence in qualitative terms, for example, "strong scientific evidence" or "scientifically supported strong suspicions." The development and calibration of such specifications would be an important contribution to the operationalization of the precautionary principle (Levin et al., 2004).

The third dimension is the *type of action* that is triggered by the precautionary principle. Several formulations of the principle emphasize that there is a wide range of alternative actions that it can trigger. For instance, the European Commission's communication from 2000 says:

> "In some cases, the right answer may be not to act or at least not to introduce a binding legal measure. A wide range of initiatives is available in the case of action, going from a legally binding measure to a research project or a recommendation."
>
> (Communication from the Commission, 2000)

In the same document it is emphasized that measures should be proportionate to the potential danger, and based on an examination of the "potential benefits and costs of action or lack of action (including, where appropriate and feasible, an economic cost/benefit analysis)." Other specifications of the principle have gone even further in their endorsement of cost–benefit analysis. For instance, the Rio Declaration refers to "cost-effective measures to prevent environmental degradation," and in 1990 the British Government issued a White Paper, *This Common Inheritance*, stating that it

> "... will be prepared to take precautionary action to limit the use of potentially dangerous materials or the spread of potentially dangerous pollutants, even where scientific knowledge is not conclusive, *if the balance of likely costs and benefits justifies it.*"
>
> (O'Riordan and Jordan, 1995, p. 197, italics added)

On the other hand, there are interpretations of the precautionary principle, mostly by NGOs and NGO-affiliated authors, stating expressly that the principle requires protection of the environment even when a cost–benefit analysis would not support such protection (Burgos and Defeo, 2004; O'Riordan and Cameron, 1994; O'Riordan et al., 2001). Obviously, this is an important difference. The relation between the precautionary principle and cost–benefit optimization will be further discussed in Section 12.7.

The fourth dimension is the *normative status* of the principle. Variants of the principle that assign a high normative status to its recommendations are often expressed with words like "should" or "must"; thus the Wingspread Statement says that precautionary actions "should be taken" and the European Commission's White Paper from 2001 that such actions "must be taken" under the specified circumstances. The legal status of the principle is an important parameter, since experience shows that attempts to apply it are often challenged in court by corporate entities that have something to lose from its application.

12.4 UNDERLYING THEORY

The precautionary principle refers to cases with decision-relevant scientific uncertainty. In order to understand how the principle relates to science it is important to distinguish between the use of scientific data on the one hand for scientific judgments and on the other hand for practical decisions. For our present purposes we can focus on judgments and decisions concerning how much we should rely on various factual statements that fall within the purview of science.

In science, this is essentially a matter of how much evidence scientists should require for accepting a scientific claim made by colleagues. Science has a long tradition of putting the onus of proof on those who make a new claim. Before a new hypothesis or idea is accepted in science, quite convincing evidence in its favor has to be presented. Those who claim the existence of an as yet unproven effect—such as a toxic effect of a chemical substance—have the burden of proof. There are good reasons for this. If we accept something as a scientific fact, then it will be taken for given in future investigations, and both theoretical and experimental work will be based on the assumption that it is correct. If false statements are accepted as scientific facts, they can block scientific progress and lead to the pursuit of all sorts of blind alleys. Although this cannot be completely avoided, it is essential for the progress of science that strict criteria of proof are applied so that it only happens very seldom.

It must also be emphasized that decisions what to accept as "scientific facts" are always provisional. In science nothing is accepted once and for all. Instead, scientific statements are accepted provisionally, that is, until reasons to doubt them arise. The provisionality of scientific beliefs, and the continuous search for new information that may us give reasons to change what was previously taken for given, constitute much of the "secret" of science, the reason why it is able to improve ceaselessly and to provide us with better and better information. This stands in stark contrast to various "alternatives" to science. It is, for instance, interesting to compare scientific pharmacology with the pseudoscience of homeopathy. From the

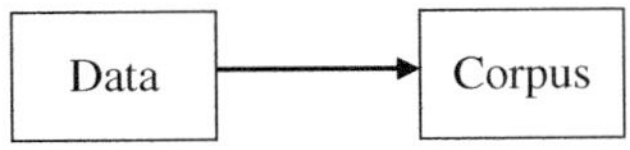

Figure 12.1. The scientific corpus is based on scientific data.

1790s to today, scientific pharmacology has gone through several radical changes in its methodology, including the introduction of quantitative chemical analysis, the concept of an active ingredient, the importance of the administered dose of that component, and the use of blinded or double-blinded clinical trials to determine the effects of pharmacological treatments. It should therefore be no surprise that modern toxicology and ecotoxicology, the sciences that provide us with information on the adverse effects of chemical substances in humans and in the environment, are based to a large extent on the methods developed in pharmacology. In contrast, due to its inability to learn, homeopathy is still based on ideas from the 1790s that were thoroughly disproved a few decades later. Unsurprisingly, it has made no contribution whatsoever to toxicology or ecotoxicology.

The statements that are taken to be scientific facts—provisionally, until we have reasons to revise or reject them—are said to comprise the scientific *corpus*, or body of current scientific beliefs. Figure 12.1 illustrates how it is constructed (Hansson, 2004). Scientific knowledge is based on *data* that originate in laboratory experiments, field studies, and other observations. Through a process of critical assessment, these data give rise to the scientific corpus. The corpus is of course not defined with perfect precision. Sometimes it is difficult to determine whether a particular statement belongs to the corpus. However, these cases are few in comparison to the immensity of the corpus. In other words, the fuzzy edge of the corpus is thin.

Since the entry requirements for the corpus are rather strict, the information that it contains is usually reliable enough when we need scientific information as a basis for practical decisions. For many (probably most) practical decisions it is sufficient to just use the information that is available in the corpus, that is, the scientific information that scientists see no reason to doubt. However, this does not hold without exceptions. In some practical decisions we wish to apply standards of proof that differ from those of science. Almost always when this happens, the scientific standards of proof are stricter than those that we wish to apply in our practical decision.[3] This is illustrated in the following example:

> *The Baby Food Example*
> New scientific evidence indicates that a common preservative agent in baby food may have a small negative effect on the child's brain development. According to the best available scientific expertise, the question is far from settled but the evidence weighs somewhat in the direction of there being such an effect. A committee of respected scientists unanimously concluded that although the evidence is not conclusive it is

[3] On situations in which it is the other way around, see Hansson (2014a).

more probable that the effect exists than that it does not. The food safety agency has received a petition whose signatories request the immediate prohibition of the preservative.

In this case, a claim that the substance impairs human brain development would *not* be treated as a (provisional) scientific fact and entered into the corpus. But on the other hand, it is fairly obvious that the vast majority of parents would like the substance to be removed from the food that their children eat. This need not be due to any particular standpoint that they have about chemical substances. Presumably, they would react similarly in the following example:

The Volcano Example
A group of children are tenting close to the top of an old volcano that has not been active for thousands of years. While they are there, seismographs and gas detectors suddenly indicate that a major eruption may be on its way. A committee of respected volcanologists immediately convene to evaluate the findings. They conclude that the evidence is uncertain but weighs somewhat in the direction that a major eruption will take place in the next few days. They unanimously conclude that although the evidence is not conclusive it is more probable that an eruption is imminent than that it is not.

It would be surprising if any of the parents of these children opposed to their evacuation with the argument that it was not scientifically settled whether an eruption would take place or not. The situation is analogous to the baby food example. In both cases, scientific evidence that is too weak to constitute a (provisional) scientific fact is nevertheless considered to be sufficient for a practical decision.

These examples are hopefully sufficient to show beyond reasonable doubt that there are occasions when we wish to base a practical decision on scientific information, but we also wish to apply a less strict standard of evidence for action than the scientific standard of evidence for treating a statement as a (provisional) scientific fact. At least in the baby food case we can also describe this as an application of the precautionary principle.

Figure 12.2 illustrates the use of scientific information for policy purposes. The most common way in which we do this is to use information from the corpus (arrow 2). As already mentioned, for many purposes, this is the only sensible thing

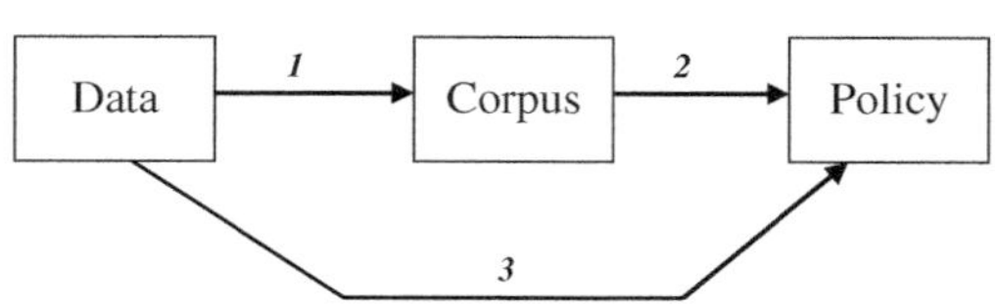

Figure 12.2. The use of science in risk management.

to do, and we need to do nothing more. However, as we have just seen there are situations when exclusive reliance on the corpus would have unwanted consequences. These are the situations where we want to use scientific information for practical purposes although the requirements for corpus entry are not satisfied, as exemplified by the baby food and volcano examples. In these cases we need to have a to have a direct way, a bypass route, from data to policy (arrow 3).

Applying the bypass route is often difficult, and it may require more detailed interaction between scientists and policy makers than the use of information from the corpus. It is essential that the bypass route does not lead to arbitrary and therefore inefficient decisions with priorities determined by the whims of uninformed opinion. To avoid this, we need to ensure that full use is made of scientific information when the bypass route is taken, although the required level of evidence will be adjusted to the practical problem at hand. The following three principles have been proposed for science-based decisions employing the bypass route:

1. *The same type of evidence should be taken into account in the policy process as in the formation of the scientific corpus. Policy decisions are not well served by the use of irrelevant data or the exclusion of relevant data.*

   Example: A decision is going to be made on whether the use of a suspected reproductive toxicant should be restricted. Then the same types of reprotoxicity studies should be used as a basis for this decision that would have been used in a purely scientific review of the effects of the substance.
2. *The assessment of how strong the evidence is should be the same in the two processes.*

   Example: If there is stronger scientific evidence that substance A accumulates in biota than that substance B does so, then this evidence should be counted as stronger in policy discussions as well.
3. *The two processes may differ in the* required *level of evidence. It is a policy issue how much evidence is needed for various practical decisions.*

   Example: Several types of measures can be considered in response to the evidence indicating that bisphenol A may be toxic to children. Examples of such measures are: support for research on the effects of the substance, a public information campaign, labelling requirements for products containing the substance, consultations with industry to achieve a voluntary withdrawal of the substance, a ban in products for children, a total ban of the substance. There are strong reasons why the required level of evidence should be different for these different types of measures. There is no reason why the required level of evidence for any of them should coincide exactly with the level of evidence that is required for scientists to consider it a (provisional) scientific fact that the substance is toxic to children (Hansson, 2008).

Practical decision-making following these principles can be described as complying with the tenets of *science-based precaution*.

12.5 RESEARCH ON USE AND EFFICIENCY

Although the precautionary principle has been the subject of much discussion, there are relatively few cases in which it has been explicitly invoked as a basis of decision-making, and even in these cases it is usually difficult to determine if and how the decision would have been different if the principle had not been applied.[4] Cautious decisions have often been made without reference to the precautionary principle (in particular, in the United States), and even when the principle was referred to it may not have made much difference. Therefore, it is difficult to determine empirically what effects the application of the principle has had.

Proponents of the principle have provided long lists of historical examples of decisions in environmental policy that were insufficiently protective (Harremoës et al., 2001). Many of these examples are quite convincing in retrospect, but of course it is difficult to determine to what extent an explicitly adopted precautionary principle would have led to a different decision. As was noted by John Mendeloff (1988, p. 7), even when government agencies have been explicitly authorized to act upon reasonable suspicions of harmful effects, their actual regulatory activities have been predominantly devoted to substances with well-documented adverse effects in humans. A plausible reason for this may be that there have mostly been a sufficient number of cases of the latter type to keep the agency busy. There are often good reasons for giving known hazards higher priority than suspected ones.

Opponents of the principle have provided examples in which its application seems to have led to overregulation, that is, unnecessarily strict (and expensive) preventive measures (Durodié, 2003; Nilsson, 2004). Much of this criticism does not take into account the impossibility of knowing beforehand exactly in which cases precautions are needed. If we want to avoid volcano disasters we will have to evacuate neighborhoods of volcanoes also on some occasions when we will learn afterward that the evacuation was unnecessary. A criticism of a disaster prevention program based on a list of such "useless" evacuations would be meaningless unless it also took into account the cases in which evacuations saved humans from disaster. Similarly, an unbiased appraisal of precautionary measures would have to systematically evaluate both cases in which the suspected danger was real and cases in which it was not. No such evaluation seems to have been performed.

12.6 WEAKNESSES, LIMITATIONS, AND CRITICISM

12.6.1 Is the Principle Asymmetric?

The precautionary principle has been promoted as a special principle for decision-making on potential adverse environmental and health effects, in particular, adverse

[4] But see Foster (2011) for an investigation of how the principle is applied in international courts and tribunals.

effects of chemicals. This presentation of the principle is arguably not quite accurate since it is, as we saw in Section 12.4, actually an application of a decision-making approach that is widely used in other decision areas. However, given how the principle is usually conceived and often applied, its purpose is to avoid *detrimental effects on human health and the environment on using particular chemical substances*. Given this, at least two types of symmetry issues will emerge:

1. There may also be (certain or uncertain) *detrimental effects on human health and the environment of not using the substance in question*. If the precautionary principle is not applied to these effects, then the outcome of applying the principle may be increased risk taking with human health and the environment.
2. There may also be (certain or uncertain) *detrimental effects of other types associated either with the use or the non-use of the substance in question,* for instance, losses of cultural heritage, increased unemployment, and economic costs.

The first type of asymmetry is the one most often discussed. One of the most commonly used examples is precautionary reduction in the use of pesticides in developing countries. Pesticide use in the third world causes considerable environmental damage, and due to unsafe working conditions it also frequently gives rise to severe occupational disease (Hansson, 2014b). But on the other hand, in some cases a decision not to use these substances may lead to an increased risk of crop failure or to increased spread of diseases such as malaria. It has been argued that if the precautionary principle is applied to these other risks, then it may lend support to the use of pesticides (Attaran et al., 2000; Gray and Hammitt, 2000).

We do not have to delve into the complex details of these particular examples. On a basic, principled level there can hardly be any doubt that asymmetrical applications of the precautionary principle are not defensible. However, it is important to note that this is not a problem with the precautionary principle itself but with the way in which it is applied. Any decision rule will yield strange outcomes if it is applied non-uniformly to the different options or potential outcomes in a decision. For instance, similar problems can arise if we apply expected utility maximization but exclude some of the effects of a decision from the calculations. Therefore the symmetry problems are not arguments against the precautionary principle as such, but rather against certain ways to apply it.

This being said, it must also be noted that the precautionary principle loses much of its bite when scientific uncertainties can be invoked in arguments both for and against a proposed decision. In such a case, the precautionary principle tells us to take the uncertainties on both sides into account. However, it does not provide us with any advice on how to weigh them against each other, and it is far from self-evident whether and in that case how its application will affect the decision.

12.6.2 Strawman Criticism

The precautionary principle has been subject to more criticism than most other safety principles. Unfortunately, much of this criticism commits the strawman fallacy, that is, it is aimed at some easily refutable version of the precautionary principle that is very different from the authoritative definitions of the principle issued by the European Commission and other authorities that subscribe to the principle and promote its application. Some authors have claimed that the precautionary principle requires that even highly implausible suspicions of danger should lead to costly precautionary measures. For instance, Whelan (2000) claimed that the principle requires that "we act on all the remote possibilities in identifying causes of human disease", letting "the distraction of purely hypothetical threats cause us to lose sight of the known or highly probable ones" (for similar views, see: Manson, 1999; McKinney, 1996; Nollkaemper, 1996). Others have claimed that the principle is anti-scientific; for instance Charnley (2000) maintained that proponents of the principle are "challenging the legitimacy of science in general and risk analysis in particular" and that science is threatened by "the belief that instead of science, the precautionary principle will somehow solve all our problems." These critics all construe the principle quite differently from how it has been defined and explained in international treaties, EU documents, etc., where it is made very clear that the principle is only triggered when there are scientifically respectable indications of danger.[5]

12.7 RELATION TO EXPECTED UTILITY AND PROBABILISTIC RISK ASSESSMENT

It is important to recognize that the precautionary principle is a principle for the treatment of (uncertain) scientific information in decision-making. It is *not* a decision-making principle on a par with expected utility maximization, the maximin principle, and others that prescribe what decision to make, given certain options, (values of) outcomes, and probabilities. However, the precautionary principle can be combined with or explicated in terms of some of these principles. Its relationship with expected utility maximization is particularly important.

Expected utility maximization is the dominating approach to decision-making with known probabilities (decision-making under risk). It is indeed "the major

[5] Admittedly, there is one international declaration that supports an interpretation of the precautionary principle that does not require any scientific evidence to trigger its application, namely the declaration of the North Sea conference in The Hague in 1990 that is quoted above in Section 12.2. However, this is a rather unique statement. It is gainsaid in numerous other declarations and in the European documents on the principle that are also quoted in Section 12.2. For instance, in the declaration adopted at the 2002 North Sea conference, the ministers strongly supported the use of science in environmental protection, and emphasized that protective measures should be based on "the current state of scientific knowledge, coupled with a sound application of the precautionary principle" (Bergen Declaration, 2002).

paradigm in decision-making since the Second World War" (Schoemaker, 1982, p. 529), and most of modern risk analysis is based on it. It could more precisely be called "maximization of probability-weighted utility," since its central idea is to weigh the utilities (measures of usefulness) of outcomes according to their probabilities.

Utilities come in many variants. One common approach is to convert everything into money, and put a price on all factors that may influence how we evaluate an outcome (including human lives, lost species, etc.). This is the approach of cost–benefit analysis. Another approach, common in risk analysis, is to focus on fatalities, and use the number of lives lost as a measure of the (dis)value of possible options. A third approach, favored by utilitarian ethicists, measures utility in an arbitrary, hypothetical unit. Suppose, for instance, that you regard the preservation of the giant panda to have twice as much value as the preservation of the Egyptian pyramids. Then it does not matter for the analysis if you assign to them the values 2 and 1 or the values 5000 and 2500. It is only the proportions that matter.

The basic principle of expected utility is very simple, and perhaps best explained with an example. Suppose that you are offered an investment from which you have a probability of 1/3 of winning $30,000 and a probability of 2/3 of losing $13,500. The expected (probability-weighted) outcome in dollars of making that investment is:

$$1/3 \times 30.000 + 2/3 \times (-13.500) = 1000.$$

According to expected utility maximization, this investment is equal in value to receiving $1000.[6] Similarly, a probability of 1/1000 that 2000 people will die should be valued equally as certainty that two persons will die, etc.

Expected utility maximization is generally taken to be the *risk-neutral* decision rule. Decision rules that put more emphasis on avoiding the worst possible outcomes are called risk averse. Decision rules that deviate in the opposite direction may be called risk seeking, risk taking, or risk prone (Arrow, 1971; Hurley, 1993; Kahneman and Tversky, 1984). It is commonly assumed that standard risk analysis is risk neutral and that the precautionary principle deviates from it in the direction of risk aversion. However, although this may be true of some variants of the precautionary principle it is not true in general. A major reason for this is that risk analysts usually base their analysis exclusively on the best available expert assessment rather than using a weighted average of the various assessments that are available. For a simple, schematic example, consider two substances A and B that are alternatives for being used in an application where they will spread into the aquatic environment (Hansson, 2008). B is highly persistent and bioaccumulative, whereas A will readily degrade into water and carbon dioxide in the environment. A has been extensively tested for ecotoxicity, and there are strong reasons to assume that it is ecotoxic. B has not been tested for ecotoxicity. However, it belongs to a group of chemically related substance. Forty of the substances in that group have been tested for ecotoxicity. Of these 2 were

[6] For simplicity, we assume here that the value of money is proportionate to its amount.

TABLE 12.1. The Scientific Information About Two Hypothetical Substances A and B. "PB" means "persistent and bioaccumulative"

Substance	Environmental fate	Effect on biota	Disvalue if toxic	Probability of toxicity
A	Non-PB	Ecotoxic	−10	1
B	PB	Ecotoxicity unknown	−500	0.05

ecotoxic (both to about the same degree as substance A) whereas the other 38 were not. Therefore, the best possible estimate of the probability that B is ecotoxic is 5%. For decision-theoretical purposes, tentative numerical values have to be assigned to possible outcomes. In this example, the (dis)value of spreading a toxic substance that is persistent and bioaccumulating is −500 units whereas that of spreading an equally toxic substance that degrades readily is −10 units. (Of course, the unit is arbitrary, and it is the ratio that matters.) This information is summarized in Table 12.1.

In standard risk analysis, no conclusion would be drawn from the fact that substance B belongs to a group in which 5% of the substances are ecotoxic. This means that the analysis will proceed in the same way as if we knew that B is not harmful to the environment. The outcome of such an analysis is shown in Table 12.2. It supports a decision to use substance B rather than substance A.

In Table 12.3, standard decision theory is applied to the same problem. In other words, the best available probability assignments are used. The outcome of this calculation supports a decision to prefer substance A to substance B. This analysis supports what would usually be regarded as a "precautionary" decision. Informally, the argument for preferring A to B is that although it is not known if B causes environmental damage, if it does so then the damage may be very serious. If the severity of a possible danger is large, then it may be rational to take action against that danger even if the probability is relatively small that the danger will materialize.

As this example shows, the common practice to disregard all but the most probable hypothesis is typically not risk neutral. To the contrary, if the competing hypotheses indicate a larger risk, then this practice is risk seeking. Perhaps surprisingly, the risk-neutral ideal may be better approximated by decision rules constructed to comply with the precautionary principle (Hansson, 2006; Hansson and Johannsson, 1997). According to that principle, unproven but scientifically credible dangers should be

TABLE 12.2. Standard Risk Assessment of the Two Hypothetical Substances

Substance	Environmental fate	Effect on biota	Disvalue if toxic	Probability of toxicity	Risk assessment
A	Non-PB	Ecotoxic	−10	1	−10
B	PB	Ecotoxicity unknown	−500	0 *(reduced)*	0

TABLE 12.3. Expected Utility Applied to the Same Problem

Substance	Environmental fate	Effect on biota	Disvalue if toxic	Probability of toxicity	Risk assessment
A	Non-PB	Ecotoxic	−10	1	−10
B	PB	Ecotoxicity unknown	−500	0.05	−25

taken into account in decision-making. If such dangers are assigned weights in accordance with their estimated probabilities, then this variant of the precautionary principle comes closer than standard risk analysis to what is traditionally called a risk-neutral approach.

Currently, chemical substances with unknown properties are in practice treated in much the same way as substances for which there is information strongly indicating that they are harmless. For instance, whereas substances with known hazardous properties have to be provided with labelling specifying the dangers and how to avert them, no labelling is required for substances whose effects on human health and the environment are unknown. This is clearly not a risk-neutral approach. It would be more risk neutral to label these substances as "insufficiently investigated" and to treat them as if they had the average toxicity of chemical substances in use. (For an estimate of this "default toxicity," see Hansson and Rudén, 2008.) Needless to say, such a default level would have to be replaced by rulings based on the actual properties of the substance as soon as these become known. For our present purposes we can leave out the technical details. The important conclusion is that treating substances with unknown properties as if they were harmless is a risk prone, not a risk-neutral practice. If the precautionary principle is applied in such cases in the way described above it will result in a movement away from an ongoing risk-seeking practice in the direction of risk neutrality.

12.8 RELATIONS TO OTHER SAFETY PRINCIPLES

There are many other safety principles that tend to support the same practical measures as the precautionary principle. This applies for instance to the substitution principle and inherent safety. (See Chapters 17 and 26.) In this section the focus will be on two principles that are often seen as closely related to the precautionary principle, or even identical to it, namely the maximin rule, and a reversed burden of proof, and finally to a principle that tends to contradict it, namely "sound science."

12.8.1 Maximin

The *maximin* strategy can be summarized as the precept "Behave as if we live in the worst of all possible worlds." More precisely, it can be formulated as follows:

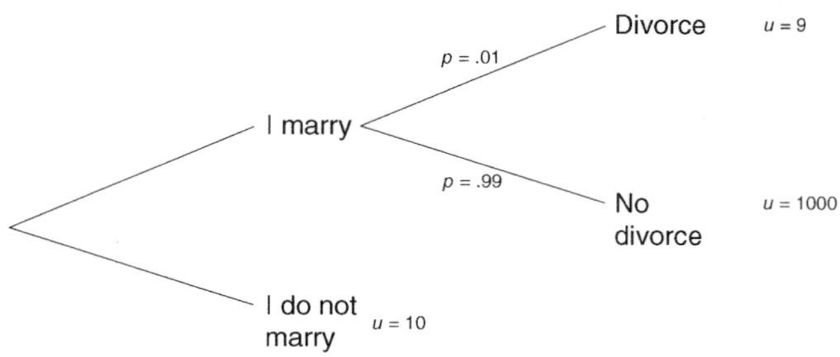

Figure 12.3. The maximin rule exemplified in a decision tree.

> The maximin rule:
> Consider only the worst possible outcome associated with each option. It is called the security level of that outcome. Choose the option (one of the options) whose security level is not worse than that of any other options.

For an example, suppose that you consider whether or not to marry the person you love. You consider living alone to be pretty depressing, but you are convinced that a lifelong marriage with this person would make your life wonderful. The worst thing that could happen would be that your future spouse leaves you, but you consider that to be very improbable. Putting numbers on this, we can assume that you consider unmarried life to have the value 10, lifelong marriage the value 1000 and marriage followed by a divorce the value 9. Furthermore you believe the probability of divorce to be 1%. This is illustrated in a decision tree in Figure 12.3.

The security level of marrying is 9 and that of not marrying is 10, so if you apply the maximin rule you will decide not to marry. (In contrast, the expected utility of not marrying is 10 whereas that of marrying is just above 990, so the expected utility rule would recommend marriage.) As this example illustrates, the maximin rule is implausibly risk averse, and it can be seen as an expression of uncompromising cautiousness.

The maximin rule also has another problem: It is extremely sensitive to how we choose to delimit a decision. We normally exclude very implausible outcomes from our considerations. For instance, when deciding whether to accept an invitation to Sicily I do not worry about the possibility that Mount Etna may have a sudden unpredicted eruption. The inclusion or exclusion of outcomes with extremely low probabilities has a much larger impact on decision-making with the maximin rule than for many other rules such as expected utility maximization. In order to avoid a total focus on unrealistic scenarios the maximin rule will have to be combined with some formal or informal procedure that excludes at least some of the extremely improbable consequences from consideration. The precautionary principle does not have the corresponding problem, at least not in its science-based variants, since only scientifically plausible events are taken into account.

12.8.2 A Reversed Burden of Proof

It has often been proclaimed that in the regulation of chemical hazards, the onus of proof should fall to those who claim that a substance can be used with impunity, not those who wish to restrict its use. However, although much may seem to speak in favor of such a principle, in practice it cannot be implemented. Although it can often be proved beyond reasonable doubt that a substance has a particular adverse effect, proving the opposite is much more difficult. It is difficult enough to prove beyond reasonable doubt that a substance does not have a particular adverse effect. Proving that it has no adverse effect at all is impossible in the vast majority of cases.

One major reason for this is that, with respect to serious health effects, we care about risks that are small in terms of experimental statistics. If we only cared about whether or not a substance causes a disease in more than one-tenth of the exposed population, then this problem would not arise. For ethical reasons, we wish to detect much lower frequencies of adverse effects. However, for statistical reasons it is very difficult to exclude the existence of low-frequency effects. As a rough rule of thumb, even under favorable conditions an epidemiological study cannot detect an excess relative risk reliably if it is smaller than 10%. For the more common types of disease, such as lung cancer and ischemic heart disease, lifetime risks are between 1% and 10%. Therefore, even in the most sensitive studies, lifetime risks smaller than 10^{-2} or 10^{-3} cannot be observed (Vainio and Tomatis, 1985). Animal experiments have the same type of statistical problems. In order to detect a small effect, very large groups of animals would have to be used. For instance, in order to detect an increase of the mutation rate by 0.5%, about 8,000,000,000 mice could be required (Weinberg, 1972, p. 210). Furthermore, due to inter-species differences, the absence of effects in animals is no proof of safety in humans. In conclusion, the "reversed burden of proof" is a chimera. After a substance has been thoroughly investigated and no signs of adverse effects have been found, we can conclude that chances are much higher than for many other substances that it will not cause any serious problems. Such thorough investigations are also what can be demanded of producers of a substance. Requiring that they prove that the substance is non-hazardous is (unfortunately) tantamount to requiring the impossible (Hansson, 1997).

12.8.3 Sound Science

According to proponents of the so-called "sound science," only well-established scientific facts should be used in decision-making. This means, in practice, that the bypass route to policy decisions described in Section 12.4 will be closed. When there are indications but no full proof that a substance is harmful, it will be treated as innocuous, that is, as if the probability of danger is zero.

The promotion of "sound science" began in 1993, when the tobacco company Philip Morris initiated and funded an ostensibly independent organization called The Advancement of Sound Science Coalition (TASSC). Its major purpose was to discredit the then rapidly accumulating evidence that passive smoking has negative health effects. The term "sound science" was launched by TASSC as a slogan for the position that no action should be taken against a possible danger until full scientific proof of that danger has been obtained (Ong and Glantz, 2001).

The "sound science" campaign was based on a remarkably unsophisticated approach to the relationship between science and policy. As we saw in Section 12.4, science cannot tell us what levels of evidence we should require for various policy measures. That has to be decided by policy makers, based on their mandate, goals, and values. Just like decision-makers in all other parts of the society, those who make decisions on safety, health, and the environment often have to act against a potential danger at a stage when it is still not know for sure whether it is also a real danger. Not doing so invites disaster.

ACKNOWLEDGMENT

I would like to thank Marko Ahteensuu, Anna Beronius, and Niklas Möller for valuable comments on an earlier version of this chapter.

REFERENCES

Ågerstrand, M., Edvardsson, L., and Rudén, C. (2014). Bad reporting or bad science? Systematic data evaluation as a means to improve the use of peer-reviewed studies in risk assessments of chemicals. *Human and Ecological Risk Assessment*, 20, 1427–1445.

Ahteensuu, M. (2008). *In Dubio Pro Natura? A Philosophical Analysis of the Precautionary Principle in Environmental and Health Risk Governance*. Turku: University of Turku.

Anses (2013). *Évaluation Des Risques du Bisphénol A (BPA) Pour La Santé Humaine. Tome 1. Avis De L'ANSES. Rapport D'expertise Collective*. Paris: Agence nationale de sécurité sanitaire de l'alimentation, de l'environnement et du travail.

Arrow, K. J. (1971). *Essays on the Theory of Risk-Bearing*. Amsterdam: North-Holland.

Attaran, A., Liroff, R., and Maharaj, R. (2000). Doctoring malaria, badly: The global campaign to ban DDT. *British Medical Journal*, 321, 1403–1404.

Bergen Declaration. (2002). In: Fifth international conference on the protection of the North Sea, Bergen, Norway, March 20–21, 2002. Available at: http://www.ospar.org/site/assets/files/1239/5nsc-2002_bergen_declaration_english.pdf

Beronius, A., Rudén, C., Håkansson, H., and Hanberg, A. (2010). Risk to all or none? A comparative analysis of controversies in the health risk assessment of Bisphenol A. *Reproductive Toxicology*, 29, 132–146.

Boehmer-Christiansen, S. (1994). The precautionary principle in Germany—enabling government. In: T. O'Riordan, and J. Cameron (Eds.), *Interpreting the Precautionary Principle* (pp. 31–59). London: Cameron May.

Bridges, J. W., and Bridges, O. (2001). Hormones as growth promoters: The precautionary principle or a political risk assessment? In: P. Harremoës et al. (Eds.), *Late Lessons from Early Warnings: The Precautionary Principle 1896–2000* (pp. 149–156). Luxembourg: Office for Official Publications of the European Communities.

Burgos, R., and Defeo, O. (2004). Long-term population structure, mortality and modeling of a tropical multi-fleet fishery: The red grouper *Epinephelus morio* of the Campeche Bank, Gulf of Mexico. *Fisheries Research*, 66, 325–335.

Cameron, J., and Abouchar, J. (1996). The status of the precautionary principle in international law. In: D. Freestone, and E. Hey (Eds.), *The Precautionary Principle and International Law: The Challenge of Implementation. International Environmental Law and Policy Series* (Vol. 31, pp. 29–52). The Hague: Kluwer Law International.

Charnley, G. (2000). 1999 annual meeting: Past president's message: Risk analysis under fire. *RISK Newsletter*, 20(1), 3.

Commission White Paper on the Strategy for a Future Chemicals Policy. (2001, February 27). Available at: http://eur-lex.europa.eu/legal-content/EN/TXT/HTML/?uri=URISERV:l21275&from=EN

Communication from the Commission on the Precautionary Principle. (2000, February 2). Available at: http://ec.europa.eu/dgs/health_consumer/library/pub/pub07_en.pdf

Consolidated version of the Treaty on the Functioning of the European Union. (2012, October 26). Available at: http://eur-lex.europa.eu/legal-content/EN/TXT/PDF/?uri=CELEX:12012E/TXT&from=EN

Durodié, B. (2003). The true cost of precautionary chemicals regulation. *Risk Analysis*, 23, 389–398.

Foster, C. E. (2011). *Science and the Precautionary Principle in International Courts and Tribunals: Expert Evidence, Burden of Proof and Finality*. Cambridge, MA: Cambridge University Press.

Freestone, D. (1991). The precautionary principle. In: R. Churchill, and D. Freestone (Eds.), *International Law and Global Climate Change* (pp. 21–39). London: Graham & Trotman/Martinus Nijhoff.

Freestone, D., and Ellen, H. (1996). Origins and development of the precautionary principle. In: D. Freestone, and E. Hey (Eds.), *The Precautionary Principle and International Law: The Challenge of Implementation. International Environmental Law and Policy Series* (Vol. 31, pp. 3–15). The Hague: Kluwer Law International.

Friedberger, M. (1994). Cattlemen, consumers, and beef. *Environmental History Review*, 18(3), 37–57.

Gray, G. M., and Hammitt, J. K., (2000). Risk/risk trade-offs in pesticide regulation: An exploratory analysis of the public health effects of a ban on organophosphate and carbamate pesticides. *Risk Analysis*, 20, 665–680.

Haigh, N. (1994). The introduction of the precautionary principle into the UK. In: T. O'Riordan, and J. Cameron (Eds.), *Interpreting the Precautionary Principle*(pp. 229–251). London: Cameron May.

Hansson, S. O. (1997). Can we reverse the burden of proof? *Toxicology Letters*, 90, 223–228.

Hansson, S. O. (2004). Philosophical perspectives on risk. *Techne*, 8(1), 10–35.

Hansson, S. O. (2006). Economic (ir)rationality in risk analysis. *Economics and Philosophy*, 22,231–241.

Hansson, S. O. (2008). Regulating BFRs—from science to policy. *Chemosphere*, 73, 144–147.

Hansson, S. O. (2014a). How context dependent is scientific knowledge?. In: F. Lihoreau, and M. Rebuschi (Eds.), *Epistemology, Context and Formalism* (pp. 127–140). Synthese Library, Springer.

Hansson, S. O. (2014b). Occupational risks in agriculture In: P. B. Thompson, and D. M. Kaplan (Eds.), *Encyclopedia of Food and Agricultural Ethics* (pp. 1461–1467). Dordrecht: Springer.

Hansson, S. O. (2016). How to be cautious but open to learning: Time to update biotechnology and GMO legislation. *Risk Analysis*, 36, 1513–1517.

Hansson, S. O., and Johannesson, M. (1997). Decision-theoretic approaches to global climate change In: G. Fermann (Ed.), *International Politics of Climate Change* (pp. 153–178). Oslo: Scandinavian University Press.

Hansson, S. O., and Rudén, C. (2008). A risk-neutral default for chemical risk management. *American Journal of Industrial Medicine*, 51, 964–967.

Harremoës, P., Gee, D., MacGarvin, M., Stirling, A., Keys, J., Wynne, B., and Guedes Vaz, S. (2001). *Late Lessons from Early Warnings: The Precautionary Principle 1896–2000.* Luxembourg: Office for Official Publications of the European Communities.

Hileman, B. (1998). Precautionary principle. *Chemical and Engineering News,* 76(6), 16–18.

Hurley, S. L. (1993). Justice without constitutive luck. *Philosophy*, 35(Suppl.), 179–212.

JoU 1984/85:30. Jordbruksutskottets betänkande om kemikaliekontroll. Stockholm: Sveriges Riksdag.

Kahneman, D., and Tversky, A. (1984). Choices, values, and frames, *American Psychologist*, 39,341–350.

Karlsson, M. (2006). The precautionary principle, Swedish chemicals policy and sustainable development. *Journal of Risk Research*, 9, 337–360.

Levin, R., Hansson, S. O., and Rudén, C. (2004). Indicators of uncertainty in chemical risk assessments. *Regulatory Toxicology and Pharmacology*, 39, 33–43.

Manson, N. (1999). The precautionary principle, the catastrophe argument, and Pascal's wager. *Ends and Means*, 4(1),12–16.

McKinney, W. J. (1996). Prediction and Rolston's environmental ethics: Lessons from the philosophy of science. *Science and Engineering Ethics*, 2,429–440.

Mendeloff, J. M. (1988). *The Dilemma of Toxic Substance Regulation.* Cambridge, MA: MIT Press.

Ministerial Declaration of the Third International Conference on the Protection of the North Sea, The Hague, March 8, 1990. Available at: http://www.ospar.org/site/assets/files/1239/3nsc-1990-hague_declaration.pdf

Ministerial Declaration on the Protection of the North Sea (1987). *Environmental Conservation*, 14, 357–361.

Morris, J. (2000). Defining the precautionary principle. In: J. Morris (Ed.), *Rethinking Risk and the Precautionary Principle* (pp. 1–21). Oxford: Butterworth-Heinemann.

Nilsson, R. (2004). Control of chemicals in Sweden: An example of misuse of the 'precautionary principle', *Ecotoxicology and Environmental Safety*, 57,107–117.

Nollkaemper, A. (1996). 'What you risk reveals what you value', and other dilemmas encountered in the legal assaults on risks. In: D. Freestone, and E. Hey (Eds.), *The Precautionary Principle and International Law* (pp. 73–94). Dordrecht: Kluwer Law International.

Ong, E. K., and Glantz, A. S. (2001). Constructing 'sound science' and 'good epidemiology': Tobacco, lawyers, and public relations firms. *American Journal of Public Health*, 91, 1749–1757.

O'Riordan, T., and Jordan, A. (1995). The precautionary principle in contemporary environmental politics. *Environmental Values*, 4, 191–212.

O'Riordan, T., and Cameron, J. (Eds.). (1994). *Interpreting the Precautionary Principle*. London: Earthscan.

O'Riordan, T., Cameron, J., and Jordan, A. (Eds.). (2001). *Reinterpreting the Precautionary Principle*. London: Cameron May.

Ozonoff, D. (1999). The precautionary principle as a screening device. In: C. Raffensperger, and J. Tickner (Eds.), *Protecting Public Health and the Environment: Implementing the Precautionary Principle* (pp. 100–105). Washington, DC: Island Press.

Proposition 1969:28. (1969). Om miljöskyddslag [Governmental Bill]. Stockholm: Regeringskansliet.

Pyhälä, M., Brusendorff, A. C., and Paulomäki, H. (2010). The precautionary principle. In: M. Fitzmaurice, D. M. Ong, and P. Merkouris (Eds.), *Research Handbook on International Environmental Law* (pp. 203–226). Cheltenham, UK: Edward Elgar.

Raffensperger, C., and Tickner, J. (Eds.). (1999). *Protecting Public Health and the Environment: Implementing the Precautionary Principle*. Washington, DC: Island Press.

Report of the United Nations Conference on Environment and Development, Rio de Janeiro, June3–14, 1992. Available at: http://www.un.org/documents/ga/conf151/aconf15126-1annex1.htm.

Rudén, C., and Hansson, S. O. (2010). REACH is but the first step—how far will it take us? Six further steps to improve the european chemicals legislation. *Environmental Health Perspectives*, 118(1),6–10.

Sandin, P. (1999). Dimensions of the precautionary principle. *Human and Ecological Risk Assessment*, 5,889–907.

Schefczyk, M. (2016). Financial markets: Applying argument analysis to the stabilisation task. to be published In: S. O. Hansson, and G. H. Hadorn (Eds.), *The Argumentative Turn in Policy Analysis: Reasoning about Uncertainty* (pp. 265–290), Springer.

Schoemaker, P. J. H. (1982). The expected utility model: Its variants, purposes, evidence and limitations. *Journal of Economic Literature*, 20,529–563.

Smith, B. (1998). Ethics of Du Pont's CFC strategy 1975–1995. *Journal of Business Ethics*, 17, 557–568.

Vainio, H., and Tomatis, L. (1985). Exposure to carcinogens: Scientific and regulatory aspects. *Annals of the American Conference of Governmental Industrial Hygienists*, 12,135–143.

Weinberg, A. M. (1972). Science and trans-science. *Minerva*, 10, 209–222.

Whelan, E. M. (2000). Can too much safety be hazardous? A critical look at the 'Precautionary Principle.' Available at: http://www.virginiabiosolids.com/pdf/Precautionary_Principle.pdf.

Wiener, J. B., and Rogers, M. D. (2002), Comparing precaution in the United States and Europe. *Journal of Risk Research*, 5,317–349.

Wingspread Statement on the Precautionary Principle. (1998). Available at: http://www.sehn.org/emandeh.html.

Wood, H. W., Jr. (1985). The United Nations World Charter for Nature: The developing nations' initiative to establish protections for the environment. *Ecology Law Quarterly*, 12,977–996.

FURTHER READING

Ahteensuu, M. (2007). Defending the precautionary principle against three criticisms. *Trames*, 11, 366–381. [A defence of the principle against claims that it is vague, incoherent, and counterproductive.]

Hansson, S. O., and Rudén, C. (2004). A science-based strategy for chemicals control. *Industry and Environment*, 27(2–3), 12–18. [Applies the principle to chemicals policy.]

Steel, D. (2015). *Philosophy and the Precautionary Principle. Science, Evidence, and Environmental Policy*. Cambridge, MA: Cambridge University Press. [Very good coverage of the literature on the interpretation of the principle.]

Sunstein, C. R. (2005). *Laws of Fear. Beyond the Precautionary Principle*. Cambridge, MA: Cambridge University Press. [A highly critical appraisal of the principle.]

13

OPERATING PROCEDURE

Jinkyun Park

13.1 INTRODUCTION

According to operational experience of complex socio-technical systems such as nuclear power plants (NPPs), it has been shown that human performance is a key factor in securing their operational safety (Disaster Database, 2017; Khalaquzzaman et al., 2010; Preischl and Hellmich, 2013; Ruckart and Burgess, 2007). For example, Kim et al. (2009) claimed that about 60% of unexpected reactor trip events that have occurred in the domestic NPPs of Republic of Korea during full-power operations were attributable to human errors. Although unexpected reactor trip events did not cause any serious consequence, it is evident that all kinds of events in complex socio-technical systems have a potential to trigger catastrophic failures with a dramatic casualty toll, such as massive deaths and/or injuries, significant environmental damage and tremendous financial losses. Consequently, diverse countermeasures to reduce and/or prevent human errors have been evolved for many decades. One of the typical countermeasures is to develop an operating procedure.

According to available online dictionaries, the term procedure can be understood in several ways. For example: "The act, method, or manner of proceeding in some

Handbook of Safety Principles, First Edition. Edited by Niklas Möller, Sven Ove Hansson, Jan-Erik Holmberg, and Carl Rollenhagen.

action; especially, the sequence of steps to be followed" (Your Dictionary, 2014); "A series of actions conducted in a certain order or manner" (Oxford Dictionary, 2014); "A fixed, step-by-step sequence of activities or course of action (with definite start and end points) that must be followed in the same order to correctly perform a task" (Business Dictionary, 2014). Although there are slight differences in describing what the *procedure* is, the abovementioned explanations allow us to imagine that operating procedures can support human operators through providing detailed instructions that specify what to be done and how to do it (Blackmore, 2004; Hale and Borys, 2013). More precisely, a *good* procedure can be defined as a procedure that provides crucial contents (e.g., information, detailed action specifications, and the sequence of actions) so that people, even novices, are able to properly perform the required actions to achieve their goal or purpose in real life (Park, 2009, p. 5). If so, it seems to be apparent that the possibility of human errors will be reduced in effect if all the contents of instructions being prescribed in a given procedure are properly performed.

In this regard, the bright side of a good procedure could be at least threefold: (1) reducing physical and/or mental load, (2) preventing forgetting and/or skipping a crucial task, and (3) minimizing the variability of human operators in conducting a required task (Bullemer and Hajdukiewicz, 2004; HSE, 2007; Park and Jung, 2008). Actually, operational experience of complex socio-technical systems for several decades has clearly demonstrated that operating procedures are very practical for reducing the possibility of human errors (DOE, 1992; EPA, 2001; HSE, 2005; HSE, 2007; Inaba et al., 2004; Praino and Sharit, 2016; Wieringa et al., 1998).

At the same time, however, it is revealed that a large portion of human errors still has been caused by procedure problems (Bullemer and Hajdukiewicz, 2004; Hale, 1990; HSE, 2007; Reason et al., 1998). For example, Hale and Borys stated that: "Reports of accidents, such as Challenger point to the normalization of deviance from rules as a primary cause of such accidents, whilst the enquiry into the Deepwater Horizon disaster castigates the company and the regulator for not having explicit procedures to govern changes in the well-drilling, –capping and –testing methods as used in that case. A Dutch study, analyzing incidents of loss of containment in the chemical process industry, found 50% related to procedures, of which 10% where there were no or unclear procedures, 12% where the procedure was wrong and 28% where a correct procedure was not followed (p. 208)." These contradictory observations may indicate that we probably missed a couple of critical principles that should be thoroughly implemented during the development of *good* operating procedures. From this concern, based on the results of recent studies, an additional principle is issued in this chapter with two kinds of promising rules, which could be considered for developing the procedure of socio-technical systems.

Before the explanation of the additional principle being suggested in this chapter, however, it is indispensable to understand required aspects to be incorporated in an operating procedure. In other words, although explanations shown above are helpful for defining the role of the procedure, they are not sufficient for clarifying essential information, such as *who is the expected user?, what are the minimal contents to be described in the instruction of a procedure?*, and *who are able to develop the procedure?* In this regard, it would be better to start this chapter with the comparison of working definitions on three similar documents (i.e., manual, guideline, and procedure).

13.2 MANUAL, GUIDELINE, AND PROCEDURE

Although the nature of a manual, guideline, and procedure are quite different, it is true that there are times when we use them interchangeably without clear distinctions. For example, it is perhaps not easy for us to draw a sharp line among *maintenance manual*, *maintenance guideline*, and *maintenance procedure*. In this light, it is very interesting to compare their working definitions as summarized in Figure 13.1.

From Figure 13.1, it is possible to list up interesting features for each document from different perspectives, such as *scope*, *purpose*, and *contents* to be included. For example, the *scope* of the manual (or handbook) focuses on a special subject, while that of the guideline as well as procedure should cover a specific policy. Accordingly, the *purpose* of the manual is to provide ready-to-use references pertaining to the given subject as much as possible, while that of the guideline is to provide a statement for determining a course of action that are able to materialize the given policy. In addition, the *purpose* of the procedure seems to be more circumstantial than that of the guideline because it contains required tasks to implement the given policy, which describe more detailed instructions to be followed by an end user. Accordingly, it is strongly expected that the *contents* of procedures should include instructions describing a fixed, step-by-step sequence of activities or a course of actions with definite start and end points, which must be followed to correctly perform a task. This alludes to the fact that any kind of deviation from prescribed instructions is strictly prohibited (i.e., it is mandatory for the end user to follow instructions institutionalized in a procedure).

To this end, however, the end user of a given procedure has sufficient knowledge and/or experience in a specific domain. That is, for example, it is absurd for a high school student to use the maintenance procedure of a commercial aircraft. In contrast, since a manual contains all kinds of ready-to-use references (e.g., figures, tables, charts, and even the telephone numbers or official homepages of related organizations), it is anticipated that anyone who is interested in a given subject (e.g., a smart phone or laptop computer) can properly use it. In this regard, Table 13.1 shows the results of comparisons among three types of documents with respect to the associated features.

From Table 13.1, it should be noted that the characteristics of the last two perspectives (i.e., *Expected developer* and *Sensitivity to environment*) are deduced from other characteristics. For example, in the case of a procedure, it is strongly anticipated that the developer of a certain procedure should be a person who has sufficient knowledge and/or experience in a specific domain, if it contains detailed instructions to be

Type	Working Definition	Reference
Manual (handbook)	• *Handbook* is sometimes applied to documents that are produced within an organization but not designed for publication. In this case, the term is used nearly synonymously with *manual*. • A handbook is a type of reference work, or other collection of instructions, that is intended to provide ready reference. • A handbook is a treatise on a special subject. • Handbooks may deal with any topic, and are generally compendiums of information in a particular field or about a particular technique. They are designed to be easily consulted and provide quick answers in a certain area.	Wikipedia, 2014a
Guideline	• A guideline is a statement by which to determine a course of action. • A guideline aims to streamline particular processes according to a set routine or sound practice. • By definition, following a guideline is never mandatory. Guidelines are not binding and are not enforced. • Guidelines may be issued by and used by any organization (governmental or private) to make the actions of its employees or divisions more predictable, and presumably of higher quality.	Wikipedia, 2014b
Procedure	• A procedure is a document written to support a *policy directive*. • A procedure is designed to describe *who, what, where, when,* and *why* by means of establishing corporate accountability in support of the implementation of a policy. • For example, a manufacturing facility established a policy that all overtime shall be approved. A procedure can be created to establish *who* can approve overtime (ranks, roles, and responsibilities), *what* forms/systems need to be used, *where* they are located, *when* overtime is applicable. And the *why* refers to the management directive established via a *policy*.	Wikipedia, 2014c

Figure 13.1. Comparing working definition of three document types.

TABLE 13.1. Comparing the Features of a Manual, Guideline, and Procedure

Perspective	Manual (handbook)	Guideline	Procedure
Purpose	Providing ready-to-use references pertaining to a given subject	Providing a statement for determining a course of action to implement a given policy (e.g., criteria)	Providing circumstantial tasks to implement a given policy
Contents	Compendium of ready-to-use information • Tables • Processes to do something • Figures • Models, etc.	The list of particular processes being identified based on rationales such as a set routine or sound practice	A fixed, step-by-step sequence of activities or a course of actions with definite start and end points that must be followed to correctly perform a task
Expected end user	Anyone who has an interest in a given subject	Personnel related to operation (e.g., qualified operators, process supervisors, decision-makers)	Qualified operators who are responsible for manipulating components or equipment
Expected developer	Anyone who has sufficient knowledge related to a given subject	Qualified personnel with sufficient knowledge and experience in a specific domain	Qualified personnel with sufficient knowledge and experience in a specific domain
Sensitivity to environment	Less sensitive or insensitive	Less or slightly sensitive	Very sensitive

conducted by qualified operators. In addition, it is very natural to presume that the more the detailed instructions provided the more their applicability largely depends on the nature of task environments such as the variation of process variables (e.g., the fast vs. slow variation of a temperature), weather, the time of a day, location, and available resources. This characteristic will be reconsidered in Section 13.3.

13.3 EXISTING PRINCIPLES FOR DEVELOPING A GOOD PROCEDURE

From the characteristics of a procedure summarized in Table 13.1, it is possible to say that we have a *good* procedure if it contains all the necessary instructions to accomplish required tasks in an optimal way. In other words, if we have such a procedure, then it is believed that every situation can be soundly controlled by conducting prescribed instructions in a step-by-step and timely manner. In this light, many guidelines as well as principles have been suggested for several decades, which seem to mainly focus on how to describe necessary instructions so that end users can easily and clearly recognize what should be done from a given procedure (i.e., an *easy-to-read* procedure). For example, in the nuclear industry, NUREG-0700 (O'Hara et al., 2002) and NUREG-0711 (O'Hara et al., 2012) give specific guidelines on how to design a procedure (e.g., human factors principles) and how to validate it (i.e., required tests and results). In addition, more practical principles are provided in the form of a procedure writer's manual dealing with the format of a procedure, font sizes to enhance the readability of a procedure, and writing styles (e.g., do not use a negative expression) with sentence structures (e.g., do not overuse conditional expressions including AND, OR, and NOT) to describe necessary instructions (Brune and Weinstein, 1983; DOE, 1992; EPA, 2001; Fuchs et al., 1981; Inaba et al., 2004; Wieringa et al., 1998).

With these guidelines and principles, however, it is continuously reported that various kinds of performance deficiencies still resulted from procedure-related problems (Boy and Schmitt, 2013; Hale and Borys, 2013; Praino and Sharit, 2016). Table 13.2 shows a couple of observations reported from diverse industrial sectors, which are reproduced from Park and Jung (2003). From these observations, it is

TABLE 13.2. Non-compliance Behaviors Observed from Diverse Industrial Sectors

Most airline employees (about 77%) felt that written materials were available when needed. But 57% of airline employees pointed out that written materials are not presented in understandable format	Rankin et al. (2000)
Pilots' knowledge, expertise, and know-how significantly influence the following of written procedures. These cognitive functions enable them to evaluate the situation, to categorize information presented, to evaluate the relevance and the feasibility of information presented, to plan, and to execute adequate actions at the proper time	Brito (2002)
The operators reported that working with procedures made work much less rewarding and the job more difficult than it would otherwise be. Consequently, the operators expressed a desire to see more concise forms of procedures which described tasks only in terms of important details	Marsden (1996)
It is assumed that all plant personnel act in a manner they believe to be in the best interests of the plant. Any intentional deviation from standard operating procedures is made because the employee believes their method of operation to be safer, more economical, or more efficient, or because they believe performance as stated in the procedure to be unnecessary	Fullwood and Hall (1988)

Source: Park and Jung 2003. Reproduced with permission of Elsevier.

possible to make an assumption specifying why people (e.g., operating personnel) frequently do not follow a procedure as written. That is, if people thought that the contents of procedures are inaccurate, incomplete, and obsolete so that the required purposes of goals cannot be accomplished in an effective and understandable manner, then they are susceptible to deviating from procedures.

Actually, the results of several studies directly support this assumption (Hale and Borys, 2013; Praino and Sharit, 2016; NRC, 1997). For example, Reason et al. (1998) describe 10 forms of rule-related behaviors with the associated characteristics: (1) mistake, (2) correct improvisation, (3) correct but unrewarding violation, (4) correct violation, (5) mispliance, (6) incorrect but rewarding compliance, (7) misvention, (8) incorrect but rewarding violation, (9) correct but unrewarding compliance, and (10) correct compliance. Accordingly, it is promising to expect that a couple of important principles could not be properly incorporated in the development of an operating procedure. In this regard, we are able to identify two kinds of radical problems from procedure characteristics summarized in Table 13.1.

The first problem is the level of details caused by the difference of domain knowledge and/or experience between the developer and end user of an operating procedure. For example, Bovair and Kieras (1991) claimed that good and bad instructions could be identified not from text characteristics pertaining to reading comprehension (e.g., complex conditional expressions including OR, AND, and NOT) but from the quality of their contents. In other words, even though there is an instruction described by a simple conditional expression, it is a poor instruction if critical information is omitted (e.g., the orientation of a part to be assembled). In addition, DOE (1992) emphasized the importance of level of details as: "Writing at the appropriate level of detail is the key to successful communication with procedure users. For ease of use and to reduce distraction and confusion, only include information in the procedure that relates directly to completing the task (p. 10)."

In this regard, it is very important to point out that the developer of an operating procedure is usually highly experienced personnel who have sufficient knowledge and/or experience in a specific domain. In other words, the amount of knowledge being possessed by the highly experienced personnel would be much plentiful compared to that of less experienced personnel, although the latter have passed required qualifications.

For example, let us consider an arbitrary system comprised of four valves with a large reservoir (Figure 13.2). In this configuration, in order to isolate the reservoir, a highly experienced operator (i.e., a procedure developer) can describe the corresponding instruction in several ways, such as (1) "Isolate the reservoir," (2) "Close all the inlet and outlet valves linked to the reservoir," (3) "Close valves 101, 102, 103, and 104," and (4) "Close valves 101, 102, 103, and 104. Note that switch knobs are located in Panel A with distinctive labels."

Of them, the procedure developer probably does not want to adopt the last description style because it seems to be too inquisitive as an instruction to be conducted by qualified operators. Actually, this expectation (i.e., evading the provision of an instruction with detailed descriptions) is true if this instruction is faced with a

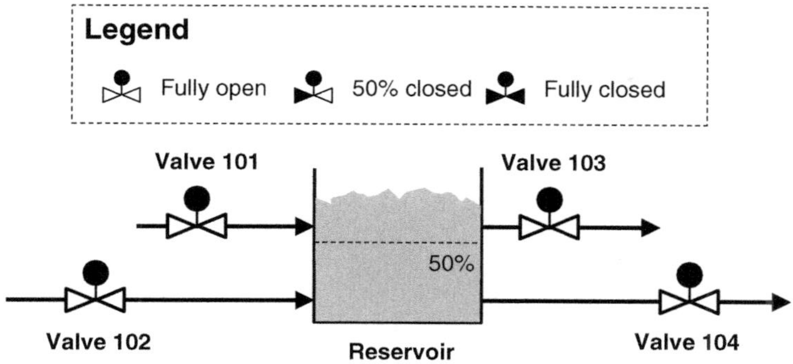

Figure 13.2. An arbitrary system with four valves and one reservoir.

qualified operator whose knowledge and experience are comparable to those of the procedure developer. Unfortunately, it seems that less experienced (but still qualified) operators are apt to feel a distraction when they have to conduct an instruction described in a succinct style (e.g., "Isolate the reservoir"). One promising explanation is that, in order to properly isolate the reservoir, additional domain-specific knowledge such as the number and name of valves linked to the reservoir may be indispensable. That is, since there is a certain knowledge gap between the less and highly experienced operators, it is likely that the latter frequently omit important information during the development of operating procedures, which is very essential for the former.

It is true that, although instructions should be described in detail as much as possible, there are times when the provision of detailed instructions is not possible or even adequate. For example, let us recall Figure 13.2 with an instruction specifying how to maintain the level of the reservoir at less than 50%. At a glance, it is possible to provide very detailed instructions, such as "Close valves 101 and 102," and "Open valves 103 and 104." Unfortunately, this job is difficult than it seems because the contents of instructions can become different with respect to the nature of an ongoing status (e.g., the water level of the reservoir) and the actual configuration of the corresponding components. Figure 13.3 compares the contents of hypothetical instructions along with two kinds of the postulated levels of the reservoir.

For example, since the level of the reservoir in the first situation of Figure 13.3 is less than 50%, that is, the suction limit of the valve 103, it is not necessary to consider its closure in describing the corresponding instructions. In contrast, when the level of the reservoir is greater than 50%, the close of valves 103 and 104 should be described as the contents of the corresponding instructions. Therefore, it is apparent that the amount of instructions to be included in an operating procedure will be avalanched if a procedure developer wants to prescribe all the plausible situations. In this light, it is inevitable to provide a less detailed instruction, such as "Maintain the level of the reservoir less than 50% by using valves 101, 102, 103, and 104," in which end users (e.g., qualified operators) have to decide appropriate responses (e.g., which

Situation	Component Configuration	Instruction
1	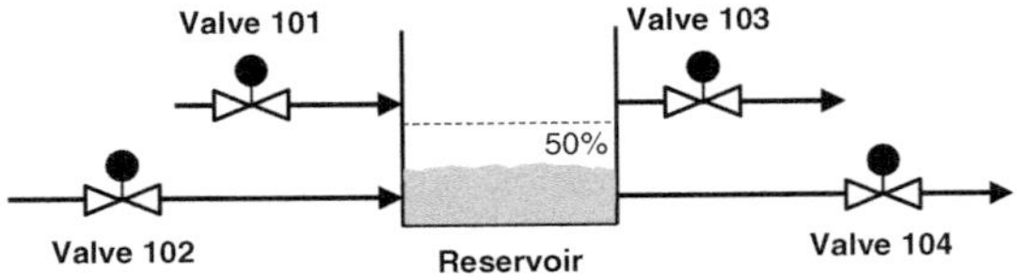	• Open valves 101 and 102 • Close valve 104
2	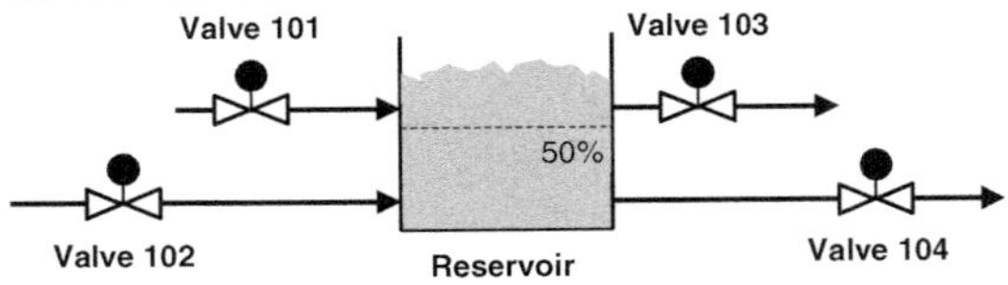	• Open valves 101 and 102 • Close valves 103 and 104

Figure 13.3. Two kinds of instructions along with the level of the reservoir.

valves should be manipulated in a situation at hand?) to accomplish the required goal.

Actually, as already mentioned at the end of Section 13.2, this is why the performance of operating procedures is very sensitive to task environments being faced by end users. In addition, this sensitivity alludes to the fact that the complexity of a task being accomplished by the performance of instructions should be taken into account for the development of a procedure. This is because instructions that are sensitive to task environments probably demand extensive cognitive resources of end users, which are necessary to conduct high-level cognitive activities including situation assessment (e.g., the current level of the reservoir with its trend) and decision-making (e.g., which valves are effective for increasing or reducing the level of the reservoir in this situation?). Unfortunately, as the second problem, it seems to be very rare to consider the complexity of instructions in the course of the development of operating procedures.

It is to be noted that these two problems are common, independent of the type of procedures. For example, in the nuclear industry, most of the operating procedures can be classified into two: event-based (or event-oriented) procedures and symptom-based (or symptom-oriented) procedures. In brief, the technical contents of an event-based procedure are dedicated to coping with a specific event, while those of symptom-based procedures provide how to respond to the change of important symptoms, such as pressure, temperature, and flow rate. More detailed information can be found from Park et al. (1995), Park (2009, p. 51), and IAEA (1985, 1998).

13.4 ADDITIONAL PRINCIPLE TO DEVELOP A GOOD PROCEDURE

In Section 13.3, two kinds of critical problems are identified (i.e., the level of details and the complexity of instructions included in an operating procedure) which are necessary to be carefully considered in developing a good procedure. Accordingly, as an additional principle to develop a good procedure, we are able to state that "Each instruction should be described so that the expected end users can conduct it without challenging their cognitive abilities at any situation." In other words, in order to develop a good procedure, we have to seriously consider not only how to provide *easy-to-read* instructions but also how to provide *easy-to-conduct* instructions. To this end, it is inevitable to solve the abovementioned two problems.

Here, it is anticipated that the first problem can be resolved if we are able to distinguish the preferred level of descriptions along with end users. In addition, one of the promising solutions to unravel the second problem is to determine the edge of a complexity, from which the end users start to feel a burden to conduct given instructions. From these concerns, in this section, two kinds of implementation rules that are helpful to provide *easy-to-conduct* instructions are suggested based on the results of recent studies.

13.4.1 Tailoring the Level of Details

As explained earlier, it is highly recommended that instructions included in an operating procedure should be described with the appropriate level of details that are supposed to be preferred to by expected end users. For example, let us assume that the expected end users of an operating procedure are qualified operators who experienced the operation of NPPs less than 5 years. In this case, the corresponding instructions should be written so that they can be soundly conducted by the expected users. From this request, the challenging question may be "How we are able to characterize the level of details to be preferred to by qualified operators who experienced the operation of NPPs less than 5 years?" In this regard, the study of Zach (1980) provided an important starting point.

For example, it is revealed that highly experienced operators generally preferred a procedure prescribed by a brief instruction such as "Isolate the reservoir." Meanwhile, most operators who have experienced plant operations less than 10 years preferred an instruction containing somewhat detailed information, such as "Close valves 101, 102, 103, and 104." Accordingly, if we are able to systematically distinguish the differences between these two instructions, then it is possible to establish a rule determining the proper level of details along with the operational experience of expected end users. To this end, it is indispensable to consider the key attributes of instructions first.

According to existing literature, each instruction can be described based on six kinds of key attributes (DOE, 1992; DOE, 1996; DOD, 1999). They are: (1) action verb, (2) object, (3) acceptance criterion, (4) means, (5) constraint, and (6) peculiarity. Table 13.3 summarizes the meaning of each attribute in brief (Park et al., 2010).

It is to be noted that the last attribute of Table 13.3 is very unusual because instructions in an operating procedure are supposed to be followed in a step-by-step manner. However, there are times when some instructions should be performed

TABLE 13.3. Six Key Attributes to Describe an Instruction

Attribute	Meaning
Action verb	Specifying what is to be accomplished in a given instruction
Object	Designating what is to be acted on to accomplish a given instruction
Acceptance criterion	A value or a worth that can be used to decide the achievement of a given instruction
Means	Explicit and/or implicit methods, tools, or devices which specify how to actually achieve the acceptance criterion of a given instruction
Constraint	A restriction (e.g., specific conditions or timing) that should be satisfied to accomplish a given instruction
Peculiarity	Any instruction that can be performed in a non-sequential manner

in a non-sequential manner. For example, according to DOE (1992), non-sequential instructions can be used if it is beneficial for end users to be provided with two or more plausible options. In addition, some instructions could impose time requirements on end users by specifying a period of time. In these cases, we need to mark the corresponding instructions because, from the point of view of the performance of instructions, end users probably spend more cognitive resources than ordinary instructions. Figure 13.4 exemplifies two kinds of non-sequential instructions (DOE, 1992).

With the six key attributes shown in Table 13.3, each instruction can be analyzed and compared in a formalized way. For example, Figure 13.5 depicts two kinds of instructions from the viewpoint of six key attributes.

From Figure 13.5, it is possible to identify several differences between two instructions. For example, we can recognize that the instruction of "Isolate the reservoir" does not include any acceptance criterion. As can be seen from Table 13.2, the role of an acceptance criterion is to determine whether or not the goal of a given instruction is achieved. In this light, there is no *overt* value clarifying the success of the reservoir isolation. Although highly experienced operators could identify surrogate indications to confirm the isolation of the reservoir (e.g., the fixation of the reservoir's level after completing the isolation), it is not easy for less experienced operators to come up with such surrogates. This strongly indicates that less experienced operators do not prefer an instruction without an acceptance criterion. In contrast, since the success of a valve closure can be directly recognized by reading the status indication of the corresponding valve (i.e., *OPEN* or *CLOSE*), it is reasonable to assume that most experienced operators are likely to prefer this kind of instruction.

Another difference is the existence of designated means. That is, similar to the acceptance criterion, it is anticipated that highly experienced operators prefer the instruction "Isolate the reservoir" even though it does not state how to actually isolate the reservoir (i.e., designated means). Meanwhile, less experienced operators probably prefer the instruction "Close valves 101, 102, 103, and 104," which manifests how to isolate the reservoir (i.e., specific valves to be closed).

[1] Perform one of the following actions: Check action taken

- Set Switch S-7 to "ON" ___
- Set Switch S-9 to "ON" ___

NOTE *The following action step starts a time-limited sequence. Action steps **[2]** through **[5]** are to be performed within 20 minutes.*

[1] Calculate the time by which action step **[5]** is to be completed by the following steps:

[a] Record the time that action step **[2]** is to be started. ___________

[b] Add 20 minutes. + 20 minutes

[c] Record the time that action step **[5]** is to be completed. ___________ Complete by

[2] Place the Incident Action switch to TEST. ___

[3] Verify the amber *ALARM TEST* light on the AIA graphic panel is LIT. ___ IV

[4] Place the *Incident Action* switch to OFF. ___

[5] Push the *RESET* pushbutton on the AIA logic tester and record the time. ______

[6] IF the time recorded in action step **[5]** is later the time recorded in action step **[1] [c]**, THEN REPEAT action steps **[1]** through **[5]**.

Figure 13.4. Examples of non-sequential instructions. *Source:* Department of Energy 1992. Public domain.

Instruction	Isolate the reservoir	Close valves 101, 102, 103, and 104
Action verb	Isolate	Close
Object	The reservoir	Valves 101, 102, 103, 104
Acceptance criterion	Not specified	Dichotomous (e.g., open–close)
Means	Not specified	Designated means
Constraint	No constraint	No constraint
Peculiarity	No peculiarity	No peculiarity

Figure 13.5. Analyzing two kinds of instructions.

TABLE 13.4. Properties of the Acceptance Criterion and Means

Key Attribute	Property	Meaning
Acceptance criterion	Objective criterion	Providing invariant value, such as *pump on/off*, and *temperature less than 300° C*
	Reference information	Providing reference tables, graphs, or charts
	Subjective criterion	Providing vague criterion that can be interpreted differently, such as *sufficient flow*, *stable temperature*, and *pressure decreasing in uncontrollable manner*
	No criterion	No description on the acceptance criterion
Means	Designated means	Overt means to be manipulated, such as *valve 101*
	Inherent means	Covert means to be manipulated, such as missing the name of a unique component
	No means	No description on the means

If we focus on the abovementioned differences, in theory, it is promising that the level of details in each instruction can be classified by the combination of the six key attributes. In this regard, Park et al. (2010) suggested 12 categories based on the properties of the acceptance criterion and means (Table 13.4). More detailed information on the properties of these key attributes can be found from Park (2009).

Based on these 12 categories, Park et al. (2013) carried out a series of experiments with seven kinds of different procedures and 98 human operators working in NPPs. To this end, as summarized in Table 13.5, they subdivided the 12 categories into

TABLE 13.5. Levels of Details Categorized by Three Groups with Respect to the Properties of the Acceptance Criterion and Means

Group	Acceptance Criterion	Means
Detail level description	Objective criterion	Designated means
Intermediate level description	Objective criterion	Inherent means
	Objective criterion	No means
	Reference information	Designated means
	Subjective criterion	Designated means
	No criterion	Designated means
Problematic level description	Reference information	Inherent means
	Reference information	No means
	Subjective criterion	Inherent means
	Subjective criterion	No means
	No criterion	Inherent means
	No criterion	No means

TABLE 13.6. Fraction of *Detail Level Descriptions* with Respect to the Years of Plant Operations (subjective difficulty is fixed as 2.5)

Year	Fraction (%)	Corresponding Correlation
Less than 5	72.1	Subjective difficulty = −0.026·Fraction + 4.375
5–10	71.7	Subjective difficulty = −0.027·Fraction + 4.438
10–15	55.2	Subjective difficulty = −0.018·Fraction + 3.494
Greater than 15	62.9	Subjective difficulty = −0.021·Fraction + 3.821

three groups (i.e., *Problematic level, Intermediate level*, and *Detail level description*), which represent distinct levels of details. After that, they asked all the participants to rate the subjective difficulty of each procedure based on a 5-point Likert scale (1 represents the most easy procedure, while 5 denotes the most difficult procedure to be performed).

As a result, Park et al. (2013) suggested a series of correlations that can be used to determine the proper level of details in providing instructions with respect to the years of plant operations. For example, let us assume that we need to develop an operating procedure to be used by qualified operators who have experienced plant operations for less than 5 years. In this case, if we want to develop an operating procedure, of which the subjective difficulty is about 2.5, then it is anticipated that the fraction of *Detail level description* included in it should be greater than at least 70%. In contrast, if the target of the subjective difficulty is 2.0, then it is strongly recommended that the fraction of *Detail level description* included in the operating procedure has to be increased up to 90%. Table 13.5 summarizes the fraction of *Detail level description* with respect to the years of plant operations, when the target of the subjective difficulty is fixed as 2.5. It should be noted that the fraction of *Detail level description* for another target of the subjective difficulty can be easily calculated by using formulas shown in the last column of Table 13.6 (Park et al., 2013).

13.4.2 Tailoring the Complexity of Instructions

As briefly explained in Section 13.3, end users need a lot of cognitive resources to conduct an operating procedure because the performance of prescribed instructions could be affected by the nature of an ongoing situation. In other words, since dynamic situations should be covered by static instructions, it is not easy to develop an operating procedure that contains precise instructions applicable to each and every situation. From the point of view of the six key attributes in describing instructions, this limitation can be understood such that it is not possible to develop an operating procedure, of which all the instructions belong to *Detail level description* (refer to Table 13.5). Accordingly, it is inevitable for end users to be exposed to various kinds of instructions that probably have a wide spectrum of task complexity. This strongly implies that the complexity of instructions being faced by end users should be properly managed from the early phase of a procedure development. Otherwise, there are times when the end users have to conduct a series of instructions that demand huge amount of cognitive resources exceeding their cognitive capability (Park, 2009). One promising solution to unravel this problem is to systematically evaluate the complexity of instructions. To this end, Park (2009) suggested the TACOM (*ta*sk *com*plexity) measure.

In brief, the TACOM measure quantifies the complexity of a given task that consists of one or more instructions based on a weighted Euclidean norm in a complexity space. According to Harvey and Koubek (2000), one of the plausible complexity spaces is comprised of three dimensions such as TS (task scope), TS (task structurability), and TU (task uncertainty). Basically, TS represents the breadth, extent, range, or general size of a task being considered, and TR denotes whether or not the sequence and the relationship among instructions are well structured. In addition, TU specifies the degree of predictability or confidence of the task.

In order to calculate the TACOM score of a task, each dimension includes one or two sub-measures that characterize significant factors making the performance of the task complicated (see Figure 13.6). More detailed explanations on the quantification of TACOM scores can be found from Park (2009), Podofillini et al. (2013), and Park (2014) with a couple of demonstration examples.

$$\text{TACOM} = \{0.621 \cdot (\text{TS})^2 + 0.239 \cdot (\text{TR})^2 + 0.140 \cdot (\text{TU})^2\}^{1/2}$$

$$\text{TS} = 0.716 \cdot \text{SIC} + 0.284 \cdot \text{SSC} \qquad \text{TU} = \text{EDC}$$

$$\text{TR} = 0.891 \cdot \text{SLC} + 0.109 \cdot \text{AHC}$$

Sub-measure	Description
SIC (step information complexity)	Complexity due to the amount of information to be processed by human operators
SSC (step size complexity)	Complexity caused by the number of actions to be conducted by human operators
SLC (step logic complexity)	Complexity originated from the sequences of actions to be followed by human operators
AHC (abstraction hierarchy complexity)	Complexity resulted from the amount of domain knowledge to be required by human operators
EDC (engineering decision complexity)	Complexity related to the amount of cognitive resources to establish an appropriate decision criterion

Figure 13.6. TACOM quantification formula with the associated sub-measures.

One of the most interesting results identified from the comparison between TACOM scores and the associated human performance data is that the TACOM measure seems to be a general tool for quantifying the complexity of a task. For example, Podofillini et al. (2013) pointed out that the probability of human errors that were observed from the full scope simulator of NPPs is increased in proportion to the increase of TACOM scores. In addition, even though qualified operators have to accomplish a series of tasks under different environments, Park (2014) claimed that the task completion times being collected from the full scope simulator of NPPs are not quite different if they are exposed to tasks, of which the TACOM scores are similar.

The abovementioned results strongly allude to the fact that the TACOM measure can specify a certain border that allows us to identify a complex task resulting in the degradation of human performance. From this concern, Park and Jung (2015) investigated the effect of TACOM scores on the variation of human performance data that were collected from railway industries. As a result, they claimed that human operators could feel a moderate difficulty in conducting a task, of which the TACOM score is greater than 4.3. They also pointed out that qualified operators probably feel a high difficulty if they are exposed to a task, of which the TACOM score exceeds 5.3.

13.5 CONCLUDING REMARKS

Although there are many advantages in using procedures, most existing literature has frequently emphasized at least three benefits for many decades. Accordingly, from the point of view of safety-critical systems, it is very natural to expect that their safe and sustainable operations largely depend on the development of good operating procedures. In this regard, many kinds of principles to develop good operating procedures have been established till today. Unfortunately, it seems that most existing principles are focusing on how to develop *easy-to-read* procedures instead of *easy-to-conduct* procedures especially under a dynamically changing situation. In this regard, an additional principle is issued in this chapter, such that "Each instruction should be described so that the expected end users can conduct it without challenging their cognitive abilities at any situation." Moreover, two kinds of implementation rules that are helpful to implement the additional principle are suggested based on the results of recent studies that deal with the performance variation of human operators who use *easy-to-read* procedures.

For example, in the case of NPPs, each and every operating procedure should be developed along with specific procedure writer's guidelines being prepared on the basis of existing principles. Therefore, if we scrutinize the performance variation of qualified operators working in NPPs, it is anticipated that useful insights or clues can be identified, those which are not properly covered by existing principles. As a result, the fraction of *Detail level description* to be included in an operating procedure is distinguished with respect to the years of plant operations (refer to Table 13.6). In addition, as explained at the end of Section 13.4.2, the distinction of a complicated task is outlined by the TACOM measure.

It is true that the appropriateness of the implementation rules is still an open question because they largely depend on empirical observations collected from the full scope simulators of NPPs. In other words, without sufficient evidence gathered from other industries, it seems to be premature to say that the validity of the implementation rules. This is because human behaviors being observed from simulated environments are apt to be biased to some extent to those of a real world (Criscione et al., 2012; IAEA, 1989; NEA, 1988). In addition, since it is still uncertain which amount of the bias exists between two kinds of human behaviors (i.e., observed from simulated environments and a real world), it is not possible to estimate the effect of the implementation rules on reducing the operational risk of socio-technical systems in a consistent manner.

At the same time, however, it is worth saying that the implementation rules should keep investigating because they allow us to identify effective countermeasures to enhance the performance of qualified operators who have to conduct procedures under dynamic situations. In other words, if we apply the first rule (i.e., the fraction of *Detail level description*) to an existing procedure, it is possible to clarify whether or not its level of details is fit to end users. This indicates that the list of instructions to be modified can be systematically identified from the existing procedure; or to some extent, it could provide a useful probe in identifying the characteristics of operating personnel because their behaviors are apt to be largely affected by the contents of procedures (Woods et al., 1990).

As Reason et al. (1998) pointed out, it is expected that the likelihood of rule violations is proportional to the experience level of operating personnel because they regard the ability of rule bending as the hallmark of capability or prominence for their job. Accordingly, if we are able to clarify a context in which the operating personnel are likely to bend a rule, then it would be a good starting point to investigate their characteristics on the safety culture (refer Chapter 28) or behavior-based safety (BBS; refer Chapter 29). For example, two organizations that use procedures sharing a similar description level showed different responses (e.g., correct compliance vs. non-compliance), it is reasonable to assume that there should be another factor making the responses of similar organizations distinctive.

A more direct application is to distinguish complicated tasks demanding significant burden on operating personnel based on the second rule (e.g., tasks of which TACOM scores are greater than 4.3 or 5.3). This is because most human reliability analysis (detailed information can be found in Chapter 25) considers the complexity of a task as one of the most important contributors increasing the possibility of human error. For example, in the case of SPAR-H (Standardized Plant Analysis Risk—HRA) method developed by US NRC (Nuclear Regulatory Commission), the possibility of human error can be affected by three levels of a task complexity as shown below (Gertman et al., 2004, p. 29).

- "Highly complex—Very difficult to perform. There is much ambiguity in what needs to be diagnosed or executed. Many variables are involved, with concurrent diagnoses or actions (i.e., unfamiliar maintenance task requiring high skill)."
- "Moderately complex—Somewhat difficult to perform. There is some ambiguity in what needs to be diagnosed or executed. Several variables are involved, perhaps with some concurrent diagnoses or actions (i.e., evolution performed periodically with many steps)."
- "Nominal—Not difficult to perform. There is little ambiguity. Single or few variables are involved."

With these levels, the TACOM measure can give objective criteria for identifying a moderately complex task (e.g., a task of which the TACOM score is between 4.3 and 5.3) and highly complex task (e.g., a task of which the TACOM score is greater than 5.3).

Another application of the TACOM measure is to identify effective countermeasures along with the characteristics of a complexity factor. That is, since the TACOM measure consists of five sub-measures, it is possible to come up with dedicated countermeasures with respect to the nature of a complexity factor (such as the amount of actions to be conducted by qualified operators, or the amount of domain knowledge to be required by qualified operators) attributable to a high TACOM score. Consequently, under the current status in which it is not easy to develop an *easy-to-conduct* procedure, the implementation rules will contribute to resolve this problem.

REFERENCES

Blackmore, A. (2004). The myth of the SOP. *The Quality Assurance Journal*, 8, 3–12.

Bovair, S., and Kieras, D. E. (1991). Toward a model of acquiring procedures from text. In: R. Barr, M. L. Kamil, P. Mosenthal, and P. D. Pearson (Eds.). *Handbook of Reading Research* (Vol. II, pp. 206–229). New York.

Boy, G. A., and Schmitt, K. A. (2013). Design for safety: A cognitive engineering approach to the control and management of nuclear power plants. *Annals of Nuclear Energy*, 52, 125–136.

Brito, G. (2002). Towards a model for the study of written procedure following in dynamic environments. *Reliability Engineering and System Safety*, 75, 233–244.

Brune, R. L, and Weinstein, M. (1983). Checklist for evaluating emergency operating procedure used in nuclear power plants. NUREGCR-2005. US Nuclear Regulatory Commission, Washington, DC.

Bullemer, P. T., and Hajdukiewicz, J. R. (2004). A study of effective procedural practices in refining and chemical operations. *Proceedings of the Human Factors and Ergonomics Society Annual Meeting*, 48(20), 2401–2405.

Business Dictionary. (2014). Available at: http://www.businessdictionary.com/definition/procedure.html

Criscione, L., Shen, S., Nowell, R., Egli, R., Chang, Y., and Koonc, A. (2012). Overview of licensed operator simulator training data and use for HRA. *Proceedings on Probabilistic Safety Assessment and Management (PSAM)*. Helsinki: Finland.

Department of Defense. (1999). Human factors engineering program process and procedures. MIL-HDBK-46855A. Washington, DC.

Department of Energy. (1992). Writer's guide for technical procedures. DOE-STD-1029-92. Washington, DC.

Department of Energy. (1996). Human factors engineering. GPG-FM-027. Washington, DC.

Disaster Database. (2017). Available at: https://www.publicsafety.gc.ca/cnt/rsrcs/cndn-dsstr-dtbs/rfrnc-tbl-smbls-dfntns-en.aspx

Environmental Protection Agency (2001). Guidance for preparing standard operating procedures. EPA240B-01004. Washington, DC.

Fuchs, F., Engelschall, J., and Imlay, G. (1981). Evaluation of emergency operating procedures for nuclear power plants. NUREGCR-1875. US Nuclear Regulatory Commission, Washington, DC.

Fullwood, R. R., and Hall, R. E. (1988). *Probabilistic Risk Assessment in the Nuclear Power Industry: Fundamentals and Applications*. Oxford: Pergamon Press.

Gertman, G., Blackman, H., Marbles, J., Byers, J., and Smith, C. (2004). The SPAR-H human reliability analysis method. NUREGCR-6883. US Nuclear Regulatory Commission, Washington, DC.

Hale, A. R. (1990). Safety rules OK? *Journal of Occupational Accidents*, 12, 3–20

Hale, A., and Borys, D. (2013). Working to rule, or working safely? Part 1: A state of the art review. *Safety Science*, 55, 207–221

Harvey, C. M., and Koubek, R. J. (2000). Cognitive, social, and environmental attributes of distributed engineering collaboration: A review and proposed model of collaboration. *Human Factors and Ergonomics in Manufacturing*, 10, 369–393.

Health and Safety Executive. (2005). Inspection toolkit—Human factors in the management of major accident hazards. Available at: www.hse.gov.uk/humanfactors/comah/toolkitintro.pdf

Health and Safety Executive. (2007). Revitalising procedures. Available at: www.hse.gov.uk/humanfactors/comah/procinfo.pdf

Inaba, K., Parsons, S. O., and Smillie, R. (2004). *Guidelines for Developing Instructions*. Boca Raton, FL: CRC Press.

International Atomic Energy Agency. (1985). Developments in the preparation of operating procedures for emergency conditions of nuclear power plants. IAEA-TECDOC-341. Vienna, Austria

International Atomic Energy Agency. (1998). Good practices with respect to the development and use of nuclear power plant procedures. IAEA-TECDOC-1058. Vienna, Austria.

Khalaquzzaman, M., Kang, H. G., Kim, M. C., and Seong, P. H. (2010). Quantification of unavailability caused by random failures and maintenance human errors in nuclear power plants. *Nuclear Engineering and Design*, 240, 1606–1613.

Kim, J., Park, J., Jung, W., and Kim, J. T. (2009). Characteristics of test and maintenance human errors leading to unplanned reactor trips in nuclear power plants. *Nuclear Engineering and Design*, 239, 2530–2536.

Marsden, P. (1996). Procedures in the nuclear industry. In: N. Stanton (Ed.), *Human Factors in Nuclear Safety*. London: Taylor & Francis.

Nuclear Energy Agency. (1988). The human factor in nuclear power plant operation, NEA issue brief—An analysis of principal nuclear energy issues, No. 2, Vienna: Austria

Nuclear Regulatory Commission. (1997). US Nuclear Regulatory Commission, NRC news. Available at: http://www.nrc.gov/reading-rm/doc-collections/news/1997/97-034.html

O'Hara, J. M., Brown, W. S., Leiw, P. M., and Persensky, J. J. (2002). Human-system interface design guidelines. NUREG-0700, Rev. 2. US Nuclear Regulatory Commission, Washington, DC.

O'Hara, J. M., Higgins, J. C., Fleger, S. A., and Pieringer, P. A. (2012). Human factors engineering program review model. NUREG-0711, Rev. 3. US Nuclear Regulatory Commission, Washington, DC.

Oxford Dictionary. (2014). Available at: http://www.oxforddictionaries.com/definition/english/procedure

Park, J. (2009). *The Complexity of Proceduralized Tasks*. Berlin: Springer-Verlag.

Park, J. (2014). Investigating the TACOM measure as a general tool for quantifying the complexity of procedure guided tasks. *Reliability Engineering and System Safety*, 129, 66–75.

Park, J., and Jung, W. (2003). The operator's non-compliance behavior to conduct emergency operating procedures—Comparing with the work experience and the complexity of procedural steps. *Reliability Engineering and System Safety*, 82, 115–131

Park, J., and Jung, W. (2008). A study on the validity of a task complexity measure for emergency operating procedures of nuclear power plants—Comparing task complexity scores with two sets of operator response time data obtained under a simulated SGTR. *Reliability Engineering and System Safety*, 93, 557–566.

Park, J., and Jung, W. (2015). Identifying objective criterion to determine a complicated task—A comparative study. *Annals of Nuclear Energy*, 85, 205–212.

Park, J., Jung, W., and Yang, J. E. (2013). Investigating the appropriateness of a decision chart to characterize the level of task descriptions in nuclear power plants. *Progress in Nuclear Energy*, 66, 41–51.

Park, S. H., Kwon, J. S., and Kim, S. R. (1995). Emergency procedure recommendation for Wolsong 2, 3 & 4 NPP. In: *Proceedings of the Korean Nuclear Society Autumn Meeting* (pp. 272–277). Seoul: Republic of Korea.

Park, J., Park, S., and Cho, S. (2010). Some insights to determine the proper level of descriptions about proceduralized tasks. *Progress in Nuclear Energy*, 52, 214–224.

Podofillini, L., Park, J., and Dang, V. N. (2013). Measuring the influence of task complexity on human error probability: An empirical evaluation. *Nuclear Engineering and Technology*, 45(2), 151–164.

Praino, G., and Sharit, J. (2016). Written work procedures: Identifying and understanding their risks and a proposed framework for modeling procedure risk. *Safety Science*, 82, 382–392.

Preischl, W., and Hellmich, M. (2013). Human error probabilities from operational experience of German nuclear power plants. *Reliability Engineering and System Safety*, 109, 150–159.

Rankin, W., Hibit, R., Allen, J., and Sargent, R. (2000). Development and evaluation of the Maintenance Error Decision Aid (MEDA) process. *International Journal of Industrial Ergonomics*, 26, 261–276.

Reason, J., Parker, D., and Lawton, R. (1998). Organizational controls and safety: The varieties of rule-related behavior. 71, 289–304

Ruckart, P. Z., and Burgess, P. A. (2007). Human error and time of occurrence in hazardous material events in mining and manufacturing. *Journal of Hazardous Materials*, 142(3), 747–753.

Wieringa, D. R, Moore, C., and Barnes, V. (1998). *Procedure Writing Principles and Practices* (2nd ed.). Columbus, OH: Battelle Press.

Wikipedia. (2014a). Available at: http://en.wikipedia.org/wiki/Handbook

Wikipedia. (2014b). Available at: http://en.wikipedia.org/wiki/Guideline

Wikipedia. (2014c). Available at: http://en.wikipedia.org/wiki/ Procedure_(term)

Wood, D. D., Roth, E. M., and Pople, H. E., Jr. (1990). Modeling operator performance in emergencies. In: *Proceedings on the 34th Human Factors and Ergonomics Society Annual Meeting* (pp. 1132–113).

Your Dictionary. (2014). Available at: http://www.yourdictionary.com/procedure

Zach, S. E. (1980). Control room operating procedures: Contents and format. In: *Proceedings of the Human Factors and Ergonomics Society—24th Annual Meeting* (pp. 125–127).

FURTHER READING

Dols, T., and Silyius, A. J. (2010). Exploring the influence of national cultures on non-compliance behavior. *Communications of the IIMA*, 10(3), 11–32.

Health and Safety Executive. (2016). Human factors that lead to non-compliance with standard operating procedures. Available at: http://www.hse.gov.uk/research/rrpdf/rr919.pdf

Reason, J. (Ed.). (1990). *Human Error*. New York: Cambridge University Press.

Solomon, J. N., Gavin, M. C., and Gore M. L. (2015). Detecting and understanding non-compliance with conservation rules. *Biological Conservation*, 189, 1–4.

14

HUMAN–MACHINE SYSTEM

Anna-Lisa Osvalder and Håkan Alm

The aim of this chapter is to discuss the overall conditions for process control and how the user interfaces between the operator and process should be designed to support operator supervision, operation, and control of a complex human–machine system which optimizes on safety, efficiency, and well-being.

Looking back in time can give us a useful perspective on the technological development and its impact on human–machine systems. Hollnagel (2010) notes that human use of technology before the industrial revolution was characterized by systems that were limited in size, had uncomplicated technology with linear cause and effect relations and because of that, easy to understand. It can be argued that in many cases, it was possible to see how different parts of the systems used were connected and had impact on other parts. The information an operator needed to control and maintain the system was often possible to inspect using direct perception (using visual, auditory, and other input channels). Work was mainly manual, sources of energy few (e.g., water, wind, human or animal power). Artifacts (objects designed by humans) were few and system integration more or less non-existent. Coupling between subsystems were few, if any.

Handbook of Safety Principles, First Edition. Edited by Niklas Möller, Sven Ove Hansson, Jan-Erik Holmberg, and Carl Rollenhagen.

As mentioned in Chapter 10, the development in electronics, computers, and software has resulted in a technological revolution in human–machine interaction. The shift from analog to digital technology has resulted in possibilities to increase the information flow to the operator of a system, and increasing the complexity of the system to be controlled. Today, systems are more complex and integrated. The technological revolution and the introduction of the computers in system control have changed the operators' tasks in many ways. Since information can be distributed quickly between many subsystems, and what happens within the system is not directly visible for operators, the task to control a process or system has become increasingly complex and abstract. Since processes are only possible to understand indirectly, using different user interfaces, the demands on human cognitive resources (e.g., perception, attention, interpretation, decision-making) have increased.

14.1 HUMAN–MACHINE SYSTEM

A human–machine system is a system which integrates human and machine (Chapanis, 1965). It consists of humans and technology interacting together in a certain context in order to perform given tasks. Information is in a cyclic manner continuously exchanged in the interaction influenced by the surroundings. In a broader scope, machine covers everything from very simple technical products such as a digital alarm clock or a coffee machine to complex safety-critical systems as a control room in process industry or the medical operating theatre. The user in the human–machine system is referred to as the operator, who controls and operates the system. Interfaces such as displays and instruments, present information about the status of the system. The operator attends, perceives, and interprets the presented data for decision-making of what to do next, and then operates the system by using different types of controls such as buttons, levers, and keyboards. Then the machine processes and encodes the data and displays a new status.

The part of the system that communicates with the operator is called the user interface. When incidents and accidents occur, operators in the system tend to be blamed for human errors. However it is often found that human errors happen due to poorly designed user interfaces not adapted to the situation, the task to be performed, or the operators' abilities and limitations. An efficient human–machine system offers many advantages such as increased safety, effectiveness, and operator satisfaction.

In order for a human–machine system to fulfil its purpose, it must be useful. Usefulness is divided into utility (the functionality of the system) and usability (how well the operator can use the functionality the system provides). When designing a useful human–machine system, the goal is a functioning interaction between the operator and the machine, which exploits human ability to operate the system while countering the effects of human limitations.

14.2 COMPLEX SYSTEMS

Human–machine systems are often complex systems. Complexity can be defined in different ways, but it is generally agreed (Brehmer, 1992) that complex systems have a number of properties. First of all, complexity varies with the number of system components and the number of goals and goal conflicts that exist in a system. Complexity also varies with the coupling or interdependency between different subsystems (Perrow, 1999). The causal structure of a complex system is another important factor. A system with a linear causal structure can be regarded as a simpler system compared to a system with a net structure. In a system with a linear causal structure, prediction of future events resulting from control efforts is easier to make, compared to a system with a net structure where different side effects may occur. The relation between different processes is also important. Nonlinear relations between processes may increase complexity.

A complex system may also change autonomously by its inner dynamics, factors outside the system, and by the impact of different efforts to control the system. This creates a problem to differentiate between spontaneous changes in a system, changes associated with external factors, and from changes resulting from attempts to control the system. According to Perrow (1999), it is very likely and even normal that complex systems will produce surprises, sometimes in the form of incidents or accidents.

A trend today is that systems are becoming increasingly more complex and coupled, and consequently harder to control. The goal of safety must therefore have high priority in a complex human–machine system. It is therefore of utmost importance to understand goal conflicts or challenges that may exist in a system and the goal conflicts' impact on safety. It is also important to apply a system perspective in the development of human–machine interaction and not only focus on separate components and their relation to other components.

14.3 TO CONTROL A COMPLEX SYSTEM

To control any system, a number of conditions must be fulfilled (Conant and Ashby, 1970).

1. An operator must have access to a model (descriptive and causal) of the system to be controlled. Stated in other words, knowledge about different processes and how the different processes are causally connected to each other. If this model is incorrect or incomplete, the control of the system may, in the long run, be less than optimal. This can lead to low productivity or, in the worst case, loss of control, resulting in incidents and accidents.
2. There must also be a goal for the system to be controlled, for instance, the goal of productivity. In complex systems, many goals exist and goal conflicts are common, for instance, the conflict between productivity and safety.

3. It must be possible to find valid (correct) information concerning the current status of the system.
4. There must also exist possible ways to change or control the status of the system and to receive useful feedback concerning the impact of control activities on system status. For complete control of a system, the variation in the control system must match the variation in the system to be controlled, that is, the law of requisite variety (Ashby, 1958).

14.4 OPERATOR DEMANDS

To control a complex system, the operator must have knowledge of the goal, or goals of the system. The operator must also have a model of the system, a mental model or some other kind of representation (for instance visual and auditory displays) of the system, in order to make correct decisions concerning the state of the system. The operator must have the necessary skills to observe what happens in the system. This means the knowledge of what to look for in different situations, such as start-up, normal, shut down, and critical phases, and also knowing where to find relevant information for these different phases.

The timing aspect is also important, that is, when to search for relevant information in different phases. The next step involves interpretation of information to understand the status of the system. This is a decision task where a number of cues are combined, using some rule or rules for combining cues, to a judgment of the status of the system.

In the next phase, the operator must have skills to decide whether the status of the system is normal or not, in comparison with some kind of reference model. The decision involves a criterion for when to intervene and implement some plan to change the status of the system.

The operator must have the skills necessary to implement plans to control the system and keep the system in a desired state. Finally, the operator must be motivated enough to control the system according to the plans. These skills will most likely be rule-based (if X then Y) and possible to document in different ways. Some skills will, most likely, be in the form of tacit knowledge (Polanyi, 1958). In case of unexpected events, the operator needs problem-solving skills as well as skills to communicate with support personnel.

14.4.1 Mental Models

A mental model is an internal representation of a system, where the properties of the system and their relations are stored in the long-term memory. Of importance is that the mental model is correct and corresponds to the properties of the system to be controlled. An implication of this statement is that an optimal mental model of a system should include every subsystem and thus be a correct and complete description of the system. Or, in many or most cases, a correct and complete description of the subsystem controlled by the operator.

To be useful for the control of the system, the mental model should also describe the relations between different subsystems and the causal structure of the system. For instance, if pump A is activated then changes in X, Y, Z are expected. An optimal mental model makes it possible for an operator of a system to make a mental simulation of what will happen when a certain action is performed, and be proactive.

Earlier research (Norman, 2002) has shown that in many cases, people's mental models of different technical artifacts are incomplete, and sometimes wrong. Based on this knowledge, it is important to support operators of a system with a correct description of the real system, using a description that matches the operators' capacities to process information.

To make it possible for operators to form a correct mental model of a system, it is important to describe the dynamic aspects of the system or subsystem. Knowledge of how fast different subsystems will react and give input to other subsystems is important for decisions concerning when to control the system. For instance, to control a car and overtake another car, it is important to have a correct mental model concerning the acceleration capabilities of the car. To control some chemical reactions in process control, it is of importance to have a correct mental model of how much time different processes need to produce the desired result.

As stated earlier, a complex system with a linear causal structure is easier to understand and predict, compared to a complex system with a net structure. Linear relations between subsystems make a system easier to understand compared to a system with nonlinear relations between subsystems. For a system with a linear causal structure, a good decision support system should have the capacity to show the operator how a control action will propagate through the system. For instance, if pump A is activated, this will have an impact on water level in process A.

For a system with a net structure, the decision support system should also show side effects of a control action. For instance, if pump A is activated, then the water level in process A will increase. At the same time, the level in process B may change. This is, of course, only possible if the system can be fully understood and completely described. In many cases, this is not possible, and the best possible approximation of the system must be used, maybe combined with an indication of uncertainty. For example, pump A will start and possible side effects might be changes in processes X, Y, Z. The decision support system should be able to show when knowledge is missing and when a perfect prediction is impossible.

The optimal decision support system should also make it possible for the operator to perform a simulation of what will, or may happen, given certain control actions, before implementing them in real life. For instance, if pump A produces more effect, then the water level in A and B will rise faster, but vibrations may increase as a side effect.

14.4.2 Situation Awareness

Situation awareness (SA) is a central concept in complex systems where people are interacting with technology. According to Endsley (1995), SA can be divided into three levels. The first level is perception of task-relevant elements, for instance the information sources (displays and co-workers) that are of importance for the control of a process. The second level is the interpretation of the task-relevant elements, for instance, the interpretation of the information from different displays and co-workers. The third level represents a prediction of what will happen next, for example, what will happen in the near future.

To control a system successfully, an operator or a team of operators must have a high level of situation awareness and have knowledge of task-related events for different tasks. Also be able to correctly interpret task-related events, and from them understand what will happen in the near future.

To design for situation awareness, it is necessary to make information from task-relevant events easy to notice, and separate signal from noise as much as possible. It is also necessary to make the interpretation of task-relevant elements as easy as possible. Finally, to support the operators' ability to make a correct prediction of what may happen in the near and far future.

It can be argued that what task-relevant elements are in the control of a process depends upon the strategy used for control. A proactive strategy (what may happen soon?) may make different demands on information compared to a reactive strategy (what happens now?). To be proactive, an operator may need information concerning increased variation in different parts of the system, something that later may result in unwanted consequences.

14.4.3 Decision-Making

Research in decision-making has shown that human decision-makers normally do not follow normative models when assessing or updating probabilities or expected values. High and low probabilities are often not correctly assessed. Normally, low probabilities are overestimated and higher probabilities are underestimated.

Rather than being rational and search for the outcome with the highest expected utility, human decision-makers often can be characterized as quasi-rational. Quasi-rationality means that the search for a solution to a problem may stop when the solution is considered to be good enough, probably reflecting a trade-off between efficiency and time constraints (see for instance Hollnagel, 2009). When updating probabilities, human decision-makers are generally conservative and do not follow normative models (i.e., Bayes theorem).

Human decision-making often seems to be based on simple heuristics, or rules of thumb (Kahneman, 2011). There exist a large number of heuristics and only a few will be mentioned here. When categorizing events, human decision-makers often use the heuristic called "representativeness," it looks like something experienced earlier.

The heuristic "availability" suggests that when human decision-makers estimate the frequency of something, this is done by how easy such instances can be recalled from memory. How easy something can be retrieved from memory depends upon its salience, dramatically, and personal experience.

One conclusion from the research in decision-making is that a human decision-maker can be very skilled when it comes to finding and interpreting different cues in a decision task, but not very good at combining the different cues in an optimal way. Besides that, decision-makers are generally not able to be consistent in the use of an integration rule for cues (Brehmer, 1978). When decision tasks are based on incomplete information and can be classified as probabilistic, research has shown that decision-makers perform in a non-optimal manner (trying to maximize instead of optimize). An interpretation is that they do not understand uncertainty and are trying to find the correct answer to every instance, instead of minimizing the error in the long run.

Another striking conclusion is the problems to understand and cope with nonlinear relations between variables. The relation between variables is very often described with simple linear rules, a strong preference for linearity.

Important research in dynamic decision-making (e.g., Dörner, 1980) has shown that operators of complex systems have more problems to learn to control processes with causal net structure (when side effects occur) compared to processes with causal chain structure (no side effects can occur).Dörner (1980) also identified other problems people have in dealing with complex processes. One problem was to consider the dynamic or time aspect of a process. People also had problems to deal with nonlinear relations between control actions and system responses. Other researchers (e.g., Brehmer and Allard, 1986) have also pointed out that delay in feedback from control activities may have a negative impact on operators' ability to control a system.

Decision support systems can be used to help the human decision-maker to integrate cues in an optimal manner and to be consistent in the use of the integration rule. Consistent means trying to minimize errors in the long run. Decision support systems could also help operators to correctly estimate probabilities of different events and the dynamic aspects of different events.

14.4.4 Mental Workload

Modern work life has shifted from physically demanding to mentally demanding. More and more tasks today involve cognitive functions like perceiving, coding, interpreting, remembering, decision-making, problem-solving, etc. The demands on the cognitive system and how to optimize these demands have been in focus for many researchers. Behind the idea of optimizing the demands on the cognitive system lies an assumption about limitations or bottlenecks in the cognitive system. Similar to physical load, it is assumed that our ability to process information has limitations and that it may be unwise to exceed these limitations. Stress, fatigue, and increased risks for unwanted and inappropriate actions may be the result of badly adjusted mental demands.

There exist a number of definitions of mental workload. One recent example is a definition suggested by Young and Stanton (2001).

"The mental workload of a task represents the level of attentional resources required to meet both objective and subjective performance criteria, which may be mediated by task demands, external support, and past experience".

It should, however, be noted that a formal and universally accepted definition of mental workload does not exist today. It was suggested by Gopher et al. (1989) that measures of workload can be used to describe the performance of a task relative to the operator's capacity. Task demands versus human resources. So what are the human mental resources? Today, it is generally accepted that there exist multiple mental resources (Wickens and Hollands, 2000; Baddeley, 1997) and a single measure of mental workload may be an extreme simplification. Workload should be considered as multidimensional.

In order to measure mental demands or mental workload, a number of different methods can be and has been used: subjective measures, physiological measures, performance measures on a primary and secondary task. A safe conclusion seems to be that there does not exist any single and universally accepted method to measure workload.

Hart and Staveland (1988) developed the NASA TLX method to measure mental workload. They identified six variables that, based on empirical and theoretical research, should have an impact on mental workload. These factors were

- Mental demand, defined as the demand on the cognitive system to search for and interpret information.
- Physical demand, associated with the actual performance of the task.
- Time pressure, the demands resulting from the time available to perform a task.
- Effort, how hard it was necessary to work in order to perform the task.
- Frustration level, uncertainty and stress during task performance.
- Performance, the degree to which task performance was regarded as acceptable.

The relation between levels of workload and performance is not clear. High levels of workload may make subjects change strategies and make the relation between workload and performance hard to interpret. But it seems reasonable to assume that extreme levels of workload, high as well as low, should be avoided.

There are some implications for design of systems using the six factors identified in the NASA TLX.

To avoid extreme levels of mental demand and possibly also frustration, a system should be designed in a way so that an operator easily can access important information and important controls. The information necessary to supervise or control the system should be designed to make interpretation of their meaning easy. This will have a positive impact on an operator's ability to have a high level of situation awareness.

To avoid extreme levels of physical demands, the system should be designed to make controlling as easy as possible.

To avoid extreme levels of time pressure, a system should be designed so that an operator will have time enough to access, interpret, and respond appropriately.

It seems that these principles have been used when the guidelines for the design of alarm systems have been proposed, and will be discussed later.

14.5 PERFORMANCE-SHAPING FACTORS

Within a human–machine system, there are several factors which have influence on overall performance, known as performance-shaping factors. One common definition of performance-shaping factors is anything that has an impact on human performance. A common and simplified distinction is between internal factors, something the human brings to a situation (such as physical and mental conditions), and external factors (such as latent and operational factors).

The internal performance-shaping factors are those which operators, as unique individuals, bring with them to the technical system. Physical conditions are, for example, age, physical fitness, hearing, vision, and general health, while mental conditions are, for example, personality, attitude, emotional state, motivation, stress tolerance, behavior, and group identification. Through training and education, qualities such as competence, experience, proficiency, and problem-solving capabilities can also be influenced. If there are specific requirements in terms of an operator's internal performance-shaping factors, individuals with specific qualifications may have to be selected.

The external performance-shaping factors are those which are present within our surroundings and in work conditions (latent factors), and those which are task- and process-dependent (operational factors). Latent factors are, for example, premises design, surrounding environment, working hours and breaks, shift rotation, staffing levels, organizational structure, management and leadership, pay and rewards. Operational factors are, for example, work procedures, working methods, instructions, communication opportunities, group cooperation, equipment and tools, operator interfaces.

This distinction may be useful despite the fact that it is simplified and ignores the interaction between internal and external factors. Just to mention one example, environmental factors such as noise extreme temperatures, crowding may result in an increased stress level and fatigue after some time of interaction. Another important fact is that a team of operators controls many tasks, especially in complex systems. It may therefore be necessary to add team factors to the group of performance-shaping factors. Examples of group factors may be communication problems, authority problems, group think. The internal and external factors, both latent and operational, can affect humans either individually or in combination in the form of various stressors.

14.5.1 Stressors

Stressors can have a direct or indirect influence on interaction within the human–technology system, which in turn affects decision-making and action. One example is that if an operator is exposed to strong vibrations, it can be harder to control the system and read displays. Not having enough time can restrict the opportunity to consider additional courses of action. Stressors are divided into psychological and physiological stressors.

Psychological stressors are high work load, high work pace, surprising events, threats and significant risks, distraction, dulling of the senses, lack of feedback, lifestyle stress, and domestic circumstances. Physiological stressors are long-term stress, exhaustion and sleep disorders, pain, discomfort, tiredness, hunger, thirst, extreme temperatures, inadequate ventilation, vibrations, high sound levels, and exposure to chemical substances.

Dividing stressors up in this way shows the direct relationships between stressors and internal and external performance-shaping factors. For example, latent factors in the surrounding environment are likely to include physiological stressors, whilst psychological stressors such as high workload can have more of an impact if the individual has low stress tolerance. Many different workplace phenomena can have a detrimental effect on performance, such as:

- Insufficient separation of workplaces, which affects the execution of tasks.
- Unsuitable positioning of workstations, which can lead to inadequate opportunities for communication.
- Badly adapted illumination levels may create glare and make it impossible to read information on paper and screens simultanously.
- High sound levels in the room, which make communication between the people within the system difficult.
- Disruptive non-operational staff at workstations, such as repairmen and maintenance workers.

When someone is exposed to stressors, the ability to process information is affected.

Stress often builds upon a pronounced feeling or fear of something. Internal psychological definitions of stress often relate to a perception of time pressure. Different individuals may feel different levels of stress in the same situation. This is because individuals have different abilities and experiences when it comes to dealing with situations or because the individual has exaggerated the risk or is blissfully unaware of the risk which the situation involves. Generally speaking, when we experience pressure in a situation, we start by performing better as the pressure increases. Once the pressure increases to a certain level, we reach our limit and our performance deteriorates as the pressure continues to rise. This is the Yerkes–Dodson law, and the curve is described as an inverted U. The fact that we start off with a rising performance curve may be due to the fact that we are gathering our resources and making the maximum effort. The optimum levels of stress also vary according to the task in question. Complex tasks which require a degree of innovation are generally best carried out at a lower stress level, whereas automated tasks which the individual is used to can be performed at a higher stress level.

Recent research (for instance Vischer, 1989, 1996, 2007) has focused on the physical environment and its effects on work performance. A theoretical model of workplace stress has been developed, based on several studies showing that the physical environment in which people work affects job performance as well as job satisfaction.

Research has also shown links between employee health and properties of the physical environment, such as indoor air quality, lighting, and ergonomic furniture (Milton et al., 2000; Veitch and Newsham, 2000). A mismatch or misfit between the resources an individual possesses and the demands from the situation, may generate stress and have a negative impact upon comfort (Czikszentmihalyi, 1990; Kaplan, 1983; Lawton, 1980).

Research in ergonomics has shown the importance of variables like lighting, noise and noise control, furniture, and spatial layout in offices. Concerning stress related to the physical environment, McCoy and Evans, 2005 have suggested that stress can be triggered when properties of the physical environment interfere with plans and behavior. A supportive physical environment should make it possible for operators to focus on the task to be performed without being occupied with properties in the environment that interfere with plans to be executed. Environmental comfort may, according to Vischer (1996), be a function of three hierarchically-related categories: physical, functional, and psychological. Physical comfort is defined as basic human needs, such as safety, hygiene, and accessibility. Functional comfort is defined as ergonomic support for work tasks and psychological comfort has to do with feelings of belonging and control over the workspace.

14.6 USER INTERFACE DESIGN

Operational interfaces need to be designed to support the operator's mental model of the system to be controlled, situation awareness, and decision-making.

14.6.1 Information Design

Too much information or poorly organized presentation of information must be avoided. If large amounts of information must be presented, it may be necessary to divide the responsibility for the control of the system between more than one operator, and to create effective channels and tools for the communication between members of the team.

Another task for the operator is to interpret and understand the implications of the information presented and attended to, and to make a judgment whether some actions are required now or later. If an action is required later, this will also create a need to remember the action to be performed and when to perform it, that is, to avoid a lapse, or to forget to perform the action (Reason, 1990).

14.6.2 Design for Attention

Besides having access to a correct model of the system, it is also important to know where attention to different parts of the system should be focused when different tasks are being performed (Woods, 1995). The time aspect (Hollnagel, 2002) plays an important role in the control of dynamic systems. Besides knowing what kind of information should be attended to during control of a system, it is important to know when to focus on the information.

Attention can be defined as a function that makes it possible to focus limited mental resources on certain parts of the external or internal environment. When we attend to something, working memory is involved, the ability to process information is limited in working memory.

In a complex environment, it is not possible to attend to every aspect of the environment and some choices have to be made. It is possible to make a distinction between four types of attention in everyday and work situations.

Selective Attention. In the first one, an operator must monitor a number of information sources to determine the status of a system. The operator has to select the relevant information sources (cues), or task-relevant elements, interpret them and make a decision concerning the status of the system. Mental resources of some kind are needed for these different subtasks.

To design for selective attention, it is recommended to make task-relevant elements easy to perceive and interpret. Possible ways to do that can be to highlight the information sources or task-relevant elements for different system phases (i.e., start-up, normal, and abnormal phases). Another possibility is to remove or shut down irrelevant information. If possible, avoid multiple visual channels. If multiple visual channels must be used, use as few as possible. Provide information concerning the importance of different information sources. In the case of multiple visual information sources, place them close together to minimize scanning behavior. In the case of auditory information, avoid masking information and increase the intensity of the most important information source.

Focused Attention. In this case, an operator has to focus attention on some information source and ignore other information sources. To design for focused attention, one possibility may be to increase the intensity of the information that should be in focus, or to remove or minimize other information sources.

Sustained Attention. Here the operator must attend some information source or sources over a long time period and be able to detect something that may occur infrequently. To design for sustained attention, it may be recommended to make the information source(s) easy to perceive. Visual information sources can be big, clear, and distinctive. Auditory information sources should be easy to hear and interpret, adapted to background noise, and not overlapping.

Divided Attention. In this case, an operator must share mental resources between more than one task concurrently. This may, and often will, increase the demands on the cognitive system and performance may decrease on some task. The decrease in performance is explained as a result of limited resources in the cognitive system. The resources in the working memory are distributed with separate and independent resources for spatial and verbal information. To design for divided attention, it may be possible to reduce the number of information channels, and to indicate the importance of different information channels. It is also important to design the information sources so that information can easily be picked up and to keep the mental workload optimal. It is recommended to use different input channels if possible. Information presented in the visual modality combined with information in the spatial modality will be easier to handle compared to information presented in a single modality.

14.6.3 Design for Perception

Perception is a process which organizes those stimuli which have been noticed and gives them meaning, that is, pattern recognition is used. We build up our perception by reinforcing, altering, and discarding information.

14.6.3.1 Design Legible Displays *Visual Displays*: Legibility requires high contrast, good illumination, and the correct viewing angle. Good legibility is essential, but does not necessarily mean that this is all that is needed in order to present usable information.

Auditory Displays: Design sound with good audibility. Frequency intervals need to be adapted in order to avoid masking, whereby sounds can cancel out other sounds in the same frequency range. Unfortunately, the intensity of warning sounds is often excessive. There is very rarely any benefit in warning sounds which scare and surprise the operator. It is also hard to concentrate and think when there is a cacophony of sound. The operator can wear a headset or sit close to the source of the sound; here, it is primarily information which is needed, not distracting, stressful noise. However, there are alarms—such as burglar alarms and car alarms, where the sound in itself should act as a deterrent and attract attention over long distances.

14.6.3.2 Avoid too Many Levels for Absolute Judgement of Information *Visual Displays*: It is incredibly difficult to assess, for example, the size, thickness, or color of a presented variable if the number of possible levels which the variable can assume exceeds five. It is often recommended that only three levels should be used. In terms of colors, only red or green and yellow or blue should be used in order to ensure good design for those who are color-blind.

Auditory Displays: Auditory signals are not suited to be presented at multiple levels.

Sounds will mask each other, and since sound is transient, they cannot be compared with each other side by side. It is possible to compare a signal with another signal that was heard a few seconds ago.

14.6.3.3 Avoid Exclusively Top-Down Processed Data *Visual Displays*: We often discern and interpret signals according to what we think they should signify, based on our previous experience. If a signal is presented for which the message is the opposite of what is expected—for example, information about an unexpected event—the information must be clearly reinforced physically by flashing, being illuminated in red, occupying a central position within the interface, or increasing in size. This makes it easier to notice and interpret the signal.

Auditory Displays: For normal communication, reading back the information which has just been heard may be useful. This can be reinforced through the use of the phonetic alphabet, replacing A, B, C with Alfa, Bravo, Charlie, etc.

14.6.3.4 Using Redundancy *Visual Displays*: It is highly likely that information will be interpreted more accurately if the same message is presented in more than one way. The information human-technology systems will be particularly clear if different modalities are used, for example, if a message is presented in both visual and auditive forms at the same time. Different physical forms of visual presentation also reinforce clarity, such as using both text and images or color and shape. Traffic lights are a good example of redundant presentation: both the position and the colors show which lamp is illuminated.

Auditory Displays: Different physical forms of sound such as auditory icons, beeps, and voices reinforce clarity. In particular, location can be emphasized using 3D sound, for example, information which relates to an engine placed on right-hand side may be presented with sound coming from the right. Alternatively, sound can be used exclusively to direct attention towards a visual display.

14.6.3.5 Avoid Similarity between Objects *Visual Displays*: Objects which are perceived as being similar cause both confusion and a risk of confusion and misinterpretation. It should be possible to differentiate clearly between objects, regardless of whether they should be interpreted immediately or whether the appearance of the signal should be retained actively in the short-term memory in order to be compared with forthcoming signals. If there is a significant risk of confusion, the design must enhance the differences and reduce the similarities in what is being presented. This is particularly true of written text and series of numbers. Compare the two vehicle registration numbers AJB648 and AJB658. They look more similar than

48 and 58. In both cases, only one digit has been changed, but in the latter case, it is much easier to spot the difference since the differences are emphasized.

Auditory Displays: Sound which is used as a warning must be easily associated with the function in question, or must be easy to discriminate from the other sounds within the system. Auditory icons and voices can be very effective when there are a large number of different sounds.

14.6.4 Design for Memory Functions

The human memory is sensitive, particularly the short-term memory with its limited capacity. Only a very small amount of information can be actively retained and processed at the same time. Quickly finding necessary information in the long-term memory at the right time is also difficult, particularly where the interface does not provide any cues. However, there are certain things which we remember all too well and which we persist in using when we need them.

14.6.4.1 Knowledge of the World Around Us *Visual Displays*: The design of an interface should generally not require that the operator keeps important information active in his or her short-term memory, or uses data stored in his or her long-term memory in order to carry out a task. Current information, such as combinations of figures and letters, checklists, and reference values, should therefore be presented on-screen. This means that the short-term memory does not need to be burdened with as much data, and its entire capacity can be used for problem-solving. However, if too much information is presented on-screen at the same time, this can make it hard to find and interpret information. This is not to say that systems which are dependent on the use of stored knowledge acquired by memorizing (the knowledge in our head) are necessarily a bad thing. Experts who use a computer system often want to be able to use shortcut commands which they know by heart instead of navigating using menus and searching for the various functions. A good design must achieve a balance between operators using the knowledge in the world and knowledge in the head (Norman, 1988).

Auditory Displays: Ideally, warning sounds are not presented very often and operators do not therefore become accustomed to these sounds. It is therefore important that sounds should be intuitive, and should be easy to associate with their warning function. Voice messages can be used to remind the operator or to keep the operator informed during a procedure.

14.6.4.2 Anticipate System Statuses *Visual Displays*: People are not particularly good at anticipating what will happen within a system on the basis of a large number of process parameters. This is a difficult cognitive task which requires the use of multiple sources of information simultaneously within the short-term memory. We need to think about the current system status and possible future scenarios while also processing this within our mental model of the system's functionality.

When our mental resources are also used to focus on other tasks and new stimuli, we find it hard to focus on what may happen in the future. Our behavior becomes reactive, and we concentrate on what has already happened or what just happened, instead of being proactive and acting on what we expect will happen within the system. Since proactive behavior is more effective than reactive behavior, the interface must assist the operator by providing information which clearly shows the anticipated system status based on current and previous circumstances. This type of information description takes a resource-demanding cognitive task away from the operator, replacing it with simpler perceptual information.

Auditory Displays: Sound which changes—trend sounds—can be designed to inform operators that something is increasing in speed by increasing the frequency, or to show that the number of revolutions is falling by the tone becoming lower. This is the equivalent of rising and falling graphical trend curves. Sound from different angles and 3D sound can also be used in this context.

14.6.4.3 Consistent Presentation

Visual Displays: When actions saved in the long-term memory are often used, the operator may instinctively and automatically continue to carry out learned actions which are inappropriate for the system in question. Old habits die hard. When designing new interfaces, such as for steering gear, it is important that these should correspond as closely as possible with the interfaces which the operator is used to. It is important to keep the same color coding as before when introducing new systems. Color coding should be consistent throughout the entire system.

Auditory Displays: In many systems, there is natural sound which the user is familiar with, such as warning sounds, amplified signals, or background noise. If sound is added or taken away, it is important that information is not lost and that the sound which the user is used to is taken into account. Confusion of sounds should be avoided.

14.6.5 Feedback

Feedback from control activities can be provided in different ways. A common strategy is to use outcome feedback and simply provide information whether the activity had the desired outcome or not. For instance, pump A did not start.

Cognitive feedback, on the other hand, aims at a deeper understanding of the process or task to be controlled. Research (e.g., Hammond, 1993) has shown that cognitive feedback may result in faster learning of a task compared to outcome feedback. Using outcome feedback means to provide the operator with important information concerning properties of the process being controlled. For instance, the relation between control activities and system responses (i.e., the relation between the time a pump is activated and the temperature of a process). By doing this, it is possible to help operators to form a mental model of the process being controlled. As an example, pump A did not start because the water level was correct.

14.6.6 Alarms

Alarm systems are used in many different settings (e.g., health care, process industry, energy producing industry, transportation industry) to help operators control processes of different kinds. The processes that are being controlled in these settings have a number of important characteristics. They are continuous and relatively slow. Feedback from control activities are often delayed and seldom, if ever, available for direct perception. Complex processes are common and often composed of many subprocesses organized in different ways. Subprocesses may be organized hierarchically but also cross-coupled. If subprocesses are cross-coupled, changes in one process may have an impact on other processes and produce side effects of different kinds. Furthermore, complex processes are partly unpredictable. To control a partly unpredictable process, alarm systems are often used. Alarm systems should be able to detect deviations from ideal states, especially deviations that may cause hazards for people or be very costly for process owners, or both.

The main aim of an *alarm system* is to draw the operator's attention to a deviation in the human–technology system. Alarm systems often include both visual and auditory components, that is, an auditory alarm is often accompanied by visual information in on-screen alarm lists. Alarm systems are common in complex technical systems, such as in control rooms at nuclear power plants, refineries, paper and pulp factories, traffic control centers, air traffic control centers, and surveillance centers. Operators rely on alarm systems particularly in process control rooms, since the underlying technical systems are not visible to operators. In certain types of control centers and surveillance centers, operators may receive additional information about the reason for a system alarm by being able to view the underlying technical system visually, such as in an air traffic control tower. In other technology-intensive industries, such as medical technology and aviation, the operator can also obtain information from a secondary source of information. In certain situations, a pilot can look through the windscreen and assess the situation or sense vibrations from the engines, while an intensive-care nurse can assess a patient's condition by directly observing skin color, skin temperature, pupil size, movements, and the ability to speak.

The role of an alarm system is to assist the operator in detecting deviations, correcting errors and dealing with disruptions. The alarm system should not, however, be used to draw the operator's attention to planned events; instead, it should be reserved for unexpected, unforeseen events. The aim of an optimal alarm system is to warn of an abnormal situation, provide information about what has happened, guide the operator in identifying and saving the situation, and confirm that the response carried out is correct (Thunberg and Osvalder, 2009). Alarms presented by the system should always be relevant, and the number of alarms occurring should be kept to a minimum.

By experimenting with variables such as sound level, pitch, and intensity, an auditory alarm can be used to inform the operator immediately of the type of deviation and its priority level. Auditory alarms should be prioritized at a maximum of three levels, since we find it hard to differentiate sound at more than three levels. One objective, in order for the lowest priority level of an auditory alarm to be noticed, is that the alarm should be 10 dB above the machine's normal sound level. It is important that the alarm system not only informs the operator that something has happened, but also calls upon the operator to carry out some form of physical and/or mental action. The system must also raise the alarm quickly so that there is plenty of time for the operator to be able to respond before it is too late, and must confirm within a reasonable period of time whether the operator's response has rectified the deviation. Depending on the human–machine system in question, a high-priority alarm should include a time indication which states that the deviation must be remedied within, for example, 2 minutes, otherwise damage to people, materials, or the environment may occur. Time and consequence are the parameters which should govern how the alarm is prioritized. If correct action is not taken, the alarm system should describe what happens next. Instead of—or as well as—on-screen visual information, speech can be used to guide the operator in responding to the alarm.

The qualities that characterize a good alarm are (Thunberg and Osvalder, 2009):

- *Relevant*—not a false alarm or an alarm which is of little use.
- *Unique*—the alarm should not be duplicated by another alarm.
- *Timely*—the alarm should occur at the right time: not a long time before a response is required, and not too late for the operator to be able to respond.
- *Prioritized*—the importance of the alarm should be indicated so that the operator can prioritize his or her actions if necessary.
- *Comprehensible*—the alarm should have a clear and easily-understood message.
- *Diagnostic*—the alarm should identify the problem which has arisen.
- *Advisory*—the alarm should indicate which action needs to be carried out.
- *Focusing*—the alarm should direct attention to the most important aspects.

14.7 DEMANDS ON THE ENVIRONMENT

14.7.1 Organization

The concept organization can have different meanings. One meaning refers to how a work process is organized, how different units of the work process interact to achieve their goals. In complex systems, it is common to have many goals and as a result, goal conflicts. For example, conflicts between productivity and safety. To be effective, organizations need to change when different demands, internal or external, changes. It is common that organizational changes take time and in some cases, changes slower than what could be optimal from a productivity or safety perspective.

The concept "safety culture" or "organizational culture" has been widely used in the nuclear power industry as well as in many other industries. Industrial processes are controlled by organizations, but at the sharp end, it is individuals who are handling technology and equipment. Human behavior is dependent not only on individual factors, but also on social and environmental factors. Social and environmental factors in their turn are dependent on organizational factors. Organizational factors are, in many ways, dependent upon political factors.

There exist many definitions of "safety culture" or "organizational culture," but a common theme is a focus on the attitudes and behavior of the individuals in the context. Reason (1997) suggests the following definition of organizational culture:

Shared values (what is important) and beliefs (how things work) that interact with an organizationás structures and control systems to produce behavioral norms (the way we do things around here).

It is here suggested that actual behavior is added to this definition, simply for the reason that the correlation between values, beliefs, norms, and actual behavior is far from perfect.

A safety culture should, according to Reason (1997) have a number of properties.

It should be a reporting culture, where near misses, incidents, and accidents can be reported without fear for some kind of punishment. It must have a reporting function which can be used easily, function(s) that support reporting of unwanted events, and provide feedback clearly and without long time delays.

It should be a just culture, a culture where someone making a mistake or performing an unsafe act is not punished without considering the circumstances. On the other hand, a culture where the line between acceptable and non-acceptable actions is as clear as possible is to prefer.

It should also be a flexible culture, with the ability to adapt quickly to changes in demands from the environment and the demands from new challenges. In the best of worlds, it should be capable of managing the unexpected, be prepared to cope with surprises of different kind. The ability to deal with a crisis situation is largely dependent on the structures that have been developed before chaos arrives.

Highly reliable organizations (HROs) are using five principles to guide their work.

HRO Principle 1: Preoccupation with Failure—Any lapse may be a symptom that something may be wrong with the system, something that could have severe consequences if several small errors happened to coincide. HROs encourage reporting of errors and are aware of liabilities of success, complacency, temptation to reduce safety margins, the drift into automatic processing.

HRO Principle 2: Reluctance to Simplify—The world is complex, unstable, unpredictable; take steps to create a complete and nuanced picture of what you are facing. Welcome skepticism and accept diverse experience and different opinions. To recognizing something may be dangerous; superficial similarities between present and past events could lead you in the wrong direction.

HRO Principle 3: Sensitivity to Operations—Attentive to the front line, the sharp end where the actual work is being done. The "big picture" in HROs is less strategic and more situational than is true in most other organizations.

HRO Principle 4: Commitment to Resilience—No system is perfect and we must learn from failures. The essence of resilience is the intrinsic ability of an organization (system) to maintain or regain a dynamically stable state, which allows it to continue operations after a major problem or during continuous stress. HROs develop capabilities to detect, contain, and bounce back from those inevitable errors that are a part of an undetermined world. Not error-free, but errors that do not disable it. Resilience is a combination of keeping errors small and of improvising strategies that allow the system to keep functioning. The demands are deep knowledge of technology, the system, one's co-workers, and oneself.

HRO Principle 5: Deference to Expertise—HROs prefer diversity because it helps them to notice more in complex organizations and to act more. Rigid hierarchies have their own special vulnerability to error. Errors at higher levels tend to pick up and combine with errors at lower levels, making the problem bigger, harder to understand, and more prone to escalation. HROs push decision-making down and around. Decisions are made on the front line, authority mitigates to the people with the most expertise, regardless of their rank.

Finally, a safety culture should be a learning culture, where earlier experience of mistakes and successes are stored and made available for all employees.

14.7.2 Communication

Interaction between human beings or human beings and technology to solve different problems are dependent upon communication, or exchange of information. With the help of communication, it is possible to formulate goal(s), develop action plans, implement them, check the status of action plans, coordinate them with other action plans, create an overview of action plans, send feedback to involved partners, evaluate the outcome, and learn from the mistakes.

In one-way communication, the sender (human or non-human) sends a message or many messages to the receiver (human or non-human). The message(s) must be coded in some way (verbal, visual, haptic, numeric, non-verbal, etc.) and then transmitted to the receiver.

Two-way communication means that a feedback loop is created between the sender and the receiver of information. This creates possibilities for the receiver to ask for more details, repetition of the message, etc.

Technological development has opened up new possibilities for how communication between human beings and artifacts can be improved. One important step would be to change from one-way communication to two-way communication, and make it possible for the operator of a system to ask questions to understand what happened, why it happened, and what to do.

As a real example, when the display in a certain modern car shows a symbol (a red car and a tool), it is a one-way communication where the artefact (the car) makes an attempt to transmit information meaning that it is time for service. The owner of the car, as well as the garage with repair service, may need more information in order to know what to do and how much time to allocate to the job, plus the cost for the service or repair. To get more detailed information, the owner of the car must know that s/he should activate the menu, scroll to vehicle information, and to the subsection service. This results in another one-way message to the owner of the car, the answer to the questions what, why, etc., and the answer was: B – plus two symbols whose meaning was not fully understood. Clearly this communication process could be improved.

Besides problems caused by one-way communication, it is also possible to produce very large amounts of information, not always adapted to the intended userás needs of information or their limitations to process and understand the information presented. Too much information may lead to overload and stress, lack of information, ambiguous information, conflicting information, may create uncertainty, and lead to interpretation problems.

The cognitive system of human user normally receives information through five senses. Most of the research on human perception has had a focus on the visual and auditory channels or senses. The haptic or tactile channels or senses are less well-researched, with a few exceptions. The same applies to smell and taste channels or senses. It is assumed that each input channel or sense also has a sensory memory, where the information is stored in its original form for a short time. The iconic memory (visual information) and echoic memory (auditory memory) are the most well-known sensory memories. The processing of information from different senses is assumed to take place in the working memory (Baddeley, 2000). Top-down and bottom-up processes are supposed to interact and create meaning or interpretation of whatever. Figure 14.1 shows the architecture of the working memory.

As can be seen from Figure 14.1, working memory is supposed to consist of five different and interacting subsystems. On the top is the central executive that interacts with two slave systems, the phonological loop (speech perception, written material), and the visual-spatial sketchpad (visual and spatial information). The central executive, the boss at the office, controls the cognitive resources, selects information from the different subsystems, integrates and coordinates information from them. The episodic buffer is a backup function, communicating with long-term memory and other components. The active place in the model, where the mental work is done, is in the central executive. Input from the slave systems is integrated and interpreted, to be used for further decision-making, problem-solving, development of plans, etc.

When people are interacting actively with technology, it may be to search for information, receive support of some kind, create or modify something, and to control processes. In all these cases, there is normally an intention or many intentions behind the interaction with technology. This or these intention(s) must be communicated or sent to the technological system used. A well-designed interaction between the human being and the technological system used should minimize the efforts in communicating the intention(s) to the system and avoid the "gulf of execution."

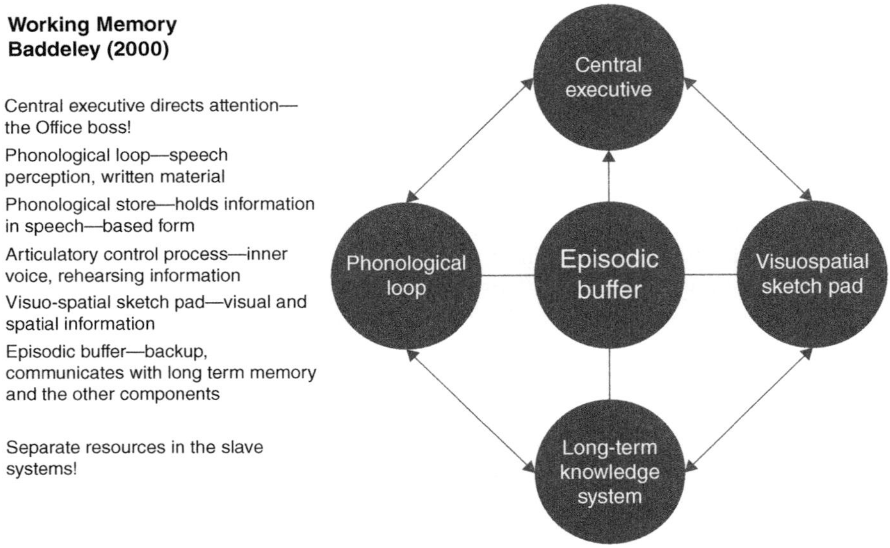

Figure 14.1. Architecture of working memory. Adapted from Baddeley, 2000.

Furthermore, a well-designed interaction between human beings and technology should minimize the efforts to evaluate the result(s) of the interaction between humans and technology and avoid the "gulf of evaluation" (Norman, 2002).

This raises two interesting questions: How can intentions from a human user be designed to minimize the gulf of execution? And how should feedback from the technological system be designed to minimize the gulf of evaluation? An important distinction is between outcome feedback and cognitive feedback (Hammond, 1993). Outcome feedback only provides information concerning whether an action succeeded or not. Cognitive feedback provides information that may be used to understand the relation between actions and system response.

One possible way to address the question concerning the communication of intention(s) is to start with human–human communication. When people are interacting face-to-face, intentions may be clarified by verbal messages and by body language, most often in combination. Verbal messages can be analyzed on different levels, for instance, what is said and how it is said (phrasing, intonation, emphasis, etc.). According to Dell (1986), it is possible to analyze the communicative process at four levels: The semantic level, the meaning in what is said; the syntactic level, or the grammatical structure; the morphological level, the morpheme or meaningful units in what is said; the phonemic level, or the basic units of sound in the speech. Body language may take many forms, from bodily posture, gestures, eye contact, facial expressions, touch, and in some cases, smell.

In a face-to-face communication process, verbal and non-verbal communication may interact and support each other, be neutral, or contradict each other.

Grice (1967) suggests that effective communication between human beings must be based on cooperation between the humans involved. This implies that the partners involved listen to each other and show mutual respect. Besides that, Grice assumes that four other aspects are of vital importance.

- The quantity aspect, meaning that the speaker should be as informative as possible
- The quality aspect, implying that the speaker should stay to the facts and not distort the message
- The relation aspect, the speaker should speak of aspects that are important for the task to be performed
- The clarity aspect, the speaker should make it easy to understand the message.

These principles should also be possible to use to improve the communication between human beings and artifacts. Team player approach!

Advances in technological development have made it possible to rethink human–machine interaction. Normally, it is assumed that the operators of a system or process are supposed to learn the properties of the system or process and construct a correct mental model of the system.

If we take human interaction as an example, then both parties in an interaction process will learn something about each other, and this will, in the best cases, make future interaction easier and, hopefully, more effective. Using human interaction as a model would imply that the system or process to be controlled can store information about operator performance, either individually if possible, or as a team or group. For instance, common errors could be registered and saved, such as responding too early or too late, or not responding at all when needed. This information could be used later, to provide operators with feed-forward information. Advances in technology should thus make it possible for the system or process to form a model of different operators' behavior and adapt to their ways of controlling the system.

14.8 HANDLING COMPLEXITY

When complexity increases in organizations, strategies to handle complexity must be developed. One strategy is to create teams with special knowledge in certain fields, technology, human resource management, safety, economy, etc. This strategy is well-known, as also problems created by using it. Fragmentation of organizational knowledge, different cultures with different conceptual frameworks, different views on goals and their importance, communication problems, just to mention a few. For a team to be a good and effective team, Cannon-Bowers (1993) suggested a number of characteristics of high-performance teams.

- Team members should have the competence needed for their own task and necessary team working skills.
- Communication, or exchange of information across all levels, is important and must be clear and concise.
- The level of motivation to perform the job and work as a team must be enough to perform well.
- Collective orientation, the team members' view of themselves in relation to other team members. Belief in the teams' ability to solve problems.
- Shared goals and mission is needed.
- Team leadership.
- Shared understanding of work tasks, own role, and the roles of other team members.

Another strategy to reduce complexity is to use automation as a solution. Processes which are possible to decompose and describe completely can be automated and removed from the operator's task(s). There are, however, some possible or likely problems associated with automation of processes. In her classical article, Bainbridge (1999) suggested a number of ironies associated with automation. First of all, automation can change the task to be performed by a human operator from an active control task to a passive supervising task. Instead of being actively involved in the performance of the task, the operator will observe the performance of the automatic system. Going from actor to observer may have consequences for situation awareness and activity level.

The "Ironies of Automation" (Bainbridge, 1999) suggests that attempts to automate out the human operator completely will probably fail. As Sheridan (1992) puts it, "we automate what we can and leave the rest to the operator." This statement is far from ideal for the entire system. By automating simple tasks, difficult tasks—that is, those tasks which are too hard to automate or not worth automating—become even more troublesome for the operator. Because these troublesome tasks only arise rarely, the operator will be unlikely to have the knowledge and experience necessary to deal with these.

Monitoring a system and checking that everything works can be understimulating, and may have a negative impact on attentiveness and vigilance. If disruption occurs which needs to be dealt with, this often requires experience. Experience is gained by carrying out tasks continuously, but disruption rarely occurs in a reliable system. All in all, this means that it is likely that human error will occur when unexpected and unforeseen events arise. This is a challenge for developers, who need to determine where the boundary lies in terms of how much automation is appropriate to introduce into a technical system.

REFERENCES

Ashby, R. W. (1958). Requisite variety and its implications for the control of complex systems. *Cybernetica*, 1(2), 83–99.

Baddeley, A. D. (1997). *Human Memory—Theory and Practice*. Psychology Press.

Baddeley, A. D. (2000). *Working Memory*. New York: Clarendon PressOxford University Press.

Bainbridge, L. (1999). Ironies of automation. *Automatica*, 19(6), 775–779.

Brehmer, B. (1978). Response consistency in probabilistic inference tasks. *Organizational Behavior and Human Performance*, 22, (103–115).

Brehmer, B. (1992). Dynamic decision making: Human control of complex systems. *Acta Psychologica*, 81(3), 211–241.

Brehmer, B., and Allard, R. (1986). Learning to control a dynamic system. In E. de Carte, et al (Eds.), *Learning and Intsruction*.

Cannon-Bowers, J., Salas, E., and Castellan, S. (1993). Shared mental models in expert team decision making. In: J. Castellan (Ed.), *Individual and Group Decision Making*. Hillsdale, NJ: Lawrence Erlbaum.

Chapanis, A. (1965). *Man–Machine Engineering*. Belmont, CA: Wadsworth.

Conant, R. C., and Ashby, W. R. (1970). Every good regulator of a system must be a model of the system. *International Journal of System Science*, 1, 89–97.

Czikszentmihalyi, M. (1990). *Flow: The Psychology of Optimal Experience*. New York: Harper Collins.

Dell, G. S. (1986). A spreading-activation theory of retrieval in sentence production. *Psychological Review*, 93, 283–321.

Dörner, D. (1980). On the problems people have in dealing with complexity. *Simulation and Games*, 11, 87–106.

Endsley, M. R. (1995). Measurement of situation awareness in dynamic systems. *Human Factors*, 37(1), 65–84.

Gopher, D., Weil, M., and Siegel, D. (1989). Practice under changing priorities: An approach to the training of complex skills. *Acta Psychologica*, 71, 147–177.

Grice, H. P. (1967). Logic and conversation. In: P. Cole and J. L. Morgan (Eds.). *Studies in Syntax* (Vol. III). New York: Seminar Press.

Hammond, K. R. (1993). Ref in G. A. Klein, J. Orasano, R. Calderwood, and C. E. Zsambok (Eds.), *Decision Making in Action: Models and Methods*. Norwood, NJ: Ablex Publishing Corporation.

Hart, S. G., Staveland, L. E. (1988). Development of NASA-TLX (Task Load Index): Results of empirical and theoretical research. P. A. Hancock and N. Meshkati (Eds.), *Human Mental Workload* (pp. 239–250). Amsterdam: Elsevier.

Hollnagel, E. (2009). *The ETTO Principle*. Ashgate.

Hollnagel, E. (Ed.). (2010). *Safer Complex Industrial Environments—A Human Factors Approach*. CRC Press.

Kahneman, D. (2011). *Thinking, Fast and Slow*. Penguin Books.

Kaplan, R. D. (1983). Person–environment fit, past, present and future. In: C. L. Cooper (Ed.), *Stress Research: Issues for the Eighties* (pp. 35–78). Hoboken, NJ: John Wiley & Sons.

Lawton, M. P. (1980). *Environment and Aging*. Monterey, CA: Brooks-Cole.

McCoy, J. M., and Evans, G. (2005). Physical work environment. In: J. Barling, E. K. Kelloway, and M. Frone (Eds.), *Handbook of Work Stress* (pp. 219–245). Thousand Oaks, CA: Sage Publications.

Milton, D. K., Glencross, P. M., and Walters, M. D. (2000). Risk of sick leave associated with outdoor air supply rate, humidification and occupant complaints. *Indoor Air*, 10(4), 212–221.

Norman, D. A. (1988). *The Design of Everyday Things*. Basic Books.

Norman, D. A. (2002). Emotion and design: Attractive things work better. *Interactions Magazine*, ix (4), 36–42. Available at: http://www.jnd.org/dn.mss/emotion_design.html

Perrow, C. (1999). *Normal Accidents. Living with High-Risk Technologies*. Princeton University Press.

Polanyi, M. (1958). *Personal Knowledge: Towards a Post-Critical Philosophy*. Chicago, IL: Routlege & Keagan Paul Ltd.

Reason, J. (1990). *Human Error*. Cambridge Press.

Reason, J. (1997). *Managing the Risks of Organizational Accidents*. Ashgate.

Sheridan, T.B. (1992). *Telerobotics, Automation, and Supervisory Control*. Cambridge, MA: MIT Press.

Thunberg, A., and Osvalder, A-L. (2009). Good practice regarding ALARM system design and alarm management strategies for complex process control settings. In: VTT Symposium 258: European Conference on Cognitive Engineering 2009, ECCE (pp. 107–110), Helsinki, Finland, September 30–October 2, 2009.

Veitch, J. A., and Newsham, G. R. (2000). Exercised control, lighting choices, and energy use: An office simulation experiment. *Journal of Environmental Psychology*, 20(3), 219–237.

Vischer, J. C. (1989). *Environmental Quality in Offices*. New York: Van Nostrand Reinhold.

Vischer, J. C. (1996). *Workspace Strategies: Environment as a Tool for Work*. New York: Chapman and Hall.

Vischer, J. C. (2007). The effects of the physical environment on job performance: Towards a theoretical model of workspace stress. *Stress and Health*, 23, 175–184.

Wickens, C. D., and Hollands, J. G (2000). *Engineering Psychology and Human Performance*. Prentice Hall.

Young, M. S., and Stanton, N. A. (2002). It's all relative: Defining mental workload in the light of Annett's paper. *Ergonomics*, 45(14), 1018–1020.

Young, M. S., and Stanton, N. A. 2004. Applying interviews to usability assessment. In: N. Stanton, A. Hedge, K. Brookhuis, E. Salas, and H. Hendrick (eds.), *Handbook of Human Factors and Ergonomics Methods*. New York: CRC Press.

Part III

DEMONSTRABILITY

15

QUALITY PRINCIPLES AND THEIR APPLICATIONS TO SAFETY

Bo Bergman

15.1 INTRODUCTION

In this section, we will provide an overview of the ideas developed in the quality movement, especially the side of the movement that is characterized by continual improvement.

After the Japanese export successes during the 1970s, the quality movement has affected management thinking and management models under different labels (see, e.g., Cole and Scott, 2000). Some of these labels are just-in-time, statistical process control, ISO 9000, balanced score cards, process management, total quality control, total quality management (TQM), quality improvement, Lean, Six Sigma, etc. Many of these concepts can be traced back to—often via Japanese adaptations—scientific management (see, e.g., Tsutsui, 1998) on the one side and the ideas put forward by Shewhart (1931, 1939) and further promoted by Deming (1993) on the other. This has spilled over to the safety management area, as seen, for example, in Chapter 7 of this book. Some illustrations of such spill-over to other areas of safety management might be found, for example, in Aven (2013, 2014), Aven and Krohn (2014), and Aven and Bergman (2012). Also, the ISO 9000 (originally in the scientific

Handbook of Safety Principles, First Edition. Edited by Niklas Möller, Sven Ove Hansson, Jan-Erik Holmberg, and Carl Rollenhagen.

management spirit) has affected thinking on safety management systems (see, e.g., Hale et al., 1997). One field that is common to safety and quality is reliability (see, e.g., Bergman, 1985, 1992; Bergman et al., 2009; and Johannesson et al., 2013). However, we will not focus on reliability issues in this chapter more than concerning some personal experiences in Section 15.5. More on reliability-related matters are found in Chapters 4, 5, 23, 24, and 29 of this book. In this chapter, we will reflect on general ideas on quality Improvement and the application of these ideas to safety, as well as their potential for future-enhanced safety principles.

Quality

Quality has been defined in many ways (see, e.g., Bergman and Klefsjö, 2010), and safety is an important dimension in most definitions. Bergman and Klefsjö (2010) defined quality as "the ability of a product or service to satisfy or preferably exceed customer needs and expectations." The customer concept is often extended to more general stakeholders, thus covering safety related to third parties.

Principles, Practices, and Tools and Improvement Knowledge

Some management and operations researchers have characterized TQM as a coherent set of principles, practices, and tools (Dean and Bowen, 1994; see also Hackman and Wageman, 1994). We will, without reference to TQM, use this framework and complement it with the basic theoretical knowledge areas suggested by Deming (1993), who called the corresponding knowledge system *profound knowledge*. These basic theoretical knowledge areas—variation understanding, psychology, knowledge theory, and system thinking—have been particularly emphasized in the health-care improvement movement (Batalden and Stoltz, 1993). Especially in health care, this knowledge system is called "improvement knowledge" (see Batalden and Stoltz, 1993; Berwick et al., 1990; and Berwick, 2008), and it has had an important influence on an emerging science of improvement in health care (see, e.g., Bergman et al., 2015; Berwick, 2008; and Perla et al., 2013). These basic theoretical disciplines provide a basic understanding that is necessary for an organizational transformation in line with the organizing principles for quality improvement. In the following, we will take these knowledge areas as our points of departure and reflect on how the quality discourse has already affected or has a potential to affect safety principles. However, we will first provide a short overview of the principles and indicate how practices and tools are aligned and in support of these principles.

Quality Principles

In the quality discourse, some use the concept *quality principles* and others use *quality values*. In quality award criteria (e.g., the Malcolm Baldrige award criteria in the United States and the Swedish Quality Award), *values* are often used, and *principles* are used in ISO 9000:2015. One way of looking at this is that values are difficult to change both at individual and group level. Much less is it possible for somebody else to decide regarding such changes. However, we can decide on principles to follow in our activities, behavior, and decisions. In the long run, we can hope that behavior will shape attitudes, and eventually values as well. Some support for this process of change from behavioral changes toward attitudes has been discussed in dissonance theory (Festinger, 1957). However, a long distance exists between expressing principles, following principles, shaping behavior, and eventually changing attitudes and values. One way to look at "Principles for Quality Improvement" is that they build on the assumption that an organization is shaped and organized by the activities and the decisions made in the organization. If the principles are accepted and followed by major players (informal and formal leaders, as well as others with high legitimacy) in the organization, this will eventually also determine how people in the organization look at the organization and how it is organized, consequently creating a culture of "this is how things are done here." For discussions on these and similar aspects, see, for example, Sztompka (1993). More insights on cultural change with special emphasis on safety culture are provided by Reiman and Rollenhag in Chapter 28.

No complete consensus exists concerning principles for Quality Improvement (or principles for quality management). As a starting point, we will use principles presented in the new ISO 9000 standard from 2015, though with some minor changes. The principles we use essentially agree with those suggested in 1990 by Bergman and Klefsjö (2010); they are as follows:

1. *Customer Focus*, meaning that the current and future customers' needs should be in focus and the organization should strive to exceed the expectations of their customers;
2. *Factual Approach to Decision-Making*, meaning that effective decisions should be based on analyses of data about customers, environment, and processes;
3. *Process Approach*, meaning that processes ("the way we do things here") that are repeated within the organization should be managed and targets for improvement activities;
4. *Continual Improvement*, meaning that small as well as large, innovative steps should be a permanent objective of the organization;
5. *Involvement of People*, a crucial issue here as people are the essence of the organization and their full commitment and empowerment is important for the organizations' benefit, improvement, and innovation; especially, the importance of intrinsic motivation should be emphasized;
6. *System Approach to Management*, as the organization itself is embedded in larger systems with interrelated processes and the ability to use this for the management of the organization contributes to its effectiveness and efficiency in achieving its objectives, and identification of the value-creating networks for and with the customers and other parties in these networks (such as suppliers) is crucial; and finally,
7. *Leadership Engagement*, the leadership principle is decisive; leaders on all levels of the organization should lead according to the other six principles.

To some extent, Principles 2–4 are already implicitly discussed in Chapter 7.

Practices—An Illustration

As a short teaser, we will illustrate how practices and tools have been tailored to support the principles above. We will use Principle 4, continual improvement. A number of practices on different organization levels have been suggested to support continual improvement. On the top-management-level, policy deployment (called Hoshin Kanri in Japanese, see, e.g., Hutchins, 2008) has been suggested as a means of aligning improvement and innovation initiatives in the organization to the organization's vision and strategy. This is achieved in an ongoing application of a learning cycle, often visualized as the PDSA cycle discussed later in this chapter (see for example Figure 15.6).

Another practice in support of the continual improvement principle is the problem-solving practice in the organization. We will use this practice, or rather, a special instantiation thereof, and some of its tools as our illustration. When problems (or opportunities) are identified in an organization, some of them are of a simple nature—just-do-it problems—while others are unclear, requiring a systematic process for advancing the topic and achieving a real improvement. It is an enormous advantage to an organization if the different steps in a problem-solving practice are well-known and routinized (see, e.g., Peng et al., 2008). Instead of discussing the problem-solving process, energy could be immediately directed to the problem at hand. One such systematic practice goes under the heading of Six Sigma (see, e.g., Lifvergren et al., 2010; Magnusson et al., 2003), which is also called the DMAIC process and which places a strong emphasis on the definition of the problem. It is also important to plan and allocate resources to the problem-solving process and to assure that if an effective solution is found, then it will also be adopted in the organization. To make that possible, it is necessary that both the management in control of resources and those who will be affected by future changes are involved in the early stage of problem definition, which dramatically increases the likelihood for acceptance of the results of the process. Within a set of tools—"the seven management tools" (Mizuno, 1988), some special tools have been suggested for involving people and finding their ideas and opinions regarding the problem at hand. One such tool is the affinity diagram, which is based on a systematic kind of brainstorming activity. This first phase in the problem-solving process is called DEFINE. The second phase, MEASURE, is to find the ways in which to measure the problem; that is, to be able to find a criterion: "How do we know that a change has produced an improvement?" In the MEASURE phase, we also gather relevant information that could elucidate the problem. It is important to gather both numerical data, using statistical tools, and verbal data (e.g., in the form of affinity diagrams and narratives) in order to increase understanding of the problem.

The third phase in the Six Sigma problem-solving process is ANALYZE. Note that no solutions are allowed in this phase, nor in the earlier ones. Experience tells us that all too often solutions are advised without a thorough analysis of ill-defined problems and without any criteria to judge if a change has provided an improvement, often leading to unsuccessful changes and lost energy in the organization. A lot of different statistical tools are available during the ANALYZE phase; these tools also involve designed experimentation (Design of Experiments (DoE)) to better understand the system being studied. Additionally, it is possible to utilize more qualitative methods such as the affinity diagram. Finding solutions only takes place after the ANALYZE phase, in the IMPROVE phase. Now improvement ideas are suggested and tested in a small scale. Not rarely, iterations are needed when efforts do not result in expected results; we have to rethink and sometimes reanalyze data in order to acquire a better understanding of the problem at hand. Sometimes, it is also necessary to gather more information, that is, going back to the MEASURE phase. Even redefinition of the problem may be advisable when new contradictory information has been gained from testing. Once good results have been achieved, the changes have to be firmly adopted in the organization and new procedures and work routines are needed. Follow-up should be made to assure that changes are made and sustained in the CONTROL phase. In parallel with the CONTROL phase (or included in this phase), it is important to reflect on the problem-solving process and to learn for future events regarding both the physical process where improvements were made and the problem-solving process itself. What could have been done better? Is it possible to make changes to the local problem-solving practice? This last phase is called LEARN in the problem-solving process at the Skaraborg Hospital Group, as discussed by Lifvergren et al. (2010).

It should be noted that the problem-solving process described above is an enhancement of the PDSA learning cycle, and it is also closely related to the structure of the "Improvement Story" utilized in Japanese reports on improvement cases.

Regarding the above discussion of the problem-solving practice, it should be noted that it is not only the "continual improvement" principle (Principle 4) that shapes the practice. Improvements are usually seen from the perspective of the customer—Principle 1; facts/data are needed—Principle 2; improvements are made to the processes—Principle 3; and the involvement of people is crucial—Principle 5. Preparedness that the current local theory of the system is wrong and the tested changes do not work should always exist—Principle 6. Finally, the support of leaders found in Principle 7 is crucial, not only in DEFINE, but also in the complete problem-solving process. Indeed, all principles are utilized in the shaping of the practice; in the same way, all organizational practices and routines should be in line with all principles so as to result in a coherent whole. It should be noted that organizations in the change process toward continual improvement should tailor their practices to fit their organization and their history.

The Rest of the Chapter

An effort to make a comprehensive study of all ideas within the quality field and their current and potential applications to safety principles would be too heavy a task for this chapter. Instead, we will look at the underlying theoretical knowledge areas; that is, *improvement knowledge*, see Figure 15.7, and what could be learned from them with respect to safety. Of course, many of the ideas put forward here will concern ideas that are probably well-known in the safety field—the interest will probably be how these theoretical ideas are taken further into feasible practices and tools. The areas that seem to stand out the most in improvement knowledge with respect to the safety discourse seem to be "understanding of variation" and, to some extent, "knowledge theory." Psychological insights and system thinking will not be treated extensively due to limited space. In the end of the chapter, first an application to patient safety will be made; it should be noted that medical errors are estimated to be the third leading cause of deaths in the United States (see Makary and Daniels, 2016). Finally, relations to other chapters in this book are given and some personal experiences will be shared.

15.2 IMPROVEMENT KNOWLEDGE AND ITS APPLICATION TO SAFETY

In this section, we will take Improvement Knowledge as our point of departure to see how insights from the quality movement might have relevance to safety and safety principles. We will primarily focus on understanding variation and knowledge theory.

15.2.1 Understanding Variation

Variation is everywhere: it has been with us since the Big Bang and is an important force in the evolution of organic life as well as knowledge development. Handling variation played an important part in the industrial revolution, with the invention of tolerance intervals. The next step of handling variation in industry was suggested by Walter A. Shewhart. But variation is also important for our understanding of the development of organizations and how people behave differently due to different mental models. These and some similar ideas have been central in the history of quality improvement, with obvious applications to safety, and will be the focus of this section. We start with the more manifest kinds of variation in industry and conclude the section with the more soft aspects of variation in organization life.

Two Kinds of Variation

The quality improvement movement originated in the 1920s with the works of Walter A. Shewhart (1931, 1939). As a physicist, Shewhart understood how physical variation could be precisely described using a statistical language with a possibility to make predictions, in a statistical sense, about future outcomes. However, in his studies of manufacturing variation, he found that the underlying mechanisms sometimes differed from those in physics, where large numbers of infinitesimally small perturbations added to the final manifest variation. In manufacturing, sometimes sources of variation gave singular large contributions, what Shewhart called assignable cause variation. The occurrence of such assignable causes of variation destroyed the statistical predictability. However, if these sources of variation could be identified and taken away or handled in some way, variation could be reduced, improving the manufacturing process. The goal became to make processes predictable producing good results from a customer perspective with only small random variation that could be described using statistical distributions. A process without assignable causes is said to be in statistical control.

The existence of assignable cause variation in safety contexts is discussed by Cutler (1997) and Aven (2014); see also Bergman (2009). In many safety contexts, especially when some types of acceptable risks are calculated, it seems to be assumed that there is always a probability distribution behind the observed variation; in fact, this is a rather crude assumption and not at all a well-founded one. In many cases, new sources of variation occur and cause trouble. Sometimes these sources of variation could be understood early in product life as wear-out phenomena, for example, or when material from a new supplier is introduced into a process/system. However, statistical properties of the introduction of these causes are missing in most cases, due to epistemological reasons or to lack of information. Sometimes, the assignable causes of variation may be completely unforeseen, such as what Taleb (2010) called Black Swans (see Aven, 2014). In most cases, it is important to be able to detect when assignable causes of variation appear; the earlier they are detected, the smaller the risk that they will cause problems to quality and safety. It should be noted that initially, in most processes, there are a lot of assignable sources of variation. With more experiences with the process, many of these assignable causes may be found, and remedies taken.

Shewhart advised a number of criteria to be utilized for detecting assignable causes of variation. In the next sub-section, we will study the most well-known criterion: the control chart.

Control Charts

The control chart is a means to identify when in time (or space) assignable cause variations occur. Natural random variations exist in any process, and a possible assignable cause has to be detected (alarmed for) in spite of that variation. Then, of course, there are two possibilities to make a faulty decision concerning the existence of an assignable cause variation: (1) either we judge an assignable cause of variation to exist but it does not exist, or (2) we let an assignable cause of variation go undetected. The control chart is a device that should not take too long to detect the appearance of an assignable cause of variation, and the rate of false detections/alarms should be low. The principles of a control chart are given in Figure 15.1.

Once an alarm situation occurs, a detective story that is either easy or complex must be solved—what is the assignable cause (if there is one), and what remedies should be taken?

We will illustrate the idea using a recently published application, Bergquist and Söderlund (2015). It considers the twist on railroads; that is, how much the height of two rails differs over a given length. In the sketch in Figure 15.2, twist is defined as cant difference in millimeter per meter.

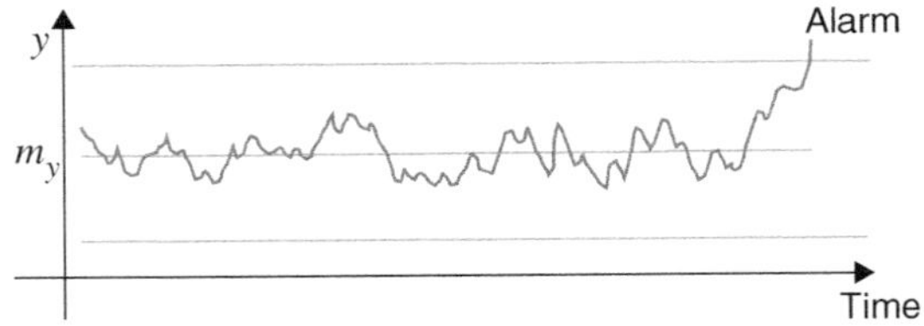

Figure 15.1. The principles of a control chart. A measure y of a process/system is followed over time and the mean value m_y of y and the alarm limits $m_y \pm ks_y$ are given lines in the chart. Here, s_y is the standard deviation (or estimated standard deviation) of y. If the process/system measure y goes beyond the alarm limits, an alarm is signaled. Often, it is found that $k = 3$ is a reasonable choice, but this may be dependent on the situation; it is important to reflect on the two risks discussed in the beginning of this sub-section.

Measurements are taken regularly, and if the twist is too large, a maintenance action has to be performed. An excessively large twist may cause a car to derail. The data shown in Figure 15.3 are retrospective data used by Bergquist and Söderlund (2015) to indicate that a use of control charts may be a much better and faster way to find assignable causes of variation, rather than relying on calculated safety margins. As illustrated in the control charts in Figure 15.3, the control chart signals significantly earlier than the traditional safety margin reasoning.

Moreover, it should be noted that the statistical properties of the variation of the process in statistical control do not apply for when the process is out of control; that is, when there is an assignable cause of variation. Rather, we should consider the process as non-predictable: if no further information is available, we cannot predict how long it will take until a dangerous deviation is reached. Probabilistic calculations based on statistical properties of the process in statistical control (where there is usually a lot of data) do not tell us anything about the variation of the process when there are assignable causes of variation.

The case at hand shows that if we want to monitor important system parameters from a safety point of view, a control chart is an obvious possibility. Of course, there is

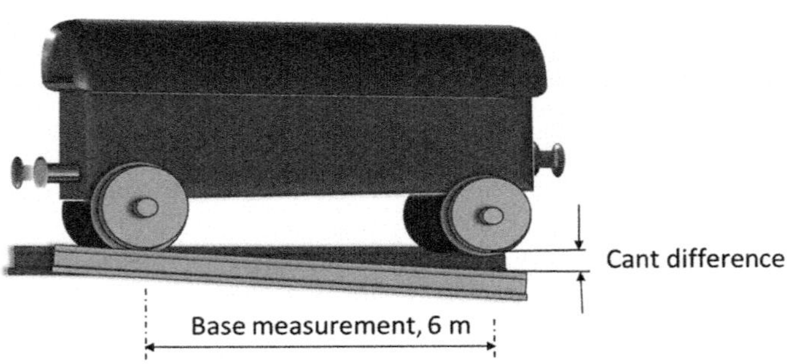

Figure 15.2. Illustration of the measure used for monitoring the twist of a railroad.

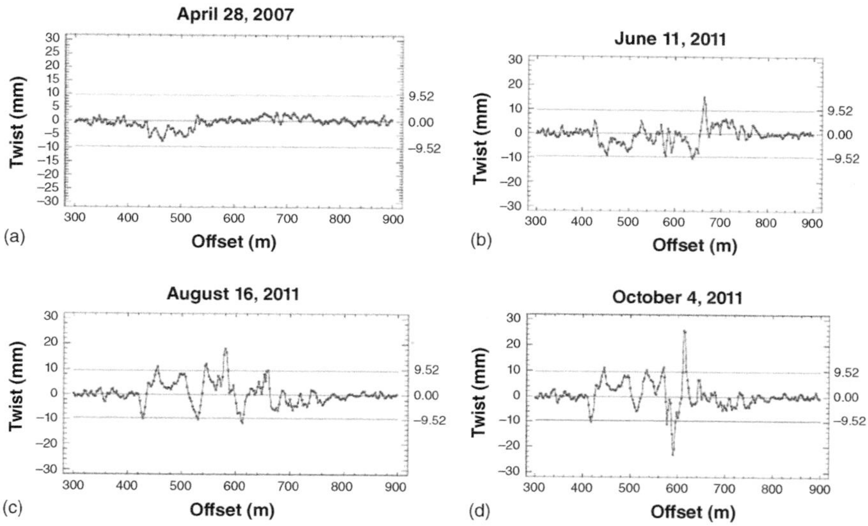

Figure 15.3. Measurements over a certain section of the railroad. Note the involvement of both a geospatial dimension and a time dimension. In the June measurements, it is already possible to detect an assignable cause of variation. The traditional way of reacting would typically be based on safety distances, and an alarm of this kind was in fact given in October, with significantly shorter time for solutions before failure. *Source*: Bergquist and Söderlund 2015. Reproduced with permission of John Wiley & Sons, Inc.

much more to control charts than what has been provided here. For example, a control chart is a good way to find out whether an intervention has really given a wanted improvement. For a deeper understanding of the subject, we recommend Bergman and Klefsjö (2010), Box and Luceño (2009), and Montgomery (2012).

The control chart provides just one type of criteria for judging if a process is in statistical control. In his discussions on manufacturing processes, Shewhart recommended a number of other criteria. These criteria were made available in simple graphical forms by Kaoru Ishikawa in the beginning of the 1960s (see, e.g., Ishikawa, 1985), in the so-called Seven QC (quality control) tools used in quality improvement projects by employees organized in quality control circles (see also Bergman and Klefsjö, 2010). These tools and ideas are central in the so-called Six Sigma problem-solving process that was described earlier.

Robust Design and Insensitivity to Variation

Variation is found in all aspects of life and in all aspects of engineering systems. In a technical system, the main sources of unwanted variation are due to manufacturing, material properties, usage environment, customer use, wear, and inner variation due to variation in different parts of the system being studied. These sources of variation are usually called noise factors. The effects of such noise factors should be taken into account and, if possible, their system effects should be reduced in the early design of the system.

It is important to take variation into consideration in the design of safety critical systems. A system will experience a lot of different sources of variation during its lifetime (and during its recycling), some of which will prove problematic to the safety of the system. Traditionally, designers have tried to catch these problematic issues by working with Failure Mode and Effects analysis, looking at components and parts of the system and trying to figure out which components are the most critical and in need of improvement. A complementary technique has been to utilize Fault Tree Analysis, starting from dangerous events and attempting to find out which events may contribute to the dangerous one. However, another way of thinking about the problem is to think about sources of variation that may be problematic to the system and its functions. A systematic way of doing that—Variation Mode and Effects Analysis—has been suggested by Johansson et al. (2006) (see also Bergman et al., 2011 and Johannesson et al., 2013). Variation Mode and Effects Analysis is a systematic process for investigating important noise factors that may affect each important function of a system (the variation of which may be critical to the system).

Many failures occur due to excessive variation in noise factors. Therefore, it is advantageous if a system is made insensitive to this variation. A conceptually interesting case is that of spherical ball bearings that were made insensitive to angular variation in 1907 by Sven Wingquist. In this case, reliability was increased dramatically. A later successor in the same tradition was Magnus Kellström with the CARB—a roller bearing that was made insensitive to both angular variations and axial variations through a creative design solution (see Figure 15.4). Here, creative means that products were made robust against disturbing noise factors already in the early, creative part of concept design.

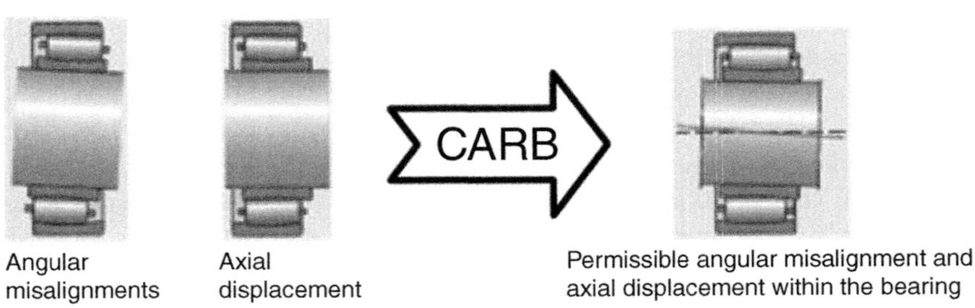

Figure 15.4. Compact aligning roller bearing combining self-aligning capability with accommodation of axial displacements; an illustration of a creative robust design making the product insensitive to two important noise factors. *Source*: SKF Publication 5102E, 2004. Reproduced with permission of SKF (U.K.) Limited.

Creative robustness is closely related to poka yoke (mistake-proofing) related to what has been called lean manufacturing, see Bergman and Klefsjö (2010). A creative robust design is determined early in the conceptual design of a system; however, it is not always possible to find such creative solutions. Later, when the concept is frozen, there are still a number of design parameters to determine the values of. In the 1960s, Japanese engineer Genichi Taguchi suggested a systematic process for deciding values of design parameters in order to reduce a system's sensitivity to variation (see Taguchi and Wu, 1979 and Phadke, 1989). A conceptual model of a system—a so-called P-diagram—is given in Figure 15.5; it illustrates that a system can be seen as a function of a number of factors. Signals are factors which are given to the system to determine its ideal response—a customer wants the system to do different things in different situations and the response should be different according to these wants. But there are also a number of sources of variation, called noise factors, which will force the system to produce responses that deviate from the ideal ones. Finally, there are a number of design parameters (also called control factors) that the designer may choose in the design process in order to achieve the best possible results. Traditionally, these parameters are set to reach the targeted response levels, but in the Taguchi approach they are first used to minimize the influence from the noise factors. Thereafter, the problem of reaching the target level is solved (usually a much easier problem supported by traditional engineering methods).

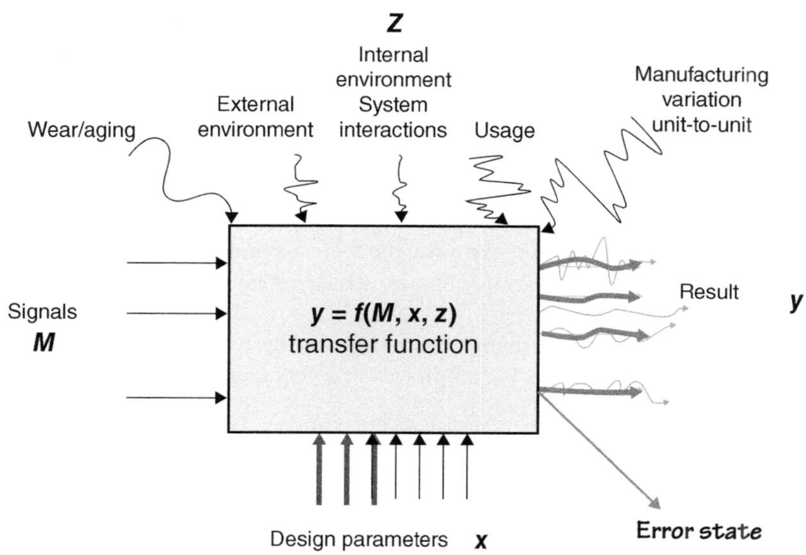

Figure 15.5. The P-diagram is a schematic picture of a system. Design parameters should be chosen so that influence from noise factors is reduced in the results and the error state avoided.

The idea behind the systematic process for determining design parameters (called Parameter Design) is that in complex systems, there are a lot of interactions; of special interest are the interactions between noise factors and design parameters. These may also be seen as nonlinearities in the system transfer function. By utilizing these interactions (nonlinearities), it is possible to reduce the influence from noise factors.

The following is a very simple and somewhat unrealistic illustration: assume that the transfer function for a system may be written as $y = b_0 + b_1x_1 + b_2x_2 + b_zz + b_{2z}x_2z$, where y is the system response that should reach a certain target T. Also, x_1 and x_2 are design parameters that could be set by the designer without any restrictions, and z is a noise factor that is not controllable during operation. Furthermore, assume that b_0, b_1, b_2, b_z, and b_{2z} are known constants. Since there are no restrictions on how x_1 and x_2 may be chosen, it is possible to find a value of x_2 making y completely independent of the noise z and a value of x_1 such that y reaches the target T; let $x_2 = -b_z/b_{2z}$. It illustrates how a simple nonlinearity, an interaction between a noise factor (here z) and a design parameter (here the parameter x_2), is used to eliminate the influence from the noise factor. Complete elimination might not be possible in more realistic cases where x_1 and x_2 may not be chosen without restrictions; however, some reduction of the influence from noise factors may still be achievable.

To find out about interactions, it is often necessary to perform designed experiments. Traditional one-factor-at-a-time experiments are not enough; in fact, they are completely inadequate, and Statistical Design of Experiment should be used instead. Unfortunately, there has been quite a lot of discussion concerning how this is done. Taguchi suggested a simple standardized way to find out where the signal-to-noise ratio is the largest; that is, where the sensitivity to noise is the smallest; an ISO standard, ISO 16336:2014 describes the procedure. However, this procedure has been refuted by some reasonable arguments based on mathematical statistics as well as on different views on the role of design engineers (see, e.g., Box et al., 1988). At the same time, these discussions should not take away the focus on making systems insensitive to sources of variation. For more information, see Bergman and Klefsjö (2010) and their cited references.

Remarkably, there are not many illustrations in the research literature where Robust Design Methodology has been utilized for safety enhancement.

A World of Variation

Variation is an unavoidable part of reality, and an important concept when we want to understand our world from the initial variation in the Big Bang, to our understanding of physical reality (quantum mechanics), organic evolution (Dawkins, 1976), and the fast-evolving social structuration of our social world (see, e.g., Sztompka, 1993). We construct our understanding of the world around us. But due to different experiences, different persons understand the same phenomenon in different ways (Lewis, 1934). Different perceptions of reality are studied in phenomenography suggested by

Ference Marton (see, e.g., Marton, 1981). Different understandings of phenomena are also conceptualized as differences in mental models by Senge (1990) in his book on learning organizations. In order to learn collectively, we need to understand each other's mental models—the dialog is a means toward that. Moreover, to understand a new concept we need variation—we cannot understand the concept color if we have only experienced the color, say, blue. Or, as in a seemingly contradictory citation given in the preface of Lo (2012): "He cannot, England know, who knows England only." This core role of variation in learning theory is emphasized by Ference Marton in his theory of variation (see, e.g., Lo, 2012). In fact, learning from variation is an important possibility for improvement (see, e.g., Bergman, 2013).

From this sketchy discussion of variation, we should understand that variation is not always possible to reduce, as has been the target for much efforts in both the safety and quality discourses. Rather, it is important to understand variation from a variety of perspectives.

15.2.2 Knowledge Theory

Knowledge theory was a starting point for Shewhart and Deming when the PDSA cycle was conceived. The PDSA cycle (called the Shewhart cycle by Deming) has proved important in the safety discourse. In this section, we will explain the relation between knowledge theory and the PDSA cycle. We will also discuss some other aspects of knowledge theory that are important in improvement work.

In his efforts to frame his theory of economic control of quality, Shewhart (1939) found the works of the pragmatic philosopher Clarence I. Lewis (1929, 1934) most helpful. In his book "Mind and World Order," Lewis discussed the creation of knowledge and how we interpret reality based on our a priori theory (understanding). Today, we would probably use the concept of "mental models" instead of a priori (see Senge, 1990). When the experiences we undergo—that is, the "sensuously" given interpreted in the light of our a priori—do not quite fit, we have to reflect, learn, and deliberately update our a priori. Essentially, Lewis talked about the individual's knowledge development. However, Shewhart (1939) put this in an industrial context, which was the starting point of the Plan-Do-Check-Act (PDCA), or as Deming later called it, the PDSA cycle or Shewhart cycle (see, e.g., Mauleón and Bergman, 2009, and the illustration in Figure 15.6). A simplistic description of the PDSA cycle with respect to improvement goes as follows: the planning of an improvement (PLAN) is based on a theory of reality; when this plan is put into action (DO), the results should be studied (STUDY); and if the theory is corroborated, then action to go further with the improvement to broader areas should be performed (ACT). However, if the result is *not* as expected, learning has to take place—meaning that the basic theory was not correct and should be revised (the second type of ACT). In the first ACT case, the theory (the a priori) became strengthened, and in the future, it is worthwhile using the same theory (mental model/a priori); if not, a revised theory has to be used and

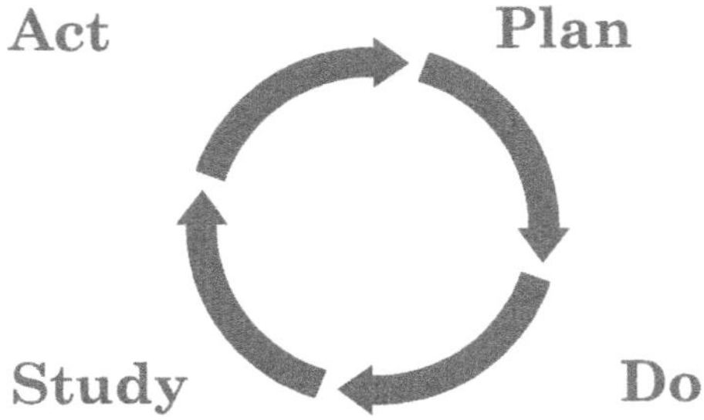

Figure 15.6. Illustration of the PDSA cycle. It should be noted that the Act part means different things if the results of Study are positive (that is, if things work out as anticipated) or if the results are negative (that is, if the test does not work out as expected). In the first case Act means that we should continue, and if necessary, further test or implement the solution. In the negative case, the theory used for the planning of the intervention might be wrong, and a revision is needed before a new test is initiated.

new improvement efforts must be made. Of course, the revision of a theory of a piece of reality is no easy thing, requiring thorough reflection, and often, new theoretical input (Deming said "Knowledge comes from the outside"; in the same spirit, Deming also said "Without a theory, there is no learning"; Deming, 1993). The PDSA cycle is applicable not only to improvement interventions, but also to normal operations. Operation has been planned based on some theory, put into operation, and continuously studied. When non-expected results occur, action should be taken as in the above-described PDSA cycle for improvements (in a Safety II context, the ACT phase should probably mean "Learning to cope with these types of deviations").

It should be noted that a similar type of learning cycle with a slightly different argumentation was earlier emphasized by John Dewey. Later, Kolb (1984) provided a thorough discussion of experiential learning.

Notably, the problem-solving practice earlier described as the DMAIC process is most probably a continuation of the same ideas. In Japan, the PDSA cycle was an important concept in the improvement activities performed in Quality Circles. However, when improvement projects were described in Quality Circle conferences, the QC circle teams used a scheme called the QC Story, which essentially contains the DMAIC steps. Motorola, the originator of Six Sigma, looked carefully at the Japanese practices when designing its Six Sigma program.

For a more detailed discussion on the use of the PDSA cycle in safety contexts, please refer to Chapter 7—Experience Feedback.

Even though the PDSA cycle is the most well-known aspect of Deming's knowledge theory, the relation between knowledge, theory, and prediction was probably the most important aspect to Deming himself. He emphasized that management is prediction, and prediction has to be based on theory that is guided by empirical results. Moreover, as we saw in the final step of the PDSA cycle, further knowledge has to be built on our theories, even though these theories may have to be revised in view of gained experiences.

The knowledge theory also has other implications apart from the PDSA cycle. As individuals have different past experiences, they also have different mental models. Thus, different individuals' experiences of the same phenomenon may be quite different—the experience is the interpretation of the sensuously given, *interpreted* via the a priori, mental model. Common action based on these different interpretations may be problematic. Senge (1990) emphasized the importance of making the mental models open to dialogue in order for the organization to learn from past history and perform common actions.

Another interesting aspect of mental models (i.e., the theories which are in use in the organization) was given by Chris Argyris (1990), who emphasized that in any organization we have to separate between "espoused theories" and "theories in use." When different aspects of the organization are discussed, there might be a consensus regarding how to interpret these aspects and what sorts of decisions or actions should be taken; that is, an *espoused theory*. But when it really comes to action, such actions could be very different; that is, *the theory in use* is different from the *espoused theory*. Assumedly, most of the readers remember situations where they know, in theory, what should have been done—but just in this situation, their actions became quite different. Similarly, Kahneman (2010) in his book *Thinking, Slow and Fast*, discussed how humans sometimes are bad decision makers due to more or less automatic responses from the fast and emotionally biased system: if the same person is making a more reflective decision, the final decision may be quite different.

Recently, this difference between fast and slow thinking has also received some support from neuroscience, which provided evidence that we are controlled in our actions by two very different cognitive systems: one corresponds to automatic, emotionally controlled processes, and the other corresponds to controlled, cognitive processes. Lieberman (2007) called these two systems reflexive (X) and the reflective (C), respectively. As discussed by Healy et al. (2015), this may sometimes have detrimental effects on teamwork. Even though the members of the team do agree in rational dialogues, they may act very differently whenever automatic and fast responses are called for. Awareness of these aspects is important in working with teams. Xu et al. (2014) discussed implications of the two-system view for safety work; their study is based on an experimental study involving 99 operators in a nuclear power plant.

Knowledge theory is, of course, also related to epistemological questions and, for example, the relations between variation (especially the two kinds of variation) and probabilities and uncertainties. Sometimes it is assumed that probabilities are always suitable to describe a phenomenon that is uncertain due to variation. However, for a phenomenon to vary according to some probabilistic model, we need to assume that it is under some kind of statistical control. It was just this aspect for which Shewhart (1939) used the control chart. Essentially, a process that is under statistical control could be described as exchangeable, a concept de Finetti (1974) took as a basic concept upon which to build subjective probability models. Of course, the assumption that a process is under statistical control is an important basic assumption also for classical statistical analyses even if that is seldom discussed. These aspects are touched upon in, for example, Chapter 20, but could have been interesting to investigate further in this context. For further discussions, see Bergman (2009) and Aven and Bergman (2012).

It should also be noted that much of the above content also could go under the heading Cognitive Psychology.

15.2.3 Psychology

The way people are willing to contribute and engage their energy in the operations of the organization and in its quality improvement is important. Research results from psychology concerning intrinsic and extrinsic motivation have a lot to offer with respect to both quality improvement work and safety.

Psychology of Motivation

Deming (1993) suggested psychology as an important theoretical domain for improvement work. Particularly, he discussed how intrinsic motivation is a strong driving factor in organizations, as well as the problems with extrinsic motivation. These aspects have been even more pronounced in later psychological research. For example, Deci and Ryan (2002), in their Self-Determination-Theory (SDT), have deepened our understanding of what drives people. Furthermore, studies in line with that of Csikszentmihalyi (1990) and his theories on flow and optimal experiences have continued with a strong emphasis on what has been called positive psychology (see Seligman and Csikszentmihalyi, 2000). These theories and their applications in organizational contexts have been described in a popular way by Pink (2009). Deming's discussions are very much in line with SDT and Positive Psychology, even though he did not have much of this research available at the time.

Social Psychology, Organizational Psychology, and Sociology

If we look at psychology from a system's perspective, we come into the areas of social psychology, organizational psychology, and sociology. Even though these areas are of importance to both quality improvement and safety, we will not go deeper into these domains here.

15.2.4 System Thinking

A system's perspective is important both for quality improvement and safety. However, in this chapter, we will only make some brief remarks on some different aspects of system thinking.

We will take Deming's views (Deming, 1993) on system thinking, or as he called it, "system appreciation." In many ways, his views may be traced back to the

Aristotelian view that "the whole is more than its parts" as well as to a teleological view. He was very much influenced by the general systems theory as discussed by Ludwig von Bertalanffy and Russel Ackoff. Deming characterized systems as having an aim ("without an aim there is no system") and as a network of interdependent parts. Some of the more important conclusions were that everyone performing at their best is not enough to guarantee the success of the system in reaching its aim. If people do not know how their work contributes to the overall aim of the system, how can we blame them for sometimes making wrong decisions?

The above discussion leads us to another very important aspect of Deming's thinking. When an unwanted event happens, the attention should be on the system instead of on the individual. This means ending a blaming culture, shifting from "who to blame" for the failure toward a focus on the system and what could be done to improve it. This has very much affected the health-care area and patient safety, which we will discuss in the next section; it has also been discussed by Cutler (1997) in a more general safety context.

Another aspect on systems thinking is that each system can be considered part of a larger system. Thus, enlarging the system borders make it possible to achieve better solutions; that which is optimal for the smaller system might not be optimal from a more universal point of view. This is of course problematic when different optimization procedures are considered—they are by necessity optimal only with respect to a limited perspective. Closely related to this is Deming's dislike of competition; he took Alfie Kohn's discussions on the subject (Kohn, 1980) as his point of departure.

We have already in Section 15.2.2 touched upon organizational learning and the importance of system thinking—the fifth discipline in the words of Senge (1990).

The complexity of social and socio-technical systems is an important aspect of system thinking. In a complex system, small changes may produce large and unforeseen consequences. This of course has far-reaching consequences for our ways of organizing for quality and safety. Awareness of these aspects is increasing (see, e.g., Plsek and Greenhalgh, 2001; Stacey, 2012; and Waldrop, 1992). However, to go deeper into this area lies outside the scope of this chapter. We refer to Chapters 3, 5, and 28 for some further insights into complexity and socio-technical systems.

15.3 HEALTH-CARE IMPROVEMENT AND PATIENT SAFETY

Improvement knowledge has been given a more prominent place in health care than in manufacturing industry from which the ideas of quality management and improvement originally emanated. As an application area of the use of quality principles and especially improvement knowledge, we will discuss health-care improvement and safety in this section. One reason for this selection is the importance of the area, it is, for example, estimated that medical errors are the third largest cause of deaths in the United States with only heart diseases and cancers larger (see Makary & Daniels, 2016). Another reason is that patient safety is only marginally discussed in other chapters of this book.

By the mid-1980s, health-care professionals had already become interested in the ideas of Deming and what was going on in industrial improvement activities. Successful interaction between companies engaging in improvement activities was an important motivator; see, for example, Berwick et al. (1990). These ideas, especially those concerning improvement knowledge, have had an important impact on how quality and safety are looked upon in health care; see, for example, Batalden and Stoltz (1993); Berwick et al. (1990); Berwick (1989, 2008); Marshall et al. (2013); and Perla et al. (2013).

Improvement knowledge has been the important frame for these endeavors; see Batalden and Stoltz (1993). Those authors emphasized the interaction between the specific domain/professional knowledge and the generic improvement knowledge, as discussed earlier in this chapter. For an illustration, see Figure 15.7 adapted from Batalden and Stoltz (1993).

Since it was established in 1990, the Institute for Health-Care Improvement (IHI) has been a strong proponent of these views. For example, IHI initiated and led a successful "100,000 lives" campaign that focused on improving patient safety. A seminal report from the Institute of Medicine (Kohn et al., 2000) suggesting that a rethinking of how patient safety issues should be dealt with spread world-wide; focus shifted away from blaming individuals toward a system improvement focus. This

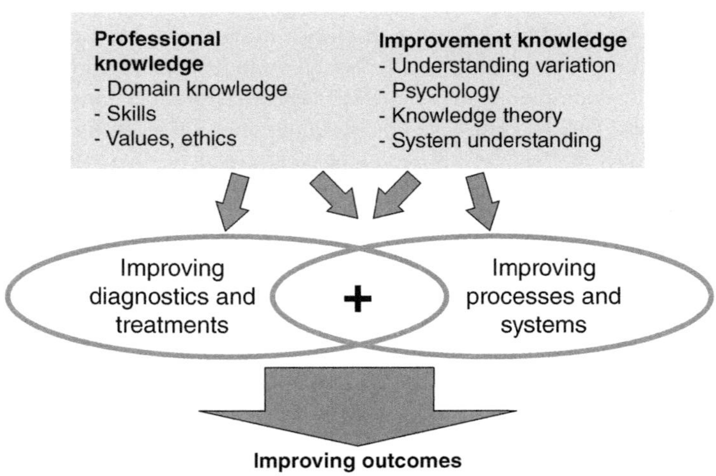

Figure 15.7. This figure is an enhanced version of the illustration by Batalden and Stoltz (1993). It illustrates how improvement knowledge should be integrated with professional knowledge for the improvement of outcomes. *Source*: Bergman et al. 2015. Reproduced with permission of Taylor & Francis.

also affected how patient safety and quality issues were viewed in countries such as Sweden, where a new regulation from 2005 reflected these points of view. However, similar points of view had earlier come from Swedish researchers in an attempt to affect legislation to promote a non-blaming culture (open letter to the Swedish Government, November 22, 1993). The letter was written as a reminder of the counterproductive treatment of the tragic dialysis accident in Linköping that took three lives. Thorough reliability studies had been performed by experienced reliability engineers, who had reached the conclusion that the accident was indeed a system failure. Still, a nurse was convicted as the sole party responsible for the accident.

In general, a number of initiatives concerning health-care improvement and patient safety were made in Sweden in the beginning of the 1990s. Some of these are still very active, such as in the Jönköping County (see, e.g., Andersson Gäre and Neuhauser, 2007), while others have disappeared. More recent activities in the Västra Götaland Region, especially at the Skaraborg Hospital Group, have been reported in, for example, Bergman et al. (2015), Hellström et al. (2010), and Lifvergren et al. (2010).

In Sweden, a large number of quality registers that report process measures, outcome measures, and patient-reported measures provide a good opportunity for successful improvement work, even though these opportunities have not yet been utilized to their full potential. An inter-professional Swedish national platform for improvement knowledge has been created and one of its aims is to promote education of improvement knowledge in health-care professional education.

15.4 WEAKNESSES, LIMITATIONS, AND CRITICISM

It can be difficult to change an organizational culture toward one following the principles and to embrace the values related to these principles; not many organizations have fully achieved this. Doing so is greatly dependent on top management. For example, an organization that won the Swedish Quality Award some years ago had indeed succeeded in this regard, according to those who investigated the organization closely. However, only a few years later, when the enthusiastic CEO proceeded toward new challenges, the quality culture seemed to have faded away. Did the organization really have a changed culture, or was the culture destroyed by the new management team? The Pulmonary Medicine clinic at the Regional Hospital in Linköping won the Swedish Quality Award in the mid-1990s. The head of the clinic quit shortly after since he felt himself pressed by his peers and not appreciated by top management; not everyone wanted change. Another illustration is a hospital that, in the mid-1990s, was very successful with a strong patient and process focus with very good results (see, e.g., Alänge and Steiber, 2011). After some years, a merger destroyed most of what had been achieved. To keep an improvement culture requires energy; otherwise, the gains are lost. See also Chapter 28 in this book.

Of course, many top management efforts have been little more than lip service. Top management often talks enthusiastically for some time and hope that the organization will continue in the pointed direction while they themselves go on to new challenges. Such tactics are doomed to failure. Some company leaders have described their organizations as Lean, Six Sigma, TQM, etc. without involving themselves. This has created cynicism within these organizations and new transformation efforts have been made ever more difficult; for further discussion, see van der Wiele et al. (2000) and Dale et al. (2001). The critical management literature also contains a lot of research results describing fads and fashion in management; see, for example, Abrahamson (1991) and Gibson and Tesone (2001). Essentially, the same management ideas have appeared repeatedly. These ideas have been preached by consultants who have new promises, but sparse results due to insufficient energy invested by top management, and perhaps also because of a lack of understanding regarding what is required (see, e.g., Weick, 2001).

The application of ideas from the quality movement has obviously had an impact on safety work, as illustrated in Chapter 7 and also in some research reports (see, e.g., Aven, 2014 and Aven and Krohn, 2014). Also, the standardization of safety management has been influenced by the corresponding standardization from a quality point of view; here also, requirements on integrated management systems in line with ISO 9000 have become stronger.

A crucial question is, how can further interaction between ideas developed in the quality movement and safety principles be supported? As indicated above, necessary change efforts are not always strong enough to achieve the desired results.

15.5 SOME PERSONAL EXPERIENCES

The safety discourse has been on the agenda for a long time. In this section, some personal experiences from the aerospace industry gained during the 1970s will be provided.

During the 1970s, the author of this chapter worked at Saab Aerospace in a department for "Design for Quality" (in Swedish called "Typkvalitet" at that time). Central issues were reliability, quality, safety, and maintainability. These different perspectives were treated simultaneously. Even though we did not know about the Japanese experiences regarding robust design, the chief reliability engineer had similar thoughts strongly in mind. Redundancies were applied, of course, but viewed skeptically—complexity increased un-proportionally and the risk of hidden failures reducing the gain from redundancies was taken very seriously (see also Chapters 4 and 5 in this book). Thus, finding creative solutions that could improve reliability was preferred over the insertion of redundancies. If necessary, redundancies using different technical solutions (called diversity in Chapter 4) were preferred over more simple ones. In many respects, the ideas now called Safety II were adhered to.

In the early 1970s, there was a lot of preventive maintenance with replacements of working units with the intention of increasing reliability and safety. Later on, it was realized that in many cases (exponential times to failure) these preventive replacements did not improve reliability; they sometimes even reduced reliability and safety due to the risk of faulty maintenance actions. Instead, condition monitoring of components and functions became an important means to increase operational reliability and safety (see, e.g., Bergman, 1977, 1978). This maintenance strategy, that is, Reliability Centerd Maintenance, was proposed by analysts from United Airlines (Nowlan and Heap, 1978) and developed into Maintenance Steering Group 3 (MSG 3), which has since been important for aerospace reliability and safety. See also Chapter 24 in this book.

Some of the research activities related to the practical reliability and safety work at the author's time at Saab Aerospace are reported in Bergman (1985). His view on safety and reliability has changed in many ways since then. The skepticism related to probabilistic reports for some kind of ALARP (as low as reasonably practicable) has increased even more (see also Chapter 26). In addition, new methods to treat reliability and safety issues have enabled new ways of handling reliability and safety in the research and development process (see, e.g., Bergman et al., 2011; Johannesson et al., 2013).

15.6 RELATIONS TO OTHER SAFETY PRINCIPLES

Ideas emanating from the quality movement could be helpful for the support of safety principles. This section starts by studying the relations between these ideas from the quality movement and the principles presented in the other chapters of this book. Finally, some comments concerning future enhancement of the quality and safety discourses.

Resilience and Quality Principles

It seems to the author of this chapter as if resilience engineering and the quality movement have a lot in common. Both focus on outcome and focus on the ordinary operation of the system. Even the definition of resilience, as described in Chapter 3, is closely related to robustness as defined earlier in this chapter. However, as described in Section 15.2.3, robustness concerns technical systems—resilience is defined for socio-technical systems and the ranges of the definitions differ. However, that does not apply to the relation between resilience engineering and the quality movement in general. The principles described in Section 15.1 of this chapter relates to the organization as a social system.

Hollnagel (Chapter 3) described the following four cornerstones of a resilient system: its ability to respond, its monitoring capacity, its learning capacity, and its capacity for anticipation. These have direct relations to different aspects of ideas put forward in this chapter. Of the four cornerstones, the *ability to respond* may be the least treated in the quality discourse but have a strong relation to Principle 5 discussed in Section 15.1 of this chapter; it is highly dependent on the operators' knowledge and understanding of the system they are operating. It is a requirement on the operators' domain/professional knowledge and ability to learn from experiences, and perhaps even on their curiosity and manipulation of the system under ordinary situations to learn more about the system and its responses to different changes in input. The ideas of Box and Draper (1965/1998) regarding evolutionary operation (EVOP) come to mind. Today, such exercises can be performed safely utilizing simulators. The *monitoring capacity* is related to statistical process control and the PDSA is a symbol for *learning* from experiences. *Anticipation* is also closely related to what Deming emphasized: "management is prediction." Here we could go deeper into how anticipation is performed. Anticipation depends on theories we have about reality and how these theories (with their time dimensions) have the possibility to bridge the gap between the past and the current on one side and the future on the other.

It seems to me that there are a lot of commonalities between resilience engineering and the quality movement presented above. It would be interesting to investigate these further and what can be learned from them. In the quality movement, a lot of practices and tools have been developed to support the ideas from resilience engineering and the philosophy of resilience engineering may give important contributions to the quality discourse. The peer-reviewed literature on resilience and quality does not indicate that much work has been done in these directions.

Professional and Improvement Knowledge

The importance of knowledge and of learning is emphasized in many chapters, and here the discussion on the relations between domain/professional knowledge and improvement knowledge may be of interest; for example, see Figure 15.7.

The Human Dimensions

From Principle 5, it is obvious that human factors as described in Chapter 9 is an important aspect in the quality discourse; see, for example, Axelsson et al (1999) and Axelsson and Bergman (1999).

Probabilistic Reasoning

Many quantitative discussions involving probabilities (see, e.g., Chapters 4, 12, 20, 23, and 24) involve implicit assumptions about processes in statistical control, that is, without assignable causes. Even though authors may be critical of these assumptions, the argumentation is not always clear. What is the relation between the epistemic uncertainty and the aleatory uncertainty? Here, the debate in the quality movement may be of some help.

Design for Safety

Many chapters (for example Chapters 6, 10, and 20) deal with design for safety, with varying degrees of implicitness. There could be an interesting dialogue involving the more general ideas on design for quality, see, for example, Part III of Bergman and Klefsjö (2010). Inherently, safe design discussed in Chapter 17 is closely related to what is called creative robustness in this chapter and to poka yoke, see Bergman and Klefsjö (2010).

Reliability Centred Maintenance

Ideas concerning condition monitoring and what can be learned from data are important for maintaining safety in the long run, as discussed, for example, in Chapters 23 and 24.

Some Thoughts for the Future

The discussions in this chapter should make it clear that issues of safety and quality should be seen together, rather than as separate issues. This is particularly important when we think of top management involvement and engagement. If too many concepts of importance are seen as different disciplines and with different background theories, how would it be possible for top management to have a coherent approach to their leadership? Of course, there is a need for specialist knowledge in different specialized areas and domains, but they need to have a common frame. The quality movement may provide such a frame. The name is less important, whether it is Total Quality Management, Improvement Science, the Deming management method, or perhaps going back to an old concept of value-based engineering (or management), perhaps with a slightly different content. Or perhaps the overall framework should be resilience? The discussions in Woods (2015) seem to be an interesting stepping stone for continued inquiries into future ways of managing complex adaptive systems to be safe and produce value with and for their customers now and in the future.

Returning to safety and reliability in a way that was necessary for writing this chapter has convinced the author that there are a lot of learning possibilities, not only from the safety discourse to that of quality, but also the other way around. In particular, the concept of adaptive resilience as discussed by Woods (2015) and implicitly by Hollnagel (described in Chapter 3) is of interest for the further development of a science of improvement. Hopefully, this chapter has provided some ideas of use in the safety discourse.

REFERENCES

Abrahamson, E. (1991). Managerial fads and fashions—The diffusion and rejection of innovations. *Academy of Management Review*, 16(3), 586–612.

Alänge, S., and Steiber, A. (2011). Diffusion of organisational innovations: An empirical test of an analytical framework. *Technology Analysis & Strategic Management*, 23(8), 881–897.

Andersson-Gäre, B., and Neuhauser, D. (2007). The health care quality journey of Jonkoping County Council, Sweden. *Quality Management in Health Care*, 16(1), 2–9.

Argyris, C. (1990). *Overcoming Organizational Defences*. New York: Prentice Hall.

Arnér, M. (2014). *Statistical Robust Design: An Industrial Perspective*. Chichester, UK: John Wiley & Sons.

Aven, T. (2013). On black swans in a risk context. *Safety Science*, 57, 44–51.

Aven, T. (2014). On the meaning of the special-cause variation concept used in the quality discourse and its link to unforeseen and surprising events in risk management. *Reliability Engineering and System Safety*, 126, 81–86.

Aven, T., and Bergman, B. (2012). A conceptualistic pragmatism in a risk assessment context. *International Journal of Performability Engineering*, 8(3), 223–232.

Aven, T., and Krohn, B. S. (2014). A new perspective on how to understand, assess and manage risk and the unforeseen. *Reliability Engineering & System Safety*, 121, 1–10.

Axelsson, J., and Bergman, B. (1999). The dynamics of quality and work life improvement—the Scandinavian heritage. In: *The Nordic School of Quality Management*. Lund, Sweden: Studentlitteratur.

Axelsson, J., Bergman, B., and Eklund, J. (1999). TQM and human factors—towards successful integration. In: Proceedings of the International Conference (Vol. I–II), Linköping, Sweden.

Batalden, P. B., and Davidoff, F. (2007). What is quality improvement and how can it transform healthcare? *Quality and Safety in Health Care*, 16, 2–3.

Batalden, P. B., and Stoltz, P. (1993). A framework for the continual improvement of health care; building and applying professional and improvement knowledge to test changes in daily work. *The Joint Commission Journal on Quality Improvement*, 19(10), 432–452.

Batalden, P., Davidoff, F., Marshall, M., Bibby, J., and Pink, C. (2011). So what? Now what? Exploring, understanding and using the epistemologies that inform the improvement of healthcare. *BMJ Quality and Safety*, 20, i99–i105.

Bendell, A. (Ed.). (1989). *Taguchi Methods: Proceedings of the 1988 European Conference*. Essex, UK: Elsevier Applied Science.

Bendell, A., Disney, J., and Pridmore, W. A. (1989). *Taguchi Methods Applied in World Industry*. A. Bendell, J. Disney, and W. A. Pridmore (Ed.). Bedford, UK: IFS Publication.

Bergman, B. (1977). Some graphical methods for maintenance planning. In *Annual Reliability and Maintainability Symposium* (pp. 467–471), Philadelphia.

Bergman, B. (1978). Optimal replacement under a general failure model. *Advances in Applied Probability*, 10, 431–451.

Bergman, B. (1985). On reliability theory and its applications (with discussion), *Scandinavian Journal of Statistics*, 12, 1–42.

Bergman, B. (1992). The development of reliability techniques: A retrospective survey. *Reliability Engineering and System Safety*, 36, 3–6.

Bergman, B. (2009). Conceptualistic pragmatism: A framework for Bayesian analysis? *IIE Transactions*, 41, 86–93.

Bergman, B. (2013). Even better care—What can we learn from the variation in open comparisons? (In *Swedish, Ännu bättre vård—vad kan vi lära från variationen i Öppna jämförelser*?). SALAR report. Stockholm, Sweden.

Bergman, B., and Klefsjö, B. (2010). *Quality from Customer Needs to Customer Satisfaction* (3rd ed.). Lund, Sweden: Studentlitteratur. [First edition published in Swedish 1990].

Bergman, B., Lorén, S., de Maré, J., and Svensson, T. (2009). *Robust Design Methodology for Reliability, Exploring the Effects of Variation and Uncertainty*. Chichester, UK: John Wiley & Sons.

Bergman, B., Hellström, A., Gustavsson, S. M., and Lifvergren, S. (2015). An emerging science of improvement in healthcare, with discussion. *Quality Engineering*, 27, 17–43.

Bergquist, B., and Söderlund, P. (2015). Data analysis for condition-based railway infrastructure. *Maintenance, Quality and Reliability Engineering International*, 31(5), 773–781.

Berwick, D. M. (1989). Continuous improvement as an ideal in health care. *New England Journal of Medicine*, 320(1), 53–56.

Berwick, D. M. (2008). The science of improvement. *JAMA*, 299(10), 1182–1184.

Berwick, D. M., Godfrey, A. B., and Roessner, J. (1990). *Curing Health Care. New Strategies for Quality Improvement* (1st ed.). San Francisco, CA: Jossey-Bass.

Box, G. E. P., and Draper, N. (1965/1998). *Evolutionary Operation: A Statistical Method for Process Improvement*. John Wiley & Sons.

Box, G. E. P., and Luceño, A. (2009). *Statistical Control by Monitoring and Adjustment*. Hoboken, NJ.: John Wiley & Sons.

Box, G. E. P., Bisgaard, S., and Fung, C. A. (1988). An explanation and criticism of Taguchi's contribution to quality engineering. *Quality and Reliability Engineering*, 4(2), 123–132.

Chakhunashvili, A., Barone, S., Johansson, P., and Bergman, B. (2009). Robust Product Development Using Variation Mode and Effect Analysis in Bergman et al. (2009).

Cole, R. E. (1999). *Managing Quality Fads: How American Business Learned to Play the Quality Game*. Milwaukee, WI.: ASQ Quality Press.

Cole, R. E., and Scott, R. (Eds.). (2000). *The Quality Movement and Organization Theory*. Sage Publications.

Csikszentmihalyi, M. (1990). *Good Business Leadership, Flow and the Making of Meaning*. New York: Penguin Books.

Csíkszentmihályi, M. (1991). *Flow: The Psychology of Optimal Experience*. New York: Harper & Row.

Csíkszentmihályi, M. (1997). *Finding Flow: The Psychology of Engagement with Everyday Life*. New York: Harper Collins.

Cutler, A. (1997). Deming's vision applied to probabilistic risk analysis. Paper presented at the second Edinburgh Conference on Risk: Analysis, Assessment and Management, Edinburgh, UK, September 1997.

Dale, B. G., Elkjaer, M. B. F., van der Wiele, A., and Williams, A. R. T. (2001). Fad, fashion and fit: An examination of quality circles, business process re-engineering and statistical process control. *International Journal of Production Economics*, 73(2), 137–152.

Dawkins, R. (1976). *The Selfish Gene*. Cambridge: Cambridge University Press.

Dean, J., and Bowen, D. (1994). Management theory and total quality: Improving research and practice through theory development. *Academy of Management Review*, 19, 392–418.

Deci, E. L., and Ryan, R. M. (2002). *Handbook of Self-Determination Theory*. Rochester, NY: University of Rochester Press.

De Finetti, B. (1974). *Theory of Probability*. New York: John Wiley & Sons.

Deming, W. E. (1993). *The New Economics for Industry, Government and Education*. MA: MIT Center for Advanced Engineering Study.

Festinger, L. (1957). *A Theory of Cognitive Dissonance*. Stanford, CA: Stanford University Press.

Fredriksson, J. J., Ebbevi, D., and Savage, C. (2015). Pseudo-understanding: An analysis of the dilution of value in healthcare. *BMJ Quality & Safety*, 24(7), 451–457

Gibson, J. W., and Tesone, D. V. (2001). Management fads: Emergence, evolution, and implications for managers. *Academy of Management Executive*, 15(4), 122–133.

Hackman, J. R., and Wageman, R. (1994). Total quality management: Empirical, conceptual and practical issues. *Administrative Science Quarterly*, 40, 309–342.

Hale, A. R., Heming, B. H. J., Catfhey, J., and Kirwan, B. (1997). Modelling of safety management systems. *Safety Science*, 26(l and 2), 121–140.

Healy, M., Vuori, T., and Hodgkinson, G. P. (2015). When teams agree while disagreeing: reflexion and reflection in shared cognition, *Academy of Management Review*, 40(3), 399–422.

Hellström, A., Lifvergren, S., and Quist, J. (2010). Process management in healthcare—Investigating why it is easier said than done. *Journal of Manufacturing Technology*, 21, 499–511.

Hutchins, D. (2008). *Hosin Kanri, the Strategic Approach to Continuous Improvement*. Aldershot, UK: Gower.

Ishikawa, K. (1985). *What is Total Quality Control? The Japanese Way*. Engelwood Cliffs, NJ: Prentice Hall.

ISO 16336. (2014). Applications of statistical and related methods to new technology and product development process—Robust parameter design (RPD), ISO TC 69, Application of Statistical Methods.

Johannesson, P., Bergman, B., Svensson, T., Arvidsson, M., Lönnqvist, Å., Barone, S., and de Maré, J. (2013). A robustness approach to reliability. *Quality and Reliability Engineering International*, 29(1), 17–32.

Johansson, P., Chakhunashvili, A., Barone, S., and Bergman, B. (2006). Variation mode and effect analysis: A practical tool for quality improvement. *Quality and Reliability Engineering International*, 22, 865–876.

Kahneman, D. (2010). *Thinking, Slow and Fast*. New York: Farrar, Straus, and Giroux.

Kohn, A. (1980). Why competition. *The Humanist*, 40, 14–15.

Kohn, L. T., Corrigan, J. M., and Donaldson, M. S. (Eds.). (2000). *To Err is Human, Building a Safer Health System*. Washington, DC: Institute of Medicine, National Academy Press.

Kolb, D. A. (1984). *Experiential Learning: Experience as the Source of Learning and Development*. Englewood Cliffs, NJ: Prentice Hall.

Lewis, C. I. (1929). *Mind and the World Order: Outline of a Theory of Knowledge*. New York: Charles Scribner's Sons.

Lewis, C. I. (1934). Experience and meaning. *The Philosophical Review*, xliii(2), 125–46.

Lieberman, M. D. (2007). Social cognitive neuroscience: A review of core processes, *Annual Review of Psychology 2007*, 58, 259–289.

Lifvergren, S., Gremyr, I., Hellström, A., Chakhunashvili, A., and Bergman, B. (2010). Lessons from Sweden's first large-scale implementation of Six Sigma in healthcare. *Operations Management Research*, 3(3–4), 117–128.

Lo, M. L. (2012). *Variation Theory and the Improvement of Teaching and Learning*. Göteborg, Sweden: Acta Universitatis Gothoburgensis.

Lönnquist, Å. (2009). Including noise factors in Design Failure Mode and Effect Analysis (D-FMEA) – A case study at Volvo Car Corporation. Chapter 4 in Bergman et al. (2009).

Magnusson, K., Kroslid, D., and Bergman, B. (2003). *Six Sigma, the Pragmatic Approach* (2nd ed.). Lund, Sweden: Studentlitteratur.

Makary, M., and Daniels, M. (2016). Medical error—The third leading cause of death in the US. *BMJ 2016*, 353, i2139.

Marshall, M., Dixon-Woods, M., and Pronovost, P. (2013). Promotion of improvement as a science. *Lancet*, 381, 419–421.

Marton, F. (1981). Phenomenography—Describing conceptions of the world around us. *Instructional Science*, 10(2), 177–200.

Mauléon, C., and Bergman, B. (2009). Exploring the epistemological origins of Shewhart's and Deming's theory of quality: Influences from C. I. Lewis' conceptualistic pragmatism. *International Journal of Quality and Service Sciences*, 4(1), 77–89.

Mizuno, S. (1988). *Management for Quality Improvement: The Seven New QC Tools*. Cambridge: Productivity Press.

Mizuno, S., and Akao, Y. (1994). *QFD: The Customer-Driven Approach to Quality Planning and Deployment*. Tokyo: Asian Productivity Organization.

Montgomery, D. C. (2012). *Statistical Quality Control: A Modern Introduction* (7th ed.). New York: John Wiley & Sons.

Nowlan, F. S., and Heap, H. F. (1978). *Reliability-Centered Maintenance*. San Francisco, CA: United Air Lines Inc.

Peng, D. X., Schroeder, R. G., and Shah R. (2008). Linking routines to operations capabilities: A new perspective, *Journal of Operations Management*, 26, 730–748.

Perla, R. J., Provost, L. P., and Parry, G. J. (2013). Seven propositions of the science of improvement: Exploring foundations. *Quality Management Health Care*, 22(3), 170–186.

Phadke, M. S. (1989). *Quality Engineering Using Robust Design*. London: Prentice-Hall.

Pink, D. (2009). *Drive—The Surprising Truth About What Motivates Us*. Riverhead Books.

Plsek, P. E., and Greenhalgh, T. (2001). Complexity science: The challenge of complexity in health care. *British Medical Journal*, 323(7313), 625–628.

Schein, E. H. (1992). *Organizational Culture and Leadership*. San Fransisco: Jossey-Bass.

Scott, R., and Cole, R. (Eds.). (2000). *The Quality Movement & Organization Theory*. Thousand Oaks, CA: Sage.

Seligman, M. E., and Csikszentmihalyi, M. (2000). Special issue: Positive psychology. *American Psychologist*, 55(1), 5–14.

Senge, P. M. (1990). *The Fifth Discipline; the Art and Practice of the Learning Organization*. New York: Dubleday Currency.

Shewhart, W. A. (1931). *Economic Control of Quality of Manufactured Product* (Republished ed.). Milwaukee, WI: 50th anniversary Commemorative Reissue by American Society for Quality control.

Shewhart, W. A. (1939). *Statistical Method from the Viewpoint of Quality Control*. Washington, DC: Dover Publications.

Sila, I., and Ebrahimpour, M. (2002). An investigation of the total quality management survey based research published between 1989 and 2002: A literature review. *International Journal of Quality & Reliability Management*, 19(7), 902–970.

Stacey, R. (2012). *Tools and Techniques of Leadership and Management: Meeting the Challenge of Complexity*. New York: Routledge.

Sztompka, P. (1993). *The Sociology of Social Change*. Wiley-Blackwell.

Taguchi, G. (1986). *Introduction to Quality Engineering*. Tokyo: Asian Productivity Center.

Taguchi, G., and Wu, Y. (1979). *Introduction to Off-line Quality Control*. Tokyo: Central Japan Quality Control Association.

Taleb, N. N. (2010). *The Black Swan:The Impact of the Highly Improbable* (2nd ed.). London: Penguin.

Tsutsui, W. M. (1998). *Manufacturing Ideology, Scientific Management in Twentieth-Century Japan*. Princeton, NJ: Princeton University Press.

Waldrop, M. W. (1992). *Complexity the Emerging Science at the Edge of Order and Chaos*. New York: Simon and Schuster.

Weick, K. E. (2000). Quality Improvement: A Sensemaking Perspective, in Cole and Scott (2000), pp 155–172.

Weick, K. E. (2001). Gapping the relevance bridge: Fashions meet fundamentals in management research. *British Journal of Management*, 12(S1), S71–S75.

van der Wiele, A., Williams, A. R. T., and Dale, B. G. (2000). Total quality management: Is it a fad, fashion or fit? *Quality Management Journal*, 7(2), 65–79.

Woods, D. D. (2015). Four concepts for resilience and the implications for the future of resilience engineering. *Reliability Engineering & System Safety*, 141, 5–9.

Xu, Y., Li, Y., Ding W., and Lu, F. (2014). *Controlled versus Automatic Processes: Which Is Dominant to Safety? The Moderating Effect of Inhibitory Control.* PLoS ONE 9(2):e87881.

FURTHER READING

Bergman, B., Hellström, A., Gustavsson, S., and Lifvergren, S. (2015). An emerging science of improvement in healthcare, with discussion. *Quality Engineering*, 27, 17–43.

Bergman, B., and Klefsjö, B. (2010). *Quality from Customer Needs to Customer Satisfaction* (3rd ed.). Lund, Sweden: Studentlitteratur.

Cole, R. E., and Scott, R. (Eds.). (2000). *The Quality Movement and Organization Theory*. Sage Publications.

Deming, W. E. (1993). *The New Economics for Industry, Government and Education*. MA: MIT Center for Advanced Engineering Study.

Montgomery, D. C. (2012). *Statistical Quality Control: A Modern Introduction* (7th ed.). New York: John Wiley & Sons.

16

SAFETY CASES

Tim Kelly

16.1 INTRODUCTION

A safety-case-based approach to safety assurance demands that the developers and operators of safety critical systems develop and present (typically for approval by a regulator) the argument and evidence to support the claim that their system is acceptably safe to operate. The origins of the approach stem from the nuclear industry as far back as the 1960s. However, this approach to regulation has gained substantial traction and has spread to an increasing number of domains, including medical devices, automotive, chemical, defense, and air traffic control.

16.2 ORIGINS AND HISTORY

In the United Kingdom at least, a number of serious accidents such as the Windscale Nuclear Incident (Arnold, 1992), Piper Alpha Offshore Oil and Gas Platform Disaster (Cullen, 1990), and Clapham Rail Disaster (Edwards, 1997) have been recognized as being instrumental in prompting a reconsideration of how safety assurance was

Handbook of Safety Principles, First Edition. Edited by Niklas Möller, Sven Ove Hansson, Jan-Erik Holmberg, and Carl Rollenhagen.

being managed in the safety-critical sector. Here, we briefly discuss how each of the accidents ushered in a safety case approach to regulation.

16.2.1 Windscale

In October 1957, a fire in the Number 1 pile at Windscale resulted in a significant release of radioactivity (20,000 Ci of Iodine-131). The reactors at Windscale used natural uranium as fuel, graphite as the moderator, and were cooled by air. The properties of graphite as a moderator were only just beginning to be understood at the time of building the Windscale reactors. The moderator was found to store energy (known as *Wigner Energy*) that could be spontaneously released in the form of heat. This energy had to be routinely released through an annealing process. The storage and release of this energy were not well understood. During one such annealing process, the energy was released too quickly, starting a fire. The fuel in the core melted, fuel cans burst and the uranium ignited, causing fission products to be released through the cooling ducts to the atmosphere (Arnold, 1992).

Following the Windscale accident, a number of actions were taken. First, the Nuclear Installations (Licensing and Insurance) Act was introduced in 1959 to regulate commercial nuclear reactor installations. As part of this Act, following recommendations from the Fleck Committee set up as a result of the enquiry into Windscale, the Nuclear Installations Inspectorate (NII) was established to regulate all land-based reactors within the United Kingdom. In order to obtain an operating license, a set of reports must be presented to the NII that justifies the safety of the design, construction, and operation of the plant. The nuclear certification process is widely cited as one of the first examples of a safety case regime, although the term safety case was not used at this time.

16.2.2 Flixborough

In 1974, an explosion occurred at the Nypro factory at Flixborough causing 28 deaths on site and extensive damage and injuries in the surrounding villages. The explosion occurred in a part of the facility involved in the production of Nylon. One of the six reactors in a process to oxidize cyclohexane developed a crack. It was removed and quickly replaced by a temporary pipe. After 2 months of operation, on June 1, a slight rise in pressure caused the pipe to rupture, resulting in 30–50 tonnes of highly pressurized cyclohexane being vented to the plant within 50 seconds. The cyclohexane then ignited causing a vapor cloud explosion that destroyed the oxidation unit, neighboring units, and a nearby office block (Kletz, 1988).

Following the Flixborough accident, an Advisory Committee on Major Hazards was established within the Health and Safety Executive. The committee recommended that regulations be established to ensure identification, assessment, and management of potential hazards in chemical installations. This recommendation resulted in the formulation of the Hazardous Installations (Notification and Survey)

Regulations. These regulations were never enacted but instead formed the basis of a European Community Directive produced in response to the Seveso accident that occurred in July 1976. The UK implementation of this directive was introduced in 1984 as the Control of Industrial Major Accident Hazards (CIMAH) Regulations (CIMAH, 1984). A key requirement of the CIMAH Regulations was the production of a Safety Report (Case) that demonstrates adequate consideration of dangerous substances, potential accidents, and provision of effective safety management systems.

16.2.3 Piper Alpha

On Piper Alpha in July 1988, a combination of poor procedures and communication meant that a pump that was out of commission for routine maintenance was re-commissioned hurriedly and switched on. The resulting gas explosion killed two men. This explosion would have been survivable had it not been for the absence of blast walls in the platform design. The blast started an oil fire. Again, this would have been controllable except that adjacent platforms in the oil field continued to pump oil and gas through the pipelines connecting the rigs to the shore, thus feeding the fire. Eventually, gas lines near the oil fire ruptured creating an uncontrollable fire fed by thousands of tonnes of pressurized gas contained within the pipelines. The crew on the platform had been given minimal training in emergency procedures. Many of the crew assembled in the accommodation block awaiting evacuation via the helipad on top of the block, following the minimal instruction they had been given. However, following the first gas explosion, this evacuation route was unworkable. No alternative procedures were communicated to the crew. The majority of the crew died waiting in the accommodation block (Cullen, 1990).

Following the Piper Alpha disaster, a public enquiry chaired by Lord Cullen was initiated. The purpose of this enquiry was both to determine the causes of the accident and to make recommendations so that similar accidents would not occur in the future. The findings of the enquiry are published in Cullen (1990). Heavily influenced by the experience of the chemical industry in its use of safety cases as required by the CIMAH Regulations, one of the main recommendations was that platform operators should be required to submit safety cases. The purpose of these documents being to present a clear and comprehensive argument of platform safety. As a direct result of this recommendation, the Offshore Installations (Safety Case) Regulations were introduced in the United Kingdom in 1992.

16.2.4 Clapham

In 1988, 35 people were killed in a collision between two trains resulting from a signaling failure. The signal failure was found to be caused by a wiring fault introduced in maintenance. A wire was improperly terminated and by-passed crucial safety interlock circuitry. The consequences of collision were particularly bad as it involved old "Mark 1" rolling stock that copes poorly with rear collisions. In such collisions, carriages of this type can easily ride over one another and slice through the passenger space.

Although the cause of the accident at Clapham was relatively straightforward to identify and eradicate in future installations, it was felt in the ensuing enquiry that the accident had been symptomatic of the whole culture (Edwards, 1997). This thinking, together with a growing concern for railway safety as a result of privatization, led to the introduction of the Railway (Safety Case) Regulations 1994 (HSE, 1994). These regulations required that the railway infrastructure controller and all train and station operators must prepare safety cases that demonstrate sufficient consideration of management of all credible hazards.

16.2.5 The Introduction of Safety Cases—A Shift in Emphasis

The four accidents described here were instrumental in prompting a reconsideration of how safety is managed in each of the respective industries. In each of these cases, there had not been a total ignorance of safety concerns, or even a complete absence of safety standards. Instead, the underlying problem was that the operator had failed to demonstrate a *systematic* and *thorough* consideration of safety. Prior to the introduction of a safety-case-based regime, assurance and regulation can be said to have focused primarily on prescriptive safety requirements, for example, construction codes as described in Chuse (1993). With such approaches, developers and operators can implicitly claim safety through satisfaction of the *regulator's* requirements. With the introduction of safety cases, the responsibility is shifted back to the operators. It is up to the operators to demonstrate that they possess adequate arguments and evidence of safety.

Since the early developments described in this section, the safety case approach has been adopted by a number of other domains. It now lies at the heart of defense acquisition practice in the United Kingdom (MoD, 2014), and other countries such as Australia (DoD, 1998). It has also recently been incorporated into the functional safety standard for the automotive safety standard ISO26262 (ISO, 2011). The US FDA has introduced the approach in the regulation of one class of devices—infusion pumps—in their 510K guidance (FDA, 2014). It has also been widely adopted as the core approach for the assurance and regulation of air traffic management (ATM) within Europe (Eurocontrol, 2006).

16.3 DEFINITIONS AND TERMINOLOGY

A safety case can be defined in the following terms:

> *A safety case is the argument and evidence supporting claims about the safety of the system in operation in a defined context.*

The following are important aspects of the above definition:

- ***Argument***—Above all, the safety case exists to communicate an *argument*. Safety can rarely be proven. Instead, a *sufficient* argument must be presented. This argument must indicate how someone can reasonably conclude that a system is acceptably safe from the evidence available.
- ***Evidence***—Arguments without evidence are unfounded and unsupported. Ultimately, the claims that a safety case argument presents need to be backed by suitable sources of information.
- ***Acceptable Safety***—Absolute safety is an unobtainable goal. Safety cases are required to justify that the system is safe *enough* (when compared against some definition or notion of tolerable risk).
- ***System***—The system to which a safety case refers can be anything from a network of pipes or a software configuration to a set of operating procedures. The concept is not limited to the consideration of conventional engineering "design."
- ***Operation***—A safety case inevitably presents arguments and evidence about a future state—that is, the operation of the system. This means that the safety case will typically need to rely on prediction, estimate, and (in some cases) assumption.
- ***Context***—Context-free safety is impossible to argue. Almost any system can be *unsafe* if used in an inappropriate or unexpected manner. The safety case must clearly state the context within which acceptable safety can be argued.

In addition to the above aspects, it is often stated that a safety case *should* be clear, comprehensive, compelling, and defensible:

- ***Clear***—A safety case is a device for *communicating* ideas and information, usually to a third party (e.g., a regulator). In order to do this convincingly, it must be as clear as possible.
- ***Comprehensive***—Safety case arguments and evidence should sufficiently address all of the aspects of system design and operation that could affect safety. The safety case should also not be *deliberately selective* in its presentation of evidence (e.g., excluding counter-evidence of safety).
- ***Compelling***—Because safety cannot be proven beyond all doubt (e.g., consider attempting to prove that all hazards have been identified), the residual issue is whether the safety arguments and evidence are *sufficient* and persuasive. This is, of course, not disjoint from issues of comprehensiveness.
- ***Defensible***—The arguments and evidence of a safety case should be capable of resisting potential challenge and attack. A strong safety case is one that considers and overcomes suggested forms of counter argument and evidence.

It is important to separate these attributes of a "good" safety case from the (simple) core definition of a safety case. This is so as to allow for both "good" and "bad" safety cases to be recognized. For example, it may be that the arguments and evidence that an operator has for the safety of their system are unclear, partial, and unconvincing. However, it may be their current safety case. Separating the core aspects of the definition from positive attributes also helps address assumptions that the production of a safety case somehow guarantees success.

16.3.1 Safety Cases vs. Safety Case Reports

Often the term "safety case" is used synonymously with the term "safety case report," or the assumption is that the term "safety case" refers to documentation. This lack of distinction is unhelpful. The safety case is a logical concept. At any given point in time during the development and operation, it is possible to ask, "What is the current safety case for this system?" This question is asking for an account of the current arguments and evidence that exist to justify the safety of the system under development and operation. The question could be answered with a verbal account of these arguments and evidence or a *documented* account. A documented account of the safety case is called a "Safety Case Report." At the heart of the use of safety cases as part of a regulatory regime is the concept that safety cases should be explicitly documented as safety case reports in order to enable structured, explicit, and formal review and acceptance. The following historical definition (taken from the first edition of the UK Ministry of Defence Ship Safety Management System Handbook JSP 430) reveals this document centric view (MoD, 1996).

> *"A safety case is a comprehensive and structured set of safety documentation which is aimed to ensure that the safety of a specific vessel or equipment can be demonstrated by reference to:*
>
> - *safety arrangements and organization*
> - *safety analyses*
> - *compliance with the standards and best practice*
> - *acceptance tests*
> - *audits*
> - *inspections*
> - *feedback*
> - *provision made for safe use including emergency arrangements"*

This definition also communicates another essential characteristic of a safety case report or document. It will naturally have to refer to, and pull together, many other forms of information (such as safety analyses) in order to show how the overall argument of safety is established.

16.3.2 Other Terminology

The concept of documenting and presenting the arguments and evidence of safety is more widespread than the usage of the term safety case. There are many examples of other forms of documentation that could be said to present the account of a safety case. For example, the terms "Safety Assessment Report" and "Safety Justification Report" are widespread in some domains (e.g., nuclear and defense domains). Where these reports clearly document the arguments and evidence of acceptable safety, they can be regarded as synonymous with the concept of a safety case report. In some cases, it may not be possible to identify a single report that neatly summarizes all of the arguments and evidence for the system. In these cases, it could be said that the safety case remains implicit and perhaps has documented elements that remain distributed across several supporting reports. For example, even a safety analysis report documenting the results of a fault tree analysis that presents some commentary and justification as to how the analysis supports claims of system safety can be said to be *partially* documenting the safety case. However, it is worth recognizing that where safety cases are a recognized element of a regulatory regime, an explicit—clearly documented—safety case report is an essential feature.

Some standards correctly recognize that the existence of safety case report (i.e., the document) should not be taken as an assumption of a clearly stated safety argument. For example, Issue 4 of the UK System Safety Defence Standard 00–56 (MoD, 2007) stated that, "*All Safety Cases contain an implicit safety argument but this Standard requires an explicit argument; this is usually expressed in terms of a defined hierarchy of safety claims and sub-claims that are supported by a body of evidence.*"

16.4 UNDERLYING THEORY

The safety case consists of three principal elements: Requirements, Argument, and Evidence. The relationship between these three elements is depicted in Figure 16.1.

The safety argument is that which communicates the relationship between the evidence and objectives. Much of the theory underlying safety case development relates to argumentation—its structuring, issues of confidence, evaluation, reuse, and maintenance.

16.4.1 Safety Case Argumentation

Safety Case arguments have historically been communicated through narrative text of a documented safety case report (such as shown in Figure 16.2).

In Figure 16.2, the narrative text describes how a safety requirement (P65) has been interpreted and achieved in the system. It also clearly provides references to where the evidence supporting the lower level statements can be found.

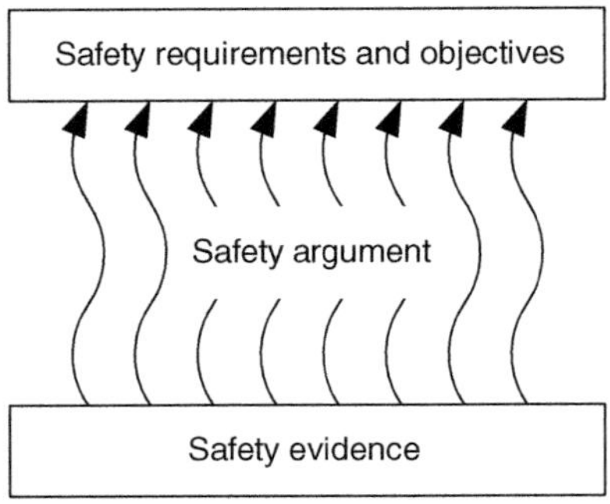

Figure 16.1. Safety requirements, argument, and evidence.

Ultimately, arguments within safety cases conform to widely accepted definitions of argument. Arguments are a connected series of propositions put forward in order to establish a conclusion. Argument in safety cases can be both deductive and inductive. A *deductive* argument is an argument where premises establish a conclusion with absolute certainty, and without any room for probabilistic treatment (Cass and Le Poidevin, 1993). An *inductive* argument is one where the truth of the conclusion is based upon the estimation of the probable truth of the premises (Cass and Le Poidevin, 1993) and the strength of the inference between the premises and conclusion. Assurance case arguments are rarely *provable* deductive arguments. Instead, they are more commonly inductive.

The presentation of safety case arguments using natural language can suffer from problems of lack of clarity, difficulty in comprehension, poor structure. As a response to these issues, structured approaches to the development and presentation of safety arguments have become popular. Examples include the Claims–Argument–Evidence notation (Bloomfield and Bishop, 2010) and the Goal Structuring Notation (GSN) (GSN, 2011).

The Goal Structuring Notation (GSN)—a graphical argumentation notation—explicitly represents the individual elements of any safety argument (requirements, claims, evidence, and context) and (perhaps more significantly) the relationships that exist between these elements (i.e., how individual requirements are supported by specific claims, how claims are supported by evidence, and the assumed context that is

> The defense-in-depth principle (P65) has been addressed in this system through the provision of the following:
> - Multiple physical barriers between hazard source and the environment (see Section X)
> - A protection system to prevent breach of these barriers and to mitigate the effects of a barrier being breached (see Section Y)

Figure 16.2. A simple textual safety argument.

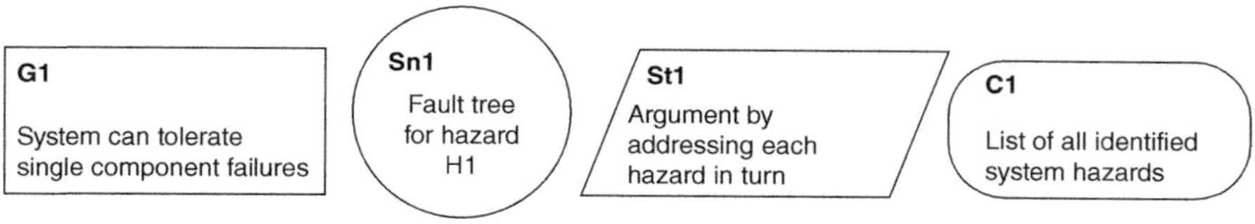

Figure 16.3. Main elements of the Goal Structuring Notation (GSN).

defined for the argument). The principal symbols of the notation are shown in Figure 16.3 (with example instances of each concept).

When the elements of GSN are connected together, they are said to form a "goal structure." Figure 16.4 shows an example goal structure. Goal structures document the asserted chain of reasoning in the argument (through the visible decomposition of claimed *goals* and the description of argument *strategies*) and indicate how this argument is supported by evidence (through *solutions*). The goal structures also clearly document the context in which the claims of the argument are being put forward.

GSN provides a means of documenting an asserted argument. The use of GSN itself does not establish the truth of that argument. The key benefit from using an explicit approach such as GSN to develop and document the arguments of any assurance case is that it can improve comprehension amongst the key stakeholders (e.g., system developers, engineers, independent assessors, and certification authorities). In turn, this improves the quality of the debate and the time taken to reach agreement on the argument approaches being adopted. For example, using the goal structure provided in Figure 16.6, it would be reasonable to question whether the allocation of SIL 4 to the primary protection system and SIL 2 to the secondary protection system had been adequately demonstrated to be appropriate to the hazards involved. This discussion could lead to a requirement for a SIL allocation justification.

GSN has been developed over the last 20 years, and now includes extensions to help represent "Safety Case Patterns" (Kelly and McDermid, 1997)—structures that document common, reusable, forms of safety case argument—and Modular Safety Case arguments (Bate and Kelly, 2003)—structures that represent how the argument and evidence of a single safety case can be partitioned into modules ("packages") of interrelated arguments and evidence. The work on safety case patterns has made it possible to represent and build up a body of knowledge on typical safety argument forms, for example, for software safety case arguments (Hawkins, 2011a).

16.4.2 Types of Safety Case Argument

Whilst the definition of the safety case argument presented earlier in the chapter is valid, it does not indicate the types of argument that typically are offered up to justify safety. Figure 16.5 shows the three main forms of argument that one would typically expect to see within a safety case.

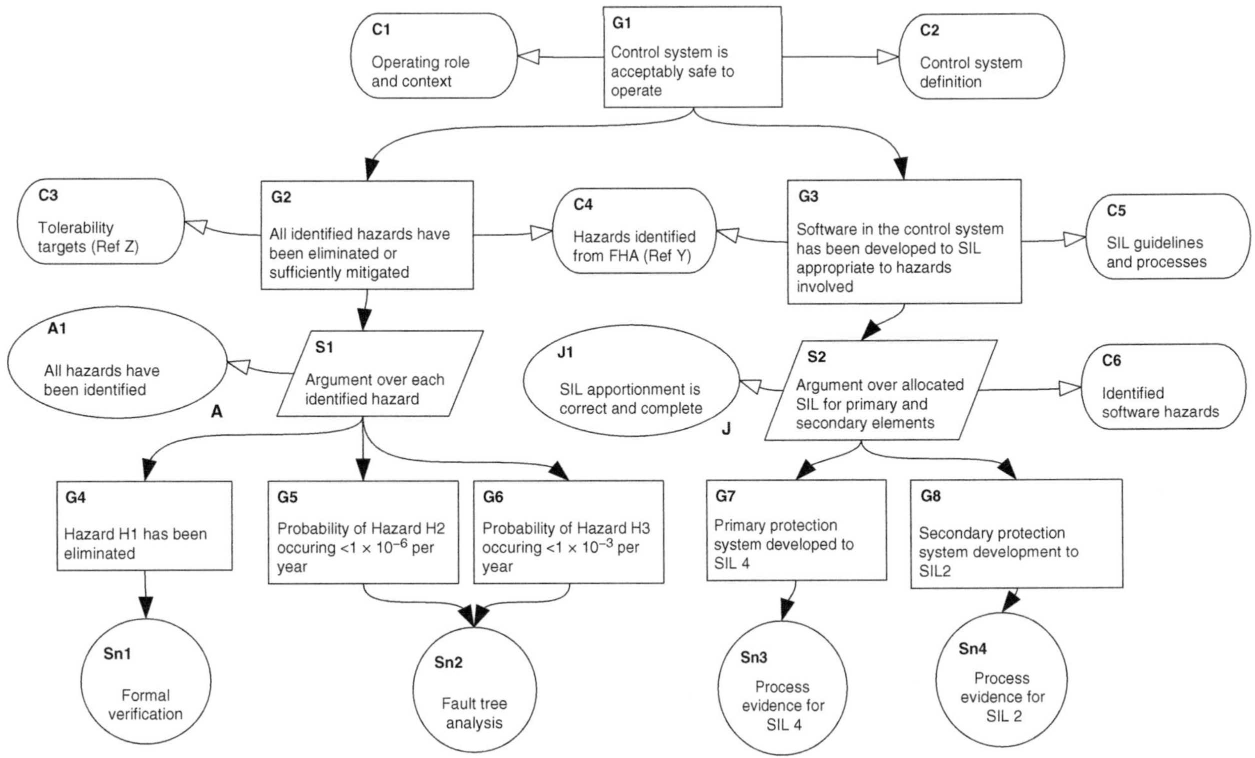

Figure 16.4. An example goal structure.

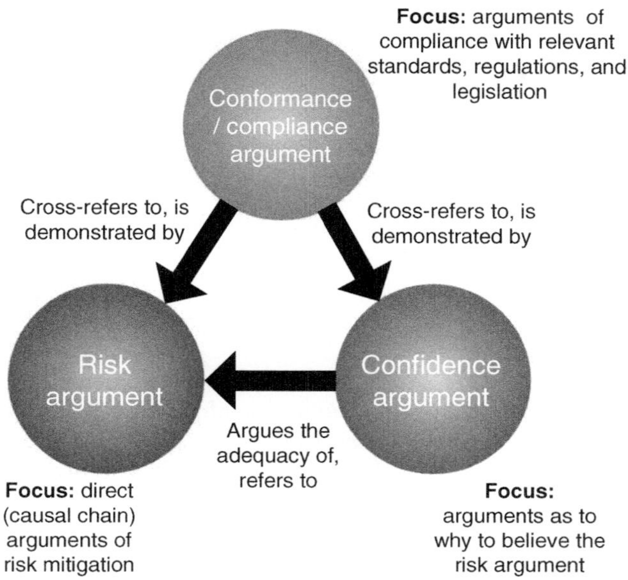

Figure 16.5. Types of safety case argument.

First, the safety case should present a *risk argument*. This is the argument that justifies safety on the basis of demonstrating an understanding of the specific potential hazards posed by a system (and their relative risk) and presenting technical arguments that these hazards have been appropriately managed (by elimination, reduction, or mitigation). These arguments should ultimately be supported by evidence that demonstrates the effectiveness of these risk management measures (e.g., test reports or operational trials).

Second, the safety case should justify the adequacy of, and the confidence held in, the risk argument. This will involve justifying attributes such as the completeness and coverage of the risk argument, the adequacy of forms of evidence being offered up (e.g., justifying the trustworthiness and appropriateness of this evidence). The confidence argument is a necessary element of the safety case given that in almost all cases, it is not possible to present a risk argument that simply *proves* safety. There will always be questions of sufficiency that need to be justified. One widely accepted principle governing the production of the confidence argument is that it should be possible to justify that the confidence in the risk argument is somehow in proportion with the level of risk under discussion—that is, the higher the risk being posed, the greater is the confidence required.

Third, the safety case will typically argue safety by demonstrating conformance or compliance with appropriate safety standards, regulations, and legislation. This is often required by the regulatory regime. In domains where there is no formal

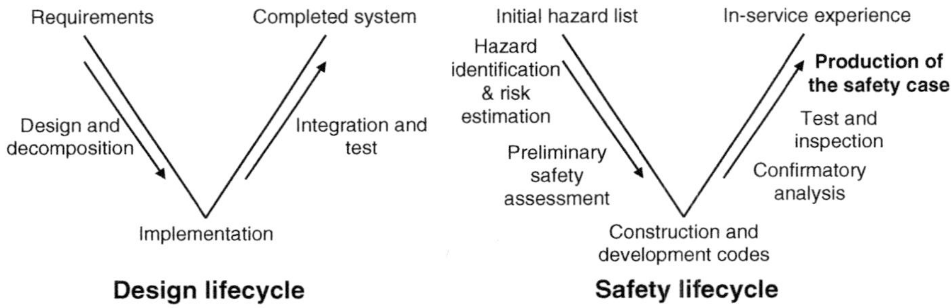

Figure 16.6. A historical view of safety case development.

regulation, this can simply be thought as arguing conformance. Where formal regulation exists, the stronger term *compliance* is more appropriate. Arguing compliance with standards can, and should overlap with the other two forms of safety case argument. Where a standard dictates specific design and operational measures to mitigate known risks, the argument will overlap with the risk argument. Where the standard provides guidance or requirements on the sufficiency of arguments and evidence (e.g., by recommending particular forms of testing), the argument will overlap with the confidence argument.

16.4.3 Safety Case Lifecycle

Contrary to what may historically have been practiced, safety case development cannot be left as an activity to be performed toward the end of the safety lifecycle. This view of safety case production being left until all analysis and development is completed is depicted in Figure 16.6.

A traditional view of the design and development lifecycle is shown on the left-hand side of Figure 16.6. Running concurrently with this, shown on the right-hand side of the diagram, is the historical view of the safety lifecycle, showing safety case development as a discrete activity to be performed following the completion of the safety assessment activities.

The problems that have been experienced with this style of safety case development can include:

- Large amounts of redesign resulting from a belated realization that a satisfactory safety argument cannot be constructed. In extreme cases, this has resulted in "finished" products having to be completely discarded and redeveloped.
- Less robust safety arguments being presented in the final safety case. Safety case developers are forced to argue over a design as it is given to them—rather than being able to influence the design in such a way as to improve safety

and improve the nature of the safety argument. This can result in, for example, probabilistic arguments being relied upon more heavily than deterministic arguments based upon explicit design features (the latter being often more convincing).

- Lost safety rationale. The rationale concerning the safety aspects of the design is best recorded at "design-time." Where capture of the safety argument is left until after design and implementation—it is possible to lose some of the safety aspects of the design decision-making process which, if available, could strengthen the final safety case.

Unfortunately, though not surprisingly, few practitioners are prepared to publicize failures of this style of safety case development. However, Cullen (Cullen, 1996) presents some of the experiences of BNFL in producing a safety case for the Sellafield Alpha Reduction Plant. For this plant, he relates that a "traditional" approach was first adopted—where "plant design has proceeded more or less independently of the production of the safety case." Design progressed to the firm proposal stage before being passed to the Safety Department. Significant safety hazards were identified with this proposal—making it impossible to produce a convincing safety case. A redesign was therefore required—resulting in great expense. The redesign was again developed into a firm proposal before the safety case was considered. However, this time, other significant problems were found with the new proposal, requiring more (expensive) redesign. It was only on the third redesign, where consideration of the safety case was integrated into the design requirements that an acceptable, arguably safe, design resulted.

16.4.4 Incremental Safety Case Development

It is now widely recognized that safety case development be treated as an evolutionary activity that is integrated with the rest of the design and safety lifecycle. This often results in the production and presentation of safety case reports at a number of stages during the development of a project. For example, it is possible to produce safety case reports at three points during the system development process:

- *Preliminary Safety Case Report*—presented after definition and review of the system requirements specification
- *Interim Safety Case Report*—presented after initial system design and preliminary validation activities
- *Operational Safety Case Report*—presented prior to in-service use, including complete evidence of satisfaction of systems requirements

The integration between the production of these safety cases and the traditional development lifecycle is depicted in Figure 16.7.

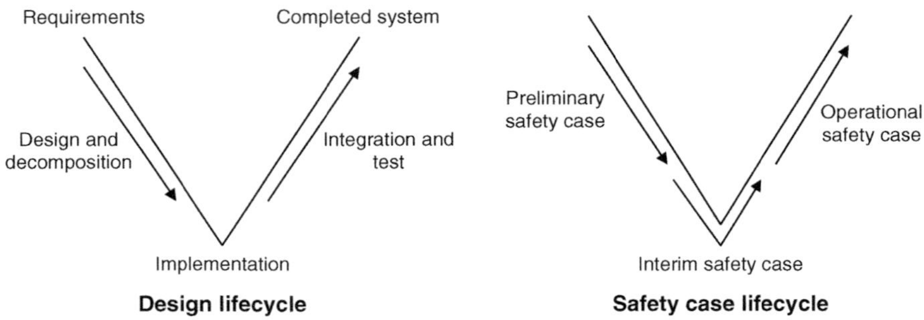

Figure 16.7. An integrated view of safety case development.

There is often some variation (according to the regulatory domain and specific needs of the project) of the exact number of safety case reports to be produced, and at what stage in the lifecycle. For example, for civil nuclear power generation in the United Kingdom, safety cases are additionally required at certain milestones in the project. In the commissioning of Sizewell "B," safety cases were presented prior to first fuel load, prior to first generation of power, and prior to being allowed to export power to the national grid (IEE, 1996). However, regardless of the specifics of the numbers of safety cases and timings of submissions, the principle of phased safety case production is increasingly being accepted as a core concept across all domains.

At the heart of the concept of phased safety case production is the presentation of an *evolving safety argument*. At the Preliminary Safety Case stage, the aim is to present an outline safety argument showing the principal objectives, approach to arguing safety, and the forms of evidence anticipated. At the Interim stage, the argument should be evolved to reflect the increased knowledge concerning the detailed design and specification of the system. At the operational stage, the argument can again be evolved further to reflect evidence concerning the system as implemented and tested.

On any safety-critical/safety-related project, it is crucial to gain an understanding of the form and expected content of the safety (or certification) argument as early as possible. By producing the safety case in step with the system development lifecycle, it enables the production of the safety case to provide feedback on the evolving state of the project. Where difficulties are encountered in making a compelling and convincing argument of safety, it means that it is still possible to influence design choices and the design process in order to strengthen the ultimate case for safety that can be made.

16.4.5 Safety Case Maintenance

In the first instance, the safety case argument will typically be constructed and presented (e.g., to a regulatory authority) prior to the system operating for the first time. The argument is often therefore based on estimated and predicted operational behavior rather than observed evidence. For this reason alone, even in the absence of changes to the system or the regulatory environment, it is almost inevitable that the safety case will require updating throughout the operational lifetime of the system. Operational experience must be reconciled with the predictions made in the initial safety argument.

The system operators, as the "owners" of the safety case, are typically responsible not only for its initial production, but also for its maintenance throughout the lifetime of the system. There is growing recognition in the standards that appropriate mechanisms must be in place for the ongoing maintenance of the safety case.

The safety case will have been produced initially to present a valid safety argument with respect to the regulations, evidence, and contextual information appropriate at the time. The difficulty in safety case maintenance is that any or all of these three elements may *change over time*. For example:

- An additional regulatory requirement may be added following an operational incident. An example of this from the civil aerospace domain would be the addition of a regulation regarding inadvertent thrust reverser deployment (JAA, 1990) following the Lauda Air thrust reverser deployment in flight accident.
- The design of a system may be changed for perfective, corrective, or adaptive maintenance reasons or through technology obsolescence. Hogberg (Hogberg, 1994) describes responding to unanticipated *problems* with the design of a class of nuclear reactors.
- Assumptions regarding the operational lifetime of a system also form an important part of the safety case context. Such assumptions may be challenged by a desire to extend plant life beyond the originally intended period. Clarke (Clarke, 1989) describes such a case for the life-extension of the UK civil Magnox nuclear reactors.
- Operational experience may challenge the evidence used as the basis of the original safety argument. For example, the safety case may estimate that a certain failure mode of a component will occur at a certain rate. This rate may be brought into question by operational data.

Having recognized a challenge to the safety case, the ongoing safety case maintenance process must evaluate the impact how, and the extent to which, this challenge affects the currently understood and recorded case. Having understood the impact of the challenge, decisions can then be made as to the natural and timescale of the action to be taken—for example, design review, further analyses, or operational limitation.

16.4.6 Safety Case Evaluation

An assurance-case-based regime requires a strong review element. Typically, one party is responsible for preparing the assurance case. Another party (the certification authority) is responsible for accepting the assurance case. Assurance cases are, by their nature, often subjective. The objective of assurance case development, therefore, is to obtain mutual acceptance of this subjective position. The move from less prescriptive standards to "goal-based" standards has both strengthened the need for assurance cases and increased the required review capability of the acceptance authorities.

The most obvious position in the system lifecycle for review of the assurance case position is "pre-operational"—that is, just prior to the system being approved to enter service. However, staged safety case review is far less risky (in terms of the project risk of a safety case that fails to gain a system clearance to operate). If there are problems with the arguments and evidence being offered up by the assurance case, it is desirable to find this out as early as possible in the lifecycle.

The most compelling staged reviews of assurance cases will involve representatives from the acceptance authority. It is often not possible to get an acceptance authority to unconditionally agree that an interim assurance case is acceptable. Instead, the concern when involving these stakeholders is to obtain the "non-negative" response—that is, to know that the assurance case, as it stands, does not contain any serious flaws in reasoning or weaknesses in evidence.

Even where it is not possible to involve the acceptance authorities in interim review activities, self-review (by the preparing organization) of an assurance case is still an extremely useful activity. (Often the most difficult people to convince of the assurance of a system are those that know it best!). Self-review either requires the involvement of people within the organization who have maintained some independence from the assurance case development, or the involvement of people capable of imaginative role-play. ("If I were the acceptance authority, what would I find unconvincing about this argument?").

16.4.7 Safety Case Confidence

Assessing or reasoning about the level of confidence in the arguments and evidence of an assurance case remains a challenge. A number of approaches to reasoning about the level of confidence have been proposed. Quantitative methods of reasoning about the possibility of perfection have been described in Littlewood (2012). Reasoning about assurance case confidence based on a Baconian philosophy in terms of the identification and mitigation of argument and evidence "defeaters" has also been proposed (Weinstock, 2013). Hawkins (2011b) outlined an approach that encourages the production of two separate (but cross-linked) arguments within any assurance case: A technical *risk* argument, with a companion *confidence* argument that justifies the adequacy of the arguments and evidence of the risk argument (as described earlier in this chapter). Such an approach allows for an *explicit* and *specific* (albeit qualitative) treatment of confidence when applied to assurance arguments. Assurance case developers are free at this point in the confidence argument to state that their confidence in the adequacy of the technical risk arguments and evidence comes from having followed the guidance of a standard (e.g., appealing to having satisfied the evidence requirements of a specific SIL). However, in Hawkins (2011a), we encourage that the confidence argument should engage in the specific certainties and uncertainties (termed "*assurance deficits*" in this approach) of the risk arguments and evidence—for example, by considering the potential impact of the uncertainty in the arguments and evidence on the technical risk claim.

16.5 EMPIRICAL RESEARCH ON USE AND EFFICIENCY

Although the use of safety cases is becoming increasingly widespread, with an ever-increasing number of domains adopting the approach as part of their accepted assurance and regulatory regimes, there is very little empirical research that evaluates how "successful" the approach has been. First of all, this may be because of the difficulty of defining "success" for a safety case approach. Should it be that systems are safer? Should it be that systems are better understood? Should it be that this is an efficient means of gaining assurance? Given that the production of safety cases does not replace system design or system safety process, but instead simply demands that the results of these processes be carefully and explicitly considered in the context of the form of safety argument being established, it can be hard to separate (in terms of evaluation) the effect of safety cases from the effect of a well-established system safety process. At the heart of the evaluation of safety cases lies the question as to whether the demand for producing and evaluating an explicit safety argument reveals flaws in system safety thinking (e.g., underestimated, undermanaged, or underexplored risks) over-and-above those that would have been identified simply through the execution of a system safety process (without a safety case). The formulation of this question suggests the nature of the controlled experiment that would be required, namely parallel execution of near identical system safety processes, one with safety case production and evaluation and one without. It is perhaps because of the difficulty of setting up such an experiment that explains why it has not been done.

In the literature, it is possible to find many anecdotal accounts of the success of adopting a safety case approach, such as Chinneck et al., (2004). It is clear from these accounts that many find there is a *face validity* in the value of being called to be explicit and transparent about the nature of the safety argument and evidence they have for their systems.

16.6 WEAKNESSES, LIMITATIONS, AND CRITICISMS

There are many ways in which the safety cases can be poorly implemented or executed as part of an assurance regime. Here we describe some examples:

The "Apologetic Safety Case"	Safety cases have little hope of adding value if they are impotent in their influence on the design and operation of the system in question. Safety cases should not be produced after the system design has

been finalized, but sometimes they are! The production of a thorough and rigorous safety case at this stage has the potential to reveal uncomfortable truths about the safety and certifiability of the system. Where problems are uncovered, developers are faced with the option of rework (often economically and politically unacceptable), operational limitation, or what could be termed the "apologetic safety case." Such a safety case is required to pick up the tatters of the safety evidence available and stitch them together to make the best possible job of the safety argument ("X doesn't quite work as intended, but it's OK because ..."). In the worst situations, such cases are made into deliberate masterworks of obfuscation, hiding the awkward truth that the system is not as safe as was hoped. This is not how things should be. For safety cases to work for us, we must give them the chance. Safety case development should be started from the earliest possible opportunity, incorporating design detail and associated safety evidence as soon as it becomes available. In such situations, when it becomes obvious that there will be difficulty in making the case of acceptable safety, this observation can be fed back into the design process and used as a driver for change.

The Document-Centric View

This is perhaps the biggest "bear-trap" in the field of safety case development. Since the inception of the concept, safety cases have often been thought of as documents. For example, Issue 1 of the UK Ministry of Defence Ship Safety Management Handbook JSP 430 stated that, "A safety case is a comprehensive and structured set of safety documentation ..." More correctly, this should have read, "A safety case is commonly presented using, and communicated through, a structured set of safety documentation." This distinction is important. We should not be reassured by paper, word-processor files, or HTML documents. Instead, we should be reassured by a structured, compelling argument with relevant and comprehensive supporting evidence. The safety case approach is not working if our goal is simply the production of a document. Safety case reports can exist even if there is no safety case, and vice versa. Unless we use the requirement to produce a safety case report as an opportunity to "dig deep" and truly gain (and then present) an understanding of what makes our system safe (or unsafe), then we are in danger of spending a lot of money to produce a document that will quite possibly have no impact on achieved safety.

The Approximation to the Truth For some, the easiest way to present a coherent, compelling safety case is to ignore some of the rough edges that exist in reality. This is perhaps particularly an issue when using a graphical notation such as the Goal Structuring Notation, or Claims-Argument-Evidence. For example, it is all-too-easy to state a claim in a GSN diagram that "All identified hazards have been acceptably mitigated" and direct the reader to the Hazard Log as the principal source of evidence when, in reality, the argument is not so straightforward. Such approximations to the truth are misleading and only serve to undermine the value of producing a safety case report. What is the point of producing a safety case report that cannot be trusted to give a full and accurate account of the position of safety?

Prescriptive Safety Cases Safety cases are not working well if they have become run-of-the-mill. The phrase "prescriptive safety case" may sound like a contradiction in terms. However, it is possible to fall in to the trap of making our safety cases routine. This can happen in a number of ways. We can become obsessed with compliance to accepted structures (e.g., the safety case report structure as required by CENELEC 50129) and lose sight of the principal challenge of creating convincing safety arguments for the system in hand. Alternatively, we can make our safety cases a parade of detail that may superficially sound compelling but, again, fails to establish a coherent argument. For example, this can happen as a result of structuring our arguments through a clause-by-clause analysis of a safety standard, or through a task-based discussion of every activity in our safety plan. Safety cases can also become routine through reuse and over-familiarity. If every safety case from a particular domain, or a particular company, starts looking the same, we should perhaps become worried. Have we thought long-enough and hard-enough about the issues that are specific to a given system? Is this safety case really revealing the arguments that need to be made or are we shoe-horning the issues into the carcass of a previously accepted structure?

Safety Case Shelf-Ware

Safety cases are not working well if, after their initial development and acceptance, they are consigned to a shelf, never again to be touched. In many cases, the development of an acceptable safety case involves substantial intellectual effort and contains detailed information concerning the safe maintenance and behavior of a system in operation. We have failed if we have managed to create an account so inaccessible or unapproachable that we are happy to never refer to it again.

Imbalance of Skills

The introduction of a safety case regime requires the existence of (at least) two skill sets. Developers must have the skills to assemble and communicate the safety case. Equally importantly, regulators must have the skills to review and critique the safety case. The former is widely recognized. The latter is sometimes overlooked, or underplayed in importance. The safety case approach cannot work effectively if there is an imbalance of skills between developers and assessors of the safety case. Whilst it is wrong to think of safety regulators as opponents, the health of a safety case regime lies in the existence of someone capable of challenging the claims and assumptions of the developer's case. A strong case will stand up to such a challenge. A weak case will be recognized as such and changes prompted.

The Illusion of Pictures

Graphical argumentation notations such as the Goal Structuring Notation and Claims-Argument-Evidence have been introduced to help people develop and present the safety arguments that should lie at the core of any safety case. However, there lies a risk with the use of these notations—overconfidence. We must not be "dazzled" by someone's use of a notation, the complexity of their diagrams, the hyperlinking of their argument structures, or the sophistication of their safety case tool support. The quality of an argument cannot be judged by a node-count (a common boast!), or on the basis of the number of colors used. (To persuade by virtue of node-count is merely a modern-day analogue of "arguing by weight of paper presented".) It is necessary to ensure that stakeholders are fully engaged in assessing the adequacy of the reasoning presented. The skills must exist to enable stakeholders to properly scrutinize and critique safety cases; safety cases must engage in accurate and in-depth analysis of the specifics of the system in question; finally, it must be remembered that the safety case message is more important than the medium.

16.6.1 Other Criticisms

The Nimrod review (Haddon-Cave, 2009) critically examined the production of safety cases for military aircraft in the United Kingdom following the explosion of a Nimrod aircraft in Afghanistan. Although this aircraft entered service at a time when safety cases were not required, a "legacy" safety case had recently been produced for this platform. The review was critical of both the production and evaluation of this legacy safety case, and claimed that it had been a missed opportunity to reveal errors. The review criticized many aspects of safety case production—such as outsourcing of safety case production to safety consultants with limited knowledge of the operational realities of safety and obsession with safety case documents as a "end" rather than a "means" to improving system safety. Perhaps most significantly, the review suggested that the safety cases were being produced with a positive outcome in mind, that is, the claim that the system is acceptably safe, and that this was blinding developers and operators to counter-arguments and evidence to this overall safety claim. The review made a radical suggestion (that has since been rejected by all parties involved in UK military aviation) that the safety case be renamed a "risk case" to ensure that developers were more focused on the negative arguments of safety.

This issue of confirmation bias has been picked up by Leveson (Leveson, 2011) as her major criticism of the safety case approach to regulation. Her suggestion is that the production of safety cases encourages developers to seek positive arguments and evidence of safety, and lead them to ignore negative evidence. This characterization of safety cases is at odds with the majority of safety case practice that encourages developers to take a hazard- and risk-based approach to constructing a safety argument, that is, positive safety is argued only by overcoming unsafe conditions. Confirmation bias remains a problem. However, it is worth recognizing that such bias is potentially a problem in any system safety process. A virtue of there being a documented safety case where the safety argument and evidence are made explicit (as opposed to remaining implicit simply through the execution of system safety processes) is that this bias can at least be exposed, evaluated, and addressed. Without explicit safety cases, it could be argued that there are many other instances of bias that would simply never be revealed.

16.7 RELATIONSHIP TO OTHER PRINCIPLES

Given that safety cases exist to document and expose whatever safety argument and evidence have been established to justify system safety, the approach can be said to be compatible with all of the other safety principles (e.g., safety margin) described in this handbook. Even new principles to system safety—for example, resilience—can be regarded as establishing a new form of safety argument—different from the form of arguments previously expected, but an argument (supported by evidence) nevertheless.

There are sometimes tensions between those advocating demonstrating compliance to standards as the principal assurance approach and those that promote the production of safety cases (Wassyng et al., 2010) (Leveson, 2011) These are often presented as opposing and alternative approaches. However, this is a false opposition.

Existing standards represent a substantial body of knowledge that it is unwise to ignore. If nothing else, existing safety standards should be used as an informative checklist of forms of argument and evidence that are commonly required within a safety case. If there is a *(product-oriented)* risk mitigation requirement within in a standard that is willfully not addressed in some way within a safety case, it could be considered necessary to justify as to why this requirement will not contribute to risk reduction in this case. Of course, the challenge with some standards is to glean the rationale behind some of their requirements. Without clearly presented rationale, standards can be accused as merely being *prescriptive*.

Conversely, the application of standards can benefit from the complementary use of safety cases. First, even in respect to the issue of *compliance* or *conformance* to standards, it can be observed that the requirements of standards can often require interpretation in their enactment. They can be ambiguous, deliberately stated at a high level of abstraction, or allow for flexibility in their application. In such cases, the concept of establishing a per-project explicit *conformance* or *compliance* argument (that justifies how the requirements of a standard have been complied with on a specific project) has the benefit of improving transparency in the application of the standard. See Graydon et al. (2012) for a fuller discussion of the use and role of conformance and compliance arguments. Second, assurance cases have a useful role to play recording the *justification* (for a specific project) of the specific (judgement-oriented) aspects of the enactment of a standard. This is discussed in the following section.

In a world where safety standards such as DO-178B/C and IEC 61508 prevail, it is sensible to ask where assurance cases should focus their attention in order to most usefully contribute to confidence. The positive aspect of such standards (and others such as ISO 26262) is that they can be thought of as providing the *template* for the technical *risk* argument needed at the *core* of any safety case. However, where confidence can be lost in the application of standards and therefore, where safety cases can potentially add most value, is in justification of the *specific* instantiation of these *template* structures and general guidance.

Safety cases are well suited to the (inevitably subjective) *justification* of the adequacy of the identification of hazards, and the derivation and implementation of safety requirements. They are also well suited to the *justification* of the adequacy of evidence (e.g., the appropriateness and trustworthiness of specific forms of evidence for requirements satisfaction). Whilst conformance to standards can, in some sense, be viewed as a "proxy" *generic* confidence argument (i.e., claiming adequate confidence because having followed the requirements of a particular Safety Integrity Level (SIL) or Development Assurance Level (DAL)), assurance cases are well suited to enabling documentation of a more *specific justification* of confidence (e.g., by considering the trustworthiness and appropriateness of evidence, or the acceptability of assurance deficits in the evidence presented). Safety cases provide an opportunity (if constructed well) to explicitly document and evaluate product and project-specific aspects (such as requirements and model validity, specific selection of evidence types and artefacts) that need to be argued for safety (and confidence). It is an artificial choice to say that safety engineers must choose between *either* a safety-case-oriented approach or a standards-oriented approach to assurance. They can be used together in highly complementary roles.

REFERENCES

Arnold, L. (1992). *Windscale 1957: Anatomy of a Nuclear Accident.* London: Macmillan.

Bate, I., and Kelly, T. (2003). Architectural considerations in the certification of modular systems. *Reliability Engineering & System Safety*, 81(3), 303–324.

Bloomfield, R., and Bishop, P. (2010). Safety and assurance cases: Past, present and possible future—An Adelard perspective. In: *Making Systems Safer* (pp. 51–67). London: Springer.

Cass, M., and Le Poidevin, R. (1993). *A Logic Primer.* Vortext Publishing.

Chinneck, P., Pumfrey, D., and Kelly, T. (2004). Turning up the HEAT on safety case construction. In: *Practical Elements of Safety* (pp. 223–240). London: Springer.

Chuse, R. (1993). *Pressure Vessels: The ASME Code Simplified* (7th ed.). New York/London: McGraw-Hill.

CIMAH. (1984). The Control of Industrial Major Accident Hazards (CIMAH) Regulations 1984, SI 1984/1902.

Clarke, A. W. (1989). Magnox safety review: Extending the life of Britain's work horses. *Nuclear Energy*, 28, 215–220.

Cullen, R. J. (1996). Safety as a design tool. Paper presented at Managing Risk in a Changing Organisation Climate—Proceedings of the Safety and Reliability Symposium, Swindon, UK.

Cullen, W. D. (1990). *The Public Enquiry into the Piper Alpha Disaster*. London: Department of Energy, HMSO.

DoD. (1998). The procurement of computer-based safety critical systems. Def(Aust) 5679, Safety Engineering for Defence Systems (Issue 2), Australian Department of Defence.

Edwards, C. (1997). Railway safety cases. Paper presented at Safety and Reliability of Software Based Systems—Twelfth Annual CSR Workshop, Bruges, Belgium.

Eurocontrol. (2006). Eurocontrol Safety Case Development Manual. Available at: https://www.eurocontrol.int/sites/default/files/article/content/documents/nm/link2000/safety-case-development-manual-v2.2-ri-13nov06.pdf (Accessed March 25, 2016).

FDA. (2014, December). Total Product Life Cycle: Infusion Pump - Premarket Notification [510(k)] Submissions—Guidance, US Food and Drug Administration.

GSN. (2011). GSN Community Standard Version 1. Available at: www.goalstructuring notation.info

Graydon, P., Habli, I., Hawkins, R., Kelly, T., and Knight, J. (2012). Arguing conformance. *IEEE Software*, 29(3), 50–57.

Haddon-Cave, C. (2009). *The Nimrod Review*. London: The Stationary Office.

Hawkins, R., Clegg, K., Alexander, R., and Kelly, T. (2011a). Using a software safety argument pattern catalogue: Two case studies. In: *Computer Safety, Reliability, and Security (SAFECOMP 2011)* (pp. 185–198). Springer.

Hawkins, R., Kelly, T., Knight, J., and Graydon, P. (2011b). A new approach to creating clear safety arguments. In: C. Dale and T. Anderson (Eds.), *Advances in System Safety: Proceedings of the Nineteenth Safety-Critical Systems Symposium*. Southampton, UK: Springer.

Hogberg, L. (1994). Shutting down 5 reactors: Reasons why and lessons learnt. *Nuclear Europe Worldscan*, 14, 42–43.

HSE. (1994). *Railways (Safety Case) Regulations 1994: Guidance on Regulations*. HSE Books.

IEE. (1996). *Proceedings of International Conference on Sizewell B: The First Cycle*. London: IEE.

ISO. (2011). 26262–Road vehicles-Functional safety. International Standard ISO/FDIS, 26262.

JAA. (1990, May). Joint Airworthiness Requirements JAR-E: Engines (Change 8), Civil Aviation Authority.

Kelly, T. P., and McDermid, J. A. (1997). Safety case construction and reuse using patterns. In: *16th International Conference on Computer Safety, Reliability and Security (SAFECOMP 1997)* (pp. 55–69). Springer.

Kletz, T. A. (1988). *Learning from Accidents in Industry*. London: Butterworths.

Leveson, N. (2011). The use of safety cases in certification and regulation. *Journal of System Safety*, 47(6), 1–5.

Littlewood, B., and Rushby, J. (2012, September–October) Reasoning about the reliability of diverse two-channel systems in which one channel is "possibly perfect." IEEE Transactions on Software Engineering, 38(5), 1178–1194. doi: 10.1109/TSE.2011.80. Available at: http://ieeexplore.ieee.org/stamp/stamp.jsp?tp=&arnumber=5975177™isnumber=6311391

MoD. (1996, January). JSP 430—*Ship Safety Management System Handbook*. Ministry of Defence.

MoD. (2007). Safety management requirements for defence systems, Defence Standard 00–56, Issue 4. United Kingdom Ministry of Defence.

MoD. (2014). Safety management requirements for defence systems. Defence Standard 00–56, Issue 5. United Kingdom Ministry of Defence.

Wassyng, A., Maibaum, T., Lawford, M., and Bherer, H. (2010). Software certification: Is there a case against safety cases?. In: Radu Calinescu and Ethan Jackson (Eds.), *Foundations of Computer Software: Modeling, Development, and Verification of Adaptive Systems* (pp. 206–227). Berlin, Heidelberg: Springer.

Weinstock, C. B., Goodenough, J. B., and Klein, A. Z. (2013) Measuring assurance case confidence using Baconian probabilities. In: 1st International Workshop on Assurance Cases for Software-Intensive Systems, ASSURE (pp. 7–11), San Francisco, CA. doi: 10.1109/ASSURE.2013.6614264. Available at: http://ieeexplore.ieee.org/stamp/stamp.jsp?tp=&arnumber=6614264&isnumber=6614259

FURTHER READING

Bishop, P., and Bloomfield, R. (1998). A methodology for safety case development. In: *Industrial Perspectives of Safety-Critical Systems* (pp. 194–203). London: Springer.

Hawkins, R., Habli, I., Kelly, T., and McDermid, J. (2013). Assurance cases and prescriptive software safety certification: A comparative study. *Safety Science*, 59, 55–71.

Kelly, T. P. (2007). Reviewing assurance arguments—A step-by-step approach. In: Proceedings of Workshop on Assurance Cases for Security—The Metrics Challenge, Dependable Systems and Networks (DSN).

Kelly, T. P., and McDermid, J. A. (2001). *A Systematic Approach to Safety Case Maintenance, Reliability Engineering and System Safety* (vol. 71, pp. 271–284). Elsevier.

Kelly, T. P., and Weaver, R. A. (2004). The goal structuring notation—A safety argument notation. In: Proceedings of the Dependable Systems and Networks 2004 Workshop on Assurance Cases.

Sujan, M., Koornneef, F., Chozos, N., Pozzi, S., and Kelly, T. (2013). Safety cases for medical devices and health information technology: Involving health-care organisations in the assurance of safety. *Health Informatics Journal*, 19(3), 165–182.

17

INHERENTLY SAFE DESIGN

Rajagopalan Srinivasan and Mohd Umair Iqbal

17.1 INTRODUCTION

In 1984, the release of about 25 tons of methyl isocyanate from Union Carbide's pesticide plant in Bhopal, India, claimed several thousand lives and left many others injured. Twenty-four years later, on August 28, 2008, in Bayer CropScience's pesticide plant in Institute, West Virginia, a runaway reaction resulted in the explosion of a process vessel. In its close vicinity was more than 17 tons of methyl isocyanate. If the shrapnel from the explosion had impinged on the methyl isocyanate storage, another disastrous event like the one at Bhopal would have unfolded. Thus, despite many advances in safety, it was only fortuitous that another Bhopal did not occur in Institute; the same hazard had bedeviled both plants.

Inherently safe design is a philosophy that seeks to unearth fundamental choices to eliminate or reduce hazards, typically in the early stages of a plant's lifecycle. The philosophy is best summarized in the words of Trevor Kletz, "What you don't have can't leak." While the traditional plant design philosophy aims to add protective cover to the existing hazards to reduce the probability of transformation of such hazards into accidents, the inherently safe philosophy seeks to use less hazardous chemicals

Handbook of Safety Principles, First Edition. Edited by Niklas Möller, Sven Ove Hansson, Jan-Erik Holmberg, and Carl Rollenhagen.

and operations to eliminate or reduce the possibility of accidents and minimize their consequences when they occur (Edwards, 2005).

Example 1

Access ways or manholes are placed at regular intervals in any sewer system to allow maintenance. The design of manholes has evolved over the course of time; the most preferred ones now being circular. The advantage of a circular manhole is that its lid cannot fall into the manhole. The same is not true for other shapes, for example, rectangular; the lid can easily fall along the diagonal of a rectangular manhole. So in the case of the circular design, the hazard is eliminated and the system becomes inherently safe without requiring additional safeguards.

Example 2

Consider the design of a nozzle for dispensing fuel in a gas station (i.e., petrol station). To protect against wrong fuel injection (e.g., diesel dispensed to a petrol vehicle), different nozzle sizes are used for the two fuels. The petrol nozzle is designed to be of smaller size so that the possibility of filling a petrol tank with diesel is eliminated. The incompatibility between the nozzle and the opening of fuel tank makes the system inherently safe.

17.2 ORIGIN AND HISTORY OF THE PRINCIPLE

The principle of inherent safety started gaining attention after a spate of major accidents (e.g., Flixborough accident of 1974, the Three Mile Island nuclear accident of 1979 and so on) and the consequent demands for higher standards of safety (Kletz and Amyotte, 2010). Historically, the trend was to first design and build the plant, and then consider safety and risk control. While these plants had extensive engineered safety features, these systems were not perfect. Analysis revealed that on-site errors were a major contributor to these accidents (Garrison, 1989) and these could not be easily rectified after the plant has been constructed. Even with sophisticated instruments, control, and technology, the risk associated with large inventories could only be reduced, not eliminated. The Flixborough accident of 1974 shook the world by claiming 28 lives. An estimated 30 tons of cyclohexane was released which upon ignition leveled the entire facility. Had the size of leak been smaller the accident would not have been as deadly.

Inherently safe design started gaining impetus after this incident. Trevor Kletz argued that the greatest potential opportunities for impacting the risk profile occur early during process design as there is a great deal of freedom in the selection of chemistry, solvents, raw materials, intermediates, unit operations, plant location, and process parameters (Srinivasan and Natarajan, 2012). Inherently safe design received further impetus after the Bhopal gas disaster; safety experts and the process industry felt a greater need to reduce hazards rather than seek to manage and control them. However, inherently safe design is a broad philosophy and an approach to design, which leads to an intrinsically safe system instead of relying on devices external to the core process to reduce risk to an acceptable level. It is not restricted to a specific domain.

It is useful to make a distinction between *safe* design and *safer* design. With safer design, there are fewer hazards, fewer causes, and fewer people to be exposed to the effects (Dalzell, 2004). Furthermore, safer emphasizes the relative nature because a process or operation may be safer with respect to a particular hazard but may be less safe with respect to another hazard.

17.3 DEFINITIONS AND TERMINOLOGY

Hazard. A chemical or physical condition that has the potential to cause damage to people, property, or environment (Crowl and Louvar, 2001).

Inherent Safety. A condition in which the hazards associated with the materials and operations used in the process have been eliminated or reduced, and this reduction or elimination is permanent and inseparable (CCPS, 2012).

Fail-Safe Design. A feature incorporated for automatically counteracting the effect of an anticipated possible source of failure. A system is fail-safe if failure of a component, signal, or utility, initiates action that returns the system to a safe condition (AIChE, 2011). For example, an elevator typically has brakes that are held off by the tension of the elevator cable. If the cable breaks, tension is lost and the breaks latch on, stopping the elevator from falling. Inherent safety has a close relationship with fail-safe. Both seek to minimize the severity of consequences, should an untoward incident happen. While the applications of the principles of inherent safety primarily do it by minimizing hazard, fail-safe design does it by minimizing unsafe consequences. Both the principles are based on the belief that accidents cannot be entirely eliminated; should one occur, it should have minimum impact.

Intrinsic Safety. Inherent safety is known as intrinsic safety in electrical and electronics engineering. Intrinsic safety is a type of protection based on restricting amount of electrical energy within any apparatus exposed to potentially explosive atmospheres to a level below that which can cause ignition (AN9003, 1989).

Intensification. Reduction of inventory of hazardous material in process, storage, and transportation is the key goal of process intensification. It is achieved by incorporating substantially smaller, cleaner, and more energy-efficient technology. Process intensification does not just need to be viewed as a technique concerning safety but it has its roots in economic side also; the economic need for process intensification is increasingly framed within the context of global competitiveness and sustainable development (Charpentier, 2007; Charpentier, 2010).

Minimization. An inherent safety principle. Minimization means to reduce the quantity of material or energy contained in a manufacturing process or plant so that the hazard is reduced. Minimization is related to intensification in a way that intensification often leads to minimization.

Moderation. An inherent safety principle. Moderation, also called attenuation, means using materials or processes under less hazardous conditions. Moderation of conditions can be accomplished by strategies that are either physical (i.e., lower temperatures) or chemical (i.e., development of a reaction chemistry which operates at less severe conditions (Srinivasan and Natarajan, 2012).

Passive Safety. It involves use of safety measures that do not rely on external stimulus to operate. Passive safety systems do not have to detect an unsafe condition or take any action to perform their protective safety function (CEP, 2010). They generally have no sensors or moving parts. They help in minimizing the consequence of accidents once they occur. Examples of passive safety equipment include helmets, crumple zone in cars in automobiles, blast-resistant structures, bursting disk in process vessels, and flame arrestors.

Risk. A measure of human injury, environmental damage, or economic loss in terms of both the incident likelihood and magnitude of the loss or injury (Crowl and Louvar, 2011).

Simplification. An inherent safety principle. Simplification means to design the process to eliminate unnecessary complexity making it robust and forgiving, thereby reducing the opportunities for error and wrong operation (Srinivasan and Natarajan, 2012).

Substitution. An inherent safety principle. Substitution means to replace a hazardous material or process with an alternative that eliminates or reduces the hazard.

17.4 USE AND IMPLEMENTATION

Inherent safety can be applied in various ways. The key principles of inherent safety include minimization, substitution, moderation, and simplification. These principles can be applied at various stages of the lifecycle of a plant. However, the effectiveness of these principles depends on the stage of the lifecycle when they are used. These principles are most effective at the early stages of design (Srinivasan and Natarajan, 2012). This is because during the early stages, changes can be incorporated easily. The application of inherently safe design begins with asking a question whether a hazard can be eliminated or not. If it cannot be eliminated, can it at least be minimized? (Mannan, 2012) This is followed by looking into the possibility of substitution of hazard and/or moderation and simplification. Although less effective, the application of these principles has also been demonstrated in existing plants (Ashford and Zwetsloot, 2000; Zwetsloot and Ashford, 2003). Some illustrations follow.

17.4.1 Examples of Minimization

In the early stages of design, process engineers had come up with a system of three knockout vessels in series (Paragon, 2011). The vessels had to contain a high-pressure gas. Using three knockout vessels meant large inventory of high-pressure gas. After serious consideration, they managed to combine these functions into a single vessel, which drastically reduced the inventory of gas, and made the system inherently safe. Additionally, it eliminated the problem of finding enough space for three knockout vessels.

Production of nitroglycerine is another example that demonstrates the minimization principle of inherently safe design (Kletz and Amyotte, 2010). Earlier nitroglycerin was produced in batch reactors by a reaction between glycerin and a mixture of concentrated nitric and sulfuric acids. The reaction is highly exothermic and if temperature is not kept in check, a runaway reaction would result. The volume of reaction mixture used to be around a ton with a residence time of about 2 hours; so that the consequences of any runaway would be devastating. The need for such large reactors was due to the requirement of long residence times, which was due to improper mixing. Engineers redesigned the reaction system and replaced the batch reactor with a well-mixed continuous reactor of a much smaller size. Due to better mixing, the reaction proceeds quickly and the residence time can be reduced. The smaller reaction mixture contained in the continuous reactor also ensures that the consequences of any runaway are comparatively mild. Thus, the continuous reactor system is inherently safer.

A criticality accident at a Uranium processing plant in Tokai village, Ibaraki Prefecture, Japan, in September 1999 led to three plant workers being exposed to high levels of radiation (Tsuchiya et al., 2001). As per the original design intent, the risk of criticality was managed by restricting the mass of Uranium in each step of the production process to less than 2.4 kg. This legally approved criticality control measure was contravened and triggered the accident. During the event, a large quantity of uranium, nearly seven times the approved limit, was fed into the precipitation tank due to economic pressures and the need to increase production (Akashi et al., 2000; Tsuchiya et al., 2001). Thus, the minimization principle of inherent safety that was the basis for safety in this case was defeated. Such maloperation would have been prevented if the precipitation tank had been designed with a suitable size that would preclude the addition of extra amount of material, and suitable geometry that prevented criticality (NRC, 2000).

17.4.2 Examples of Substitution

Filling of balloons with non-flammable helium rather than flammable hydrogen is a simple example of the principle of substitution. Similarly, substitution of normal cables with fire-resistant cables helps in preventing and minimizing fires. Biocatalysis and enzymes may allow production at a much lower temperature and pressure.

Natural gas power plants generate electricity using combustion turbines that use natural gas as fuel. Piping from the natural gas supply to the turbine must be installed as part of the construction process. When new piping is installed, some debris may be left behind in the pipeline. To prevent damage to the gas turbine, the piping has to be cleaned before startup. On February 7, 2010, six workers were fatally injured during a planned work activity to clean debris from natural gas pipelines at Kleen Energy's plant in Middletown, CT, USA. To remove the debris, workers used natural gas at high pressure (CSB, 2010). The gas that was vented out after cleaning accumulated due to poor ventilation and found an ignition source, leading to an explosion. The accident would have been prevented if, instead of natural gas, a less hazardous substitute such as compressed air, nonflammable nitrogen, or an air-driven pig had been used for cleaning the debris (CSB, 2010).

17.4.3 Examples of Simplification

Power plugs are often designed in such a way that they can be inserted into the socket in only one way. This makes incorrect assembly impossible and the design is inherently safe. Another example of simplification is gravity flow, which eliminates the need for pumps.

Consider a case in which three pumps A, B, and C are located adjacent to each other in that order, so pump A is the first pump, pump B the second one, and pump C the third. Over time, due to additional demands, a fourth pump (say pump D) may become necessary. Due to space constraints, the fourth pump might be installed between pump B (second pump) and pump C (third pump). Eventually, the labels on the pumps may be gone. If operators are instructed to open/close the third pump, there is a likelihood that they may act on pump D by mistake and cause undesired effects. A simpler less error-prone design would be to place pump D only after pump C so that third pump still means pump C only.

17.4.4 Example of Moderation

Physical characteristics have a considerable impact on the properties of a material, for example, the lower the particle size, the higher is the explosion hazard associated with a combustible dust. If these materials can be handled as granules or pellets instead of dusty powder, they will be inherently safer (Mannan, 2012).

Manufacture of phenol from cumene (Kletz and Amyotte, 2010) involves a stage where a reaction needs to be carried out very close to runaway temperature. In this case, it is usual to provide for a half-filled water tank just below the reactor base so that the reaction can be quenched by dumping the reactor contents into water by just opening a valve. Because of the absence of absolute reliability of instruments (sensors, valve, etc.), there is still a risk of a runaway reaction. Carrying out the reaction at a lower temperature (moderation) provides an inherently safer alternative although the reactor volume may need to be increased to ensure the same yield. This process will not even require the dump tank and associated complexities.

The above example illustrates that application of the principles may result in reduction of some hazard at the expense of another. In other words, modifications in design rarely result in monotonous reduction of risk because such modifications may alter chemistry, process route, operating conditions and various other parameters (Srinivasan and Natarajan, 2012). This means that the risk with respect to a particular hazard may decrease but the change may lead to escalation of another hazard. For example, consider substitution of a toxic non-volatile solvent with a non-toxic but volatile one. The risk associated with the toxicity of the material gets eliminated but it poses an alternate risk associated with volatility. Therefore, it becomes imperative to weigh the alternate designs in terms of their safety and evaluate the trade-offs among various design options.

To enable decision-making in such situations, various inherent safety indices have been proposed to determine the risks associated with a particular design. The Dow Fire and Explosion Index and Mond index are used for determining the fire hazard and toxicity levels of a process respectively. These require detailed knowledge of inventories in the process which may be difficult to estimate in the early stages of the design lifecycle. To address this shortcoming, a number of inherent safety indices have been developed including Prototype Index for Inherent Safety (Edward and Lawrence, 1993; Lawrence, 1996); Inherent safety Index (Heikkila, 1999; Heikkila et al., 1996); iSafe (Palaniappan et al., 2002a, 2002b); Intergrated Safety Index (Khan and Amyotte, 2004); and inherent benignness indicator (Srinivasan and Nhan, 2008). As an alternative to such indices, Gupta and Edwards (2003) proposed a graphical method for measurement of inherent safety.

17.5 EMPIRICAL RESEARCH ON USE AND EFFICIENCY

Several companies have reported the use of inherent safety principles including ABB, BASF, BP Exploration, Dow Chemical, DuPont, Shell Upstream, and Miox Corporation (Srinivasan and Natarajan, 2012). However, information on the specific nature of the inherent safety intervention adopted by companies is sparse given the sensitivities and confidentialities often associated with such developments. Here, we highlight a few published cases.

Beverage bottling plants are required to maintain high standards of sanitation and have to clean preparation tanks and lines to prevent contamination, flavor carryover, and corrosion. This cleaning often requires high temperatures and hazardous chemicals (such as chlorine). Miox Corporation (Boeh and Christy, 2011) reported the adoption of an electro-chemically activated clean-in-place method that produces chlorine and other oxidant species *in situ*, reducing the need for their large inventory.

With the ever-growing demand for energy, offshore oil and gas exploration and production have become very prominent. The dynamic situations that usually characterize offshore settings pose unique challenges. Therefore, new/modified design strategies have to be evolved regularly. Many companies in the offshore industry, ranging from Kellogg Brown and Root (Chia et al., 2003), to BP (Considine et al., 2010), and Shell (Gilmour and Deveney, 2010) have reported applications of inherently safe design principles in the design of new oil and gas platforms (Khan and Amyotte, 2002).

Zwetsloot and Ashford (2003) report on the application of inherently safe design in four facilities—one in the design stage of the lifecycle and three in operating plants. Many hazardous situations that could be improved through application of inherent safety principles were identified. A majority of the resulting design options were economically feasible with payback period of less than 2 years. They also report that a fourth of all the identified options were implemented in the plants. The authors also show that these principles are not restricted only to the process but they can be applied to other departments such as logistics.

17.6 WEAKNESSES, LIMITATION, AND CRITICISM

The subject of inherent safety is quite popular among safety experts and has seen some success in day-to-day usage. In a survey conducted by Gupta and Edwards (2002), 64% of the respondents were "unconvinced of benefits" of inherently safer design (Edwards, 2005). A fundamental issue is that safety is usually not given high priority in the early stages of the design lifecycle where project economics is considered primary. Safety has traditionally been taken care of at a later stage by which time the philosophy of inherently safe design cannot be applied as effectively. Furthermore, while the principles of inherently safe design are well known, tools that enable their easy application are limited. The few that are available are often complicated or inaccessible.

Since safety is inherently multi-faceted, any safety principle may not offer monotonic risk reduction on all fronts. For example, when chlorofluorocarbons first became available, and substituted older solvents, they were hailed as wonder solvents which were nonflammable and had low toxicity relative to older solvents (Kletz and Amyotte, 2010). However, now their effect on the ozone layer is understood, their use is discouraged.

Economic feasibility remains a fundamental basis for any production process. It is sometimes argued that the objectives of safety can be in conflict with costs. There is an old saying attributed to Kletz "If you think safety is expensive, try having an accident." Thus in terms of lifecycle cost, safety is rarely in conflict with economy since avoiding accidents prevents losses of life, property, business interruption, goodwill, etc. Similarly, it is sometimes observed that ensuring safety and process availability may be in conflict—production pressures may require that the plant be operated with its safety mechanisms compromised (c.f. Chernobyl, Tokiamura). A strong safety culture is a strong antidote to ensure that the safety is not seen as being in conflict to production or economics.

17.7 RELATION TO OTHER PRINCIPLES

Inherently safe design cannot be considered in isolation. It has to be viewed in relation to other principles, methodologies and imperatives such as green chemistry and sustainability. Green chemistry has been defined as the design of chemical products and processes that reduce or eliminate the use and generation of hazardous substances (Anastas et al., 2000). While the primary aim of green chemistry is to reduce the impact of pollution on environment, it also helps in the improvement of safety (Amyotte et al., 2009). Similarly, the application of the principles of inherent safety translates to environmental, social, and health benefits (Edwards, 2005)—the three pillars of sustainability.

REFERENCES

AICHE. (2011). The principle of fail-safe available: [Online], American Institute of Chemical Engineers. Available at: http://www.aiche.org/chenected/2011/02/principle-fail-safe (Accessed September 14, 2015).

Akashi, M., Aoki, H., and Endo, A. (2000, April). Lessons learned from the JCO nuclear criticality accident in Japan in 1999 [Online]. The International Atomic Energy Agency. Available at: http://www-ns.iaea.org/downloads/iec/tokaimura-report.pdf (accessed on July 11, 2017).

Amyotte, P. R., Khan, F. I., and Kletz, T. A. (2009). Inherently safer design activities over the past decade. Symposium Series No. 155, IChemE, pp. 736–734.

AN9003 (1989). *A User's Guide to Intrinsic Safety*. Norfolk, UK: Measurement Technologies Ltd.

Anastas, P. T., Heine, L. G., and Williamson, T. C. (2000). *Green Chemical Syntheses and Processes*. American Chemical Society; Distributed by Oxford University Press.

Ashford, N. A., and Zwetsloot, G. (2000). Encouraging inherently safer production in European firms: A report from the field. *Journal of Hazardous Materials*, 78(1), 123–144.

Boeh, B., and Christy, J. C. (2011). Bottling a winning: Clean-in-place solution. *Water Wastewater International*, 26, 48–50.

Center for Chemical Process Safety. (2012). *Guidelines for Engineering Design for Process Safety*. John Wiley & Sons.

Charpentier, J. C. (2007). In the frame of globalization and sustainability, process intensification, a path to the future of chemical and process engineering (molecules into money). *Chemical Engineering Journal*, 134(1), 84–92.

Charpentier, J. C. (2010). Among the trends for a modern chemical engineering, the third paradigm: The time and length multiscale approach as an efficient tool for process intensification and product design and engineering. *Chemical Engineering Research and Design*, 88(3), 248–254.

Cheremisinoff, N. P. (2014). *Dust Explosion and Fire Prevention Handbook: A Guide to Good Industry Practices*. John Wiley & Sons.

Chemical Engineering Progress. (2010, July). Passive Safety Equipment [Online]. *Process Safety Beacon*. Available at: https://www.aiche.org/sites/default/files/cep/20100767.pdf (accessed on June 1, 2017).

Chia, S., Walshe, K., and Corpuz, E. (2003). Application of inherent safety challenge to an offshore platform design for a new gas field development–Approaches and experiences. In: Institution of Chemical Engineers Symposium Series, 149 ed., Manchester, pp. 257–265.

Considine, M., Fargie, D., and Cooke, W. (2010). Enhancing offshore safety: A description of the offshore major accidental risk (OMAR) model. In: 2010 AIChE Spring Meeting and 6th Global Congress on Process Safety, San Antonio, TX.

Crowl, D. A., and Louvar, J. F. (2001). *Chemical Process Safety: Fundamentals with Applications*. Pearson Education.

Crowl, D. A., and Louvar, J. F. (2011). *Chemical Process Safety: Fundamentals with Applications* (3rd ed.). Upper Saddle River, NJ: Prentice Hall.

CSB (2010). Investigation report. Kleen Energy Natural Gas Explosion. U.S. Chemical Safety and Hazard Identification Board.

Dalzell, G. A. (2004, January). Inherently safer design; changing attitudes and relationships. In: SPE International Conference on Health, Safety, and Environment in Oil and Gas Exploration and Production. Society of Petroleum Engineers.

Edwards, D. W. (2005). Are we too risk-averse for inherent safety?: An examination of current status and barriers to adoption. *Process Safety and Environmental Protection*, 83(2), 90–100.

Edwards, D. W., and Lawrence, D. (1993). Assessing the inherent safety of chemical process routes: Is there a relation between plant costs and inherent safety? *Process Safety and Environmental Protection*, 71(B4), 252–258.

Garrison, W. G. (1989). *Large Property Damage Losses in the Hydrocarbon-Chemical Industries: A Thirty-Year Review*. M & M Protection Consultants.

Gentile, M., Rogers, W. J., and Mannan, M. S. (2003). Development of a fuzzy logic-based inherent safety index. *Process Safety and Environmental Protection*, 81(6), 444–456.

Gilmour, N., and Deveney, D. (2010). Floating LNG: Shell's recent history and current approach. In: Proceedings of the 16th International Conference on Liquefied Natural Gas.

Gupta, J. P., and Edwards, D. W. (2002). Inherently safer design—present and future. *Process Safety and Environmental Protection*, 80(3), 115–125.

Gupta, J. P., and Edwards, D. W. (2003). A simple graphical method for measuring inherent safety. *Journal of Hazardous Materials*, 104(1), 15–30.

Heikkilä, A. M. (1999). *Inherent Safety in Process Plant Design: An Index-Based Approach.* VTT Technical Research Centre of Finland.

Heikkilä, A. M., Hurme, M., and Järveläinen, M. (1996). Safety considerations in process synthesis. *Computers & Chemical Engineering*, 20, S115–S120.

Khan, F. I., and Amyotte, P. R. (2002). Inherent safety in offshore oil and gas activities: A review of the present status and future directions. *Journal of Loss Prevention in the Process Industries*, 15(4), 279–289.

Khan, F. I., and Amyotte, P. R. (2004). Integrated inherent safety index (I2SI): A tool for inherent safety evaluation. *Process Safety Progress*, 23(2), 136–148.

Kletz, T. A. (1978). *Process Plants: A Handbook for Inherently Safer Design*. Philadelphia, PA: Taylor & Francis.

Kletz, T. A., and Amyotte, P. (2010). *Process Plants: A Handbook for Inherently Safer Design.* CRC Press.

Lawrence, D. (1996). Quantifying the inherent safety of chemical process routes. Unpublished Doctoral Thesis, Loughborough University of Technology, UK.

Mannan, S. (Ed.). (2012). *Lees' Loss Prevention in the Process Industries: Hazard Identification, Assessment and Control.* Butterworth-Heinemann.

Palaniappan, C., Srinivasan, R., and Tan, R. (2002a). Expert system for the design of inherently safer processes. 1. Route selection stage. *Industrial & Engineering Chemistry Research*, 41(26), 6698–6710.

Palaniappan, C., Srinivasan, R., and Tan, R. B. (2002b). Expert system for the design of inherently safer processes. 2. Flowsheet development stage. *Industrial & Engineering Chemistry Research*, 41(26), 6711–6722.

Srinivasan, R., and Natarajan, S. (2012). Developments in inherent safety: A review of the progress during 2001–2011 and opportunities ahead. *Process Safety and Environmental Protection*, 90(5), 389–403.

Srinivasan, R., and Nhan, N. T. (2008). A statistical approach for evaluating inherent benignness of chemical process routes in early design stages. *Process Safety and Environmental Protection*, 86(3), 163–174.

Sutton, I., and Paragon, A. M. E. C. (2011). Inherent safety in front end engineering. In: American Institute of Chemical Engineers 2011 Spring Meeting and 7th Global Congress on Process Safety, Chicago, IL, March 13–16, 2011.

Tsuchiya, S., Tanabe, A., Narushima, T., Ito, K., and Yamazaki, K. (2001). An analysis of Tokaimura nuclear criticality accident: A systems approach. In: The 19th International Conference of the System Dynamics Society, System Dynamics Society, Atlanta, Georgia.

US Nuclear Regulatory Commission. (2000, April). NRC review of the Tokai-mura criticality accident.

Zwetsloot, G. I., and Ashford, N. A. (2003). The feasibility of encouraging inherently safer production in industrial firms. *Safety Science*, 41(2), 219–240.

18

MAINTENANCE, MAINTAINABILITY, AND INSPECTABILITY

Torbjörn Ylipää, Anders Skoogh, and Jon Bokrantz

18.1 INTRODUCTION

Maintenance is important for securing the reliability of the equipment in a plant. It is also a factor that promotes production efficiency, minimizing waste, and environmental effects as well as providing safer and healthier working conditions. However, the cost of maintenance is as high as 15–70% of the total production cost (Bevilacqua and Braglia, 2000) and one-third of these maintenance costs are wasted through improper maintenance management (Mobley, 2004). Apart from the costs, maintenance highly affects the safety aspect in a production system as well as the neighborhood and environment outside the factory. Consequently, the maintenance area has moved from almost nowhere to top priority for company managers Gopalakrishnan (2016).

Moreover, Reason (1998) claims that maintenance can seriously damage the production system if it is not carried out incorrectly. Studies in the campaign for safety on maintenance by the European Agency for Safety and Health at work (EU-OSHA) have shown that maintenance is a high-risk activity to which it was estimated that 10–15% of all fatal accidents and 15–20% of all accidents are related(EU-OSHA, 2010). Production disturbances often result in direct accident risk for maintenance workers

Handbook of Safety Principles, First Edition. Edited by Niklas Möller, Sven Ove Hansson, Jan-Erik Holmberg, and Carl Rollenhagen.

(Toulouse, 2002). Further, Lind et al., (2008) describe several risk-increasing factors for maintenance workers, such as operating under the pressure of time, performing independent maintenance work during night shifts, as well as poor work practices, work guidance, and risk analysis. Pintelon and Muchiri (2009) claim, despite increase in safety knowledge, there has been an increase in automation, so that the equipment is more complex to operate and maintain. In most maintenance concepts developed in recent times, a high involvement of the operators in maintenance tasks is proposed. However, as Pintelon and Muchiri claim, this will expose the operator to more potential hazards (ibid.).

Maintenance errors have been among the principal causes of several major accidents such as Piper Alpha, Texas Refinery, and the Bhopal disaster (Reason and Hobbs, 2003).

The cause of the Piper Alpha disaster, for example, was connected with preventive maintenance. Preventive maintenance aims to avoid failures as well as accidents and creates other afore-mentioned benefits. However, badly performed, it can seriously damage systems and result in a catastrophe. The Piper Alpha case description below, serves as an example of how easily things can go wrong when maintenance is poorly performed. It is worth pointing out, that in contrast to an emergency maintenance, this was routine work in a peaceful environment. Maintenance errors in these situations are not uncommon. Studies performed by Rasmussen, who categorized 200 hundred significant events during maintenance related tasks in nuclear power plants, indicated that the majority of errors were from preventive and corrective maintenance, surveillance testing, removal and restoration of equipment, checking, supervision, post-maintenance testing, and modifications (Reason, 1990).

18.1.1 The Piper Alpha Disaster

Piper Alpha was a North Sea oil production platform. The platform began production in 1976, first as an oil platform and later converted to gas production. An explosion and the resulting oil and gas fires destroyed it totally and killed 167 people. The total insured loss was about £1.7 billion. At the time of the disaster, the platform accounted for approximately 10% of North Sea oil and gas production (Mannan and Lees, 2005).

Two condensate pumps on the platform, designated A and B, compressed the gas for transport to the coast. Pump A's pressure safety valve was removed for routine maintenance. The pump's fortnightly overhaul was planned but had not started. The now open condensate pipe was temporarily sealed with a flat metal disc. Because the work could not be completed during this shift, the metal disc remained in place. The on-duty engineer filled out a permit which stated that Pump A was not ready and must not be switched on under any circumstances. As the on-duty custodian was busy, the engineer neglected to inform him of the condition of Pump A. Instead, he placed the permit in the control center and left. The permit disappeared and could not be found. Consequently, another permit was issued, when in fact it had begun but not completed and should not be switched on.

Then condensate Pump B stopped suddenly and could not been restarted. The entire power supply for offshore construction work depended on this pump. The manager had only a few minutes to bring the pump back online; otherwise, the power supply would fail completely. A search was made through the documents to determine whether Condensate Pump A could be started or not. The permit for the overhaul was found, but not the other permit stating that the pump must not be started under any circumstances due to the missing safety valve. The valve was in a different location from the pump and therefore the permits were stored in different boxes, as they were sorted by location. None of those present was aware that this safety part of the machine had been removed. The manager assumed from the existing documents that it would be safe to start pump A. The missing valve was not noticed by anyone, particularly since the metal disc replacing the safety valve was located several meters above ground level and obscured by machinery. Condensate Pump A was switched on. Gas flowed into the pump, and, due to the missing safety valve, produced an overpressure which the loosely fitted metal disc did not withstand. The gas leaked out at high pressure, ignited, and exploded, blowing through the firewall into the control room and the catastrophe was a fact (ibid.).

18.2 ORIGIN AND HISTORY

Maintenance has probably existed from the time man learned to sharpen pieces of stone into such implements as stone axes and arrows (Patton, 1994). Ljungberg (2000) describes how maintenance work has developed from this time, when the manufacturer, user, and the maintenance person could be the one and same person, to the situation we see in todays' factories, where we have divided technological specializing, such as mechanics, electricians, and instrument technicians. At the same time, the equipment has become more advanced, automated, computer controlled, and network connected. Consequently, maintenance has become more difficult to handle and demand has been put on maintainability aspects of products and for equipment in production systems. Maintainability needs to be designed for the system, so that the time required to complete scheduled and unscheduled maintenance can be reduced. Further, if system maintenance is easy to work with, it will reduce maintenance errors, reduce maintenance related injuries and accidents, minimize personnel training requirements, improve trouble-shooting performance, and lower both maintenance costs and revenue losses.

The maintainability field can be traced back to the early 1900s when the Wright brothers' airplane was developed and the US Army stated that the airplane should be made easy to operate and to maintain. After that many military documents on maintainability were published by the United States Department of Defense, see for example MIL-470 (1966). A history of the maintainability field is given in (Retterer and Kowalski, 1984). The first commercial work entitled Electronic Maintainability appeared in 1960 (Dhillon, 2008). Usually, terms like inspectability, that is, failure-finding maintenance and inspections, and supportability, that is, administrative time and logistic time, are included into maintainability together with the actual repair time Misra (2008). However, today it is common to separate these terms, see for example (Tiusanen et al., 2012).

Reliability, mainly decided during the design of a product or a process, is closely related to maintenance. If a system is well designed with high inherent reliability of the system or device, maximum achievable reliability is based on configuration and component selection. If operated correctly, maintained correctly, and inspected at appropriate intervals, it will be possible to attain the full inherent reliability. If poorly operated, poorly maintained, or allowed to develop and retain defects, reliability can be significantly less than the inherent reliability (Daley, 2008). The reliability field can be traced back to the early 1930s when probability concepts were applied to problems associated with electric power generation. However, the real development in the reliability field started during the World War II, when Germany had huge problems with the reliability of their V1 and V2 rockets (Dhillon, 2008).

18.3 UNDERLYING THEORY, THEORETICAL ASSUMPTIONS, DEFINITION, AND TERMINOLOGY

The cost of maintenance is estimated to be the major cost in many plants (Mobley, 1990). For example, Bevilacqua and Braglia (2000) point out that 15–70% of total production cost is spent on maintenance. Therefore, corrective and preventive maintenance, including inspections, must be carefully designed. According to the European Standard, SS-EN 13306:2010, maintenance is defined in as:

> *A combination of all technical, administrative, and managerial actions during the life cycle of an item, and intended to maintain it in, or restore it, to a state in which it can perform the required function.*

Maintenance is generally divided into two main categories, that is, unplanned maintenance and planned maintenance, see Figure 18.1. The first, corrective maintenance actions, cannot be planned and occur when we do not want them to. The second, preventive maintenance actions aim to control the level of failures. Of these two types, it is preventive maintenance, and especially when to do this, that has received most of the academic interest in the past. For failure maintenance, both the timing and the content involved in the maintenance task are unknown (Misra, 2008), and as Misra and many other authors point out, it is always advantageous to shift work to a more manageable category. In general, uncertainty will always mean higher risks and it is the same in the maintenance area. Moreover, studies have shown that unplanned maintenance is three times as costly as planned maintenance (Mobley, 1990). In short, breakdown maintenance is not safe, and may sometimes involve danger to life Misra (2008).

Over time, different generations of maintenance have been developed and maintenance has undoubtedly undergone major changes during the past decade, from

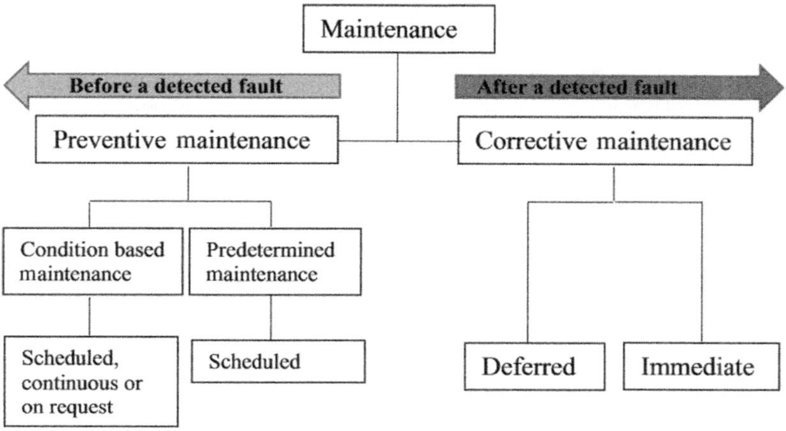

Figure 18.1. The concept of maintenance. Adapted from SS-EN 13306:2010.

being viewed as almost a "necessary evil" to a professional business function and an area of intensive academic research (Pintelon and Parodi-Herz, 2008).

Moubray (1997), and others, have described how maintenance has evolved through three-four generations of development, from reactive maintenance toward predictive and risk-based maintenance (Figure 18.2). There has also been a major shift in organizational thinking toward participation, teamwork, flexibility, and integration in maintenance (Misra, 2008; Peng, 2000). Each generation responds to the increased expectations in form of higher plant availability, better product quality, as well as greater safety and less environmental impact from equipment failures. It can be argued that we are now into the fourth generation of maintenance, that Dunn (1998) claims will have more focus on safety, a greater integration between functional requirements, equipment design and maintenance, as well as a much greater use of information technology to detect, predict, and diagnose equipment failures. This development in the maintenance area can be seen in Figure 18.2, where different generations have been placed to correspond with different times.

Reactive maintenance	Planned maintenance	Condition-based maintenance	Risk-based maintenance
Fix it when it broke	Time- or use-based replacements	Proactive & predictive maintenance	Risk-based inspections & maintenance
1940<	1960	1980	2000>
First genaration	Second genaration	Third genaration	Forth genaration

Figure 18.2. Development of maintenance types over four generations. Based on Moubray 1997, Dunn 1998, Alsyouf 2007, Arunraj and Maiti 2007, and Pinteleon and Parodi-Herz 2008.

Qingfeng et al. (2011) claim there exists a close relationship between maintenance efficiency and maintenance types. For example, preventive is better than reactive maintenance, as shown in Figure 18.3. However, this is not always true for every single failure mode, but it is a common goal that preventive maintenance should be above 80% of the total maintenance performed (Wireman, 2005).

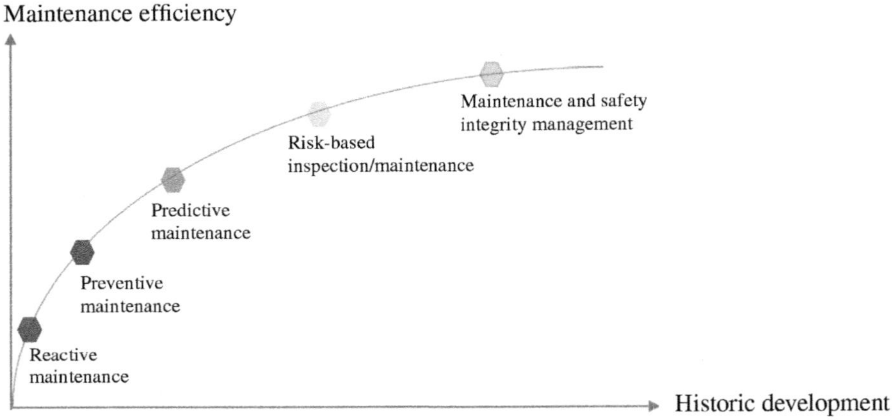

Figure 18.3. Maintenance types and their corresponding efficiency. *Source*: Qingfeng et al. 2011. Reproduced with permission of Elsevier.

Several different maintenance concepts have been developed during the generations of maintenance presented in Figures 18.1 and 18.2. It is of course, out of the scope of this chapter to cover all these concepts, but the two most common concepts that can be found in several of the other concepts are Total Productive Maintenance (TPM) and Reliability Centered Maintenance (RCM), see for example (Wireman, 2005). Related to the maintenance concepts are different types of Equipment Integrity Management and Asset Management. Equipment Integrity Management is specially developed for the process industry, where Qingfeng et al. (2011) propose a Maintenance and Safety Integrity Management System as the final step in their maintenance model's development structure, see figure 18.3. Another safety-related concept is RAMSI (Reliability, Availability, Maintainability, Safety, and Inspectability). The RAMSI concept has been focused on such industries as power plants, wind energy, space, railway transport, and oil/gas industries for decades, while other industries are only now awakening to the need for RAMSI.

Tiusanen et al. (2012) describe a RAMSI model used in a wind turbine application that consists of two main sections: safety and dependability, see figure 18.4. These are divided into elements and further into RAMSI factors. The system dependability sector is divided into four elements, and the system safety sector into three elements. Each element consists of numerous factors, not shown in the picture. Elements of dependability are defined in the dependability management standard

SS-EN 60300-1 (2014). System safety in this paper is seen from the viewpoint that failures cause hazards for people, the operational environment or nature (Tiusanen et al., 2012).

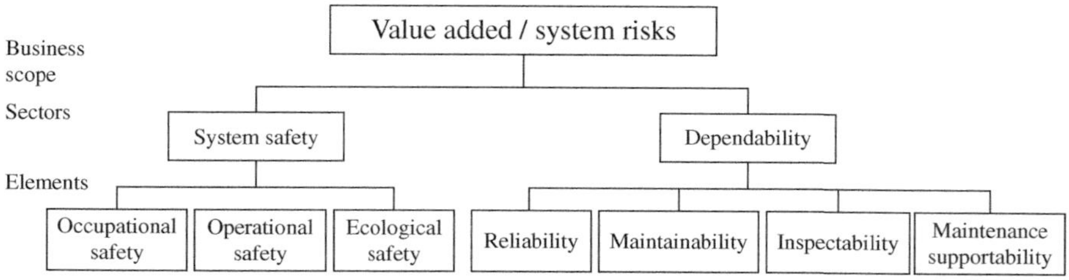

Figure 18.4. Elements related to added value and system risks simplified picture. *Source*: Tiusanen et al. 2012. Reproduced with permission of Risto Tiusanen, VTT Technical Research Centre of Finland.

Of special interest in this chapter of the book is the concept of dependability, seen to the right in Figure 18.3. In short, maintenance in industry is an important area to secure equipment reliability and other highly important and related goals in connection to the business scope. The maintenance and reliability relationship can be described by the concept of dependability as shown in Figure 18.3. All four elements are contributing to the final availability measurement that dependability is measured in. Operational availability can simply be expressed as (Brown and Porton, 2013):

$$\mathrm{A_O} = \frac{\mathrm{MOT}}{\mathrm{MOT} + \mathrm{MDT}_{NS} + \mathrm{MDT}_{S}} \tag{18.1}$$

The operational availability, $\mathbf{A_O}$, accounts for the effects of scheduled and unscheduled maintenance as well as other down time factors, for example, inspectability and supportability factors. MOT is the mean operating time, MDT_{NS} is mean not-scheduled down time for unscheduled maintenance and MDT_{S} is the mean scheduled downtime compromise of any down time associated with preventive maintenance, upgrades, etc.

Reliability is the ability of an item to perform a required function under given conditions for a given time interval. Reliability may also be defined as the probability that equipment will perform its intended function without failure for a specific time under defined conditions.

A commonly used measurement is mean time to failure (MTTF[1]) or mean time between failure (MTBF[2]). The probabilistic reliability function, can be expressed by

[1] MTTF stands for a non-repairable item.
[2] MTBF stands for a repairable item.

using the reliability function where t is the time, $R(t)$ is the reliability function, and $f(t)$ is the probability density function, for example, exponential, Weibull, normal.

$$R(t) = \int_{t}^{\infty} f(t)dt \tag{18.2}$$

The inherent reliability of equipment is the maximum achievable reliability based on configuration and component selection. If operated correctly, maintained correctly, and inspected at appropriate intervals, it will be possible to attain full inherent reliability.

However, if poorly operated, poorly maintained, or allowed to develop and retain defects, reliability can be significantly less than inherent reliability.

Maintainability is the measure of the ability of an item to be retained in or restored to a specified condition when maintenance is performed by personnel having specified skill levels, using prescribed procedures and resources, at each prescribed level of maintenance and repair. Maintainability is also a design feature, with goals such as increasing the ease of maintenance (i.e., improving the ergonomics), minimizing the time for performing preventive and corrective maintenance as well as minimizing the logistical burden through resources required for maintenance and support and reducing support cost. The mean time to repair (MTTR) is probably the most commonly used measure of maintainability. It can be calculated as shown in expression 2.

$$\text{MTTR} = \frac{\sum_{1}^{N} \text{repair time}}{N} \tag{18.3}$$

Maintainability can also be expressed as the probability that a repair is completed in a specified time interval by using the maintainability function where t is the time, $M(t)$ is the maintainability function, and f_{rp}.is the probability density function of the repair times, for example, exponential, lognormal, normal.

$$M(t) = \int_{0}^{t} f_{\text{rp}}(t)dt \tag{18.4}$$

Inspection is a conditional task, the result of which is a statement about the condition of the item, that is, the condition is satisfactory or it is unsatisfactory. During the operation of the item, inspections are performed at specified fixed intervals until the critical level is reached, when prescribed preventive maintenance tasks take place. If the item fails between inspections, corrective maintenance takes place.

Inspectability has often been an integral component of maintenance and maintainability issues (Tiusanen et al., 2011). However, over the last few years, inspection and testing have received increasing attention due to various incidents involving offshore installations and it is foreseeable that inspectability will be included in new standards and guidelines (ibid.).

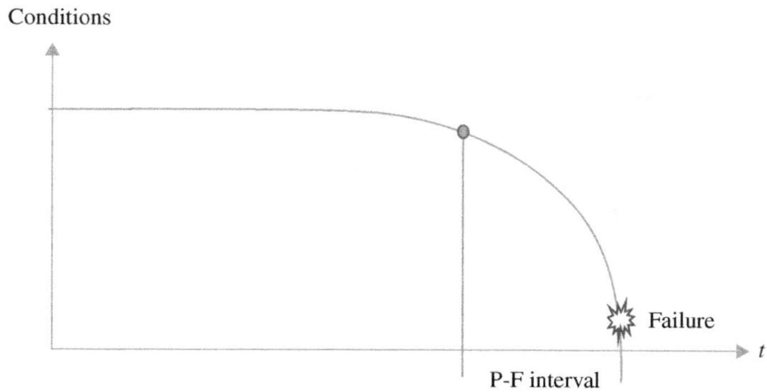

Figure 18.5. Potential failure interval. Adapted from Moubray 1997.

In for example the RCM concept, inspections of safety-related functions are seen as failure finding tasks where preventive maintenance tasks are performed at a specified interval to determine whether a hidden functional failure has occurred or not. Moreover, on condition that the task is a periodic or continuous inspection designed to detect a potential failure condition prior to functional failure, that is, potential failure interval (PF-interval), see Figure 18.5. Residual useful life (RUL) can then be calculated and, with that, the interval from detection to a failure. The optimal inspection interval proposed in RCM is then half the interval from detection to a failure, that is, PF-interval/2 (Moubray, 1997). In practice, this is not easy to decide. Different detection methods give different intervals. For example, pump bearing damage can be detected by vibration analysis, infrared heat detection, or simply an operator hearing a jarring noise or noticing an increased heat of the pump. The different detection methods will give different PF-intervals. However, the method to decide when to inspect the equipment and the cost for it, must be weighed against the negative aspect of the failure, that is, if it is small productivity loss the inspection cannot be to excessive, but if the consequences of the failure is safety or other major loss it may demand a redesign of the system.

The final element in the dependability concept is maintenance supportability, which is defined as the ability of having the right maintenance support in the necessary place to perform the required maintenance activity at a given instant of time, or during a given time interval (SS-EN 13306: 2014). Resources include human resources, support equipment, materials and spare parts, maintenance facilities, documentation and information, and maintenance information systems.

18.4 USE AND IMPLEMENTATION

The real developments in maintenance and reliability area started around World War II in the military field. After that, it was the military and space industry that stood for most of the developments in this area. Nowadays, there is also a lot of commercial work performed, mainly in high-risk industries such as nuclear power production, aviation, offshore, and the process industry. These industries mainly use some kind of risk-based maintenance such as RCM, RBM, RBI, RAMI, RAMS, or RAMSI. In more safety oriented concepts, such as RAMS and RAMSI there is more frequent use of so-called risk analytic tools and methods, for example, preliminary hazard analysis, fault tree analysis, event tree analysis, failure mode and effect analysis, and reliability block diagram simulation (Brown and Porton, 2013; Tiusanen et al., 2011). According to Cotaina et al. (2000), due to many tragic chemical accidents, these companies must follow a well-established set of laws and rules to enforce security and reliability. These tools and methods, approved by OSHA (included, but limited to), include all of the above-mentioned tools and methods. They are included in the RCM process used in chemical companies (Cotaina et al., 2000).

In discrete production, especially in the automotive industry, concepts such as TPM, Value Driven Maintenance (VDM), and Lean Maintenance are widely used. Often, there is a combination of the TPM and the RCM concept. Usually TPM is implemented before RCM. In the TPM concept, small defects are fixed along with debugging of the equipment, that is, the equipment is brought back to its original condition. After that, RCM can be applied in order to decide the right maintenance task and its appropriated interval. Put the other way around, this will result in a maintenance program dealing with a lot of small defects and short intervals (Scontrino, 1995). In this sense, RCM can be used to fine-tune TPM by an integration of the two concepts as proposed in Lean Maintenance (Smith and Hawkins, 2004).

In short, in RCM, you determine what maintenance that you really need to do, that is, you try to tailor-make a unique maintenance program for your current operating context. In TPM, you want to do this as cheaply and effectively as possible (ibid.). Moreover, in RCM you try to reduce unnecessary maintenance as much as possible, especially time-based preventive maintenance. Reduction of unnecessary maintenance will reduce the risk of introducing failures, for example, by human errors during maintenance. According to Reason (1998), is it easy to introduce failure during maintenance, especially when you have some kind of shift changes or different personnel who reassemble and assemble a system. Reason has an example of a bolt with eight nuts. According to Reason, there is only one way to take it apart but over 40 000 combinations to assemble it in the wrong order, if every nut, say nuts A to H, is to be in its right place, see figure 18.6. This situation, and also the situation with production personnel, for example, not highly maintenance trained operators, who will be carrying out more and more maintenance, can easily lead to introduction of failures, when the purpose is to prevent them.

Ways to deal with this problem, with the exception of, as in RCM, reduction of unnecessary maintenance are, for example, education, training, good instruction and design for optimal maintainability, inspectability, as well as poka-yoke design (Yang and El-Haik, 2008). A trained user can reduce the probability of failure.

Figure 18.6. The bolt-and-nuts example. *Source*: Adapted from Reason 1998.

Operators trained in TPM can reduce the level of production disturbances by up to 85% Nakajima (1989).

Per se, the dependability elements in Figure 18.3, for example, maintainability and inspectability, are the backbone in production operator maintenance in TPM. Operator maintenance, also called autonomous maintenance, is divided into seven steps where cleaning, inspections, and lubrications (CIL) are the activities focused upon in the first steps. The purpose of these steps is to discover the "minor" defects that trigger breakdowns and to correct them. If these defects are not corrected then failure will soon be a fact. Higher up the autonomous maintenance ladder, operators are trained to do repair work and quality assurance of important parameters in the equipment.

However, these activities demand high maintainability of the system and include a lot of redesign of the "standard" equipment. At the end of the traditional 12-step implementation ladder, see for example Nakajima (1988), this knowledge is used in, for example, major changes of the production system. Then, of course, the companies wish to buy equipment with good maintainability, that is, "TPM equipment," which is ready-built so that it is easy to use and maintain.

Inspection is a factor focused upon in both the safety area and the maintenance area. Both TPM and RCM stress the importance of inspections. The operator maintenance in TPM has already been mentioned, but in RCM this is called "failure finding task," and potential failure intervals are used to find "optimal" inspection intervals. Outside these concepts, authors deal with residual useful life (RUL), judgment of inspections, and maintenance actions.

To sum up these maintenance ideas and concepts, the development stems from different industries and cultures. TPM, VDM, and Lean Maintenance have their origins in the Japanese car industry, where tools and methods such as operator maintenance, improvement teams, company-wide cooperation, early equipment management, and maintenance prevention design were developed. In high-risk industries in the western world, there has been a development of more risk-based maintenance, with concepts such as RCM, RMB, RBI, and RAMSI, but there is an overspill between ideas from discrete production to process industry and vice versa.

Whatever the concept or types of maintenance, implementations are often time- and resource-consuming. In fact, most companies fail in their implementation attempts. The reported cases are often the ones who have been successful. Or, as these authors express it, implementation of improvement programs, such as TQM, JIT, DFM, and lean manufacturing have proven successful for some companies, but not for the majority (Hayes and Pisano, 1994; Oishi, 1989).

Ravishankar et al. (1992) reported from a TPM case in the semiconductor industry that training is a fundamental aspect of an improvement program, and can be the difference between success and failure that is often overlooked. Their improvement teams were trained in effective meetings and in data driven and decision-making problem solving tools. Team dynamics were stressed as being much more important than producing output (ibid.). Wireman (1992) also point out that: If management is not willing to invest in this type of training, DO NOT begin a TPM program; it will only end in failure.

In high-risk industries the many regulations and government demands force implementation. In discrete industry, these forces are not so obvious. Here the competitive factor is the main force behind the implementation of maintenance concepts.

Digitalization in society as well as in industry will raise the demand for maintenance, reliability, and safety areas. For example, Germany has started to implement "Industrie 4.0," known in the United States as "Industrial Internet Consortium." These production systems of the future will entail production systems that are highly automated, efficient, and flexible. One of the challenges in achieving these goals is to be able to handle new technology and make dependability and safety a built-in feature of the systems. This puts pressure on the production development process to use analytic tools and methods such as those included in the RAMSI concept.

18.5 EMPIRICAL RESEARCH ON USE AND EFFICIENCY

The realization of the "Internet of Things" will require effective maintenance organizations working with highly sophisticated engineering tools and methods. However, little is known about their actual extent of use in industry (Bokrantz et al., 2016). Moreover, a survey study of more than 70 Swedish manufacturing companies shows a potential for increasing the use of engineering tools and methods in both the operational and the design and development phase (ibid.). Worth noting in this study was that almost none of the participating companies used specific software to conduct risk or reliability analyses, despite several of the companies being big multinational companies within such industries as automotive (both car and truck industry), the chemical process industry, the steel industry, etc.

Generally in industry, there has been little integration between safety and maintenance areas. Both areas have been developed separately and performed by different departments in the companies. The authors and others believe that in future production systems there will need to be more integration between different organizational functions as proposed in, for example, TPM, where engineering and maintenance work together through early equipment management and maintenance prevention (Gotoh, 1991), and between operations, quality, and maintenance through autonomous maintenance and focused improvements of OEE (Nakajima, 1988). For example, the safety, health, and environment goals in the TPM concept are very ambitious, that is, zero accident, zero health damage, and zero fires (Venkatesh, 2007). The other major maintenance concept, RCM, is more oriented toward risk-based maintenance than TPM. However, the use of safety tools in both concepts are limited as are safety analyses.

Despite the theoretical developments in the maintenance, reliability, and safety areas with different maintenance generations and concept, in the industry they still spend more time on reactive repairs compared to the time they spend on preventive maintenance (Ylipää et al., 2016). In this study, based on overall equipment effectiveness (OEE) data from 98 different companies collected from over 50 industrial workshops during 2006 to 2012, the unplanned stops were 9.6% and planned stops 6.3%. The average OEE value for these 98 companies was found to be 51.5%, which shows that the manufacturing industry suffers from a good many production disturbances (ibid.).

As several authors claimed, accidents tend to occur more often in abnormal, conflicting system states than in normal. For example, Barroso and Wilson found that the accidents occurring while the operator is involved in troubleshooting or disturbance control tasks varies between one-third and two-thirds of the total accidents analyzed (Barroso and Wilson, 1999). Therefore, the bottom line is that a production system with high dependability can be assumed to provide better safety and a higher added value for the system.

18.6 WEAKNESSES, LIMITATIONS, AND CRITICISM

The TPM company-wide improvement program demands high top management attention, but not all top managements are interested in maintenance. As mentioned before, implementation is not an easy or straightforward task, especially when it comes to cultural changes and concepts such as those upon which TPM is built. Even a slight change, as compared to implementation of the whole TPM concept, reported by Sandberg et al. (2015), in shifting from a reactive to a preventive mind-set was a hard struggle, took several years, and failed to really succeed in achieving the goal of more focus on preventive than corrective maintenance.

The weakness in the RCM concept is that it is time-consuming, at least when done by the book. There have been developed different light and streamlined versions of RCM that are less time consuming and also use of ABC classification, according to the Pareto principle, for selecting the most important equipment so that the RCM analysis is focused on the most critical aspects of the production system, such as bottleneck machinery. However, drawbacks have been reported by both this ways of handle the RCM analysis (Maintenance Technology, 2001; Gopalakrishnan, 2016).

18.7 RELATIONS TO OTHER SAFETY PRINCIPLES

- Inherent reliability will increase when the inherent safety increases, especially in passive inherent safety systems.
- The principle of simplification has connections to maintainability and inspectability.
- Defense-in-depth and safety barriers are ways to increase reliability, if they are used wisely.
- Redundancy can, for example, comprise two or more pumps on stand-by. In the Piper Alpha example described in the introduction of this chapter, redundancy in terms of two pumps was used. One pump can be running and the other can be under preventive maintenance or on stand-by ready to take over if the first one brakes down. There are also some other types of redundancy.
- Fail-safe design can be used to secure the reliability of equipment.
- Foolproof design can be used to prevent human errors during maintenance, for example, mitigate failures introduced to the equipment during reactive and preventive maintenance.
- Human error assessment and human reliability prediction are methods and techniques closely related to maintenance, maintainability, and inspectability. Several studies point to up to 90% of failures being related to human errors. In many industrial applications a combination of human and equipment failure can create a total system failure.
- In the interests of the safety culture of a company, it is important to progress from reactive or pathological steps to more proactive. As mentioned before, breakdown maintenance is much more hazardous than preventive or proactive work.

REFERENCES

Alsyouf, I. (2007). The role of maintenance in improving companies' productivity and profitability. *International Journal of Production Economics*, 105, 70–78.

Arunraj, N. S., and Maiti, J. (2007). Risk-based maintenance - techniques and applications. *Journal of Hazardous Materials*, 142, 653–661.

Barroso, P. M., and Wilson, R. J. (1999). HEDOMS—Human error and disturbance occurrence in manufacturing systems. *Human Factors and Ergonomics in Manufacturing,* 9(1), 87–104. Available at: http://onlinelibrary.wiley.com/doi/10.1002/(SICI)1520-6564(199924)9:1%3C87::AID-HFM5%3E3.0.CO;2-2/abstract?systemMessage=Wiley+Online+Library+will+be+unavailable+on+Saturday+01st+July+from+03.00-09.00+EDT+and+on+Sunday+2nd+July+03.00-06.00+EDT+for+essential+maintenance.++Apologies+for+the+inconvenience

Bevilacqua, M., and Braglia, M. (2000). The analytic hierarchy process applied to maintenance strategy selection. *Reliability Engineering & System Safety*, 70(1), 71–83. Available at: https://www.journals.elsevier.com/reliability-engineering-and-system-safety

Bokrantz, J., Skoogh, A., Ylipää, T., and Stahre, J. (2016). Handling of production disturbances in the manufacturing industry. *Journal of Manufacturing Technology Management*, 27(8), 1054–1075. Available at: https://doi.org/10.1108/JMTM-02-2016-0023

Brown, R., and Porton, M. (2013). The development of a methodology to allocate RAMI requirements to DEMO: The development of dtools & methodologies for DEMO in Europe. Paper presented at the 2013 IEEE 25th Symposium on Fusion Engineering (SOFE), 2013, San Francisco, CA.

Cotaina, N., Matios, F., Chabrol, J., Djeapragache, D., Prete, P., Carretro, J., and Pena, J. M. (2000). Study of existing RCM approaches used in different industries Madrid. Department de Arquitectura y Tecnologia de Computadores.

Daley, D. T. (2008). *The Little Black Book of Maintenance Excellence*. New York: Industrial Press.

Dhillon, B. S. (2008). *Mining Equipment Reliability, Maintainability, and Safety*. Berlin, London: Springer.

Dunn, S. (1998, May). Re-inventing the maintenance process-towards zero downtime. Paper presented at the Queensland Maintenance Conference.

EU-OSHA. (2010). Safe maintenance in practice. [Online]. European Agency for Safety and Health at Work. Available at: https://osha.europa.eu/en/tools-and-publications/publications/reports/safe-maintenance-TEWE10003ENC/view (Accessed February 16, 2016).

Gopalakrishnan, M. (2016). *Towards effective maintenance planning—Utilizing continuous awareness of the critical resource* (Licentiate thesis, Chalmers University of Technology, Göteborg, Sweden).

Gotoh, F. (1991). *Equipment Planning for TPM: Maintenance Prevention Design*. Portland, Oregon: Productivity Press, Inc.

Hayes, R. H., and Pisano, G. P. (1994). Beyond world class: The new manufacturing strategy. *Harvard Business Review*, 72(1), 77–84. Available at: https://hbr.org/1994/01/beyond-world-class-the-new-manufacturing-strategy

Lind, S., Nenonen, S., and Kivistö-Rahnasto, J. (2008). Safety risk assessment in industrial maintenance. *Journal of Quality in Maintenance Engineering*, 14(2), 205–217.

Ljungberg, Ö. (2000). (In Swedish) *TPM Vägen Till Ständiga Förbättringar*. Lund, Sweden: Studentlitteratur.

Mannan, S., and Lees, F. P. (2005). *Lees' Loss Prevention in the Process Industries*. Boston, Amsterdam: Elsevier Butterworth-Heinemann.

MIL-STD-470. (1966). Maintainability program requirements, US Department of Defense, Washington, DC.

Misra, K. B. (2008). *Handbook of Performability Engineering*. London, England: Springer-Verlag.

Mobley, R. K. (1990). *An Introduction to Predictive Maintenance*. New York: Van Nostrand Reinhold.

Mobley, R. K. (2004). *Maintenance Fundamentals*. Oxford: Butterword-Heinemann.

Moubray, J. (1997). *Reliability-Centered Maintenance*. Oxford: Butterworth-Heinemann Ltd.

Nakajima, S. (1988). *Introduction to TPM. Total Productive Maintenance*. Cambridge: Productivity Press, Inc.

Nakajima, S. (Ed.) (1989). *TPM Development Program: Implementing Total Productive Maintenance*. Cambridge: Productivity Press, Inc.

Oishi, O. (1989). Building a JIT pyramid. In: Society of Engineers International Conference, Detroit, Michigan, USA.

Patton, J. D. (1994). *Maintainability and Maintenance Management*. Research Triangle Park, NC: Instrumentation, Systems, and Automation Society.

Peng, K. (2000). The post-maintenance era of complex equipment management in the semiconductor industry—The case of Intel Corporation (PhD thesis, Golden Gate University).

Pintelon, L., and Muchiri, P. N. (2009). Safety and Maintenance. In: M. Ben-Daya, S. O. Duffuaa, A. Raouf, J. Knezevic, and D. Ait-Kadi (Eds.), *Handbook of Maintenance Management and Engineering* (pp. 613–648). London, Dordrecht, Heidelberg: Springer.

Pintelon, L., and Parodi-Herz, A. (2008). *Maintenance: An Evolutionary Perspective Complex System Maintenance Handbook* (pp. 21–48). London: Springer. doi:10.1007/978-1-84800-011-7_2. Available at: https://link.springer.com/chapter/10.1007%2F978-1-84800-011-7_2

Qingfeng, W., Wenbin, L., Xin, Z., Jianfeng, Y., and Qingbin, Y. (2011). Development and application of equipment maintenance and safety integrity management system. *Journal of Loss Prevention in the Process Industries*, 24, 321–332.

Ravishankar, G., Burczak, C., DeVore, R. (1992). Competitive manufacturing through total productive maintenance. In: IEEE/SEMI Int'l Semiconductor Manufacturing Science Symposium.

Reason, J. (1990). *Human Error*. Cambridge, England: Cambridge University Press.

Reason, J. (1998). *Managing the Risks of Organizational Accidents*. Hampshire, England: Ashgate Publishing Limited.

Reason, J., and A. Hobbs (2003). *Managing Maintenance Error—A Practical Guide*. Hampshire, England: Ashgate Publishing Company.

Retterer, B. L., and Kowalski, R. A. (1984). Maintainability: A historical perspective. *IEEE Transactions on Reliability*, R-33 (1). Available at: http://ieeexplore.ieee.org/stamp/stamp.jsp?arnumber=6448275

Sandberg, U., Ylipää, T., Skoogh, A., Isacsson, M., Stieger, J., Wall, H., Andersson, M., Johansson, H., Nilsson, I., Agardtsson, J., Vikström, S., and Nyström, M. (2014). Working with forces promoting or hindering implementation of strategies for maintenance—Experiences from Swedish Industry. In: The Sixth Swedish Production Symposium, Göteborg, Sweden, September 16–18, 2014. Available at: http://publications.lib.chalmers.se/records/fulltext/210299/local_210299.pdf

Scontrino, M. P. (1995). *TPM in Process Industries* (Vol. 48). Durham: Wiley Subscription Services, Inc.

Smith, R., and Hawkins, B. (2004). *Lean Maintenance: Reduce Costs, Improve Quality, and Increase Market Share*. New York: Elsevier Butterwort-Heinemann.

SS-EN 13306. (2010). Maintenance terminology. European standard. CEN (European Committee for standardization), Brussels, Belgium. Available at: https://enav.sis.se/sv/Standard/?std=STD-74825

SS-EN 60300-1. (2014). Dependability management—Part 1: Guidance for management and application. Available at: https://enav.sis.se/sv/Standard/?std=STD-3336435

Tiusanen, R., Jännes, J., and Liyanage, J. P. (2011). From-design-to operations risk mitigation in Nordic wind energy assets: A systematic RAMS+I management model. In: Proceedings of the Sixth World Congress on Engineering Asset Management 2011.

Tiusanen, R., Jännes, J., and Liyanage, J. P. (2012). RAMSI management model and evaluation criteria for Nordic offshore wind asset. VTT Technical Research Centre of Finland.

Toulouse, G. (2002). Accident risks in disturbance recovery in an automated batch-production system. *Human Factors and Ergonomics in Manufacturing*, 12(4), 383–406.

Venkatesh J. (2007). An Introduction to Total Productive Maintenance (TPM) [Online]. Available at: http://www.plant-maintenance.com/articles/tpm_intro.shtml

Wireman, T. (2005). *Developing Performance Indicators for Managing Maintenance* (2nd ed.). Industrial Press.

Wireman, T. (1992). *Inspection and Training for TPM*. New York, Industrial Press, cop.

Yang, K., and El-Haik, B. S. (2008). *Design for Six Sigma*. McGraw Hill Professional.

Ylipää, T., Gopalakrishnan, M., Bokrantz, J., and Skoogh, A. (2016). Identification of maintenance improvement potential using OEE assessment. *International Journal of Productivity and Performance Management*, 66(1), 126–143. Available at: https://doi.org/10.1108/IJPPM-01-2016-0028

FURTHER READING

Hollnagel, E. (2005). Human reliability assessment in context. *Nuclear Engineering and Technology*, 37(2).

Part IV

OPTIMIZATION

19

ON THE RISK-INFORMED REGULATION FOR THE SAFETY AGAINST EXTERNAL HAZARDS

Pieter van Gelder

19.1 INTRODUCTION

Countries are becoming more densely populated, not only in terms of population size, but also in terms of industrial activities, which include the use and the transport of hazardous substances and the storage of large amounts of energy, and in terms of critical infrastructures, which include ICT networks, energy and transport networks, and flood defenses. Many risks are associated with this for which optimal regulation on local, regional, and (inter)state level is needed.

Our society is safer than ever, as can immediately be seen from the increase of our life expectancy. The average lifetime nowadays in the Netherlands is 78 years and still increasing. Some 200 years ago, life expectancy was only 40 years. Better food, clean drinking water, and healthcare, and also better construction of buildings and sewerage systems have contributed significantly to the increase of life expectancy.

The Netherlands has a population of 16.8 million people. In 2012, a total of 140,000 people died. Hence, the overall probability of dying in the Netherlands is 8.3×10^{-3} per year; in other words, 83 out of 10,000 people die each year. The

Handbook of Safety Principles, First Edition. Edited by Niklas Möller, Sven Ove Hansson, Jan-Erik Holmberg, and Carl Rollenhagen.

TABLE 19.1. Reduction of Life Expectancy Due to Increase in Exposed Risk

Extra Probability of Decease in a Year	Expected Age	Decrease of the Life Span in Years	Decrease of the Expected Life Span in Days
10^{-3}	74.97	3.16	1153
10^{-4}	77.81	0.32	117
10^{-5}	78.10	0.03	11
10^{-6}	78.13	0.00	1
10^{-7}	78.13	0.00	0

majority of these 83 people die of natural causes. Only 3.3 of these 83 people die of non-natural causes, which amount to about 6000 casualties per year.

Of these 6000 casualties, 2800 took place by accidents at home, 1600 suicide, 650 traffic, 150 murder and assault, 60 work-related, and 750 other. Delayed deaths because of asbestos and the preventable deaths in hospitals because of errors in surgery are left out in this overview.

Mortality rate also changes over time. The average mortality rate is, as given above, 8.3×10^{-3} per year. The lowest rates go as low as 10^{-4} per year, for children in between 4 and 16 years. If they could stay in that condition, they would become 10,000 years old. The influence of increasing the mortality rate with small probabilities, ranging from 10^{-3} per year (the probability of dying because of smoking) to 10^{-7} per year (the probability of dying after a bee sting), on the overall life expectancy is shown in Table 19.1.

The individual risk is the probability per year that a person residing continuously and unprotected at a location, dies as a result of an incident at that location.

It is very difficult to determine what is an acceptable individual risk, because every death is still one too many. Nevertheless, in order to arrive at a norm, it is examined what the probability is that a random person, under "normal" circumstances, dies. The standard for the individual risk, in the Netherlands, is derived from the Dutch death rate (the number of people of a certain age group who die annually). Because the premise is that the death rate may be hardly affected by economic interests, it is determined that the individual risk must be within 1% of the lowest death rate, that of 4–16 year olds. This results in a chance of one in a million per year. Within the area with risk levels higher than one in a million, no new occupancy is allowed. However, because this standard would cause major problems in existing situations, one has since lowered the standard to 1 in 100,000 years.

According to Table 19.1, this corresponds to a decrease of expected lifetime of 1 day for the 10^{-6} level and 11 days for the 10^{-5} level. Although this may seem not very much, we have to note that not just one life is involved in a disaster, but usually a large number. Therefore also criteria for group risk have been formulated by the Ministry of Infrastructure and Environment. The group risk is described by the probability that a large group of people might get killed due to a hazardous event, and modeled by a so-called FN curve, which denotes the probability that at least n people lose their lives $P(N > n)$.

The need for a group risk criterion, apart from an individual risk criterion, can be well illustrated in a following case, related to the risk of an avalanche that is transverse to a descent. Passing the 50 m avalanche track is possible in less than 3 s, resulting in a negligible individual risk. However, at a constant flow of skiers, there are at least 10 skiers entrained as the avalanche slides off. For the elevator company, the group risk is too large. The skiers find it excessive given the small individual risk. The question which risk criterion should be applied, in other words "which safety level is safe enough?," is a very important research question in safety sciences. And a very difficult one for that matter, since it also involves moral considerations.

Van Dantzig (1960) showed in the 1950s that the problem of the acceptable level of risk can be formulated as an economic decision problem (Figure 19.1). The expenditure for a safer system is equated with the gain made by the decreasing present value of the risk. In this way, optimal risk criteria could be derived for the flood protection system of the Netherlands, resulting in safety levels from once in 10,000 years for the coastal-flood-prone areas (shown by the dark grey areas in Figure 19.2) to once in 1250 years for the river-flood-prone areas (shown by the light grey areas in Figure 19.2).

Risk-informed regulation in nuclear engineering started in the 1990s with probabilistic risk assessment (PRA) principles. In the early 2000s, the risk-informed regulation came into force (IAEA 2005; USNRC 2002). IAEA established regulatory requirements for reactor, materials, and waste applications to ensure that "no

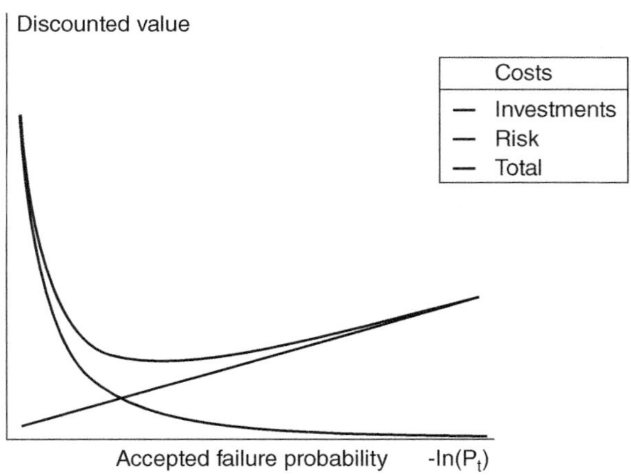

Figure 19.1. The economic decision problem concerning safety.

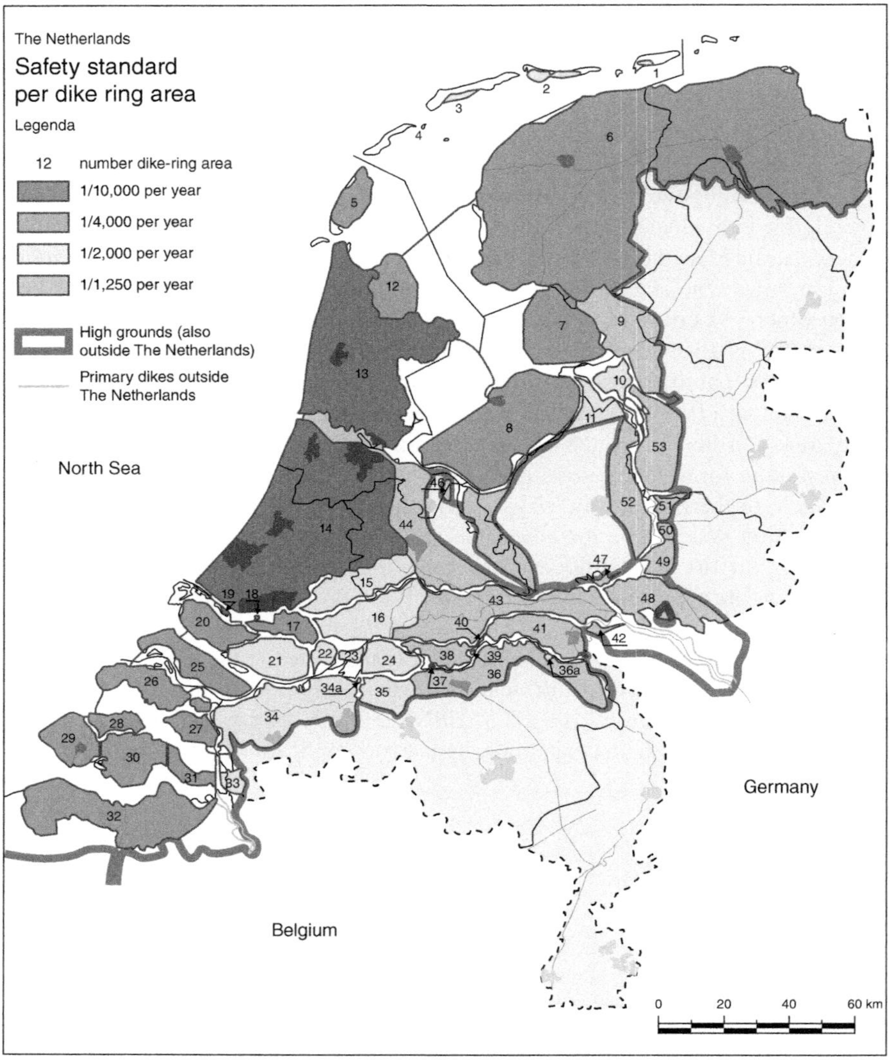

Figure 19.2. Safety standard against flooding in the Netherlands.

undue risk to public health and safety" results from licensed uses of facilities and materials.

A recent overview of risk-informed regulation in the chemical—and oil and gas sector is given by Pasman (2015), and risk-informed regulation for critical infrastructures, in general, is described by Stapelberg (2008).

19.2 RISK-REGULATION IN SAFETY AGAINST ENVIRONMENTAL RISKS

There has been in recent years a development of a deterministic calculation of embankment height (highest observed water level + 1 meter) to an exceedance probability calculation (height of dam such that the crest height is exceeded by the water level once in 10,000 years) to a failure probability calculation (all failure mechanisms are included; not only overflow caused by exceedance of water levels) and since 2014, a transition to a death probability calculation (strength of the embankment such that the risk of drowning behind the dike is less than 10^{-5} per year). In this sense, the standardization is in line with the individual risk safety standard, which indicates that nowhere in the Netherlands someone may be at greater risk than 10^{-6} per year and for existing situations, a temporary standard of 10^{-5} per year. This would mean that the safety of the dike is still to become a factor of 10 more stringent over the years.

We thus need computational models which describe the tails of the distributions up to odds of 10^{-5}–10^{-6} per year. A dike of 5 meters + NAP at Hoek van Holland has a chance of being exceeded on average once every 10,000 years. However, there is a factor of 10 uncertainty around it. The chance could also be once in 1000 years, or the other way once in around 100,000 years. The reason for this uncertainty is the enormous extrapolation from the relatively limited data set (about 120 years of water-level data are available along the Dutch coast and rivers). This results in wide confidence intervals around the mean estimate.

When calculating the probability of failure, the same issues play a role, but also the uncertainties in the strength variables come into play. A dike is more than just a height; also the breadth, soil parameters, presence of sheet pile walls, slope protection, etc. should be considered. The width of the uncertainty interval therefore further increases.

For the death probability calculation, even further increase of the uncertainty takes place by taking into account the uncertainties in the flood propagation, uncertainties in evacuation capabilities, and uncertainties in self-reliance of people. Especially, the latter factor is very uncertain. So even larger confidence intervals will emerge.

So very many uncertainties, in which the government must decide about the protection levels for the people. The consequences of a wrong decision are huge (>10% of GNP if dike-ring 14 would be flooded), and therefore it is better to take protective measures on the safe side of the uncertainty intervals, but the question is how safe. Let us look at the cost–benefit analysis of the Delta Committee (1960). Investment costs for higher dikes can be weighed against the risk (probability x consequence) of the area to be protected. This approach has led to a safety standard. In the 1960s, a failure probability of 10^{-4} per year resulted. If we multiply this by the conditional probability of drowning given a flood, often taken equal to 1%, then an allowable chance of drowning of 10^{-6} per year appears. Since 1960, the economic value and the demographics changed dramatically; especially if we think of the cascading effects that we have created in our modern society. This would mean that the allowable drowning probability should even be lower than 10^{-6} per year, which is a more stringent value than is presented by the current Delta Commission. Group risk has not been accounted yet in this argumentation. Floods are, after all, hazards that may simultaneously cause large numbers of casualties. We speak of dependent failure. If the house of your neighbor is flooded, it is very likely that your home is also under water. The same reason can be used to argue that insurance against floods is not desired. The government should itself organize the safety against flooding and not to pass on to insurance companies, which will bring to it high premiums in charge.

19.3 DEALING WITH UNCERTAINTIES IN RISK-INFORMED REGULATION

A rational way to deal with uncertainty in a decision model is to integrate this uncertainty, according to a Bayesian approach. If the uncertainty is represented by the variable б and its probability density function f(б), then integrating over б will result in a б-independent outcome function $F(X)$ where $F(X) = \int F(X \mid б) f(б) dб$.

Figure 19.3 shows the influence of intrinsic uncertainty $б_1$, statistical uncertainty $б_2$, and model uncertainty $б_3$ on the tails of the crest height $F(X)$ of the IJsselmeer

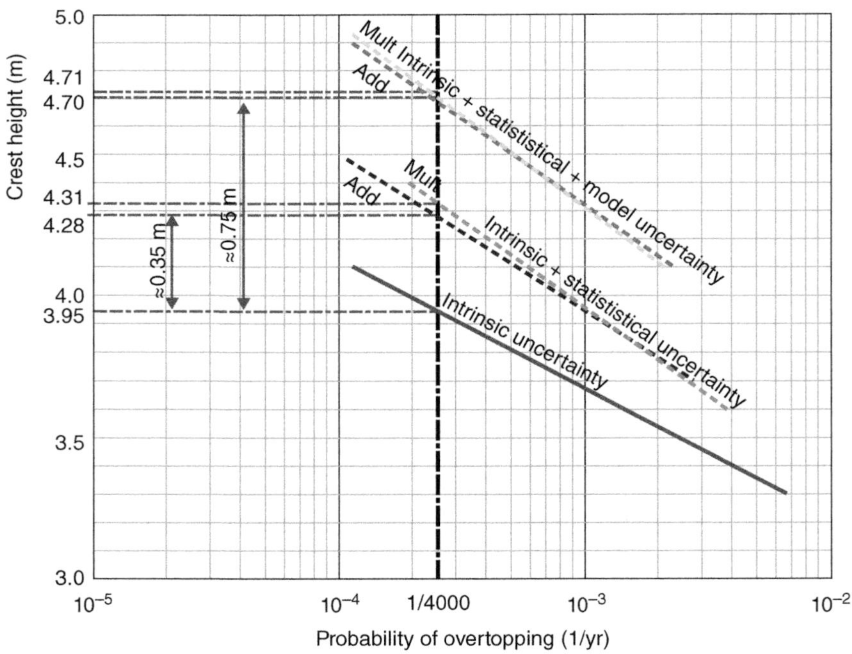

Figure 19.3. Tails of water level distributions.

Figure 19.4. Pressure vessel in steel.

dikes (Vrijling et al., 1999). At a fixed failure rate of 1/4000 per year, we see an increase of up to 75 cm.

The approach in regulation, based on integrating uncertainties, can be very well illustrated in the following setting, adapted from Vrouwenvelder and Vrijling (2000). Let us assume that there are 10,000 high pressure vessels (Figure 19.4) in the Netherlands containing liquefied petroleum gas (LPG, a mixture of propane C3 and butane C4), of which 10% might have a weak spot in the relief valve. The other 90% of the tanks are considered safe.

A random owner of the LPG tank does not know if he has a safe or unsafe tank. He can perform nondestructive testing, such as ultrasonic testing, radiography, and pressure tests, usually involving water, also known as a hydro test, to his tank. If the test shows that the relief valve is unsafe, it can be replaced, resulting in a safe tank again. A rightly cautious decision has been made to inspect the tank.

If the test on another tank does not detect the weak spot, no replacements have to be made, and retrospectively, this second tank had been judged, unfairly, as unsafe. The intrinsic safety of the tank has not been changed, by the inspection.

The above discussion of the safety of the tank is based on the intrinsic property of a high-pressure tank, but this is only known after the nondestructive inspection has taken place. In retrospect, after the realization of the inspection has "removed" the uncertainty, the situation has become simplified. Before that, both situations were less safe, even though one tank was intrinsically unsafe and the other one was intrinsically safe. The knowledge uncertainty (does my tank have a weak spot?) cannot be disregarded. It is an integral part of the safety consideration and of the related decisions.

The rational way to take away the knowledge uncertainty in the safety consideration is determining the average probability of failure. If the probability of failure of the weak tank equals 10^{-2} per year and the probability of failure of the strong tank equals 10^{-6} per year, then the probability of failure of a random tank (for which it is unknown yet which type it is), is equal to $0.1 \times 10^{-2} + 0.9 \times 10^{-6} \sim 10^{-3}$ per year.

This procedure is quite similar to the above presented approach of integrating the (knowledge) uncertainty. One can speak of the average probability of failure: the intrinsic failure rates have been averaged over their chances of occurrence. The result can be described as "the best estimate of the probability of failure." Yet, for many tank owners, this approach may feel unsatisfactory, perhaps because a frequentist interpretation is not possible. The idea remains that the tank has a failure rate of 10^{-2} or 10^{-6}, but never 10^{-3}. In itself, the question of the value of the probability of failure in the above situation is actually less interesting. A probability is an intermediate result and not an end result. What matters is which decision is ultimately taken.

Suppose the tank owner is given a choice to (19.1) directly inspect the tank, repair if necessary, and then use the tank in operation for 1 year, and (19.2) use the tank in operation for 1 year (without inspection). Alternative (19.1) seems the wisest, but there are costs involved for an investigation, equal to I. The costs of possible failure are equal to S. The tank owner now faces a choice problem. The cost balancing of the two alternatives is as follows:

$$I + 10^{-6}\, S \tag{19.1}$$

$$10^{-3}\, S \tag{19.2}$$

Depending on the values of I and S, an optimal choice can be made.

The tank has a failure rate of 10^{-2} or 10^{-6}, but never 10^{-3}. Similar as an accident happens or it does not happen. But, there is no 10^{-6} accident. Therefore, the question whether "average probability" should be regarded as a "feature of the tank" is not relevant. It is also worth to realize that the problem of interpretation of the individual plant owner is completely different if we put us in the position of the government. For the government, the reality is a population with 10% bad high pressure tanks. If $N = 10{,}000$ tanks, then there are $1000 \times 10^{-2} = 10$ explosions to be expected every year.

19.4 LIMITATIONS OF THE CURRENT RISK MEASURES

Risk is referred to as the probability of an accident associated with the consequences. The effects of a disaster are strongly dependent on the type of accident. Too often, schematization of the consequences to one dimension takes place. Accidents in the chemical industry with emissions of toxic gases focused in the past usually on the number of casualties and not so much on the damage, because the latter in these cases was probably subordinate. In considering the damage caused by flooding, much attention was given to the material damage and much less to the number of deaths. Material damage appeared in greater or lesser degree decisive for the choice of the most economically acceptable flooding probability. The Delta Committee included in their economic considerations a valuation of 100,000 Dutch Guilders per human life (about 50,000 Euro) and showed that this would result in an increase of the dike with only 3 cm (Delta Report Part 3, 1960). Subsequent policymakers paid therefore little attention to the multivariate character of the consequences.

In general, however, the damage of an accident is very diverse (following Ravenzwaaij, 1994):

(I) Effects on People
Direct death
Direct wounded, sick, disabled
Delayed deaths as a result of short exposure
Delayed wounded as a result of short exposure
Delayed deaths as a result of prolonged exposure
Delayed wounded as a result of prolonged exposure
Evacuation
Socio-psychological effects

(II) Effects on Goods and Territories
Damage to capital
Damage to infrastructure (roads, water, gas, etc.)
Indirect damage: loss of income
Multiplier effects
Loss of confidence

(III) Effects on Plants and Animals
Damage to ecosystems

In order to achieve a view of the extent of the damage in the event of an accident, the total effect is usually outlined in one dimension. In one case, the damage is expressed in monetary terms and in other cases, as the number of deaths. The advantage of this schematization is the simplicity of the quantitative analysis. Of course, the disadvantage is the lack of representation of the overall effect of a disaster. Easily, there may arise differences between the model-based quantitative approach to the acceptability of the risk and the actual social judgment. A good compromise is to indicate the impact on at least the above specified three dimensions; numbers of deaths, damage (euros), and the environment (to be determined measure).

The individual risk and the group risk are common quantities in describing risks in terms of human lives only. However, both of these descriptions should also be expressed for the two other dimensions. For the material damage, the FN curve would then be converted to an FD curve in which the damage D along the horizontal axis is measured in euros.

There is another aspect in which the perception of the consequence of an accident can differ from the usual FN curve measure. The consequences of an accident are limited to the number of external casualties c.q. local residents. Employees, users, and visitors will not be counted in the construction of the FN curve. For conventional plants with hazardous substances, this might not be a large problem, but in the FN curve for airports such as Schiphol, it is indicated that approximately 50 people at the site of impact will die because of a crash of a plane. In reality, the consequences will be much larger because apart from local residents at the site of impact, also the passengers and crew will perish. Although these two categories are not linked to the issues of spatial planning, the difference between the considered number 50 and the actual number of victims of 200–300 is enormous. The motorists who are stuck in traffic on the motorway along the airport Schiphol are also not included in the risk calculations, although they can be counted to the spatial planning domain. Another example is a temporary populated, by a large crowd of people, sports facility like a swimming pool or a stadium near the hazard source. The additional risk of such a number of temporary people present should be included in an FN curve calculation.

19.5 SPATIAL RISK

Another gap in the current method of determining the risk is the limitation to one single plant or one kilometer transport axis. While each of the hazard sources themselves comply with the requirements, the overall threat of these sources together becomes unacceptable. It seems wise for a city or region to determine the sum of the risks to get the overall risk of an activity. This should eventually result in a regional or national standard, instead of a site-related standard. Risk accumulation over space according to $P(\text{A or B}) = P(\text{A}) + P(\text{B})$ if A and B are mutually exclusive is relatively straightforward. Difficulties will occur in case of dependency or cascading effects.

In case of dependent effects, the correlation between events should be assessed. If events are positively correlated, say with correlation coefficient ϱ, the uncertainty in the accumulation will be underestimated when just summing up the individual standard deviations, by the amount of $2\varrho\sigma_1\sigma_2$, where σ_1 stands for the standard deviation of the first event and σ_2 the standard deviation of the second event. Negative correlations would be beneficial in this case; exceedances to the right would then be compensated by exceedances to the left.

Probabilities of dependent events should be determined based on conditional probabilities, $P(\text{A or B}) = P(\text{A}) + P(\text{B}) - P(\text{A}) \times P(\text{B} \mid \text{A})$ where $P(\text{B} \mid \text{A})$ represents the probability of the event B given that the event A has occurred. Bayes theorem will enable to determine $P(\text{B} \mid \text{A}) = P(\text{A} \mid \text{B}) \times P(\text{A})/\text{P}(\text{B})$. Bayesian networks are applied to propagate incidents as (partial) cascading effects through networks.

Let us consider two locations in a spatial environment, which are described by an uncertain capacity R. We use the notation R_1 to model the capacity or resistance

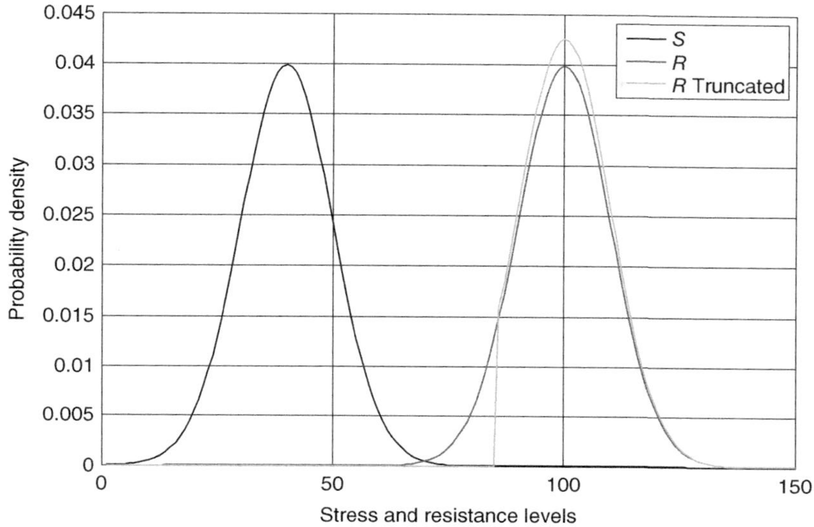

Figure 19.5. Probability density functions for stress S and resistance R, and truncated resistance.

at the first location R_1 and R_2 for the second location. The locations are at risk if the environmental stress S at these locations exceeds the capacities R_1 and/or R_2. Stresses and capacities are uncertain and should therefore be modeled as probability density functions, as shown in Figure 19.5. The uncertainty in a capacity is reduced if we observe a survival of an extreme stress level. If a structure, system, or process survives a stress level with an intensity of R_0, then the resistance distribution of that system can be left truncated at R_0 (Figure 19.5). The actual strength is stronger than R_0. The uncertainty in the resistance distribution has reduced (smaller standard deviation) and the mean value of the resistance distribution has increased.

Let us introduce the notation R'' for the updated variable $\{R \mid R > R_0\}$.

We will start from the U''_R value and calculate the corresponding U_S-value. Given U''_R, the value of the complement of the Probability Distribution Function for R'' is given by:

$$P(R'' > r) = 1 - F''_R(r) = \Phi\left(-U''_R\right)$$

$$\begin{aligned} \text{As } P(R'' > r) &= P(R > r \mid R > R_o) \\ &= P(R > r \text{ and } R > R_o)/P(R > R_o) \\ &= P(R > r)/P(R > R_o) \text{ for } r > R_o \end{aligned}$$

we have

$$R'' = \mu_R + U_R\sigma_R$$

with U_R following from $\Phi(-U_R) = P(R'' > r)\ P(R > R_o)$.

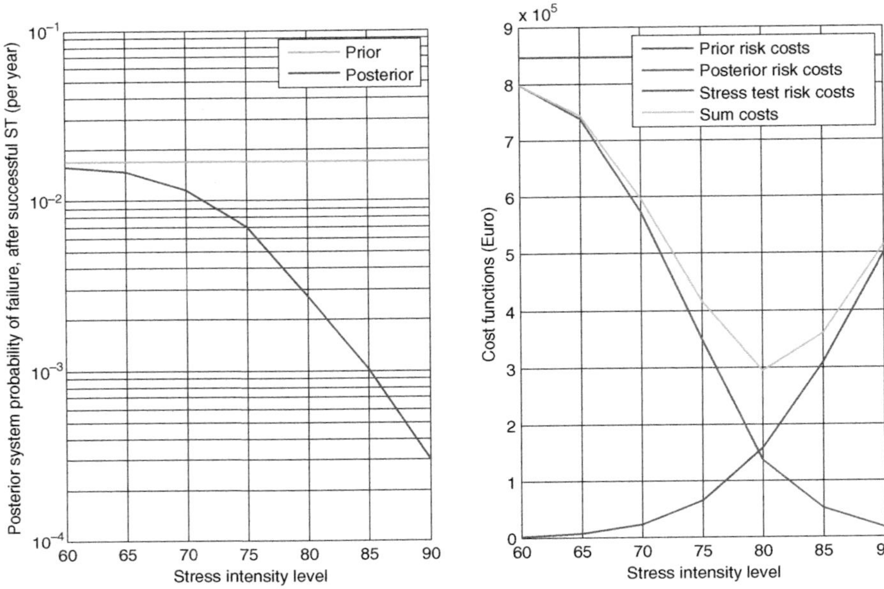

Figure 19.6. Risk reduction curves after observing a range of stress intensity levels.

Note that the mean and standard deviation of the posterior will not be the prior values. The posterior distribution of the resistance U_R is not a standard normal distributed variable anymore.

Figure 19.6 (left) shows the posterior probability (dark grey line) after successfully observing a positive stress test (ST) without failure. The horizontal light grey line is the prior probability of failure (prior to the ST). Assuming the costs involved with conducting a ST, as well as the costs after the failure of a ST, an optimization can be made with respect to the stress intensity level (Figure 19.8, right). The minimum of the total costs function leads to the optimal stress intensity level (all based on hypothetical data).

Let us assume a strong spatial correlation structure for the capacities at locations 1 and 2, with $\varrho = 0.9$. The contour lines of equal probability after a successful ST at location 1 are shown in Figure 19.7. Marginalizing the joint probability density function over component 1 results in Figure 19.8 (left) and over component 2 results in Figure 19.8 (right). The posterior of component 1 typically shows the discontinuous truncated behavior, whereas component 2 shows a continuous skewed distribution to the right, after having observed a result of "its neighbor."

The above "toy example" calculations show how risks in a spatial geographical area can be mathematically described, and how the observations of extreme stress levels and possible survivals or failures can be used to update the risk levels.

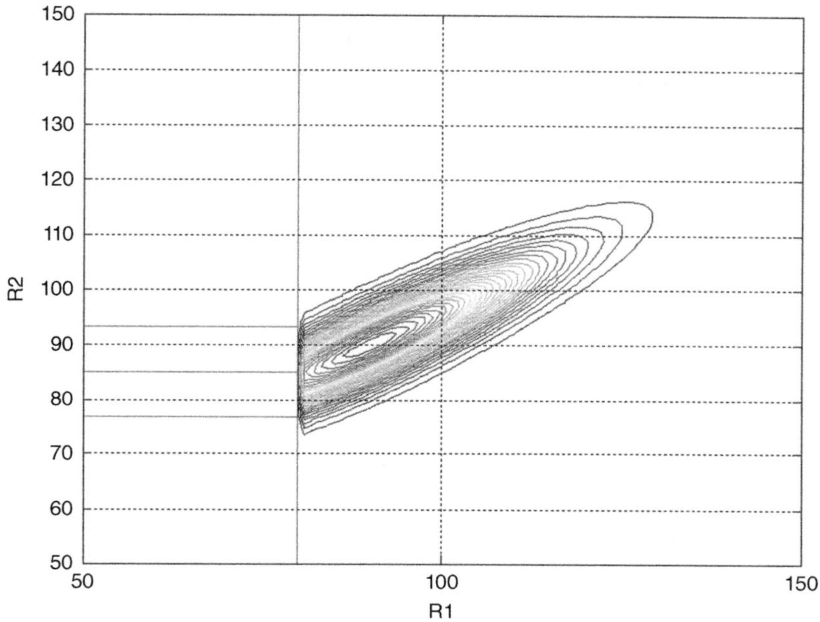

Figure 19.7. Contour lines of equal probability to model spatial risk.

19.6 TEMPORAL RISK

Individual risk (IR) refers to risk which is randomly distributed over a period of time and more or less constant during this time period. When we have a risk which is only present on, let us say, 1 day during a year, and the other 364 days the risk is completely

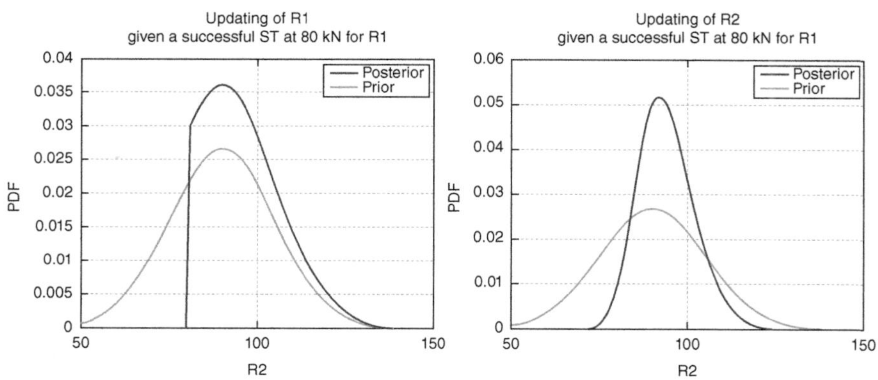

Figure 19.8. Bayesian updating of the spatial risks after observing extreme stress levels.

absent, can we say that an IR of 365×10^{-6}/year during that one day (3.65×10^{-4}) is acceptable, because the average still leads at 10^{-6} per year over a year? If we divide the year in seconds (31×10^{6}), we can see that this argument is invalid.

However, in the construction work of large hydraulic structures in the Oosterschelde (the Netherlands), the designers counted with an elevated probability of failure of 10^{-2}–10^{-3} per year during the construction phases, as repair materials and equipment were at hand. Loss of lives barely played a role. Starting from a permitted level of 10^{-6} per year, the failure opportunity for 100 days (the typical duration of a dangerous construction activity) would be 3.65×10^{-2}, corresponding with the rates above.

Of course it is true that the mortality rate does not rise much in a year by a short-term high-risk, but this is a dangerous argument if everyone applies it.

A crane accident in Alphen aan den Rijn (the Netherlands) in August 2015 occurred. During the construction work on a bridge, two cranes had to perform some dangerous activities during a time period of a few hours. There was a danger that the cranes would collapse on a populated area, which actually occurred (with surprisingly no human casualties). The question was which risk level would have been acceptable during this short period of time. A pure cost–benefit analysis can provide partial insight. Suppose the potential damage of the crane accident is estimated at 10 lives. Assuming a value of statistical human life (VSHL) of 10 million euros, the damage is 10^{7} euros. If the probability of an accident during the construction phase would be 10^{-2}, then the risk is 10^{5} euros. If preventive evacuation of the people in the risk-prone area would have been less than 10^{5} euros, the evacuation is advisable. Risk aversion factors can be applied to further refine the analysis, although the cost–benefit framework is in principle applicable. It may serve as a guideline to the question whether one should temporarily evacuate people in an inhabited area during hazardous construction work.

A similar discussion is also observed in the safety of dikes against flooding, but on a longer time scale. The probability of failure of a dike is the highest immediately after the commissioning of the dike. If the dike survives the first year, it has a proven strength and it will probably survive the rest of the years. In other words, the probability of failure, as time continues, decreases to 0 (until aging mechanisms come into play). This applies to situations of regulated water levels (when the uncertainty in the strength dominates the uncertainty in the load). For flood defenses with a non-regulated water level, the above is in general not true, because every year a new maximum load is "drawn" from the inherent uncertain load distribution, which dominates the uncertainty in the strength distribution.

The 10^{-4} standard of the Dutch Water Act can be interpreted as a requirement which applies separately for each year and which has to be met separately in each year. We can outline the following dilemma:

- Structure A has a failure rate in year 1 which is two times the norm and a failure rate of 0 in the 49 years ahead;
- Structure B has a failure rate for 50 years which equals the annual norm.

In total, therefore, the failure probability of structure A is equivalent to 2× norm and structure B has a failure probability which equals 50× norm. Structure B is allowed, whereas A is not in accordance, with a strict interpretation, of the Water Act.

The above result was considered by the Ministry of Water Management illogical. They chose, quite arbitrarily, for an averaging period of 10 years. The background is that the averaging period should be relatively small compared to a human lifetime. It seems better to conceive the number in the law for the individual risk as a maximum acceptable average at a permitted averaging period. However, a more fundamental basis for this choice is desired, and recommended for future research.

The first essential question is what the requirement $IR < 10^{-6}$ really means:

A. A death rate of 10^{-6} per year in which no count may be exceeded (similar to the speed limit for a car).
B. A summed probability over a year of 10^{-6}.
C. An average of 10^{-6} per year for an averaging period other than 1 year (1 month, 10 years, etc.).

Furthermore, there is a choice whether the requirement is for a permanently located hypothetical person at a fixed location or for any relevant person whose place and activity can change (real individual risk).

Consider for the purpose of the discussion, a person who is half of his time in an office and the other half of his time spending at home. As the home and office both yield a death rate of 10^{-6} per year during continuous presence, this person has a death rate of $0.5 \times 10^{-6} + 0.5 \times 10^{-6} = 10^{-6}$ per year. In short, the shorter presence at home and at the office is no reason to adjust the safety requirements for the home and office. Both must be 10^{-6} per year to achieve a correct total aggregate. So we have to be careful with reductions in norms through shorter residence times.

Compensation of safety levels might be allowed in case a (significantly) lower death rate in another time or activity can be derived, such as the above 10-year averaging period proposed by the Ministry of Water Management. The idea is now also being applied to failure of bridges in Groningen (the Netherlands) due to human-induced earthquake risks. Who passes a high-risk bridge runs by definition an earlier or later less dangerous (due to earthquakes) highway.

Finally, the element of voluntariness should also not be ruled out. Sometimes, one is undergoing medical treatment or stays in the hospital, increasing the individual risk to temporarily high levels, with the aim to lower them again. This element also plays in construction works such as Alphen aan den Rijn, because the traffic situation for all increases with the new bridge, willing to accept temporarily a higher risk.

19.7 CONCLUSIONS AND RECOMMENDATIONS

Based on the above critical analysis of the individual and group risk concepts, we conclude that risk regulation should take place on an aggregated (regional or national) level and should be expressed not only for the category human lives (immediate and delayed), but also for economy and environment, as well as the categories employees, users, and third parties.

Furthermore, a risk criterion for a short period of time should be further elaborated. Not only new perspectives for the individual risk (IR) in time are desired. Also new perspectives for the IR in space are desired, such as taking the sum of the IRs over several risk sources in space, in which dependencies in the failure of the sources of danger are taken into account. A "riskometer" which, on an individual level, monitors the risk continuously in time and space, would be the ideal image. In the medical world, the Stroke Riskometer is an innovative example in this respect (Feigin et al., 2015).

After observations of survival or failure of systems, structures or processes under extreme stress levels, risks can be updated with the Bayesian paradigm. A toy example for designing and optimizing STs in systems has been presented in this chapter, incorporating temporal and spatial uncertainties in loading and resistance variables. It has been shown that elevated stress levels can be applied to individual components and/or a combination of components in order to reduce the probability of failure (after observing a positive outcome of the test), and to cut off the tails of the distributions.

Regarding the question of an objective standard for risk regulation, a pure operational-research type of cost–benefit analysis under boundary conditions regarding minimum safety levels, is a rational way to come up with objective criteria, avoiding so-called risk control reflexes. Risk aversion factors, incorporating uncertainties in the probability of events, its consequences, and the effectivity of safety measures, can readily be embedded in such overall framework.

ACKNOWLEDGMENT

The author acknowledges the fruitful discussions which he had on this topic with TU Delft Professors Vrijling, Vrouwenvelder, Steenbergen, and Jonkman.

REFERENCES

Van Dantzig, D. (1960). The economic decision problem concerning the safety of the Netherlands against storm surges. Report of the Delta Commission, Contribution II.2, pp. 59–110.

Delta Committee Report. (1960). Final report and interim advice, Part 3.

Feigin, V. L., Wang, W., Fu, H., Liu, L., Krishnamurthi, R., Bhattacharjee, R., Parmar, P., Hussein, T., and Barker-Collo, S. (2015). Primary stroke prevention in China—A new approach. *Neurological Research*, 37(5).

IAEA. (2005). Risk informed regulation of nuclear facilities: Overview of the current status. IAEA-TECDOC-1436. International Atomic Energy Agency, Vienna.

Pasman, H. J. (2015). *Risk Analysis and Control for Industrial Processes—Gas, Oil and Chemicals* (pp. 461). Elsevier.

Van Ravenzwaaij. (1994). Risk-information in safety policy-making (in Dutch: Risico-informatie in het veiligheidsbeleid) (Ph.D. thesis, Universiteit Utrecht).

Stapelberg, R. F. (2008). Infrastructure systems interdependencies and risk informed decision making (RIDM): Impact scenario analysis of infrastructure risks induced by natural, technological and intentional hazards. *Journal of Systemics, Cybernetics and Informatics*, 6(5), 21–27.

USNRC. (2002). An approach for using probabilistic risk assessment in risk-informed decisions on plant-specific changes to the licensing basis. Regulatory Guide 1.174. US Nuclear Regulatory Commission.

Vrijling, J. K., van Gelder, P. H. A. J. M., van Asperen, L., van de Paverd, M., Westphal, R., Berger, H., and Voortman, H. G. (1999). Uncertainty analysis of water levels on Lake IJssel in the Netherlands. In: Proceedings ICASP8, Sydney.

Vrouwenvelder, A., and Vrijling, J. K. (2000). Probabilities, uncertainties and its interpretations. TNO report, CON-DYN/M2107.

20

QUANTITATIVE RISK ANALYSIS

Jan-Erik Holmberg

20.1 INTRODUCTION

Risk analysis is a broad area of methods and applications. Society for Risk Analysis[1] defines risk analysis "to include risk assessment, risk characterization, risk communication, risk management, and policy relating to risk, in the context of risks of concern to individuals, to public- and private-sector organizations, and to society at a local, regional, national, or global level." Risk analysis has been described as addressing the three questions (Apostolakis, 2004; Kaplan and Garrick, 1981):

1. What can go wrong?
2. How likely is it?
3. What would be the consequences?

[1] http://www.sra.org/about-society-risk-analysis

Handbook of Safety Principles, First Edition. Edited by Niklas Möller, Sven Ove Hansson, Jan-Erik Holmberg, and Carl Rollenhagen.

A simple division of methods for risk analysis can be made between quantitative and qualitative risk analysis. Quantitative risk analysis (QRA) includes a calculation of the two components of risk:

- The magnitude of the potential loss
- The probability that the loss will occur

Definition of risk has thus these two components: a metric for losses and a probability distribution of losses. In the qualitative risk analysis, the probabilities are not explicitly assessed, and even the losses may be characterized by qualitative categories. Qualitative and quantitative risk analyses should, however, not be seen as two complementary approaches. On the contrary, it is a matter of spectrum of approaches to the risk analysis where the role quantification and use of probabilities vary. Quantification may sometimes be regarded as a means to give a structure for a qualitative risk analysis. Vice versa, qualitative risk analysis is often a preceding step before the development of a quantitative risk model. Qualitative risk analysis is discussed in Chapter 21 of the handbook.

This chapter discusses the principles and methods for quantitative risk analysis as generally applied in the engineering context. Engineering QRA covers a risk assessment of hazardous socio-technological processes, where the losses can be described in terms of human, environmental, or economic losses. One of the objectives is to show compliance with safety objectives expressed by numerical probabilistic criteria, called acceptable risk. Further objectives are to identify weak points, search for design improvements, and to optimize the system. In the engineering risk analysis, there is usually a regulatory framework, including a regulator (safety authority) who stipulates the requirements for the QRA.

Quantitative risk analysis is a synonym to probabilistic risk analysis (PRA) and probabilistic safety assessment (PSA), which terms are, for example, used in the nuclear safety applications.

Another significant area of application of quantitative risk analysis is the financial sector. Even though there are methodological similarities with the engineering risk analysis, the financial risk analysis is dedicated to the optimization of financial investments such as portfolios. It is usually much more driven by data and use of stochastic models than the engineering risk analysis, where the analyzed objects (technical systems) are nearly always in some way unique. See, for example, Luenberger (1998) for methods and theory for the financial risk analysis.

20.2 ORIGIN AND HISTORY

Concept of risk was already known in ancient times. In Egypt, measures were taken to manage flooding and dry periods of the Nile river, which had a great impact on the success or failure of agriculture. Egyptians kept record on the periods and built storehouses of grain to prepare against catastrophic periods.

The word "risk" is said to come from Greek word "rhize" meaning "hazards of sailing too near to the cliffs: contrary winds, turbulent downdraughts, swirling tides." The origin of the word "risk" is however disputed, and there are a number of other explanations for it, see, for example, Althaus (2005).

More professional risk analysis became necessary in the expansion of the colonialization in the seventeenth century. Each journey to the colonies far away was risk taking where rewards could be high but also the probability to lose the investment (ship) was significant. Risk could be managed by having more stakeholders and more ships. An insurance market was developed in order to handle the risk.

A precondition for the development of risk analysis was the development of the mathematical theory of probability and statistics. In short, the insurance and gambling problems discussed in the seventeenth and eighteenth centuries raised the need to develop theories and concepts. Many of the commonly applied probability models and statistical estimation methods that are used even today in risk analysis and quality control had been invented by the beginning of twentieth century. Important progress from the risk analysis point of view included axiomatic foundation of probability concepts (Kolmogorov axioms, subjective probability) and application of stochastic models in operations research (decision theory, reliability theory, renewal theory, and Monte Carlo).

The way risk analysis is performed for hazardous technological installations has its roots in the development of reliability analysis, operations research, statistical quality control, and decision analysis methods since the Second World War. First application areas were military, nuclear, space technology, and aviation sectors in the 1950s and 60s.

For the development of risk acceptance criteria, Farmer's (1967) proposal to use of probabilistic criteria in the regulatory framework was an important step. It follows the mathematical representation of the risk as a curve, which plots $F(C)$ against C, where $F(C)$ is the frequency of events with consequences greater than C (see Figure 20.1). A safety criterion could be defined in a similar manner (see Figure 20.2).

In the nuclear side, the Reactor Safety Study published 1975 became a cornerstone (U.S. NRC, 1975), and the same event tree–fault tree modeling approach has been applied after that. In the chemical industry, the Canvey Island Study can be mentioned as an early quantitative risk assessment which evaluated the risks associated with an integrated chemical/petrochemical processing complex in the UK (HSE, 1978).

Since the Seveso directive (EEC, 1982), risk analysis has been standardized as a methodology to control all kinds of industrial hazards within the European Community. Since the 1980s, probabilistic safety criteria and ALARP (as low as reasonably practicable) principles have been widely applied in the nuclear safety field, and are used in the regulatory process even in other industries, for instance, in

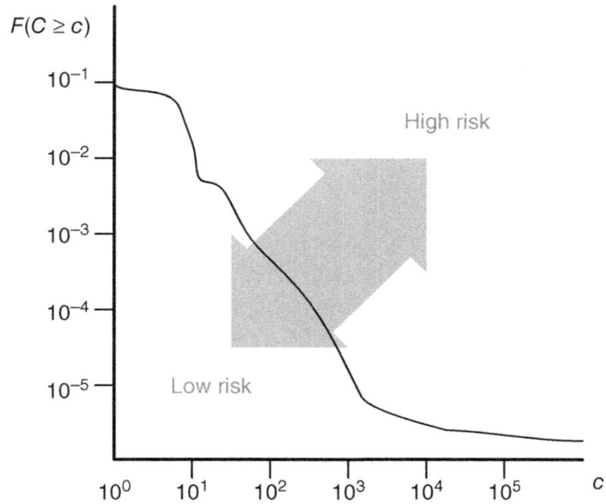

Figure 20.1. Complementary probability distribution of consequences (presented in log-log scale).

the Netherlands, Norway, United Kingdom, Switzerland, United States, see further discussion below.

Reviews of the history of risk analysis can be found, for example, in Ballard (1993), Cooke (2009), and Covello and Mumpower (1985).

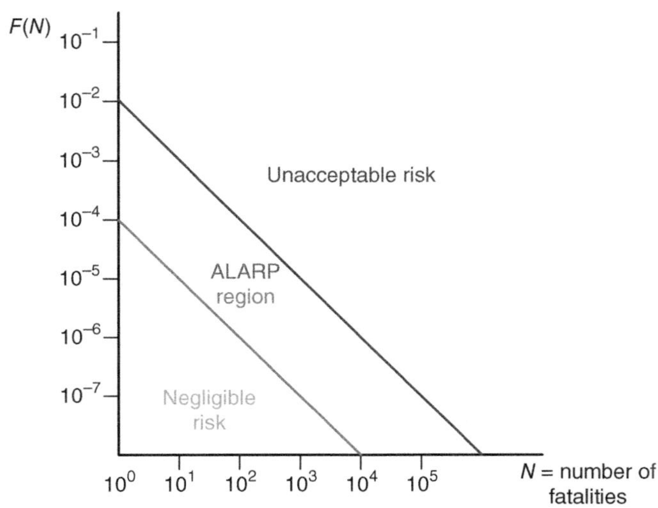

Figure 20.2. ALARP regions for risk curves (HSE).

20.3 UNDERLYING THEORY AND THEORETICAL ASSUMPTIONS

The underlying theory and related assumptions of QRA have two parts: (1) how risk analysis is related to decision-making and (2) how the probabilistic model is made. Both aspects are discussed in this section.

20.3.1 Risk

Risk can be defined in many ways. Aven (2012) presents the history of definitions of risk and compares different definitions. In this chapter, we choose the definition where risk is a combination of hazards measured by a probability. This is often expressed with the formula

$$R = \sum_i P_i C_i, \tag{20.1}$$

where P_i is the probability of the scenario i and C_i is a metric for the consequences. The expected value of the consequences, like defined in the formula above is not the only way to characterize the risk. The complementary probability distribution of the consequences, also called risk curve or Farmer's curve, $F(C \geq c)$ is a popular way of presenting results from a risk. Figure 20.1 depicts a Farmer's curve for a fictive example.

Kaplan and Garrick (1981) defines risk as a set of triplets

$$R = \{\langle S_i, L_i, X_i \rangle\}, \tag{20.2}$$

where S_i denotes the scenario, L_i denotes the likelihood of the scenario, and X_i denotes the resulting consequences. This definition corresponds with the QRA, where a set of scenarios are identified and quantified.

20.3.2 Probability

Probability is a mathematically formalized concept which is used to measure the likeliness that an event will occur. The two major interpretations of probability are the frequentist interpretation and the subjective interpretation. In the frequentist interpretation, true probabilities exist and could be assessed if we had an infinite sample of similar situations to the one analyzed. In the subjective interpretation, the probability express the degree of belief of the assessor. The subjective interpretation is usually applied in a Bayesian way, where expert knowledge is combined with observations to obtain probability estimates.

In the quantitative risk analysis, there is a need to estimate large number of various probability parameters. The Bayesian approach has generally become the state-of-the art. As an example, the risk analysis model may include a failure event caused by a component failure. The probability (or frequency) of the event is quantified using an appropriate reliability model, for instance, a model for a repairable component. The average unavailability of the component is

$$U = \frac{\text{MTTR}}{\text{MTTF} + \text{MTTR}}, \tag{20.3}$$

where MTTF is the mean-to-time-failure and MTTR is the mean-to-time-repair.

Assuming that repair and failure times are exponentially distributed with intensities λ and μ, respectively, the mean values can be estimated from the data. In the Bayesian analysis, prior distributions are first defined for λ and μ, for example, the non-informative Jeffrey's prior, and given representative data from similar components, a posteriori distributions can be derived for the parameters using Bayes' formula (Gelman et al., 1995).

As an example, failure statistics have been collected for a group of homogeneous components, having totally T hours operating time and experiencing n failures. Assuming that the failure process follows the homogeneous Poisson process, the failure rate λ is often estimated choosing first an improper gamma distribution

$$\lambda_{\text{prior}} \sim \Gamma(0.5;\, T), \tag{20.4}$$

as the prior distribution, which yields a posterior distribution

$$\lambda_{\text{post}} \sim \Gamma(0.5 + n; T), \tag{20.5}$$

The posterior expected value of the failure rate is

$$E[\lambda|n,\, T] = \frac{0.5 + n}{T}. \tag{20.6}$$

QRA typically relies heavily on probability assessments described above. In addition, the models usually include events for which representative data lacks. Examples are human errors, extreme weather conditions, and new technologies. Expert judgements have to be used, and validity of such assessments is often a debatable issue (e.g., the discussion in the chapter on human reliability analysis).

20.3.3 Uncertainty

Uncertainty is an important concept in risk analysis, and the probability is a measure of uncertainty. Assessment of uncertainties in the context of QRA is, however, a broader issue than simply an application of probability theory (Apostolakis, 1988). A thorough discussion on uncertainties in risk analysis can be found, for example, in *Reliability Engineering and System Safety*, 23(4), 1988.

When dealing with uncertainties, there is thus on the one hand the question of how uncertainties related to the scenarios are covered and captured within QRA—in terms of probabilistic modeling—and on the other hand how uncertainties of the ORA itself (the probabilistic model) are addressed and taken into account, for example, in decision-making. Uncertainties can be thus approached from the philosophical basis of probabilities and risk concepts as well as from the practical QRA performance point of view.

Common characterizations of uncertainty in risk analysis are (EPRI, 2004):

- Uncertainty versus randomness
- Aleatory versus epistemic
- Parametric, modeling, and completeness
- Sensitivity analysis versus uncertainty analysis

Uncertainty and randomness are not always distinguished in the QRA terminology, but there is a clear difference between them. Randomness is a fundamental variability of an outcome, while uncertainty refers to lack of knowledge regarding the modeled phenomena. Gaining more knowledge can decrease uncertainty but not randomness.

Terms "aleatory" and "epistemic" uncertainties are synonymous to the terms above, and are often used in the uncertainty discussions. In reality, it is difficult to distinguish between the two categories of uncertainty and it can be difficult to apply the concepts when addressing uncertainties in QRA.

Division of uncertainties into parametric uncertainty, modeling uncertainty, and incompleteness is a practical way of addressing uncertainties of QRA. Parametric uncertainties refer to lack of evidence to properly estimate the parameters of the risk analysis model such as the reliability parameters (λ and μ above). Model uncertainties comprise analysis assumptions-related uncertainties, such as the logic model uncertainties.

Incompleteness is a sub-category of modeling uncertainties. In principle, it includes both "known unknowns" and "unknown unknowns." The first category can be someway captured by a proper description of the scope and assumptions of ORA and if possible an assessment of impacts of limitations from the conclusions and decision-making point of view.

Sensitivity and uncertainty analyses are sometimes used interchangeably, but a distinction can be made by defining sensitivity analysis as a what-if analysis where a selection of modifications are made to the model structure, its input parameters, and assumptions, with the purpose of identifying which of these are more important. Uncertainty analysis refers to the application of mathematical techniques to propagate the uncertainty of the individual input variables through the model quantification in order to produce an output result in the form of a probability distribution (EPRI, 2004).

Techniques for analyzing uncertainties of QRA in nuclear power plant context can be found in EPRI (2004). The outlined approach is, however, applicable to any industrial area, that is, how parametric and modeling can be analyzed. It can be also used as an example of coupling uncertainties to the overall requirements for QRA.

20.3.4 Expected Value and Utility Principle

Formula (20.1) relates the definition of risk to the expected value of the consequences. It is the probability-weighted average of all possible outcomes. In the decision theory, if we can choose between lotteries, the optimal choice is to take alternative which minimizes the expected value of losses (or vice versa maximizes the expected value of revenues).

From the decision-making point of view, the main principles applied are the expected utility principle and acceptable risk principle. The two principles are not fully compatible with each other, but the acceptable risk—if such is applied—may be seen as a boundary condition, and the expected utility principle gives guidelines to choose between options as long as the risk is acceptable.

Given decision options $a_1, \ldots, a_n$, and associated probability distributions for the consequences $P_i(C)$, $i = 1, \ldots, n$, where consequences are measured as negative losses, such as human lives or monetary losses, the optimal option yields the minimal expected loss, that is,

$$i^* = \arg\min_i E[C|a_i]. \tag{20.7}$$

It should be noted that expected utility principle is seldom strictly applied in the engineering applications, but the consideration between safety improvements is made by a combination of qualitative arguments and cost–benefit evaluations. This is further discussed in the chapter on cost–benefit analysis.

One of the reasons not to apply expected value principle is the uncertainty of risk analysis. In particular, it is questionable to perform strict optimization when the result is dependent on scenarios which have low probability and high consequence. There is a large uncertainty about the absolute expected value. On the other hand, we have better confidence when comparing decision options for which the space of outcomes and associated probability distributions do not differ too much. The expected value principle is, however, the normative reference for risk decision-making.

One weakness of the expected value principle is that it does take into account risk aversion for large losses. This could be overcome by replacing the value function by a utility function. The use of expected utility principle can be, however, even more controversial than the risk neutral expected value function. It should be noted that risk aversion is accounted in the definition for the acceptable risk criterion; see the next section.

20.3.5 Risk Criteria

Risk criteria can be generally understood as references used in the comparison of numerical results of a risk analysis against some targets values. Typical application of risk criteria is to show the compliance of a hazardous installation against a risk target in order to be convinced that the risk is at an acceptable level. Two main risk metrics applied are the individual fatality risk and societal risk or group fatality risk, but there are many other possible risk criteria as is discussed in Section 20.4.

The individual risk is usually defined for a hypothetical most exposed person in the vicinity of an installation, sometimes for an average person. The individual risk criterion, p^*, can be expressed like

$$p < p^*. \tag{20.8}$$

As a reference for the criterion, the general accidental death, which is about 1×10^{-4} per year, can be used. Using the factor of 100 or 1000, the safety goal for individual risk from an industrial accident should be 1×10^{-6} or 1×10^{-7} per year, meaning *no significant additional accident risk to an individual*. Different criteria may be set for a worker and non-worker.

The group fatality risk criterion can be expressed like

$$f(n) < f^*(n), \tag{20.9}$$

That is, the frequency of a single accident causing n or more fatalities shall be less than $f^*(n)$. Examples for $f^*(n)$ are $1 \times 10^{-3}/n^2$ per year used by Dutch authorities for hazardous installations (Ale, 2002), $1 \times 10^{-3}/n$ per year, used by Australian authorities for existing dams ($1 \times 10^{-4}/n$ for new dams) and "total risk of 100 or more fatalities (ANCOLD, 2003)," limit 1×10^{-5} per year, objective 1×10^{-7} per year used by U.K. HSE (2001).

Societal risk as a concept is sometimes interpreted as a synonym for group fatality risk, but it may also cover other negative consequences for the society, such as risk of evacuation and restrictions on land use. For simplification, group fatality risk is often used as the surrogate of the societal risk.

20.3.6 ALARP

A numerical safety goal can be a mandatory criterion (limit), a desired target (an objective), a compensatory criterion, or an informal goal. In mandatory use, the value must be strictly met. An objective is a desired target that should be aimed at, but where violations can be accepted and justified.

The ALARP (as low as reasonably practicable) principle adopted by the U.K. HSE includes both a limit and a target. In the ALARP framework, risk is classified into three regions: unacceptable, tolerable, and broadly acceptable (see Figure 20.2). The boundary between unacceptable and tolerable risk is a limiting risk criterion,

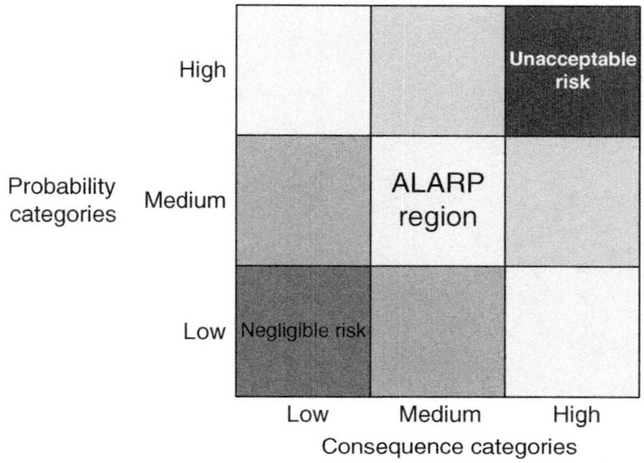

Figure 20.3. Risk matrix.

and the boundary between tolerable and broadly acceptable is a target. The tolerable region is also called ALARP region, since the fundamental principle of ALARP is to ensure that risk has been reduced as low as reasonably practicable. Formal cost–benefit analysis may be applied to make the judgement (see also Chapter 22).

ALARP principle matches well with the use of a risk matrix for the presentation of results of a risk analysis. In a risk matrix, various hazards are plotted into different cells with respect to the magnitude (category) for the probability versus consequence. Right-upper corner of the matrix corresponds with unacceptable risks while left-down corner is the area of negligible risk. Figure 20.3 presents an example with three categories for consequences and probabilities. It should be noted that this results presentation and interpretation approach is common to both QRA and qualitative risk analysis (Chapter 21).

As an alternative to a risk matrix, results can be plotted in a scatter diagram (Figure 20.4). In this representation, ALARP regions can be defined as a function of the product $P \times C$. In the ALARP region, safety improvements can be considered, and a further ranking can be done by the Pareto efficiency principle (here applied in a risk analysis context): a hazard is Pareto efficient (i.e., kind of highly critical) if there is no other hazard which has higher probability and higher consequence. If the number of hazards is high, the decision-maker should first pay attention to the most critical (Pareto efficient) hazards, assuming that the unacceptable risks have been first eliminated.

20.3.7 Subsidiary Risk Criteria

A "subsidiary criterion" is a criterion on a lower-technical level to assess in a simplified way the consequences on a higher level. Subsidiary criteria are commonly

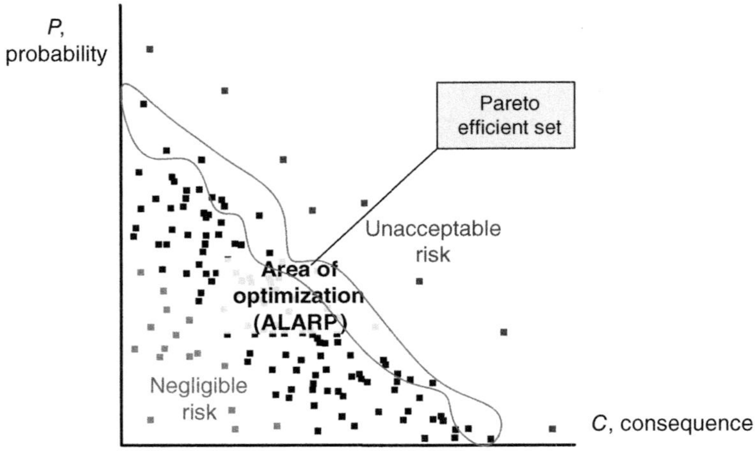

Figure 20.4. Scatter diagram for the presentation of results of QRA.

applied in many industries, but the way they are defined is specific to the industry (Figure 20.5). As an example, in the nuclear power plant risk assessment the core damage frequency and large release frequency are used as subsidiary criteria for risk of offsite consequences in countries where level 3 PSA is not required. In some documents, the term "surrogate criterion" is used instead. In the railway sector, safety system failure rate, called "tolerable hazard rate," is used as a surrogate for the individual passenger risk frequency.

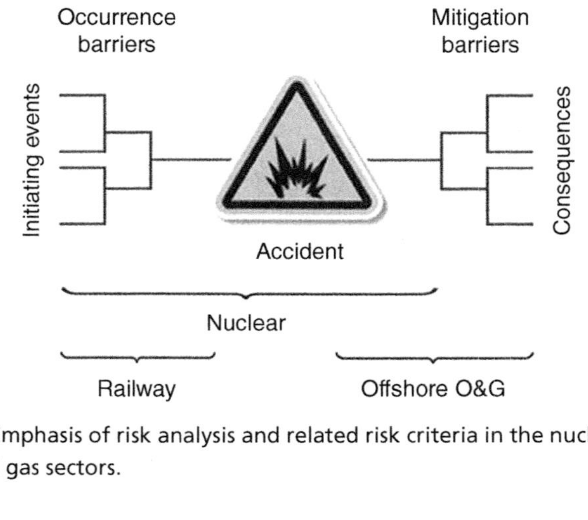

Figure 20.5. Emphasis of risk analysis and related risk criteria in the nuclear, railway, and offshore oil and gas sectors.

Subsidiary criteria are advocated for several reasons (Bengtsson et al., 2011):

- To perform a full-scope QRA is a resource demanding effort, which can be avoided if the safety can be demonstrated by a lower-level safety system analysis.
- The uncertainties in the risk assessment of societal and individual risk are considerably larger than in the assessment of reliability of safety systems. There are also less uncertainties in the assessment of compliance with subsidiary risk criteria.
- Subsidiary risk criteria put focus on defense-in-depth, in particular attention is paid to accident prevention and mitigation.
- Subsidiary risk criteria can be used as a basis for the definition of safety function or system level reliability requirements, providing better support than higher-level criteria to the actual design of safety functions and systems.
- Subsidiary risk criteria are closer to day-to-day operational safety management concerns of the utility, and they are closer to risk-informed applications.

The following concerns may be expressed in relation to the use of subsidiary criteria:

- The metric of different subsidiary risk criteria typically differ a lot, which complicates any tries to verify the assumed correspondence with higher-level safety criteria.
- Technology dependency and site dependency can be difficult to take into account in subsidiary criteria.
- Subsidiary criteria can be difficult to compare to other risks of the society, which are typically expressed on a higher level (degree of damage to individuals or groups).
- In the communication with the public, subsidiary criteria may be seen as more abstract and harder to understand than top level risks (like off-site consequences).

20.3.8 Event Tree–Fault Tree Modeling

Event trees and fault trees are complementary methods for risk and reliability analysis, and they can be used in an integrated manner as is done, for example, for nuclear power plant risk assessments. Both apply a graphical representation of cause–consequence relationship between events, and the model is based on Boolean logic (multi-valued logic may be possible, too) and a probability metric for the quantification part.

Event tree analysis is a forward, bottom-up modeling method. An event tree starts from an initiating event followed by (typically in a chronological order) intermediate

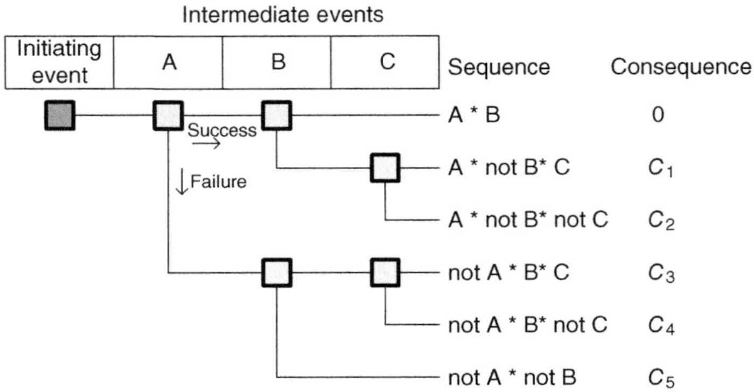

Figure 20.6. Event tree example.

events, which represent successes or failures of, for example, various barriers mitigating the consequences of the initiating event. Each combination of successes and failures of the intermediate events corresponds with an event sequence, which can have a specific consequence. In a simple analysis, an end state is either a system success or a system failure.

Figure 20.6 shows an event tree for a system with three intermediate events and six sequences. Quantification of sequence probabilities can be carried out in several manners. The branch probabilities can be direct probability estimate or there can be complex reliability models (e.g., a fault tree) behind them.

Fault tree analysis is a top-down modeling method. A fault tree starts from a top event representing the system failure. The top event is broken down into sub events explaining conditions which can lead to the top event. Logical operators AND, OR, exclusive OR, k-out-of-n and NOT are applied in the definition of failure conditions. As long as an event is broken down into sub events it is called a gate. An event which is not broken down anymore is a basic event, and it needs to be associated with a probability for the quantification purposes. A fault tree is usually solved by searching minimal cut sets[2] which is a unique representation of failure conditions for a coherent system,[3] and is very applicable formula for the quantification of the top event probability. For large fault trees, approximation methods are applied for solving minimal cut sets and quantifying the top event probability.

Figure 20.7 shows an example fault tree and a corresponding reliability block diagram, which is an alternative way of representing a system's reliability structure. A fault tree follows a failure logic; that is, it represents conditions for a system failure as a function of component failures. Reliability block diagram follows a success

[2] A cut sets is combination of basic events causing the top event. A cut set is minimal if it loses its cut set property when any of its basic events is removed.

[3] A system is coherent if all of its components are beneficial for its reliability at any configuration.

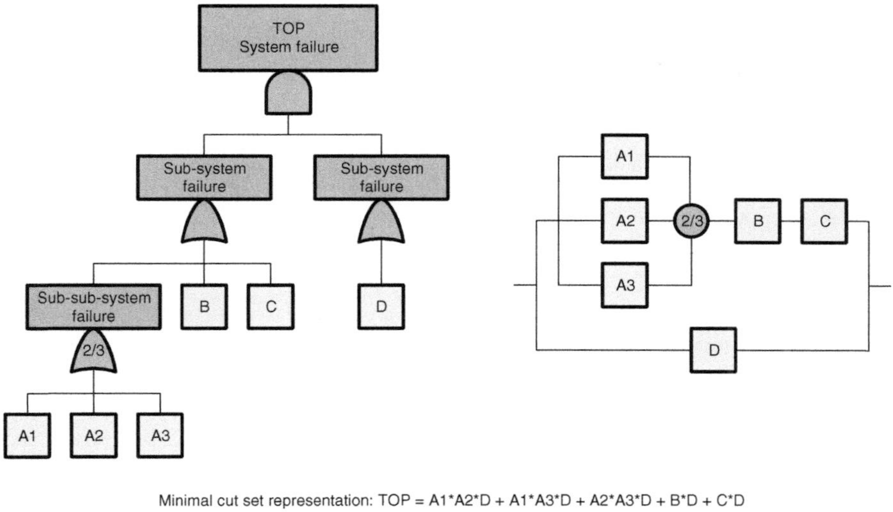

Figure 20.7. Corresponding fault tree and reliability block diagram.

logic; that is, it represents conditions for a system success as a function of component successes. From the reliability theory point of view, system failure and system success representations are duals, and it depends on the modeling convention which representation is applied.

Event tree is more applicable for a scenario type of modeling, while fault trees enable very detailed modeling of system dependences. These features are combined when the two modeling techniques are integrated, that is, event trees are applied as high-level modeling structures governing how detailed fault trees are combined with each other for the quantification of sequences and consequences.

Besides quantifying top event probabilities or consequence probabilities, it is common to analyze relative importance of basic events by means of reliability importance measures. Commonly used importance measures are

- Fractional contribution or Fussell–Vesely importance (these two give practically the same value)
- Risk increase factor or risk achievement worth
- Birnbaum.

Fractional contribution, *FC*, is the relative share of minimal cut sets containing the basic event *X*,

$$\mathrm{FC}(X) = 1 - \frac{P(\mathrm{TOP}|X = 0)}{P(\mathrm{TOP})}. \tag{20.10}$$

It gets values between 0% and100%. It presents how much risk (or top event probability) can be decreased if the basic event is false ($X = 0$); that is, the component is made fully reliable. For components with high-fractional contribution, means to improve reliability should be considered.

Risk Increase Factor, RIF, expresses how much the risk (or top event probability) increases when the basic event is true ($X = 1$); that is, the component is made unavailable,

$$\mathrm{RIF}(X) = \frac{P(\mathrm{TOP}|X = 1)}{P(\mathrm{TOP})}. \tag{20.11}$$

For components with high-risk increase factor, means to implement a backup should be considered.

Birnbaum measure is the derivative of the top event probability with respect to the component's failure probability

$$B(X) = P(\mathrm{TOP}|X = 1) - P(\mathrm{TOP}|X = 0). \tag{20.12}$$

Birnbaum is a sensitivity metric, that is, it shows the influence of the stochastic behavior of the component on the system reliability.

The benefit of using risk importance measures is that they provide a relative ranking of basic events, which makes them less sensitive to uncertainties as absolute numbers. They are especially used for the identification of weak points of the system.

Event tree and fault tree models are limited by the static structure of the model and static probabilities. There are techniques to implement dynamic properties in the model, but ultimately these models are static. Markov models and Petri nets are examples of probability models which allows the modeling of infinite event sequences, which can be practical when accounting context dependent system recoveries.

20.3.9 Bayesian Belief Network

Bayesian belief network (BBN) is becoming a popular modeling for representing causal dependences in a network format. Each node represents an event or condition, and causal relationships are defined by arcs between nodes. A conditional probability function (a probability table when node states are discrete) is associated with each node.

BBN is a seemingly attractive modeling approach for many risk analysis applications (Fenton and Neil, 2012; Trucco and Leva, 2012). The model can represent both probabilistic and causal dependences, and expresses the role of prior knowledge and data in a Bayesian manner. Compared to the fault tree models the size of models are limited. It may be also hard to validate a BBN-model, which is a prerequisite when the risk analysis is used in a regulatory context. The challenge can be both when using extensively expert judgements and when applying larger models with a complex structure.

20.3.10 Bow-Tie Method

Bow-tie diagram is a visualization of accident prevention and mitigation related events (Figure 20.5). Mathematically bow-tie model can be constructed similarly to event tree–fault tree approach or BBNs. Bow-tie gives an overview of multiple scenarios in one diagram. It requires that the meaning of an accident (also called hazardous event or top event) is clear so that the prevention versus mitigation barrier can be distinguished (Salvi and Debray, 2006).

20.3.11 Monte Carlo Simulation

Monte Carlo simulation is a multi-purpose computer algorithm where (pseudo-) random samples are drawn to produce numerical results such as probability distributions of complex probability models.

In risk analysis applications, Monte Carlo method is widely applied as a quantification method. Monte Carlo is used (1) to evaluate top event probabilities for a complex models, for which the probability function cannot be easily computed (quantification of P), (2) to evaluate effect of parametric uncertainties in the models (uncertainty distribution of P), and (3) to derive a probability distribution over quantifiable consequences (probability distribution of C). Monte Carlo can be also utilized in the parameter estimation, for example, in the Bayesian models, and in the risk-based optimization problems.

20.4 USE AND IMPLEMENTATION

Choice of QRA methods is an application area specific issue. Each industry has historically started doing risk analyses in certain manner with different emphasis on the role of quantification and use of numerical risk criteria. This chapter summarizes use of risk criteria and discusses a few industries which regularly use QRA for safety management. Emphasis of the discussion is in the decision-making and risk criteria.

20.4.1 National Risk Criteria

Many societal necessary projects involve risks of fatal accidents. Therefore, some sort of regulation is required to ascertain that the risks are not unfairly distributed. Typically the probabilistic safety goals used consider loss of life and economic damage as a consequence. Different probabilistic safety goals are categorised according to the consequences they consider (Jonkman et al., 2003):

- Fatalities
    - Individual risk
    - Societal risk
- Economic damage
- Environmental damage
- Integrated safety goals
- Potential damage

Examples of national criteria for individual and societal (group) risk are summarized in Tables 20.1 and 20.2.

20.4.2 IEC 61508 and Safety Integrity Levels

The concept of Safety Integrity Levels (SIL) is introduced in the increasingly important standard IEC 61508, which deals with the functional safety of electrical, electronic and programmable electronic safety-related systems (IEC, 2010). The standard applies quantitative requirements to systems operating on demand and to system operating continuously in order to maintain a safe state. IEC 61508 is a general standard that forms a basis for different industry specific standards.

A requirement on a SIL sets the requirements on the process how the equipment is manufactured, tested etc. The IEC 61508 standard sets requirements on several aspects that together form the basis for the safety integrity level (Figure 20.8).

Different functions with safety requirements need to meet different Safety Integrity Levels (SIL). To meet a SIL level one has to, for example,

- Use hardware that meets the reliability requirements of the specific SIL level.
- Architectural design for safe failure fractions.
- Develop software according to quality graded requirements (high SIL means high requirements on the software process).

Table 20.3 shows the SIL levels and their corresponding assumed failure probability and failure rate. Assignment of SIL requires a risk analysis where the risk associated with a specific hazard is calculated without the beneficial risk reduction effect of the safety system. The amount of required risk reduction is compared with the SIL target. Each order of magnitude of risk reduction that is required correlates with an increase in one of the required SIL numbers. Methods to assess SIL is not fixed.

TABLE 20.1. Comparison of National Criteria of Individual Risk

Country	Application	Maximum Tolerable Risk	Negligible Level of Risk	Comment
The Netherlands	Established plants or combined plants	10^{-5} ALARA principle applies	Not applied	
	New plants	10^{-6} ALARA principle applies	Not applied	
UK	Existing hazardous industries	10^{-4} ALARP principle applies	Broadly accepted limit 10^{-6}	Negligible limit 10^{-7}
	Existing dangerous goods transportation	10^{-4}	10^{-6}	
	New housing areas near existing plants	10^{-5}	10^{-6}	
Czech Republic	Existing installations	10^{-5}		Risk reduction must be carried out
	New installations	10^{-6}		
Hungary	Hazardous facilities	10^{-5} Upper limit	3×10^{-6}–10^{-6} Lower limit	
Hong Kong	New plants	10^{-5}	Not used	
Australia (New South Wales)	New plants and housing	10^{-5}	Not used	
Australia (Victoria)	Existing installations	10^{-5}	Acceptable limit 10^{-7}	
USA, California	New plants	10^{-5}	10^{-6}	
Germany	Transportation	10^{-5}		

TABLE 20.2. Comparison of National Criteria of Societal risk (N = Number of Fatalities)

Country	Application	Maximum Tolerable Risk	Negligible Level of Risk	Comment
The Netherlands	Established and new plants	$10^{-3}/N^2$	Not applied	
UK	Hazardous installations	$10^{-2}/N$		
	Existing harbors	$10^{-1}/N$	$10^{-4}/N$	
Hong Kong	Hazardous installations	$10^{-3}/N$	$10^{-5}/N$	Limit for maximum N = 1000
USA, California	On-site risk	$10^{-1}/N^2$	$10^{-3}/N^2$	
	Off-site risk	$10^{-3}/N^2$	$10^{-5}/N^2$	
Australia (Victoria)	Hazardous industries	$10^{-2}/N^2$	$10^{-4}/N^2$	
Switzerland	Hazardous installations	$10^{-5}/N^2$ (for $N > 10$)	$10^{-7}/N^2$ (for $N>10$)	Limit for maximum N = 1000. $N < 10$ domain of no serious damage
Denmark	Hazardous installations	$10^{-2}/N^2$		

20.4.3 Nuclear Power Plants

Performance of QRA is practically obligatory in all countries having commercial nuclear power plants. In nuclear context QRA is called probabilistic risk analysis (PRA) or probabilistic safety assessment (PSA). Methodology is almost the same, and has been the same since the pioneering WASH-1400 study (U.S. NRC, 1975). PRAs for nuclear power plants are extensive models using event trees for the modeling of accident scenarios and fault trees for the assessment of systems' reliability.

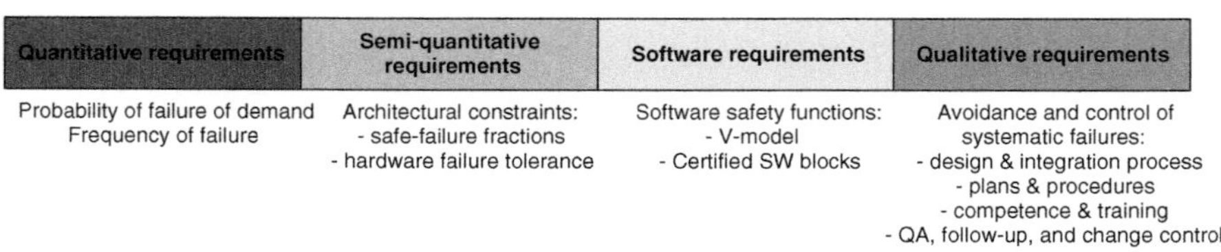

Figure 20.8. Types of requirements set by the safety integrity levels.

TABLE 20.3. Safety Integrity Level and Their Corresponding Reliability Targets

Safety Integrity Level (SIL)	Average Probability of a Dangerous Failure on Demand of the Safety Function
4	10^{-5} to 10^{-4}
3	10^{-4} to 10^{-3}
2	10^{-3} to 10^{-2}
1	10^{-2} to 10^{-1}

Safety Integrity Level (SIL)	Average Frequency of a Dangerous Failure of the Safety Function [h^{-1}]
4	10^{-9} to 10^{-8}
3	10^{-8} to 10^{-7}
2	10^{-7} to 10^{-6}
1	10^{-6} to 10^{-5}

Nuclear power plant PRAs are divided into three levels according to the consequence assessed:

- Level 1 PRA: assessment of reactor core damage risk
- Level 2 PRA: assessment of the risk of external radioactive release
- Level 3 PRA: assessment of the health and environmental consequences

For most nuclear power plants, there are level 1 and 2 PRAs, and the state-of-the-art follow closely the IAEA guidelines (IAEA, 2010a, 2010b), although use of national guidelines and topic specific guidelines (e.g., human reliability analysis, fire risk analysis) are also common. Level 3 PRA is required only in a few countries (Caldwell et al., 2014).

There is a large difference in the use of numerical criteria (OECD, 2009). Some countries apply numerical criteria in the licensing of new plants, others do not apply them in regulatory context. There are also differences on what type of numerical criteria is applied: fatality risk level criteria or lower-level surrogate criteria.

20.4.4 Oil and Gas Industry in Europe

In the European oil and gas industry, risk acceptance criteria (RAC) are used to express a risk level with respect to a defined period of time or a phase of the activity. RAC may be qualitative or quantitative. RAC are also known variously in the oil and gas industry as "risk criteria," "decision criteria," "screening criteria," or "tolerability criteria."

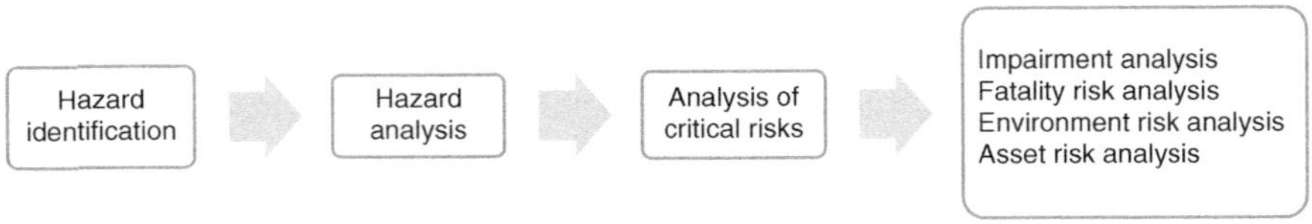

Figure 20.9. Main elements of the offshore QRA.

An overview of regulation and methods applied for offshore risk assessment can be found in Vinnem (2011). Modeling of hazards is related to the most prominent hazards for offshore installations, which include fire, explosion, collision and marine hazards. Risk to personnel is addressed most thoroughly, and risk to the environment and material damage risk are also covered. Main elements of QRA include

- identification of initiating events
- cause analysis to identify and quantify causes for an initiating event
- consequence analysis to estimate loads, responses, probabilities and damages of accidents

The bow-tie metaphor discussed in Section 20.3.10 corresponds well for this QRA approach.

Figure 20.9 describes the main elements of the offshore QRA in a flow chart (adapted from Vinnem, 2011). Hazards are the various initiating events which can form a threat to humans, environment or properties. Various consequence analyses are only made to "critical risk," which those risks that are shown to make a significant contribution to the total risk.

There is a broad set of alternatives for the qualitative analysis methods used in the identification of initiating events and their causes such as hazard and operability studies (HAZOP) and failure mode and effects analyses (FMEA). For the quantitative analysis combinations of event tree and fault tree models can be applied as well as Monte Carlo based methods. Consequence calculations require dedicated models for, for example, assessment impacts from fires and explosions.

As another example, in the Netherlands, it is required to perform QRA for all hazardous installations and transport systems. The Committee for the Prevention of Disasters (CPR) has issued reports describing the methods for QRA. The "purple book" describes starting points and data needed for QRA (VROM, 2005). Issues covered by the book are:

- selection of installation for QRA (screening criteria for installations)
- loss of containment events to be included (list of initiating events and their frequencies)
- consequence analysis models
  - modeling source term and dispersion
  - modeling exposure and damage

- calculation and presentation of results, including comparison with health risk criteria
- tools and models for quantitative environmental risk analysis
- uncertainty assessment

Compared to the nuclear PRAs, in these analyses, there is relatively seen more emphasis in the consequence assessment side (corresponding level 2 and 3 PRA in nuclear PRA terminology) while the analysis of hazards (offshore) or loss of containment events (chemical industry) is more straightforward compared to nuclear level 1 PRA models.

20.4.5 Railway Safety in Europe

The railway safety analysis is focused on analyzing the train control system's failure probability. Reliability criteria have been derived from the acceptable fatality risk level, and the attempt is to have common European target to facilitate border-crossing railway operation.

The responsibility for establishing safety targets for railway systems is described in the standard EN-50129 (CENELEC, 2003) and is divided between each railway authority and each supplier. The principle is that a tolerable hazard rate (THR) is allocated by the railway authority to the supplier for a specific defined hazard. Each hazard and THR is then by the supplier apportioned within their system to each relevant subsystem. This means that the overall risk analysis is mainly the responsibility of the railway authority, and that the supplier is responsible for hazard control and to verify their results against the safety target set by the railway authority.

The verification against the THR is done by the manufacturer of the system at different levels. Usually it is analyzed using a fault tree analysis. There is a conceptual fault tree specified by UNISIG (2008) that qualitatively analyses the top hazard. The fault tree will be adapted to the specific system being analyzed and to the mode of operation. The verification of the safety target will be by comparing the result of the fault tree analysis to the THR. If satisfactory results are not achieved, then a redesign would be considered.

20.4.6 Other Industries

Some other safety goals used for various technologies: (adapted from Kafka, 1999 and Pfitzer et al., 2004):

- Marine structures: failure probability for different accident classes 10^{-6} to 10^{-3}
- Aviation, air planes: catastrophic failure per flight hour, less than 10^{-9}
- Space vehicles: catastrophic consequence for crew transfer vehicle less than 1 in 500 missions.

- Missile range criteria for falling debris: for example, maximum acceptable probability for individual fatality (for the public) during one mission 10^{-7} and during one year 10^{-6}.

20.5 EMPIRICAL RESEARCH ON USE AND EFFICIENCY

Quantitative risk analysis is an extremely wide area to be empirically studied from the efficiency point of view. There are experience from problem areas, which relate to problems of human beings deal with probabilities. Cognitive biases with probabilities and deviations from the expected utility based decision-making (Kahneman and Tversky, 1979) are well known phenomena. Problems experienced with risk communication are also common, as discussed in Chapter 11.

To know the efficiency of risk analysis-based safety management, a comparison should be made against non-probabilistic (deterministic) safety management framework. History of nuclear industry could be used as an example. However, it is impossible say that thanks to QRA "safety has been improved this much" or that "operation of an installation is this much more cost-effective." Cases exist but a generalization can be difficult to draw.

20.6 WEAKNESSES, LIMITATIONS, AND CRITICISM

Uncertainties of QRA make the application of risk criteria problematic, as there are uncertainties in QRA which are clearly identifiable but difficult to quantify. Uncertainties should be accounted for in decision-making but there is no formal method, within the QRA methodology, how to do it. In principle, to deal with uncertainties we could use fuzzy numbers or apply some second order metric to represent uncertainties, but these approaches do not eliminate the problem of uncertainties. As long as safety goals have an informal role, uncertainties can be handled by discussion. However, a mandatory risk limit requires a strict comparison of two numbers.

Interpretation of the probabilities is a well-known problem. It is not only the general problem of understanding probabilities, which makes the communication hard. In many safety-critical applications, it is much harder to communicate very small probabilities (low-probability-high-consequence dilemma). In this context we have, for example, the following hard assessment tasks:

- An event has a low probability, but it is hard to estimate it due to scarcity of data
- An event is practically impossible, but we cannot fully exclude it

Risk analysis raises the question of comparing different risks in the society, for example, when applying numerical risk criteria for decision-making. Some risk comparisons are more likely than others to be perceived as an effort to pre-empt judgments about the acceptability of a risk. On the basis of this experience, different kinds of risk comparisons have been ranked in terms of their acceptability to people in the community (Covello et al., 1988). Examples of high-ranking comparisons include comparisons of the same risk at different times, comparisons with a standard, and comparisons with different estimates of the same risk. On the other hand, the comparisons of unrelated risks is very low-ranking, meaning that one should be very cautious when comparing, for example, traffic risks with nuclear risks. Low-ranking comparisons can be misleading and regarded as manipulative. This complicating nature of risk comparisons is a fact to be accounted, when justifications for safety goals are looked upon other contexts.

Normative risk decision-making is based on the theory of expected utility. The applicability of the theory of expected utility as a practical guideline in decision-making under risk can be questioned for several reasons. The theory supposes that possible outcomes and associated probabilities can be fully assessed, which is a very hard requirement for real world cases. Both the assessment of outcomes and of probabilities are demanding exercises. Secondly, real world cases are usually diffuse and complex, so that any model only reflects a small piece of the decision-making problem. Thirdly, the points of view of multiple stakeholders should be accounted for. If stakeholders do not agree on the probabilities and outcomes, the problem is outside of the theory of expected utility. Finally, in practical decision-making, people do not behave according to the axioms of the theory (Kahneman and Tversky, 1979). These situations deal more with decision-making under *uncertainty*, where the probabilities of outcomes are not well explicated

It is attempting to include risk analysis in the safety regulation. The regulator's perspective is to supervise and regulate all risks to human beings and environment. The decision-making on acceptable risk is culminated in the licensing process of a process (installation) or a product. The regulator would like to see a demonstration that the risk from a system is small enough compared to other technological risks, which is a comparative risk acceptance principle:

- comparison and valuation of risks in society to be used as reference
- justification of results and conclusions of the risk assessment

Accident statistics exist for the determination of the overall risk level for human beings. The assessment of different risks is a more complex issue since, many factors affect people's risk perception, as discussed before. The assessment of an acceptable risk level can, thus, is a many faceted decision-making situation:

- It is a political problem in the sense that society should decide what kind of risks are accepted and carry the responsibility of this decision.
- It is a juridical problem in the sense that the legal role of a safety goal needs to be clarified.

- It is a research problem in the sense that objective knowledge needs to be gained about various risks in society.
- It is a systems engineering problem; first to design a plant that fulfils the requirement, and then to demonstrate the fulfilment of the requirement.

Commercially and politically, a workable risk-based safety regulation should further include harmonization of requirements between countries and industries. Otherwise, industry and also citizens are not treated equally. While harmonization is discussed a lot, there is still a long way to a harmonized regulation.

20.7 RELATIONS TO OTHER SAFETY PRINCIPLES

QRA can be seen as one part of the safety assessment or safety cases for safety-critical technological systems. QRA is usually compared with a deterministic safety assessment, which is based on the assessment of postulated initiating events together with some failure and acceptance criteria.

QRA and qualitative risk analysis form the whole category of various risk analytic methods where the use of probabilities and associated risk metric vary. Qualitative risk analysis is discussed in Chapter 21.

Defense-in-depth is overall design principle for safety management, as discussed in Chapter 2. Defense-in-depth may regarded as a deterministic principle to cope with uncertainties of risk-based design. In defense-in-depth multiple accident prevention and mitigation barriers are required regardless of their reliability (in simple terms). From risk point of view this may be non-optimal, but it can be nevertheless a reasonable safety management strategy due to uncertainties of QRA.

QRA is prerequisite for risk-based optimization (Chapter 23) and cost–benefit analyses (Chapter 24). Therefore, same principles and challenges apply to those principles.

Chapter 25 discusses human reliability analysis (HRA), which is essential part of QRA applications involving human interactions. Same overall principles apply for HRA as for QRA although HRA includes several challenging aspects such as definitions of "failure modes" for human interactions and quantification of human error probabilities.

REFERENCES

Ale, B. J. M. (2002). Risk assessment practices in The Netherlands. *Safety Science*, 40(1), 105–126.

Althaus, C. E. (2005). A disciplinary perspective on the epistemological status of risk. *Risk Analysis*, 25(3), 567–88.

ANCOLD. (2003). Guidelines on risk assessment. Australian National Committee on Large Dams Incorporated, Australia.

Apostolakis, G. E. (2004). How useful is quantitative risk assessment? *Risk Analysis*, 24(3), 515–520.

Aven, T. (2011). Interpretations of alternative uncertainty representations in a reliability and risk analysis context. *Reliability Engineering and System Safety*, 96(3), 353–360.

Aven, T. (2012). The risk concept—historical and recent development trends. *Reliability Engineering and System Safety*, 99, 33–44.

Bengtsson, L., Holmberg, J.-E., Knochenhauer, M., and Rossi, J. (2011). Probabilistic safety goals for nuclear power plants; phases 2–4. Final report. NKS-226. Nordic Nuclear Safety Research (NKS), Roskilde.

Caldwell, A., Olsson, A., Nordqvist, M., Johanson, G. Holmberg, J.-E., Sunde, C., and Karanta, I. (2014). Addressing off-site consequence criteria using Level 3 PSA—Phase 1 status report, NKS-303. Nordic Nuclear Safety Research (NKS), Roskilde.

CENELEC. (2003). EN 50129: Railway applications—Communication, signaling and processing systems—Safety related electronic systems for signalling. CENELEC, Brussels.

Cooke, R. (2009). A brief history of quantitative risk assessment. *Resources*, 172, 8–9.

Covello, V. T. and Mumpower, J. (1985). Risk analysis and risk management: A historical perspective. *Risk Analysis*, 5(2), 103–20.

Covello, V. T., Sandman, P. M., and Slovic, P. (1988). Risk communication, risk statistics, and risk comparisons: A manual for plant managers. Chemical Manufacturers Association, Washington, DC.

EEC. (1982). Seveso directive. Council directive of 24 June 1982 on the major-accident hazards of certain industrial activities (82/501/EEC). Office Journal of the European Communities No L 230/2.

EPRI. (2004). Guideline for the treatment of uncertainty in risk-informed applications. Technical Basis Document. 1009652, Electric Power Research Institute, Palo Alto.

Farmer, F. R. (1967). Siting criteria—A new approach. IAEA SM-89/34, 1967. *Reprinted in Nuclear Safety*, 8, 539–48.

Fenton, N., and Neil, M. (2012). *Risk Assessment and Decision Analysis with Bayesian Networks*. CRC Press.

Gelman, A., Carlin, J. B., Stern, H. S., and Rubin, D. B. (1995). *Bayesian Data Analysis* (1st ed.). Chapman & Hall.

HSE. (1978). *Canvey: An Investigation of Potential Hazards from Operations in the Canvey Island/Thurrock Area*, London: H.M. Stationary Office.

HSE. (2001). *Reducing Risks, Protecting People. HSE's Decision Making Process*. Norwich: H.M. Stationary Office.

IAEA. (2010a). Development and application of level 1 probabilistic safety assessment for nuclear power plants. Specific Safety Guide Series No. SSG-3, IAEA, Vienna.

IAEA. (2010b). Development and application of level 2 probabilistic safety assessment for nuclear power plants specific safety guide no. SSG-4. IAEA, Vienna.

IEC. (2010). Functional safety of electrical/electronic/programmable electronic safety-related systems (E/E/PES). IEC 61508, ed. 2.0, International Electrotechnical Commission, Geneva.

Jonkman, S. N., van Gelder, P. H. A. J. M., and Vrijling, J. K. (2003). An overview of quantitative risk measures for loss of life and economic damage. *Journal of Hazardous Materials*, A99, 1–30.

Kafka, P. (1999). How safe is safe enough? An unresolved issue for all technologies; safety and reliability. In: Proceedings of ESREL99, Rotterdam.

Kahneman, D., and Tversky, A. (1979). Prospect theory: An analysis of decision under risk. *Econometrica*, XVLII, 263–291.

Kaplan, S., and Garrick, B. J. (1981). On the quantitative definition of risk. *Risk Analysis*, 1(1), 11–27.

Luenberger, D. G. (1998). *Investment Science*. New York: Oxford University Press.

OECD. (2009). Probabilistic risk criteria and safety goals. NEA/CSNI/R(2009)16, OECD, Nuclear Energy Agency, Committee on the Safety of Nuclear Installations, Paris.

Pfitzer, B., Hardwick, M., and Pfitzer, T. (2004). A comparison of QRA methods used by DOD for explosives and range safety with methods used by NRC and EPA. Presentation at the 22nd International System Safety Conference, August 2004.

Salvi, O., and Debray, B. (2006). A global view on ARAMIS, a risk assessment methodology for industries in the framework of the SEVESO II directive. *Journal of Hazardous Materials*, 130, 187–199.

Trucco, P., and Leva, M. C. (2012). BN applications in operational risk analysis: Scope, limitations and methodological requirements. In: P. Premchaiswadi (Ed.) *Bayesian Networks*. InTech.

UNISIG. (2008). ETCS application levels 1 & 2—Safety analysis, SUBSET-088, UNISIG.

U.S. NRC. (1975). Reactor safety study. WASH-1400, U.S. Regulatory Commission, Washington, DC.

U.S. NRC. (1990). Severe accident risks: An assessment for five U.S. nuclear power plants. Final Summary Report, NUREG-1150, Volume 1, U.S. Regulatory Commission, Washington, DC.

Vinnem, J.-E. (2011). *Offshore Risk Assessment: Principles, Modeling and Applications of QRA Studies* (2nd ed.) London: Springer.

VROM. (2005). Guidelines for quantitative risk assessment. Publication Series on Dangerous Substances (PGS 3), The State Secretary of Housing Spatial Planning and the Environment (VROM).

FURTHER READING

Aldrich, J. (2012). Figures from the history of probability and statistics [Online]. Available at: http://www.economics.soton.ac.uk/staff/aldrich/Figures.htm.

Apeland, S., Aven, T., and Nilsen, T. (2002). Quantifying uncertainty under predictive, epistemic approach to risk analysis. *Reliability Engineering and System Safety*, 75(1), 93–102.

Apostolakis, G. (2012). A proposed risk management regulatory framework. NUREG-2150, U.S. Regulatory Commission, Washington, DC.

Apostolakis, G. E. (1988). The interpretation of probability in probabilistic safety assessments. *Reliability Engineering and System Safety*, 23(4), 247–252.

Apostolakis, G., Soares, G., and Kondo, S. (Eds.). (1996). Special issue on treatment of aleatory and epistemic uncertainty. *Reliability Engineering and System Safety*, 54(2–3), 91–262.

Aven, T. (1992). *Reliability and Risk Analysis*. London: Elsevier Applied Science.

Aven, T. (2003). *Foundations of Risk Analysis: A Knowledge and Decision-Oriented Perspective*. John Wiley & Sons.

Aven, T. (2011). On different types of uncertainties in the context of the precautionary principle. *Risk Analysis*, 31(10), 1515–1525.

Aven, T., and Pörn, K. (1998). Expressing and interpreting the results of quantitative risk analyses: Review and discussion. *Reliability Engineering and System Safety*, 61(1), 3–10.

Aven, T., and Vinnen, J. E. (2005). On the use of risk acceptance criteria in the offshore oil and gas industry. *Reliability Engineering and System Safety*, 90(1), 15–24.

Ayyub, B. M. (2003). *Risk Analysis in Engineering and Economics*. Chapham & Hall.

Ballard, G. (1993). Guest editorial: Societal risk—Progress since Farmer. *Reliability Engineering and System Safety*, 39(2), 123–127.

Bäckman, J. (2002). Railway safety—Risks and economics (PhD thesis, Royal Institute of Technology, Stockholm).

Beckjord, E. S., Cunningham, M. A., and Murphy, J. A. (1993). Probabilistic safety assessment development in the United States 1972–1990. *Reliability Engineering and System Safety*, 39(2), 159–170.

Bottleberghs, P. H. (2000). Risk analysis and safety policy developments in the Netherlands. *Journal of Hazardous Materials*, 71(1), 59–84.

Fishburn, P. C. (1970). *Utility Theory for Decision Making*. New York: John Wiley & Sons.

French, S. (1986). *Decision Theory: An Introduction to the Mathematics of Rationality*. Chichester: Ellis Horwood Limited.

HSE. (2002). *Marine Risk Assessment, Offshore Technology Report 2001/063*. Norwich: H.M. Stationary Office.

Kirchsteiger, C. (1999). On the use of probabilistic and deterministic methods in risk analysis. *Journal of Loss Prevention in the Process Industries*, 12(5), 399–419.

Modarres, M. (2006). *Risk Analysis in Engineering: Techniques, Tools, and Trends*. Taylor & Francis.

Norsk Olje & Gass. (2004). Application of IEC 61508 and IEC 61511 in the Norwegian petroleum industry. OLF 070, Norwegian Oil and Gas Association.

NORSOK. (2001). Risk and emergency preparedness analysis, NORSOK Standard Z-013, Rev.2. Norwegian Technology Centre, Oslo.

Otway, H. J., and Pahner, P. D. (1976). Risk assessment. *Futures*, 8(2), 122–134.

Paté-Cornell, M. E. (1996). Uncertainties in risk analysis: Six levels of treatment. *Reliability Engineering and System Safety*, 54(2–3), 95–111.

Rausand, M., and Høyland, A. (2004). *System Reliability Theory Models, Statistical Methods, and Applications* (2nd ed.). New York: John Wiley & Sons.

Smith, D. J., and Simpson, K. G. L. (2010). *Safety Critical Systems Handbook. A Straightforward Guide to Functional Safety: IEC 61508 (2010 Edition) and Related Standards* (3rd ed.). Elsevier.

Tait, N. R. S. (1993). The use of probability in engineering design—An historical survey. *Reliability Engineering and System Safety*, 40(2), 119–132.

Trbojevic, V. M. (2005). Risk criteria in EU. In: *Advances in Safety and Reliability*. Two volume set: Proceedings of the European Safety and Reliability Conference (ESREL 2005), Tri City (Gdynia–Sopot–Gdansk), Poland, June 27–30, 2005. CRC Press.

U.S. NRC. (1981). Fault tree handbook. NUREG-0492, U.S. Regulatory Commission, Washington, DC.

Vatn, J. (1998). A discussion of the acceptable risk problem. *Reliability Engineering and System Safety*, 61(1), 11–19.

Vose, D. (2000). *Risk Analysis: A Quantitative Guide*. John Wiley & Sons.

21

QUALITATIVE RISK ANALYSIS

Risto Tiusanen

21.1 INTRODUCTION

All activities in society involve risks. Organizations manage risk by identifying it, analyzing it and then evaluating the risk by considering the need for risk treatment actions in order to reduce the risk to an acceptable level. Generally speaking, the aim of risk assessment is to support decision-making. The goal of risk assessment is to identify and describe the risks associated with a decision problem and to analyze the potential impacts of the risks. Risk analysis is part of the risk assessment process which again is part of the overall risk management framework in companies and organizations. Risk assessment is applied to entire organizations, at their operating areas and levels, as well as to specific functions, projects and activities in the organization at any time.

Comprehensive identification of risks is critical, because a risk that is not identified will not be included in further analysis. As well as identifying hazards and what might happen, it is important to consider possible causes and scenarios of what consequences can occur. All significant causes and consequences associated with the risk should be considered.

Handbook of Safety Principles, First Edition. Edited by Niklas Möller, Sven Ove Hansson, Jan-Erik Holmberg, and Carl Rollenhagen.

New digital technology increases the complexity of the systems and introduces new potential causal factors. In complex systems, accidents occur on account of the interaction of perfectly functioning components. In practice, when one is designing an innovative multi-technology solution, there are no failure data or user experiences available, or data are very limited, to certain specific applications. It can be argued that current accident models and safety engineering techniques do not cover all of the new technological and operational aspects. This implies that proactive analysis and control of system hazards is growing increasingly important. Qualitative risk analysis techniques, which have long been used in industries that use dangerous processes and for other hazardous systems, can identify the causes of accidents that have never occurred before. In unique, new technology systems, analysis should begin with identification of all potential hazardous events and situations and then involve assessment of whether they are possible or not.

The concept of risk and history of risk analysis principles are covered in previous Chapter 20 on quantitative risk analysis. In this chapter, the theoretical basis for qualitative risk analysis principles, typical qualitative methods and experiences of their use will be presented. In addition to that overall systems engineering viewpoints to systematic risk assessment and need for qualitative risk analysis methods will be presented in this chapter.

21.2 ORIGIN AND HISTORY OF THE PRINCIPLE

Risk assessment is used to support decision-making in all system life cycle phases from conceptual design to operation and maintenance and finally to reuse or decommissioning. Risk assessment is an overall process of risk identification, risk analysis and risk evaluation.

Methods used in analyzing risks can be qualitative, semi-quantitative, or quantitative. The degree of detail required will depend upon the particular application, the availability of reliable data and the decision-making needs of the organization. Some methods and the degree of detail of the analysis may be prescribed by legislation in certain industrial sectors or applications. Performing qualitative and performing quantitative risk analysis are two processes within the risk management process dedicated to specific needs in a company. While qualitative risk analysis should generally be performed on all risks, for all projects, quantitative risk analysis has a more limited use, based on the type of project, the project risks, and the availability of data to be used to conduct the quantitative analysis.

Quantitative analysis estimates practical values for consequences and their probabilities, and produces values of the level of risk in specific units defined when developing the context. Full quantitative analysis may not always be possible or desirable due to insufficient information about the system or activity being analyzed, lack of data, influence of human factors, etc., or because the effort of quantitative analysis is not warranted or required. In such circumstances, a comparative semi-quantitative or qualitative ranking of risks by specialists, knowledgeable in their respective field, may still be effective. In cases where the analysis is qualitative, there should be a clear explanation of all the terms employed and the basis for all criteria should be recorded (IEC/ISO 31010, 2009).

This chapter deals with qualitative risk analysis principles, methods, and techniques. Qualitative risk analysis is a management technique concerned with examining the probability of a risky event and the impact of the risk if it occurs. All risks have both probability and impact that characterizes the risk. Probability is the likelihood that a risk event will occur, and impact is the significance of the consequences of the risk event.

21.3 DEFINITIONS

Definitions and the basic theory of risk analysis principles are also well described in Chapter 20 on quantitative risk analysis. The concept of risk is complex and the term "risk" has various definitions, depending on the context in which the term is used (Rausand, 2011). In literature, the word "risk" is used in many different senses and many kinds of risk are discussed: business risk, social risk, economic risk, safety risk, investment risk, military risk, political risk, etc. The different points of view related to the concept of risk, risk perception, quantification of risk, risk analysis, and risk assessment are discussed widely in literature among others in Kaplan and Garrick (1981) and Renn (1998). Renn (1998) expresses that there is no commonly accepted definition for the term risk, however, all risk concepts have one element in common—the distinction between reality and possibility. Kaplan and Garrick (1981) simplify the idea of risk analysis to be an answer to the following three questions: What can happen? How likely is that it happens? If it does happen, what are the consequences? From a safety engineering point of view, ISO/IEC Guide 51 (1999) defines risk as a combination of the probability of occurrence of harm and the severity of that harm. This definition has been adopted also in the basic machinery-safety standard ISO 12100 (2010).

A qualitative risk analysis prioritizes risks using a pre-defined rating scale. Risks will be scored based on their probability or likelihood of occurring and the impact on project objectives if they occur.

Qualitative assessment defines consequence, probability, and level of risk by significance levels such as "high," "medium," and "low," may combine consequence and probability, and evaluate the resultant level of risk against qualitative criteria. The impact scale is organizationally defined (e.g., a 1–5 scale, with 5 being the highest impact on project objectives—such as budget, schedule, or quality). In a safety engineering example, a qualitative analysis would use a scale of "low," "medium," and "high" to indicate the likelihood of occurrence of a harm, and scale of "fatal, significant injury, slight injury" to indicate the severity of the harm.

What are qualitative factors or measures used in qualitative risk analysis? Qualitative factors or measures are observed and can generally not be measured with a numerical result. Some engineering and scientific measures are qualitative. An expert judgement can be a qualitative measure. Human factors are probably one of the most important issues that deal with qualitative properties as well as environmental issues.

Semi-quantitative methods use numerical rating scales for consequence and probability and combine them to produce a level of risk using a formula. Scales may be linear or logarithmic, or have some other relationship; formulae used can also vary.

The quantitative risk analysis methods have been introduced and discussed in Chapter 20. The main difference between qualitative and quantitative risk analysis is that the former uses a relative or descriptive scale to measure the probability of occurrence whereas quantitative analysis uses a numerical scale.

21.4 UNDERLYING THEORY AND THEORETICAL ASSUMPTIONS

Qualitative risk analysis is a process of assessing the impact and likelihood of identified risks. This process prioritizes risks according to their potential effect on the objectives. Qualitative risk analysis is one way to determine the importance of identified risks and guiding the risk management actions. The two primary functions of a qualitative risk analysis are:

- Risk identification
- Risk characterization and estimation

According to ISO Guide 73 (2009), risk assessment is an overall process of risk identification, risk analysis, and risk evaluation. The risk-assessment standard ISO 31000 (2009) describes the general risk-assessment process, its phases, and its connections to the overall risk-management framework in detail. Figure 21.1 illustrates the

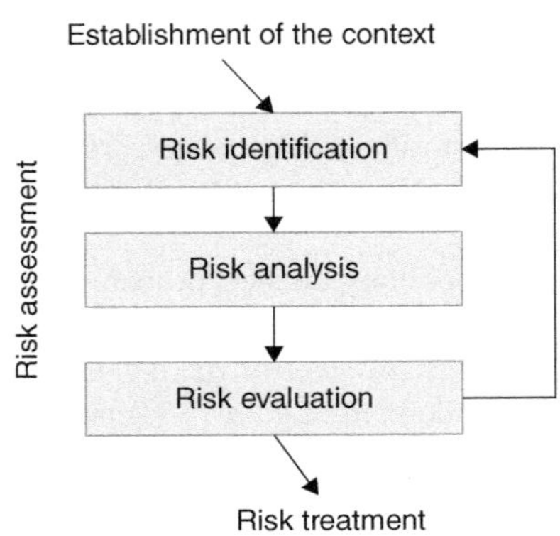

Figure 21.1. The general risk-assessment process phases.

general risk-assessment process phases. The other risk-assessment standard (IEC/ISO 31010:2009) describes a wide variety of general risk-assessment tools and techniques, categorizes them, and evaluates their applicability for risks' identification, analysis, and evaluation.

Risk identification is the process of finding, recognizing, and describing risks in a narrative fashion. In the safety engineering context, we talk about hazard identification. The hazard identification answers qualitatively the questions of how likely the identified hazard is to cause harm and how severe the consequences could be.

There are several techniques for qualitative risk analysis to identify hazards and hazardous events and determine the probability and impact of risks, including the following:

- Brainstorming, interviewing, check lists, Delphi technique
- Historical data
- Preliminary hazard analysis (PHA)
- Operating hazard analysis (OHA)
- Hazard and operability study (HAZOP)
- Risk matrixes

Some of them are introduced briefly in the following chapters. More information about the techniques can be found, for example, from IEC/ISO 31010 (2009).

21.4.1 Brainstorming

The term "brainstorming" is often used very loosely to mean any type of group discussion. Brainstorming involves stimulating and encouraging free-flowing conversation amongst a group of knowledgeable people to identify potential failure modes and associated hazards, risks, criteria for decisions and/or options for treatment.

Brainstorming involves particular techniques to try to ensure that people's imagination is triggered by the thoughts and statements of others in the group. Brainstorming can be used in conjunction with other risk assessment methods described below or may stand alone as a technique to encourage imaginative thinking at any stage of the risk management process and any stage of the life cycle of a system. It may be used for high-level discussions where issues are identified, for more detailed review or at a detailed level for particular problems.

Brainstorming places a heavy emphasis on imagination. It is therefore particularly useful when identifying risks of new technology, where there are no data or where novel solutions to problems are needed. Brainstorming may be formal or informal. Formal brainstorming is more structured with participants prepared in advance and the session has a defined purpose and outcome with a means of evaluating ideas put forward. Informal brainstorming is less structured and often more *ad hoc* (IEC/ISO 31010, 2009).

21.4.2 Preliminary Hazard Analysis

Preliminary hazard analysis (PHA) is a simple, inductive method of analysis whose objective is to identify the hazards and hazardous situations and events that can cause harm for a given activity, facility, or system. It is most commonly carried out early in the development of a project when there is little information on design details or operating procedures and can often be a precursor to further studies or to provide information for specification of the design of a system. It can also be useful when analyzing existing systems for prioritizing hazards and risks for further analysis or where circumstances prevent a more extensive technique from being used (IEC/ISO 31010, 2009).

The PHA method is described, for example, in ISO 12100 (2010), IEC/ISO 31010 (2009), and Vincoli (2006). A PHA type method can be applied for the overall production-area analysis and in analysis of system operation and maintenance concepts in the conceptual design and system-definition phases. In machinery applications, the PHA covers conceptual work site level issues, the machinery system under study and its operation environment, and interfaces to other systems and activities in that environment. The aim of the PHA is to cover all phases of the system life cycle, including construction, testing, and commissioning of the system.

21.4.3 Scenario Analysis

In scenario analysis, descriptive models are developed to describe how the future might turn out. It can be used to identify risks by considering possible future developments and exploring their implications. Sets of scenarios reflecting best, worst, and expected cases can be used to analyze potential consequences and their probabilities. This is close to sensitivity analysis when analyzing risks.

Scenario analysis can be used to assist decisions making and planning future strategies as well as to consider existing activities. Because scenarios are only defined views of possible future happenings, it is important that the probability of a particular scenario is considered. For example, where best-case, worst-case and expected-case scenarios are used, some attempt should be made to qualify, or express the probability of each scenario occurring (IEC/ISO 31010:2009).

21.4.4 Operating Hazard Analysis

The operating hazard analysis (OHA) focuses on system operation and maintenance procedures, aspects of human–technology interaction, and system functions that extend throughout the machinery application. The objective in the OHA is to identify potential hazards and hazardous events in the operation procedures for the system in the selected phases in the life cycle, in view of both human error and technical failures. OHA can be used to analyze operational risks both in new system concepts and existing processes, production plants, or work places. The OHA method is described, for example, in Stephans (2004) and Rausand (2011).

To be able to conduct the hazard analysis, the operation and maintenance procedures and the work tasks need to be described and analyzed first. Task analysis is a detailed examination of the activities associated with the execution of a required task or work. A commonly used method for this purpose is a hierarchical task analysis, which uses top-down approach to systematically go through operations and actions people have to do to meet the goals of the task or system operation.

The scope of OHA should cover all kinds of operations, not only the "normal operation," but also installation, testing, maintenance, repair and emergency operations. Human factors, human errors, and human technology interaction (HTI) aspects are important factors in system operations and successful execution of work tasks. Among others, the following types of operator errors are used to guide the analysis:

- Neglect to perform required actions
- Perform actions that are not required
- Fail to recognize needed action
- Respond improperly (early, late, or inappropriately)
- Engage in poor communication
- Make a maintenance error

Human factors engineering, core-task analysis and related methods are described and discussed more specifically in Chapter 9, "Principles of Human Factors Engineering," in this book.

When designing new production systems or processes, OHA should be conducted in early phases of the system concept's development or at the beginning of the customer-specific application project, depending on the stage in the system's development. The aim is to estimate the operational risks, to evaluate the safety measures designed or planned, and to specify possible additional safety measures for this application. In such cases, OHA should be updated in the system's implementation and commissioning phase for purposes of validating the safety measures designed and implemented to control the operations' risks and to identify any new site-specific risks (or potential risks) that were not identified in the system development phase.

21.4.5 HAZOP Studies

HAZOP is the acronym for HAZard and OPerability study. HAZOP is a structured and systematic examination of a planned or existing product, process, procedure, or system. It is a technique to identify risks to people, equipment, environment and/or organizational objectives. The study team is also expected, where possible, to provide a solution for treating the risk. The HAZOP process is a qualitative technique based on use of guide words which question how the design intention or operating conditions might not be achieved at each step in the design, process, procedure, or system. It is generally carried out by a multi-disciplinary team during a set of meetings.

HAZOP is similar to FMEA because it identifies failure modes of a process, system, or procedure and their causes and consequences. HAZOP differs from FMEA to the extent that the analysis team considers unwanted outcomes and deviations from intended outcomes and conditions and works back to possible causes and failure modes, whereas FMEA starts by identifying failure modes (IEC/ISO 31010, 2009).

In safety engineering, the objective in HAZOP studies is the identification and assessment of functional safety risks related to possible technical failures, software errors, and human errors. The HAZOP studies are typically supported by system architecture modeling, function-level drawings, and the use-case descriptions. These supporting documents can be utilized for sharing of information within the analysis team.

21.4.6 Risk Matrixes

Risk matrix, also called a consequence–probability matrix, is used to rank risks on the basis of the level of risk. Risk matrices are commonly used as a screening tool together with hazard identification methods when many risks have been identified, for example, to define which risks need further or more detailed analysis, which risks need treatment first, or which need to be referred to a higher level of management.

The risk-estimation method utilizing risk matrix is described, for example, in ISO 12100 (2010). The probability of the harm and the severity of the consequences can be estimated, with a scheme involving, for example, five categories, and the final rating of risk level employed, for example, three categories: low, medium, and high risk (see Table 21.1). To assist in the probability estimation, the categories can be concretized, for example, with the following hints:

1	=	Definite	Harm occurs continuously when the system is operated in the manner specified
2	=	Very possible	Harm can easily occur in normal operation conditions
3	=	Possible	Harm can occur in normal operating conditions
4	=	Remotely possible	Harm can occur only in certain operation conditions
5	=	Very unlikely	Harm can occur only if several errors or failures occur at the same time.

TABLE 21.1. An Example of a Risk Rating Matrix for Estimation of Risks to Personal Safety

	Probability				
Severity	Definitely A	Very possible B	Possible C	Remotely possible D	Very unlikely E
Multiple-fatality 1	1	2	4	7	11
Fatality/paralysis 2	3	5	8	12	16
Reportable injury 3	6	9	13	17	20
Lost-time injury 4	10	14	18	21	23
Minor/no-loss injury 5	15	19	22	24	25

The severity of the personal-safety consequences and physical damage or loss of production can be estimated, for example, with the following hints:

1	=	Multiple-fatality	Death of more than one person
2	=	Fatality	One person dying or being paralyzed
3	=	Reportable injury	One person being seriously injured
4	=	Lost-time injury	One person being injured (>3 days' absence)
5	=	Minor or no injury	A maximum of 3 days' absence

When designing new systems, risk estimation can be typically done in two phases: first, without any specific safety measures and, second, in light of the existing, planned, and proposed safety measures. First, the severity of the consequences and the likelihood of the harm are estimated in the scenario of the system and its environment without any specific risk-reduction measures. Second, the risk-reduction measures designed and built into the system concept by the system supplier are recorded and remaining risks are estimated again.

21.5 USE AND IMPLEMENTATION

In this section, different views to the use of qualitative risk analysis principles and reasoning of the use of qualitative methods will be introduced and discussed. The use and implementation of qualitative risk analysis methods will be described first from the overall systems engineering point of view and then from different safety engineering perspectives. Safety engineering practices here mean approaches and normative guidelines developed to support the designers, manufacturers, or end users in development and maintaining of safety in industrial applications. Safety engineering practices have been developed in light of needs and interest in various sectors of industry. Between diverse domains such as manufacturing, the process industry, the nuclear power sector, civil aviation, the space industry, and defense-sector engineering, efforts aimed at reaching safety differ considerably. Four general approaches and practices for safety engineering will be introduced:

- Industrial safety engineering
- System-safety engineering
- Machinery-safety engineering
- Functional safety engineering

21.5.1 Systems Engineering Approach to Risk Assessment

This section reviews the systems engineering approach (or "systems engineering management") and the systems engineering process in brief. The aim is to introduce the basic ideas of the systems engineering approach and identify risk-management activities in the general systems engineering life cycle model.

The systems engineering discipline is thought to have begun in the 1950s amidst the development of the first complex military applications, such as ballistic missile systems. In the 1960s, the Apollo program brought the systems engineering approach to non-military applications. This century's literature has produced several definitions for systems engineering. For example, system-engineering guidelines from the U.S. Department of Defense (DoD DAU 2001) highlight that systems engineering involves two significant disciplines: the technical knowledge domain in which the engineer operates and systems engineering management. According to the SE Handbook (2011), systems engineering is based on systems thinking. The systems engineering approach has a horizontal orientation and includes both technical and management processes. The SE Handbook (2011) states also that decisions made in early phases of the system life cycle, when consequences are not understood, can have enormous implications later in the life cycle.

International systems engineering standards are quite new. All of them have been published in this century. ISO/IEC 15288 (2008) introduces and specifies various system life cycle processes, such as agreement processes, organizational project enabling processes, project processes, and technical processes). ISO/IEC 26702 (2007) specifies and gives guidelines for the applications and management of the systems engineering process. ISO/IEC 16085 (2006) standard introduces and describes a process for the management of risk during systems or software acquisition, supply, development, operation, and maintenance.

The systems engineering process is an essential element of systems engineering management. According to the DoD's DAU (2001), this process is an iterative and recursive problem-solving process to be applied sequentially throughout all stages of development. The aim of the systems engineering process is to transform needs and requirements into a set of system descriptions, to generate information for

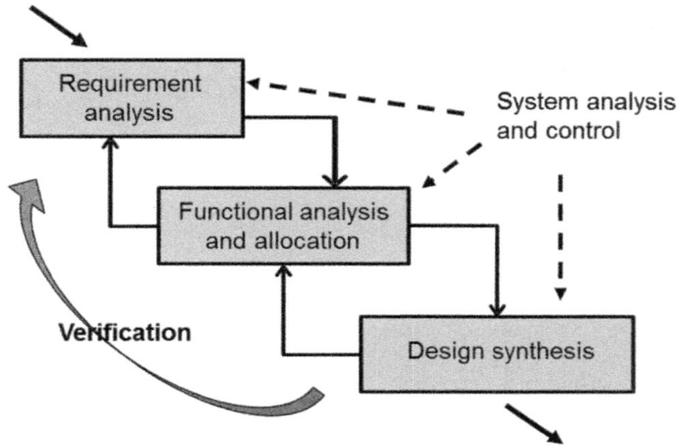

Figure 21.2. Problem solving process phases and activities.

decision-makers, and to provide input for the next level of development. The main activities in the process are requirements' analysis, functional analysis and allocation, and design synthesis (see Figure 21.2).

From a systems engineering management perspective, the risk-management process is a continuous process to identify, analyze, treat, and monitor risks related to the acquisition, development, maintenance, or operation of a system. The general risk-assessment process is described in ISO/IEC 15288 (2008) and specified in more detail in the specific risk-assessment process standard ISO/IEC 16085 (2006). The integration of risk-management activities into the systems engineering process application in each phase in the system life cycle is described in ISO/IEC 26702 (2007). The standard procedure for systems engineering process application described in ISO/IEC 26702 (2007) includes safety engineering activities, which are systematically integrated into each phase of the process and system-analysis efforts (see Figure 21.3).

21.5.2 System-Safety Engineering

The system-safety approach was developed first for complex military, aviation, and space-industry systems, and nowadays it is also applied in the process industry. System safety has been defined as a "sub-discipline of systems engineering that applies scientific, engineering, and management principles to ensure adequate safety, the timely identification of hazard risk, and initiation of actions to prevent or control those hazards throughout the life cycle and within the constraints of operational effectiveness, time and cost" (Vincoli 2006).

System safety is concerned primarily with new systems and has the goal of reducing the risk to an acceptable level. Vincoli (2006) points out that following

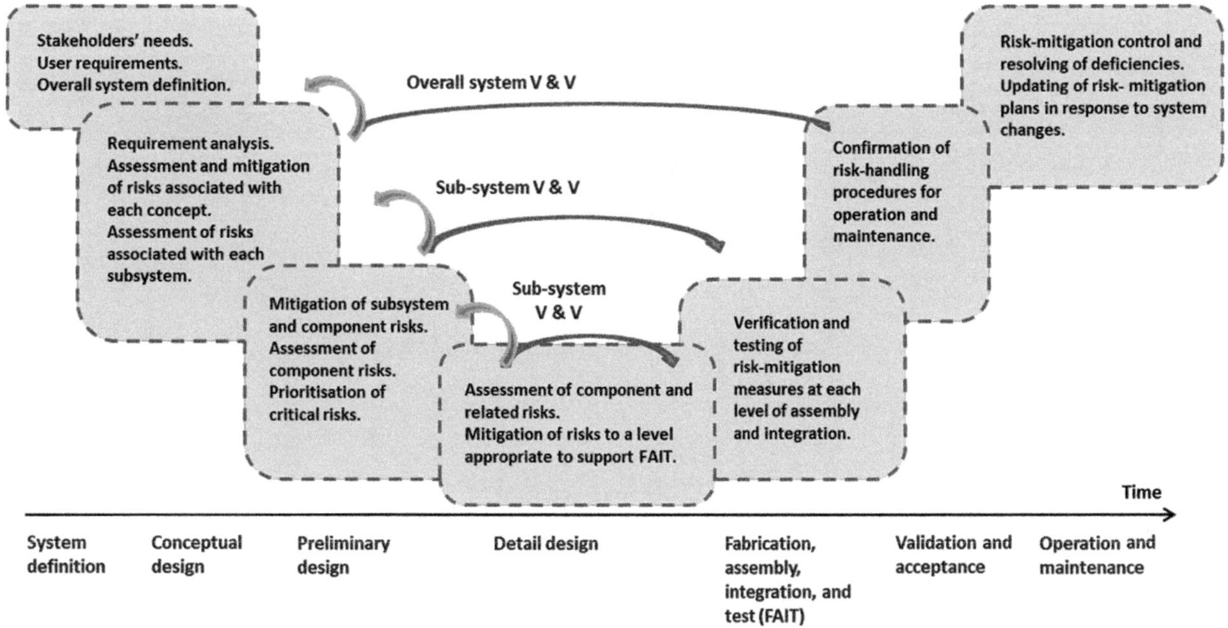

Figure 21.3. General risk-assessment activities related to the system's development and operation phases. Adapted from ISO/IEC 26702:2007 and ISO/IEC 16085:2006.

safety regulations, standards, and written codes is aimed only at meeting minimum safety requirements. The risk-management work associated with system safety is an attempt to exceed these minimum compliance standards and provide the highest level of safety (the lowest level of acceptable risk) achievable for the target system with acceptable cost implications (see also ALARP principle in Chapter 24 and ALARA principle in Chapter 26). It is important to consider direct costs and indirect implications—such as operation restrictions, system performance, operation schedules, and downtime—related to alternative risk-reduction solutions. In complex systems, the human–system interface and, more generally, human-technology interaction issues and design requirements are getting more and more important.

The system-safety standard MIL-STD-882D (2000) does not give any guidance for the selection of risk-analysis methods but does refer to the publications of the System-Safety Society. It is also worth mentioning that the system-safety literature does not refer to the widely applied general functional safety standard IEC 61508-1:2010. This is one indication of how differentiated safety engineering development work has become over the years.

According to Stephans (2004), the main stages in a system's life cycle are the concept, definition, development, production, and system operation. The three basic system-safety analysis methods are preliminary hazard analysis (PHA), fault hazard analysis (FHA), and fault tree analysis (FTA). These differ fundamentally in their concepts and practical execution. Distinct analyses methods have been developed for certain purposes to support decision-making in certain phases of the system life cycle (see Figure 21.4). A list of various methods and more detailed descriptions of them can be found, for example, in Stephans (2004), and Vincoli (2006). The system-safety literature and the standard MIL-STD-882D (2000) introduce a risk-matrix method for the estimation of risks. For example, Vincoli (2006) proposes four categories for hazard severity and five categories for the occurrence of harm (mishaps). Risks are classed into four categories, with the following criteria: unacceptable (changes must be made), unacceptable (make changes if possible), acceptable with management review, and acceptable without review.

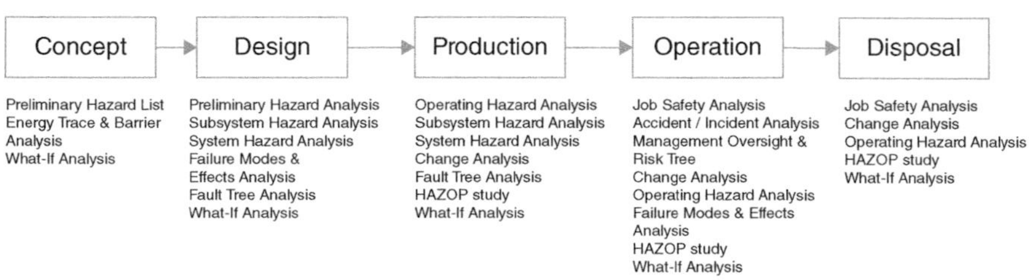

Figure 21.4. Phases in the life cycle and the primary system-safety tasks for a "one-of-a-kind" project or product. *Source*: Vincoli 2006. Reproduced with permission of John Wiley & Sons, Inc.

21.5.3 Industrial Safety Engineering

Traditional occupational safety and health work in industry focuses on improving the safety, health, and welfare of people at work. The work aimed at improving existing work sites and workplaces and at investigating individual past accidents is called industrial safety engineering by, among others, Leveson (2003). The legislation pertaining to work-environment safety requirements in Europe is based on Directive 89/391/EEC (1989), called the Occupational Safety and Health Framework Directive, which sets the general objectives for occupational safety and health work in the workplace and imposes obligations for both employer and employees.

Where the hazards presented by the undertaking are few or simple, for example, in many small businesses, it is appropriate to just carry out a simple qualitative risk assessment. This can be a very straightforward process based on informed judgement and reference to appropriate guidance. Where the hazards and risks are obvious they can be addressed directly, and no complicated process or skills will be required (MHSWR, 1999). In a qualitative risk assessment at work places, it is appropriate to complete the following steps:

- Identify the hazards
- Identify the possible consequences (decide who might be harmed and how)
- Evaluate risk (and decide whether the existing precautions are adequate or whether more should be done)
- Record the findings

In many cases where the hazards are neither few and simple, nor numerous and complex, for example, if there are some hazards that require specialist knowledge, such as a particular complex process or technique, it may be appropriate to supplement the simple qualitative approach with a semi-quantitative assessment. In carrying out semi-quantitative risk assessments, simple qualitative techniques, supplemented by, for example, measurements to identify the presence of hazards from chemicals or machinery, or the use of simple modeling techniques may be appropriate. Simple modeling techniques may also be used to estimate the severity of the consequences and likelihood of realization of hazards.

One of the latest internationally well-known guide for management of occupational health and safety at work is BS 18004 (2008), which gives companies and other organizations guidance on how to build occupational health and safety management elements for their overall management system to manage their occupational health and safety risks and improve their occupational health and safety performance. As the guidelines emphasize, it is quite obvious that there is no single method of hazard identification and risk assessment that can suit all organizations. BS 18004 (2008) introduces some risk-assessment tools and methods, and it points out that in many cases occupational health and safety risks can be addressed via simple methods and the assessment can be qualitative. Methods such as checklists and questionnaires, risk matrices, ranking and voting tables, failure mode and effect analysis (FMEA), hazard and operability (HAZOP) studies, and computer modeling are cited as examples of applicable methods.

According to the standard, an acceptable risk is a risk that has been reduced to a level that can be tolerated by the organization with regard to its legal obligations and policies. The evaluation of risks' acceptability can be based, for example, on a five-band structure reflecting use of the "as low as reasonably practicable" (ALARP) principle, which is described, for example, in IEC/ISO 31010 (2009).

21.5.4 Machinery-Safety Engineering

Machinery-safety guidelines are developed mainly to help the machine-manufacturing industry build safe manual stand-alone machines. The standard ISO 12100 (2010) introduces risk-assessment and risk-reduction processes for machine manufacturers and system designers. Risk assessment is described as a series of logical steps to enable analysis and evaluation of the risks associated with machinery. It is followed by risk reduction. Iteration of this process may be necessary for minimization of hazards or at least to reduce risks adequately via the implementation of protective measures (see Figure 21.5). The objective of risk assessment is the best practicable risk reduction. The risk-assessment process is iterative, and several applications of it might be necessary for reducing the risk and making the best use of the available technology. In carrying out this process, it is necessary to take into account the following four factors, in decreasing order of priority (ISO 12100:2010).

- The safety of the machine over all phases in its life cycle
- The ability of the machine to perform its function
- Usability of the machine
- The manufacturing, operation, and dismantling costs of the machine

Hazard identification in this approach involves systematic identification of reasonably foreseeable hazards (constant hazards and hazards that can appear unexpectedly), hazardous situations, and/or hazardous events in all phases of the machine life cycle. *Risk estimation* is carried out for each hazard via determining of two factors: severity of harm and the probability of occurrence of that harm. The probability factor is presented as a function of three parameters: people's exposure to the hazard, the probability of occurrence of a hazardous event, and the technical and human possibilities for avoiding or limiting the harm (ISO 12100:2010).

After risk estimation, risk evaluation is carried out to determine whether risk-reduction measures are required. The adequacy of the risk reduction shall be determined after each step in the risk reduction until sufficient reduction in risk has been achieved. Risk reduction in this approach is described as a three-step process. The

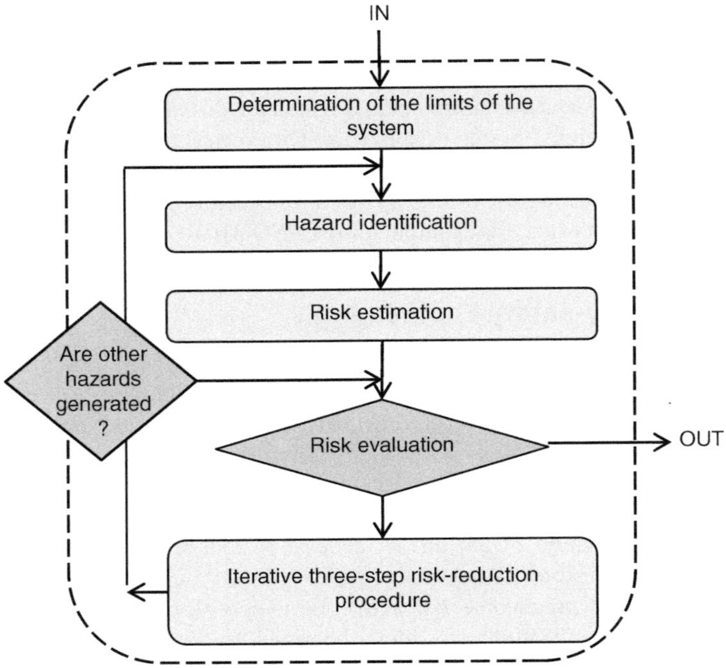

Figure 21.5. The risk-assessment and risk-reduction model modified and simplified. Adapted from ISO 12100:2010.

three steps in risk reduction are inherently safe design measures, safeguarding and/or complementary protective measures, and information for use. This approach also includes guidelines for the consideration of protective measures implemented by the end user such as safe work procedures, use of personal protective equipment, and training (ISO 12100:2010).

ISO TR 14121-2 (2007) introduces two types of hazard-identification approaches for machine design: a checklist-based top-down approach, which starts with potential consequences and examines the possible causes, and a bottom-up approach, which identifies all possible hazards, causes, and consequences. Both ISO 12100 (2010) and ISO TR 14121-2 (2007) can be applied for complicated machinery applications, but they do not primarily offer support for system-level hazard identification and risk estimation for complex automated machinery applications.

21.5.5 Functional Safety Engineering

The functional safety approach has been established to ensure safety of automation in various fields of industry. In general, the functional safety engineering approach is aimed at ensuring safety by eliminating the risks, reducing them to an acceptable level, or rendering them as low as is reasonably practical for reaching a tolerable risk level.

The IEC 61508-1 (2010) introduces and specifies a generic approach to safety engineering that covers all activities in the safety life cycle of systems utilizing E/E/PE components to perform safety functions. Although functional safety is a perspective aimed at ensuring overall safety of the system, the SFS EN 61508 standards (2010) focus only on that portion of the overall risk reduction that is allocated to the safety-related E/E/PE parts of the control system. According to IEC 61508-1 (2010), the objectives of the hazard- and risk analysis task in the overall safety life cycle approach are to determine the hazards, hazardous events, and hazardous situations related to the equipment under control (EUC) and its control system in all modes of operation, in all reasonably foreseeable circumstances (including fault conditions and reasonably foreseeable misuse); to determine the sequences of events leading to the hazardous events; and to determine the EUC risks associated with the hazardous events. Although hazard and risk analysis is introduced as one particular phase in the safety life cycle model, it may be necessary to conduct more than one hazard and risk analysis during the parts of the overall safety life cycle devoted to requirement specifications. If decisions taken in later parts of the safety life cycle change the basis for the earlier decisions, a further hazard and risk analysis should be carried out. IEC 61508-1 (2010) describes the outline of the hazard- and risk-analysis procedure. Hazard and risk analysis can be conducted via application of qualitative or quantitative methods and techniques. A qualitative risk graph or risk matrix can be used for risks' estimation.

Several application standards based on the main functional safety standard have been developed for diverse domains employing safety-related software and programmable electronic control systems, such as avionics, medical devices, railways, road vehicles (ISO 26262-2, 2011), and machinery. The machinery-sector application standard IEC 62061 (2005) has been created to provide guidelines for the requirement specification, design, and verifications of safety-related control systems in machinery applications. The IEC 62061 (2005) standard does not deal with the general risk-assessment process in machinery applications; neither does it give guidance on how the safety-related functions should be identified. It refers to ISO 12100 (2010). It states that the safety-related functions will be specified in accordance with the risk-assessment and risk-reduction processes described in those standards.

An application standard for the process industry (IEC 61511-1:2003) has been established to specify guidelines and practices for functional safety engineering for safety-instrumented systems designed to ensure the safety of an industrial process. There is an important difference between the basic functional safety standard and this application standard. While SFS EN 61508-1 (2010) has been developed for manufacturers and suppliers of safety-related devices, IEC 61511-1 (2003) was developed for designers of safety-instrumented systems, integrators of these systems, and the end users. According to IEC 61511-1 (2003), the objectives of the hazard and risk analysis in process-industry applications include:

- to identify the hazards and hazardous events of the process and associated equipment;
- to estimate the process risks associated with the hazardous event in terms of consequences and likelihood of the event, to determine the sequence of events leading to the hazardous event (in consideration of the various operation modes);
- to determine any requirements for risk reduction, to determine the safety functions required for the necessary risk reduction; and
- to determine whether any of the safety functions are safety-instrumented functions.

IEC 61511-2 (2004) emphasizes that in the process industry, a preliminary hazard and risk assessment should be carried out early, in the project's basic process design phase. At this stage, the objective should be to try to eliminate hazards or reduce the risks as far as is reasonably practicable by applying inherent safety principles and good engineering practice. It is important to start the hazard- and risk-assessment work as early as possible because the assessment results serve as input for system-architecture design and, at the same time, designing and implementing a safety-instrumented system can take a long time. The process-design and system-architecture information is needed before the process and instrumentation diagrams can be finalized.

A final hazard and risk assessment should be completed once the process and instrumentation design and diagrams have been completed. The so-called final analysis should use a formal, fully documented procedure such as hazard and operability studies (HAZOP). The objective of the final assessment should be to confirm that the safety layers designed are adequate for guaranteed safety of the process plant (IEC 61511-2:2004). To emphasize the importance of distinct safety layers and of both technological risk-reduction measures and the role and actions of human operators, IEC 61511-1 (2003) introduces examples of typical risk-reduction methods in the context of a process plant.

21.6 STRENGTHS, WEAKNESSES, LIMITATIONS AND CRITICISM

Strengths, uncertainties, limitations of the qualitative risk analysis methods, and analysis techniques are discussed widely in literature. In this section, some of those views, guidelines, experiences and criticism are shortly described.

Health & Safety Laboratory in England has studied risk assessment practices in industry (Gadd et al., 2003). The HSE report sets out examples of good practice in relation to risk assessment, and contrasts these with pitfalls encountered by HSE, illustrated by case study examples.

According to HSE's report in cases where the hazards are few or simple, for example, in many small businesses, it is appropriate to just carry out a simple qualitative risk assessment. This can be a very straightforward process based on informed judgement and reference to appropriate guidance. Where the hazards and risks are obvious they can be addressed directly, and no complicated process or skills will be required. A qualitative risk assessment should be a systematic examination of what in the workplace could cause harm to people, so that decisions can be made as to whether existing precautions or control measures are adequate or whether more needs to be done to prevent harm.

An outline of good practice in the use of risk assessment is presented in the report, and common industry pitfalls are illustrated throughout this section of the report by the inclusion of 26 case study examples. The identified pitfalls were as follows (Gadd et al., 2003):

- Carrying out a risk assessment to attempt to justify a decision that has already been made using a generic assessment when a site-specific assessment is needed
- Carrying out a detailed quantified risk assessment without first considering whether any relevant good practice was applicable, or when relevant good practice exists
- Carrying out a risk assessment using inappropriate good practice
- Making decisions on the basis of individual risk estimates when societal risk is the appropriate measure
- Only considering the risk from one activity
- Dividing the time spent on the hazardous activity between several individuals—the "salami slicing" approach to risk estimation
- Not involving a team of people in the assessment or not including employees with practical knowledge of the process/activity being assessed
- Ineffective use of consultants
- Failure to identify all hazards associated with a particular activity
- Failure to fully consider all possible outcomes
- Inappropriate use of data
- Inappropriate definition of a representative sample of events
- Inappropriate use of risk criteria
- No consideration of ALARP or further measures that could be taken
- Inappropriate use of cost benefit analysis
- Using "Reverse ALARP" arguments (i.e., using cost benefit analysis to attempt to argue that it is acceptable to reduce existing safety standards)
- Not doing anything with the results of the assessment
- Not linking hazards with risk controls

21.7 EXPERIENCES OF PRELIMINARY HAZARD IDENTIFICATION METHODS

Systems engineering guidelines do not name specific analysis methods for system-safety activities, but it is strongly emphasized that the initial hazard analysis for the system of interest should be started as early as possible in the conceptual design phase and should be continued throughout the system life cycle (SE Handbook, 2011).

Leveson (2003) and Vincoli (2006) have described the relationship and differences between industrial safety engineering and system-safety analyses and engineering practices. They state that PHA methodology using brainstorming sessions and systematic job-safety analysis suite well to the overall-level (i.e., work-site-level) risk analysis. Vincoli (2006) also claims that system-safety analysis (including methods such as PHA, SHA, and OHA) provides an excellent way for industrial-safety experts to achieve an accident-free work environment.

According to the system-safety literature, the bulk of the system-safety activity in the conceptual phase consists of the preparation of a preliminary hazard list (PHL). The preliminary analysis of hazards should be carried out and updated in the system's design phase. Stephans (2004) emphasizes that, even though the primary purpose of PHA in the design phase is to analyze previously identified hazards and to propose safety measures to reduce the risks, the hazard analysis should be continued throughout the design phase. The focus should be on the identification of new hazards and hazardous events, especially in relation to system interfaces and changes in system design. In large hierarchical systems with several levels of subsystems or in large distributed systems, analysis methods such as subsystem hazard analysis (SSHA) or system hazard analysis (SHA) can be used to supplement and update the PHA.

According to Vincoli (2006) PHA is especially useful in the concept design phase. The system-safety experts Leveson (2003) and Vincoli (2006) also state that PHA is an applicable method not only for new complex system concepts but also for such systems in actual use. The commonly expressed view is that OHA should be done as early as possible in the design phase (Vincoli, 2006). It is also important to consider and analyze human factors with a wider scope and more extensively in complex human-machine-system applications. Comparing OHA with HAZOP study, one can find that OHA provides support for the creation of new views of operation situations and human factors.

21.8 EXPERIENCES OF HAZOP STUDIES

Redmill et al. (1999) have studied HAZOP extensively from the system-safety perspective, also as a method to identify hazards in human-centered systems, where the human operator is critical or even central to system functionality or safety. They declare that HAZOP study is an effective way to identify hazards not only in technical systems but also in human-centered systems in looking at deviations from the design intent of the processes and tasks and in the information flowing between them. The HAZOP method can be used to explore humans' deviations from the design intent just as well as deviations on the part of the technical system. The interpretations of the guide words used must be modified to be appropriate for the human perspective. On the other hand, system-safety guidelines propose a dedicated methodology, distinct from HAZOP study, called operating hazard analysis (OHA) that would be more suitable for the identification and analysis of hazards connected with operation and maintenance procedures in the early system definition and development phases. OHA is an attempt to identify hazards resulting from tasks, activities, or operation of system functions. The analysis approach is similar to PHA but focuses the analysis on the level of operation events and activities. The approach of HAZOP study is limited to the designed, intended use of the system, and in this sense OHA is a PHA-type hazard-identification and analysis method. It seems that the analysis of operator errors should be broadened to cover the factors related to the operation situations, operation environment, and factors enabling and supporting the intended, correct, and safe operation.

21.9 EXPERIENCES OF RISK ESTIMATION METHODS

Author's practical experiences and observations of risk estimation difficulties are in line with the conclusions Cox (2008) has come to with respect to the utility of risk matrices in risk management decision-making. He claims that risk matrices do not necessarily support good risk management decisions or effective allocation of risk reduction resources. According to Cox (2008), risk management decisions cannot be based in principle only on mapping of ordered category ratings of severity and probability factors to recommendations for actions or priorities. Risk matrices should be used with caution. The reasoning and judgement behind the risk level, just as much as the information related to each matrix element, should be carefully explained for those interpreting the risk estimation results. This reflects the approach that is emphasized in BS 18004 (2008), according to which the ALARP ("as low as reasonably practical") safety engineering principle and cost-effectiveness analysis should be applied for risks of between acceptable and unacceptable level, for finding appropriate and effective safety solutions.

Vincoli (2006) has developed the risk-estimation matrix further for the system-safety process by applying hazard severity categories and hazard probability levels from MIL-STD-882D (2000) (see Table 21.2). In that matrix, there are four severity categories, five probability categories, and four risk levels. The probability represents qualitative estimation of the likelihood of occurrence of harm caused by an uncontrolled or uncorrected hazard relative to the lifetime of the item or system in question. Risk levels are expressed thus: "unacceptable risk" (changes must be made),

TABLE 21.2. Risk-Estimation Matrix Modified from Vincoli (2006).

	Severity			
Probability	Catastrophic I	Critical II	Marginal III	Negligible IV
Frequent (A)				
Probable (B)				
Occasional (C)				
Remote (D)				
Improbable (E)				

"undesirable risk" (make changes if possible), "acceptable with management review," and "acceptable without review." The risk-classification principle clearly directs the risk-evaluation decision-making and connects it to the correct level of project management or company management.

21.10 SUMMARY OF STRENGTHS AND LIMITATIONS

The following table summarizes the strengths and limitations of the selected techniques and methods. Table 21.3 is modified from the risk assessment standard IEC/ISO 31010 (2009)

21.11 EXPERIENCES FROM COMPLEX MACHINERY APPLICATIONS

In this section, the author introduces and discusses results from two case studies on the evaluation of usefulness of risk assessment methods. Both cases are related to early life cycle phases of an automated mobile work machine system. The cases represent different industrial sectors and automation technologies: underground mining machinery and container handling systems in port terminals.

21.11.1 Change from Machines to Automated Machine Systems

From a machine manufacturer's point of view, single industrial machines and automated machine systems differ greatly considering their whole life cycle. Industrial machines are products that are placed on the market, but automated machine systems are unique customer-specific projects. The systems are built and commissioned at the work site in the final production environment. From both of these perspectives, such automated machine systems can be compared with process-automation applications (Figure 21.6). The change from single machines to automated machine systems transforms the design and engineering problems from machine design and manufacturing

TABLE 21.3. Summary of Strengths and Limitations of Qualitative Methods and Techniques

Technique or Method	Strengths	Limitations
Brainstorming technique	– Encourages imagination which helps identify new risks and novel solutions; – Involves key stakeholders and hence aids communication overall; – Is relatively quick and easy to set up.	– Participants may lack the skill and knowledge to be effective contributors; – Since it is relatively unstructured, it is difficult to demonstrate that the process has been comprehensive, for example, that all potential risks have been identified; – There may be particular group dynamics where some people with valuable ideas stay quiet while others dominate the discussion.
Structured or semi-structured interviews	– Structured interviews allow people time for considered thought about an issue; – One-to-one communication may allow more in-depth consideration of issues; – Structured interviews enable involvement of a larger number of stakeholders than brainstorming which uses a relatively small group.	– Is time-consuming for the facilitator to obtain multiple opinions in this way; – Bias in results is tolerated and not removed through group discussion; – The activation of imagination which is a feature of brainstorming may not be achieved.
Check-lists	– Check-lists may be used by persons who are not expert on the matter under discussion; – When well designed, they combine wide ranging expertise into an easy to use system; – Check-lists can help ensure common problems are not forgotten.	– Check-lists tend to inhibit imagination in the identification of risks; – Check-lists address the "known knowns," not the "known unknown's" or the "unknown unknowns"; – Check-lists encourage "tick the box" type behavior; – Check-lists tend to be observation based, so miss problems that are not readily seen.

(*continued*)

TABLE 21.3. (*Continued*)

Technique or Method	Strengths	Limitations
Preliminary hazard analysis (PHA)	– PHA is able to be used when there is limited information; – PHA allows risks to be considered very early in the system lifecycle.	– PHA provides only preliminary information; it is not comprehensive, neither does it provide detailed information on risks and how they can best be prevented.
HAZOP	– HAZOP provides the means to systematically and thoroughly examine a system, process or procedure; – It involves a multidisciplinary team including those with real-life operational experience and those who may have to carry out treatment actions; – HAZOP generates solutions and risk treatment actions; – It is applicable to a wide range of systems, processes and procedures; – It allows explicit consideration of the causes and consequences of human error; – It creates a written record of the process which can be used to demonstrate due diligence.	– A detailed HAZOP analysis can be very time-consuming and therefore expensive; – A detailed HAZOP analysis requires a high level of documentation or system/ process and procedure specification; – HAZOP can focus on finding detailed solutions rather than on challenging fundamental assumptions (however, this can be mitigated by a phased approach); – The discussion can be focused on detail issues of design, and not on wider or external issues; – HAZOP study is constrained by the (draft) design and design intent, and the scope and objectives given to the team; – HAZOP process relies heavily on the expertise of the designers who may find it difficult to be sufficiently objective to seek problems in their designs.

TABLE 21.3. (*Continued*)

Technique or Method	Strengths	Limitations
Scenario analysis	– Scenario analysis takes account of a range of possible futures which may be preferable to the traditional approach of relying on high-medium-low forecasts that assume, through the use of historical data, that future events will probably continue to follow past trends. This is important for situations where there is little current knowledge on which to base predictions or where risks are being considered in the longer term future.	– This strength, however, has an associated weakness which is that where there is high uncertainty some of the scenarios may be unrealistic; – The main difficulties in using scenario analysis are associated with the availability of data, and the ability of the analysts and decision makers to be able to develop realistic scenarios that are amenable to probing of possible outcomes; – The dangers of using scenario analysis as a decision-making tool are that the scenarios used may not have an adequate foundation; that data may be speculative; and that unrealistic results may not be recognized as such.
Risk matrix (consequence/ probability matrix)	– It is relatively easy to use; – Risk matrix provides a rapid ranking of risks into different significance levels.	– A risk matrix should be designed to be appropriate for the circumstances so it may be difficult to have a common system applying across a range of circumstances relevant to an organization; – It is difficult to define the scales unambiguously; – The use of risk matrix is very subjective and there tends to be significant variation between experts; – Risks cannot be aggregated (i.e., one cannot define that a particular number of low risks or a low risk identified a particular number of times is equivalent to a medium risk); – It is difficult to combine or compare the level of risk for different categories of consequences.

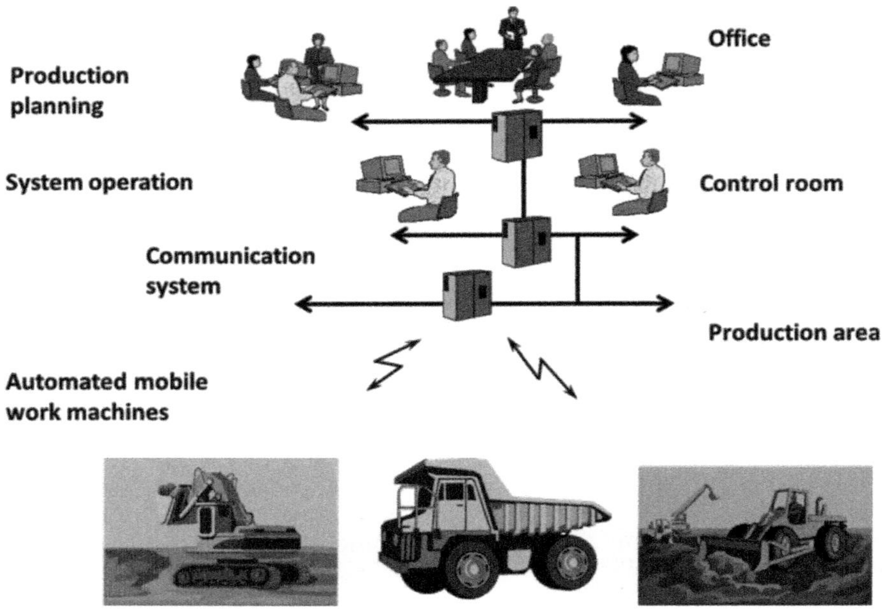

Figure 21.6. The main elements of an automated mobile work-machine system.

issues into system design, systems engineering and system integration ones (Tiusanen, 2014).

The change toward automated complex machine systems takes risk management considerations to a new, system risk level. According to Leveson (2011a), new digital technology increases the complexity of the systems and introduces new potential causal factors. In complex systems, accidents occur on account of the interaction of perfectly functioning components. New safety threats due to system complexity in automated machinery systems are seen in complex human–machine interactions, complex system operations and maintenance situations, systematic or random system failures in control systems, and system interfaces within the operation environment at the work sites. According to Leveson (2011b), current accident models and safety engineering techniques do not cover all of the new technological and operational aspects. Experiences from various sectors of industry confirm these concerns. Among others, Rasmussen (1997), Leveson (2011a), and Rausand (2011) have pointed out that the system complexity, increased amount of software, automated functions, and automatic operation bring out new safety issues and design problems for system designers and safety engineers.

This development implies that proactive analysis and control of system hazards are growing increasingly important. The case with new technology applications is even more demanding as comprehensive data are not available for modeling, simulations and qualitative analysis to help support risk estimation, evaluation and assessment in the early conceptual phases of the system lifecycle. Moreover, risk analysis methods based on a systems approach become essential when identifying and analyzing potential dependability and safety issues from technical and operational perspectives. From the systems engineering view, there is lack of knowledge and experience of approaches and practical risk assessment methods applicable for system availability, maintainability, and safety issues in early conceptual design phases.

21.11.2 Case Studies on Qualitative Methods

The usefulness of qualitative risk assessment methods in automated mobile work-machine applications was examined through evaluation of benefits of the results and impacts of the risk-assessment work in light of the risk-assessment results achieved in the projects. The risk-analysis and risk-estimation methods were analyzed through examination of the pros and cons of the selected methods and work practices in the projects. The evaluation based on project material, documented experiences and observations were supplemented afterward with expert interviews. The case studies and the evaluation method are presented in more details in Tiusanen (2014). In case 1, the usefulness of qualitative risk analysis and risk-estimation methods (Preliminary Hazard Analysis (PHA), HAZOP study and risk matrix) in system design phase were examined. In case 2, the usefulness of the PHA, Operating Hazard Analysis (OHA), and risk matrix methods in conceptual design and in system design phases were examined.

In case 1, the automated ore-transportation system in question was unique—the first of its kind in the underground mining industry. The target system uses autonomously operating dump trucks that transport ore from the transfer points (loading points) to the crusher pin (dump point). The main subsystems in this complex application are a production planning system, a mission-control system and operator stations at the surface control-room level, and automated dump trucks, and the local safety system in the underground production area. A wireless communication system connects the machines to a mine-wide high-speed communication system. The autonomous dump trucks are loaded with manually operated LHDs at the transfer points.

The other case system was related to the development of a semi-automatic container terminal utilizing automatic crane systems for containers' stacking and handling. The target system was an automatic stacking crane (ASC) system that handles containers in one block with three automatic cranes. The complexity of the system can be understood by its multi-machine nature and the control systems and data transmission systems needed for automatic operation. In this application, two identical "inner cranes" use the same tracks, while one "outer crane" uses a separate track. The outer crane is able to pass both inner cranes (Tiusanen, 2014).

21.11.3 Case Study Results

According to the case study results of the mining automation case the analysis method utilizing brainstorming sessions and team discussions was easy to learn and suitable for hazard identification and analysis for an automated mobile work machine system in its conceptual design phase. In the container handling automation case the PHA method, utilizing brainstorming sessions for hazard identification and creation of accident scenarios in combination with team discussions for specifying causes and consequences, worked out well. The hazards identified in the conceptual design phase and the conceptual safety solutions were used later as a baseline in customer site-specific OHA. The system supplier stated that the scope of the risk analysis should be broadened to cover also system installation, building, and testing phases on the site.

Use-case descriptions covering various human–human and human–technology interactions were defined for the specified system operations in OHA. In the container handling case the OHA report including worksheets was considered a useful and practicable work document, describing the system operations and functions at such a level that all system designers, with diverse technological backgrounds, can take part in the analysis and discussion. Comparing OHA with HAZOP study for upper system level risk analysis aimed at analysis of system operations, one finds that OHA provides support for the creation of new views of operation situations and human factors. The approach of HAZOP study is limited to the designed, intended use of the system, and in this sense, OHA is a PHA-type hazard-identification and analysis method.

The results show that the risk estimation methods and risk-evaluation practices utilizing risk matrices need to be developed to be more appropriate for the specific needs of risk assessment activities at the various levels of systems engineering and in the individual phases in the system life cycle. It was also noted that deficiencies in the risk analysis methods applied at the time caused problems near the end of the risk assessment process in both cases. It was difficult to perceive and decide upon the appropriate level of the safety measures. The case study results emphasize that the interpretation and explanation of the implications of the final risk-level results should be developed such that they give better support to the decision-making in risk evaluation and trade studies in requirement, functional, and design analyses in systems engineering processes.

Systematic analysis of system operations and system functions via PHA, OHA, and HAZOP studies in both case studies brought out a great deal of information that was not directly related to safety but did have links to system availability, system usability, or system reliability. Applying a bottom-up analysis method in reliability engineering later in the life cycle may not bring out these issues, because the objectives and analysis method are not aligned with each other. To allow systematic utilization of this valuable information, the companies could establish and maintain an overall RAMS (reliability, availability, maintainability and safety) management process that brings system-availability and system-safety information together (Tiusanen, 2014).

21.12 RELATIONS TO OTHER SAFETY PRINCIPLES

Qualitative risk analysis has relations to many of the safety principles introduced and discussed in this book. Generally speaking qualitative risk analysis is an essential part of risk management framework and risk assessment process in all organizations. Qualitative analysis and estimation of risks are needed together with quantitative risk analysis (Chapter 20). To some extent, qualitative analysis is needed for the specification and design of risk management principles (Chapters 4 and 5); specification of RAMS risk reduction and control measures (Chapter 23); ALARA, BAT, and substitution principle (Chapter 26); specific safety barriers (Chapter 5) and design of safety-related automation (Chapter 10).

Industrial systems are always complex socio-technical systems. Consideration and understanding of human factors, human behavior and human capabilities/limitations are essential in systems engineering and safety engineering. Qualitative risk analysis methods and risk estimation principles are needed when analyzing human factors (Chapter 9) and human reliability issues (Chapter 25). Qualitative risk identification and risk estimation results give also essential input for the development and evaluation of system operation procedures (Chapter 13) as well as input for the development of maintenance optimization principles (Chapter 24).

REFERENCES

BS 18004:2008, Guide to achieving effective occupational health and safety performance.

Cox, L. A. Jr. (2008). What's wrong with risk matrices? *Risk Analysis*, 28(2), 497–511.

Directive 89/391/EEC. (1989). Council Directive 89/391/EEC of 12 June 1989 on the introduction of measures to encourage improvements in the safety and health of workers at work. European Commission. Official Journal of the European Union, L 183, p. 1–8.

DoD DAU. (2001). *Systems Engineering Fundamentals. Department of Defence.* Defence Acquisition University Press.

Gadd, S., Keeley, D., and Balmforth, H. (2003). *Good practice and pitfalls in risk assessment.* Health & Safety Laboratory. Health and Safety Executive. Research report 151. Sheffield. ISBN 0 7176 2732 2

IEC 61508-1:2010, Functional safety of electrical/electronic/programmable electronic safety-related systems—Part 1: General requirements. IEC.

IEC 61511-1. (2003). Functional safety—Safety instrumented systems for the process industry sector—Part 1: Framework, definitions, system, hardware and software requirements.

IEC 61511-2. (2004). Functional safety—Safety instrumented systems for the process industry sector—Part 2: Guidelines for the application of IEC 61511-1.

IEC 62061. (2005). Safety of machinery: Functional safety of electrical, electronic and programmable electronic control systems.

ISO/IEC Guide 51:1999, Safety aspects—Guidelines for their inclusion in standards.

IEC ISO 31010. (2009). Risk management—Risk management techniques. International Organization for Standardization (ISO).

ISO 12100. (2010). Safety of machinery. General principles for design. Risk assessment and risk reduction.

ISO 26262-2. (2011). Road vehicles—Functional safety—Part 2: Management of functional safety. ISO.

ISO 31000. (2009). Risk Management—Principles and guidelines.

ISO Guide 73. (2009). Risk management—Vocabulary.

ISO IEC 15288. (2008). Systems and software engineering—System life cycle processes. IEEE Std 15288-2008, 2nd ed.

ISO IEC 16085. (2006). Systems and software engineering – Life cycle processes – Risk management. IEEE Std 16085-2006.

ISO IEC 26702. (2007). Systems engineering—Application and management of the systems engineering process. IEEE Std 1220-2005, 1st ed.

ISO TR 14121-2. (2007). Safety of machinery. Risk assessment. Part 2: Practical guidance and examples of methods.

Kaplan, S., and Garrick, J. (1981). On the quantitative definition of risk. *Risk Analysis*, 1(1), 11–27.

Leveson, N. (2003). White paper on approaches to safety engineering [Online]. Available at: http://sunnyday.mit.edu/caib/concepts.pdf (accessed on November 2, 2015).

Leveson, N. (2011a). *Engineering a Safer World. System Thinking Applied to Safety*. Cambridge: The MIT Press.

Leveson, N. (2011b). Applying systems thinking to analyse and learn from events. *Safety Science*, 49(2011), 55–64.

MHSWR. (1999). Management of Health and Safety at Work Regulations, Approved Code of Practice and Guidance. HSC 2000, HSE L21. ISBN 0 7176 2488 9.

MIL-STD-882D. (2000). *Standard Practice for System Safety*. Department of Defense.

Rasmussen, J. (1997). Risk management in a dynamic society: A modelling problem. *Safety Science*, 27(2), 183–213.

Rausand, M. (2011). *Risk Assessment: Theory, Methods, and Applications*. John Wiley & Sons.

Redmill, F., Chudleigh, M., and Catmur, J. (1999). *System Safety: HAZOP and Software HAZOP*. Chichester: John Wiley & Sons.

Renn, O. (1998). The role of risk perception for risk management. *Reliability Engineering and System Safety*, 59, 49–62.

SE Handbook. (2011). *Systems Engineering Handbook A Guide for System Life Cycle Processes and Activities*. San Diego, CA: INCOSE.

Stephans, R. (2004). *System Safety for the 21st Century*. New Jersey: John Wiley & Sons.

Tiusanen, R. (2014). An approach for the assessment of safety risks in automated mobile work-machine systems. VTT Science: 69. Dissertation. 2014. VTT, Espoo, 200 p. + app. 6 p. ISBN 978-951-38-8172-6; 978-951-38-8173-3 (PDF).

Vincoli, J. W. (2006). *Basic Guide to System Safety*. John Wiley & Sons.

22

PRINCIPLES AND LIMITATIONS OF COST–BENEFIT ANALYSIS FOR SAFETY INVESTMENTS

Genserik Reniers and Luca Talarico

22.1 INTRODUCTION

One of the definitions of safety is that it is a *dynamic non-event* consistent with the understanding of safety as *the freedom from unacceptable risk* (Hollnagel, 2014). If safety is regarded as a dynamic non-event, the question raises how to count or detect the non-events in time, since this is actually what safety represents or can be regarded as a proxy for. A non-event, is by definition, something that did not happen or does not happen, and therefore, relatively difficult to measure. At the end of a working day, a factory worker, after having come home safely, may ask himself/herself "how many times was I not injured at work today?" or "how many accidents did I not encounter today?" These are all legitimate questions, yet very hard to answer. Nonetheless, this is what companies pay for: for dynamic non-events, or in other words, for events not to happen! However, statistics of non-events within companies do not exist. There are no statistical data or there is no information on non-events.

Therefore, it is obviously very difficult to prove how costly, or maybe how cost-efficient, safety really is. It is only possible to use non-safety information, such as accidents, incidents, and the like, to verify the cost and benefit of non-safety. A related

Handbook of Safety Principles, First Edition. Edited by Niklas Möller, Sven Ove Hansson, Jan-Erik Holmberg, and Carl Rollenhagen.

problem is that non-safety information can only be collected for events with reasonable frequencies, and not for extremely rare events or disasters, since it is evidently not possible to have information based on a number of disasters that happened within a firm and/or an industrial area. A disaster may strike only once. Hence, as can be seen from the above, the economics of operational safety is not a straightforward subject.

Roughly, three types of risks can be distinguished: risks where a lot of historical data are available (type I), risks where little or extremely little historical data are available (type II), and risks where no historical data are available (type III). Whereas type I risks usually lead to low-impact high-probability events, such as most work-related accidents (e.g., falling, little fires, slipping), type II negative risks can result in catastrophes with major consequences and often with multiple fatalities; such risks can lead to so-called high-impact low-probability events. Type II accidents do occur on a (semi-)regular basis from a worldwide perspective, and large fires, large releases, explosions, toxic clouds, etc. belong to this class of accidents. Type III risks (they can be considered as an extreme limit of type II risks) may transpire into "true disasters" in terms of the loss of lives and/or in terms of economic devastation. These accidents often become part of the collective memory of humankind. Examples include disasters such as Seveso (Italy, 1976), Bhopal (India, 1985), Chernobyl (USSR, 1986), Piper Alpha (North Sea, 1988), 9/11 terrorist attacks (USA, 2001), and more recently, Deepwater Horizon (Gulf of Mexico, 2010), and Fukuchima (Japan, 2011). Remark that once type III risks have turned from the theoretical phase into reality, they become type II. Type III events should not be looked at from an economic analysis perspective. However, they should be part of assessments, in the light of taking precautionary- and emergency measures.

For each of these categories of risk, a discussion is possible. For example, what does the company consider to be a "very small risk," or what is considered a "very large risk?" The definitions of these categories are usually decided and discussed within the organization and they can differ from company to company. The risks, using their likelihood and consequences, are usually displayed on a so-called (likelihood, consequences) risk-assessment decision matrix, or abbreviated a risk matrix, and the need for further reduction of risk (or not) is usually determined by the position of the risk within the risk matrix. Depending on their position in the matrix, the risks either should merely be monitored (negligible risk region), or they have to be reduced immediately (unacceptable risk region), or they have to be reduced to the lowest level practicable (tolerable risk region), bearing in mind the benefits of further risk reduction and taking into account the costs of that risk reduction. A company usually has two choices when the risk is located in the tolerable region: either take further risk reduction measures or show that additional risk reduction is not reasonably practicable. "Not reasonably practicable" usually means that the risk reduction costs are higher than the accompanying benefits (the reader is referred to Rushton, 2006 for more details).

However, such an approach implies that a company has sufficient knowledge of the risk to assign it to a certain risk matrix cell. This is a first problem: companies often do not possess adequate information on all risks to be able to assign them unambiguously to one risk matrix cell. Unknown risks will never occur in risk matrices for example. The more the information about a risk, the easier it is to assign it to one specific risk matrix cell. Another disadvantage is that the risk matrix does not distinguish between different types of risks and thus all risks are treated in the same way. This is very dangerous, as it has led toward "blindness for disaster" in the past such as the BP Texas City disaster in 2005 (Hopkins, 2011). Risks of different types should not be mixed especially if safety investment decisions are to be made.

For both type I and type II events which are situated in the tolerable region, different economic considerations should be made and different kinds of economic analyses carried out. In some cases, decisions may be straightforward, but in the majority of the cases, this is not true. Especially in the process industries, where there are obviously important type I as well as type II risks that should be managed and controlled, risk management options may be difficult and not at all obvious. Certainly, cost–benefit analysis (CBA) represents a valid support for decision-making. However, cost–benefit analysis is not a pure science and sometimes models and quantitative methods used within CBA need to employ debatable concepts such as the value of human life, the value of body parts, the question who pays the prevention costs, and who receives the safety benefits.

Nevertheless for some specific problems such as deciding how to secure a chemical plant, a pragmatic approach would be needed by adopting a scientific and sound financial approach. This would require a clear definition of the problem using assumptions to model reality as close as possible, specifying well-defined goals, constraints, and boundaries. Since safety is largely the product of human actions consequent upon prior choices, the economics of operational safety consist in pricing it, or compensating its absence, as to produce the economically optimal amount of safety at socially optimal cost. Depending on the type of accident and the type of risk under consideration, companies carry out cost–benefit analysis to assess potential safety investments resulting in different investment recommendations. In the remainder of the chapter, two groups of CBA methods, respectively, for type I and type II safety investments are analyzed. For each group, an overview of the main CBA approaches is provided describing conditions for applicability, principles, weakness, and advantages.

22.2 PRINCIPLES OF COST–BENEFIT ANALYSIS

CBA represents a powerful methodology to compare, in a systematic way, the total costs, associated with an investment decision, with the benefits that might result from such intervention. In addition, CBA can be used to assess the desirability of an investment whether the benefits outweigh the costs. In case of several investment options, a structured CBA can support decision-makers to rank the alternatives and compare them with the aim of selecting the most promising option.

When dealing with safety investments, decision-makers should consider some important limitations such as maximum budget expenditure, technical constraints, legal frameworks, while maximizing the overall level of safety. In the context of safety investments, costs are generally associated with the sacrifices directly or indirectly related to the safety interventions and expressed in quantifiable and monetized terms. Conversely, benefits are quantified considering the potential savings that are achievable due to the security measures to be implemented compared to a non-investment scenario.

Beside the savings obtained by avoiding potential accidents, one might consider the direct effects of safety interventions such as the costs reduction due to financial disruption avoidance as a consequence of enhanced levels of safety, lower level of absenteeism/illness/fatalities due to improved working conditions, lower level of defects, and better quality of the final products/services. In other words, the safety measures aimed at preventing the risk of accidents can enhance the overall company performance by increasing the positive factors (e.g., sales, margin) and improve the economic net result of any organization. In order to quantify costs and benefits associated with a safety intervention, standard CBA methodology should take into account the following issues:

- A base case (the "do-nothing" scenario) which is used as a reference benchmark to which the marginal outcome of a safety investment needs to be compared.
- All differences between the intervention under analysis and the base case are to be considered and somehow measured. These values should be consistent, and possibly expressed in market prices, and using the same time, but also using appropriate quantitative scales;
- The impact of each investment should be assessed over a clearly defined time horizon depending on the duration of the assets or technologies being involved;
- Sensitivity analysis should be performed by decision-makers to understand whether certain criteria have a higher impact on the outcomes.

A quantitative CBA can be used in two different ways. *Ex ante* CBA could be effectively adopted to support decisions related to a safety budget allocation in order to maximize the benefits from prevention investments. Once the investment has already been made, *ex post* CBAs can be used to assess both the validity of previous decisions and to evaluate the overall impact of safety investments. A method based on a CBA assumes that all costs and benefits associated with a safety investment can be quantified and measured in monetary terms including tangible and intangible impacts to other people and infrastructure in addition to those immediately impacted. Therefore, decision-makers need to explicitly and systematically consider the various factors which should influence strategic choices. This represents one of the major advantages of a quantitative CBA.

22.3 CBA METHODOLOGIES

One of the elements which characterize the concept of risk is, without any doubt, the notion of uncertainty. Unexpected events may happen at different times triggering economic losses, human injuries and human losses, material damages and/or environmental detriments. However, the level of uncertainty may differ from event to event. As mentioned before, the level of uncertainty significantly affects both the likelihood that the unexpected event will occur and its consequences considering both financial and human aspects. In many cases, not all hazards can be known in advance. Moreover, not everything is known about the recognized hazards and there might be a lack of information about possible exposures. Finally, an exact quantification of the losses might be unknown, not knowing a priori all potential losses that may be triggered by an accident scenario. Hence the different domains belonging to the risk triangle hazards–exposure–losses (Meyer and Reniers, 2013) are accompanied by high uncertainties.

Taking into account these aspects, accidents can be classified according to three groups depending on the availability of data and therefore the associated level of uncertainty: accidents for which historical data are available (type I), accident with little available historical data (type II), and events where no historical data are available (type III), see introductory section.

In order to prevent type I accidents many companies adopt risk management techniques which include statistical and mathematical models to predict the likelihood of occurrence of such accidents. Sometimes, quantitative CBA is used to support risk management decisions regarding the selection of the appropriate measures to prevent or mitigate undesired events. In case of Type I accidents, CBA makes use of well-known economic indicators used to evaluate investments such as the net present value, the payback period, and the internal rate of return. These economic indicators can indeed be effectively used to support decision-making under the following conditions: the availability of data, low data variability, the easiness of predicting such types of accidents, and the possibility to mathematically describe clear relationships between prevention measures and levels of accident reduction.

A study conducted in 2010 by OPPBTP (2015) analyzed 27 small, medium, and large companies located in France, assessing 101 safety prevention measures for type I accidents. An economic CBA methodology has been adopted to evaluate the economic impact of each safety intervention. Conversely, type II accidents are extremely difficult to be predicted since the frequency with which these events happen in a single company is very low. Therefore, the scarcity of historical data does not allow the use of statistical models. Extrapolation techniques, assumptions, and expert opinions are used to estimate the likelihood of such accident scenarios happening. However, one might expect that the level of uncertainty and the margin of errors are much higher than in case of type I accidents.

In addition to data availability, data variability might play a crucial role in the decision-making pushing organizations to take different economic considerations

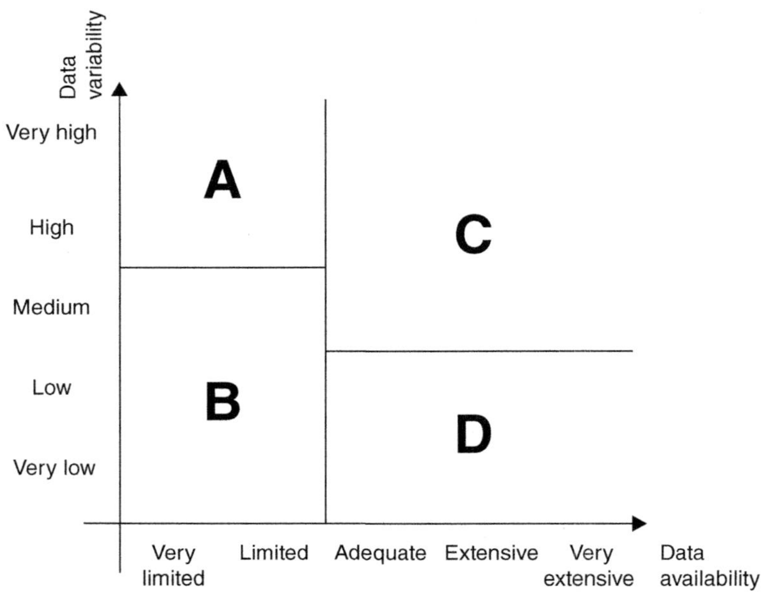

Figure 22.1. CBA clusters depending on data availability and data variability.

(Aven, 2013). Based on these considerations, different types of economic CBAs should be carried out as described later in this chapter. Reniers (2016) proposes a risk management matrix, shown in Figure 22.1, to select the most appropriate CBA decision model depending on two dimensions: data variability and data availability. Each organization can customize the parameters of the matrix and define the economic CBA approaches which should be used for safety investment decisions based on the specific characteristics of the company.

In general, the level of safety which is considered "reasonable" for the decision-makers should be preliminary defined considering a combination of physical limits, economic constraints, and moral aspects (Wilson and Crouch, 2001). For areas classified as D (see Figure 22.1) characterized by an extensive data availability and low data variability (which is typical of a type I accident), the mostly adopted CBA techniques are based on quantitative methods such as the net present value, the return on investment, and the cost–benefit ratio. In case of scarce data availability and high variability (cluster A) which is typical for type II accidents, CBA approaches should certainly rely on risk assessments performed by risk experts. Therefore, a CBA based on a multi-criteria decision-making is to be recommended after a quantitative preliminary risk assessment and prioritization. The remaining clusters (B and C) can make use of a wide spectrum of CBA methodologies depending from case to case.

When the level of uncertainty is low, data are sufficiently available and the data volatility is limited, techniques borrowed from the economic theory based on accurate estimation of costs and benefits may be applied. More specifically, a financial evaluation based on the net present value can be applied as well as the methodologies which rely on the calculation of payback period, the cost–benefit ratio, the internal rate of return as discussed in Section 3.1.

When the level of uncertainty is relatively high and due to the lack of data, not all the economic measures involved in the decision can be calculated with an acceptable level of accuracy, and the notions of cost and benefit need to be broadened. On the one side, costs should include not only the financial sacrifice demanded from an organization to purchase a safety measure, but also criticalities and negative side effects (both quantitatively measurable) correlated to the safety measure are to be considered. On the other side, benefits not necessarily imply monetary return of the investment, but include the concepts of advantages, reputation, respect of legal frameworks, and policies. These latter represent elements difficult to be quantified. Nonetheless, these elements can be assessed by using disproportion factors and multi-criteria approach (see Section 3.2). These CBAs are aimed at ranking alternative investment options providing recommendations to make investment decisions.

22.3.1 CBA for Type I Accidents

For events classified as type I (e.g., operational work-related accidents), an economic evaluation of the safety measures, aimed at preventing or mitigating the consequences of potential accidents, can be carried out within a CBA by using some of the well-known economic and financial indicators. For simple investments where costs and benefits occur in a single identical point of time, a benefit–cost ratio can be adopted as an indicator to determine which options provide the greater benefit for which investment. Where costs and benefits associated with the investments will materialize in different points in time during a fixed time horizon, a CBA would typically use a net present value (NPV) calculation. If the sum of the actualized costs are lower than total actualized benefits, the value of NPV is positive and thus the investment is profitable. Other indicators such as the payback period or the internal rate of return can be used to provide decision-makers with a more comprehensive overview of the safety investments under investigation. Some of the most common quantitative CBA methodologies are reported in the following paragraphs.

22.3.1.1 Cost–Benefit Ratio For some relatively simple safety interventions, it is possible to determine whether the costs of a safety measure outweighs—or not—its benefits. In other words, an index is associated with each measure by comparing the costs of the safety measure with its benefits. As mentioned in the previous sections, the direct costs of a safety measure are relatively easy to determine, while the expected hypothetical benefits are much more difficult to calculate. In case of type I risks, the benefits can be expressed as the "reduced risk," taking into account the costs of accidents with and without the safety measure implementation. The following equation may be used for this exercise (see OGP, 2000 for more details):

$$[(C_{\text{without}} \cdot F_{\text{without}}) - (C_{\text{with}} \cdot F_{\text{with}})] \cdot Pr_{\text{control}} > \text{Safety measure cost} \qquad (1)$$

Or, if no sufficient information regarding the initiating events' frequencies is available for using Eq. (1) :

$$(C_{\text{without}} - C_{\text{with}}) \cdot F_{\text{accident}} \cdot Pr_{\text{control}} > \text{Safety measure cost} \qquad (2)$$

with

$$\begin{cases} C_{\text{without}} = \text{Cost of accident without safety measure} \\ C_{\text{with}} = \text{Cost of accident with safety measure} \\ F_{\text{without}} = \text{Statistical frequency of initiating event if the safety measure is not implemented} \\ F_{\text{with}} = \text{Statistical frequency of initiating event if the safety measure is implemented} \\ F_{\text{accident}} = \text{Statistical frequency of the accident} \\ Pr_{\text{control}} = \text{Probability that the safety measure will perform as required} \end{cases}$$

The formulas above show immediately why this approach may only be carried out for type I risks where sufficient data are available. If not, the required "statistical frequencies" are not known, the probabilities may not be known, and rough estimates should be used, leading to highly unreliable results. If sufficient information is available, results from using these equations for determining the cost–benefit of a safety measure, are reliable.

In case of type I risks, certain economic concepts that are linked with the costs and the benefits could support investment analysis in order to steer a recommendation for the safety investment. Some examples of these economic concepts are the "Internal Rate of Return" (IRR), "Payback Period" (PBP), and other indices, described in the following paragraphs.

22.3.1.2 Net Present Value When a safety measure presents benefits and costs which materialize in different points in time within a fixed time horizon, the NPV associated with the measure can be computed and used to rank alternative measures on the same basis. The main concepts behind the application of the NPV calculation is that all costs and benefits associated with the safety investments can be measured and expressed in monetary terms. Moreover, the effect of time on the positive and negative cash flows (benefits and sacrifices of the measure, respectively) are considered by using a discount factor to quantify and compare all actualized costs and benefits on the same point of time, which is usually year 0 when the investment takes place.

More specifically, a benefit occurring in year 1 needs to be discounted for one period to calculate its present value in year 0. Similar to the calculation of a benefit occurring in year 1, the present value of benefits occurring in year 2 and year 3 are obtained by discounting them respectively 2 and 3 periods. Similar to the previous calculations, the present value (PV) of a benefit (B_t) occurring in year t is obtained by discounting it t periods (Campbell and Brown, 2003). The same applies to costs (C_t). These calculations can be found in the following range:

$$\begin{aligned}
\mathrm{PV}(B_1) &= B_1/\,(1+r)\\
\mathrm{PV}(B_2) &= B_2/\,[(1+r)(1+r)] \;= B_2/(1+r)^2\\
\mathrm{PV}(B_3) &= B_3/\,[(1+r)\;(1+r)\;(1+r)] = B_3/(1+r)^3\\
&\qquad\qquad\ldots\\
\mathrm{PV}(B_n) &= B_n/\,[(1+r)\;(1+r)\;\cdots\;(1+r)] = B_n/(1+r)^n
\end{aligned}$$

The term r represents the discount factor used to actualize costs and benefits. The higher the discount rate, the lower the present values of future cash flows. In this way, the net present value NPV associated with a safety intervention can be calculated by summing up all the PVs related to costs and benefits of the investment as shown in Eq. (3).

$$NPV = \sum_{t=0}^{T} \frac{F_t}{(1+r)^t} \tag{3}$$

The term F_t represents the net cash flow (the difference between the actualized benefits and costs $\mathrm{PV}(B_t) - \mathrm{PV}(C_t)$ in year t, T is the time horizon in which evaluating the investment (expressed in years and usually associated to the duration of the safety investment), and r is the discount rate. An investment is recommended when its NPV is positive, while the safety intervention is not economically convenient and thus often rejected when the NPV is negative. Applied to operational safety, a positive NPV for a given safety investment indicates that the project benefits are larger than its costs.

22.3.1.3 Internal Rate of Return The IRR represents the discount factor that makes the NPV of all future cash flows (or monetized expected hypothetical benefits) equal to zero or, in other words, the value of interest rate at which an investment reaches a break-even point. It can be used to measure and compare the profitability of investments. Generally speaking, the higher an investment's internal rate of return, the more desirable it is to carry on with the investment. As such, the IRR can be used to rank several possible investment options an organization is considering. Assuming all other factors are equal among the various investments, the safety investment with the highest IRR would be recommended to have priority. IRR is sometimes referred to as "economic rate of return" (ERR). An organization should, in theory, undertake all safety investments available with IRRs that exceed a minimum acceptable rate of return predetermined by the company which most often represents the cost of the capital. Investments may of course be limited by the availability of funds to the company. Because the internal rate of return is a rate quantity, it is an indicator of the efficiency, quality, or yield of an investment. This is in contrast with the NPV, which is an indicator of the value or magnitude of an investment. The internal rate of return r^* can be computed as follows:

$$\mathrm{NPV}(r^*) = \sum_{n=0}^{N} \frac{F_n}{(1 + r^*)^n} = 0 \tag{4}$$

Given the same initial investment in a fixed time horizon, the higher the IRR, the more promising the investment for the organization, which is thus more confident about its profitability, also in case of fluctuation of the interest rate. More specifically, if the discount rate materializing in the future (after that the investment took place and during the duration of the project) is lower than the IRR, the expected return of the investment will be higher than expected. In other words, the higher the IRR, the higher the confident zone of the discount rate for which the investment will produce positive cash flows.

When two safety measures present the same IRR, the higher the NPV, the better the investment. In cases a safety investment presents a lower IRR but a higher NPV compared to an alternative safety investment, the first investment should be accepted over the second one. It should be noted that the IRR should not be used to compare investments of different duration. For example, an NPV associated with an investment with longer duration but lower IRR could be higher than an NPV of a similar investment (in terms of total net cash flows) with shorter duration and higher IRR. In Figure 22.2, a graphical representation of the IRR equal to 24.93% is shown for

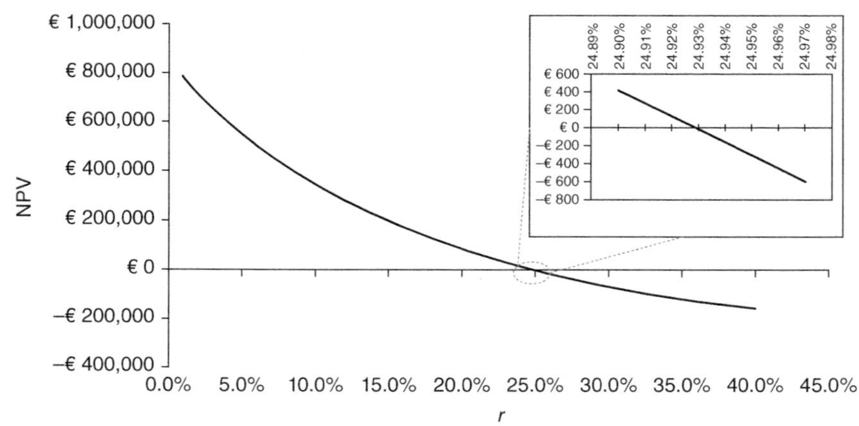

Figure 22.2. IRR rapresentation.

a safety intervention whose duration is 20 years and the initial investment equal to €0.48 million.

22.3.1.4 Payback Period The PBP represents the time needed (usually expressed in years) to recover the total cost of the investment. In other words, PBP simply computes how fast a company will reach a break-even point by recovering the negative cash flows of the safety investment. The PBP of a certain safety investment is used within a CBA to determine whether to proceed with the safety project or not, as longer payback periods are typically not desirable for some companies. It should be noted that PBP ignores any benefits that occur after the fixed time horizon in which the investment is evaluated and it does not measure the investment's profitability. Moreover, neither the time value of money nor opportunity costs are taken into account in the concept of payback period. PBP may be calculated as the cost of safety investment divided by the annual benefit inflows. It is worth noticing that PBP calculation uses cash flows, not the net income.

The shorter the payback period, the better the investments since the company will recover faster its investment on safety. However, together with the payback period other metrics (NPV, IRR, etc.) should be used to avoid myopic operational safety focused on short term. Figure 22.3 shows a graphical representation of the PBP associated with the safety investment having a value of IRR equal to 24.93% as described before.

22.3.1.5 Weaknesses and Limitations of CBA for Type I Accidents CBA can effectively be used within a more robust risk management phase to support decision-makers. Several methods can be used to evaluate safety investments such as the cost–benefit ratio, the net present value, and the internal rate of return. These approaches borrowed from the economic theory require reliable estimations of

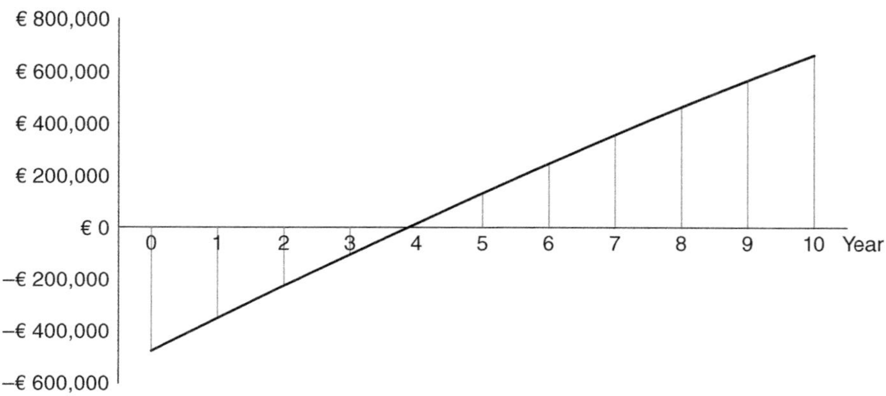

Figure 22.3. PBP representation.

all costs and benefits associated with the safety investment to be assessed. Even though uncertainty may be embedded in the discount factor that is employed, in the NPV calculation, an evaluation of all costs and benefits associated with the safety measures should be conducted in a preliminary stage by risk experts. In general, these approaches are suitable for "soft and standard" safety interventions mainly grouped in the cluster of health and safety measures such as training, simple equipment, small changes to work organization, which are characterized by low investment costs, easiness to assess direct costs, and low incidence of indirect costs which are more difficult to estimate. In addition, these safety solutions present limited time horizons in which the safety investment is evaluated, low levels of uncertainty, and high level of standardization. Therefore, such economic investment analyses of safety investment options are mainly recommendable for type I risks.

Additional drawbacks of CBA methodologies, described before, are mainly due to the accuracy of the source data used to support decision-making. As mentioned before, type I risks are characterized by higher frequencies and higher predictability due to a larger amount of historical data. This allows the use of statistical methods to quantify accurate estimation of the probability of an accident which is used to assess the beneficial effects of safety investments compared to a base non-investment case.

Other limitations are due to the difficult quantification of benefits and due to the possible lack of effectiveness of the safety measure that needs to be put in place. In fact, a wrong assumption may lead to an overestimation of the degree of safety that should be guaranteed by the safety investment. Other limitations are due to the methodological CBA approach that is used. In many cases, this latter approach may be too simplistic and/or it might include many elements of uncertainty such as the discount factor for the NPV methodology that can significantly influence final decisions. Existing cost–benefit software tools mainly focus on type I accidents having thus a rather limited scope. Some examples include CEOccAcc tool (Vandekerckhoven, 2008), Safety pays, Prevention Matrix, and Tyta (see Reniers and Brijs, 2014 for more details). These tools provide useful support in assessing the direct and indirect costs of potential accidents.

22.3.2 CBA for Type II Safety Investments

In case of type II accidents, the possible consequences of an unwanted event/scenario might be significant (e.g., many fatalities and huge economic losses.) In these cases, advanced safety measures might be required to mitigate and/or prevent the consequences of a major accident scenario. Sometimes the required safety investments might be substantial for a firm/organization.

In addition, such events are very rare with probabilities of happening in the range $[10^{-10}–10^{-4}]$ depending on the accident scenario. Nevertheless, in case of the accident, the consequences might cost many millions of euros. Since the probabilities of a major accident happening are extremely low, the risk of a disaster is most often perceived by any decision-maker as a really remote event that will not happen in his/her company. In light of this, safety investments are mainly seen as a cost to be avoided or postponed in the future simply because the (real) costs of safety measures grossly exceed the value of the perceived (hypothetical) benefits.

Considering type II accidents for which only very scarce historical data are available about probabilities or frequencies of accidents and on the correlation between prevention measures and risk reduction, traditional CBA methodologies are not a viable option. Therefore, an alternative quantitative methodology based on a CBA can be applied within a more risk-informed decision-making process to make sure that safety is properly addressed and prioritized during the decision-making. A disproportion factor (DF) is adopted in these cases as a bias in favor of safety assuming quantitative estimates about costs and benefits involved in decision-making. When comprehensive quantitative data are not available or when critical elements to be considered in the decision process are not easy to be quantified and expressed in a monetary form, a multi-criteria decision-making approach can be used to rank and assess safety measures. In the following paragraphs, more discussion about these alternative CBA methods is provided.

22.3.3 Disproportion Factor

Type II accidents are related to extremely low frequencies and a high level of uncertainty. To take this into account, CBA preferably involves a so-called disproportion factor in order to reflect an intended bias in favor of safety above costs. This safety mechanism is vital in the calculation to determine the adequate level of investment in prevention measures, as on the one hand, the probability influences the hypothetical benefits substantially through the number of years over which the total accident costs can be spread out, and on the other hand, the uncertainty regarding the consequences is high (see Goose, 2006 for more details).

Traditional CBAs discourage safety investments when the costs are higher than the benefits. If however a disproportion factor is included, an investment in safety is reasonably practicable unless its costs are grossly disproportionate to the benefits. If Eq. (5) is true, then the safety measure under consideration is not reasonably practicable, as the costs of the safety measure are disproportionate to its benefits (Goose, 2006).

$$\frac{\text{Costs}}{\text{Benefits}} > \text{DF} \rightarrow \text{Costs} > \text{Benefits} \times \text{DF} \tag{5}$$

In order to determine the size of the disproportion factor, some guidelines and rules of thumb are available. Disproportion factors are rarely greater than 10. Moreover, the higher the risk is, the higher the value of disproportion factor to be employed in order to emphasize the magnitude of those risks in the CBA. Therefore, if the risk of accident is very high, it might be acceptable to use a disproportion factor greater than 10 (Goose, 2006). However, Rushton (2006) strongly advices not to use a disproportion factor greater than 30.

As mentioned, for type II safety investments, a disproportion factor is used to increase the risk awareness of the decision-maker and make the benefits comparable with the costs. Thomas and Jones (2010) presented an approach that is widely used within companies to assess alternative safety measures based on the maximum budget that is reasonable investing in safety to prevent major accidents. They referred to this maximum budget as the "maximum justifiable spend" (MJS) which is defined as follows:

$$\text{MJS} = \text{Cost of failure} \cdot \text{Probability of failure} \cdot \text{DF} \tag{6}$$

In other words, MJS represents the amount of money it is worth spending to reduce the risk to "acceptable" levels based on the magnitude of the risk. It provides a high level criterion useful to identify what additional risk reduction measures are worth considering. According to this threshold, the safety investments being higher (more costly) than the MJS are to be discarded. Conversely, those interventions which cost less than the MJS should be implemented. However, this approach requires preliminary assessments aimed at determining a reasonable level of the disproportion factor. Goose (2006) suggests calculating the disproportion factor by multiplying three "how" factors and adding 3 (dimensionless number) to this product as follows:

$$\text{DF} = \text{“How bad”} \cdot \text{“How risky”} \cdot \text{“How variable”} + 3 \tag{7}$$

where the "how bad" factor is computed considering the effects of the average number of fatalities (N_{av}) per event on the disproportion factor as in Eq. (8).

$$\text{“How bad”} = \log_{10}(N_{av}) \tag{8}$$

The "how risky" factor represents the effect on the DF of the potential loss of life (i.e., the average number of casualties expected per year EV):

$$\text{“How risky”} = \log_{10}(10^5 \times \text{EV}) \tag{9}$$

The "how variable" factor represents the effect on the DF of the ratio between the maximum potential fatalities (N_{max}) and the average number of fatalities (N_{av}) per event:

$$\text{“How variable”} = \log_{10}\left(\frac{N_{max}x}{N_{av}}\right) \tag{10}$$

The formula in Eq. (6) represents an alternative to the NPV which can be used within a CBA to shortlist alternative safety investments whether costs and benefits associated with these safety measures can easily be estimated.

22.3.3.1 Risk Experts' Evaluation This methodology is mainly based on a CBA carried out by risk experts. In case of complex investments such as those involving sophisticated technical equipment or infrastructure or high level of uncertainty and variability of data, a quantitative assessment performed by risk experts represents the only available option. Especially for type II accidents with very low probability of happening, risk experts may simulate several scenarios, assessing for each of them, possible consequences and hypothetical benefits due to the application of safety measures.

In many cases, costs and benefits associated with the investments are difficult to be estimated. In addition, the probability associated with an accident scenario may be so low that a quantitative assessment based on the well-known net present value might not be applicable, without using a disproportion factor as a bias in favor of safety. In these cases, several criteria different from costs and benefits need to be used to rank and classify alternative safety investments. A quantitative cost–benefit analysis is still possible if costs and benefits are envisioned from a wider perspective where costs are associated with the disadvantages (i.e., points of weakness, criticalities, and so on) and benefits represent advantages (i.e., points of strengths) associated with the safety measure. In light of this, a quantitative approach can still be used assigning a score to each of the criteria (i.e., advantages and disadvantages) used for the evaluation. To underline the relative importance of one criterion over another, a weight can be used. A total score is thus computed for each safety investment and used as a measure to rank and compare alternative safety solutions. More complex multi-criteria approaches can be used depending on the situations and the expertise of the decision-maker. No matter the specific multi-criteria approach used, one should bear in mind that a quantitative and general CBA may be available to support decision-makers enhancing the quality and the effectiveness of the safety management process.

CBAs based on multi-criteria analysis have a well-established record of providing robust and effective support to decision-makers considering a wide range of problems in different industrial sectors (Connolly and Ordóñez, 2003). This methodology can be applied also for type I accidents when indirect costs are not easily computable and/or benefits are not accurate.

The MCA and the NPV are both analytical approaches of comparing alternative investments. Using the NPV calculation (as shown in previous sections), all inputs and outputs are converted into a common monetized set of values. When it is impossible or impractical to quantify, in a monetary way, costs and benefits associated with an investment, the NPV cannot be applied. In this case, a valid quantitative CBA can make use of a multi-criteria analysis (MCA) which enables the comparison of alternatives considering factors which are not necessarily measured in monetary terms. In other words, MCA can be used to rank several alternatives of safety investments presenting different scores for the criteria which are considered important by decision-makers.

The main benefit of the MCA is to enable decision-makers to consider a number of different criteria, based on a set of complex information, in a consistent way. In

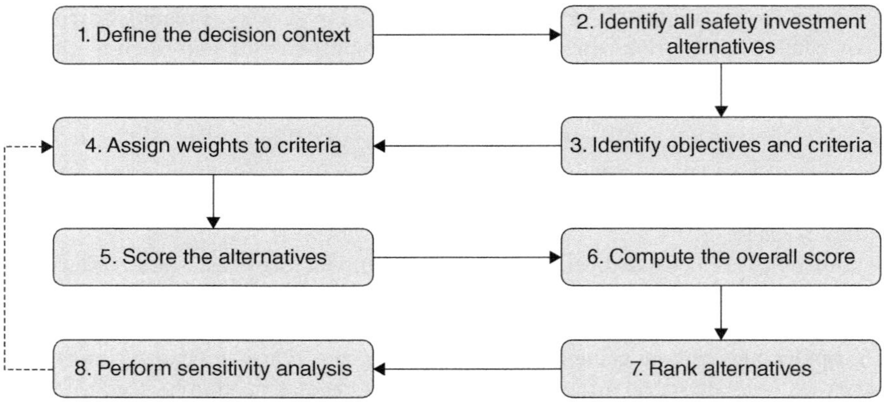

Figure 22.4. MCA scheme.

case quantified values are available or sufficient information exists to estimate or extrapolate appropriate quantified values for inputs, these will be used. The MCA can be used by decision-makers for many purposes: ranking options, identifying the most suitable investment, identifying a limited number of options for more detailed appraisal. The reader is referred to De Keyser and Springael (2010) and Figueira et al. (2005) for more details about multi-criteria decision-making.

The indicators considered by decision-makers should be somehow significantly correlated to the goals that should be achieved capturing important features of the safety investment that could have an influence on the decision that is to be made. Therefore, establishing the most appropriate evaluation criteria within a MCA represents the foundation of a more general CBA. A structured MCA method for cost–benefit analysis should be characterized by the steps shown in Figure 22.4.

22.3.3.2 Weakness and Limitations of CBA for Type II Accidents

Generally speaking, all efforts put in place to reduce or mitigate risks of accidents present a cost that might vary depending on the adopted measures. Nevertheless, a risk reduction measure presents some beneficial effects mainly reducing the likelihood of major accidents from happening or mitigating their consequences. In other words, benefits are mainly linked to the expected reduction of both financial and human aspects. In order to assess safety investments within a structured CBA model, decision-makers are demanded to compare pros and cons and evaluate them on a quantitative and comparable (often monetary) basis.

When evaluating the costs of an accident, two types of losses (financial loss f and human loss h) are usually considered using a multi-attribute utility function in Eq. (11) to quantify the potential consequences of an accident expressed in a monetary form. The financial losses might be high; nevertheless, in many cases, they are not comparable with the tragic human losses due to major accidents. In fact, human losses resulting from major accidents represent a critical factor and the main driver for the estimation of the total costs associated with type II events.

$$v = f + c \cdot h \tag{11}$$

The term c is also known as the value of a statistical life (VSL) which quantifies in a monetary form, the loss of a single human life. It plays a crucial role for any CBA analysis providing a reference point to decision-makers in order to assess the benefits of risk reduction efforts. Behind this approach many ethical issues are hidden. For example, one might find not moral or extremely difficult to quantify and measure in monetary terms the value of human life.

The value associated with human life might depend on the demographic characteristics of the groups of population that is under investigation. Since the characteristics of population groups are most often different from country to county and from group to groups, the values of VSL used in each country and industry may present relevant differences which are, moreover, in constant evolution over time. In other words, the value of VSL to be used within a CBA for a specific country and considering a specific group of population today is largely different from the one adopted 10 years ago. Moreover, depending on the decision-maker (e.g., public institution, government and/or private investors), the selected VSL may differ depending on the assumptions, the qualitative methods used by risk experts, or the quantitative statistic methods to quantify the VSL within hedonic market studies.

Although CBA is a useful tool to support strategic decisions, its use has received severe criticisms by many authors (Anderson, 1995; Sagoff, 1988) mainly because of the controversial assignment of a monetary price to (the loss of) human life. In Ackerman and Heinzerling (2002), some criticisms about CBA and the arguments in defense on why it is wrong to put a price tag on human life are addressed. Ten different classes of philosophical problems that might affect a quantitative CBA are described such as moral aspects, bias in the delimitation of consequences, incommensurability of consequences, and difficulties in defending the essential requirement of transferability across contexts. The so-called "human capital" approach was refused in part because it simply translated greater worldly success into a higher personal "worth" or value, but now the same results appear in the contingent valuation approach. Yet this should not be surprising, since any attempt to derive values from consumer demand in the marketplace will attach greater weight to the most affluent consumers. In other words, highly paid workers have a higher perceived value of life.

While the CBA approaches based on the evaluation of human costs have been increasingly used from an organization point of view in a relative way to compare and rank alternative safety investments, it has been criticized when used from a societal perspective for decisions that may affect large groups of individuals. As mentioned, the main argument against this approach is that human lives do not have a monetary price. In reality, a pragmatic approach should be used. Human life is incommensurable; however, the VSL included in a CBA is for calculation purposes,

TABLE 22.1. VSL in Different Countries

Country	VSL[a] (million €)
Australia	3
The Netherlands	5.8
United Kingdom	2.4
USA	7.8

[a]All amounts are calculated in 2014 euros, for easy comparison.

only to make the impact of an accident on people somehow measurable and support the assessment of safety interventions. Moreover, since the benefits of a safety intervention are always compared to a base non-intervention case, and assuming that the costs of human life are much greater than the financial consequences, the value itself assigned to the VSL has only a relative impact useful to rank and compare alternative safety investments.

In Viscusi and Aldy (2003), a survey is presented in which the cost of human life is analyzed considering ten different countries. Moreover, the impact of several factors such as econometric issues, the role of unionization in risk premiums, and the effects of age are considered while measuring the VSL. Several authors have also studied the relationship between the value of a statistical life and mortality risk showing that there is a negative correlation between the VSL and the annual occupational mortality. The higheer the mortality the lower the VS (Miller, 2000; Sunstein, 2014). Table 22.1 provides some VSL values used in different countries.

Even if the human life represents an incommensurable value from an ethical and/or philosophical stand point, many researches in the past years conducted studies and came up with ranges to be used in a pragmatic way to cope with safety investment decisions. Several approaches can be used to estimate the VSL based on risky behavioral choices. A first approach to estimate the VSL consists in quantifying the additional amount of money earned by a worker who performs a risky job compared to the average salary of a safer administrative duty in the same sector. A second approach can be based on human behavior in decision-making. How much one is willing to pay in safety features to increase its overall safety level? There is no perfect methodology, but over the past years, researchers have reached a consensus that $1.5 million is much too low. Viscusi and Aldy (2003) used different methodologies and techniques to assess the value of human life in the United States which is estimated in the range $7–9 million. In the past decades, several studies have been carried out to quantify and harmonize the different values of VSL in different countries. In Sunstein (2014), the values of VLS ranges between 45,000€ and 28,000,000€ depending on the country. Although the values are so different, Miller (2000) showed that across countries, the average VSL increases almost linearly with income.

Assuming that the value for a human life is worth $9 million, will this figure remain constant in the future also? Since safety decisions generally assume long time horizons, it might happen that a safety measure might prevent a major accident and thus save several human lives in 40–50 years from the moment at which the decision is made. A discount factor might be used to actualize this hypothetical future benefit. However, an accurate and realistic value for the discount factor should be defined as it needs to be used for the NPV methodology and several sensitivity analyses may be carried out to produce optimistic, pessimistic, and average scenarios to support decision-makers.

The second main critic of using a quantitative CBA for type II risks is the following. It should be noticed that for type II accidents, higher levels of uncertainty may affect measurements of hypothetical benefits and costs scenarios. In addition, the potential consequences of a rare accident are much harder to identify and to quantify than for type I accidents. For example, especially in case of major accidents, indirect costs largely exceed the direct costs and are not always easy to be estimated. In many cases, due to a lack of historical data, not all consequences can be identified. Moreover, even if these consequences are identified, their quantifications presents come criticalities. Nevertheless, a CBA model for type II accident prevention, although representing reality with reasonable levels of approximation, is certainly advisable to help the decision-maker.

Lastly, the dependence of the CBA on the decision perspective might significantly affect the results. Due to the rarity of these events and the crucial role played by many parameters (e.g., DF, VSL) within a CBA to select and/or rank safety investments, assumptions and models used by the decision-makers might have a significant influence on final decisions. In addition, the assessment of the probability of a rare event may not be accurate. All these problems may be overcome by doing appropriate sensitivity analysis to assess the robustness of the decisions. As in many other scientific problems, one might consider that there is no simple recipe to evaluate and manage the risks and uncertainties that are associated with decision-making (Klinke and Renn, 2002). Therefore, especially for major accidents the results of a CBA should always be used with great caution and the recommendations of risk experts might be always carefully evaluated in the final decisions. However, a CBA represents a rational and scientific sound perspective to assess safety decisions from a decision-maker's point of view.

22.4 CONCLUSIONS

Cost–benefit analyses in case of safety investments constitute much more than calculating the costs of actual accidents, or determining the costs of prevention. Hypothetical benefits, that is, the benefits gained from accidents that have never occurred, should be considered, and type I as well as type II risks should be taken into account when dealing with prevention investment choices. Decisions concerning safety investments make up for a complex decision problem.

The net present value can be defined as the net value on a given date of a payment or series of payments made at other times. If the payments are made in the future, they are discounted to reflect the time value of money and other factors such as investment risk. Net present value calculations are widely used in business and economics to provide a means to compare cash flows at different times on a meaningful "like-to-like" basis. Discounted values reflect the reality that a sum of money is worth more today than the same sum of money at some time in the future. Therefore, in cost–benefit analyses, prevention costs incurred today should be compared with hypothetical benefits obtained at some time in the future, but equated to today's values.

REFERENCES

Ackerman, F., and Heinzerling, L. (2002). Pricing the priceless: Cost–benefit analysis of environmental protection. *University of Pennsylvania Law Review*, 1553–1584.

Anderson, E. (1995). *Value in Ethics and Economics*. Harvard University Press.

Aven, T. (2013). The concepts of risk and probability: An editorial. *Health, Risk & Society*, 15(2), 117–122.

Campbell, H. F., and Brown, R. P. C. (2003). *Benefit–Cost Analysis: Financial and Economic Appraisal Using Spreadsheets*. Cambridge University Press.

Connolly, T., and Ordóñez, L. (2003). *Judgment and Decision Making*. Wiley Online Library.

De Keyser, W., and Springael, J. (2010). *Why Don't We KISS!?: A Contribution to Close the Gap Between Real-World Decision Makers and Theoretical Decision-Model Builders*. ASP/VUBPRESS/UPA.

Figueira, J., Greco, S., and Ehrgott, M. (Eds.). (2005). *Multiple Criteria Decision Analysis: State of the Art Surveys* (Vol. 78). Springer Science & Business Media.

Goose, M. H. (2006). Gross disproportion, step by step—A possible approach to evaluating additional measures at COMAH sites. In: *Institution of Chemical Engineers Symposium Series* (Vol. 151, p. 952). Institution of Chemical Engineers.

Hollnagel, E. (2014). *Safety-I and Safety-II: The Past and Future of Safety Management*. Ashgate Publishing.

Hopkins, A. (2011). Risk-management and rule-compliance: Decision-making in hazardous industries. *Safety Science*, 49(2), 110–120.

Klinke, A., and Renn, O. (2002). A new approach to risk evaluation and management: Risk-based, precaution-based, and discourse-based strategies. *Risk Analysis*, 22(6), 1071–1094.

Meyer, T., and Reniers, G. (2013). *Engineering Risk Management*. Walter de Gruyter.

Miller, T. R. (2000). Variations between countries in values of statistical life. *Journal of Transport Economics and Policy*, 169–188.

OGP. (2000). *Fire system integrity assurance*. International Association of Oil and Gas Producers, London, UK.

OPPBTP. (2015). *An economic approach to prevention—2014 consolidated results*. The French Professional Agency for Risk Prevention in Building and Civil Engineering. Available at: www.preventionbtp.fr

Reniers, G. (2016). *Operational Safety Economics within Organizations. Fundamentals and Practical Approaches*: John Wiley & Sons.

Reniers, G., and Brijs, T. (2014). Major accident management in the process industry: An expert tool called CESMA for intelligent allocation of prevention investments. *Process Safety and Environmental Protection*, 92(6), 779–788.

Rushton, A. (2006). CBA, ALARP and industrial safety in the United Kingdom.

Sagoff, M. (1988). Some problems with environmental economics. *Environmental Ethics*, 10(1), 55–74.

Sunstein, C. R. (2014). The real world of cost–benefit analysis: Thirty-six questions (and almost as many answers). *Columbia Law Review*, 167–211.

Thomas, P. J., and Jones, R. D. (2010). Extending the J-value framework for safety analysis to include the environmental costs of a large accident. *Process Safety and Environmental Protection*, 88(5), 297–317.

Vandekerckhoven, D. (2008). A practical tool for the estimation of the direct and indirect costs of occupational accidents: CEOccAcc (Thesis, KU Leuven, Belgium).

Viscusi, W. K., and Aldy, J. E. (2003). The value of a statistical life: A critical review of market estimates throughout the world. *Journal of Risk and Uncertainty* [Online]. Available at: http://doi.org/10.1023/A:1025598106257

Wilson, R., and Crouch, E. A. C. (2001). *Risk–Benefit Analysis*. Harvard University Press.

23

RAMS OPTIMIZATION PRINCIPLES

Yan-Fu Li and Enrico Zio

LIST OF ACRONYMS

A	Availability
C	Cost
EA	Evolutionary algorithm
M	Maintainability
MOEA	Multi-objective EA
MOO	Multi-objective optimization
R	Reliability
RAMS	Reliability, availability, maintainability, and safety
RAMS+C	RAMS and cost
S	Safety

Handbook of Safety Principles, First Edition. Edited by Niklas Möller, Sven Ove Hansson, Jan-Erik Holmberg, and Carl Rollenhagen.

23.1 INTRODUCTION TO RELIABILITY, AVAILABILITY, MAINTAINABILITY, AND SAFETY (RAMS) OPTIMIZATION

Reliability optimization aims at maximizing system reliability, and related metrics, while minimizing the cost associated with the reliability choices. It has been an active research domain since the 1960s. Various optimization problems have been formulated and various solution techniques proposed. In general, the decision variables of the optimization problems encode the parameters that drive system reliability properties, for example, the inherent component reliability (e.g., failure probability, failure rate), the system logic configuration (e.g., number of redundant components, component assignments), and those relevant to testing and maintenance activities, which govern the system availability and maintainability characteristics (e.g., test intervals, maintenance periodicities).

In terms of the objectives, system reliability optimization is typically based on the quantification of the effects that the design choices (e.g., component reliability, system logic configuration), and testing and maintenance activities have on reliability-related attributes: reliability, maintainability, and availability (RAM) (Martorell et al., 2005). For hazardous systems, risk attributes must also be accounted for, that is, simultaneous consideration of RAM and Safety criteria (RAMS) (Frank, 1995). Moreover, any design, testing, and maintenance activity is associated with certain costs. In conclusion, system reliability optimization has essentially a multi-objective formulation, which aims at finding the appropriate choices of reliability design, testing, and maintenance procedures that optimally balance the conflicting RAMS and costs (RAMS+C) attributes (Marseguerra and Zio, 2006). To this aim, the decision variable vector $\boldsymbol{x}$ is evaluated with respect to numerical objectives regarding RAMS+C attributes: $R(\boldsymbol{x})$ = system reliability; $A(\boldsymbol{x})$ = system availability; $M(\boldsymbol{x})$ = system maintainability, that is, the unavailability contribution due to test and maintenance; $S(\boldsymbol{x})$ = system safety, normally quantified in terms of the system risk measure $Risk(\boldsymbol{x})$ (e.g., as assessed from a probabilistic risk analysis); $C(\boldsymbol{x})$ = cost required to implement the vector choice $\boldsymbol{x}$.

Many researchers convert the multi-objective optimization (MOO) problem into a single-objective one, by for example, regarding one of the RAMS attributes or cost as the single objective and the other attributes as constraints, or by aggregating all attributes into a single objective. The solution techniques to single-objective RAMS+C optimization problems have been well documented in the surveys by Kuo and Prasad (2000) and by Kuo and Wan (2007). The main problem types surveyed are redundancy allocation and component assignment. They are combinatorial and difficult to solve by mathematical programming approaches. The methods surveyed are evolutionary algorithms (EAs), for example, genetic algorithm, simulated annealing, and ant colony optimization; exact methods, for example, dynamic programming and branch-and-bound. The surveys reveal a rising trend in the applications of EAs and hybridizations which combine local search methods with different EAs to improve computational efficiency or with exact methods to reduce the search space.

Tutorials on multi-objective evolutionary algorithms (MOEAs) have been published by Marseguerra and Zio (2006) and Konak et al. (2006).

Optimization is a key step in the decision-making process involved in RAMS management, which often considers multiple conflicting objectives, for example, safety and cost. The standard approach of multiple criteria decision-making typically specifies an aggregated single objective of the optimization, often in terms of monetary return, such that the best action (i.e., the set of decision variable values) that minimize the aggregated objective can be eventually identified. However, often the decision-maker (DM) is not able to provide sufficient preference information for defining the objectives aggregation. Efforts are required to assist DM in systematically probing his/her true feelings for the relevant trade-offs between incommensurable objectives (Keeney and Raiffa, 1993). This can be performed before, after, or within the optimization step: if the DM is able to provide beforehand the information required for objectives aggregation then the MOO problem can be transformed into single objective and the corresponding solution techniques can be applied; if the DM wishes to examine all efficient (i.e., Pareto-optimal) solutions, then, the solution techniques that simultaneously optimize all objectives can be applied and finally the DM selects his/her most preferred one from a set of optimal solutions which are non-dominated to each other; if the DM is driving the search in an iterative manner, then, the phase of solving the optimization problem alternates with the phase of interaction with the DM, for the dynamic and iterative definition of his/her preferences and trade-offs.

To the knowledge of the authors, in the literature there is no reference synthesis of the optimization principles for decision-making in RAMS engineering and management. This chapter fills the gap by presenting an overview of the MOO process, including the various MOO principles, performance measures, and preferred-solution selection methods.

The rest of this chapter is organized as follows. Section 23.2 presents the general formulation and the concepts of MOO. Section 23.3 presents the methods for solving MOO problems. Section 23.4 presents the performance measures. Section 23.5 presents the methods of selecting preferred solution(s). Based on the materials in Sections 23.3–23.5, and the research experience and the knowledge of the authors, Section 23.6 synthesizes the guidelines for implementing the MOO methods. Section 23.7 illustrates the implementation on one redundancy allocation case study. Section 23.8 discusses weaknesses, limitations, and criticism about the MOO methods. Section 23.9 presents the relations to other safety principles.

23.2 MULTI-OBJECTIVE OPTIMIZATION

Reliability of a certain component or system measures its capability to sustain operation without failure under specified conditions during a given period of time. It is an intrinsic property that directly depends on the component or system physical characteristics and its design, and not on its maintenance. Maintenance relates to all activities performed on the component or system during the operational lifetime, in order to sustain or restore its functional capabilities. In spite of the positive effects on component or system functionality, maintenance downtime puts the component or system out of service due to maintenance activities. In the end, a component or system does not perform its intended function either because of failure or maintenance downtime. Maintainability measures the capability of the system to be maintained under specified conditions during a given period of time. Availability measures the effects on the component or system functionality of both the failure process and the maintenance. For quantitative analysis, the above-mentioned metrics are typically all defined in probabilistic terms.

Safety is defined as the capability to prevent or mitigate the consequences of postulated accidents on specified targets (e.g., workers, public, and environment). Risk is often adopted as the quantitative metric of interest, in relation to scenarios, probabilities of occurrence, and consequences.

With the given definitions of RAMS metrics, the MOO problem formulation is introduced in the following section.

23.2.1 Problem Formulation

In general, a MOO problem can be formulated as follows:

$$\text{Minimize} \quad f_i(\boldsymbol{x}), \quad i = 1, \ldots, M \tag{23.1a}$$

$$\text{Subject to} \quad \begin{cases} g_j(\boldsymbol{x}) = 0, & j = 1, \ldots, J \\ h_k(\boldsymbol{x}) \leq 0, & k = 1, \ldots, K \end{cases} \tag{23.1b}$$

where f_i is the ith of the M objective functions, $\boldsymbol{x} = (x_1, x_2, \ldots, x_N)$ is the vector that represents a solution in the solution space $\mathbf{R}^N$, g_j is the jth of the J equality constraints, and h_k is the kth of the K inequality constraints. Let $z_i = f_i(\boldsymbol{x}), \ \forall i$, then $z = (z_1, z_2, \ldots, z_M)$ is the objective vector and z is inside $\mathbf{R}^M$ the objective space.

For simplicity of illustration, we assume that all objective functions are to be minimized: if one $f_i(\boldsymbol{x})$ were to be maximized, it can be converted into $-f_i(\boldsymbol{x})$ for minimization. Adapting the general definition for RAMS+C optimization, the MOO problem has the following formulation:

$$\text{Minimize} \quad (1 - R(\boldsymbol{x}), 1 - A(\boldsymbol{x}), M(\boldsymbol{x}), Risk(\boldsymbol{x}), C(\boldsymbol{x})) \tag{23.2a}$$

$$\text{Subject to} \quad R(\boldsymbol{x}) \geq R_L \tag{23.2b}$$

$$A(\boldsymbol{x}) \geq A_L \tag{23.2c}$$

$$M(\boldsymbol{x}) \leq M_U \tag{23.2d}$$

$$Risk(\boldsymbol{x}) \leq R_U \tag{23.2e}$$

$$C(\boldsymbol{x}) \leq C_U \tag{23.2f}$$

$$\boldsymbol{x} = \left(y_1, \ldots, y_{N_d}\right) \in \mathbf{R}^{N_d} \tag{23.2g}$$

The quantities R_L, A_L, M_U, R_U, C_U represent the constraining threshold values for the reliability, unavailability, maintainability, risk, and cost objectives, respectively. Martorell et al. (2005) gives typical threshold values for RAMS+C objectives. N_d is the number of decision variables.

MOO requires minimizing all objectives simultaneously. If there is no conflict between any pair of the objectives, then one would find a single solution that minimizes all objectives at the same time. In this case, solving the MOO problem is equivalent to minimizing one of the objectives. The MOO methods are worth applying only when there exist conflicts among the objectives: in this case, due to the contradiction and possible incommensurability of the objective functions, MOO methods identify a set of optimal solutions $\boldsymbol{x}_l^*$, $l = 1, 2, \ldots, L$, instead of a single optimal solution.

23.2.2 Pareto Optimality

In the set of optimal solutions of a MOO problem, no one can be regarded as better than any other with respect to all the objective functions. The identification of this set of solutions can be achieved in terms of the concepts of Pareto optimality and dominance (Sawaragy et al., 1985): in the case of a minimization problem, solution $\boldsymbol{x}_a$ is regarded to dominate solution $\boldsymbol{x}_b$ ($\boldsymbol{x}_a \succ \boldsymbol{x}_b$) if both the following conditions are satisfied:

$$\forall i \in \{1, 2, \ldots, M\},\ f_i(\boldsymbol{x}_a) \leq f_i(\boldsymbol{x}_b) \tag{23.3a}$$

$$\exists j \in \{1, 2, \ldots, M\},\ f_j(\boldsymbol{x}_a) < f_j(\boldsymbol{x}_b) \tag{23.3b}$$

If one or both of the above conditions are violated, $\boldsymbol{x}_b$ is said to be non-dominated by $\boldsymbol{x}_a$ in Figure 23.1. Within the entire search space, the solutions non-dominated by

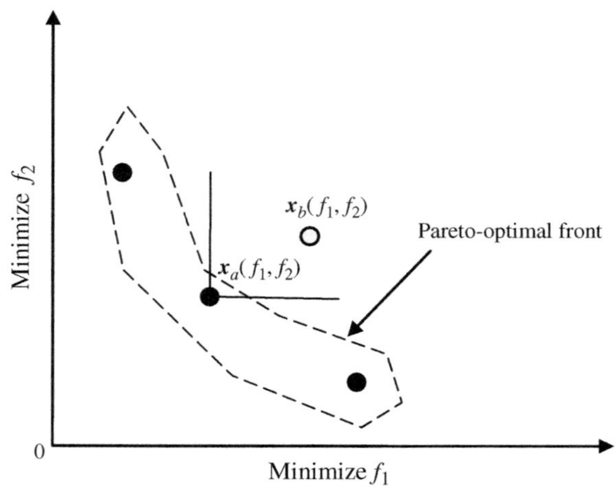

Figure 23.1. Pareto dominance and Pareto optimality.

any others are *Pareto optimal* and constitute the *Pareto-optimal set*; the corresponding z values form the *Pareto-optimal front* in $\mathbf{R}^M$. The goal of a MOO method is to search for solutions in the Pareto-optimal set, while maintaining diversity so as to cover well the Pareto-optimal front and, thus, allowing flexibility in the final decisions on the solutions to be actually implemented.

23.3 SOLUTION METHODS

In the past decades, a number of methods have been developed for solving generic MOO problems. Miettinen (1999) has classified them into the following four groups: no-preference methods, a priori methods, posteriori methods, and interactive methods. No-preference methods use heuristics to find a single optimal solution without any information about the preference of each objective. One example of such kind is the method of global criterion (Zeleny, 1973). Note that the no-preference methods generally do not attempt to find multiple Pareto-optimal solutions. A priori methods require sufficient information about the preference of each objective before the solution process starts. Usually only one preferred Pareto-optimal solution is found. The well-known examples include weighted-sum method and goal programming. Posteriori methods attempt to find all Pareto-optimal solutions or a representative subset of them. Posteriori methods can be further classified into mathematical-programming-based methods (e.g., ε-constraint approach), where each run of the algorithm produces one Pareto-optimal solution, and evolutionary methods, where each run of the algorithm might produce multiple Pareto-optimal solutions. Interactive methods involve the decision-maker (DM)'s continuous interactions with the method, in the search for the most preferred solution.

Multi-objective approaches are relatively new for solving RAMS+C problems. The rest of this section reviews four different approaches from recent literature. The advantages and drawbacks of each method are also discussed.

23.3.1 Weighted-Sum Approach

The weighted-sum approach aggregates multiple objectives into one single objective using a weighting vector $\boldsymbol{w} = (w_1, w_2, \ldots, w_M)$, as the following:

$$\text{Minimize} \quad \sum_{i=1}^{M} w_i f_i(\boldsymbol{x}) \tag{23.4a}$$

$$\text{Subject to} \quad \begin{cases} g_j(\boldsymbol{x}) = 0, & j = 1, \ldots, J \\ h_k(\boldsymbol{x}) \leq 0, & k = 1, \ldots, K \end{cases} \tag{23.4b}$$

where each $w_i \geq 0$ and $\sum_{i=1}^{M} w_i = 1$. Consequently, the solution to equation (23.4) is also a solution to equation (23.1). The coefficients are selected dependent on the DM's preferences. In multiple criteria decision analysis theory, there are a number of methods, for example, AHP and TOPSIS, developed for the quantification of the

DM's preferences. Interested readers could refer to Keeney and Raiffa (1993) for detailed information.

This approach is the most straightforward to solve a MOO problem. For problems that have a convex Pareto-optimal front, it guarantees finding the solutions on the entire Pareto-optimal set by varying the values of the weight vector. Elegbede and Adjallah (2003), Villanueva et al. (2008), Wattanapongskorn and Coit (2007), Hadavi (2008), Martorell et al. (2010), Pereira et al. (2010), Guo et al. (2011), Kancev et al. (2011) implemented this method for RAMS+C optimization. However, this method has several disadvantages. To name a few: different weight vectors do not necessarily result in different Pareto-optimal solutions; uniformly distributed sets of weight vectors do not necessarily result in uniformly distributed Pareto-optimal solutions; there are difficulties to handle objectives of different numerical scales; there are difficulties to find certain Pareto-optimal solutions in a non-convex objective space.

23.3.2 ε-Constraint Approach

To alleviate the difficulties faced by the weighted-sum approach, the ε-constraint approach was proposed in 1971 by Haimes et al. (1971). It reformulates the MOO problem by keeping one objective and transforming the others into constraints bounded by user-specific values. The transformed problem has the following expression:

$$\text{Minimize} \quad f_\theta(\boldsymbol{x}) \tag{23.5a}$$

$$\text{Subject to} \quad \begin{cases} f_i(\boldsymbol{x}) \leq \varepsilon_i, & i = 1, \ldots, M \ \ and\ i \neq \theta \\ g_j(\boldsymbol{x}) = 0, & j = 1, \ldots, J \\ h_k(\boldsymbol{x}) \leq 0, & k = 1, \ldots, K \end{cases} \tag{23.5b}$$

where ε_i is the upper bound of $f_i(\boldsymbol{x})$ and it need not necessarily be a small value close to zero. By iteratively increasing/decreasing the value of ε_i, the DM can obtain different Pareto-optimal solutions on the entire Pareto-optimal set. This method has been recently applied for RAMS+C optimization. Khalili-Damghani and Amiri (2012) applied an augmented version of the ε-constraint approach to solve binary-state multi-objective redundancy allocation series–parallel problems. Cao et al. (2013) utilized an enhanced adaptive ε-constraint approach that has a changing step size at each iteration to increase the efficiency of the ε-constraint method. The disadvantages of this method lie in the difficulties to determine the ranges of the objectives being constrained and the values of ε_i, when many objectives are involved. The lower limit of each converted objective can be obtained by solving the individual single-objective optimization using the same objective. However, computing the upper limit is not straightforward. A payoff table is usually implemented (Isermann and Steuer, 1988).

23.3.3 Goal Programming

First introduced by Charnes et al. (1955) in 1955 in an application of a single-objective linear programming problem, goal programming now is widely used for solving MOO problems. The main idea of goal programming is to find solutions that achieve predefined targets at one or more objectives. If there is no solution achieving the predefined targets in all objectives, then the task is to find the solutions that minimize the deviations from the targets. On the other hand, if there exist solutions within the desired targets, then, the task is to identify those solutions. For MOO problem, the simplest version of goal programming requires the user to set the target and relative weight for each objective function. An optimal solution $\boldsymbol{x}^*$ is defined as the one that minimizes the deviation from the set targets. Goal programming generally takes the following form:

$$\text{Minimize} \quad \sum_{i=1}^{M} c_i \left(d_i^+ + d_i^-\right) \tag{23.6a}$$

$$\text{Subject to} \quad \begin{cases} f_i(\boldsymbol{x}) + d_i^+ - d_i^- = f_i^0 \\ d_i^+ d_i^- = 0 \\ d_i^+, d_i^- \geq 0 \quad i = 1, \ldots, M \end{cases} \tag{23.6b}$$

where c_i is the weight of the deviation of each objective, d_i^+ and d_i^- are the deviations, and f_i^0 is the predefined target for the ith objective. Some researchers have used it for the RAMS+C problems: Gen et al. (1993) proposed a fuzzy goal programming and applied it to the redundancy optimization problem; Azaron et al. (2009) considered a goal attainment formulation together with a single-objective genetic algorithm (SOGA) to solve the component assignment problem. The disadvantages of this method are similar to those of weighted-sum and ε-constraint methods: the DM has to provide targets and weights for each of the objective functions.

23.3.4 Evolutionary Algorithms

The above-mentioned approaches for solving MOO problems are often referred to as "classical." They all suggest certain way of converting a MOO problem into a single-objective optimization problem. They have some common difficulties, only one Pareto-optimal solution can be found in one simulation run, and certain problem knowledge, such as w_i, ε_i, c_i, and f_i^0 are required from the DM.

The evolutionary algorithms (EAs), for example, genetic algorithms (GAs) (Holand, 1975), are stochastic optimization methods mimicking the biological evolutions on a group of individuals (solutions). The parallelization and evolution operations of EAs are well-suited to the characteristics of MOO problem: parallelization helps to identify multiple solutions on the Pareto front in one simulation run, without soliciting problem knowledge from the DM; the evolution operators have the capability to avoid trapping into the local minima (which is common in non-convex objective spaces). These properties render the EAs by far the most popular method implemented for RAMS+C MOO. The standard operation procedures of a single-objective GA (SOGA) are presented as follows:

```
Set t ← 0
Initialize B_t
Evaluate B_t
While t < G_max do
   t ← t + 1
   Select B_t from B_{t-1}
   Crossover B_t
   Mutate B_t
   Evaluate B_t
   Apply elitist strategy to B_t (given B_{t-1})
End While
```

where $\boldsymbol{B}_t$ is the population of solutions (i.e., individuals), t is the generation counter (i.e., the population iteration index) and $G_{\max}$ is the maximum number of generations. $\boldsymbol{B}_0$ consists of a group of encoded individuals $\boldsymbol{x}$ randomly generated at the initialization step. The evaluation of the population means computing the value of the objective function in correspondence of each individual in the population and converting it into the fitness value, to measure the quality of the corresponding individual. The selection step determines the group of individuals entering the evolution processes that drive the optimality search, based on probabilities related to the fitness values of the individuals. Usually, the individuals with higher fitness values have larger probabilities to survive and continue to evolve the search. The crossover and mutation are the evolutionary operators of GA, which allow generating new individuals in the search for the optimal solution. In crossover, generally two individuals, named parents, are paired to produce new individuals, named offsprings, by exchanging some parts of the encoded solutions. Crossover allows parts of the good solutions to be retained and copied in the population, so that the algorithm can eventually converge to an overall good solution. On the other hand, the mutation operator randomly changes the coding of the individuals. The perturbations brought by mutation introduce diversity into the population and assist the search in escaping from local minima. The elitist strategy maintains the fittest individual of the population; it ensures that the best solution will not be lost during the stochastic search process.

Based on a group of solutions, GA is well-suited for solving MOO problems. The multi-objective GAs (MOGAs) aim at modifying the SOGA in order to find a set of non-dominated solutions in a single run. The well-known MOGAs include vector evaluated GA (VEGA) (Schaffer, 1985), niched Pareto GA (NPGA) (Horn et al., 1994), non-dominated sorting GAs (NSGAs) (Deb et al., 2002; Srinivas and Deb, 1994), strength Pareto EAs (SPEAs) (Zitzler and Thiele, 1999; Zitzler et al., 2001).

The details about MOGAs for RAMS+C optimization are well documented in the tutorials by Marseguerra and Zio (2006) and Konak et al. (2006).

Despite the popularity of the multi-objective EAs (MOEAs), they have the following disadvantages: there is no guarantee to find a true Pareto-optimal solution; there is no guarantee to identify all Pareto-optimal solutions; they are computationally expensive with large population size.

23.4 PERFORMANCE MEASURES

To quantify the performance of MOO methods, a number of measures have been proposed. For a comprehensive review and comparative study, please refer to Zitzler et al. (2003). In this chapter, we introduce three representative ones. The first, named as generational distance (GD) (van Veldhuizen, 1999), measures the distance of an approximation to the true Pareto-optimal front. It is defined as

$$G = \frac{\sqrt{\sum_{i=1}^{m} d_i^2}}{m} \tag{23.7}$$

where m is the number of points (solutions) on the approximate front and d_i is the Euclidean distance between the point i and its nearest neighbor on the true Pareto front. This metric needs a reference or the true Pareto front to be known.

Another measure, called hypervolume (Zitzler and Thiele, 1998), does not have such requirement. It is the area (volume or hypervolume) under the dominated region defined by the approximate front. In Figure 23.2, we sketch the dominated region by

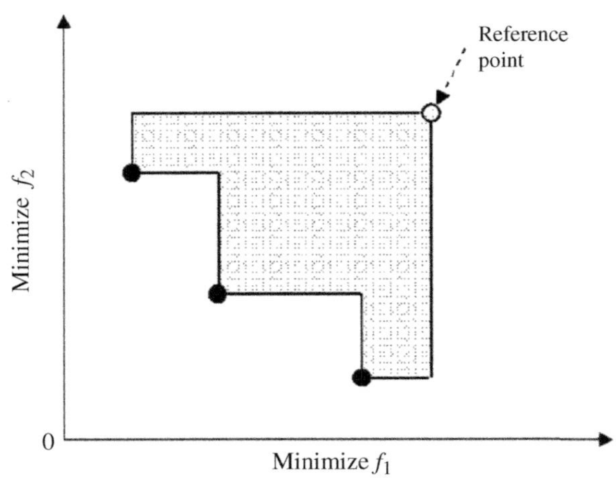

Figure 23.2. Hypervolume of the set of three non-dominated solutions.

three non-dominated solutions in a two-dimensional objective space. The hypervolume is the shaded region in the figure. The reference point for calculating the hypervolume is set to be a point with the two objective values larger than the maximum objective values of the non-dominated solutions, individually. Details of computing the hypervolume measure can be found in Deb (2001).

The two measures above evaluate the quality of a single approximate front. Differently, the coverage function (Zitzler and Thiele, 1998) compares two approximate fronts (A and B) and gives the fraction of solutions in B that are weakly dominated by any solution in A. It is defined as

$$C(A, B) = \frac{|\{b \in B | \exists a \in A : a \succcurlyeq b\}|}{|B|} \tag{23.8}$$

where $|B|$ is the cardinality of set B and $a \succcurlyeq b$ means solution a weakly dominates solution b, that is, only the condition (23.3a) is satisfied.

23.5 SELECTION OF PREFERRED SOLUTIONS

Once the Pareto-optimal solution set is obtained, higher-level decision-making might be necessary to choose one or more preferred solutions according to different application backgrounds and specific preferences. The methods for selecting the best compromise solution are called post-optimal techniques and the methods for selecting a preferred Pareto-optimal region are called optimization-level techniques (Deb, 2001). In this chapter, we introduce one popular method for selecting the best compromise solution. For detailed information about other methods, please refer to Deb (2001).

23.5.1 "Min–Max" Method

It is a widely used approach for defining a single best compromise solution (Belegundu and Chandrupatla, 1999). Let $z = (z_i),\ i = 1, \ldots, M$ denote one solution in the M-dimensional Pareto front and z_i^{max} denote the maximum value of the ith objective function on such fronts. The relative deviation of each objective of each solution is calculated as $r_i = (z_i^{max} - z_i)/z_i^{max}$ and $r_z = \min_i\{r_i\}$ is taken as the representative value of each solution. The solution z^* with the maximum r_z is selected to be the best compromise solution. In practice, the DM should adopt this method when he/she desires a solution that is representative of the "center" of the Pareto front. Figure 23.3 illustrates one example of using the "min–max" method to choose the best compromise solution from a bi-objective optimization problem.

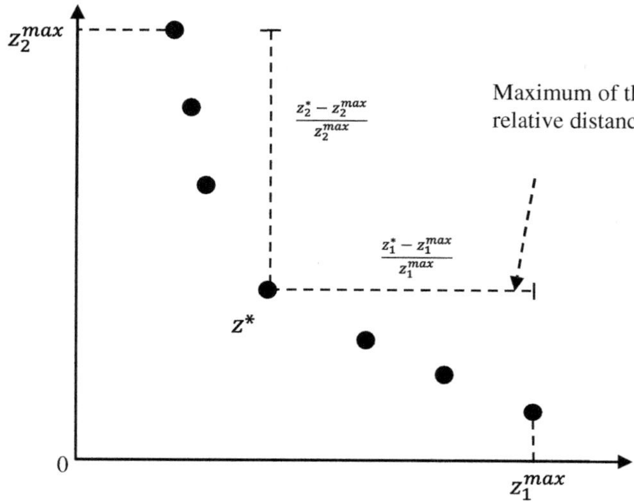

Figure 23.3. Best compromise solution z^* selected from the Pareto front by the "min–max" method, for a bi-objective optimization problem.

23.6 GUIDELINES FOR IMPLEMENTATION AND USE

The complete framework to deal with RAMS+C optimization problems includes problem formulation, solution method selection, and preferred solution(s) selection. The implementation guidelines are presented as the following steps, together with the main points of attention:

1. *Formulate the RAMS+C Optimization Problem.* As stated in the previous sections, RAMS+C optimization is essentially multi-objective; the following aspects have to be taken into account in the problem formulation:
   a. All objectives need to be first analyzed to reveal the relations between them. For example, the generic unavailability $U(\boldsymbol{x}) = 1 - A(\boldsymbol{x})$ might contain maintainability $M(\boldsymbol{x})$ to describe the unavailability due to test and maintenance activities. The MOO methods are worth applying only when there exist conflicts between at least one pair of objectives, otherwise a single-objective method can be used to find one solution that optimizes all objectives.
   b. The DMs need to be solicited for their preferences on each objective. The preferences can be represented by weights or converted into utility functions. If there is sufficient information about the preferences, then, in the next step a priori solution methods need to be selected; otherwise the posteriori solution methods have to be chosen. In addition, if the DMs need

to intervene during the optimization process, then the interactive methods could be considered.

2. *Select Appropriate Optimization Methods.* The choice of optimization method depends on the formulation of the problem and can largely impact the optimization results. The following aspects need to be considered in this step:
   a. If each individual objective with the constraints can be analytically solved by a single-objective mathematical programming method in polynomial time, then it is recommended to combine the mathematical programming and the classical MOO problem solution approaches, for example, weighted-sum (a priori) and ε-constraint (posteriori) approaches. If the DM provides preference weights (and targets), then weighted sum (or goal programming) can be used, whereas if the DM wishes to obtain a complete Pareto-optimal front then the ε-constraint needs to be used. It is not recommended to use weighted sum or goal programming to produce the complete Pareto-optimal front, due to their disadvantages presented in Sections 23.3.1 and 23.3.3.
   b. If one of the objectives exhibits difficult characteristics, for example, non-linear, non-convex, NP-hard, then it is recommended to use MOEAs or the classical MOO problem solution approaches combined with single-objective EAs (SOEAs). Note that the original problem can also be relaxed and, then, solved approximately by mathematical programming techniques. This approach is recommended if the practitioner has good knowledge about advanced mathematical optimization theory.
   c. Test more than one method especially for difficult problems because no one is the best for all cases and each has its own advantages and drawbacks.
3. *Solve the Optimization Problem.* Depending on the optimization methods selected, the following aspects need to be taken into account:
   a. In case the exact solutions to all individual objectives can be found via mathematical programming in polynomial time, if a priori methods are used, then, the whole problem will be solved once and one exact solution will be obtained, whereas if posteriori methods are used, then, the whole problem needs to be solved multiple times, each under a different setting of the parameter, for example, ε, and a set of the solutions on the Pareto-optimal front will be obtained.
   b. If MOEAs (or ε-constraint + SOEAs) are used, then multiple simulation runs will be required due to the stochastic nature of these algorithms. The parameters of EAs need to be tuned and the convergences of the EAs ensured. Typically, each simulation run delivers one approximate Pareto front. To obtain the best results across all runs, all fronts need to be combined and, then, a final “front of fronts” is selected out of them. Note that for ε-constraint + SOEAs, each simulation run delivers one single solution

and different ε values need to be explored to obtain the fronts. The final front can be selected from all results.

c. In case a priori MOO problem solution methods are used together with SOEAs, multiple simulation runs, parameter tuning, and convergence insurance are also necessary.

d. The performance measures are, then, used to evaluate the quality of the obtained results. If the results are fronts, then, the measures presented in Section 23.5 will be useful, whereas if the results are single solution points, then, statistics such as min, mean, standard deviations will be used.

e. The optimization results are recommended to be presented, compared, analyzed, and validated at this step.

4. *Select the Preferred Solution(s).* This step is necessary when the posteriori methods are used. First, it is recommended to select the best front from all the final fronts of different methods. The DMs are, then, solicited to determine the preferred solution(s) or the method that selects the preferred solution(s).

5. *Validate the Results and the Procedures.* All results and procedures need to be thoroughly checked to ensure the correctness of the implementation and the meaningfulness/usefulness of the results.

23.7 NUMERICAL CASE STUDY

The implementation guidelines are illustrated on a benchmark problem of redundancy allocation for multi-state series–parallel system (MSSPS), which is actually a well-known RAMS+C optimization problem in literature (Kuo and Wan, 2007). To illustrate the problem, Figure 23.4 depicts a simple flow transmission system made of two subsystems in series, S_1 and S_2, each one containing a number of pumps in parallel. Between the two subsystems there is a reservoir, here represented as a connecting line. The two subsystems must deliver the flow from the source to meet the demand

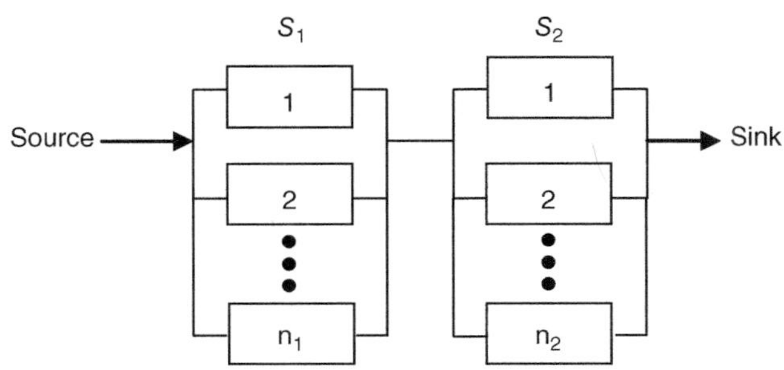

Figure 23.4. A simple MSSPS.

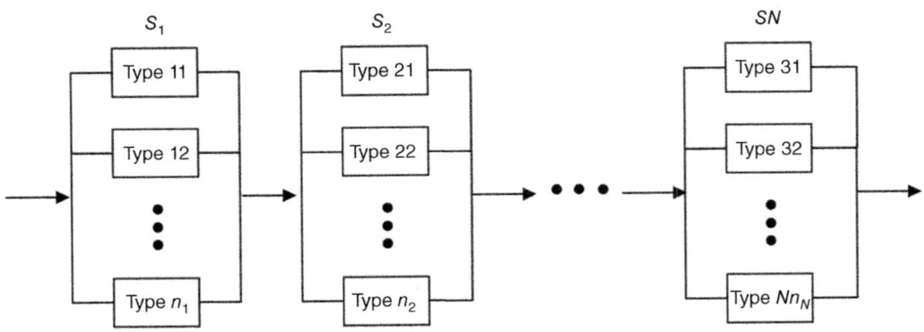

Figure 23.5. The generic MSSPS.

at sink. Each pump has a number of pumping levels, indicating the amounts of flow delivered per time unit. In simple words, the redundancy allocation problem concerns the determination of the (optimal) type and number of pumps to be installed in each subsystem, so that the total cost is minimized and system demand (flow) is satisfied.

In general terms, the diagram of an MSSPS is shown in Figure 23.5.

The MSSPS typically consists of N subsystems connected in series. For one subsystem i, it has $n_i > 0$ component types. For one component type j of subsystem i ("component type ij" in short), its state (or performance level) is a discrete random variable (RV) G_{ij} taking values from the finite set $\{g_{k_{ij}} | g_{k_{ij}} \in \mathbb{R}_{\geq 0},\ k_{ij} \in \mathbb{N}_{\geq 0},\ k_{ij} \leq m_{ij}\}$, where m_{ij} is the highest state of component type ij. The state probability $p_{k_{ij}}$ is defined as $p_{k_{ij}} = \Pr(G_{ij} = g_{k_{ij}})$. The following assumptions are typically made for MSSPS: (1) all components are mutually statistically independent; (2) for component type ij, $g_{k_{ij}-1} < g_{k_{ij}}$ holds for any $1 \leq k_{ij} \leq m_{ij}$, g_0 is the complete failure state and $g_{m_{ij}}$ is the perfect functioning state; (3) for component type ij, it is associated with a cost $c_{ij} > 0$; (4) the state of subsystem i is a discrete RV: $G_i = \sum_{j=1}^{n_i} G_{ij}$, the quantity that sums up all its component states; (4) the state of the whole system is a discrete RV: $G = \min(G_1, \dots, G_N)$, the minimal value of its subsystem states; (5) the demand is a discrete RV, W, following an assigned distribution.

The benchmark (Levitin et al., 1998) consists of four subsystems connected in series. For each subsystem, there are four to six different component types available. The parameters of each component type are presented in Table 23.1. Note that all components are binary stated. The $g_{1_{ij}}$s and $p_{1_{ij}}$s denote the performance and probability of the perfect functional state of component type ij, respectively.

The system demand follows the distribution presented in Table 23.2.

Step 1: Formulate the RAMS+C Optimization Problem

Given the above definitions and assumptions, the reliability (or availability, if the components are repairable) of the MSSPS is computed as $R = \Pr(G \geq W)$. The decision vector of a generic redundancy allocation problem on MSSPS can be written as $\boldsymbol{x} = (x_{11}, \dots, x_{1n_1}, \dots, x_{N1}, \dots, x_{Nn_N})$, where the element x_{ij} is a non-negative integer

TABLE 23.1. Parameters of the Component in the Benchmark

Subsystem	Type	c_{ij} (million \$)	$g_{1_{ij}}$ (%)	$p_{1_{ij}}$
1	1	1.020	150	0.9600
	2	0.890	100	0.9600
	3	0.720	80	0.9800
	4	0.620	80	0.9600
	5	0.520	50	0.9700
2	1	1.367	75	0.9530
	2	0.967	50	0.9600
	3	0.916	50	0.9140
	4	0.516	20	0.9670
3	1	0.813	240	0.9600
	2	0.783	200	0.9700
	3	0.614	200	0.9600
	4	0.534	180	0.9590
	5	0.384	90	0.9700
	6	0.214	60	0.9590
4	1	1.260	70	0.9800
	2	1.190	70	0.9600
	3	0.697	30	0.9800
	4	0.683	25	0.9890
	5	0.645	25	0.9700

representing the number of type ij components. The objectives of the optimization are minimizing the total cost and maximizing the system reliability. Obviously, they are conflicting with each other.

$$\min \quad C(\boldsymbol{x}) = \sum_{i=1}^{N} \sum_{j=1}^{n_i} c_{ij} x_{ij} \tag{23.9a}$$

$$\min \quad 1 - R(\boldsymbol{x}) \tag{23.9b}$$

The constraints are

$$R(\boldsymbol{x}) \geq R_0 \tag{23.9c}$$

$$1 \leq \sum_{j=1}^{n_i} x_{ij} \leq u_i \tag{23.9d}$$

TABLE 23.2. The System Demand W

Demand (%)	100	80	40
Probability	0.2	0.3	0.5

where u_i is the maximum number of components allowable in subsystem i and R_0 is the minimal reliability required for the system design. To emulate different decision-making scenarios, in this case study we consider both a priori and posteriori methods.

Step 2: Select Appropriate Optimization Methods

Since RAP on MSSPS is an NP-hard problem, there is no mathematical programming approach capable of finding the exact solution in polynomial time. To solve this problem, most existing studies (single-objective optimization) apply EAs. Therefore, in this case study two types of solution methods are considered. The first combines classical MOO methods and SOGA. The second directly applies NSGA-II (Deb et al., 2002). For illustration purposes, weighted sum and goal programming are included and different a priori preferences are simulated.

Step 3: Solve the Optimization Problem

The parameters of SOGA are: crossover probability = 0.5, mutation probability = 0.06, population size = 5, and maximum number of generations = 2200. The details about the SOGA version and the parameter tuning can be found in Li and Zio (2014). All performance measures presented in Section 23.4 are used for the experiments.

For the weighted-sum method, we select the following 11 weights pairs $\{(0, 1), (0.1, 0.9), (0.2, 0.8), (0.3, 0.7), (0.4, 0.6), (0.5, 0.5), (0.6, 0.4), (0.7, 0.3), (0.8, 0.2), (0.9, 0.1), (1, 0)\}$, representing different preferences of the DMs. The objective (23.9a) is scaled into the range [0, 1] by $\left(\frac{C(\boldsymbol{x})-1.895}{44.6-1.895}\right)$ where 44.6 and 1.895 are the respective maximal and minimal costs, satisfying the constraints in (23.9d). For goal programming, the weights are the same as in the weighted-sum method. The goals of the two objectives are set to 4.8 (million $) and 0, respectively. Note that the goal of the cost objective is set below the best solution 5.423 (million $) (with $R_0 = 0.9$) found in the literature. To apply the ε-constraint method, the second objective is converted to a constraint and the range of this constraint is set as [0, 0.1]. Because the maximal value of reliability is 1 and the minimum required by many systems is 0.9, we divide the whole range into 10 equally sized intervals, which means that there are 11 different levels of ε including the two extremes. Note that the lowest value of ε is set to 0.0001 instead of 0, because the latter is usually difficult to achieve.

In this case study, NSGA-II inherits the same parameter settings from SOGA, except for the population size which is set to $5 \times 11 = 55$. The reason is to ensure the same number of fitness evaluations for both classical MOO methods and MOEA methods.

All experiments are realized on different CUP cores of a supercomputer at the home university of the authors. Under each setting of the classical method, the SOGA has been run 50 times. NSGA-II has also been run 50 times.

Figure 23.6 depicts the best and average results of each classical approach + SOGA under 11 different settings. It can be seen that weighted sum + SOGA and

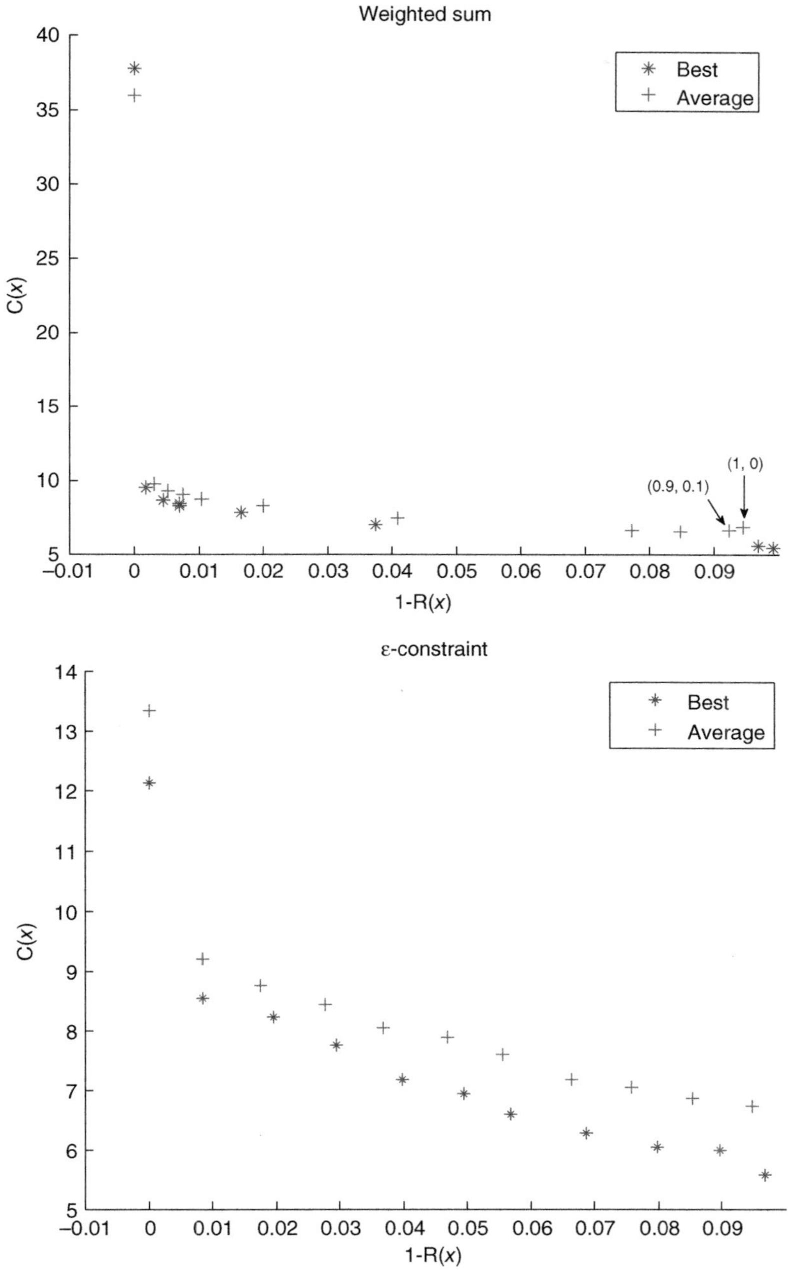

Figure 23.6. The best and average results of each classical approach + GA under the 11 different settings.

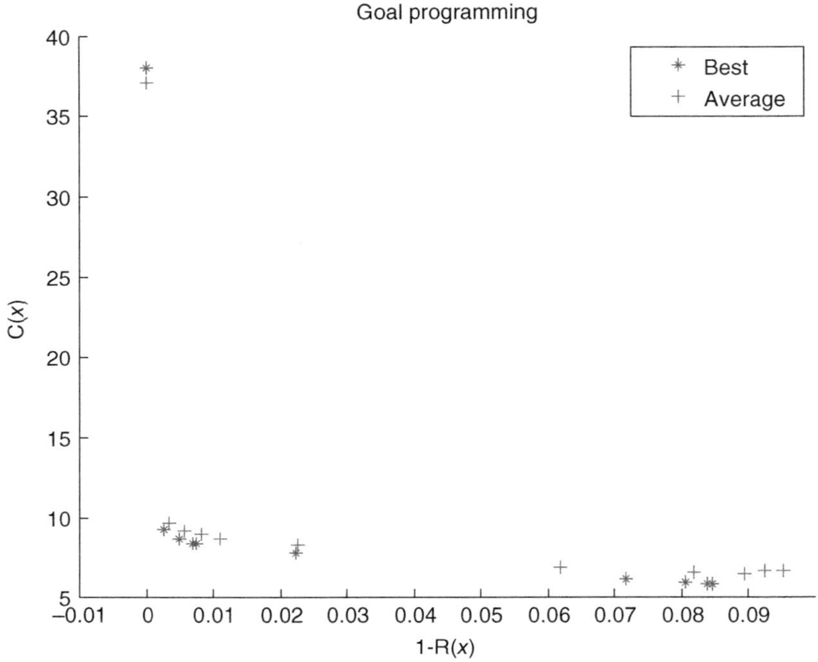

Figure 23.6. *(Continued)*

goal programming + SOGA generate similar results, which cluster mainly into three regions of the entire front with certain best results being overlapped, whereas ε-constraint + SOGA produces more evenly distributed results.

The convergence of the EAs must be checked. The plots of the average cost, reliability, and the weighted-sum objective function values over 50 simulation runs are shown in Figure 23.7 to verify this observation. It is seen that both weight settings converge at least to certain local minima as indicated by the curves of objective function values.

Different from the classical methods, NSGA-II obtains 50 approximate fronts. To aggregate all solutions, the final front is selected, for each method, from its final results of all experiments. Figure 23.8 shows that NSGA-II obviously achieves the best front in terms of the quality of the individual solutions, the number of solutions and the uniformity of the distribution of the solutions. The rightmost solution, (0.0993, 5.423), found by NSGA-II and weighted sum + SOGA is the best published result found by single-objective methods.

The quantitative evaluations are realized using the performance measures of Section 23.4. Table 23.3 presents the results. Note that the GD values are computed between the classical methods and NSGA-II, whose final front appears to be the best.

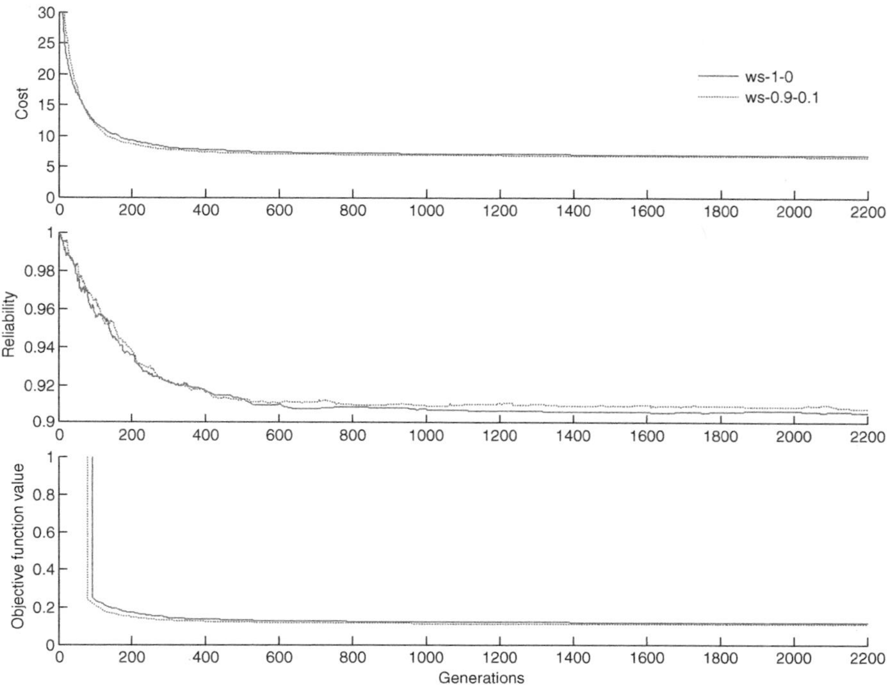

Figure 23.7. Convergence plots of the average of 50 simulation runs for weighted sum at (1, 0) and (0.9, 0.1).

The overall ranks of the methods are: NSGA-II, the first, followed by weighted sum and goal programming, and ε-constraint, the last. There is one exception in the GD value of ε-constraint which is lower than other GDs. It is mainly due to the scale of the cost being much higher than that of $1 - R(\boldsymbol{x})$, as shown in Figure 23.8. This also implies that GD might not be a very robust performance measure.

Step 4: Select the Preferred Solution(s)

The number of solutions on the respective final fronts of the individual methods are: 959 on the NSGA-II front, 61 on the goal programming front, 52 on the weighted-sum front, and 29 on the ε-constraint front. To ensure the quality of the selected preferred solutions, the overall best front is chosen from the four final fronts. There are 1007 solutions on the overall best front (shown in Figure 23.9). Among them, 948 come from NSGA-II, 27 from goal programming, 23 from weighted sum, and 9 from ε-constraint. The respective percentages for the solutions on the final fronts also being on the best front are: 98.9% for NSGA-II solutions, 44.3% for goal programming solutions, 44.2% for weighed-sum solutions, and 31.0% for ε-constraint solutions. This observation confirms the findings of the performance measures.

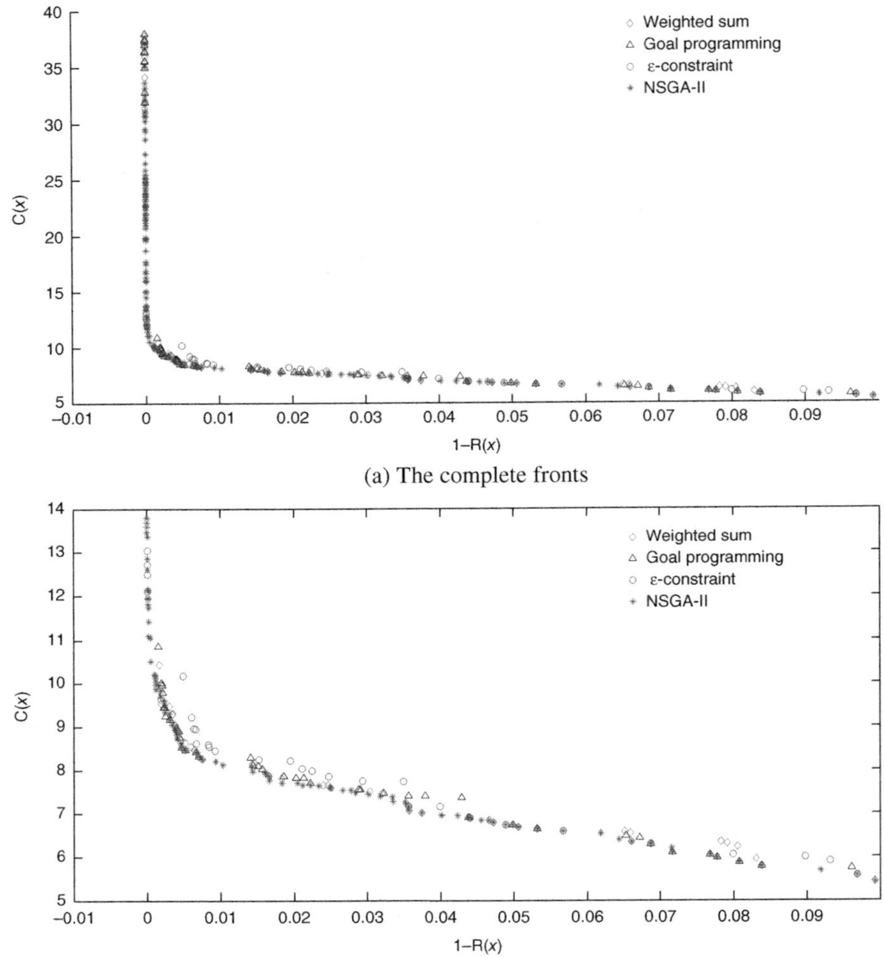

(a) The complete fronts

(b) Fronts in the rectangle region (0, 14), (1, 14), (1, 5), (0, 5)

Figure 23.8. The final fronts.

Without loss of generality, the "min–max" method presented in Section 23.5.1 is, then, used to select the best compromise solution whose objective values are (0.0184, 7.706) marked by the square in Figure 23.9.

Step 5: Validate the Results and the Procedures

All the results and procedures presented above have been thoroughly checked. For the best compromise solution, it is presented as "1(3);2(1),3(2);1(3);3(2),5(1),"

TABLE 23.3. The Numerical Results

Measures	HV	GD		CF			
Methods			Methods	WS	EC	GP	NSGA
WS	0.8160	0.0217	WS	1.0000	0.7241	0.3934	0.3107
EC	0.8181	0.0091	EC	0.1346	1.0000	0.0820	0.1043
GP	0.8157	0.0227	GP	0.3462	0.6552	1.0000	0.2336
NSGA	0.8266	NA	NSGA	0.9231	0.8966	0.8689	1.0000

HV, hypervolume; GD, generational distance; CF, coverage function; WS, weighted sum; EC, ε-constraint; GP, goal programming.

where the semicolons separate the subsystems and the term "$j(x_{ij})$" indicates the number of type j components in subsystem i: for example, 2(1) between the first and the second semicolons means two type 1 components in subsystem 2. The reliability and cost values of this solution are 0.9816 and 7.706 (million \$), respectively.

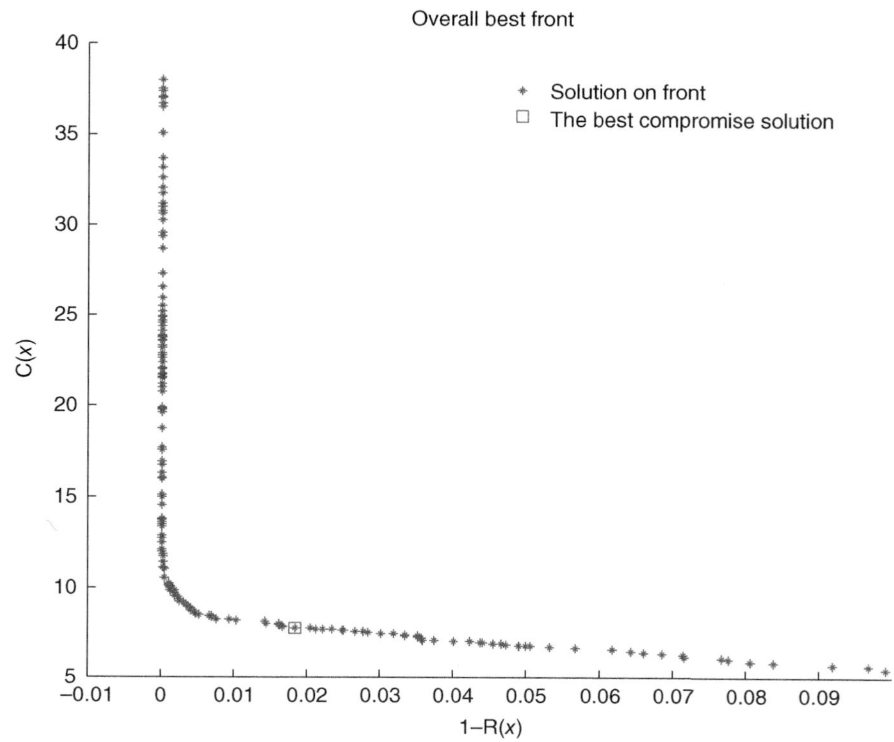

Figure 23.9. The best front and the selected preferred solution.

23.8 DISCUSSION

Often RAMS+C optimization problems are treated as single objective. The weighted-sum method (or prior method) is most often used to aggregate the different objectives of the problem into one. Based on specialized research and practice experience, in this chapter we keep with the multi-objective nature of RAMS+C optimization problems and present a holistic framework to deal with them, providing guidelines for the implementation of solution methods. The exemplifying redundancy allocation problem taken as reference generally shows the disadvantages of using the weighted-sum approach and the MOEA superiority over other methods. Nevertheless, the classical MOO methods are still preferred when the problem is solvable through mathematical programming techniques, due to the disadvantages of MOEAs. The ε-constraint is able to produce evenly distributed fronts whereas weighted-sum-based approaches are more suitable when the DM provides clear preferences via weights assignment.

On a more general level, the main drawbacks of posteriori methods are the following: there is no guarantee that the true global optimal solution can be found; there are difficulties in handling large problems because of the high computational demands; the large size of the derived Pareto-optimal solution set could cause "information overload" to the DM, impairing his/her comfortable decision. By far, the priori methods are most popular. Their disadvantage is mainly the requirement of precise and sufficient information from the DM, impairing his/her comfortable decision. Interactive methods are questioned when the frequent interactions with the DM are difficult to manage or when multiple DMs need to be solicited to guide the search.

23.9 RELATIONS TO OTHER PRINCIPLES

The relations to other safety principles are demonstrated mainly through the literature examples of practical optimization problems dealt with in different industries. In general, other safety principles such as safety margin and safety barrier can be incorporated, in the form of decision variables, objective functions, or constraints, into the optimization formulation such that the final optimal solutions are changed, reflecting safety aspects of the system that is being optimized. In Ibrahim et al. (2012), the safety margin, which is defined as the difference between the supply of an engineering system and its demand, has been considered in the design optimization of nine highway cross-sections in British Columbia, Canada. A safety factor is defined to take into account uncertainty in the margin. By integrating this factor into the objective function, the reduction in collisions is realized.

Safety barriers refer to the physical and/or non-physical means planned to prevent, control, or mitigate undesired events or accidents (Sklet, 2006). The well-being of safety barrier is necessary for system safety. In Vatn and Aven (2010), the maintenance interval of the signal red light bulbs in a Norwegian railway system is considered as decision variable; the expected safety cost due to a small fraction of bulbs failures is incorporated into the objective function.

REFERENCES

Azaron, A., Perkgoz, C., Katagiri, H., Kato, K., and Sakawa, M. (2009). Multi-objective reliability optimization for dissimilar-unit cold-standby systems using a genetic algorithm. *Computers and Operations Research*, 36(5), 1562–1571.

Belegundu, A. D., and Chandrupatla, T. R. (1999). *Optimization Concepts and Applications in Engineering*. Englewood Cliffs, NJ: Prentice-Hall Editions.

Cao, D. Z., Murat, A., and Chinnam, R. B. (2013). Efficient exact optimization of multi-objective redundancy allocation problems in series-parallel systems. *Reliability Engineering and System Safety*, 111, 154–163.

Charnes, A., Cooper, W. W., and Ferguson, R. O. (1955). Optimal estimation of executive compensation by linear programming. *Management Science*, 1(2), 138–151.

Deb, K. (2001). *Multi-Objective Optimization Using Evolutionary Algorithms* (Vol. 16). John Wiley & Sons.

Deb, K., Pratap, A., Agarwal, S., and Meyarivan, T. (2002). A fast and elitist multiobjective genetic algorithm: NSGA-II. *IEEE Transactions on Evolutionary Computation*, 6(2), 182–197.

Elegbede, C., and Adjallah, K. (2003). Availability allocation to repairable systems with genetic algorithms: A multi-objective formulation. *Reliability Engineering and System Safety*, 82(3), 319–330.

Frank, M. V. (1995). Choosing among safety improvement strategies: A discussion with example of risk assessment and multi-criteria decision approaches for NASA. *Reliability Engineering and System Safety*, 49(3), 311–324.

Gen, M., Ida, K., Tsujimura, Y., and Kim, C. E. (1993). Large-scale 0–1 fuzzy goal programming and its application to reliability optimization problem. *Computers and Industrial Engineering*, 24(4), 539–549.

Guo, S. C., Huang, H. Z., Wang, Z. L., and Xie, M. (2011). Grid service reliability modeling and optimal task scheduling considering fault recovery. *IEEE Transactions on Reliability*, 60(1), 263–274.

Hadavi, S. M. H. (2008). Risk-based, genetic algorithm approach to optimize outage maintenance schedule. *Annals of Nuclear Energy*, 35(4), 601–609.

Haimes, Y. Y., Lasdon, L. S., and Wismer, D. A. (1971). On a bicriterion formulation of the problems of integrated system identification and system optimization. *IEEE Transactions on Systems Man and Cybernetics*, SMC-1(3), 296–297.

Holand, J. H. (1975). *Adaptation in Natural and Artificial Systems*. Ann Arbor, MI: The University of Michigan Press.

Horn, J., Nafpliotis, N., and Goldberg, D. E. (1994). A niched Pareto genetic algorithm for multiobjective optimization. Evolutionary Computation. IEEE World Congress on Computational Intelligence., In: Proceedings of the First IEEE Conference on, 1994. IEEE, 82–87.

Ibrahim, S. E., Sayed, T., and Ismail, K. (2012). Methodology for safety optimization of highway cross-sections for horizontal curves with restricted sight distance. *Accident Analysis and Prevention*, 49, 476–485.

Isermann, H., and Steuer, R. E. (1988). Computational experience concerning payoff tables and minimum criterion values over the efficient set. *European Journal of Operational Research*, 33(1), 91–97.

Kancev, D., Gjorgiev, B., and Cepin, M. (2011). Optimization of test interval for ageing equipment: A multi-objective genetic algorithm approach. *Journal of Loss Prevention in the Process Industries*, 24(4), 397–404.

Keeney, R. L., and Raiffa, H. (1993). *Decisions with Multiple Objectives: Preferences and Value Trade-Offs*. Cambridge University Press.

Khalili-Damghani, K., and Amiri, M. (2012). Solving binary-state multi-objective reliability redundancy allocation series-parallel problem using efficient epsilon-constraint, multi-start partial bound enumeration algorithm, and DEA. *Reliability Engineering and System Safety*, 103, 35–44.

Konak, A., Coit, D. W., and Smith, A. E. (2006). Multi-objective optimization using genetic algorithms: A tutorial. *Reliability Engineering and System Safety*, 91(9), 992–1007.

Kuo, W., and Prasad, V. R. (2000). An annotated overview of system-reliability optimization. *IEEE Transactions on Reliability*, 49(2), 176–187.

Kuo, W., and Wan, R. (2007). Recent advances in optimal reliability allocation. *IEEE Transactions on Systems, Man and Cybernetics, Part A: Systems and Humans*, 37(2), 143–156.

Levitin, G., Lisnianski, A., Ben-Haim, H., and Elmakis, D. (1998). Redundancy optimization for series-parallel multi-state systems. *IEEE Transactions on Reliability*, 47(2), 165–172.

Li, Y. F., and Zio, E. (2014). A quantum-inspired evolutionary approach for non-homogeneous redundancy allocation in series-parallel multi-state systems. ICRMS 2014, Guangzhou, China. IEEE, 1–6.

Marseguerra, M., and Zio, E. (2006). Basics of genetic algorithms optimization for RAMS applications. *Reliability Engineering and System Safety*, 91(9), 977–991.

Martorell, S., Villamizar, M., Carlos, S., and Sanchez, A. (2010). Maintenance modeling and optimization integrating human and material resources. *Reliability Engineering and System Safety*, 95(12), 1293–1299.

Martorell, S., Villanueva, J. F., Carlos, S., Nebot, Y., Sanchez, A., Pitarch, J. L., and Serradell, V. (2005). RAMS+C informed decision-making with application to multi-objective optimization of technical specifications and maintenance using genetic algorithms. *Reliability Engineering and System Safety*, 87(1), 65–75.

Miettinen, K. (1999). *Nonlinear Multiobjective Optimization* (Vol. 12). Springer Science+Business Media.

Pereira, C. M. N. A., Lapa, C. M. F., Mol, A. C. A., and Da Luz, A. F. (2010). A particle swarm optimization (PSO) approach for non-periodic preventive maintenance scheduling programming. *Progress in Nuclear Energy*, 52(8), 710–714.

Sawaragy, Y., Nakayama, H., and Tanino, T. (1985). *Theory of Multiobjective Optimization*. Orlando, FL: Academic Press.

Schaffer, J. D. (1985). Multiple objective optimization with vector evaluated genetic algorithms. In: Proceedings of the 1st International Conference on Genetic Algorithms, L. Erlbaum Associates Inc., 93–100.

Sklet, S. (2006). Safety barriers: Definition, classification, and performance. *Journal of Loss Prevention in the Process Industries*, 19(5), 494–506.

Srinivas, N., and Deb, K. (1994). Multiobjective optimization using nondominated sorting in genetic algorithms. *Evolutionary Computation*, 2(3), 221–248.

Van Veldhuizen, D. A. (1999). *Multiobjective Evolutionary Algorithms: Classifications, Analyses, and New Innovations*. Air Force Institute of Technology, Air University.

Vatn, J., and Aven, T. (2010). An approach to maintenance optimization where safety issues are important. *Reliability Engineering and System Safety*, 95(1), 58–63.

Villanueva, J. F., Sanchez, A. I., Carlos, S., and Martorell, S. (2008). Genetic algorithm-based optimization of testing and maintenance under uncertain unavailability and cost estimation: A survey of strategies for harmonizing evolution and accuracy. *Reliability Engineering and System Safety*, 93(12), 1830–1841.

Wattanapongskorn, N., and Coit, D. W. (2007). Fault-tolerant embedded system design and optimization considering reliability estimation uncertainty. *Reliability Engineering and System Safety*, 92(4), 395–407.

Zeleny, M. (1973). Compromise programming. In: Cochrane, J. L., and Zeleny, M. (Eds.), *Multiple Criteria Decision Making*. Columbia, SC: University of South Carolina Press.

Zitzler, E., Laumanns, M., and Thiele, L. (2001). *SPEA2: Improving the Strength Pareto Evolutionary Algorithm*. Zurich, Switzerland: Swiss Federal Institute Technology.

Zitzler, E., and Thiele, L. (1998). Multiobjective optimization using evolutionary algorithms—A comparative case study. In: Eiben, A. E., Back, T., Schoenauer, M., and Schwefel, H. P. (Eds.), *Parallel Problem Solving from Nature—PPSN V*. Heidelberg: Springer.

Zitzler, E., and Thiele, L. (1999). Multiobjective evolutionary algorithms: A comparative case study and the strength Pareto approach. *IEEE Transactions on Evolutionary Computation*, 3(4), 257–271.

Zitzler, E., Thiele, L., Laumanns, M., Fonseca, C. M., and Da Fonseca, V. G. (2003). Performance assessment of multiobjective optimizers: An analysis and review. *IEEE Transactions on Evolutionary Computation*, 7(2), 117–132.

FURTHER READING

Kuo, W., Prasad, V. R., Tillman, F. A., and Hwang, C. L. (2006). *Optimal Reliability Design: Fundamentals and Applications*. Cambridge University Press.

Lisnianski, A., and Levitin, G. (2003) *Multi-State System Reliability: Assessment, Optimization and Applications* (No. 6). World Scientific.

Yang, X. S. (2010) *Engineering Optimization: An Introduction with Metaheuristic Applications*. John Wiley & Sons.

Zio, E. (2009). *Computational Methods for Reliability and Risk Analysis* (Vol. 14). World Scientific.

24

MAINTENANCE OPTIMIZATION AND ITS RELATION TO SAFETY

Roger Flage

24.1 INTRODUCTION

System safety is influenced by the system reliability and hence by the maintenance performed on the system in question. The relation between maintenance and safety can be summarized by the following points (Vatn and Aven, 2010):

- Maintenance will typically increase system reliability and thereby the system safety level.
- Maintenance actions could induce failures.
- Accidents could occur during maintenance.

Hence, maintenance optimization often needs to be seen in relation to (safety) risk management and its methods and principles, such risk assessment methodology and risk reduction principles.

Maintenance optimization refers to the development and use of theory, principles, methods, and models to find maintenance policies that are in some sense optimal. Mathematical models for determining optimal maintenance policies are usually based on a classical cost–benefit rationale, where all effects studied (typically including

Handbook of Safety Principles, First Edition. Edited by Niklas Möller, Sven Ove Hansson, Jan-Erik Holmberg, and Carl Rollenhagen.

material costs, loss of assets, and safety) are transformed into monetary value and different concerns are balanced through expected value-based calculations. Recently, the implementation of risk acceptance criteria in the context of maintenance optimization modeling has been studied, leading to constrained optimization. Furthermore, some have argued a need for seeing beyond expected values in situations with high safety impacts.

These are some key characteristics and issues when considering maintenance optimization in the context of safety, and in this chapter, we review and discuss this context. The chapter is to a large extent based on Flage (2013) but with some modifications and extensions. The review ranges from pure cost–benefit optimization, via quantitatively accounting for safety beyond the expected value calculations of cost–benefit optimization through constrained optimization, to a broader decision basis, as addressed above.

The remainder of this chapter is organized as follows: Section 24.2 provides an overview of some principles and terms that are related to maintenance optimization from the point of view of safety. Section 24.3 then reviews maintenance optimization principles, and Section 24.4 provides a discussion and some conclusions. Finally, Section 24.5 provides some references for further reading.

24.2 RELATED PRINCIPLES AND TERMS

In this section, we look into some principles related to maintenance optimization when viewed in the context of safety. These are cost–benefit analysis, risk assessment, and risk management, as well as the ALARP principle and risk acceptance criteria. Before doing so, we introduce some key terms.

24.2.1 Key Terms

In this subsection, we introduce some key terms related to maintenance, based on Rausand and Høyland (2004), reference to which is made for more details and further terms.

Maintenance can be defined as "the combination of all technical and corresponding administrative actions, including supervision actions, intended to retain an entity in, or restore it to, a state in which it can perform its required function" (IEC 50(191), in Rausand and Høyland, 2004 p. 361).

Different types of maintenance can be distinguished (Rausand and Høyland, 2004): Preventive maintenance refers to planned maintenance carried out while an item is still in the functioning state, with the purpose of preventing future failures by reducing the probability of such failures. Preventive maintenance includes age-based maintenance, performed when the item reaches a specified age; clock-based maintenance, performed at specified calendar times; condition-based maintenance (also sometimes called predictive maintenance), performed based on measurement of one or more quantities giving information about the item; and opportunistic maintenance, applicable to multi-component systems and then performed on a component when an opportunity to do so arises either due to maintenance of other components in the system or due to a system shutdown. Preventive maintenance also includes failure-finding maintenance, performed to reveal hidden failures, for example, in items performing off-line functions. Corrective maintenance (or repair), on the other hand, refers to maintenance that is performed after the occurrence of a failure, with the purpose of bringing the item back to a functioning state.

Finally, the availability (or dependability) of a system is often seen as determined by its inherent reliability, maintainability, and maintenance support (Rausand and Høyland, 2004). Availability can be defined as "the ability of an item (under combined aspects of its reliability, maintainability, and maintenance support) to perform its required function at a stated instant of time or over a stated period of time" (BS 4778, in Rausand and Høyland, 2004 p. 362). Reliability can be defined as "the ability of an item to perform a required function, under given environmental and operational conditions and for a stated period of time" (ISO 8402, in Rausand and Høyland, 2004 p. 603). The maintainability of an item can be defined as "the ability of the item, under stated conditions of use, to be retained in, or restored to, a state in which it can perform its required functions, when maintenance is performed under stated conditions and using prescribed procedures and resources" (BS 4778, in Rausand and Høyland, 2004 p. 362). The maintainability of a system is hence influenced by design factors such as accessibility and how easy it is to dismantle and reinstall the item, whereas maintenance support is influenced by the personnel performing the maintenance, their availability, skill level and tools, as well as spare part availability and quality (Rausand and Høyland, 2004).

24.2.2 Maintenance Optimization Models as Special Types of Cost–Benefit Analysis

As described by Vatn and Aven (2010), maintenance optimization tends to be treated in the scientific literature as a problem of optimizing some object function. Accordingly, research in this field to a large extent deals with developing mathematical models that produce optimal maintenance decisions that balance costs and benefits. Maintenance optimization models, then, can be seen as a type of cost–benefit analysis (CBA). The object function typically expresses the expected cost per unit of time or the expected discounted cost, and could reflect various cost dimensions, such as the cost of maintenance, the cost of production loss and safety-related costs; see, for example, Vatn et al. (1996). Incorporation of safety-related costs is typically done according to a traditional CBA approach: A so-called "value of preventing a fatality" (VPF) is chosen and multiplied with the expected number of fatalities, creating a weight for the importance and "cost" of safety; cf. for example, Vatn and Aven (2010) and see Section 24.3.1.

24.2.3 Risk Assessment and Risk Management

Risk management can be defined as "the coordinated activities to direct and control an organization with regard to risk" (ISO, 2009). The purpose of risk management can be described as to achieve an appropriate balance between realizing opportunities to achieve benefits on the one hand, and minimizing losses on the other (Aven and Vinnem, 2007). This balancing act could go beyond conventional principles of CBA, as will be seen in the coming subsection on the ALARP principle as well as in the discussion in Section 24.4.

A key part of the risk management process is risk assessment, which comprises several steps: Identify hazards/threats, assess potential causes and consequences of these, and characterize the uncertainties concerning the occurrence of these hazards/threats, causes, and consequences. Various categories of risk assessment methods can be distinguished, for example, (Aven, 2015): Simplified, standard, and model-based. The category of standard risk assessment methods comprises methods such as failure mode, effects and criticality analysis (FMECA) and Hazard and Operability analysis (HAZOP), whereas model-based risk assessment methods makes use of formal and typically quantitative techniques, such as fault tree analysis (FTA) for cause analysis, and event tree analysis (ETA) for consequence analysis.

The so-called bow-tie diagram constitutes a basic conceptual model of what is known as the causal chain approach to risk assessment. In the middle of the diagram can be found the initiating event (IE); on the left-hand side potential causal event sequences leading up to the IE, including the performance of barriers in place to avoid that the IE occurs; and on the right-hand side potential event sequences developing into consequences that may result from the IE, including the performance of the barriers in place to avoid the IE developing into harmful consequences. Figure 24.1 shows a bow-tie diagram adapted to the maintenance optimization setting. Characteristic IEs in this setting could be "system failure" and "preventive maintenance" (cf. the links between maintenance and safety mentioned in Section 24.1), and Figure 24.1 illustrates both the use of event trees to analyze whether or not an accident occurs following these IEs, as well as the distinction into the consequence dimensions loss of life and economic loss.

Quantitative risk assessment (QRA), also known as probabilistic risk assessment (PRA), is a type of model-based risk analysis. A QRA/PRA gives the decision-maker a quantitative description of the risk associated with a set of decision options. According to Apostolakis (2004), the purpose of a QRA should be to support risk-informed decision-making. However, it is not the only basis for decision (Apostolakis, 2004, p. 518): "I wish to make one thing very clear: QRA results are never the sole basis for decision-making by responsible groups. In other words, safety-related decision-making is risk-informed, not risk-based." The point is that a risk-informed decision does not rely solely on calculated risk numbers; also elements of traditional engineering approaches are emphasized, including safety margins and the principle of defense

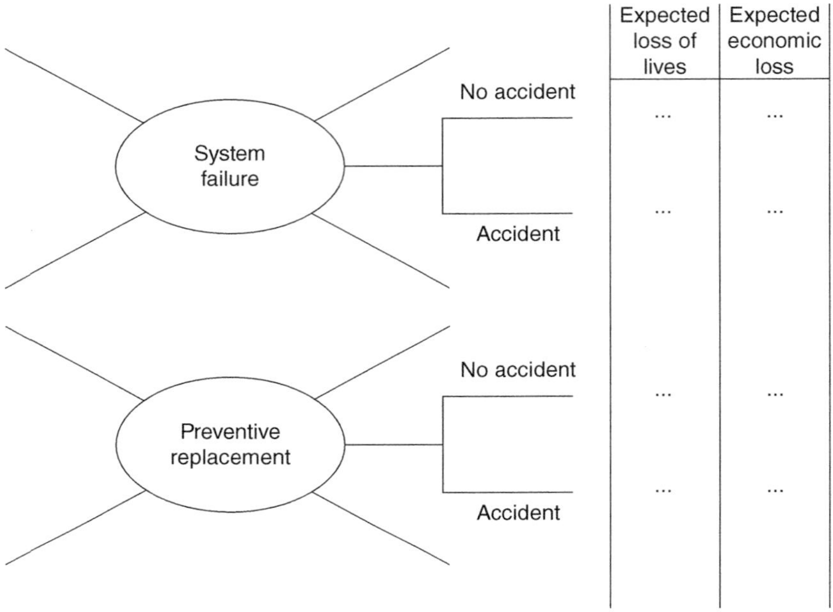

Figure 24.1. Bow-tie diagrams with "system failure" and "preventive replacement" as initiating events (IEs) and considering loss of lives and economic loss as consequence dimensions.

in depth (see, e.g., NRC, 2002); and there will usually also be other concerns involved in a risk-related decision besides risk itself.

Risk management is applied across different types of industries. The term asset risk management may refer to risk management of financial assets (see, e.g., Esch et al., 2005), but it may also refer to risk management of industrial assets. Industrial asset risk management, then, addresses a variety of decision problems, ranging from problems with potentially high safety impact and characterized by substantial epistemic uncertainty (i.e., lack of knowledge about phenomena, usually translating into model uncertainty or uncertainty about the parameters of models used to describe random variation), to problems more straightforwardly subject to pure economic cost–benefit optimization and mainly characterized by so-called aleatory uncertainty (i.e., variation in populations, in a maintenance optimization setting typically variation in the lifetimes of populations of similar systems).

Various risk management principles can be adopted for treating risk, including the (pre-)cautionary principle and the ALARP principle—the latter typically in conjunction with risk acceptance criteria. The following subsection reviews the ALARP principle and its often-seen link to risk acceptance criteria.

24.2.4 The ALARP Principle and Risk Acceptance Criteria

The ALARP principle is a key risk management principle that prescribes risk reduction to the extent that the remaining risk is "as low as reasonably practicable" (ALARP) (e.g., Aven and Vinnem, 2007). The "reasonably practicable" criterion can be interpreted in various ways; one common interpretation being that a risk reducing measure should be implemented unless it can be demonstrated that there is a "gross disproportion" between the burdens (costs, in a wide sense) and benefits of doing so. Hence, this is a principle of reversed burden of proof, since the prescription is that a risk reducing measure is to be implemented unless it can be shown that the burdens of doing so is grossly disproportionate to the gains. The "gross disproportion" formulation is also open to interpretation; however, as pointed out by Jones-Lee and Aven (2011, p. 878):

> "... it would seem to follow as a matter of logical necessity that since costs can naturally be viewed as being "disproportionate" to benefits whenever they exceed the latter, then the stipulation that a safety improvement should be undertaken provided its costs are not grossly disproportionate to expected benefits is effectively to require that safety expenditure should be undertaken to a point well in excess of that which would be justified if the aim was to ensure a socially optimal use of scarce resources."

The ALARP principle is often implemented with reference to three regions of risk (Aven and Vinnem, 2007):

1. Intolerable risk
2. Neither intolerable nor negligible risk—the ALARP principle applies
3. Negligible or broadly acceptable risk

The UK Health and Safety Executive (HSE) Tolerability of Risk (TOR) framework (HSE, 2001) is an example of such a partitioning. In the Norwegian petroleum regulations only the regions 1 and 2 apply; that is, there is no level of risk that is "negligible" or "broadly acceptable" and so the ALARP principle applies in all cases whenever the risk is not intolerable.

Risk acceptance (tolerability) criteria (RAC) are used to delineate such regions of risk. The effect of imposing such criteria in the context of maintenance optimization modeling is essentially to restrict the set of permissible maintenance policies. Clearly, the result could be sub-optimal scheduling of maintenance from a pure CBA point of view.

Both risk acceptance criteria and the ALARP principle are often imposed as regulatory requirements. As an example, oil and gas companies operating on the Norwegian Continental Shelf are required to define acceptance criteria for major accident risk and environmental risk (Management regulations, 2011). The Petroleum Safety Authority Norway also defines some minimum requirements. The ALARP principle is incorporated into Section 11 of the framework regulations (2011b) of the Norwegian petroleum regulatory system, although the term ALARP is not used. The principal regulatory references when it comes to maintenance planning in the Norwegian petroleum industry are the activities regulations (2011c) for offshore facilities and the technical and operational regulations (2011d) for onshore facilities. The primary focus of these regulations is health, safety, and environmental (HSE) aspects. The regulations require a classification to be performed of systems and equipment with respect to the HSE consequences of failure and, furthermore, that this classification should a basis for maintenance planning (activities regulations Section 46, 2011c):

> "Facilities' systems and equipment shall be classified as regards the health, safety and environment consequences of potential functional failures. [...] The classification shall be used as a basis in choosing maintenance activities and maintenance frequencies, in prioritizing between different maintenance activities and in evaluating the need for spare parts."

The guidelines to the activities regulations (2011c) refer to the NORSOK Z-008 standard (NORSOK, 2011) in relation to the HSE consequence classification.

The quantity introduced as VPF in Section 24.2.2 is sometimes also referred to as the value of a statistical life (VSL) or the implied cost of averting a fatality (ICAF). Different values are typically used in different contexts; common values used in CBA range from \$1 to \$10 million (Aven and Vinnem, 2007). As an example, the UK HSE refers to a VPF equal to \$1 million, but for the offshore industry an ICAF value of \$6 million is considered to be the minimum level, corresponding to a disproportion factor of 6 (HSE, 2006). The disproportion factor value defines what is judged as "grossly disproportionate" in relation to the ALARP principle. The UK HSE overall attitude toward ALARP evaluation appears to be that the disproportion factor should at least be greater than 1 (HSE, 2001 p. 67):

> "In comparing cost against risks HSE, when regulating, will be governed by the principles that: there should be a transparent bias on the side of health and safety. For duty holders, the test of "gross disproportion" implies that, at least, there is a need to err on the side of safety in the computation of health and safety costs and benefits."

Some plausible justifications for the "gross disproportion" interpretation of ALARP is offered by Jones-Lee and Aven (2011); these can be summarized as:

a. Accounting for other possible benefits than those considered
b. Making allowances for uncertainty
c. Accounting for greater willingness to pay for safety improvements in cases of abnormally high levels of baseline risk
d. Incentivizing cost-minimization and "best practice" engineering

Here (a) refer to indirect benefits, for example, societal benefits that may not be properly accounted for if only the direct benefits of the individuals who benefit directly from a safety improvement is included. Moreover, making allowances for uncertainty (b) means to see beyond expected value calculations for future benefits and costs. Furthermore, (c) refers to accounting for aversion to higher levels of risk by increasing the VPF as the risk level increases. Finally, (d) is to reduce the probability that overstating of costs is done to avoid implementation of a safety improvement.

In Section 24.3.2.1, we study how the ALARP principle and RAC could be implemented in the context of maintenance optimization modeling. Particular attention is there paid to justification (c) described above. Then, in Section 24.4, justification (d) and a general ALARP verification procedure extending beyond CBA is considered. First, however, we introduce the setting of maintenance optimization.

24.3 MAINTENANCE OPTIMIZATION

This section reviews maintenance optimization modeling. We start by considering pure cost–benefit optimization models, including the incorporation of safety costs into such models. Then a framework for and numerical example of implementing risk acceptance criteria (RAC) and the ALARP principle in the context of maintenance optimization modeling is described. Finally, some comments on empirical research on use of maintenance optimization models are given.

24.3.1 Theory

Before applying mathematical maintenance optimization models the decision problem needs to be structured. More qualitative techniques such as reliability-centered maintenance (RCM) and total productive maintenance (TPM) can be used for this purpose. RCM is a common methodology for preventive maintenance planning that involves a two-step procedure (Selvik and Aven, 2010):

- Inductive analysis to reveal potential failures, typically using a variant of failure mode, effects and criticality analysis (FMECA); the purpose being to determine critical system components.
- Application of the so-called RCM logic (using logical decision diagrams) to specify suitable preventive maintenance categories, for example, predictive maintenance or replacement.

See, for example, Rausand (1998), Rausand and Vatn (2008), and Bloom (2006) for detailed methodological descriptions of RCM.

A general framework for maintenance of large complex systems is described by Labib (1998), wherein all elements of the system (machines, units, components) are seen as having failure characteristics which can be summarized in terms of a small number of measures, for example, "average downtime per failure (downtime)" and "number of failures per unit of time (frequency)" (Scarf, 2007); Depending on the failure characteristics of the system in terms of these measures, different types of maintenance is prescribed. For example, for low frequency/low consequence failures, an operate-to-failure strategy is prescribed, whereas for low frequency/high consequence, failures condition-based maintenance is prescribed.

A general framework for reliability analysis and maintenance optimization can be obtained based on semi-martingale and point process theory. According to Aven and Jensen (2013), the use of point processes in the study of reliability and maintenance optimization was initiated by Arjas (1981a, 1981b).

A general lifetime/failure model and set-up for maintenance optimization based on such a general framework is presented by Jensen (1996). In this general failure model, the system failure time evolution is represented in terms of a counting process on a smooth semi-martingale form, and the maintenance optimization is framed as an optimal stopping time problem. There are two particularly attractive features of this approach, namely the ability to incorporate time dynamics and different information levels. This is done by conditioning on different stochastic process histories, by means of the concept of filtration (Heinrich and Jensen, 1992; Jensen, 1996). We will not go further into this general framework here but instead refer to the references mentioned above.

Many maintenance optimization models can be formulated on a general format, for example, the format described by Aven and Dekker (1997) and Aven and Bergman (1986). In the former case, the object function $B(T)$ is written as

$$B(T) = \frac{C(T)}{D(T)},$$

where T is the decision parameter in question and $C(T)$ and $D(T)$ denote, respectively, the expected costs incurred in a replacement cycle and the expected cycle length. Aven and Dekker (1997) show that by making certain assumptions about the properties of the functions $C(T)$ and $D(T)$, some general results can be derived, for example, the shape of the optimization criterion and algorithms for minimizing $B(T)$. Frameworks like the one that arises from the above formulation is useful considering future model development, since it illustrates that there is often at play a common underlying optimization mechanism. In the following, an example is given illustrating this optimization framework as well as the incorporation of safety costs in such models.

The so-called age replacement model is an example of a basic preventive maintenance optimization model. We consider a system with lifetime X having distribution function F, that is, $F(t) = P(X \leq t)$ is the probability that the system fails at or before time t. The system is replaced T time units after being put into operation, or upon failure, whichever occurs first. The former type of replacement is referred to as preventive replacement and the latter as corrective replacement. Let b denote the expected cost of replacing the system and k the added expected cost of a system failure. Then the expected long-run cost per unit of time can be written as

$$B(T) = \frac{b + kP(X \leq T)}{E \min\{X, T\}} = \frac{b + kF(T)}{\int_0^T (1 - F(x))dx}.$$

If system failure could have safety effects, a typical way to include these in the modeling is to consider loss of life as the potential consequences of an accident and to determine the "safety costs" as (Vatn and Aven, 2010)

$$k_S = p \times \text{PLL} \times v, \tag{24.1}$$

where p is the probability of an accident given that the system has failed, PLL the potential loss of life (defined as the expected number of fatalities) given an accident, and v the so-called value of a prevented fatality (VPF) (cf. Sections 24.2.1 and 24.2.3). The safety cost k_S is hence a cost incurred if a system failure occurs and then leads to an accident. This safety cost may be incorporated in the basic age replacement model by decomposing the corrective maintenance cost as follows:

$$k = k' + k_S,$$

where k′ is a fixed cost incurred when the system fails, regardless of whether or not the failure leads to an accident. Determining the quantities p and PLL falls within the domain of quantitative risk assessment (QRA); cf. Section 24.2.3.

We now extend the above model formulation by considering a system maintained according to the age replacement policy and subject to two types of accidents: Type 1 accidents are failure-induced accidents that may occur when the system fails. Type 2 accidents, on the other hand, are maintenance-induced accidents that may occur when preventive maintenance (i.e., replacement) is performed on a system that is still in the functioning state. Four scenarios of a replacement cycle are then possible:

- Scenario 1: The system fails at time $X < T$, without a failure-induced (type 1) accident occurring, and is then correctively replaced.
- Scenario 2: The system operates without failure until time $T < X$ and is then preventively replaced, without a maintenance-induced (type 2) accident occurring.
- Scenario 3: The system fails at time $X < T$, and a type 1 accident occurs.
- Scenario 4: The system operates until time $T < X$ and upon an attempted preventive maintenance, a type 2 accident occurs.

A type 2 accident is assumed not to be able to occur in Scenario 1, as the system has failed and in is in a non-operating state.

Let, as above, b denote the expected cost of a preventive replacement and k the expected cost of a corrective replacement; in addition, let a_1 and a_2 denote the expected costs of a type 1 accident and a type 2 accident, respectively. The conditional probabilities of type 1 and type 2 accidents, given system failure and preventive replacement, are denoted p_1 and p_2, respectively. The expected long-run cost per unit of time then becomes

$$B(T) = \frac{b + (k + p_1 a_1) F(T) + p_2 a_2 (1 - F(T))}{\int_0^T (1 - F(t))\, dt} = \frac{b + p_2 a_2 + (k + p_1 a_1 - p_2 a_2) F(T)}{\int_0^T (1 - F(t))\, dt}.$$

Now define $c = b + p_2 a_2$ and $h(t) = 1 - F(t)$, and also $m(t) = (k + p_1 a_1 - p_2 a_2) z(t)$, where $z(t) = F'(t)/(1 - F(t))$. We see that c is a constant, $h(t)$ the probability that the system fails before time t, and $m(t)$ is the hazard rate function $z(t)$ multiplied by a constant. The hazard rate function expresses the proneness of the system to fail at time t, since $z(t)\, \Delta t \approx P(X \leq t + \Delta t \,|\, X > t)$ for small Δt; that is, since the hazard rate at a given time t multiplied by the length of a small time increment is approximately equal to the probability of failure in the next time interval of that length. We see that the above expression can be written as

$$B(T) = \frac{c + \int_0^T m(t) h(t) dt}{\int_0^T h(t) dt},$$

and is hence on the general form studied by Aven and Dekker (1997). Hence, Theorem 1 therein applies and so if it can be assumed that $m(t)$ is strictly increasing for $T > T_0 = 0$ and that $\lim_{T \to \infty} m(T) = \infty$, then $g(T_0) = m(T_0)\, D(T_0) - \int_0^{T0} m(t)\, h(t)\, dt = 0 <$ c for $T_0 = 0$, and it then follows that $g(T_1) >$ c for some $T_1 > T_0$ and hence that $B(T)$ has a minimum B^* in T^*, which is unique on $[T_0, \infty)$. In terms of the age replacement model, the preceding essentially says that if the system ages in such a way that at any time the probability of failure in the next time instant is always greater than before, and the hazard rate eventually becomes infinitely large, such as to ensure the inevitable failure of the system, then there is an optimal positive finite T^* at which the expected long-run cost per unit of time is minimized.

The description above mainly covers single system optimization (local optimization). A global optimization problem would be to select a portfolio of maintenance policies. For example, Mild et al. (2015) describe how portfolio decision analysis (PDA) in the form of robust portfolio modeling (RPM) has been used for bridge maintenance planning. In the remainder of the present chapter, we will focus on the single system setting.

24.3.2 Use and Implementation

24.3.2.1 Implementing Risk Acceptance Criteria and the ALARP Principle in the Context of Maintenance Optimization Modeling From a mathematical optimization point of view, one way of implementing the TOR framework is by dividing the risk level into three regions (cf. Section 24.2.4) as

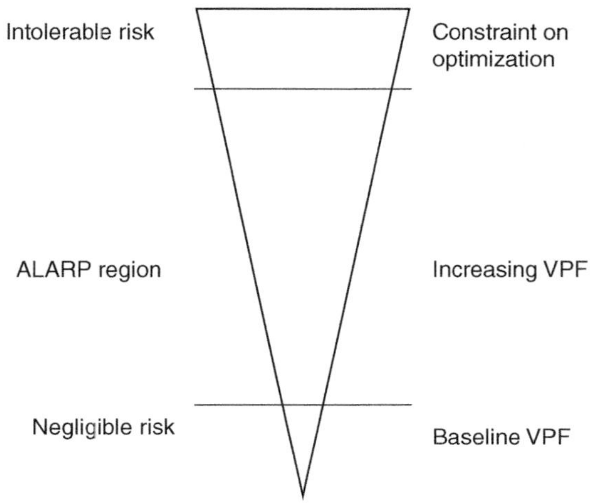

Figure 24.2. Risk regions and associated VPF.

shown in Figure 24.2: For decision parameter values with an associated risk level so low that it is considered broadly acceptable, a baseline VPF equal to v is used in the optimization criterion. For decision parameter values with an associated risk level in the ALARP region, a VPF that is a factor d times higher than the baseline VPF is used, where $d \geq 1$ is a factor intended to reflect the "gross disproportion" criterion. The disproportion factor could be an increasing function of the risk level, where the VPF value used in the ALARP area increases from the baseline level $1 \times v$ up to a level $d_1 \times v$, where d_1 is the "gross disproportion" factor when the risk level is just below the region of unacceptable risk; cf. justification (c) in Section 24.2.4. Decision parameter values leading to a risk level in the region of intolerable risk are not permitted.

In the next three subsections, we look into the problem of delineating the region of unacceptable risk and of implementing the "gross disproportion" criterion (factor), and lastly present a numerical example.

24.3.2.1.1 Maintenance Optimization Subject to a Risk Acceptance Criterion One way to manage risk in the context of maintenance optimization and system safety is to impose quantitative performance requirements on safety-critical systems. Safety constraints applied to maintenance optimization was studied by Aven and Castro (2008a, b). The idea is to first perform pure economic optimization and then check the resulting economically optimal maintenance policy against various safety constraints; if the economically optimal solution satisfies the safety constraint(s) it will be chosen, otherwise the economically optimal solution that just satisfies the safety constraint(s) is chosen. An example of such a safety constraint is that the probability of one or more failures in a fixed time interval should not exceed a fixed value.

In the present chapter, we are, for the most part, assuming that the system deterioration process is known (or well characterized). On the other hand, if the process is not known, changes in system state or condition (or lack of such) between observations of the system will also be of interest, since it increases the knowledge we have about the process. This way of reasoning leads naturally to an adaptive Bayesian model, that is, a model that incorporates new information as and when it becomes available. In Bayesian statistics, the way to handle lack of knowledge (i.e., uncertainty) about a stochastic process is to assign a subjective probability distribution on the probability models used to characterize the process (in practice on the parameters of such models), and then to incorporate new information using Bayes' theorem; see, for example, Bernardo and Smith (1994). In a Bayesian setting, a limiting relative frequency (in an exchangeable sequence) is sometimes referred to as a "chance," to distinguish it from the term probability which then refers to a degree of belief; see, for example, Singpurwalla (2006) and Singpurwalla and Wilson (2008). Following this terminology, two types of constraints can be imposed; one in terms of "probability of chance" and one in terms of probability, the latter understood as expected chance. For example, Flage et al. (2012) study these types of constraints, focusing on the following two specific ones:

1. With a certain probability (degree of belief), the chance (limiting relative frequency) of one or more failures during a specified time interval shall not exceed a pre-defined limit.
2. The probability (i.e., expected chance) of one or more failures during a specified time interval shall not exceed a pre-defined limit.

Reliability constraints are related and sometimes technically equivalent to safety constraints; for an example of the former, see, for example, Scarf et al. (2005). In the examples given above, the safety constraints are formulated purely in terms of system reliability characteristics; then safety constraint can be seen as another name for reliability constraint, but are referred to as safety constraint as a system failure could have safety effects. Safety/reliability constraint models are applicable for delineating the limit(s) of the ALARP region in terms of the decision parameters of maintenance optimization models.

As opposed to reliability constraints restricting, say, the probability of failure or another reliability characteristic, risk acceptance criteria are typically formulated to include a loss dimension, where loss is not restricted to production loss but also includes harm to humans and the environment. As an example, expected loss, suitably normalized, is a common format for risk acceptance criteria. Risk acceptance criteria are typically not expressed in terms of PLL values, which is the (non-normalized) quantity that typically enters the maintenance optimization criterion (cf. Section 24.3.1). They are rather formulated in terms of risk indices such as individual risk (IR) figures or as an F–N curve. If n individuals are under exposure to the risk associated with an activity, each one having an individual risk

IR_i defined as the probability of death by person i as a result of the activity, $i = 1, 2, \ldots, n$, then the PLL value is given as the sum of the individual risk values. Hence, it is clear that if only the PLL value enters the optimization criterion, there is no guarantee that the individual risk of any single person does not exceed an RAC formulated in terms of individual risk. Exceptions include the cases where $n = 1$ as well as when homogeneous risk in the exposed group results in the relation IR = PLL/n. The optimal solution to a maintenance optimization problem constrained by the requirement that the individual risk of all individuals exposed should not exceed a fixed number r_1 is then to let the optimal replacement time be the value of T that minimizes the cost criterion $B(T)$ while fulfilling the risk acceptance criterion that the individual risk should be less than r_1 for all individuals.

Another common format for risk acceptance criteria is as mentioned above that of *F–N* curves. Here N refers to the number of fatalities, and imposing such criteria in terms of an *F–N* curve amounts to placing a restriction on the value of $H(i) = P(N \geq i)$ for each $i = 0, 1, 2, \ldots, n$; that is, on the probability of having i or more fatalities. Adhering to the so-called Farmer principle of diminishing probability of occurrence with increasing consequence (see, e.g., Andrews and Moss, 2002), a risk acceptance criterion formulated in terms of an *F–N* curve may, for example, take a linearly decreasing format. Maintaining the assumption from above that n persons are under exposure, it follows from the well-known relation between the expected value of a non-negative unknown quantity and its complementary cumulative distribution function, that the PLL value is given as the sum of all the $H(i)$ values. Again, it is clear that using only the PLL value as input to the optimization process, in the way described in Section 24.3.1, a risk acceptance criterion (here if formulated by specifying a bound on $H(i)$ for all i is not necessarily satisfied.

With some level of generality, then, we can formulate the problem of maintenance optimization subject to risk acceptance criteria as follows: Let $R(T)$ be a risk index taking a maintenance optimization decision parameter T as its function (both possibly vector-valued), and suppose that the risk index is required to be less than r_1 (also possibly vector-valued). For example, we may define $R(T)$ as a vector of IR values or $H(i)$ values as a function of T. The optimal replacement time T^*_{RAC} is then the value of T that minimizes the cost criterion $B(T)$ while fulfilling the risk acceptance criterion that $R(T)$ should be less than r_1.

24.3.2.1.2 Maintenance Optimization Subject to the ALARP Principle

One way to implement the "gross disproportion" criterion of the ALARP principle is to introduce a function d taking the risk index $R(T)$ as its argument and let the safety cost first defined by Equation (24.1) now be given by

$$k_{\mathrm{S}}(T) = p \times \mathrm{PLL} \times v \times d(R(T)),$$

which hence becomes a function of the decision parameter T. We could, for example, define $d(y)$ to be equal to 1 if y is smaller than the upper limit of broadly acceptable risk r_0 and equal to a non-decreasing function greater than 1 between r_0 and the upper limit of tolerable risk r_1. The definition of $d(y)$ for y greater than the upper limit of tolerable risk r_1 is arbitrary, since maintenance policies for such values of the risk index are not permissible and, hence, will not be considered.

24.3.2.1.3 Numerical Example We consider the extended age replacement model introduced in Section 24.3.1. Let X be the time to system failure, assumed to follow a Weibull distribution F with scale parameter $\alpha = 5$ and shape parameter $\beta = 1.5$. The following cost structure is assumed: A preventive replacement costs $b = 0.1$, a corrective replacement incurs a fixed costs $k = 5$, and the baseline value of preventing a fatality is $v = 15$. Furthermore, we assume that $p_1 = 0.1, p_2 = 0.02$, $\text{PLL}_1 = 0.2, \text{PLL}_2 = 0.1$. Finally, the function d is assumed to increase linearly from $d_0 = 1$ to $d_1 = 5$ on the interval from $r_0 = 1 \times 10^{-5}$ to $r_1 = 5 \times 10^{-3}$, where r_0 and r_1 are, respectively, the upper limits of broadly acceptable and tolerable risk. The linear form of g is arbitrarily chosen. This form implies, as an example, that increasing the risk index, taking 1×10^{-5} as a starting point, by a factor of 25, 50, or 75 leads to increasing the disproportion factor by a of factor 2, 3, and 4, respectively.

Let $\text{PLL}_T(t, t+s)$ denote the PLL in the time interval $(t, t+s)$ when the preventive replacement time is T. Then, using Blackwell's theorem for renewal reward processes (e.g., Ross, 1996, p. 159) and considering loss of life as the "reward," we have that for $t \to \infty$,

$$\text{PLL}_T(t, t+s) \to s \times \frac{(p_1 \text{PLL}_1 F(T) + p_2 \text{PLL}_2 (1 - F(T)))}{\int_0^T \bar{F}(x)\, dx}.$$

Now let $\overline{\text{PLL}}_T(u)$ denote the above limit and $\overline{IR}_T(u)$ the associated asymptotic individual risk when assuming equal distribution of risk in the population of exposed individuals. Suppose that only one individual is exposed to the risk associated with performing maintenance on a system, so that $\overline{IR}_T(u) = \overline{PLL}_T(u)$. We may, for example, consider the most risk-exposed individual in a group, thus ensuring that the RAC is satisfied for all individuals in the group. Nevertheless, it should be pointed out that $\overline{IR}_T(u)$ is the asymptotic individual risk, while the risk acceptance criterion will typically be formulated to apply to any time interval of length s. Hence, it is important to be aware that $\overline{IR}_T(u)$ may be lower than the applicable RAC without actually satisfying the RAC.

The principle for finding the optimal replacement time T is constrained optimization, that is, to find the most economically favorable value of T that does not violate the RAC. This is analogous to the optimization performed by Aven and Castro (2008a, b); however, here we have incorporated the disproportion factor as described in Section 24.3.2.1 and above, which depends on the risk level. In Figure 24.3 is shown the value of the object function $B(T)$ based on the fixed baseline VPF for all T (solid line; cf. the left-hand axis), and based on a baseline and increasing VPF

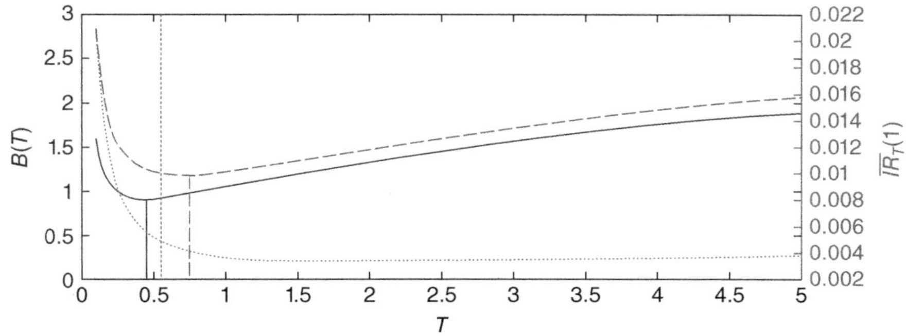

Figure 24.3. Expected long-run cost *B*(*T*) (solid and dashed lines) and asymptotic IR (dotted line) versus *T*.

depending on *R*(*T*) (dashed line; cf. the left-hand axis), as well as the asymptotic IR (dotted line; cf. the right-hand axis) for different values of *T*. When based on a fixed baseline VPF, the optimal replacement time *T* is 0.45, corresponding to a value of *B*(*T*) equal to 0.91; however, as indicated by the dotted horizontal line, for values of *T* less than 0.55 the asymptotic IR exceeds the risk acceptance criterion of 5×10^{-3}. Accordingly, we would in this case choose *T* equal to 0.55 as the "optimal" replacement time, corresponding to *B*(*T*) equal to 0.92. On the other hand, when based on a baseline and increasing VPF, the optimal replacement time is 0.75, which satisfies the RAC and corresponds to an expected long-run cost per unit of time equal to 1.2.

24.3.2.2 Empirical Research on Use and Efficiency The industrial and commercial incentive to optimize reliability and maintenance is reflected by several case studies, for example, Arthur (2005), Castanier and Rausand (2006), Hokstad et al. (1994) and Aven (1989). The body of literature proposing maintenance optimization models is extensive and rapidly increasing. Surveys and reviews in this area include Dekker (1996), Cho and Parlar (1991), Valdez-Flores and Feldman (1989) and Wang (2002). These surveys and reviews illustrate that developing maintenance optimization models is paid a considerable amount of attention from researchers who see opportunities when it comes to using applied mathematics to solve maintenance optimization problems.

Nevertheless, despite the great number of optimization models developed, there appears however to be some challenges in applying these to real problems. Rausand and Høyland (2004 p. 362) note that:

> "It is often recommended [...] to establish mathematical models that can be used to assess the impacts of maintenance decisions. This approach seems to give promising results but has not yet been sufficiently developed in an industrial context. By using

mathematical/stochastic models it may be possible to "simulate" maintenance strategies and to reveal the associated effects and maintenance costs and operational performance. The simulation may, in some cases, be used to determine the best maintenance strategy to implement."

As pointed out by Scarf (1997), many applications and case studies appear to have been for the most part motivated by the need to find applications for a particular model that has been developed, rather than finding solutions to real maintenance-related problems. The need for more application oriented research is echoed also by, for example, Dekker (1995). Dekker and Scarf (1998) present several arguments in favor of more qualitative approaches such as RCM and TPM, and condition-monitoring approaches; at the same time they point out that optimization models can offer more than such qualitative approaches, and case studies and decision support systems are presented making the case that "there is certainly a need for optimization."

24.4 DISCUSSION AND CONCLUSIONS

In this section, we discuss some limitations and weaknesses of maintenance optimization modeling, centering on the following points where the first two relate to such modeling in general while the last three relate to its use to support risk/safety management:

- The stochastic nature of many maintenance optimization models limits their applicability when the system being analyzed is unique.
- A large number of maintenance optimization models have been developed, yet examples of their application in practice are relatively scarce.
- It has been argued that it is necessary to look beyond expected values (losses) in situations with potentially high safety impacts.
- There are other ways of achieving risk reduction than through adjustment of the decision parameter(s) of a given maintenance optimization model.
- The philosophy and practice of the ALARP principle goes beyond a more or less "mechanical" cost–benefit optimization.

A maintenance optimization model can be described as "a mathematical (stochastic) model in which costs (in a wide sense) are quantified and optimized" (Aven and Sandve, 1999, p. 141). The stochastic nature which characterize many models means that probability models are used. A probability model is a model of variation and its interpretation is the fraction of times the event in question occurs if the situation were repeated an infinite number of times under similar conditions. Such models hence require the introduction of a (hypothetical) infinite number of similar systems. Firstly, this may be difficult to grasp conceptually (Aven and Sandve, 1999). Secondly, if the system in question is unique, and data are scarce or non-existent, such models become difficult to use. Apeland and Scarf (2003a, 2003b) describe a "fully subjective approach" to inspection maintenance and capital equipment replacement, where such infinite populations, and non-observable parameters defined from these, are not introduced. Instead subjective (Bayesian) probability is used only to express uncertainty about observable quantities only. In between the infinite and unique cases, Barlow (1998), giving as examples a "special order" of only a few items, and the existence of only five space shuttles in 1996, describes a finite population model, noting that (Barlow, 1998 p. 3):

> "All populations are actually finite, although it is often useful to consider unbounded populations. It is usual to start with the assumption that random quantities (also called random variables) are *i*ndependent, *i*dentically *d*istributed (*iid*). However, this cannot be justified for finite populations."

For the issue of lack of practical implementations of maintenance optimization models, reference is made to Section 24.3.2.2.

As regards the third point in the list at the beginning of this section, we have seen that maintenance optimization models can be characterized as formal tools for balancing costs and benefits, typically based on expected value calculations and accounting for safety using VPF figures. Vatn and Aven (2010) question the rationale of such an approach, asking for the meaning of a VPF figure and questioning whether it is sufficient to reflect safety aspects using expected values (losses). Their conclusion is that there is a need to see beyond expected values in settings where the safety impacts may be high. Furthermore, they present a framework covering a broader decision, including considerations on the potential for major accidents, the nature of the uncertainties involved, as well as lack of knowledge about key risk influencing factors (RIFs). In particular, decisions with a high safety impact should, they argue, not be delegated to the maintenance department but rather moved to a "Safety Board" where they can be subject to broader consideration and discussion.

Considering the second point above, we see that for a given (fixed) maintenance optimization model – take the extended age replacement model described in Section 24.3.1 as an example – risk reduction can be brought about in essentially two ways:

I. By changing T, that is, the decision parameter in the maintenance optimization model, in order to change the probability $F(T)$ of system failure
II. By implementing some risk reducing measure(s) M in order to
   a. for a fixed T, change the probability of system failure $F(T)$; and/or
   b. reduce the conditional expected loss $p \times \text{PLL}$.

Here the strategies I and IIa change the probability of system failure and can hence be seen as preventive measures, whereas IIb reduces the vulnerability given an initiating event and can be seen as a consequence-reducing measure. Strategy I presumes a fixed maintenance model and changes the probability $F(T)$ of system failure through adjustments of the decision parameter T. This as opposed to strategy IIa, which also reduces the same probability but instead by changing the structure of the model F. Moreover, for a given system, several maintenance policies could be relevant. Hence, one risk reducing measure could be to alter the maintenance policy itself. For example, a condition-based policy may be overall more expensive than a time-based policy and would hence not be considered in a conventional cost–benefit regime; however, in an ALARP regime such a modification would at least be considered since the ALARP principle places a strong focus on risk reduction. Another risk reducing measure, also outside the typical maintenance optimization context where both the system and the maintenance policy are given (fixed), could be to change the type of system subject to maintenance optimization.

From the above we see that risk reduction for a system subject to maintenance optimization goes beyond optimization with respect to a model decision parameter T. One way to handle this is to modify the set-up by extending the decision parameter to the set $T' = (T, M)$ and then optimize the optimization criterion with respect to T'. Then it is worth mentioning that the cost of a risk reducing measure M to reduce $p \times$ PLL will often comprise both an initial investment cost as well as a periodic operating cost. In this case a cost criterion expressing the expected discounted costs would be more appropriate than the more common expected long-run cost per unit of time criterion, since the former allows for easy integration of the expected discounted costs of M with the standard optimization criterion $B(T)$. In the case that the operating cost of the risk reducing measure is incurred on an on-demand basis, the integration of this cost component into a long rung average cost criterion is straightforward. An alternative approach to introducing T′ is to introduce a function such as $h(T) = F(T) \times p \times$ PLL as an additional objective function, to be minimized jointly with $B(T)$ in a multi-objective optimization procedure.

The discussion has so far been framed in a mathematical optimization setting as presented in Section 24.3.1, referring to optimization models, optimization criteria and decision parameters. However, both the philosophy and practice of the ALARP principle can be said to go beyond more or less "mechanical" cost–benefit optimization. This argument is in line with justification d) in Section 24.2.4, that is, incentivizing cost-minimization and engineering "best practice." The underlying objective of the ALARP principle is risk reduction, at least if the requirement for reversed burden of proof and the "gross disproportion" interpretations are adopted. On the other hand, the underlying objective of maintenance optimization modeling is to balance burdens and benefits in a (quantitatively) optimal manner. The outcome of the maintenance planning for a safety-critical system would in practice typically be the result of a combination of:

- Adherence to codes and standards
- Application of engineering judgement
- Decision support from various risk assessments and cost–benefit/cost-effectiveness analyses

More cautionary approaches may also be seen; for example, if there are considerable uncertainties concerning the performance of a new system, maintenance actions such as inspection or replacement would most likely be performed more often in an early phase and then less seldom as more knowledge of the system performance is gained.

Aven (2015) presents a three-step procedure for ALARP evaluations (inspired by Aven and Vinnem, 2007) of proposed risk reducing measures:

1. Crude analysis
2. More detailed (cost–benefit/cost-effectiveness) analysis
3. Assessment of other concerns

Measures where a crude analysis demonstrates obvious effects at low costs, are implemented without further detailed analysis. If the costs are high, or the risk reducing effect unclear, more detailed analyses in the form of cost–benefit/cost-effectiveness analyses are performed to give decision support. If these analyses suggest that the measure is economically favorable (positive expected net present value, or sufficiently low ICAF), the measure is implemented; otherwise an assessment is performed covering other concerns, such as consideration of uncertainty (cf. item b) in Section 24.2.4) or strategic considerations.

This chapter has addressed the relation between maintenance optimization and safety. We have seen that models for determining optimal maintenance policies commonly rely on a classical cost–benefit rationale, based on transformation of all effects studied into monetary value and balancing of different concerns based on expected value calculations. If accidents due to system failure or maintenance actions could have negative safety effects, then a potential conflicts exists between the former type of optimization approach and the ALARP principle; at least so if an interpretation of the latter is adopted where verification of ALARP extends beyond CBA as seen above. If (also) risk acceptance criteria are imposed, a constraint may be placed on the set of permissible maintenance policies resulting in sub-optimal maintenance scheduling from an economic optimization modeling point of view. The latter is illustrated in a numerical example, where also the effect of incorporating a "gross disproportion" factor into the safety effects modeling is shown. Finally, the three key points at the beginning of this section have been discussed.

FURTHER READING

Maintenance optimization considering safety brings together system reliability theory, mathematical optimization as well as risk assessment and risk management. A brief historical perspective on reliability and risk analysis is presented in Aven (1992), where references to further literature on reliability and risk analysis in a historical perspective are offered.

Numerous textbooks deal with methods and models for reliability analysis and/or risk assessment, for example, Andrews and Moss (2002), Aven (2015, 1992), Aven and Jensen (2013), Barlow and Proschan (1996), Bedford and Cooke (2001), Cox (2002), Rausand and Høyland (2004), Singpurwalla (2006), Vinnem (2007), Vose (2008), and Zio (2007, 2009). Application of risk assessment methodology to support maintenance-related decision-making is described by Aven (2008).

As mentioned in Section 24.3.2.1.1, in the present paper we are largely assuming that the system deterioration process is known (or well characterized). The Bayesian approach was mentioned as a way to handle the setting where this is not the case. Several of textbooks in the above paragraph address Bayesian reliability and Bayesian failure data analysis; additional references are, for example, Barlow (1998) and Hamada et al. (2008). The same holds for textbooks on statistical methods (also non-Bayesian) for reliability data; additional references are Crowder et al. (1991) and Meeker and Escobar (2014).

Several authors have discussed the relationship between CBA and the ALARP principle in general; including Aven and Abrahamsen (2007), French et al. (2005) and Jones-Lee and Aven (2011). On the other hand, there is an apparently limited body of literature that addresses in particular and in detail the relation between maintenance optimization and the ALARP principle and risk acceptance criteria. French et al. (2005), in a discussion of the use of discounting to trade-off risk through time, point at the fact that maintenance policies could lead to a non-uniform risk within a fixed time interval; cf. the comment about the asymptotic IR in Section 24.3.2.1.3 in the present chapter. Finally, as detailed in this chapter, Vatn and Aven (2010) discuss maintenance optimization when safety issues are important, including the ALARP principle.

The present chapter has links to several of the topics covered by other chapters in this handbook. Some of these links have been discussed in some level of detail in Section 24.2, namely maintenance optimization models as a type of cost–benefit analysis (cf. Chapter 24), the influence of maintenance on safety and hence the link to (quantitative) risk assessment (cf. Chapter 22) and risk management (cf. Chapter 30), and the implementation of the ALARP principle (cf. Chapter 25) in a maintenance optimization setting. In addition, the concepts of maintainability and maintenance support described in Section 24.2.1 are influenced by, for example, design and operator skills, respectively, and hence there is a link to fail safe design (cf. Chapter 5), human factors engineering (cf. Chapter 7), inherently safe design (cf. Chapter 14), proven design and proven technology (cf. Chapter 15), and inspectability/maintainability (cf. Chapter 19). Furthermore, as mentioned in Section 24.3.1, frequent system failures with low downtimes would probably be handled as a quality management issue (cf. Chapter 16). Finally, maintenance optimization is a special case of optimization (cf. Chapter 20) and, for example, reliability constraints are essentially constrained optimization which is discussed in the optimization chapter.

REFERENCES

Andrews, J. D., and Moss, T. R. (2002). *Reliability and Risk Assessment* (2nd ed.). London: Professional Engineering Publishing.

Apeland, S., and Scarf, P. A. (2003a). A fully subjective approach to modelling inspection maintenance. *European Journal of Operational Research*, 148, 410–425.

Apeland, S., and Scarf, P. A. (2003b). A fully subjective approach to capital equipment replacement. *The Journal of the Operational Research Society*, 54(4), 371–378.

Apostolakis, G. E. (2004). How useful is quantitative risk assessment? *Risk Analysis*, 24(3), 515–520.

Arjas, E. (1981a). A stochastic process approach to multivariate reliability systems: Notions based on conditional stochastic order. *Mathematics of Operations Research*, 6(2), 263–276.

Arjas, E. (1981b). The failure and hazard processes in multivariate reliability systems. *Mathematics of Operations Research*, 6(4), 551–562.

Arthur, N. (2005). Optimization of vibration analysis inspection intervals for an offshore oil and gas water injection pumping system. *Journal of Process Mechanical Engineering*, 219(3), 251–259.

Aven, T. (Ed.). (1989). *Reliability Achievement: The Commercial Incentive*. Elsevier.

Aven, T. (1992). *Reliability and Risk Analysis*. Elsevier.

Aven, T. (2008). Risk analysis in maintenance. In: K. Kobbacy and D. N. P. Murthy (Eds.), *Complex System Maintenance Handbook* (pp. 437–458). London: Springer.

Aven, T. (2015). *Risk Analysis* (2nd ed.). Chichester: John Wiley & Sons.

Aven, T., and Abrahamsen, E. B. (2007). On the use of cost–benefit analysis in ALARP processes. *International Journal of Performability Engineering*, 3(3), 345–353.

Aven, T., and Bergman, B. (1986). Optimal replacement times—A general set-up. *Journal of Applied Probability*, 23, 432–442.

Aven, T., and Castro, I. T. (2008). A minimal repair replacement model with two types of failure and a safety constraint. *European Journal of Operational Research*, 188(2), 506–515.

Aven, T., and Castro, I. T. (2009). A delay-time model with safety constraint. *Reliability Engineering and System Safety*, 94(2), 261–267.

Aven, T., and Dekker, R. (1997). A useful framework for optimal replacement models. *Reliability Engineering and System Safety*, 58, 61–67.

Aven, T., and Jensen, U. (2013). *Stochastic Models in Reliability* (2nd ed.). New York: Springer.

Aven, T., and Sandve, K. (1999). A note on how we should express and interpret the results of stochastic maintenance optimization models. *Journal of Quality in Maintenance Engineering*, 5(2), 141–146.

Aven, T., and Vinnem, J. E. (2007). *Risk Management: With Applications from the Offshore Petroleum Industry*. London: Springer.

Barlow, R. E. (1998). *Engineering Reliability*. Philadelphia, PA: SIAM.

Barlow, R. E., and Proschan, F. (1996). *Mathematical Theory of Reliability. Classics in Applied Mathematics*(Vol. 17). Philadelphia, PA: SIAM.

Bedford, T., and Cooke, R. M. (2001). *Probabilistic Risk Analysis: Foundations and Methods*. Cambridge: Cambridge University Press.

Bernardo, J. M., and Smith, A. F. M. (1994). *Bayesian Theory*. Chichester: John Wiley & Sons.

Bloom, N. (2006). *Reliability Centered Maintenance: Implementation Made Simple*. New York: McGraw-Hill Companies.

Castanier, B., and Rausand, M. (2006). Maintenance optimization for subsea pipelines. *International Journal of Pressure Vessels and Piping*, 83, 236–243.

Cho, D. I., and Parlar, M. (1991). A survey of maintenance models for multi-unit systems. *European Journal of Operational Research*, 51, 1–23.

Cox, L. A. (2002). *Risk Analysis: Foundations, Models, and Methods*. Boston: Kluwer Academic Publishers.

Crowder, M. J., Kimber, A. C., Smith, R. L., and Sweeting, T. J. (1991). *Statistical Analysis of Reliability Data*. London: Chapman & Hall.

Dekker, R. (1995). On the use of operations research models for maintenance decision making. *Microelectronics Reliability*, 35(9–10), 1321–1331.

Dekker, R. (1996). Applications of maintenance optimization models: A review and analysis. *Reliability Engineering and System Safety*, 51, 229–240.

Dekker, R., and Scarf, P. A. (1998).On the impact of optimization models in maintenance decision making: The state of the art. *Reliability Engineering and System Safety*, 60, 111–119.

Esch, L., Kieffer, R., and Lopez, T. (2005). *Asset and Risk Management: Risk Oriented Finance*. Chichester: John Wiley & Sons, Ltd.

Flage, R. (2013). On risk reduction principles in the context of maintenance optimization modelling. *Journal of Risk and Reliability*, 227(3), 241–250.

Flage, R., Coit, D. W., Luxhøj, J. T., and Aven, T. (2012). Safety constraints applied to an adaptive Bayesian maintenance optimization model. *Reliability Engineering and System Safety*, 102, 16–26.

French, S., Bedford, T., and Atherton, E. (2005). Supporting ALARP decision making by cost–benefit analysis and multiattribute utility theory. *Journal of Risk Research*, 8(3), 207–223.

Hamada, M. S., Wilson, A. G., Reese, C. S., and Martz, H. F. (2008). *Bayesian Reliability*. Dordrecht: Springer.

Health and Safety Executive (2001). *Reducing Risks, Protecting People: HSE's Decision-Making Process (R2P2)*. London: HSE Books.

Health and Safety Executive (2006). Offshore Installations (Safety Case) Regulations 2005. Regulation 12 Demonstrating compliance with the relevant statutory provisions. Offshore Information Sheet No. 2/2006. Available at: http://www.hse.gov.uk/offshore/is2-2006.pdf

Heinrich, G., and Jensen, U. (1992). Optimal replacement rules based on different information levels. *Naval Research Logistics*, 39(7), 937–955.

Hokstad, P., Fløtten, P., Holmstrøm, S., McKenna, F., and Onshus, T. (1994). A reliability model for optimization of test schemes for fire and gas detectors. *Reliability Engineering and System Safety*, 47, 15–25.

International Organization for Standardization (2009) ISO Guide 73:2009. Risk management—Vocabulary.

Jensen, U. (1996). Stochastic models of reliability and maintenance: An overview. In: S. Özekici (Ed.), *Reliability and Maintenance of Complex Systems*. Berlin: Springer.

Jones-Lee, M., and Aven, T. (2011). ALARP – What does it really mean? *Reliability Engineering and System Safety*, 96(8), 877–882.

Labib, A. W. (1998). World-class maintenance using a computerised maintenance management system. *Journal of Quality in Maintenance Engineering*, 4(1), 66–75.

Meeker, W. Q., and Escobar, L. A. (2014). *Statistical Methods for Reliability Data*. John Wiley & Sons.

Mild, P., Liesiö, J., and Salo, A. (2015). Selecting infrastructure maintenance projects with Robust Portfolio Modeling. *Decision Support Systems*, 77, 21–30.

NORSOK (2011). NORSOK Standard Z-008 Risk based maintenance and consequence classification. Rev. 3. Available at: https://www.standard.no/en/sectors/energi-og-klima/petroleum/norsok-standard-categories/z-regularity–criticality/z-0082/

NRC (2002). Guide 1.174 – An Approach for using probabilistic risk assessment in risk-informed decisions on plant-specific changes to the current licensing basis. Rev 1.

Petroleum Safety Authority Norway (2011a). Regulations relating to management and the duty to provide information in the petroleum activities and at certain onshore facilities (The management regulations). Available at: http://www.ptil.no/management/category401.html

Petroleum Safety Authority Norway (2011b). Regulations relating to health, safety and the environment in the petroleum activities and at certain onshore facilities (The framework regulations). Available at: http://www.ptil.no/framework-hse/category403.html

Petroleum Safety Authority Norway (2011c). Regulations relating to conducting petroleum activities (The activities regulations). Available at: http://www.ptil.no/activities/category399.html

Petroleum Safety Authority Norway (2011d). Regulations relating to technical and operational matters at onshore facilities in the petroleum activities, etc. (Technical and operational regulations). Available at: http://www.ptil.no/technical and operational regulations/category635.html

Rausand, M. (1998). Reliability centered maintenance. *Reliability Engineering and System Safety*, 60, 121–132.

Rausand, M., and Høyland, A. (2004). *System Reliability Analysis: Models, Statistical Methods, and Applications* (2nd ed.). Hoboken, NJ: John Wiley & Sons.

Rausand, M., and Vatn, J. (2008). Reliability centered maintenance. In: K. Kobbacy and D. N. P. Murthy (Eds.), *Complex System Maintenance Handbook* (pp. 437–458). London: Springer.

Ross, S. M. (1996). *Stochastic Processes* (2nd ed.). New York: John Wiley & Sons.

Scarf, P. A. (1997). On the application of mathematical models in maintenance. *European Journal of Operational Research*, 99, 493–506.

Scarf, P. A., Dwight, R., and Al-Musrati, A. (2005). On reliability criteria and the implied cost of failure for a maintained component. *Reliability Engineering and System Safety*, 89, 199–207.

Selvik, J. T., and Aven, T. (2010). A framework for reliability and risk centered maintenance. *Reliability Engineering and System Safety*, 96, 324–331.

Singpurwalla, N. D. (2006). *Reliability and Risk: A Bayesian Perspective*. Chichester: John Wiley & Sons, Ltd.

Singpurwalla, N. D., and Wilson, A. G. (2008). Probability, chance and the probability of chance. *IIE Transactions*, 41(1), 12–22.

Valdez-Flores, C., and Feldman, R. M. (1989). A survey of preventive maintenance models for stochastically deteriorating single-unit systems. *Naval Research Logistics*, 36, 419–446.

Vatn, J., and Aven, T. (2010). An approach to maintenance optimization where safety issues are important. *Reliability Engineering and System Safety*, 95(1), 58–63.

Vatn, J, Hokstad, P., and Bodsberg, L. (1996). An overall model for maintenance optimization. *Reliability Engineering and System Safety*, 51, 241–257.

Vinnem, J. E. (2007). *Offshore Risk Assessment: Principles, Modelling and Applications of QRA Studies* (2nd ed.). London: Springer.

Vose, D. (2008). *Risk Analysis: A Quantitative Guide* (3rd ed.). Chichester: John Wiley & Sons, Ltd.

Wang, H. (2002). A survey of maintenance policies of deteriorating systems. *European Journal of Operational Research*, 139, 469–489.

Zio, E. (2007). *An Introduction to the Basics of Reliability and Risk Analysis*. Hackensack, NJ: World Scientific.

Zio, E. (2009). *Computational Methods for Reliability and Risk Analysis*. Hackensack, NJ: World Scientific.

25

HUMAN RELIABILITY ANALYSIS

Luca Podofillini

25.1 INTRODUCTION WITH EXAMPLES

Human reliability analysis (HRA) is intended as a set of analyses addressing human failures to perform safety-relevant tasks, generally when interacting with a technical system. The typical personnel tasks considered are either during normal system operation, or shutdown, or in response to abnormal situations. Also, maintenance and test tasks are in the scope of HRA. The most typical applications are for the nuclear power, aerospace, oil and gas, transportation industries.

The key questions that an HRA addresses are:

- How may personnel (e.g., plant operating crew, pilots, train drivers) fail to perform a task?
- What influences their performance to accomplish the task?
- How likely the failure is (what is the probability of failure)?

Handbook of Safety Principles, First Edition. Edited by Niklas Möller, Sven Ove Hansson, Jan-Erik Holmberg, and Carl Rollenhagen.

Depending on the scope of the analysis, the identification of measures to reduce the probability of failure, or more generally to improve the performance conditions may be an explicit goal of the analysis.

The most representative elements of an HRA are:

- Task analysis and human error identification, to characterize how the task is performed and how it may fail (which actions with undesired consequences may be carried out)
- Qualitative analysis of the performance conditions as these may influence possible failure
- Quantification of the failure probability

These three elements are intended to address the three questions above. Depending on the methods used for the analysis, the three elements overlap; also, some iterations are generally required. Complete overviews of the whole HRA process can be found in the literature (e.g., Forester et al., 2007; Kirwan, 1994; Spurgin, 2010).

HRA is very often associated with the term "human error." Indeed, reference to "errors" was pervasive in the early developments of the field and it is still common among practitioners. The use of the word "error" was revisited around the 1990s to avoid any connotation of blame to the personnel. Also, evidence shows that human failures involved in serious accidents are generally driven by contextual factors, as opposed to by pure human fallacy.[1] Terms such as human interactions, erroneous actions, unsafe actions, human failure events have been proposed (Forester et al., 2007; Hollnagel, 1998; USNRC, 2000) and are generally now preferred, depending on the modeling aspect to emphasize. The present chapter will generally refer to human failures: for a deeper discussion on this issue, see Reason (1990), USNRC (2000), Hollnagel (1998), Sträter (2000), Dekker (2014).

An HRA can be carried out following two different perspectives, retrospective and prospective. A retrospective HRA refers to the analysis of a past event, aiming at identifying the factors contributing to the failure. A prospective HRA relates to potential failures that have not necessarily happened; this is the most common perspective for HRA applications. Another distinction that is generally made relates to whether the HRA is part of a probabilistic safety assessment (PSA) or it is a stand-alone analysis. When it is part of the PSA, the result of the HRA (i.e., the failure probability) is then integrated into the PSA fault trees and event trees to get the full risk profile. Also, in general, when embedded into a PSA, the failure events of interest are defined on the basis of the accident sequence models, the event trees, and fault trees—generally as part of an iterative process. When HRA is conducted as a stand-alone analysis, the use of methods for task analysis and failure identification (Kirwan, 1998a, 1998b; Kirwan and Ainsworth, 2003) is comparatively more extensive.

[1] The implication of fault or blame in human error was never the intention of HRA from its early developments; see definition of "human error" by Swain and Guttman (1983).

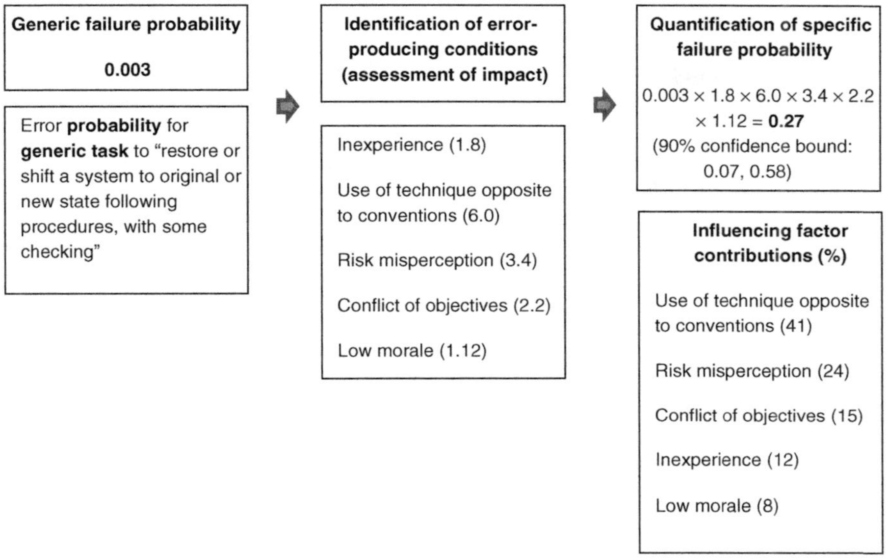

Figure 25.1. Example of quantification of human failure probability via an HRA method. Data from Kirwan 1994, using the HEART method.

The rest of this chapter presents two examples. The first shows the general principle of how HRA quantifies failure probabilities. The second one shows the type of insights that may be produced when the HRA results are embedded into the PSA model. The examples are intended also to show how quantification of a human failure allows getting concrete insights on safety.

The first example relates to the assessment of the probability of an operator's failure to isolate a bypass line route of a plant following procedures (taken from Kirwan, 1994). Figure 25.1 summarizes the analysis and quantification. As the figure shows, the quantification process starts from a generic failure rate (for the task "restore or shift a system to original or new state following procedures, with some checking," which envelops the specific operator task to be analyzed). Then, scenario and task-specific factors are identified and assessed: their influence modifies the generic failure probability to account for the specific performance conditions. Each HRA method has its own specificity, but the common idea is that the failure probability is determined by a characterization of the task to be performed and the performance conditions. In general, HRA methods support the characterization of the tasks, the analysis of the performance factors, and the quantification of their influence.

The example also shows the relative importance of the influencing factors to determine the resulting failure probability. This may suggest and prioritize possible performance improvements: from the point of view of the failure probability, the analysis suggests that changes in the technique used for isolating the system would be more effective than improving the worker's morale. Of course, the reduction of the failure probability on a specific task is only one piece of information in the identification of possible improvements, the effect of which has to be analyzed at the system level (e.g., improving the worker's morale may have a smaller impact on this specific task, but it would influence all tasks, as opposed to the isolation technique). Also, the performance issues resulting from the HRA have to be evaluated in light of more fundamental organizational issues—for example, many influencing factors identified, shown in Figure 25.1, may point to organizational deficiencies in the facility safety culture and policies for allocation of personnel.

It is worth adding a comment on the probability value of 0.27 produced by the analysis. The exact value itself is of relatively low importance. Generally, the uncertainty in the value is quite large (see confidence bounds reported in Figure 25.1). It is important to bear in mind that an HRA method is intended to be a tool to allow analysts to judge the likelihood of a failure, in practice to assess the order of magnitude of the failure probability—not an exact value. The decimals in the probability values are generally maintained for traceability of the calculation steps.

The second example shows how HRA contributes to the characterization of the overall facility risk profile. Figure 25.2 is taken from the results of the PSA of one unit of the Oconee nuclear power plant in the United States presented in Vesely et al.

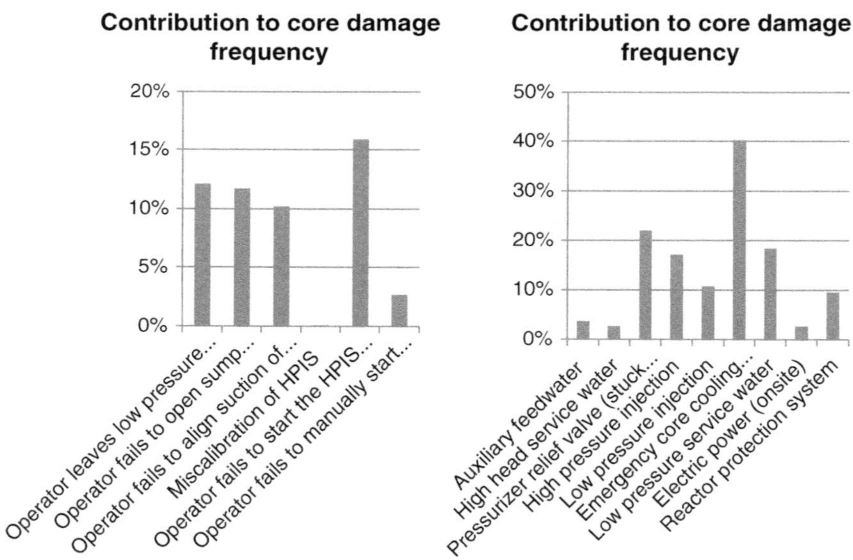

Figure 25.2. Example of quantification of contribution to risk profile: operator actions (left); plant systems (right). The contribution relates to the fraction (%) of the core damage frequency including accident sequences involving the actions or the systems (the Fussell–Vesely importance measure). *Source*: Vesely et al. 1986. Public domain.

(1986). It shows that the failures of operator actions contribute differently to the risk figure-of-merit (the core damage frequency as typical in nuclear PSA); their contribution depends not only on the assessed failure probability, but also on the frequency of the accident sequences they are involved in and on the relative importance of the plant systems whose unavailability they lead to (the contribution of the latter to the plant risk is presented on the right of the figure). Indeed, the need to consider the human element to have a complete picture of the risk profile was clear from the early developments of PSA (USNRC, 1975). The contribution of failure events (of any type), components, systems, accident categories is quantified by the so-called importance measures, which basically can be related to the degree of protection of the plant and to the potential for further risk reduction (see Fullwood, 2000, for further reading on PSA methods and applications). The risk profile needs to be investigated to evaluate which improvements will be most effective for reducing risk. In case of HRA, these generally relate to improvements in the performance conditions, for example, by enhancing the procedural guidance or modifying the human–machine interface to make the indications more salient, or altering the response strategy to increase the time available for actions. As already mentioned, the HRA results are only one element to suggest improvements, which needs to be evaluated in a broader context (e.g., improving procedure for one specific task may not be good for other tasks). Indeed, the HRA results may alternatively point to the need for relying less on a human action and more on automation to bring the facility to a safe state.

25.2 ORIGIN AND HISTORY OF THE PRINCIPLE

This chapter will present some of the key milestones for the HRA field; comprehensive and insightful historical overviews of the field can be found in Spurgin (2010), Boring (2012), Kirwan (1994). The earliest systematic methodology for conducting HRA is the technique for human error rate prediction (THERP; Swain and Guttman, 1983). The main elements of THERP, detailed task analysis, task decomposition, and the underlying HRA data, started to be developed in the 1960s, primarily for military applications in the United States. THERP's application to nuclear power plants matured for use in the WASH-1400 study (USNRC, 1975), which is recognized as the first systematic and comprehensive PSA study for nuclear power plant application. In 1979, the Three Mile Island accident in the United States shook the nuclear industry, underscoring the fundamental role of human factors in plant safety (Osif et al., 2004). Among the lessons learned were the value of PSA analyses to prioritize actual contributors to the risk and the key part of operator actions among them. THERP was finally released in 1983, incorporating lessons learned from the Three Mile Island accident as well as addressing other limitations of earlier versions (mostly related to excessive emphasis on execution errors and missing consideration of diagnosis and decision-making errors, Spurgin, 2010). THERP's handbook represents one of the pillars of HRA, not just because of the method proposed, but also because of the extensive analysis of human performance issues documented throughout the whole handbook. For some types of personnel tasks (mostly related to executing procedural steps), THERP's data has recognized acceptance by the HRA community (Kirwan, 1997a, 1997b) and its validation is still subject of current interest (Preischl and Hellmich 2013; Shirley et al., 2015).

The 1980s saw the establishment of much of the foundation of the HRA field: few highlights are given here. The systematic human error reduction and prediction approach (SHERPA) was developed (Embrey, 1986), aimed at systematically identifying possible human failures, connecting psychological mechanisms and failure modes. The basis for SHARP is the work on cognitive science and human error by Rasmussen et al. (1987), Rasmussen et al. (1981) and Reason (1987). The attempt to connect HRA methods to models of failure mechanisms and modes is also present in more recent HRA methods and is intended to provide stronger foundation of the methods on cognitive science (Moieni et al., 1994, and, up to very recently, Whaley et al., 2016). Generally, during the 1980s, most of the work related to developing methods for task analysis and error identification was also made (Kirwan, 1998a, 1998b; Kirwan and Ainsworth 2003).

Several studies in the early 1980s focused on the development of time reliability curves as a means to capture the reliability of nuclear power plant crews to respond to emergency conditions, based on data obtained from simulator studies. Probably the most influential studies were the ones behind the human cognitive reliability (HCR) model (Hannaman et al., 1984, 1985) and the operator reliability experiment (ORE), which resulted in a revision of the HCR model, into HCR/ORE (Moieni et al., 1994). The HCR/ORE curves are featured in the EPRI HRA Calculator® (Julius et al., 2005), currently the most widely used HRA tool in the United States. The insights from the ORE experiment were also the basis of the development of the cause-based decision trees (CBDTs), Parry et al. (1992), also featured in the EPRI HRA Calculator. Such trees are also featured in the very recently proposed HRA method IDHEAS (Integrated Decision-Tree Human Event Analysis System; Xing et al., 2013).

The intrinsic complexity in the collection of HRA data led to the development of a number of methods aiming at eliciting knowledge about failure probabilities based on expert judgment. The most prominent method of this type is the Success Likelihood Index Method (SLIM; Embrey et al., 1984), most often implemented in the version developed by Chien et al. (1988) (known as PLG's SLIM from Pickard, Lowe, Garrick, Inc. or FLIM from Failure Likelihood Index Method). In SLIM methods, the performance conditions are assessed with judgments on a set of weighted factors. Factor ratings are then combined into a success (or failure) likelihood index, which is then calibrated on a probability scale.

Another method that strongly influenced the HRA field is the Human Error Assessment and Reduction Technique (HEART; Williams, 1986)—used in the example of Figure 25.1. Compared to THERP, in HEART the personnel tasks are analyzed at a higher level, thus avoiding one of the criticisms of THERP: an excessively decomposed analysis would miss performance issues related to the overall task complexity. HEART was developed to cover a range of industries. This favored wide application of the method, but also led to difficulties to match specific industry characteristics to the generic task and factor definitions of HEART (see Kirwan, 1994 and Spurgin, 2010 for a deeper discussion). Recently, sector-specific versions of HEART have been developed, the Nuclear Action Reliability Analysis (NARA; Kirwan et al., 2005), the Railway Action Reliability Analysis (RARA; Gibson, 2012), and the Controller Action Reliability Analysis (CARA; Kirwan and Gibson, 2007, for the air traffic control domain). These sector-specific variants are expected to be used more consistently. They also build on the latest advances in HRA, a systematic use of a human failure database, CORE-DATA (Kirwan et al., 1997), and on literature human factor studies.

The 1990s marked a generational change for the field, resulting in the development of the so-called second-generation methods, generally more focused on addressing the influence of the context on the operator performance and the decision-making aspects of the operator response. It was clear from its early developments that the models adopted to address decision-making aspects in the operator response were a pragmatic simplification (Dougherty, 1990; Swain and Guttman, 1983;). Also, analyses of accidents and operational experience as well as advances in behavioral sciences contributed to a better understanding of the relationship between human performance and the context in which it occurs (Reason, 1990; Sträter, 2000; USNRC, 2000). In particular, the available HRA methods were perceived as not being able to represent adequately the driving influence of contextual factors (e.g., inadequate procedures, misleading human–machine interfaces, training biases) in the occurrence of unsafe events. The need to extend the scope of the human interactions modeled in the PSA was underscored: from failures to perform required actions (typically referred to as "errors of omissions") to erroneous actions which, unintentionally, aggravate the state of the plant (typically referred to as "errors of commission") (NEA/CSNI, 2002; Sträter et al., 2004). The most relevant methods belonging to this second generation are: ATHEANA (Forester et al., 2007; USNRC, 2000), MERMOS (Le Bot, 2004), CAHR (Sträter, 2000), CREAM (Hollnagel, 1998), CESA (Reer et al., 2004). Concerning the issue of errors of commissions, it is important to note that despite their recognized importance, very few attempts have been made to systematically include these in PSAs (Julius et al., 1995; Podofillini et al., 2013; Reer 2008a, 2008b; Reer et al., 2004).

The latest developments, from the year 2000 on, have been characterized by the continuous increase in use of PSA to inform decisions, operational and regulatory. On the one hand, this led to the need for assessing and evaluating the HRA methods, to enhance the confidence on the decisions they inform (Bell and Holroyd, 2009; Forester et al., 2014, 2015; Johanson, 2015a, 2015b; NEA/CSNI, 2015; Oxstrand, 2010; Pyy, 2000). On the other hand, in the nuclear field, HRA methods have been extended to cover areas of less established application, but which have become of key interest for the characterization of the risk profile: accidents initiated by events such as fires, floods internal and external to the plant, seismic events, and situations of post-core damage accident management. Section 25.5 will present more details on these recent developments.

Recent years have also seen advances in HRA applications outside the nuclear power domain. In industries for which PSA is an established practice, for example, aerospace and oil and gas, the interest in HRA methods is generally driven by the need to improve the level of detail of the qualitative and quantitative analyses supporting the PSA. Indeed, in these industries, the practice has mostly been to quantify operator failures with bounding values (e.g., 0.01 or 0.001), without performing the detailed analyses underlying a typical HRA. In the aerospace domain, in 2005, realizing the need for a systematic choice of an HRA approach, the US National Aeronautics and Space Administration (NASA) reviewed the applicability of a number of HRA methods to their mission PSA (Chandler et al., 2006). The 2011 edition of NASA's PSA guide includes guidance (NASA, 2011) for the application of four HRA methods: THERP, CREAM, NARA, and SPAR-H. A human failure data collection program is also underway to empirically evaluate the values produced by such techniques as well as to support the human factor analyses program (Calhoun et al., 2013; Chandler et al., 2010). For what concerns the oil and gas industry, the insufficient incorporation of human and organizational factors in quantitative risk assessment has been recently pointed out (Skogdalen and Vinnem, 2011). The Barrier and Operational Risk Analysis (BORA; Seljelid et al., 2006) has been developed to integrate quantitative (i.e., probabilistic) physical and non-physical barrier analyses, but its application in industrial cases is very limited (Skogdalen and Vinnem, 2011). A very recent initiative is ongoing to develop an HRA method specific to the oil and gas industry based on an adaptation of SPAR-H (Laumann et al., 2014; Rasmussen et al., 2015). Concerning other industries, the development of HEART-based methods RARA and CARA for application to the railway and air traffic control has been already cited. Efforts in the healthcare domain are also worth mentioning (Lyons et al., 2004).

Another event with significant consequences on the nuclear history has been the accident at the Fukushima Daiichi plant, hit by an earthquake followed by a tsunami on March 11, 2011. The nuclear community (industry, regulators, and research institutions) reacted worldwide to analyze the event, derive lessons learned, and implement improvements at the technical as well as organizational levels (EU, 2012; USRNC, 2011). The implications for PSA (and HRA specifically) have been analyzed by different institutions (e.g., Kumar et al., 2015; Siu et al., 2013). For HRA specifically, lessons learned relate to the need to improve models for action performed for mitigation of accidents (post-core damage, as opposed to the preventive phase) and further emphasize the importance of including errors of commission in the scope of the analysis—more details will be presented in Section 25.5.

25.3 UNDERLYING THEORY AND THEORETICAL ASSUMPTIONS

This chapter presents the underlying quantification models of the mostly used methods. Four types of models are presented: decomposition based (e.g., THERP), error mechanisms based (e.g., CBDTs), factor based (e.g., HEART, CREAM, SPAR-H), narrative based (ATHEANA, MERMOS). Other grouping criteria have been used, for example, by method generation and by sponsoring agency, see Spurgin (2010), Boring (2012), Bell and Holroyd (2009). A note on the validation of HRA methods closes the chapter.

A common assumption of HRA methods is that it is possible to develop a model that allows quantifying the probability of failure for a specific task,[2] by specifying some parameters or elements of the model representing the relevant performance conditions. The premises for this assumption have been discussed since the foundation of the field, in the THERP handbook. The authors of the THERP handbook recognize the inherent difficulties in trying to derive models of human performance in a complex technological system. Indeed, the aim is to model the role of the human only as it concerns the purpose of a PSA—quantification of a failure probability. Also, human performance is represented as a general statement: the failure probability on specific tasks carried out under specific performance conditions. The different models developed for HRA in part represent different notions and operationalizations of what influences human performance and how.

The modeling approach in THERP is based on task decomposition. This consists in decomposing the human interactions with the technical system "into small units, find data (or use expert judgment to obtain appropriate estimates) that fits these subdivisions and then recombine them to derive estimates of failure probabilities for the tasks performed in NPPs" (Swain and Guttman, 1983). The task decomposition and recombination is the distinctive element of THERP. This approach has much in common with typical system reliability modeling in which the system reliability is calculated from the reliability of its constituting components. THERP's small units are, for example, related to failures while manipulating manual controls, inadvertently skipping a procedural step, or reading displays. Opportunities to recover from the failure via self-checks or by other personnel can also be modeled. THERP also addresses cognitive aspects of the operating crew response: the model provides bounding values for the probability of no diagnosis by the operating crew as a function of the time after the abnormal event. Of course, this diagnosis model is not intended to represent the complexity of the cognitive process: it was intended as an ad interim solution to suit the PSA needs, given the cognitive science knowledge and data at that time. The data underlying THERP's model is mostly coming from human factor experiments and field studies, much from military situations. As explained in the handbook, the probability values are generally "derived data," in the sense that they contain much

[2] As mentioned earlier (see the introduction of this chapter), this chapter will refer to (human) failures, intended as failures to perform specifically defined tasks, with specific criteria defining their success (e.g., avoiding the system parameters reach some thresholds). The term "failure" does not necessarily imply "error": it just refers to not meeting the task success criteria, which are typically defined based on the overall system risk model (e.g., by event trees and fault trees). In this sense, failure probability is intended as the probability that the task success criteria are not met. For a discussion on this issue, see Reason (1990), USNRC (2000), Hollnagel (1998), Sträter (2000), Dekker (2014).

extrapolation and judgment. Concerning the diagnosis model, it is based on expert judgment; it was developed based on group consensus values from a workshop of experts (Swain and Guttman, 1983).

Modeling efforts subsequent to THERP aimed at a better characterization of the cognitive part of the operator response, resulting in the development of the HCR/ORE time reliability correlations, combined with the CBDTs (Moieni et al., 1994). The common element of time reliability correlation models is the representation of the probability of failure in the response as a decreasing function of the time since the abnormal event starts. Of course, these models are not intended to represent the complexity of the cognitive process: they are used as a surrogate model for quantifying the probability of non-response. The use of decision trees was an important step ahead to provide a systematic structure for the identification of failure modes connected with the cognitive/decision-making element of the crew response and the quantification of their probability. Eight error mechanisms connected with the cognitive response are identified and the probability of each is assessed with the support of a decision tree (see Chapter 26 for an example implementation). CBDT relationships are quantified with data from both THERP and results from the ORE experiment (Moieni et al., 1994).

A different model for quantification is taken by HEART (Williams, 1986). HEART's model for the calculation of probabilities is driven by the specific task to be analyzed. Compared to THERP, the tasks are defined at a higher level, to avoid the potential for overlooking dependencies when decomposing tasks. A probability of failure is associated to each of these generic tasks, assessed from data (Williams, 1986). The fundamental assumption of HEART is that it is possible to quantify the failure probability of a specific task enveloped by the task type by applying the effect of selected error-producing conditions. Each error-producing condition acts as a multiplier to the task failure probability: the multiplicative factor depends on the error-producing condition itself and on its strength, to be assessed by the analyst.

Besides its use in HEART and its successors (NARA; Kirwan et al., 2005; CARA, Kirwan and Gibson, 2007; RARA, Gibson, 2012), the idea of quantifying the failure probability by progressively applying multiplicative factors to a base HEP is common to a number of other methods currently in use, for example, SPAR-H (Gertman et al., 2005), CREAM (Hollnagel, 1998). These methods are conceptually very different in the "starting point" for the modification; this is the type of task for HEART-based methods, a generic decision versus execution characterization for SPAR-H, a surrogate characterization of the cognitive effort for CREAM. The underlying data for the development of the SPAR-H relationships is mostly THERP (Boring and Blackman, 2007). For CREAM, data from THERP, HEART, and other HRA method databases were used (Hollnagel, 1998). A common element of these methods is the structured analysis of the performance conditions, factor by factor. This makes the quantification framework also quite flexible: probably this is the reason why this type of methods is the most used for adaptation to different domains.

In addition, it is also worth mentioning that in the model for calculation, each factor affects the failure probability independently of the others, in the sense that the value of one multiplicative factor does not depend on the assessment of the other factors. This largely simplifies the HRA model and the quantification of its parameters, but may lead to inadequate representation of factor interactions: multiple factors acting simultaneously can amplify or compensate their effects. Recent research is addressing the use of modeling frameworks allowing better representation of these effects (Groth and Mosleh, 2012a; Mkrtchyan et al., 2015).

Quantification, in the most recently developed methods, is generally based on a rich characterization of the context and on the explicit analysis of the possible scenario variants in which human failures may unfold (ATHEANA: USNRC, 2000; Forester et al., 2007; MERMOS: Le Bot, 2004; CESA: Reer, 2009). The context is not static: different variants may have different performance influences, also changing over the accident progression. Each scenario variant is separately analyzed and quantified. Also, for each variant, the analysis perspective is strongly placed on the personnel, following how the context is perceived over time and how different decisions are made (this is the reason why these methods are generally referred to as narrative based). Compared with the other methods, the quantification of the failure probability is made for each possible context variant, combining the probability of the variant with the probability of failure given the specific context variant. The advances allowed by the application of these methods, especially concerning the modeling of errors of commission have been presented in NEA/CSNI (2002). Also, the recent HRA Empirical Studies underscored the stronger modeling potential of these techniques, mostly due to the rich underlying qualitative analyses (Forester et al., 2014). A common challenge for these models is to translate the rich qualitative analysis into a modeling framework which would make the analyses more reproducible and transparent—currently, the dominant approach for quantification is expert judgment which may have reproducibility and transparency issues. Work is indeed ongoing to develop quantification models suitable to these methods (Ekanem et al., 2015; Xing et al., 2013).

A final note relates to the validation of HRA methods. Naturally, many attempts have been made to assess the accuracy and validity of the methods. Much of these attempts were carried out in the 1980s and 1990s and have addressed first-generation methods. As a result, there is general consensus that the methods THERP and HEART provide valid probability values for certain types of tasks (typically involving execution activities as opposed to cognitive activities such as decision-making). Kirwan (1997a, 1997b) provides a thorough discussion of validation issues and results. As mentioned earlier, validation of first-generation techniques is still the subject of activities (Preischl and Hellmich, 2013; Shirley et al., 2015). Validation of second-generation methods is a more challenging endeavor. Second-generation methods are based on the characterization of error-forcing conditions, rather than the human error concept. Data are easier to collect for isolated actions from either operating experience or simulator experiments. As for second-generation HRA methods, the assessment of failure probabilities is based on context-specific information which is difficult to extrapolate from operational experience or to reproduce in simulation experiments. A recent major effort also including second-generation methods has been performed (Forester et al., 2014, 2015). More on these studies will be presented in Section 25.5.

25.4 USE AND IMPLEMENTATION

This chapter briefly presents the use and implementation of selected HRA methods. The focus of the chapter is on the-state-of-practice: the selection of methods reflects those with the most use in the industry. The chapter aims at highlighting the most representative features of the methods. For details, the reader should refer to HRA books (Kirwan, 1994; Spurgin, 2010) and to the specific method references.

Concerning task analysis and error identification, methods are presented and reviewed in Kirwan (1994), Kirwan and Ainsworth (2003), Kirwan (1998a, 1998b). The basic goal of a task analysis is to describe and analyze what needs to be performed to accomplish a task and how. The most known technique is the Hierarchical Task Analysis (HTA); an example of HTA is shown in Figure 25.3. The HTA describes the task to be performed with a top-down decomposition approach (see Figure 25.3).

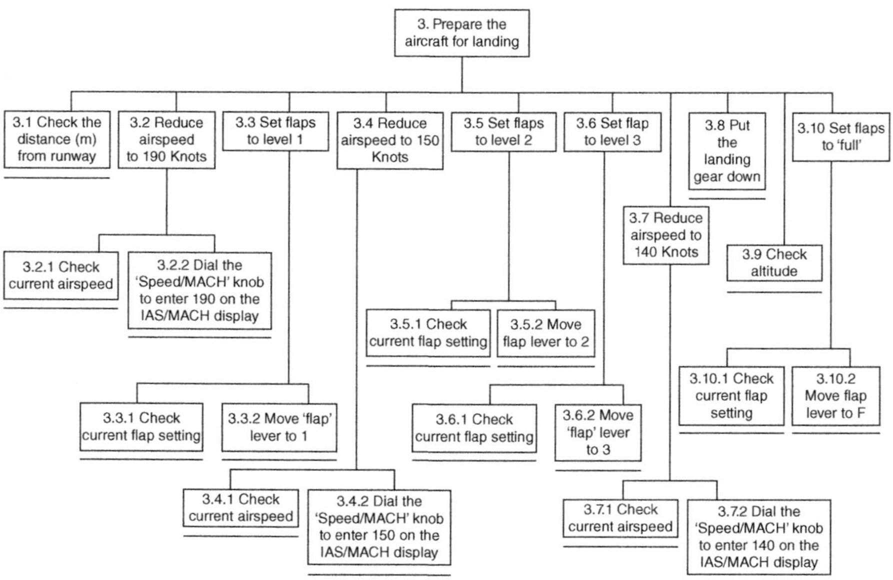

Figure 25.3. Aircraft landing Hierarchical Task Analysis, extract of full analysis. *Source*: Stanton et al. 2009. Reproduced with permission of Elsevier.

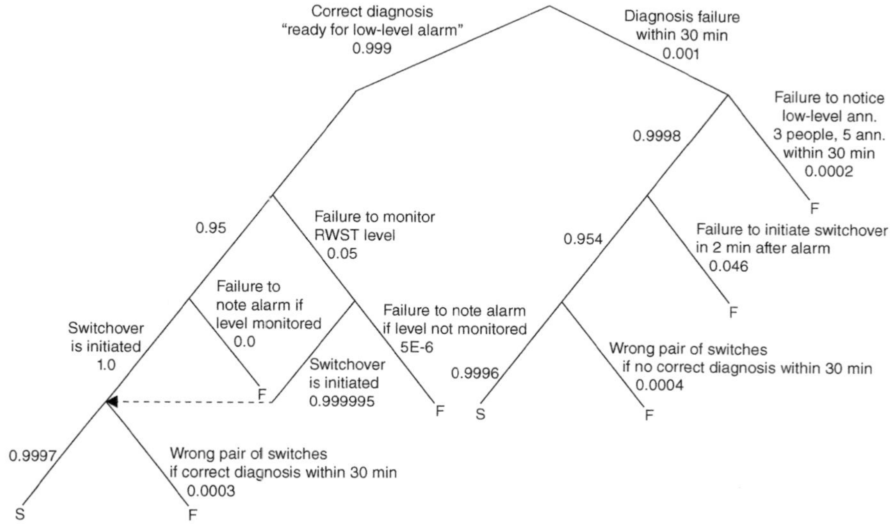

Figure 25.4. Example of THERP event tree. S, success; F, failure. *Source*: NEW CSNI 1998. Public domain. Adapted from Swain and Guttman 1983. Public domain.

Other techniques are available, which also allow to represent other interactions with the environment (e.g., timing aspects, resources needed, information flow; Kirwan and Ainsworth, 2003). The methods for the identification of human errors include modifications of the hazard and operability analysis (HAZOP, see Chapter 21), much frequently used in risk analysis, SHERPA (Embrey et al., 1984), and different taxonomies of error mechanisms and error modes (Kirwan, 1998a, 1998b). Stanton et al. (2009) and the associated references present examples of more recent work on the development and use of error taxonomies.

Concerning methods for quantification, Figure 25.4 shows an example of "HRA event tree," the tool used by THERP (Swain and Guttman, 1983) to guide the task decomposition and the calculation of the human failure probability. Depending on the nature of the items, either the diagnosis model is used or the tabled failure probabilities, possibly modified to represent specific performance conditions. The elementary probabilities are then recombined into the task failure probabilities following Boolean logic rules.

Figure 25.5 presents the model for failure probability calculation underlying the combination of the HRC/ORE curves and CBDTs (Moieni, 1994). The combination is used to quantify the cognitive part of the response (P1 and P2 in Figure 25.5); for the execution part (P3), Moieni (1994) recommends the use of THERP. CBDTs are intended for the assessment of the probability of the failure to formulate the correct response (P1). The failure to initiate timely the response (P2) is quantified with the HCR/ORE time reliability correlations. Eight error mechanisms are identified

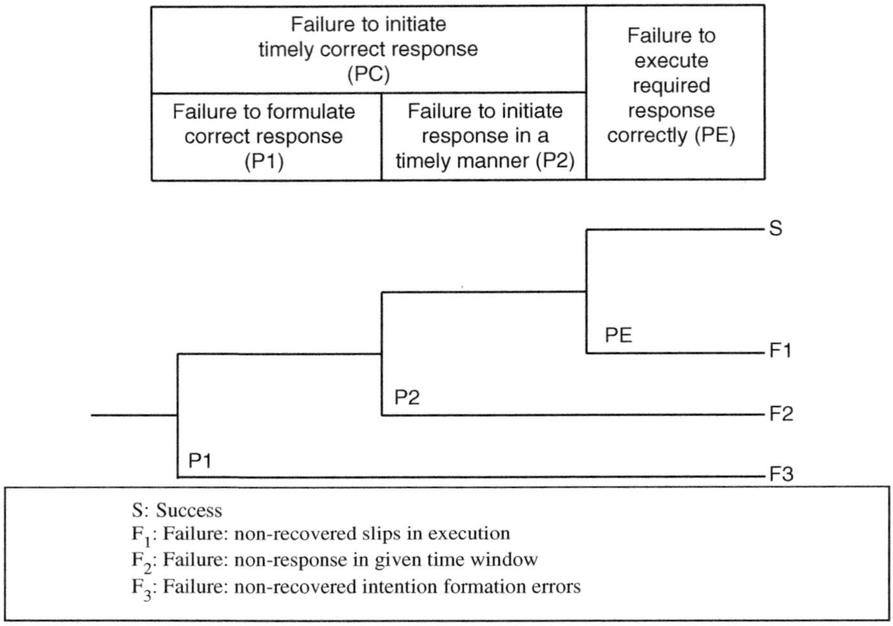

Figure 25.5. Generalized event tree for modeling procedure-driven operator actions in an accident. Source: Moieni et al. 1994. Reproduced with permission of Elsevier.

for the failure type P1, for example, problems due to available information, failure of attention, misreading/miscommunication, misleading information. Decision trees are provided for each failure mechanism. Figure 25.6 shows the decision tree related to misinterpretation of diagnostic logic in procedure. The method includes guidance to support the analyst in the selection of the appropriate tree branch.

Some of the key elements of HEART have already been presented, see Section 25.1 and Figure 25.1. Figure 25.7 presents examples of the generic task types and of the factors representing the error-producing conditions. The analyst has to assess the proportion of affect of each of the applicable conditions. Each affect is then converted into a multiplier that is proportionally applied to the generic failure probability to represent the task-specific conditions.

25.5 EMPIRICAL RESEARCH ON USE AND EFFICIENCY

When addressing HRA research, many publications stress the large number of methods that have been proposed over about 50 years of HRA (Bell and Holroyd, 2009; Lyons et al., 2004). This is generally understood as the result of some interrelated factors: inherent complexity of the field, large subjectivity involved, and difficulty to

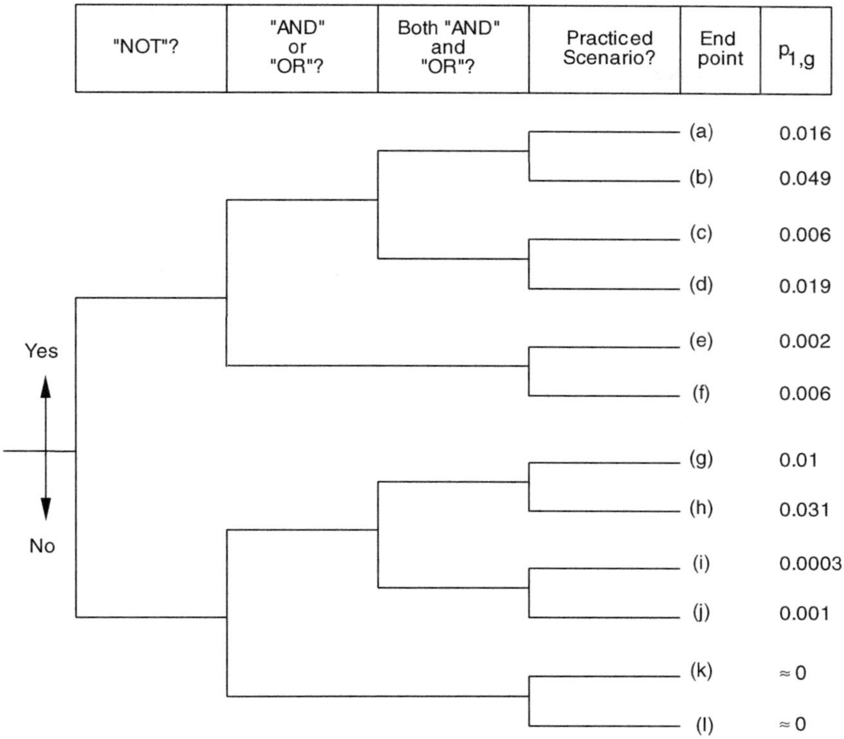

"NOT": Does the procedure step contain the word "not"?

"AND" or "OR": Does the procedure step present diagnostic logic in which more than one condition is combined to determine the outcome?

Both "AND" and "OR": Does the procedure step contain a complex logic involving a combination of ANDed and ORed terms?

"Practiced Scenario": Has the crew practiced executing the step in a scenario similar to this one in a simulator?

$p_{1,g}$ is a contribution to failure in decision-making; see Figure 25.5, P1.
The probabilities are based on data from THERP

Figure 25.6. Decision tree for determining the probability that a diagnostic logic in a procedure is misinterpreted. *Source*: Moieni et al. 1994. Reproduced with permission of Elsevier.

fully validate the model underlying the methods. However, in contrast to the large number of methods having been proposed, the number of methods that are actually used in practice is much smaller: THERP (Swain and Guttman, 1983) and its modified versions, the EPRI HRA Calculator (i.e., basically a combination of HCR/ORE, CBDTs, THERP; Julius et al., 2005), MERMOS (Le Bot, 2004), HEART (Williams,

	Generic Task Types (selected)	Nominal Error probability (5th-95th percentile bounds)
A	Totally unfamiliar, performed at speed with no real idea of the likely consequences	0.55 (0.35-0.97)
B	Shift or restore system to a new or original state on a single attempt without supervision or procedures	0.26 (0.14-0.42)
...		
E	Routine, highly practiced, rapid task involving a relatively low level of skill	0.02 (0.007-0.045)
F	Restore or shift a system to an original or new state following procedures, with some checking	0.003 (0.0008-0.007)
...		

	Error-producing condition, EPC	Multiplier for maximum affect
1	Unfamiliarity with a novel or infrequent situation that is potentially important	17
2	Shortage of time for error detection and correction	11
3	Noisy or confused signals	10
4	A means of suppressing or overriding information or features that is too easily accessible	9
5	No means of conveying spatial and functional information to an operator in a form they can readily assimilate	9
6	Poor system or human user interface. A mismatch between an operator's model of the world and that imagined by a designer	8
...		

Figure 25.7. Examples of HEART generic task types and error-producing conditions (Williams 1986, 2015).

1986) and its modified versions, SLIM (Chen et al., 1988), and CREAM (Hollnagel, 1998) cover much of the PSAs all over the world. Also, very few new methods have been proposed in recent years: instead of proposing new methods, the HRA field is now assessing and strengthening its foundations, and leveraging and building on them. Indeed, the latest major developments in the field concern: the assessment of the available HRA methods and corresponding incremental improvements; the collection of data from simulator environment; the development of alternative quantification frameworks to make better use of data and cognitive models; the adaptation of methods to less established application areas (e.g., external event PSA, non-nuclear applications).

The assessment of available methods is motivated by the continuously increasing use of PSA to inform decisions, operational and regulatory. The credibility of these decisions is subject to the credibility of the methods informing them, and therefore of HRA, and PSA methods in general. Being one of the areas of largest uncertainties for PSA, HRA is often in the spotlight. The most recent assessments of HRA methods can be found in Forester et al. (2006), Bell and Holroyd (2009), Oxstrand (2010), Johanson et al. (2015a, 2015b), Forester et al. (2014, 2015), NEA/CSNI (2015). The assessments have different focus depending on their underlying aim; typically, either assessing the acceptability of methods against regulatory requirements and good practices (Bell and Holroyd, 2009; Forester et al., 2006; Johanson et al., 2015a, 2015b), or identification of strengths and weakness to inform future developments (Forester et al., 2014, 2015; Oxstrand, 2010). The present chapter will provide some details on selected studies: the two HRA Empirical Studies, because of their strong research impact, and the EXAM-HRA project, because of the somehow different perspective it took from other, more typical assessments of the HRA practice.

The two HRA Empirical Studies, the International (Forester et al., 2014) and United States (Forester et al., 2015) have been a recent landmark for the field. The studies aimed at empirically assessing strengths and weaknesses of HRA methods. A number of HRA analyst teams applied the most commonly used HRA methods to a set of operator actions relevant for PSA. The analyses were assessed against evidence from simulator experiments involving operating crews, carried out at the HAMMLAB research simulator in Halden in Norway (for the International HRA Study, Skjerve and Bye, 2011), and at US nuclear power plant simulator (for the US study). The studies underscored the key importance of a rich qualitative analysis of the performance conditions to reach valuable results. The needs that emerged relate to a more comprehensive representation of the cognitive aspects of the operator response and of the contextual factors, as well as to improved guidance for conducting qualitative analysis and for the application of the specific methods (refer to Forester et al., 2014, 2015). These studies sparkled a number of initiatives, currently ongoing, related to the use of simulator data for HRA (Chang et al., 2014), and development of new methods. In particular, concerning the last point, the US Nuclear Regulatory Commission instituted a research effort to develop a new HRA method that addresses the identified weaknesses, while building on the strengths of the available methods: the IDHEAS method (Xing et al., 2013). IDHEAS is expected to be more solidly rooted on cognitive science: the method builds on a structured cognitive framework (Whaley et al., 2016). It features "Crew Response Trees" (Ekanem et al., 2015) to capture the context associated to the human tasks as well as the key crew decision points over the accident evolution, possibly leading to both errors of omissions and commission.

The variability of HRA results has also been recently investigated in the EXAM-HRA project (Johanson et al., 2015a, 2015b). The project analyzed the HRAs from six PSA studies of Nordic, Swiss, and German nuclear power plants, with the aim to understand how much of the differences in the results could be attributed to differences in the plants, HRA method, or analysis assumptions. The study shows that some of the differences in the HRA results could be traced to differences in the plant design and operation. This is a strongly encouraging insight as it shows how HRA models are able to reflect plant differences and, on the other hand, can be used to identify specific weaknesses and potential improvements. The project published a practical guide on the application of HRA methods, with the aim to decrease the subjectivity in the analyses and, therefore improve consistency in the application of the HRA methods (Johanson et al., 2015b). The need to address, more systematically, the errors of commission is again highlighted (Johanson et al., 2015a).

Besides the evaluation of HRA methods, another currently active direction of developments relates to data: collection, interpretation, and use. Efforts to collect and use data to support HRA have been part of the field since its beginning, over the past decades (for overviews see Hallbert et al., 2004; NEA/CSNI, 2012; Prvakova and Dang, 2013; Spurgin, 2010). As discussed in Prvakova and Dang (2013), the data needs have evolved as the HRA methods have become more sophisticated. Earlier data collection efforts were predominantly aimed at collecting performance measures (typically, failure probabilities and time performances) for different types of operator tasks (e.g., Kirwan, 1997a, 1997b; Kirwan et al., 1997; Preischl and Hellmich, 2013). Emerging as well as currently used HRA methods give more emphasis to the context, the decision-making elements of the response, and narratives of crew behavior and failure mechanisms: this puts more requirements on the data to be collected, which need to become richer in the description of the context as well as of the observed crew behavior. A challenge for future simulator studies is therefore to identify which scenarios to simulate, balancing between complexities of the scenarios (such that failures can be stimulated) and their plausibility, such that results are relevant for PSA scenarios. Recently proposed guidelines for data collection are going in this direction (Chang et al., 2014; Hallbert et al., 2013; KAERI, 2013).

Concerning HRA models, a recent direction is the development of causal models, more sophisticated than the quantification models presented in the earlier sections, potentially more suitable to represent the complex relationships among the influencing factors. The use of Bayesian belief networks (BBNs) has shown promising results in this direction: besides their ability to represent complex relationships, their strengths relate to the potential to formally aggregate different sources of information (data, cognitive models, expert judgment). The idea is to develop more "science-based" models: built based on cognitive models, learnt from data, eventually complemented with expert judgment. Recent advances can be found in Groth and Mosleh (2012a), Sundarmurthi and Smidts (2013), Mkrtchyan et al. (2015). Note that, finally, due to their ability to model soft factors and relationships, the use of BBNs is also getting strong attention to incorporate organizational factors in the PSA framework (see also discussion in Section 25.6).

Still related to the development of advanced modeling frameworks for HRA, an important role is played by dynamic PSA. The dynamic PSA framework embeds plant physical models (typically thermo-hydraulic codes in the nuclear power industry) and operating crew models into stochastic simulation engines. Variants of accident scenarios are directly and automatically produced by the simulation engine based on the interactions of the plant systems, components, and operating crew over time. Reviews of the available frameworks can be found in Siu (1994). For HRA, the interest is in the possibility to explore variants of scenarios as well as directly simulate operating crew decisions and their impact on the accident evolution (Acosta, 1991; Chang and Mosleh, 1999; Kloos and Peschke, 2008). This framework potentially integrates well with narrative-based methods, which also rely on the identification of multiple scenario variants and with the development of causal models, which could be embedded as part of the operating crew models (Groth and Mosleh, 2012a; Sundarmurthi and Smidts, 2013).

Further, industrial nuclear PSAs across the world are tending toward covering a broad scope of analyses: accidents initiated by internal events as well as external, at all plant operational modes, covering from accident initiation to the final consequences on the population. Also, the state-of-the-art and related needs for PSA methodologies have been recently scrutinized in many initiatives, following the Fukushima Daiichi accident, in particular addressing the models concerning natural hazards, multi-units plants, and post-core damage models (see, e.g., Kumar et al., 2015; Siu et al., 2013). HRA is consistently mentioned as one of the area with significant technical and methodological challenges. Indeed, most HRA methods have been developed focusing on pre-core damage PSA, for internal initiating events; generally, applications to other areas of PSA have been made with simplified, bounding analyses. The increasing requirements of completeness and realism call for advances by HRA in models and identification of good practices beyond their most typical application. The challenges for HRA methods in these situations are the representation of increased difficulty in the operator response, due to possibly significant instrumentation failures; increased workload; unfavorable environment; as well as the larger uncertainty in the plant conditions (EPRI/USNRC, 2012).

Finally, as mentioned earlier, an important area of current research is the development or adaptation of HRA methods for applications in domains other than nuclear power. As discussed in Section 25.2, these advances are driven by the need to increase the level of details of the available methods as well as to support the qualitative risk analysis to prioritize risk contributors and possible improvement measures. Section 25.2 has already presented recent activities on this.

25.6 WEAKNESSES, LIMITATIONS, AND CRITICISM

HRA results are used to inform decisions with safety-relevant implications. This motivates the continuous efforts by the community to systematically assess their quality. Earlier efforts were mostly aimed at identifying the "best" among the numerous available methods (Kirwan 1997a, 1997b). More recent ones aim at identifying weaknesses smaller selection of methods, with focus on those mostly used in practice (Bell and Holroyd, 2009; Forester et al., 2014, 2015; Johanson et al., 2015a, 2015b; Oxstrand, 2010).

A recurrent concern is about the variability of the results, across analysts and across methods. The sources of this variability are different: different scopes and aims of the methods; imprecise guidance to apply the methods and limited empirical foundation of the methods; reliance on subjective assessment (which, to some extent, is always inevitable for many risk analysis applications). As presented in Section 25.5 much is ongoing to characterize and reduce this variability: better guidance and definitions (Forester et al., 2006; Groth and Mosleh, 2012b; Johanson et al., 2015b); stronger links of the models with data, with cognitive and human factor sciences, and with more formal ways to incorporate expert judgment (Groth and Mosleh, 2012b; Groth et al., 2014; Mkrtchyan et al., 2015; Whaley et al., 2016); better characterization of the method scopes and aims (Bell and Holroyd, 2009; Forester et al., 2014, 2015; Johanson et al., 2015a, 2015b; Oxstrand, 2010); and extensions of the current scope of HRA analyses in PSA (EPRI/USNRC, 2012).

More at the technical level, the areas of HRA that are generally highlighted as open issues are:

- Limited empirical data available to support quantification considering context influences and decision-related failures (open issues relate to how to collect, interpret, and use the data, mostly from simulator studies). Related to this is also the need for data applicable to modern digital interfaces.
- HRA for operator task carried out in degraded performance conditions, for example, due to fire, flood, seismic events as well as post-accidental situations. These situations potentially pose increased challenges to the personnel (possibly unreliable instrumentation, increased workload, unclear overall picture of the damage condition of the plant) that need to be identified and modeled.
- Treatment of errors of commission, or more generally, decision failures. Besides the mentioned data problem, open issues relate to their systematic incorporation in the PSA framework and the quantification of their probability.
- Dependence analysis, which relates to models for the treatment of multiple, subsequent human failures. Related is the issue of human performance limiting values, that is, the lowest probability values that can be credibly defended.

In the rest of this section, a more fundamental issue is discussed. An often-mentioned limitation of HRA and PSA concerns the missing treatment of organizational factors (Mosleh, 2014; Mohaghegh and Mosleh, 2009; Mohaghegh et al., 2009; Øien, 2001;). The HRA models tend to focus at the "sharp end," at the technical interference with the system. Instead, investigations of major industrial accidents point to management and organizational factors as the important contributing causes, the "blunt end" (Dekker, 2014; French et al., 2011; Hollnagel, 2012). It can be argued that management and organizational factors are implicitly included in HRA: the quality of influencing factors such as the procedural guidance, training, work processes reflects choices at the managerial level. Instead, what the current PSA and HRA framework do not reflect are the interdependencies that management and organizational issue create across different safety barriers—human and hardware. Safety barriers are designed and maintained by the plant organization and its culture and attitude are key contributors to safety. Their quantification is very challenging, because of their "soft" nature, difficult to operationalize. Their influence is long term, difficult to monitor. A more subtle issue is the fact that organizational attitudes toward safety are driving the quality of the quantitative analyses themselves—PSAs are often developed internally by the organization so that deficiencies in the organization safety culture would not reflect into higher values of the quantified risk, but into inadequate risk models. For a more detailed discussion of the issue, the reader can refer to Mohaghegh and Mosleh (2009), Mohaghegh et al. (2009), Øien (2001), Groth et al. (2010). Recognizing the challenges for quantification of these influences, innovative qualitative approaches have also being developed to address the complexity of modern systems in analysis frameworks alternative to PSA (Hollnagel, 2012; Leveson, 2012).

25.7 RELATIONSHIP WITH OTHER PRINCIPLES

Among the principles addressed in the present book, HRA has the closest relationship with quantitative risk assessment and human factors.

Quantitative risk assessment is the natural framework for HRA. Indeed, HRA has been devoted to support quantification of human failures for quantitative risk assessment from its early beginning. Much of the HRA elements are influenced by the relationship with the PSA framework: the definition of the human failure events (as basic events of a fault tree); the definition of the scenarios in which the personnel actions take place; the focus on the omission of actions, as opposed to errors of commission (in analogy with the system failures included in the PSA). Discussions on how HRA integrates with the classical PSA framework are presented extensively in the literature (e.g., Kirwan, 1994; Spurgin, 2010); alternative integration frameworks are also presented in Groth et al. (2010), Mohaghegh et al. (2009), Sundarmurthi and Smidts (2013) to model the dependency introduced by human and organizational factors across the different safety barriers.

Concerning the human factors discipline, most HRA methods build on it, more or less explicitly. Human factor studies provide inputs to the definition of the factors incorporated in HRA models and methods. The human factor foundations of HRA are evident in the handbooks of THERP and ATHEANA, for example. Most of the methods for task analysis and error identification are also part of human factor analyses.

However, the two disciplines have somewhat diverged over the years (e.g., the main scientific journals and conferences rarely overlap, as noted by Boring, 2012). HRA is closely connected with the risk analysis field: it focusses on undesired, generally rare, events (rare human failures during rare accident scenarios); also, it emphasizes quantification of probabilities as a means to prioritize analyst insights. HRA is not generally intended as a design tool, whereas much of the human factor analyses are intended to support design (Spurgin, 2010). Human factor engineering is generally devoted to improving human performance during normal working conditions, not necessarily related to rare situations. The result is that the research questions, methods, and data needs are often different (Boring, 2012; Oxstrand, 2010).

The need to maintain close interaction between the two disciplines is largely recognized. Many references empathize that HRA analysts seek involvement of human factor specialists, especially to support the qualitative analysis (ATHEANA; Kolaczkowski et al., 2005; Johanson et al., 2015a, 2015b). Also, HRA research can benefit from the rigorous experimental designs and factor operationalization typical of the human factor discipline. Other areas of possible mutual benefits are discussed in Boring (2012).

REFERENCES

Acosta, C. (1991). Dynamic event tree for accident sequence analysis (Ph.D. thesis, Massachusetts Institute of Technology).

Bell, J., and Holroyd, J. (2009). Review of human reliability assessment methods. Research Report RR697. UK Health and Safety Executive.

Boring, R. L. (2012). Fifty years of THERP and human reliability analysis. In: Joint Probabilistic Safety Assessment and Management and European Safety and Reliability Conference, PSAM11/ESREL 2012, Helsinki, Finland, June 25–29, 2012.

Boring, R. L., and Blackman, H. S. (2007). The origins of the SPAR-H method's performance shaping factor multipliers. In: Joint IEEE HFPP and HPRCT Conference, Monterey, CA, August 26–31, 2007.

Calhoun, J., Savoie, C., Randolph-Gips, M., and Bozkurt, I. (2013). Human reliability analysis in spaceflight applications. *Quality and Reliability Engineering International*, 29, 869–882.

Chandler, F. T., Chang, Y. H., Mosleh, A., Marble, J., Boring, R. L., and Gertman, D. I. (2006). Human reliability analysis methods: Selection guidance for NASA. NASA Technical Report.

Chandler, F., Heard, I. A., Presley, M., Burg, A., Mideen, E., and Mongon, P. (2010). NASA human error analysis—Final report. NASA Technical Report.

Chang, Y. H. J., and Mosleh A. (1999). Cognitive modeling and dynamic probabilistic simulation of operating crew response to complex system accidents (ADS-IDACrew). Center for Technology Risk Studies, University of Maryland, College Park, MD.

Chang, J. Y., Bley, D., Criscione, L., Kirwan, B., Mosleh, A., Madary, T., Nowell, R., Richards, R., Roth, E. M., Sieben, S., and Zoulis, A. (2014). The SACADA database for human reliability and human performance. *Reliability Engineering and System Safety*, 125, 117–133.

Chien, S. H., Dykes, A. A., Stetkar, J. W., and Bley, D. C. (1988). Quantification of human error rates using a SLIM-based approach. In: Proceedings of the 1988 IEEE Conference on Human Factors and Power Plants (pp. 297–302), Monterey, CA, June 5–9, 1988.

Dekker, S. (2014). *The Field Guide to Understanding Human Error*. Ashgate.

Dougherty, E. M. (1990). Human reliability analysis—Where shouldst thou turn? *Reliability Engineering and System Safety*, 29, 283–299.

Ekanem, N. J., Mosleh, A., and Shen, S.-H. (2015). Phoenix—A model-based human reliability analysis methodology: Qualitative analysis procedure. *Reliability Engineering and System Safety*, 145, 301–315.

Embrey, D. E. (1986). SHERPA: Systematic human error reduction and prediction approach. In: International Topical Meeting on Advances in Human Factors in Nuclear Power Plants, Knoxville, TN.

Embrey, D. E., Humphrey, P., Rosa, E. A., Kirwan, B., and Rea, K. (1984). SLIM-MAUD: An approach to assessing human error probabilities using structured expert judgment. NUREG/CR-3518. US Nuclear Regulatory Commission, Washington, DC.

EPRI/USNRC. (2012). Fire human reliability analysis guidelines—Final report. NUREG-1921. Electric Power Research Institute (EPRI), Palo Alto, CA and US Nuclear Regulatory Commission, Washington, DC.

EU. (2012). Communication from the Commission to the Council and the European Parliament on the comprehensive risk and safety assessments ("stress tests") of nuclear power plants in the European Union and related activities. COM/2012/0571 final.

Forester, J., Kolaczkowski, A., Lois, E., and Kelly, D. (2006). Evaluation of human reliability analysis methods against good practices. NUREG-1842. US Nuclear Regulatory Commission, Washington, DC.

Forester, J., Kolaczkowski, A., Cooper, S., Bley, D., and Lois, E. (2007). ATHEANA User's Guide. NUREG-1880. US Nuclear Regulatory Commission, Washington, DC.

Forester, J., Dang, V. N., Bye, A., Lois, E., Massaiu, S., Bromberg, H., Braarud, P. Ø., Boring, R., Männistö, I., Liao, H., Julius, J., Parry, G., and Nelson, P. (2014). The international HRA empirical study lessons learned from comparing HRA methods predictions to HAMMLAB simulator data. NUREG-2127. US Nuclear Regulatory Commission, Washington, DC.

Forester, J., Liao, H., Dang, V. N., Bye, A., Lois, E., Presley, M., Marble, J., Nowell, R., Broberg, H., Hildebrandt, M., Hallbert, B., and Morgan, T. (2015). The US HRA empirical study—Assessment of HRA method predictions against operating crew performance on a US nuclear power plant simulator. Draft report of the US Nuclear Regulatory Commission and Electric Power Research Institute (EPRI). Available at: http://pbadupws.nrc.gov/docs/ML1322/ML13225A516.pdf

French, S., Bedford, T., Pollard, S. J. T., and Soane, E. (2011). Human reliability analysis: A critique and review for managers. *Safety Science*, 49, 753–763.

Fullwood, R. R. (2000). *Probabilistic Safety Assessment in the Chemical and Nuclear Industries*. Woburn, MA: Butterworth-Heinemann.

Gertman, D., Blackman, H., Marble, J., Byers, J., and Smith, C. (2005). The SPAR-H human reliability method. NUREG/CR-6883. US Nuclear Regulatory Commission, Washington, DC.

Gibson, H. (2012). Railway Action Reliability Assessment User Manual. UK Rail Safety and Standards Board Ltd.

Groth, K. M., Wang, C., and Mosleh, A. (2010). Hybrid causal methodology and software platform for probabilistic risk assessment and safety monitoring of socio-technical systems. *Reliability Engineering and System Safety*, 95(12), 1276–1285.

Groth, K. M., and Mosleh, A. (2012a). Deriving causal Bayesian networks from human reliability analysis data: A methodology and example mode. *Proceedings of the Institution of Mechanical Engineers, Part O: Journal of Risk and Reliability*, 226(4), 361–379.

Groth, K. M., and Mosleh, A. (2012b). A data-informed PIF hierarchy for model-based human reliability analysis. *Reliability Engineering and System Safety*, 108, 154–174.

Groth, K. M., Smith, C. L., and Swiler, L. P. (2014). A Bayesian method for using simulator data to enhance human error probabilities assigned by existing HRA methods. *Reliability Engineering and System Safety*, 128, 32–40.

Hallbert, B., Gertman, D., Lois, E., Marble, J., Blackman, H., and Byers, J. (2004). The use of empirical data sources in HRA. *Reliability Engineering and System Safety*, 83, 139–143.

Hallbert, B., Morgan, T., Hugo, J., Oxstrand, J., and Persensky, J. J. (2013). A formalized approach for the collection of HRA data from nuclear power plant simulators. NUREG/CR-7163, INL/EXT-12-26327. US Nuclear Regulatory Commission, Washington, DC and Idaho National Laboratories, Idaho.

Hannaman, G., Spurgin, A., and Lukic, Y. (1984). Human cognitive reliability model for PRA analysis. NUS-4531. Electric Power Research Institute, Palo Alto, CA.

Hannaman, G. W., Spurgin, A. J., and Lukic, Y. (1985). A model for assessing human cognitive reliability in PRA studies. In: IEEE Third Conference on Human Factors in Nuclear Power Plants, Monterey, CA, June 23–27, 1985.

Hollnagel, E. (1998). *Cognitive Reliability and Error Analysis Method (CREAM)*. Oxford: Elsevier Science Ltd.

Hollnagel, E. (2012). *FRAM: The Functional Resonance Analysis Method*. Farnham, UK: Ashgate.

Johanson, G., Jonsson, S., Bladh, K., Iseland, T., Karlsson, K.-H., Karlsson, A., Ljungbjörk, J., Becker, G., Tunturivouri, L., Porthin, M., Olsson, A., and Böhm, J. (2015a). Exam-HRA summary report. Report 11-004-01. Nordic PSA Group (NPSAG).

Johanson, G., Jonsson, S., Bladh, K., Iseland, T., Karlsson, K.-H., Karlsson, A., Ljungbjörk, J., Becker, G., Tunturivouri, L., Porthin, M., Olsson, A., and Böhm, J. (2015b). Evaluation of existing applications and guidance on methods for HRA—EXAM-HRA. A practical guide to HRA. Report 11-004-02. Nordic PSA Group (NPSAG).

Julius, J. A., Jorgenson, E. J., Parry, G. W., and Mosleh, A. (1995). A procedure for the analysis of errors of commission in a probabilistic safety assessment of a nuclear power plant at full power. *Reliability Engineering and System Safety*, 50, 189–201.

Julius, J. A., Grobbelaar, J., and Rahn, F. (2005). EPRI HRA Calculator®, Version 3. In: ANS Topical Conference on PRA, San Francisco, CA, September 2005.

KAERI. (2013). A guideline to collect HRA data in the simulator of nuclear power plants. KAERI/TR-5206. Korean Atomic Energy Research Institute, South Korea.

Kirwan, B. (1994). *A Guide to Practical Human Reliability Assessment*. Boca Raton, FL: CRC Press.

Kirwan, B. (1997a). Validation of human reliability assessment techniques: Part 1—Validation issues. *Safety Science*, 27, 25–41.

Kirwan, B. (1997b). Validation of human reliability assessment techniques: Part 2—Validation results. *Safety Science*, 27, 43–75.

Kirwan, B., Basra, G., and Taylor-Adams, S. E. (1997). CORE-DATA: A computerized human error database for human reliability support. In: Global Perspectives of Human Factors in Power Generation—Proceedings of the 1997 IEEE Sixth Conference on Human Factors and Power Plants, Orlando, FL, June 8–13, 1997.

Kirwan, B. (1998a). Human error identification techniques for risk assessment of high risk systems—Part 1: Review and evaluation of techniques. *Applied Ergonomics*, 29, 157–177.

Kirwan, B. (1998b). Human error identification techniques for risk assessment of high risk systems—Part 2: Towards a framework approach. *Applied Ergonomics*, 29, 299–318.

Kirwan, B., and Ainsworth, L. K. (2003). *A Guide to Task Analysis: The Task Analysis Working Group*. CRC Press.

Kirwan, B., and Gibson, H. (2007). CARA: A human reliability assessment tool for air traffic safety management—Technical basis and preliminary architecture. In: The Safety of Systems—Proceedings of the Fifteenth Safety-Critical Systems Symposium (pp. 197–214), Bristol, UK, February 13–15, 2007.

Kirwan, B., Gibson, H., Kennedy, R., Edmunds, J., Cooksley, G., and Umbers, I. (2005). Nuclear action reliability assessment (NARA): A data-based HRA tool. *Safety & Reliability*, 25(2), 38–45.

Kloos, M., and Peschke, J. (2008). Consideration of human actions in combination with the probabilistic dynamics method Monte Carlo dynamic event tree. *Journal of Risk and Reliability*, 222, 303–313.

Kolaczkowski, A., Forester, J., Lois, E., and Cooper, S. (2005). Good practices for implementing human reliability analysis. NUREG-1792. US Nuclear Regulatory Commission, Washington, DC.

Kumar, M., Wielenberg, A., and Raimond, E. (2015). Post Fukushima lesson learned for probabilistic safety assessment. In: Proceedings of the 25th European Safety and Reliability Conference, ESREL 2015, Zürich, September 7–10, 2015.

Laumann, K., Øien, K., Taylor, C., Boring, R. L., and Rasmussen, M. (2014). Analysis of human actions as barriers in major accidents in the petroleum industry, applicability of human reliability analysis methods. In: The Proceedings of the Probabilistic Safety Assessment and Management (PSAM12), Honolulu, HI, June 2015.

Le Bot, P. (2004). Human reliability data, human error and accident models—Illustration through Three Mile Island accident analysis. *Reliability Engineering and System Safety*, 83, 153–167.

Leveson, N. (2012). *Engineering a Safer World: Applying Systems Thinking to Safety*. MIT Press.

Lyons, M., Adams, S., Woloshynowych, M., and Vincent, C. (2004). Human reliability analysis in healthcare: A review of techniques. *International Journal of Risk & Safety in Medicine*, 16, 223–237.

Mkrtchyan, L., Podofillini, L., and Dang, V. N. (2015). Bayesian belief networks for human reliability analysis: A review of applications and gaps. *Reliability Engineering and System Safety*, 139, 1–16.

Mohaghegh, Z., and Mosleh, A. (2009). Incorporating organizational factors into probabilistic risk assessment of complex socio-technical systems: Principles and theoretical foundations. *Safety Science*, 47(8), 1139–1158.

Mohaghegh, Z., Kazemi, R., and Mosleh, A. (2009). Incorporating organizational factors into probabilistic risk assessment (PRA) of complex socio-technical systems: A hybrid technique formalization. *Reliability Engineering and System Safety*, 94(5), 1000–1018.

Molseh, A. (2014). PSA: A perspective on strengths, current limitations, and possible improvements. *Nuclear Engineering and Technology*, 46, 1–10.

Moieni, P., Spurgin, J., and Singh, A. (1994). Advances in human reliability analysis methodology. Part I: Frameworks, models and data. *Reliability Engineering and System Safety*, 44, 27–55.

NASA. (2011). Probabilistic risk assessment procedures guide for NASA managers and practitioners. NASA/SP-2011-3421. December 2011.

NEA/CSNI. (1998). Critical operator actions: Human reliability modeling and data issues. Principal working group no. 5—Task 94-1, final task report. NEA/CSNI/R(98)1. Nuclear Energy Agency.

NEA/CSNI. (2002). Proceedings of the International Workshop—Building the New HRA: Errors of Commission from Research to Application. NEA/CSNI/R(2002)3. Nuclear Energy Agency.

NEA/CSNI. (2012). Simulator Studies for HRA Purposes, Workshop Proceedings, Budapest, Hungary, November 4–6, 2009. NEA/CSNI/R(2012)1, Nuclear Energy Agency.

NEA/CSNI. (2015). Establishing the Appropriate Attributes in Current Human Reliability Assessment Techniques for Nuclear Safety. NEA/CSNI/R(2015)1, Nuclear Energy Agency.

Øien, K. (2001). Risk indicators as a tool for risk control. *Reliability Engineering and System Safety*, 74(2), 129–145.

Osif, B. A., Baratta, A. J., and Conkling, T. W. (2004). *TMI 25 Years Later: The Three Mile Island Nuclear Power Plant Accident and Its Impact*. University Park, PA: Pennsylvania State University Press.

Oxstrand, J. (2010). Human Reliability Guidance—How to Increase the Synergies between Human Reliability, Human Factors, and System Design and Engineering. NKS-229, Nordic Nuclear Safety Research Council.

Parry, G. W., Lydell, B. O. Y., Spurgin, A. J., Moieni, P., and Beare, A. N. (1992). An approach to analysis of operator actions in probabilistic risk assessment. EPRI report TR100259, Electric Power Research Institute, Palo Alto, CA.

Podofillini, L., Dang, V. N., Nusbaumer, O., and Dres, D. (2013). A pilot study for errors of commission for a boiling water reactor using the CESA method. *Reliability Engineering and System Safety*, 109, 86–98.

Preischl, W., and Hellmich, M. (2013). Human error probabilities from operational experience of German nuclear power plants. *Reliability Engineering and System Safety*, 109, 150–159.

Prvakova, S., and Dang, V. N. (2013). A review of the current status of HRA data. In: European Safety and Reliability Conference, ESREL 2013, Amsterdam, September 29–October 2, 2013.

Pyy, P. (2000). Human reliability analysis methods for probabilistic safety assessment. VTT publications 422, VTT Technical Research Center of Finland.

Rasmussen, J., Pedersen, O. M., Carnino, A., Griffon, M., Mancini, C., and Gagnolet, P. (1981). Classification system for reporting events involving human malfunctions. Report Riso-M-2240, DK-4000, Riso National Laboratories, Roskilde, Denmark.

Rasmussen, J., Duncan, K., and Leplat, J. (Eds.). (1987). *New Technology and Human Error*. Chichester: John Wiley & Sons.

Rasmussen, M., Standal, M. I., and Laumann, K. (2015). Task complexity as a performance shaping factor: A review and recommendations in standardized plant analysis risk-human reliability analysis (SPAR-H) adaption. *Safety Science*, 76, 228–238.

Reason, J. (1987). Generic error modelling system: A cognitive framework for locating common human error forms. In: J. Rasmussen, K. Duncan, and J. Leplat (Eds.), *New Technology and Human Error*. Chichester: John Wiley & Sons.

Reason, J. (1990). *Human Error*. Cambridge University Press.

Reer, B., Dang, V. N., and Hirschberg, S. (2004). The CESA method and its application in a plant-specific pilot study on errors of commission. *Reliability Engineering and System Safety*, 83, 187–205.

Reer, B. (2008a). Review of advances in human reliability analysis of errors of commission—Part 1: EOC identification. *Reliability Engineering and System Safety*, 93(8), 1991–1104.

Reer, B. (2008b). Review of advances in human reliability analysis of errors of commission—Part 2: EOC quantification. *Reliability Engineering and System Safety*, 93(8), 1105–1122.

Reer, B. (2009). Outline of a method for quantifying errors of commission. LEA 09-302, Laboratory for Energy Systems Analysis, Paul Scherrer Institute, Switzerland.

Seljelid, J., Haugen, S., Sklet, S., and Vinnem, J. E. (2006). Operational risk analysis—Total analysis of physical and non-physical barriers. BORA Handbook, Rev. 00.

Shirley, R. B., Smidts, C., Li, M., and Gupta, A. (2015). Validating THERP: Assessing the scope of a full-scale validation of the technique for human error rate prediction. *Annals of Nuclear Energy*, 77, 194–211.

Siu, N. (1994). Risk assessment for dynamic systems: An overview. *Reliability Engineering and Systems Safety*, 43, 43–73.

Siu, N., Marksberry, D., Cooper, S., Coyne, K., and Stutzke, M. (2013). PSA technology challenges revealed by the Great East Japan Earthquake. In: PSAM Topical Conference in Light of the Fukushima Daiichi Accident, Tokyo, Japan, April 15–17, 2013.

Skjerve, A. B., and Bye, A. (Eds.). (2011). *Simulator-Based Human Factors Studies Across 25 Years*. London: Springer.

Skogdalen, J. E., and Vinnem, J. E. (2011). Quantitative risk analysis offshore—Human and organizational factors. *Reliability Engineering and System Safety*, 96, 468–479.

Spurgin, A. J. (2010). *Human Reliability Assessment—Theory and Practice*. Boca Raton, FL: CRC Press.

Stanton, N. A., Salmon, P., Harris, D., Marshall, D., Demagalski, J., Young, M. S., Waldmann, T., and Dekker, S. (2009). Predicting pilot error: Testing a new methodology and a multi-methods and analysts approach. *Applied Ergonomics*, 40, 464–471.

Sträter, O. (2000). Evaluation of human reliability on the basis of operational experience. Gesellschaft für Anlagen- und Reaktorsicherheit (GRS) report GRS-170.

Sträter, O., Dang, V. N., Kaufer, B., and Daniels, A. (2004). On the way to assess errors of commission. *Reliability Engineering and System Safety*, 83(2), 129–138.

Swain, A. D., and Guttman, H. E. (1983). Handbook of human reliability analysis with emphasis on nuclear power plant applications. NUREG/CR-1278, U.S. Nuclear Regulatory Commission, Washington, DC.

Sundarmurthi, R., and Smidts, C. (2013). Human reliability modelling for next generation system code. *Annals of Nuclear Energy*, 2013, 137–156.

USNRC. (1975). Reactor safety study: An assessment of accident risks in US commercial nuclear power plants. WASH-1400, NUREG-75/014, U.S. Nuclear Regulatory Commission, Washington, DC.

USNRC. (2000). Technical basis and implementation guidelines for A Technique for Human Event Analysis (ATHEANA). NUREG-1624, U.S. Nuclear Regulatory Commission, Washington, DC.

USRNC. (2011). Recommendations for enhancing reactor safety in the 21st century—The near term task force review of insights from the Fukushima Daiichi accident. U.S. Nuclear Regulatory Commission, Washington, DC.

Vesely W. E., Davis, T. C., Denning, R. S., and Saltos, N. (1986). Measures of risk importance and their applications. NUREG/CR-3385, U.S. Nuclear Regulatory Commission, Washington, DC.

Whaley, A. M., Xing, J., Boring, R. L., Hendrickson, S. M. L., Joe, J. C., Le Blanc, K. L., and Morrow, S. L. (2016). Cognitive basis for human reliability analysis. NUREG-2114, U.S. Nuclear Regulatory Commission, Washington, DC.

Williams, J. C. (1986). A proposed method for assessing and reducing human error. In: Proceedings of the 9th Advances in Reliability Technology Symposium, University of Bradford.

Williams, J. C. (2015). Heart—A Proposed Method for Achieving High Reliability in Process Operation by Means of Human Factors Engineering Technology, Safety and Reliability, 35:3, 5–25.

Xing, J., Presley, M., Parry, G., Forester, J., Hendrickson, S., and Dang, V. (2013). IDHEAS—A new approach for human reliability analysis: An Integrated Decision-Tree Human Event Analysis System (IDHEAS) method for NPP internal at-power operation. USNRC-EPRI draft report.

26

ALARA, BAT, AND THE SUBSTITUTION PRINCIPLE

Sven Ove Hansson

26.1 INTRODUCTION

- In an American nuclear plant, a special team prepares maintenance work by setting up temporary shielding before the maintenance team arrives. Shielding is set up even when the predicted radiation exposure is so low that it is not required in the regulations. This is done since the plant puts much emphasis on the *as low as reasonably achievable* (ALARA) principle. By consistently applying this principle, they have set a national record: the lowest ever personnel radiation dose in an American nuclear plant (Fallon, 2012).
- Researchers have performed an in-depth investigation of how textile industries can best reduce their wastewater emissions of decabromodiphenyl ether (decaBDE), a toxic flame retardant that is used to make textiles less flammable. A list of 17 methods to reduce the emissions was compiled (it included replacement of the substance, reuse of rinse water, and sending process baths and rinse water to a specialized waste processing plant). The use of these methods was called *best available technology (BAT)*. The researchers proposed that future

Handbook of Safety Principles, First Edition. Edited by Niklas Möller, Sven Ove Hansson, Jan-Erik Holmberg, and Carl Rollenhagen.

regulations should require that all European textile plants reduce their waste emissions to the levels that had been achieved with these methods (Derden and Huybrechts, 2013).

- Since 1996, a major Swedish truck and bus manufacturer has published lists of chemicals used in their production that they strive to replace with alternatives that are less hazardous to the environment and to human health. These lists are publicly available and they are an important source of information for the company's suppliers. A considerable number of problematic chemicals have been replaced with the help of these lists. The company sees this as an application of the *substitution principle* (Kemikalieinspektionen, 2008).

Three principles are invoked in these examples: ALARA, BAT, and the substitution principle. The three principles have different origins, and they are usually applied in different contexts. But they have at least one important feature in common: They all tell us to improve safety when we have a chance to do so. We can call them *improvement principles* and contrast them for instance with cost–benefit analysis that tells us always to weigh safety against other values.[1]

In Sections 26.2–26.4, each of these principles will be discussed in some detail. This is followed by a concluding section in which they are compared to each other and to a couple of other safety principles that also focus on improvements.[2]

26.2 ALARA

26.2.1 History and Current Use

In the first few years after Röntgen's discovery of X-rays in 1895, the risks of radiation exposure were unknown. Consequently, researchers worked without protection, not knowing what dangers they exposed themselves to. Beginning about 1900, it was recognized that high exposures give rise to acute effects such as skin burns. Efforts were made to keep doses well below the levels that had given rise to these effects. It was believed that doses below those levels were innocuous (Kathren and Ziemer, 1980). However, some physicians were worried that X-rays might have unknown side effects. As early as 1912, a German textbook in pediatric radiology advised physicians to use only as much X-rays as was necessary for the diagnostic or therapeutic purpose of the intervention (Oestreich, 2014).

[1] It was pointed out to me by Marko Ahteensuu that as they are usually interpreted, these principles require improvements up to the level of the best available technology, but they do not require technological developments leading to improvements going beyond that level.

[2] The section on ALARA is in part based on Hansson (2013b) and that on the substitution principle on Hansson et al. (2011).

In the Manhattan project, the WW2 project that developed the first nuclear weapons, Dr. Robert S. Stone in the Health Division of the Metallurgical Laboratory in Chicago was commissioned to set "tolerance levels" for radiation in order to protect the project's own personnel. He responded to this by pointing out that there was no known absolutely safe level. Therefore, he proposed that radiation exposures should be kept on the lowest level that was practically possible to achieve. This principle was accepted (although some wartime exposures were very high, judged by modern standards). After the war, it was realized that accumulated exposures to ionizing radiation increase the risk of cancer, in particular leukemia. These effects appeared to be stochastic, and there did not seem to be any threshold dose below which they disappear completely (Lindell, 1996). This gave support to Stone's principle, and it was adopted by the United States' National Committee on Radiation Protection (NCRT). In a 1954 statement, they declared that radiation exposures should "be kept at the lowest practical level" (Auxier and Dickson, 1983). The International Commission on Radiological Protection (ICRP) made a similar recommendation in 1959:

> "The mechanism of leukemia induction by radiation is not known. It may be postulated that if the dose is lower than a certain threshold value no leukemia is produced. In this case it would be necessary to estimate the threshold dose and to make allowances for recovery, if any. There is not sufficient information to do this, but caution would suggest that an accumulated dose of 750 rems might exceed the threshold. The most conservative approach would be to assume that there is no threshold and no recovery, in which case even low accumulated doses would induce leukemia in some susceptible individuals, and the incidence might be proportional to the accumulated dose. The same situation exists with respect to the induction of bone tumors by bone-seeking radioactive substances....
>
> [T]he Commission recommends that all doses be kept as low as practicable, and that any unnecessary exposure be avoided."
>
> (ICRP, 1959, p. 4 and p. 11)

This recommendation has repeatedly been confirmed by the ICRP. In its 1977 recommendations, it was rephrased as a requirement that "all exposures shall be kept as low as reasonably achievable, economic and social factors being taken into account" (ICRP, 1977, p. 3).

ALARA is still accepted worldwide as a principle for radiation protection. In most countries, it is not much used outside of radiation protection, but in Britain, it is also used in general worker's health and safety. There are also international differences in the legal status of the ALARA principle. In Britain, an operator who satisfies the legal dose limits may still be under an ALARA-based legal obligation to take further action to reduce exposures. In other countries such as the United States, the ALARA principle does not have such legal force (French et al., 2005; Lierman and Veuchelen, 2006).

26.2.2 Definitions and Terminology

Over the years, the principle has been known under several different names:

> as low as practicable (ALAP)
> as low as reasonably achievable (ALARA)
> as low as reasonably attainable (ALARA)
> as low as reasonably practicable (ALARP)
> so far as is reasonably practicable (SFAIRP)
>
> (HSE, 2001a, p. 92)

ALAP appears to have been the first of these abbreviations. In the early 1970s, it was replaced by ALARA (mostly in the United States) and ALARP (mostly in Europe) that were thought to be less imprecise (Wilson, 2002). Some authors have tried to make distinctions between these terms:

> "What is achievable is a function of current science. But what is practicable is a function of economics and benefit-cost analysis… If ICRP were really serious about making its recommendations less tied to utilitarian and benefit-cost criteria, then it would recommend exposures that were as low as possible/achievable, not those that are as low as practicable."
>
> (Shrader-Frechette and Persson, 2002, p. 155)

In practice, however, ALARA and ALARP are taken as synonyms. The major difference is one of usage: In Britain where the principle is called ALARP, it has a general use in workplace health and safety. In countries where it is called ALARA, it tends to be restricted to radiation protection.

26.2.3 Theory and Interpretation

There have been extensive debates on the interpretation of ALARA. Three of the most debated issues are the meaning of "reasonable" (the "R" of the ALARA acronym), how the principle is related to optimization of safety measures, and whether ALARA applies to all dose levels or an exception should be made for very low ones.

26.2.3.1 What Is "Reasonable"? All non-zero radiation doses can in principle be further reduced. When is a reduction of radiation doses "reasonably" achievable or attainable? A very strict interpretation of ALARA would require that a dose is reduced whenever that is at all technically possible, irrespectively of what it would cost. According to a very weak interpretation of the principle, it would only require such measures if they can easily be afforded. Most practically influential interpretations lie somewhere between these two extremes.

The term "reasonable" is common in legal discourse, where it is usually a means to legitimate the introduction of a variety of practical considerations, including considerations that are not explicitly mentioned in the legislation. This was confirmed in a safety context by the House of Lords in a much quoted case (Marshall vs. Gotham Co Ltd, 1954):

> "The test of what is [reasonably practicable] is not simply what is practicable as a matter of engineering, but depends on the consideration, in the light of the whole circumstances at the time of the accident, whether the time, trouble and expense of the precautions suggested are or are not disproportionate to the risk involved, and also an assessment of the degree of security which the measures suggested may be expected to afford."
>
> (Jones-Lee and Aven, 2011)

In a penetrating analysis, Corten (1999) drew attention to two major functions that this word has in legal discourse. First, it makes regulations adaptable, allowing them to be applied in different circumstances. The term is used to "introduce a degree of flexibility" in legal instruments in order to resolve "a contradiction between the essentially static character of legal texts and the dynamic character of the reality to which they apply" (p. 615). The term "reasonable" provides a legitimation for judges and administrators to interpret regulations flexibly, taking into account conditions and circumstances that the lawmakers did not foresee. Second, the notion of reasonableness can provide legitimacy to a legal order "by presenting an image of a closed, coherent, and complete legal system." The notion "masks persistent contradictions regarding the meaning of a rule, behind a formula which leaves open the possibility of divergent interpretations" (p. 618).

The reasonableness incorporated into the ALARA principle (the "R" in the acronym) appears to have both these functions. First, it provides the flexibility needed to solve potential conflicts with economic and practical constraints and objectives. The degree of risk that is accepted in practice can be "balanced against time, trouble, cost, and physical difficulty of its risk reduction measures" (Melchers, 2001). Second, the notion of reasonableness creates the impression of a coherent and comprehensive system that is capable of solving all potential conflicts. Divergences between economic demands and the demands of health protection are presented as solvable within the system of radiation protection, rather than having to be treated as conflicts between radiation protection and external, economic interests. This "internalization" of economic compromises can of course be seen both as a way to resolve conflicts and as a way to hide them from view.

26.2.3.2 ALARA and Optimization In 1977, the ICRP introduced a system of three basic principles for radiation protection that are still in use. They are:

> "(a) no practice shall be adopted unless its introduction produces a positive net benefit;
>
> (b) all exposures shall be kept as low as reasonably achievable, economic and social factors being taken into account; and
>
> (c) the dose equivalent to individuals shall not exceed the limits recommended for the appropriate circumstances by the Commission."
>
> (ICRP, 1977, p. 3)

These principles are known under the names justification, optimization, and (individual) dose limits. Their combination is a "regulatory mix" intended to provide better protection than what each of them could do alone (Lierman and Veuchelen, 2006). According to the justification principle, all radiation exposures have to be justified, which is usually interpreted as meaning that the benefits (advantages) should outweigh the risks (disadvantages). The third principle (dose limits) restricts the permitted exposure of each individual. But according to the second principle (ALARA), keeping these limits is not enough. Even if exposures are below the dose limits, efforts should be made to reduce them further. These further reductions are usually conceived as referring to collective rather than individual doses, and therefore ALARA is often seen as a utilitarian principle (Hansson, 2007a, 2013b).

The term "optimization" for ALARA indicates that some sort of compromise is aimed at. To optimize two or more factors (here: dose and costs) means to find the best balance between them. To optimize radiation doses while taking costs into account would therefore seem to mean that one finds a balance from which it would be undesirable to deviate in either direction. Higher doses should be avoided since they can cost-efficiently be reduced, whereas lower doses should be avoided since they would involve indefensible costs. It is not uncommon to interpret ALARA as an "optimal balance" principle. However, there is also another interpretation, namely that the principle requires lowering of doses as far as possible. According to the "as low as possible" interpretation, further reduction of the doses can never conflict with the ALARA principle. The "as low as possible" view is arguably supported by the actual wordings of the acronym; it reads "as low as reasonably achievable," not "at the exact level that is reasonably achievable" (Hansson, 2013b).

26.2.3.3 Does ALARA Apply at All Dose Levels? According to some explications of ALARA, it should not be applied to all radiation doses but only to the middle part of the dose range. If the doses are high, then they have to be reduced irrespective of the costs, and there is no need for considerations of what is "reasonable." On the other hand, if the doses are sufficiently low, then—according to this view—there is no need for action. It is in the intermediate region, with doses that are neither negligible nor utterly unacceptable, that the ALARA principle will be applied. The upper limit of the ALARA region in this model has been called the "action level" and its lower limit the "inaction level" (Hendee and Edwards, 1986). Doses below the inaction level have also been referred to as "de minimis" doses. This term is derived from the legal phrase "de minimis non curat lex" (law does not concern itself with trifles). In the context of risk assessment, it expresses the view that some risks are so small that we have no reason to take action against them even if such action can be taken at no or negligible cost (Bicevskis, 1982; Hansson, 2013a, pp. 97–98; Pearce et al., 1981). This three-leveled interpretation of ALARA is illustrated in Figure 26.1.

According to one specification of this three-leveled interpretation, in the intermediate dose range (the "ALARA region"), "the value of a person-rem", that is, the price

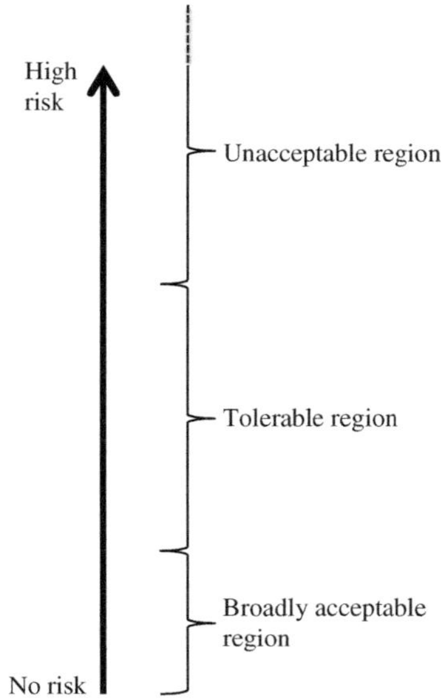

Figure 26.1. The three-leveled model of radiation protection.

that should be paid for such a reduction of the collective dose, "remains constant."[3] Below that range, the value of a person-rem "tends to zero," whereas above it that value "increases sharply" in the range (Kathren et al., 1984). As noted by Jones-Lee and Aven, this would reduce the ALARA principle to "simply a restatement of the standard cost-benefit criterion" for the dose range in which it is applied (Jones-Lee and Aven, 2011).

The three-leveled approach has been adopted by the Health and Safety Executive (HSE) in Great Britain. In their version, it is applied not only to radiation doses, but also to risk exposures in general, and therefore they define the three regions in terms of risk levels rather than dose levels. The HSE has specified tentative limits between the three regions in terms of probabilities of fatalities. The limit between the broadly acceptable and the tolerable region has been set at an individual risk of death of one in a million per year (both for occupational and non-occupational risk exposure). Hence, risks below that level would typically be taken to be negligible and therefore not in need of ALARA-based activities (HSE, 2001, p. 45). The limit between tolerable and unacceptable risks is tentatively set at one in one thousand per year for occupational risks and one in ten thousand per year for non-occupational risks (HSE, 2001, p. 46).

[3] rem (roentgen equivalent in man) is an older unit of ionizing radiation dose. 1 rem is equal to 0.01 Sv (sievert) or 10 mSv (millisievert).

The idea that ALARA is only applicable to risks above a certain level has been criticized on the grounds that unnecessary or frivolous risks should not be accepted. If a very small risk can be averted at almost no cost or effort, why not do so? In the early 1980s, the applicability of ALARA to very small risks was discussed among three major contributors to the ICRP's work, namely the American Harald H. Rossi (1917–2000), the Swede Bo Lindell (b. 1922), and the Argentinian Dan J. Beninson (1931–1994). The debate was started by Rossi who defended the "de minimis" principle and wanted it to overrule ALARA at low risk levels:

> "One stops when one has reached a level where the radiation hazard has become *de minimis*... It follows that a more accurate formulation of the ALARA philosophy is that: 'It is the obligation of those responsible for radiation protection to see to it that personal exposures are below maximum permissible limits and reduced as near as practicable to *de minimis* levels. Further reductions are not required.'"
>
> (Rossi, 1980)

In a reply, Lindell and Beninson claimed that the level below which no further reductions are needed is not the same in all contexts but has to be determined from the specific circumstances:

> "[I]n each situation, there is a level of dose below which it would not be reasonable to go because the cost of further dose reduction would not be justified by the additional eliminated detriment. That level of dose, however, is not a de minimis level below which there is no need of concern, nor can it be determined once and for all for general application. It is the outcome of an optimization assessment which involves marginal cost-benefit considerations... It is not reasonable to pay more than a certain amount of money per unit of collective dose reduction, but if dose reduction can be achieved at a lesser cost even at very low individual doses, the reduction is, by definition, reasonable."
>
> (Lindell and Beninson, 1981)

26.2.4 Effects of Applying the Principle

There is abundant experience from the nuclear industry showing that large dose reductions can be achieved through safety work based on the ALARA principle. Reportedly, the establishment of "a strong ALARA culture" is essential for efficient safety work (Fallon, 2012). There is similar experience from other areas of radiation protection. For instance, large reductions in the doses in diagnostic radiology in children have been achieved in campaigns based on the ALARA principle (Newman and Callahan, 2011). However, it must be emphasized that these experiences have not been subject to a comparative evaluation. We do not seem to have any data on how ALARA-based preventive measures fare in comparison to preventive measures based on other safety principles.

26.2.5 Weaknesses and Criticism

The ALARA principle is primarily intended for cases where there is a conflict between, on the one hand, safety or risk reduction and, on the other hand, some other type of goal, typically cost containment. The application of ALARA is less straightforward when there is a conflict between two types of risk. Such conflicts are common in healthcare since most beneficial medical interventions come with risks of side effects. A balance has to be struck between the chances of positive effects and the risk of negative ones. For instance, if too low radiation doses are used in diagnostic radiology, the quality of the information obtained will be lower, with potential negative effects for the patient. This was expressed rather drastically by one radiologist:

> "We can actually lower the radiation dose such that the correct diagnosis is not made from the imaging procedure. The patient receives radiation, albeit a low dose, but as the benefit to the patient is now nulled, the risk-to-benefit ratio becomes very high."
>
> (Cohen, 2012)

Interestingly, this is a problem that ALARA shares with the precautionary principle (and with related ideas such as the maxim "better safe than sorry"). Cautiousness can be interpreted as requiring that we do not take the risks of serious side effects of a treatment, but it can also be interpreted as requiring that we do not take the risks of a serious development of the disease caused by refraining from treatments. In such symmetric situations, we have to find a balance between different risks and then neither ALARA nor principles requiring cautiousness can help us.

26.3 BEST AVAILABLE TECHNOLOGY

26.3.1 History and Current Use

The notion of BAT has its origins in emissions control. When environmental regulation was introduced in the 1960s and 1970s, it was soon discovered that the so-called "technology-forcing" statutes that require the implementation of specific technological solutions had considerable disadvantages. Such regulations are inflexible since they do not allow industry to achieve the same effect with different means, and they also tend be become outdated when new and better technological solutions become available (Sunstein, 1991, pp. 627–628). Another alternative was regulations specifying maximal allowed emissions. Such regulations are more flexible, but they are sometimes inefficient since they do not adjust to technological development that makes the attainment of lower emission levels practically achievable (Ranken, 1982, p. 162). Legislation based on the BAT was introduced as a solution to these problems in earlier legislative methods. Industries can be required to use the BAT to reduce emissions. Such a rule is technology neutral; if there are alternative ways to reach the best result, then each industry can make its own choice among these alternatives. Furthermore, such statutes can stimulate improvements in ways that the other types of statutes do not. If new technology becomes available that allows the reduction of emissions to lower levels than what was previously possible, then the new technology becomes the new BAT standard that industry will have to adjust to.

In the United States, BAT strategies were introduced in most environmental legislations, and today they are since long "a defining characteristic of the regulation of the air, water, and workplace conditions" (Sunstein, 1991, pp. 627–628). The same applies to most European countries and to legislation on the European level. For instance, the European Directive on industrial emissions stipulates that certain large industrial installations may not be operated without a permit. Such a permit is conditional on adherence to emission limit values (ELV) that are based on BATs. The BATs are in their turn based on Best Available Techniques REFerence Documents (BREFs) that synthesize academic knowledge and industrial experience of different types of technological operations. The drafting of such a document typically involves some 40–100 experts. The BREFs are not binding but they have a large impact on European decisions on ELV and other environmental conditions for industrial activities (Merkouris, 2012; Schoenberger, 2011). The United States has a somewhat similar system in the form of a clearinghouse for information on the best available air pollution technologies. The information in that clearinghouse is based on information provided by states and local permitting agencies (Stensvaag, 2011).

The BAT concept is also used in several international treaties. For instance, the 1992 Convention on the Protection of the Marine Environment of the Baltic Sea Area (the Helsinki Convention) contains several provisions for pollution control that are based on BATs. The same applies to the Convention for the Protection of the Marine Environment of the North-East Atlantic (the OSPAR Convention) from the same year (Merkouris, 2012).

The implementation of BATs has essentially been restricted to emissions control. There is also one major category of emissions for which it is not much used: Legislation on the limitation and reduction of greenhouse gas emissions has in most cases been based on other regulatory principles, including standards based on current technologies and tradable emission permits. A major difficulty with tradable permits is that it is often not clear who is responsible for a particular emission. For instance, if a device is produced in China and then exported to the United States where it is used, should the greenhouse gas emissions that its production gave rise to be assigned to the Chinese or the American quota for greenhouse gas emissions? Recently, an interesting proposal has been made to solve this problem with the notion of BAT:

> "Hence, we propose allocating to the consumer only the emissions generated when the BAT along the whole product chain is used and allocating the difference between the real and BAT emissions to the respective producers of non-optimally performed steps."
> (Berzosa et al., 2014, p. 98)

Interestingly, the BAT concept has not been used, at least not systematically, in safety legislation. For instance, the type approval of motor vehicles, aircraft, and some types of machines is based on well-defined technical standards, not on a requirement to use the BAT. Obviously, the development of new technology can lead to stricter standards (and this certainly happens in practice), but the adoption of stricter standards may be a more complex process than that of most BAT regulations. Fixed (albeit revisable) standards do not provide the incentives to improvement and innovation that at least some BAT regulations seem to provide.

The BAT concept may also be applicable in other areas in which adaptability to technology development is important. For instance, it has been proposed that the issue of copyright liability for webhosts can be solved with a "best availability technology" principle. According to this proposal, a webhost who uses the best filtering technology should be immune from liability for copyright infringements committed by its customers (Helman and Parchomovsky, 2011).

26.3.2 Definitions and Terminology

A large number of synonyms and near-synonyms of BAT have been used in different legislations, including the following:

best available control technology (BACT)
best available techniques (BAT)
best available technology not entailing excessive costs (BATNEEC)
best environmental practice (BEP)
best practicable control technology (BPT)
best practicable environmental option (BPEO)
best practicable means (BPM)
lowest achievable emissions rate (LAER)
maximum achievable control technology (MACT)
reasonably achievable control technology (RACT)
(Merkouris, 2012; Ranken, 1982; Vandenbergh, 1996).

The difference between these terms is in most cases unclear. They are used in different legislations, but there is no obvious connection between what the different legislations demand and their choice of wordings (for instance the choice between the words "available," "achievable," and "practicable").

26.3.3 Theory and Interpretation

The major issues in the interpretation of BAT can be divided into three categories, namely the interpretations of the three words "best," "available," and "technology."

26.3.3.1 What is Best? If a plant has only one important environmental problem, then it is usually easy to determine the degree to which that problem has been solved. For instance, if sulfur emissions into the atmosphere is the only problem we have to worry about, then the technology that leads to the lowest concentrations of sulfur is the environmentally "best" technology. However, in most industries, there are several environmental problems, and consequently, BAT has to be determined according to several criteria of environmental performance. Therefore, the identification of the environmentally "best" technology is essentially the same problem as that of determining total environmental impact. Several methodologies for that purpose have been developed, including life-cycle analysis (Finnveden et al., 2009) and ecological footprint (Wiedmann and Barrett, 2010). Specialized multi-criteria methodology (for instance, with weights assigned to different factors) has also been developed for the purpose of identifying BATs (Ibáñez-Forés et al., 2015).

26.3.3.2 What is Available? The BAT principle is often conceived as less constrained by economic considerations than other ways of managing environmental requirements. For instance, in American legislation from the 1970s, "best practicable control technology" (BPT) represented lower demands on emissions control than BAT that was only to be achieved later. In determining a BPT, the authority had to perform a cost–benefit analysis. For a BAT, this was not required, although the economic impact of proposed measures had to be considered (Bush, 2012, pp. 550–551; Ranken, 1982, pp. 163–164). The lack of strict cost–benefit principles has often been seen as a major justification of BAT regulations. For instance, it has been argued that its application to offshore technology makes it possible for regulators to "ensure that deepwater exploration regulatory reforms truly implement the best available technology as opposed to technology that is only economically convenient" (Bush, 2012, p. 564).

However, BAT requirements are usually not intended to disregard economic considerations altogether. The European system of BREFs takes economic feasibility into account (but with a focus on what is feasible in a branch of industry in general, rather than what a particular company can afford). In the United States, the Water Pollution Control Act allows the EPA to make exceptions, called "variances," from BAT-based regulations. There has been considerable conflict over the use of this legal instrument. According to the EPA's interpretation, variances should only be granted under special circumstances, and a discharger's inability to pay is not sufficient argument for a variance. Unsurprisingly, this interpretation has been challenged by industry (Ranken, 1982, p. 161).

If the "available" of "best available technology" is interpreted as requiring economic feasibility, then there can be technologies that are better than the "best" available technology (but too expensive). Such technologies have been called "beyond BAT." In many cases, they can (and arguably should) be further developed and made available at a lower price that will make them into BAT (Schoenberger, 2011).

26.3.3.3 What is a Technology? The word "technology" can be interpreted as referring primarily to machines and other physical equipment. Alternatively, it can refer to the broader system of social practices in which such equipment is used.

Both these interpretations have been used in discussions on BAT. In some cases, the word "technology" has been replaced by some other word in order to express this difference. The European Directive on industrial emissions (Directive 2010/75) distinguishes carefully between two readings of BAT, "Best Available Technology" and "Best Available Techniques." "Best Available Techniques" includes not only the technological equipment, but also "the way in which the installation is designed, built, maintained, operated, and decommissioned." Thus, in this directive "best available techniques" has a wider scope than "best available technology." However, the two terms are not always conceived differently. The OSPAR convention consistently uses the term "best available techniques" whereas the Helsinki convention uses "best available technology." There is no important difference in the respective definitions of these two terms, so we can conclude that they are used here as synonyms. In addition to "best available technology" (BAT), the Helsinki convention also makes use of the term "best environmental practice" (BEP). This is a much wider notion that includes measures such as information and education in environmental matters, lawmaking, etc. (Merkouris, 2012).

26.3.4 Effects of Applying the Principle

It has often been pointed out that significant improvements in environmental practice have been achieved both in Europe and in the United States during the decades when BAT-based regulations have been used (Bugallo et al., 2013, p. 254; Bush, 2012; Vandenbergh, 1996). However, it must be recognized that we do not know what the outcomes would have been if some other legislative methodology had been chosen, such as technology-forcing or economic incentives. Generally speaking, there is a remarkable lack of comparative studies of the efficiency of different instruments in environmental policies, such as economic instruments, different types of regulations, voluntary agreements, and information measures (Edvardsson Björnberg et al., 2015). Until such studies have been performed, the current status of our knowledge is that BAT strategies are efficient in emissions control, but we do not know much about how efficient they are in other areas, and neither do we know whether there are other strategies that surpass them in efficiency.

26.3.5 Weaknesses and Criticism

BAT methodology has been criticized from different points of view. Some have argued that BAT standards attempt to achieve unattainable goals and that some of the measures they require of industry are unnecessary and too expensive (Rimmele, 1994, p. 256). It has also been claimed that BAT-based regulations are inefficient since they require the same level of emission control in all areas, "regardless of whether they are polluted or clean, populated or empty, or expensive or cheap to clean up" (Sunstein, 1991, p. 628). Other worries are that the BAT concept is difficult to apply to long-term investments and that some BAT technologies may have economies of scale that make it impossible for small businesses to afford them (Helman and Parchomovsky, 2011, p. 1215). In the cases when BAT technology is only required for new entrants, it can stifle innovation and give undue advantages to existing plants (Sunstein, 1991, p. 628).

On the other hand it has been pointed out that in some cases, BAT strategies are not sufficiently protective since the best technology is not good enough to protect the environment. (In some cases, regulatory agencies have been authorized to impose requirements stricter than the BATs in order to achieve sufficient protection of the environment; Vandenbergh, 1996, pp. 837–838 and p. 841.). Some authors claim that BAT regulations give rise to unproductive discussions on peripheral issues:

> "BAT strategies ensure that citizens and representatives will focus their attention on largely incidental and nearly impenetrable questions about currently available technologies, rather than on the appropriate levels of reduction. Technological debates are singularly ill-suited for democratic resolution. They also distract attention from the central issue of determining the appropriate degree and nature of regulatory protection."
> (Sunstein, 1991, p. 629. Cf. Ackerman and Stewart, 1988, pp. 189–190)

It has also been argued that due to their focus on the activities of individual industries, BAT strategies put focus on "superficial symptoms rather than underlying causes of pollution." For instance, they do not directly encourage measures such as energy conservation and efficiency, or shifts to cleaner, renewable fuels (Sunstein, 1991, p. 628).

There are cases when BAT technologies cannot be used due to conflicts with other goals or restrictions for the activity in question. One example is the energy refurbishment of historically valuable buildings. Typically, BAT technology would do damage to the building's historical authenticity. Instead, renovators have to apply the most energy-efficient technology that is compatible with the requirements of architectural conservation. The term "allowed best technology" (ABT) has been used for the technology that is chosen based on these criteria. A significant reduction in energy consumption is achievable with ABT, but usually not as large a reduction as BAT technology would have provided (Milone et al., 2015).

26.4 THE SUBSTITUTION PRINCIPLE

26.4.1 History and Current Use

The substitution principle can simplistically be described as the principle that a toxic chemical should be replaced by a less toxic alternative when that is possible. Such substitutions have a much longer history than the principle itself. The replacement of arsenic with tin as an alloying metal in the early Bronze Age is often cited as an example, but it is not known whether this replacement was actually due to the toxic effects of the substance (Charles, 1967; Parascandola, 2012). More recent examples are much better documented. Dynamite that was invented in 1866 by Alfred Nobel replaced other explosives due to its safety properties (Meyers and Shanley, 1990). Another example is the replacement of white (yellow) phosphorous in matches that was agreed internationally in the Berne Convention of 1906 (Crass, 1941; Marx, 2008). Already in 1949, a general requirement to replace dangerous substances with less dangerous ones was introduced into the Swedish legislation on workplace health and safety:

> "A poisonous or otherwise noxious substance shall be replaced by a non-toxic or less harmful one whenever this can reasonably be done considering the circumstances."
> (Svensk författningssamling, 1949, p. 401)

The chemical industry has a long tradition of replacing dangerous substances by less dangerous ones. For instance, one of the major advantages of dynamite was that there were much fewer major accidents in its production than in that of previous nitroglycerin products (Meyers and Shanley, 1990). In the 1970s, the disastrous explosion in a chemical plant in Flixborough gave rise to extensive discussions within the chemical industry on how accident risks could be reduced. In these discussions, Trevor Kletz pointed out that it is always better to eliminate a hazard altogether than to retain and manage it. He introduced the term "inherent safety" for this way of thinking (Kletz, 1978; see also Chapter 17). One of the major components of inherent safety is substitution, which has been defined in this context simply as "using a safer material" (Khan and Abbasi, 1998, p. 364). Engineers in one major chemical company have reported how their company has used "substitution, a strategy to replace one material with a less hazardous substance if the chemistry of the process will allow." The company has, for instance, replaced benzene with a less toxic substance, replaced a highly flammable solvent with another less hazardous solvent, and replaced chlorine gas with sodium hypochlorite as a water purification chemical (Overton and King, 2006). The basic idea behind such substitutions is that even if a dangerous material is safely contained in the ordinary process, there is always a risk of unwanted events such as fires, leakages, and uncontrolled chemical reactions. When that happens, it is always better to have as small quantities of hazardous material as possible (Hansson, 2010).

In the debates that preceded the European chemicals legislation (REACH) that came into force in 2007, one of the key conflicts, up to the last phase of the negotiations, was the proposal to include a substitution principle into the legislation. It was introduced into the legislative process in the European Commission's 2001 White Paper that recommended "the substitution of dangerous by less dangerous substances where suitable alternatives are available" (European Commission, 2001). This is a formulation that could have been taken from a paper by the experts on inherent safety of any of the major chemical industries. In spite of this, the principle was strongly contested by the European chemicals industry.

Today, the substitution principle is integrated in the REACH legislation. A substance can be classified as a substance of very high concern (SVHC) if it is

(1) carcinogenic, (2) mutagenic, (3) toxic to reproduction, (4) a persistent, bioaccumulative, and toxic substance, or (5) has some other property giving rise to an equivalent level of concern. In August 2017, there were 174 substances on the candidate list of SVHCs. After a regulatory process, these substances may be included in the Authorization List, which means that companies need an authorization to use them. In August 2017, there were 43 substances on the Authorization List. This process is a major driver of substitution in Europe.

Many methods have been used to encourage or enforce substitution. The following list of such methods is summarized from (Hansson et al., 2011):

1. *Increasing the Availability of Toxicity and Ecotoxicity Data.* In order to perform risk-reducing substitutions, companies need reliable and comparable data not only about the substances they use, but also about alternative chemical substances, mixtures, and materials. Unfortunately, such data is still scarce (Rudén and Hansson, 2006, 2010).
2. *Increasing Availability of Data on the Chemical Composition of Materials.* Even large companies often have difficulties in obtaining this information from their suppliers (Hall, 2001). By providing buyers with adequate information about what they buy, manufacturers can substantially improve their chances to make informed product choices.
3. *Increasing the Availability of Information About Technical Functionality.* According to Lohse and coworkers, "the most significant barrier for substitution appears to be the guarantee of technical properties of a product" (Lohse et al., 2003). This is, of course, a particularly pressing problem in cases when the technical functionality of a chemical can only be seen after some time, such as marine anti-fouling coatings, wood preservation agents, etc. To alleviate this problem, several authorities and organizations have developed databases with information about successful substitutions.
4. *Developing Green Chemistry.* Research is needed that provides industry with new chemical substances, or new uses of known substances, that can reduce risks in production and use (Clark et al., 2014; Thorpe and Rossi, 2007).
5. *Helpdesk Functions.* The Toxic Use Reduction Act of Massachusetts has provided companies with technical help supplied by university and government experts. This is reported to have led to substantial decreases in the use of hazardous substances (Onasch et al., 2011).
6. *Lists of Unwanted Substances.* This is one of the most common strategies to encourage substitution (Lohse et al., 2003). A large number of such lists is now available, and some of them play a major role in the substitution of hazardous chemicals. As noted by Løkke (2006), the efficiency of such lists "is highly dependent on the response of producers believing that listed substances may possibly be regulated, in case the voluntary phase out languishes."

7. *Positive Lists*. These are lists of substances whose toxicity is so low that they can often be used to substitute other, more harmful substances. Such lists are much less in vogue than lists of hazardous substances, probably because of the uncertainty associated with claims that a substance is not hazardous (Hansson, 1997).
8. *Ban of Dangerous Substances*. This is the strongest way in which a jurisdiction can require that a substance be replaced by less dangerous alternatives. A successful example of substitution by prohibition is the phase-out of ozone-depleting CFCs (Oosterhuis, 2006). However, for practical reasons, prohibition is a method that can only be used for relatively few substances or groups of substances.
9. *Required Substitution Plans*. The state of Massachusetts' Toxic Use Reduction Act requires companies using large amounts of listed substances to develop a plan that evaluates possible options for reductions in use (Ellenbecker and Geiser, 2011; Onasch et al., 2011). A similar but even more ambitious system of substitution plans has been proposed by Thorpe and Rossi (2007).
10. *Economic Incentives*. Economic incentives can be employed to discourage the use of hazardous or untested substances. An interesting example is the Norwegian environmental tax for pesticides, calculated so that the "basic fee" for each pesticide product is multiplied with a factor between 1 and 8 depending on the estimated health and environmental risk (Lohse et al., 2003).

26.4.2 Definitions and Terminology

To substitute means to replace, or to put one thing in place of another. In chemicals regulation and risk management, the substitution principle is a policy principle that requires the replacement of hazardous (or potentially hazardous) chemical substances by less hazardous alternatives. The substitution principle has also often been called the "product choice principle."

It is important to distinguish between substitution, that is, replacement in general, and applications of the substitution principle. A hazardous substance can be replaced for other reasons than hazard reduction. There are cases when manufacturers have knowingly replaced a hazardous substance with an even more hazardous one in order to avoid a legal or public relations problem with the first substance. In 2008, American authorities took measures against the high contents of lead in children's jewelry imported from China. Some manufacturers responded by replacing lead with cadmium. Cadmium is more toxic than lead, but whereas the use of lead in jewelry was prohibited, there were no legal restrictions on cadmium in jewelry (Mead, 2010; Pritchard, 2010). This type of substitution should not be seen as an application of the substitution principle.

The more precise definition of the substitution principle is controversial. The major definitional issues can be summarized under the following four headings:

- Which are the candidates for substitution?
- What can be the substitute?
- What can the purpose of substitution be?
- How functional must the substitute be?

26.4.2.1 Which Are the Candidates for Substitution? Originally, the substitution principle referred to chemical substances. Today, it is mostly conceived more broadly. In addition to chemical substances, it is applied to mixtures (i.e., chemical products), materials, and products containing potentially dangerous chemicals. A major reason for this is that there are cases when a mixture or a product gives rise to a larger hazard than any of its components. A mixture may, for instance, constitute a much larger fire or reaction hazard than any of its components, and the components of a mixture may have synergistic toxic or ecotoxic effects. It is not uncommon for consumer products with similar functional properties to differ in their environmental and health impact without this difference being due to the presence or absence of a single substance (Ingre-Khans et al., 2010). In such cases, substitution of a mixture or a product can be an appropriate measure.

26.4.2.2 What Can Be the Substitute? In some statements of the substitution principle, its application is restricted to the replacement of a chemical by another, less hazardous chemical. However, most formulations of the principle also cover the replacement of chemicals by non-chemical methods. This has been described as a distinction between chemical and functional substitution. Not surprisingly, the chemical industry has issued statements that define the substitution principle as referring to chemical substitution only:

> "Substitution is the replacement of one substance by another with the aim of achieving a lower level of risk."
>
> (CEFIC, 2005, p. 1)

However, several authors have pointed out that from the viewpoint of environmental protection, "the hazardous substance does not necessarily have to be replaced by another substance. It can also be substituted by other means of fulfilling the function it had. Thus, a hazardous cleaning agent (e.g., a chlorinated solvent) can be replaced by a less harmful one, but [it] is also conceivable that the product or production process is redesigned in such a way that the cleaning step can be omitted" (Oosterhuis, 2006). For another example, brominated flame retardants in plastic casings for electronic equipment can be replaced by other, less hazardous flame retardants for the same casings. Alternatively, the plastic casings can be replaced by metal casings. As pointed out by Jennifer Hall (2001), it makes a difference if one says "we need a substance to replace BFR" or "we need an environmentally better way to protect a material from catching fire." Most definitions of the substitution principle take this into account, and allow for substitution by non-chemical means. For instance, Charlie Auer, former director of the US Environmental Protection Agency's Office of Pollution Prevention and Toxics, defined the substitution principle as follows:

> "Informed substitution is the considered transition from a chemical of particular concern to safer chemicals or non-chemical alternatives."
>
> (Auer, 2006)

26.4.2.3 What Can the Purpose of Substitution Be? There can be many grounds for substituting a chemical by another, for instance, various economic and technical reasons. In most of the literature on the substitution principle, the justification of substitution is taken to be that the substance is "hazardous," without further specification. In practice (although this is not always clearly said), the substitution principle is taken to refer to purposes related to the environment and to human health. Substitution can be based on the toxic and ecotoxic properties of substances as well as properties such as explosiveness, inflammability, and dangerous reactivity. It is also generally accepted that the substitution principle can be applied to deal with properties related to the climate effects of producing, using, or disposing of substances and products.

In most discussions on substitution, the focus has been on normal conditions in the production and use of the substance. As was pointed out by Cordella and coworkers, a large part of the severe chemical accidents have resulted from the release of decomposition products that are more hazardous than the substance itself. (The Seveso accident is among the most well-known examples.) Therefore, unwanted products that are formed as a consequence of deviations from normal operating conditions should be included in deliberations on substitution (Cordella et al., 2009).

The different reasons for substitution can sometimes come into conflict with each other. Hence, substance A may be preferable to substance B in terms of toxicity, whereas B is preferable to A in terms of inflammability (Hansson et al., 2011). This can be taken as an argument to base the application of the substitution principle on the totality of hazards, rather than on the particular hazard that started the process.

26.4.2.4 How Functional Must the Substitute Be? Some hazardous substances are difficult to replace because it is difficult to find other materials or methods that are equally functional. There are influential definitions of the substitution principle that require the substitute to satisfy the functional requirements to the same degree as the original chemical:

> "Substitution means the replacement or reduction of hazardous substances in products and processes by less hazardous or nonhazardous substances, or by achieving an equivalent functionality via technological or organisational measures."
>
> (Lohse et al., 2003. Cf. Oosterhuis, 2006, p. 1)

> "Substitution is the replacement of a substance, process, product, or service by another that maintains the same functionality."
>
> (UK Chemicals Stakeholder Forum, 2010)

As was noted by Hansson et al. (2011), this approach entails that environmental objectives should always be given lower priority than the objectives of material production. In their view, it is not appropriate to exclude the need for compromises between the different objectives involved in such a conflict:

> "[T]he substitution principle should be seen as a principle that requires the replacement of dangerous substances by less dangerous alternatives, while attempting to (i) reduce the hazard as much as possible, (ii) retain (or in principle increase) the functionality that the original substance had as much as possible, and (iii) keep costs as low as possible. The priority between these three objectives is a matter for negotiation and adjustment in each particular case, and cannot be settled beforehand by requiring that one of these objectives always has absolute priority over the others."
>
> (Hansson et al., 2011, pp. 455–456)

26.4.3 Theory and Interpretation

In chemicals regulation, it is customary to distinguish between a hazard and a risk. A hazard can be defined as a potential for damage. As applied to chemicals, a hazard pertains to the inherent properties of the substance as such, whereas a risk also depends on how the substance is used and handled, that is, it takes exposure into account. Hazard is mostly treated as a non-quantitative concept, whereas risk is described in quantitative terms, for instance, using probabilities or statistically expected damage (Hansson, 2007b).

Public authorities have often emphasized the application of the substitution principle to chemical hazards. Substitution based on hazard can be general, that is, apply to many or all uses of a substance, and it does not require detailed assessments of the particular circumstances under which the substance is used in specific companies or by individual consumers. In contrast, representatives of the chemical industry have claimed that substitution should be performed on the level of risk, not hazard:

> "All chemical management decisions should be based on risk… Substitution is only acceptable provided the suitability of the alternative has been demonstrated through socio-economic costs/benefits analysis and a risk assessment."
>
> (CEFIC, 2005, pp. 1 and 3)

In much the same vein, one opponent of hazard-based substitutions has described the issue as follows:

> "The key issue, however, is whether regulatory decision-making should be based on hazard classification alone, or whether one should conduct a full risk analysis… Substitution is a complex process and should be based on risk analysis and not on hazard classifications."
>
> (Löfstedt, 2014, pp. 556–557)

However, the information needed for a full risk analysis is often not available. The cases when it is are not controversial. No one seems to propose that risk information should be disregarded when it is available. The controversial cases are those in which a hazard assessment is available, but a risk assessment is not available and cannot be made based on the available knowledge (and these are by far the most common situations). The key issue is therefore whether hazard-based substitutions should be promoted in cases where risk information is lacking. The following is a typical such case:

> The two substances A and B can both be used as degreasers (or for some other well-defined purpose). A dominates the market, and very few companies use B. However, a hazard assessment shows that the hazard associated with B is much smaller than that associated with A. We do not have access to the quantitative information about exposures and dose/response relationships that would be required to perform a risk assessment (in the conventional sense) (Hansson et al., 2011).

According to the opponents of hazard-based substitution, there is insufficient basis for substituting B for A since we do not have a risk assessment. However, the information that we have in such a case may be sufficient to show that a substitution would reduce the risks. Since the exposure is approximately the same for both substances, but B has less dangerous inherent properties, we can conclude that the probability of dangerous effects will decrease if we use B rather than A. The example is not untypical. To the contrary, if we replace a hazardous substance with a less hazardous one without increasing the exposure, then the risks will be reduced. Therefore, a substitution does not have to be risk *based* (based on risk information) in order to be risk *reducing*. This is, of course, the reason why most proponents of the substitution principle advocate hazard-based substitutions when full risk assessments are not available.

As noted by Aven, adversaries of hazard-based substitution ignore "uncertainties and the potential for surprises" (Aven, 2014, p. 570). There is a long history, not least from the chemical industry, showing that accidents do in fact happen and that unplanned exposures are almost sure to occur. When they happen, we are in a much better situation if hazardous substances have been replaced by less hazardous ones. This is the insight expressed in the principle of inherent safety (see Chapter 17). It is generally accepted in safety engineering that safety work must have a strong focus on the possibility of unplanned adverse events. If the same focus is applied to the choice of chemical products, then variants of the substitution principle that do not endorse risk-reducing measures based on hazard information will have to be replaced by versions that recommend risk reductions more generally.

26.4.4 Effects of Applying the Principle

There is well-documented evidence showing that significant environmental improvements can be achieved by substitution of hazardous materials. To mention just one example, the introduction of a plan for substitution of hazardous chemicals in the offshore industry in the United Kingdom resulted in a reduction of hazardous discharges. The total quantity of discharged chemicals did not decrease much, but a significant shift to less hazardous chemicals was achieved (La Védrine et al., 2015). For additional examples, see Onasch et al. (2011) and Ozturk et al. (2009).

The available evidence also gives clear indications that in order to ensure that the substitution principle is efficiently implemented, measures by regulators and public authorities are necessary. Vorschoor and Reijnders (2001) investigated five cases in which companies reduced their use of toxic chemicals. All five companies mentioned environmental legislation as a reason for doing this. A report containing 10 case studies of substitution confirmed that (actual or expected) legislation was a powerful driver (Lohse et al., 2003). It should be noted that the actions by government agencies that were influential in these examples did not have the form of detailed regulations on what substances to substitute. Instead, these agencies put pressure on companies to take initiatives themselves.

Not much is known about how the substitution principle fares in comparison to safety work based on other principles. We also lack information about the relative efficiency of different methods to achieve substitution (such as the 10 types of measures mentioned in Section 26.4.1.).

26.4.5 Weaknesses and Criticism

Substitution decisions can never be better than the available knowledge about the substances in use and the available alternatives to them. Unfortunately, incomplete information has sometimes led to counterproductive substitutions, that is, an unsafe product is replaced by another product that is incorrectly believed to be safer. In some cases, a whole series of failed substitutions has taken place:

> "The chemical trichloroethylene (TCE), a volatile organic chemical, was widely used as a degreaser in the manufacture of electronic circuits and components until concerns about TCE's environmental effects led the industry to replace it with trichloroethane (TCA), which has similar chemical structure. TCE and TCA were among the most widely used industrial degreasers, and they are now found in many of the hazardous cleanup sites listed on the National Priorities List. TCA, in turn, was replaced as a degreaser by chlorofluorocarbons such as Freon when ozone depletion concerns were raised about TCA in the 1990s. The use of Freon as a chemical degreaser was eventually phased out due to its own health and environmental concerns. Now, new mixtures of solvents are being used in vapor degreasing."
>
> (Bent, 2012, pp. 1402–1403)

Such failed substitutions depend in part on lack of information, but often also in part on the all-too-common assumption that if a substance has no known hazardous properties, then it has no hazardous properties. The former problem has to be solved with systematic investigation programs to obtain toxicity and ecotoxicity information about substances that are used or planned to be used (Rudén and Hansson, 2010). To solve the second problem, a less risk-taking attitude to toxicological uncertainty would have to be adopted. Currently, substances with unknown toxicity are still often treated in the same manner as substances that can reasonably be assumed to have no such harmful properties. A more defensible approach would be to treat substances with unknown properties as if they had (at least) the average toxicity among tested substances in the same chemical group (Hansson and Rudén, 2008).

The European Chemical Industry Council (CEFIC) and some other major business organizations have accused the substitution principle of being too costly (CEFIC, 2006). One critic has even claimed that the substitution principle implies that "almost any chemical product that has certain properties can be subjected to a ban or severe restriction, irrespective of the actual or projected level of risk" (Nilsson, 2004, p. 113). In practice, however, only few substances have been banned due to the substitution principle, and these substances have been selected because of their unusually hazardous properties (Öberg, 2014). Voluntary decisions by industry to replace a problematic substance or product are much more common than bans and other far-reaching legal restrictions.

It should also be emphasized that for most of the chemicals currently in use, we do not have access to the data needed to assess whether or not they are harmful to human health or the environment. The substitution principle does not provide a solution to that problem.[4]

> "The problem with restricted substance use is not confined to a fixed number of chemicals with a negative impact. The problem is the use of a large number of poorly understood substances. Thus, substitution of one substance with another may not actually constitute a solution."
>
> (Hall, 2001)

26.5 COMPARATIVE DISCUSSION

26.5.1 Comparisons Between the Three Principles

The three principles have their origins in different areas of activity: ALARA in radiation protection, BAT in emissions control, and the substitution principle in chemicals control. In spite of their different origins and the fact that the three principles are seldom discussed together, they are so close in meaning that it is often difficult to judge

[4] Solutions to that problem will have to include a combination of (i) the systematic creation of toxicological and ecotoxicological laboratory data, and (ii) safety measures to prevent human exposure and environmental release of substances with unknown properties.

if a particular statement of a management goal is based on one or the other of the three principles.

> "The employer shall reduce the use of a carcinogen or mutagen at the place of work, in particular by replacing it, in so far as is technically possible, by a substance, preparation, or process which, under its conditions of use, is not dangerous or is less dangerous to workers' health or safety, as the case may be."
>
> (European Union, 2004)

We can read this as an expression of ALARA since it requires reduction of the hazard "as far as… possible," as an expression of BAT since the possibility it refers to is technological feasibility, or as an expression of the substitution principle since the aim is to replace hazardous substances and preparations. As already indicated, what makes the three principles so close to each other is their focus on *improvement*. Although at least one of them (ALARA) has sometimes been described as a principle of optimization, they are all principles of improvement rather than (just) optimization. A common theme in literature promoting these principles is that we should not be satisfied with what we have achieved in terms of safety and the protection of human health and the environment. When improvement is possible, it is desirable.

But none of the three principles prescribe that all improvements should be implemented immediately, regardless of costs. The principles all allow economic and other competing considerations to have a role. However, the precise nature of that role is not well defined. It is indeed a major issue in the interpretation of each of the three principles. For ALARA, this discussion has centered around the interpretation of the word "reasonable," for BAT on what we (should) mean with "available." For the substitution principle, the focus has been on what degree of functionality a substitute must have. There is a scope for interpretation of each of these principles, but it should be noted that none of them survives as an independent principle if it is conceived as just an application of standard cost–benefit analysis.

26.5.2 Comparisons with Other Principles

Several other safety principles and general management principles promote the same basic commitment to improvements as the three principles we have studied in this chapter. For instance, experience feedback aims at improving safety performance by making systematic use of previous experiences (Chapter 7), and safety culture principles aim at creating an organization that improves its safety record (Chapter 28). The following principles have interesting relationships with the three improvement principles discussed above and are therefore in need of a somewhat more detailed treatment.

Vision Zero originated as a goal for traffic safety that was adopted by the Swedish Parliament in 1997. The Bill states that "the long-term goal of traffic safety is that nobody shall be killed or seriously injured as a consequence of traffic accidents" and that "the design and function of the transport system shall be adapted accordingly" (Government Bill, 1996/1997:137). Similar goals for traffic safety have been adopted in other countries and by many states in the United States. The Danish government has formulated its vision succinctly in the slogan "every accident is one too many" (Færdselssikkerhedskommissionen, 2000). Vision Zero has also been proposed or adopted in other areas such as industrial waste (Curran and Williams, 2012), industrial accidents (Zwetsloot et al., 2013), avalanche safety (Atkins and Granhed, 2012), premature death from ischemic heart disease (Lingfors et al., 2002), and suicide prevention (Government Bill, 2008). The basic idea in all these cases is that a far-reaching visionary goal can provide better direction and guidance than a less ambitious short-term goal, and therefore also lead to better results (Rosencrantz et al., 2007; Edvardsson and Hansson, 2005). Contrary to many other goals, Vision Zero does not tell us when to stop and say that we are finished. To the contrary, it tells us to continue improving traffic safety as long as serious accidents still happen. This is very close to the common core of ALARA, BAT, and the substitution principle that was discussed above, and we can therefore describe Vision Zero as another improvement principle. Its underlying ideas were very clearly expressed by Harvey Fineberg:

> "Too often, many people accept avoidable harm as normal. Nightly newscasts commonly report another murder and another deadly automobile collision. These unfortunate events happen. Most used to think it was normal to smoke, but today smoking is generally socially unacceptable. Thinking things are normal when they are avoidable is a psychological obstacle to implementing strategies of prevention. Safer cars and safer highways can be built; indeed, Sweden has adopted a goal it calls Vision Zero to eliminate highway fatalities, and some US jurisdictions are on the same path. And surely the right norm for the number of murders is zero."
>
> (Fineberg, 2013)

Continuous improvement (CI) is a concept that arose out of quality management programs where it gained impetus in the 1980s. Its main idea has been summarized as follows:

> "In its simplest form, CI can be defined as a company-wide process of focused and continuous incremental innovation. Its mainspring is incremental innovation–small step, high frequency, short cycles of change which taken alone have little impact but in cumulative form can make a significant contribution to performance."
>
> (Bessant et al., 1994, p. 18)

Continuous improvement has been adopted in healthcare (Berwick, 1989) and in safety management (Granerud and Rocha, 2011). One of its advantages is that it can often be integrated with other management processes in ways that strengthen the organization's commitment to safety. At least in some of its guises, continuous improvement is closely related to the three safety principles that have been at focus in this chapter.

The *precautionary principle* has often been compared to ALARA, and it has been claimed that the two principles "have a common denominator: they are general rules for decision makers on risk assessment and risk management under scientific uncertainty" (Lierman and Veuchelen, 2006). However, the similarity between the two principles should not be exaggerated. Although they are both intended to ensure that environmental and health concerns are taken seriously, they do this in different ways and at different stages in the decision-making process.

There is no consensus on what the precautionary principle means, but most versions of it refer to situations in which there is a valid (scientifically based) suspicion of a potential danger, but not full evidence of its existence. The principle requires or allows that in such a situation, measures are taken to protect the environment and/or human health (Hansson, 2008; see also Chapter 12). There are two major types of definitions of the principle: argumentative and prescriptive definitions. An argumentative version of the precautionary principle is found in Principle 15 of the Rio Declaration (UNCED, 1993). It requires that "lack of full scientific certainty shall not be used as a reason for postponing cost-effective measures to prevent environmental degradation." Prescriptive versions of the principle prescribe actions. Perhaps the most famous of these is the so-called Wingspread Statement: "When an activity raises threats to the environment or human health, precautionary measures should be taken, even if some cause-and-effect relationships are not fully established scientifically" (Raffensperger and Tickner, 1999, pp. 354–355).

Hence, the precautionary principle deals with how we should react to uncertain information. If there is no uncertainty, then there is no need to apply the precautionary principle, although there may be reasons to be cautious in other ways. (It is important not to confuse the precautionary principle with cautious decision-making in general.) In contrast, ALARA, BAT, and the substitution principle are applicable both in cases with considerable uncertainty and in cases when the risks we are dealing with are well known and well characterized. Arguably, these principles are more suitable for the latter type of situation. The precautionary principle has its focus on what factual assumptions we should base our decisions on, whereas the improvement principles pertain to our choice of goals and procedures for making those decisions.

ACKNOWLEDGMENT

I would like to thank Marko Ahteensuu, Jan-Erik Holmberg, and Christina Rudén for useful comments on a draft for this chapter.

REFERENCES

Ackerman, B. A., and Stewart, R. B. (1988). Reforming environmental law: The democratic case for market incentives. *Columbia Journal of Environmental Law*, 13, 171–199.

Atkins, D., and Granhed, M. (2012). Vision zero: Applying road safety to avalanche safety. In: Proceedings, 2012 International Snow Science Workshop (pp. 16–22), Anchorage, AK.

Auer, C. (2006). US experience in applying 'Informed Substitution' as a component in risk reduction and alternatives analyses. Transcript of an oral presentation given at the Chemicals, Health, and the Environment Conference Ottawa, Ontario, Canada, October 2006. BASTA, 2010. BASTAonline AB. Available at: http://www.bastaonline.se/english/bastaonline.4.3d9ff17111f6fef70e9800039672.html.

Auxier, J. A., and Dickson, H. W. (1983). Guest editorial: Concern over recent use of the ALARA philosophy. *Health Physics*, 44, 595–600.

Aven, T. (2014). Comment. *Journal of Risk Research*, 17(5), 569–571.

Bent, J. R. (2012). An incentive-based approach to regulating workplace chemicals. *Ohio State Law Journal*, 73, 1389–1455.

Berwick, D. M. (1989). Continuous improvement as an ideal in health care. *New England Journal of Medicine*, 320, 53–56.

Berzosa, Á., Barandica, J. M., and Fernández-Sánchez, G. (2014). A new proposal for greenhouse gas emissions responsibility allocation: Best available technologies approach. *Integrated Environmental Assessment and Management*, 10(1), 95–101.

Bessant, J., Caffyn, S., Gilbert, J., Harding, R., and Webb, S. (1994). Rediscovering continuous improvement. *Technovation*, 14(1), 17–29.

Bicevskis, A. (1982). Unacceptability of acceptable risk. *Search*, 13(1–2), 31–34.

Bugallo, P. M. B., Cristóbal Andrade, L., Magán Iglesias, A., Torres López, R. (2013). Integrated environmental permit through best available techniques: Evaluation of the fish and seafood canning industry. *Journal of Cleaner Production*, 47, 253–264.

Bush, B. J. (2012)."Addressing the regulatory collapse behind the deepwater horizon oil spill: Implementing a 'Best Available Technology' regulatory regime for deepwater oil exploration safety and cleanup technology. *Journal of Environmental Law and Litigation*, 26, 535–568.

CEFIC (2005). CEFIC paper on substitution and authorisation under REACH, pp. 1–3.

CEFIC (2006). Implementation of REACH: A demanding challenge for industry, Joint Press Release, December 6, 2006.

Charles, J. A. (1967). Early arsenical bronzes—A metallurgical view. *American Journal of Archaeology*, 71, 21–26.

Clark, J., Sheldon, R., Raston, C., Poliakoff, M., and Leitner, W. (2014). 15 years of green chemistry. *Green Chemistry*, 16(1), 18–23.

Cohen, M. D. (2012). CT radiation dose reduction: Can we do harm by doing good? *Pediatric Radiology*, 42, 397–398.

Cordella, M., Tugnoli, A., Barontini, F., Spadoni, G., and Cozzani, V. (2009). Inherent safety of substances: Identification of accidental scenarios due to decomposition products. *Journal of Loss Prevention in the Process Industries*, 22, 455–462.

Corten, O. (1999). The notion of 'reasonable' in international law: Legal discourse, reason and contradictions. *International and Comparative Law Quarterly*, 48, 613–625.

Crass, M. F. (1941). A history of the match industry. *Journal of Chemical Education*, 18, 316–319.

Curran, T., and Williams, I. D. (2012). A zero waste vision for industrial networks in Europe. *Journal of Hazardous Materials*, 207, 3–7.

Derden, A., and Huybrechts, D. (2013). Brominated flame retardants in textile wastewater: Reducing Deca-BDE using best available techniques. *Journal of Cleaner Production*, 53, 167–175.

Edvardsson, K. and Hansson, S. O. (2005). When is a goal rational?. *Social Choice and Welfare*, 24, 343–361.

Edvardsson Björnberg, K., Hansson, S. O., and Vredin-Johansson, M. (2015). Making climate policy efficient: Implementing a model for environmental policy efficiency. *International Journal of Sustainable Society*, 7, 1–21.

Ellenbecker, M., and Geiser, K. (2011). At the source: The origins of the Massachusetts toxics use reduction program and an overview of this special issue. *Journal of Cleaner Production*, 19(5), 389–396.

European Commission (2001). Strategy for a Future Chemicals Policy, White Paper, European Commission, February 27, 2001. Available at: http://europa.eu.int/eurlex/en/com/wpr/2001/com2001 0088en01.pdf

European Union (2004). Directive on the Protection of Workers from the Risks Related to Exposure to Carcinogens or Mutagens at Work (Directive 2004/37/EC), April 29, 2004.

Færdselssikkerhedskommissionen [Danish Road Safety Commission] (2000). *Hver Ulykke Er en for Meget – Trafiksikkerhed Starter med Dig* (*Every Accident is One Too Many – Traffic Safety Starts with You*). Copenhagen: Trafikministeriet.

Fallon, M. (April 2012). Palo Verde's outage ALARA success: Is it repeatable and beatable? *Nuclear News*, pp. 44–52.

Fineberg, H. V. (2013). The paradox of disease prevention: Celebrated in principle, resisted in practice. *JAMA*, 310(1), 85–90.

Finnveden, G., Hauschild, M. Z., Ekvall, T., Guinée, J., Heijungs, R., Hellweg, S., Koehler, A., Pennington, D., and Suh, S. (2009). Recent developments in life cycle assessment. *Journal of Environmental Management*, 91, 1–21.

French, S., Bedford, T., and Atherton, E. (2005). Supporting ALARP decision making by cost benefit analysis and multiattribute utility theory. *Journal of Risk Research*, 8, 207–223.

Government Bill 1996/1997:137 *Nollvisionen och det Trafiksäkra Samhället* (Vision Zero and the Traffic Safe Society) Stockholm.

Government Bill (2008). En förnyad folkhälsopolitik, prop. (A renewed policy of public health, bill) 2007/08:110, 13 Mars. Government Offices of Sweden, Stockholm.

Granerud, L., and Rocha, R. S. (2011). Organisational learning and continuous improvement of health and safety in certified manufacturers. *Safety Science*, 49, 1030–1039.

Hall, J. C. (2001). Product design to reduce restricted substances, IIIEE Reports 2001:2, Lund.

Hansson, S. O. (1997). Can we reverse the burden of proof?. *Toxicology Letters*, 90, 223–228.

Hansson, S. O. (2007a). Ethics and radiation protection. *Journal of Radiological Protection*, 27, 147–156.

Hansson, S. O. (2007b). "Risk" Entry in the Stanford Encyclopedia of Philosophy. Available at: http://plato.stanford.edu/entries/risk/

Hansson, S. O. (2008). Regulating BFRs—From science to policy. *Chemosphere*, 73, 144–147.

Hansson, S. O. (2010). Promoting inherent safety. *Process Safety and Environmental Protection*, 88, 168–172.

Hansson, S. O. (2013a). *The Ethics of Risk. Ethical Analysis in an Uncertain World*. New York: Palgrave Macmillan.

Hansson, S. O. (2013b). ALARA: What is reasonably achievable? In: D. Oughton, and S. O. Hansson (Eds.), *Social and Ethical Aspects of Radiation Risk Management* (pp. 143–156). Elsevier Science.

Hansson, S. O., Molander, L., and Rudén, C. (2011). The substitution principle. *Regulatory Toxicology and Pharmacology*, 59, 454–460.

Hansson, S. O. and Rudén, C. (2008). A risk-neutral default for chemical risk management. *American Journal of Industrial Medicine*, 51, 964–967.

Health and Safety Executive (2001). *Reducing Risks, Protecting People. HSE's Decision-Making Process*. Norwich: Her Majesty's Stationery Office. Available at: http://www.hse.gov.uk/risk/theory/r2p2.pdf

Helman, L., and Parchomovsky, G. (2011). The best available technology standard. *Columbia Law Review*, 111(6), 1194–1243.

Hendee, W. R., and Edwards, F. M. (1986). ALARA and an integrated approach to radiation protection. *Seminars in Nuclear Medicine*, 16, 142–150.

Ibáñez-Forés, V., Aragonés-Beltrán, P., and Bovea, M. D. (2015). Multicriteria decision making methodologies applied to the selection of best available techniques in the ceramic industry: Equalitarian vs prioritised weighting. In: J. L. Ayuso Muñoz, J. L. Yagüe Blanco and S. F. Capaz-Rizo (Eds.), *Project Management and Engineering, Lecture Notes in Management and Industrial Engineering* (pp. 117–129). Springer.

Ingre-Khans, E., Rudén, C., and Breitholtz, M. (2010). Chemical risks and consumer products: The toxicity of shoe soles. *Ecotoxicology and Environmental Safety*, 73, 1633–1640.

International Commission on Radiological Protection (1959). *Recommendations of the ICRP: ICRP Publication No. 1*. New York: Pergamon.

International Commission on Radiological Protection (1977). *Recommendations of the ICRP: ICRP Publication No. 26. Annals of the ICRP* 1(3), (pp. 1–53). Oxford: Pergamon.

Jones-Lee, M., and Aven, T. (2011). ALARP—What does it really mean? *Reliability Engineering and System Safety*, 96, 877–882.

Kathren, R. L., Munson, L. H., and Higby, D. P. (1984). Application of risk-cost benefit techniques to ALARA and de-minimis. *Health Physics*, 47, 195.

Kathren, R. L., and Ziemer, P. L. (1980). The first fifty years of radiation protection – A brief sketch. In: R. L. Kathren, and P. L. Ziemer (Eds.), *Health Physics: A Backward Glance* (pp. 1–9). Elmsford, NY: Pergamon Press.

Kemikalieinspektionen (2008). Att arbeta med produktval i praktiken, Rapport 2008:2. Kemikalieinspektionen, Sundbyberg.

Khan, F. I., and Abbasi, S. A. (1998). Inherently safer design based on rapid risk analysis. *Journal of Loss Prevention in the Process Industries*, 11, 361–372.

Kletz, T. (1978). What you don't have, can't leak. *Chemistry and Industry*, 287–292.

La Védrine, M. A. G., Sheahan, D. A., Gioia, R., Rowles, B., Kroeger, S., Phillips, C., Kirby, M. F. (2015). Substitution of hazardous offshore chemicals in UK waters: An evaluation of their use and discharge from 2000 to 2012. *Journal of Cleaner Production*, 87, 675–682.

Lierman, S., and Veuchelen, L. (2006). The optimisation approach of ALARA in nuclear practice: An early application of the precautionary principle? Scientific uncertainty versus legal uncertainty and its role in tort law. *European Environmental Law Review*, 15(4), 98–107.

Lindell, B. (1996). The history of radiation protection. *Radiation Protection Dosimetry*, 68, 83–95.

Lindell, B., and Beninson, D. J. (1981). ALARA defines its own limit. *Health Physics*, 41, 684–685.

Lingfors, H., Persson, L., Lindstrom, K., and Bengtsson, C. (2002). Time for a 'vision zero' concerning premature death from ischaemic heart disease? *Scandinavian Journal of Primary Health Care*, 20(1), 28–32.

Löfstedt, R. (2014). The substitution principle in chemical regulation: A constructive critique. *Journal of Risk Research*, 17, 543–564.

Lohse, J., Wirts, M., Ahrens, A., Heitmann, K., Lundie, S., Lißner, L., and Wagner, A., (2003). Substitution of hazardous chemicals in products and processes, Final report. Report compiled for the Directorate General Environment. Nuclear Safety and Civil Protection of the Commission of the European Communities, Hamburg.

Løkke, S. (2006). The precautionary principle and chemicals regulation. *Environmental Science and Pollution Research*, 13, 342–349.

Marx, R. E. (2008). Uncovering the cause of 'phossy jaw' circa 1858 to 1906: Oral and maxillofacial surgery closed case files–Case closed. *Journal of Oral and Maxillofacial Surgery*, 66, 2356–2363.

Mead, M. N. (2010). Cadmium confusion: Do consumers need protection? *Environmental Health Perspectives*, 118(12), A528–A534.

Melchers, R. E. (2001). On the ALARP approach to risk management. *Reliability Engineering and System Safety*, 71, 201–208.

Merkouris, P. (2012). Sustainable development and best available techniques in international and European Law. In: K. E. Makuch, and R. Pereira (Eds.), *Environmental and Energy Law* (1st ed.) (pp. 37–60). Blackwell.

Meyers, S., and Shanley, E. S. (1990). Industrial explosives—A brief history of their development and use. *Journal of Hazardous Materials*, 23, 183–201.

Milone, D., Peri, G., Pitruzzella, S., and Rizzo, G. (2015). Are the best available technologies the only viable for energy interventions in historical buildings? *Energy and Buildings*, 95, 39–46.

Newman, B., and Callahan, M. J. (2011). ALARA (as low as reasonably achievable) CT 2011—Executive summary. *Pediatric Radiology*, 41(Suppl 2), S453–S455.

Nilsson, R. (2004). Control of chemicals in Sweden: An example of misuse of the 'precautionary principle.' *Ecotoxicology and Environmental Safety*, 57, 107–117.

Öberg, T. (2014). Substitution of chemicals based on assessment of hazard, risk and impact. *Journal of Risk Research*, 17, 565–568.

Oestreich, A. E. (2014). ALARA 1912: 'As Low a Dose as Possible' a century ago. *RadioGraphics*, 34, 1457–1460.

Onasch, J., Shoemaker, P., Nguyen, H. M., and Roelofs, C. (2011). Helping small businesses implement toxics use reduction techniques: Dry cleaners, auto shops, and floor finishers assisted in creating safer and healthier work places. *Journal of Cleaner Production*, 19, 408–413.

Oosterhuis, F. (2006). Substitution of hazardous substances. A case study in the framework of the project ‘Assessing innovation dynamics induced by environmental policy’. Institute for Environmental Studies, Vrije Universiteit, Amsterdam 2006.

Overton, T., and King, G. M. (2006). Inherently safer technology: An evolutionary approach. *Process Safety Progress*, 25, 116–119.

Ozturk, E., Yetis, U., Dilek, F. B., and Demirer, G. N. (2009). A chemical substitution study for a wet processing textile mill in Turkey. *Journal of Cleaner Production*, 17, 239–247.

Parascandola, J. (2012). *King of Poisons: A History of Arsenic*. Dulles, VA: Potomac Book.

Pearce, D. W., Russell, S., and Griffiths, R. F. (1981). Risk assessment: Use and misuse. *Proceedings of the Royal Society of London. Series A, Mathematical and Physical Sciences*, 376(1764), 181–192.

Pritchard, J. (2010). *Perfectly Legal: Toxic Cadmium, up to 91%, Added to Kids' Jewelry as China Gets Lead Out*. Associated Press, January 2010. Available at: http://www.cleveland.com/world/index.ssf/2010/01/perfectly_legal_toxic_cadmium.html

Raffensperger, C., and Tickner, J. (Eds.). (1999). *Protecting Public Health and the Environment: Implementing the Precautionary Principle*. Washington, DC: Island Press.

Ranken, A. (1982). EPA v. National Crushed Stone Association. *Ecology Law Quarterly*, 10, 161–189.

Rimmele, K. M. (1994). Do removal credits deserve credit—An analysis of POTWs and the CWA Removal Credit Program. *Tulane Environmental Law Journal*, 8, 223–264.

Rosencrantz, H., Edvardsson, K., and Hansson, S. O. (2007). Vision zero – Is it irrational? *Transportation Research Part A: Policy and Practice*, 41, 559–567.

Rossi, H. H. (1980). What are the limits of ALARA? *Health Physics*, 39, 370–371.

Rudén, C., and Hansson, S. O. (2006). Improving REACH. *Regulatory Toxicology and Pharmacology*, 44, 33–42.

Rudén, C., and Hansson, S. O. (2010). REACH is but the First step – How far will it take us? Six further steps to improve the European chemicals legislation. *Environmental Health Perspectives*, 118(1), 6–10.

Schoenberger, H. (2011). Lignite coke moving bed adsorber for cement plants – BAT or beyond BAT? *Journal of Cleaner Production*, 19, 1057–1065.

Shrader-Frechette, K., and Persson, L. (2002). Ethical, logical and scientific problems with the new ICRP proposals. *Journal of Radiological Protection*, 22, 149–161.

Stensvaag, J.-M. (2011). Preventing significant deterioration under the Clean Air Act: The BACT determination-Part I. *Environmental Law Reporter*, 41, 11101.

Sunstein, C. R. (1991). Administrative substance. *Duke Law Journal*, 40(3), 607–646.

Svensk författningssamling (1949). Kungl. Maj:ts kungörelse med föreskrifter angående tillämpning av arbetarskyddslagen (arbetarskyddskungörelsen). Svensk författningssamling 1949: 208. Norstedt & Söner, Stockholm. [Swedish Code of Statutes]

Thorpe, B., and Rossi, M. (2007). Require safer substitutes and solutions: Making the substitution principle the cornerstone of sustainable chemical policies. *New Solutions*, 17, 177–192.

UK Chemicals Stakeholder Forum (2010). A guide to substitution. An information note from the UK Chemicals Stakeholder Forum. Final version, August 20, 2010.

United Nations Conference on Environment and Development (UNCED) (1993). *The Earth Summit: The United Nations Conference on Environment and Development (UNCED), Rio De Janeiro 1992*. Introduction and commentary by Stanley P. Johnson. London: Graham & Trotman.

Vandenbergh, M. P. (1996). An alternative to ready, fire, aim: A new framework to link environmental targets in environmental law. *Kentucky Law Journal*, 85, 803–918.

Vorschoor, A. H., and Reijnders, L. (2001). Toxic reduction in processes. Some practical examples. *Journal of Cleaner Production*, 9, 277–286.

Wiedmann, T., and Barrett, J. (2010). A review of the ecological footprint indicator—Perceptions and methods. *Sustainability*, 2, 1645–1693.

Wilson, R. (2002). Precautionary principles and risk analysis. *Technology and Society Magazine, IEEE*, 21(4), 40–44.

Zwetsloot, G. I. J. M., Aaltonen, M., Wybo, J.-L., Saari, J., Kines, P., and De Beeck, R. O. (2013). The case for research into the zero accident vision. *Safety Science*, 58, 41–48.

FURTHER READING

Hansson, S. O., Molander, L., and Rudén, C. (2011). The substitution principle. *Regulatory Toxicology and Pharmacology*, 59, 454–460.

Jones-Lee, M., and Aven, T. (2011). ALARP–What does it really mean? *Reliability Engineering and System Safety*, 96, 877–882.

Merkouris, P. (2012). Sustainable development and best available techniques in international and European Law. In: K. E. Makuch, and R. Pereira (Eds.), *Environmental and Energy Law* (1st ed.) (pp. 37–60). Blackwell.

Oughton, D., and Hansson, S. O. (Eds.). (2013). *Social and Ethical Aspects of Radiation Risk Management*. Elsevier Science.

Part V

ORGANIZATIONAL PRINCIPLES AND PRACTICES

27

SAFETY MANAGEMENT PRINCIPLES

Gudela Grote

27.1 INTRODUCTION

Wherever the safety of people and the environment is at stake, the responsible actors will aim to keep hazards at bay through a systematic assessment of risks and measures for their control. These activities, subsumed under the heading of safety management, are mostly carried out by organizations that either run hazardous work processes themselves or monitor and sanction these processes in the role of auditors or regulators. However, they are also relevant for individuals, both in their work roles and in relation to leisure time occupations such as rock climbing or scuba diving.

Large organizations tend to have quite elaborate safety management systems that support a comprehensive and integral treatment of risk. For instance, all national states that have organizations involved in civil aviation (airlines, air traffic control, etc.) are required by the ICAO to have a safety management system (see Box 27.1). This example also illustrates the linkage between operating organizations and regulators, where the regulators define requirements for what they consider to be adequate safety management.

Handbook of Safety Principles, First Edition. Edited by Niklas Möller, Sven Ove Hansson, Jan-Erik Holmberg, and Carl Rollenhagen.

Box 27.1 Safety Management in Commercial Aviation

A State Safety Program (SSP) is a management system for the regulation and administration of safety by the State. The implementation of an SSP is commensurate with the size and complexity of the State's civil aviation system and requires coordination among multiple authorities responsible for the aviation functions of the State. The objectives of the SSP are to:

a. ensure that a State has the minimum required regulatory framework in place;
b. ensure harmonization amongst the State's regulatory and administrative organizations in their respective safety risk management roles;
c. facilitate monitoring and measurement of the aggregate safety performance of the State's aviation industry;
d. coordinate and continuously improve the State's safety management functions; and
e. support effective implementation and interaction with the service provider's SMS.

Source: ICAO 2013.

A simpler example is the safety management required of mountaineers (see Box 27.2). This example illustrates that safety management is ubiquitous and not restricted to large formal organizations. Rather, we all engage in some form of safety management for a number of our activities, all the way to decisions on whether to drive our car home when we have been drinking alcohol. In this chapter, the focus will be on safety management by organizations, but of course successful safety management by organizations is always linked to adequate safety management by individuals as well, as they have to identify and evaluate risks, follow procedures, or learn from incidents.

Box 27.2 Safety Management in Mountaineering

Mountains for Growth's safety management plan emphasizes personal responsibility. Under instruction and guidance, participants are accountable for their own safety and the safety of others in the group. That said, many of the environments we travel in may be new to participants. Mountains for Growth guides draw from significant experience to teach participants to read hazards in these environments. It is the participants' responsibility to maintain the instructed safety standards throughout the program. Participants are also encouraged to call attention to anything they perceive as unsafe (no matter who it involves), and learn from the situation. Owning our journey is the only way we can possibly grow and develop as individuals.

Source: Website of Mountains for Growth. http://www.mountainsforgrowth.com/coaching/safety/

27.2 ORIGIN AND HISTORY OF THE PRINCIPLE

Safety management is as old as the first human activity that required the handling of hazards. What has changed is the degree of formalization of safety management along with growing legal requirements and a shift from individual to institutional responsibility. These changes have happened in response to our reliance on ever more complex technological systems and citizens' growing expectations to be protected against hazards by their governments, building on scientific advances in the understanding of risks (Covello and Mumpower, 1985). Along with growing knowledge both in technical and organizational domains relevant to safety management, there has been a shift in the general approach taken, from what has been termed the technical age to the human factors age to today's focus on organizational safety management (Hale and Hovden, 1998). With the more integral understanding of organizations as sociotechnical systems, there has come a broadening of relevant components of safety management, such that at this stage safety management can be considered a superordinate principle that integrates many other safety principles, such as risk analysis, accident investigation, and safety training, whose origins and definitions are treated in other chapters of this book.

In recent decades, the concept of safety management system has been proposed to systematically develop, implement, monitor, and adapt the different components of safety management within an integral approach. This concept borrows from quality management, which can be considered the first domain where comprehensive management systems have been advocated (Mangelsdorf, 1999). As a consequence, the relationship between quality management and safety management is much debated and frequently an integrated system is recommended (Beckmerhagen et al., 2003; Robson et al., 2007). Today, virtually all industries are required to have safety management systems in place. The growing emphasis on safety management systems can also been seen in conjunction with a change in regulatory practice from prescriptive and often technically driven requirements organizations had to meet to increasing self-regulation which obligates organizations to demonstrate appropriate management of safety without prescribing in much detail how that should be achieved (Hopkins and Hale, 2002; May, 2007; Reason, 1997). As a consequence, the purpose of introducing safety management systems is not only, and some would argue not even primarily, the optimization of system design and operation, but rather the defendability of organizational processes in the face of external agencies, such as regulators or insurance companies.

27.3 DEFINITIONS

Ever since industrial operations have been created that imply considerable risks for people, the environment, and the business itself, there have been attempts to mitigate these risks in order to achieve "safe" operations, that is operations with acceptable risk. The methods used for this purpose are usually subsumed under the terms of risk management or safety management. Depending on industrial sector and professional field one term is preferred over the other, so that one may conclude that they basically mean the same thing (e.g., Harms-Ringdahl, 2004). One can argue, though, that the term risk management is broader because any kind of risk is included. The respective ISO standard defines risk management as "coordinated activities to direct and control an organization with regard to risk" (ISO 31000, 2009, p. 2). Safety management tends to focus on "safety risks," that is health and environmental risks, with less explicit consideration of general business and financial risks. For this book, the term safety management was chosen due to its focus on health and environmental risks. However, in this chapter some literature will be used also that favors the term risk management.

As for most anything, many definitions exist for safety management. Two examples may suffice to show the range of different conceptions of safety management:

- Arrangements made by the organization for the management of safety in order to promote a strong safety culture and achieve good safety performance (INSAG, 1999);
- Organized approach to managing safety, including the necessary organizational structures, accountabilities, policies, and procedures (ICAO, 2013).

While in the first definition, safety management is seen as a means to several ends, specifically promotion of safety culture and good safety performance, in the second definition, components of safety management are described. Both definitions taken together cover most of the content found in the multitude of definitions of safety management. In most basic terms, following the ISO definition of risk management, one could state that safety management concerns all coordinated activities to direct and control an organization with regard to health and environmental risk.

Along with the many definitions, many different descriptions of components of safety management exist. The components listed in Table 27.1 can be considered a general standard across different literatures (e.g., Amyotte et al., 2007; ICAO, 2013; INSAG, 1999; Robson et al., 2007). The specific ways of designing and implementing them in a particular organization will depend on a number of attributes of that organization and its environment, though. These will be discussed in later sections of this chapter. When implemented and operated in an integral fashion, the components constitute a safety management system.

27.4 UNDERLYING THEORY AND THEORETICAL ASSUMPTIONS

Safety management builds on a large array of theories as it covers a broad spectrum of organizational activity. While the theories underlying specific elements of safety

TABLE 27.1. Components of Safety Management

Component	Description
Safety policy	Statement of an organization's objectives, commitments, and management principles with respect to safety
Safety resources and responsibilities	Resources (financial and human) devoted to safety and definition of roles including specific competencies and accountabilities
Risk identification and mitigation	Instruments and processes for risk assessment and for decision-making on appropriate measures to handle the risks
Safety standards and procedures	Rules for safe operations and processes for their design, implementation, monitoring, and improvement
Human factors based system design	Expertise and instruments for system design based on human factors principles
Safety training	Training for individuals and teams including evaluations of training effectiveness
Safety performance monitoring	Definition of safety performance indicators and processes for their continuous measurement and assessment
Incident reporting and investigation	Recording and analysis of safety-relevant incidents and accidents and processes for learning from the results
Auditing	Internal and external assessments of safety management practices
Continuous improvement	Instruments and processes for the continuous improvement of the safety management system
Emergency preparedness	Plans and processes for managing and recovering from emergency situations
Management of change	Instruments and processes for managing the risks of technical and organizational change

management, such as learning theories relevant for safety training or continuous improvement, probability theories used in risk analysis, or cognitive and motivational theories informing human factors based system design, are treated in other chapters that discuss these elements as safety principles in their own right, here theory is introduced as it relates to the overall approach chosen for dealing with risk in organizations. This is, on the one hand, theory on which management systems are built in general, and on the other hand, theory concerning the management of risk and uncertainty.

Inherent in management systems is the assumption that managing is done well when it follows a cycle of continuous assessment and adaptation of all operations in an organization (often called plan–do–check–act or plan–do–study–act cycle). Originating from quality control and made popular especially by Deming's work (e.g., Deming, 1986), this cycle is akin to fundamental descriptions of human behavior as developed by cognitive psychologists in the 1960s. Following general principles of control theory, Miller et al. (1960) argued that the most basic component underlying human behavior is the test–operate–test–exit (TOTE) unit, meaning that there is an initial evaluation of whether or not a current state is satisfactory, if it is not, an operation is planned and executed to move toward a more satisfactory state, ending the iteration between these two phases only if no further change is deemed necessary. Taken to the level of management systems, this thinking entails that instruments and processes need to be in place that ensure continuous monitoring and improvement of all organizational activities within iterative feedback loops.

The foremost goal of safety management is to keep all processes in an organization within the bounds of what has been determined—by the organization itself and/or by external agents such as regulatory bodies—as acceptable risk. This raises the basic questions of how acceptable risk is defined and what are considered successful actions toward achieving and maintaining acceptable risk. In order to answer these questions, first of all there needs to be a theory of risk, which currently is hotly debated, especially with respect to the relationship between risk and uncertainty and the necessity to expand quantitative accounts of risk and uncertainty by including methods for capturing and integrating qualitative uncertainty as well (e.g., Aven and Renn, 2009). What is considered acceptable risk is foremost the result of a political process, which has been aptly described for instance in the risk governance model of the IRGC (2005; see also Renn, 2008). Keeping risk within the set limits is a matter of risk mitigation for which usually four groups of instruments are discussed that aim at risk avoidance, risk reduction, risk transfer, or risk retention, respectively (Renn, 2008). The predominant focus is usually on risk reduction, with risks maintained only reluctantly in acknowledgement of the constraints highly complex and uncertain organizations and environments place on the ability to control risk. However, there is increasing awareness that increasing risk may be a viable option also in particular circumstances (Grote, 2015). For instance, the ISO 31000 standard mentions "increasing risk to pursue an opportunity" as an option for risk treatment (ISO 31000, 2009, p. 6).

As Grote (2015) discusses, considering the various options for handling risks or more broadly uncertainty within one management system entails the difficulty of bridging a broad range of theoretical assumptions and belief systems about managing uncertainty (see Table 27.2). *Reducing uncertainty* to a level of acceptable risk is the main thrust in classic risk mitigation. The overall objective is to create stable systems that allow for a maximum of central control. Measures such as standardization and automation help to streamline work processes. *Maintaining uncertainty* follows from acknowledging the limits to reducing uncertainty in complex systems, which has led to the development of concepts such as "high reliability organizations" (Weick et al., 1999) and "resilience engineering" (Hollnagel et al., 2006). Flexibility as a source for resilience, that is the capability of systems to recover from perturbations, is sought. For this purpose, control capacity needs to be decentralized, for example, by means of empowering local actors. *Increasing uncertainty*, finally, aims at flexibility not only in response to perturbations, but also in support of innovation. An important conceptual basis is complexity theory (cf., e.g., Anderson, 1999) and

TABLE 27.2. Options for Managing Uncertainty and Their Conceptual Basis

	Reducing Uncertainty	Maintaining Uncertainty	Increasing Uncertainty
Objective	Stability	Flexibility	Flexibility/innovation
Conceptual approach	Classic risk mitigation	Resilience	Complexity theory
Control paradigm	Central control	Control by delegation to local actors	Shaping contexts for self-organizing agents
Examples of measures	Standardization; automation	Empowerment	Controlled experimentation

Source: Grote 2015. Reproduced with permission of Elsevier.

self-organization as one of the theory's fundamental principles. Self-organizing local agents are assumed to not be directly controllable; instead, they are indirectly influenced in their adaptive behavior by shaping contexts, for instance, through setting incentives and constraints for experimentation.

Carroll (1998) has pointed out that the different conceptions of risk control tend to be prevalent in different professional (sub)cultures within organizations (Schein, 1996). While engineers and executives believe in uncertainty reduction through design and planning, operative personnel are very aware of the need for resilience in the face of only partially controllable uncertainties. Social scientists finally will argue for openness to learning and innovation, thereby even adding uncertainty. Reducing, maintaining, and increasing uncertainty also form the core of the three models of safety postulated by Amalberti (2013) for different industries. According to him, ultra-safe systems like aviation are built on reducing uncertainty through standardization and external supervision, while high reliability organizations, for instance in health care or the chemical industry, are characterized by their ability to collectively cope with inevitable uncertainties. Ultra-resilient systems, finally, are those that seek uncertainty as part of their business models, for instance in stock market trading or in military aviation. The safety performance of these systems is generally lower compared to the other two types of organizations, but will greatly vary based on the level of competence of the highly autonomous actors that run ultra-resilient systems. Building a shared understanding of the legitimacy of all three options of reducing, maintaining, and increasing uncertainty across professional boundaries is paramount to developing a more comprehensive approach to safety management.

27.5 USE AND IMPLEMENTATION

As stated in the introduction to this chapter, safety management has been in use for as long as there have been safety concerns. More systematic and integral forms of safety management along with regulatory requirements for safety management have only emerged in the last century, though. At different times, different industries have been considered ahead of everybody else with respect to the implementation of safety management systems. For a long time, the nuclear industry seemed to be the most advanced, due also to high public pressure. In the 1980s, the chemical industry set standards through the development of process safety management (e.g., Amyotte et al., 2007), which even the nuclear industry paid close attention to (INSAG, 1999). Most recently, there has been a strong push by aviation authorities across the world to introduce safety management systems in the different domains in aviation, like airlines, air traffic control, and aircraft maintenance (ICAO, 2013). All along, the transfer of knowledge between industry domains has oscillated between "one size fits all" and "reinventing the wheel" (e.g., Amalberti et al., 2005; Hudson, 2003).

Two points need to be stressed to foster successful implementation and use of safety management systems. Firstly, introducing a safety management system or any components thereof amounts to major organizational change, which should be approached on the basis of a thorough consideration of existing knowledge on how successful organizational change can be effected (e.g., Weick and Quinn, 1999). Importantly, the change process itself entails risks, possibly including even increased safety risks—for example, due to both managers' and workers' attention being distracted from safely running the everyday operational processes—which need to be managed well (Grote, 2007).

Secondly, the safety management system and its components need to be carefully tailored to the particularities of the organization where they are to be implemented. Unfortunately, there is little systematic research to guide such comparisons and adaptations. In the following section, existing research is briefly reviewed and relevant context factors introduced. Finally, some examples of how these contextual factors should be taken into consideration are provided.

27.6 EMPIRICAL RESEARCH ON USE AND EFFICIENCY

There are no systematic reviews available on the prevalence of safety management systems in organizations. Given the ubiquity of government and industry safety standards, one can assume that every organization in developed countries with considerable health and environmental risks has some sort of safety management in place. A growing number of studies have investigated the effects of introducing safety management systems or certain safety management practices on safety performance, however. These studies are confronted with the fundamental difficulty in safety research that results from safety being a "dynamic non-event" (Weick, 1987). Indicators for good safety performance therefore are mostly the absence of events, such as no accidents and injuries. There have been efforts to define behaviors such as reporting incidents, following rules, or making suggestions for safety improvements as safety performance indicators, but in order to validate these indicators again a relationship with accidents and injuries needs to be demonstrated. Another way out of this dilemma has been to measure individuals' perception of safety performance, for instance captured in the concept of safety climate (Zohar, 1980), but the issue whether these perceptions can be linked to objective safety performance remains. Within these limitations, research has been able to demonstrate positive effects of safety management systems in a number of different industries (Robson et al., 2007; Thomas, 2012). However, the evidence is rather weak as tests of safety intervention effects against control groups are mostly lacking and much of the research relies on self-report data. Also, the evidence for the effectiveness of particular safety management practices is rather mixed. This is not astonishing given the different domains and contexts within which the studies took place. To guide practice, it is important to specify conditions that influence the effects of safety management activities in organizations. Knowing these conditions and their impact can help practitioners to decide which activities to adopt and how to adapt them to their particular circumstances. However, it has to be acknowledged that sound empirical evidence on the generalizability of safety measures is scarce. Therefore, much of what follows should also be regarded as propositions for future research.

Grote (2012) has suggested the type of risk (e.g., risks concerning process vs. personal safety), the predominant approach to managing uncertainty in the organization/the industry sector (e.g., minimizing uncertainty vs. coping with uncertainty), and the regulatory regime (e.g., external prescriptive regulation vs. self-regulation) as three important contextual factors which should be considered when adopting safety management practices and specifically when adopting them based on experience from other industry sectors. These contextual factors will be described in more detail in the following and their relevance shown for two commonly used safety management practices: safety standards and procedures and safety training. Clearly, there are many other relevant contextual factors that should be considered as well, such as the maturity of an organization or an industry sector, the ownership structure in a company, or differences in professional cultures. The three factors discussed here can provide a first indication of how context should be considered in order to develop and implement successful safety management systems. They also align with what other authors have considered to be particularly relevant factors, for example, Goodman et al. (2011) talk about hazard profile, external pressures and regulation, and organizational strategy as major factors influencing organizational learning from errors.

27.6.1 Contextual factors

Process versus Personal Safety. Obviously, a first important consideration is to specify what it is that is to be managed by a management system, which in our case requires a closer look at safety itself. A fundamental distinction to make when discussing safety is that between process safety and personal safety (Grote, 2012; Hopkins, 2009). In process safety, the risks and uncertainties to be managed are directly linked to the primary work task of the organization, such as curing patients, producing energy or transporting goods and people. Breaches of process safety do not necessarily cause harm to the human operators involved, as in the case of the death of a patient due to false medication. In personal safety, on the other hand, potential damages always concern the human operators, but in turn are not necessarily directly linked to the primary work task. In building construction, for instance, the basic process safety requirement is the stability of the structures erected, including proper calculations for building requirements and proper execution of these requirements such as putting the required steel reinforcing in concrete slabs. These requirements for the most part are unrelated to personal safety, which for a construction worker concerns incidents such as falling of a scaffold or getting hit by a falling object. Personal protection against these hazards is mostly a secondary task, sometimes even interfering with the primary task, like ear protection preventing detection of task-relevant signals such as unusual machine noises.

Personal and process safety may or may not be closely related depending on the primary work task. Pilots put their own just as much as passengers' lives at risk when operating an aircraft. In building construction, structural safety, which is the main process safety requirement, and personal safety are quite unrelated, except when structures fail during construction already. Furthermore, causal mechanisms involved in promoting personal versus process safety most likely differ due to differences in the visibility and complexity of risks, required competence for risk handling, available incentives for safe behavior, and conflicts between safety and production requirements. However, to date research has not systematically addressed these differences (Thomas, 2012).

Minimizing versus Coping with Uncertainty. A second fundamental issue concerns the way how organizations manage uncertainties more generally—that is any situation in which there is only insufficient or ambiguous information available for decision-making—as this will also influence how safety is managed (Grote, 2009). Two paradigms can be distinguished: (1) Following the principles of scientific management and bureaucratic organizations, *uncertainty may be minimized* by high levels of standardization, central planning, automation of work processes, and high levels of specialization with few degrees of freedom for employees. (2) In line with newer organization theory, which stresses the need for flexible adaptation in highly uncertain and networked business processes, all actors in the organization may be empowered to *cope with uncertainty locally* by providing them with options for action rather than fixed plans and standards, with ample opportunity for cooperation and learning, and with decision-support tools rather than algorithmic automation (e.g., Farjoun, 2010; Weick et al., 1999). As was discussed earlier, there is an emerging third paradigm, which includes not only coping with unavoidable uncertainties, but also deliberately creating uncertainties at times (Grote, 2015). In general terms, these paradigms aim to either maximize stability or flexibility. As the levels of uncertainty that companies are confronted with can change over time and also internal demands for different parts of companies may vary, recent research has addressed the need to balance stability and flexibility rather than opt for one or the other (Farjoun, 2010; Grote, 2009), which entails decisions on reducing, maintaining, or increasing uncertainty in view of own objectives and situational demands. This balance is to some extent specific to different industries, for example, emergency management organizations like police or fire departments require more flexibility and less stability than a nuclear power plant (e.g., Bigley and Roberts, 2001). But stability and flexibility demands also vary within industry and across time. For instance, periods of maintenance outage require more flexibility than routine operations in a power plant.

Safety management tends to follow the first paradigm in search for stability, but there is an increasing awareness of the need to balance stability and flexibility (Weick et al., 1999; Hollnagel et al., 2006). Also, in regulatory approaches, the balance between minimizing and coping with uncertainties is under debate. Zero-risk regulation has been proclaimed over many years, which assumes that risks can be reduced to (almost) nothing to worry about. More recently, the feasibility and appropriateness of zero-risk regulation has been questioned and possibilities for more openly admitting to risks in order to find more appropriate ways of handling them are discussed (Gilbert et al., 2007; Power, 2004).

External Regulation versus Self-Regulation. An important element of any organization's environment concerns external oversight over the organization's operation. All companies and especially those operating with major risks are subject to some degree of external regulation, usually by government agencies. Regulation concerns most basically the permit to operate and can reach into the details of how operations are carried out in order to protect employees, customers, and the society at large. Regulatory agencies define and modify rules and are also in charge of surveying rule compliance and investigating cases of—suspected—rule violation. Recent years have witnessed a tendency to move away from prescriptive regulation which specifies in great detail how an organization has to carry out its operations, toward goal-oriented legislation, which promotes self-regulation by giving the regulated companies the freedom—but also the responsibility—to decide on the management of their operations within broadly defined safety functions and goals (Hopkins and Hale, 2002; Majumdar and Markus, 2001; May, 2007). Instead of having to continuously react to changes in external demands and adapt their safety management accordingly, companies can proactively manage safety in a way that best fits their specific situation. For regulatory agencies, the new regime implies a shift from being an independent authority and representative of public interest to being an expert and even consultant. This shift requires different competencies of regulators (Reiman and Norros, 2002) and contains the danger of undue diffusion of accountability (May, 2007).

For companies, the different types of regulation regimes imply very different demands on their internal functioning and the relationship with external regulators. When the focus is on external regulation, companies have to continuously react to changes in external demands and adapt their safety management accordingly. They also have to be ready for external audits and inspections at any time. When the focus is on self-regulation, companies have to find out for themselves to a large extent what kind of safety management best fits their particular situation and be more proactive in monitoring their safety performance (Parker, 2002). Compliance is determined not by adherence to detailed prescriptions but by whether companies have acceptable plans for achieving good safety performance and by actually reaching set safety goals (Kirwan et al., 2002; May, 2007).

27.6.2 Examples for the effects of context on safety management

Safety Standards and Procedures. Safety standards and procedures are a crucial part of any safety management system. In order to use them well, it is important to understand how rules and behavior interact (Grote, 2012; Grote et al., 2009; Hale and Borys, 2013). Rules are artifacts containing the written-down formal specific actions for behavior patterns required in predefined situations. The actual behavior patterns emerging and eventually repeated with a high degree of regularity have been called routines, or more specifically routines in practice. The latter term is used to distinguish the behavior from the more abstract understanding of the reasoning behind the behavior, which has been termed routine in principle (Becker, 2005; Feldman and Pentland, 2003). Rules are usually meant to create particular routines in practice and also convey the reasoning behind the required behavior in the sense of a routine in principle. Problems arise when these three elements are not aligned, because the rule writers and the rule followers have formed different routines in principle, based, for instance, on different priorities between personal and process safety (Grote et al., 2009). This can also be a problem in very homogeneous teams due to the danger of establishing inadequate behavioral routines based on a shared but limited view of the world. Moreover, in order to employ rules in ways that support the overall objective of adequately balancing stability and flexibility, the nature of the rules themselves needs to be examined. For this purpose, a distinction suggested by Hale and Swuste (1998) between goal, process, and action rules is useful. Goal rules only define the goal to be achieved, leaving open how this is accomplished by the actors concerned. Process rules provide guidance for deciding on the right course of action for achieving certain goals. Finally, action rules prescribe detailed courses of action, possibly without even mentioning the goal to be achieved. As a rule of thumb on good rule making, action rules should be used when stability of processes is required. Goal and process rules should be used when flexibility is required. An important final check for any rule is to assess whether the assigned responsibility for uncertainty handling in the work processes matches the capability of the actor concerned (Grote, 2008, 2009).

Some considerations regarding the design of safety standards and procedures in view of the three contextual factors are provided (cf., Grote, 2012). In order to ensure *personal safety*, action rules should be employed because the priority of safety needs to be established unambiguously. For *process safety*, goal and process rules will be more important as a means to provide flexible guidance for actors. *Minimizing uncertainty* is best achieved with action rules, which leave as little scope of action as possible and thereby help establish firm behavioral routines. Recent efforts to introduce checklists into surgery (Haynes et al., 2009) may be considered an example here. On the other hand, goal and process rules promote the forming of more flexible routines and thereby support *coping with uncertainty* (Grote et al., 2009). Standards and procedures help coordinate work processes, which also should be their main function in organizational *self-regulation*. *External regulators* tend to view standards and procedures as an instrument for defining legal accountability. When both types of regulation regime co-exist, these two perspectives need to be aligned by assigning responsibility in ways that match the coordination capabilities of the respective actors. Otherwise, conflicting rules or a discrepancy between rules and routines in practice result. A negative example found in many rulebooks across industry domains is the rule stating that deviating from rules in the interest of safety may be necessary and that actors at the operational level are responsible for making this decision. Thereby legal accountability is assigned, but actors are scapegoated rather than supported in handling difficult situations.

Safety Training. Individual safety training, usually concerned mainly with personal safety, is common practice in most high-risk industries (e.g., Burke et al., 2006). In recent years, many companies in high-risk industries have started to also invest in team training in order to build up the skills needed for effective coordination, especially in emergency situations. This has happened in response to many accident analyses that convincingly show malfunctioning coordination to be a significant precursor to accidents (Salas et al., 2006). The aviation industry was the first to develop special training programs for so-called crew resource management in the cockpit (Helmreich et al., 1999), focusing on leadership, decision-making, and communication skills. Over the last decade, similar programs have been introduced in many other domains, most prominently in medicine (Baker et al., 2005). Training methods range from classical lectures to behavior training in full-scale simulators. While there is general agreement that the more engaging the teaching methods, the more profound the learning (Burke et al., 2006), many other issues are still much debated such as the specific topics that should be taught and the overall effectiveness of training (Salas et al., 2006).

Again, some considerations regarding the design of safety training in view of the three contextual factors are provided (cf., Grote, 2012). The support of team coordination, that is, the management of task interdependencies is the core element of team training. This requirement is much more pronounced with respect to *process safety* than personal safety. Accordingly, team training is more prevalent in industries with high process risks like aviation, nuclear power plants, or medicine. For *personal safety*, training is mainly aimed at increasing individual knowledge about risks and risk protection measures, which can be found in almost any industry at this stage. When *minimizing uncertainty* is the aim, training focuses on drills in order to ensure the correct execution of a predefined action path. Especially for technical simulator training in aviation or process industries, this has long been the dominant form of training because the main concern was to support people in performing certain tasks even under very high levels of stress during emergencies. In recent years, more open forms of training aimed at adaptive decision-making and problem solving have been introduced across industries, which conform with the *coping with uncertainties* paradigm (e.g., Salas et al., 2007). There has been a push by *external regulation* to comply with certain amounts and types of training, for example, a fixed number of hours of team simulator training. Depending on organizational factors such task complexity and team diversity, different kinds of training may be necessary, though, among which organizations should carefully choose, if they are allowed to do so under *self-regulation* schemes. For instance, for complex or novel tasks training should support team problem solving, while for simple or routine tasks the training purpose is shifted toward avoiding over-routinization. In homogeneous teams, training can help to avoid complacency and overreliance on shared perceptions and assumptions. Training of cockpit crews therefore stresses explicit coordination as a means to question and reestablish common ground for action. For heterogeneous teams, training is important to first of all develop a shared understanding and to cope with internal differences in personal backgrounds, values, and perspectives. Training in health care due to the highly interdisciplinary nature of that field is always confronted with these issues. Any attempt to transfer training schemes from the cockpit to medicine should therefore be conducted with great care (e.g., Zeltser and Nash, 2010).

27.7 WEAKNESSES, LIMITATIONS, AND CRITICISM

Management systems are only useful when they help continuous monitoring and improvement. With too many components managed too bureaucratically, they may turn organizations and individuals into mindless followers of routines (Reiman and Odewald, 2007). In this way, what has been said about safety standards and procedures is also true for the safety management system as a whole, which can be considered a meta-standard producing meta-routines. Weick et al.'s (1999) call for collective mindfulness is key for keeping safety management effective.

For the design of safety management practices, the main message is not to look for solutions in other, presumably more advanced, industries per se, but to compare and contrast organizational attributes and take those safety measures on board that have worked well in similar conditions (Grote, 2012). So health care may learn from aviation regarding crew resource management training for homogenous teams because to date cockpit crews and cabin crews tend to be trained separately. However, with advances in safety training in health care, which in most cases will have to be conducted in heterogeneous teams due to the highly interdisciplinary nature of most work processes there, aviation may learn from health care when designing joint cockpit and cabin crew trainings or joint trainings between pilots and air traffic control. So mindfulness is key in the design of safety management practices as well. Moreover, whatever general approach is taken toward safety management in terms of balancing stability and flexibility should be consistently applied to all components of the safety management system. The examples discussed above—safety standards and procedures and safety training—give an indication of what are necessary considerations for achieving a consistent approach.

One safety principle that deserves special attention is safety culture. The meaning and relevance of safety culture and its relationship to safety management is much debated (e.g., Guldenmund, 2010). In the INSAG (1999) definition, for instance, the goal of safety management is to develop a safety culture, while others consider safety culture to be an important prerequisite for effective safety management or describe safety culture as a component of safety management systems (Beckmerhagen et al., 2003; McDonald et al., 2000). In order to better understand the relationship between safety management and safety culture, it is helpful to consider them as separate phenomena for a moment. Safety management, as we have seen, comprises different organizational practices aimed at controlling risk and keeping it at an acceptable level. Safety culture, on the other hand, refers to the norms and beliefs in an organization that affect the way safety is dealt with. These norms and beliefs may concern safety directly, for instance, by placing a high importance on safety, but also indirectly as they relate to work motivation, cooperation, or leadership, which all are relevant for how work is done including how safely it is done. It has been extensively written about how difficult it is to measure, understand, and manage culture. Culture generally shows itself most clearly during organizational change when basic assumptions are challenged (Schein, 1992). Therefore, introducing a safety management system will meet with acceptance or resistance in part due to its compatibility with existing norms and beliefs, while in the process of organizational change triggered by the new management system some of these norms and beliefs may get changed. Explicit consideration of culture should not be an aim in itself, but should happen as part of organizational change brought about by implementing safety management practices.

When considering the relationship between safety culture and safety management, contextual factors are also relevant (cf., Grote, 2012). For instance, regarding the kinds of safety that are to be managed, safety culture plays quite a different role in each. When *personal risks* are fairly distinct from process risks, protection against personal risks creates a task separate from the primary work task. In that case the "safety first" principle needs to be stressed and firmly founded in cultural norms in order to promote personal safety. In order to promote *process safety*, alignment between task requirements and safety requirements is generally easier because they are more tightly interlinked with the primary work task itself. Therefore, cultural norms that emphasize and support good task fulfillment will in themselves be beneficial for safety. Furthermore, shared norms and assumptions are crucial for coordination and integration in flexible and adaptive work processes allowing much decision latitude to individual actors, which is typical of a *coping with uncertainty* approach

(Weick, 1987). When coordination is achieved through high levels of standardization and strict planning, which is central to the *minimizing uncertainty* approach, a culture of rule following and obedience is helpful to enforce standards and plans, but not essential to ensure organizational functioning (Grote, 2007). Finally, the different approaches to regulation affect the way in which safety culture should be addressed. From the perspective of *external regulation*, the focus should be on prescribing, monitoring, and enforcing safety management systems, *not* safety culture. From the perspective of *self-regulation*, culture should be internally monitored and reflected upon in organizations, especially as a potential source of resistance to change. These examples illustrate the intricacies in the relationship between safety management and safety culture, which need to be well understood in order to successfully introduce and run safety management systems.

27.8 RELATIONS TO OTHER SAFETY PRINCIPLES

Safety management includes a wide range of safety management practices, which constitute safety principles in their own right. Many of these are discussed in other chapters of this book, such as risk identification and mitigation (chaps. 4, 20, 21, 25), safety standards and procedures (chap. 31), human factors based system design (chaps. 9, 10), safety training (chap. 29), safety performance monitoring (chap. 8), continuous improvement (chaps. 7, 15), and emergency preparedness (chap. 30). The chapters 3 and 28 contain discussions of overarching principles and basic attitudes and norms for organizational design. These are very relevant to the implementation and monitoring of effective safety management systems in line with what has been referred to earlier, borrowing from Weick et al. (1999), as collective mindfulness.

REFERENCES

Amalberti, R. (2013). *Navigating Safety*. Dordrecht, The Netherlands: Springer.

Amalberti, R., Auroy, Y., Berwick, D., and Barach, P. (2005). Five system barriers to achieving ultrasafe health care. *Annals of Internal Medicine*, 142, 756–764.

Amyotte, P. R., Goraya, A. U., Hendershot, D. C., and Khan, F. I. (2007). Incorporation of inherent safety principles in process safety management. *Process Safety Progress*, 26, 333–346.

Anderson, P. (1999). Complexity theory and organization science. *Organization Science*, 10, 216–232.

Aven, T., and Renn, O. (2009). On risk defined as an event where the outcome is uncertain. *Journal of Risk Research*, 12, 1–11.

Baker, D. P., Gustafson, S., Beaubien, J. M., Salas, E., and Barach, P. (2005). Medical team training programs in health care. *Advances in Patient Safety*, 4, 253–267.

Becker, M. (2005). The concept of routines: Some clarifications. *Cambridge Journal of Economics*, 29, 249–262.

Beckmerhagen, I. A., Berg, H. P., Karapetrovic, S. V., and Willborn, W. O. (2003). Integration of management systems: Focus on safety in the nuclear industry. *International Journal of Quality and Reliability Management*, 20, 210–228.

Bigley, G., and Roberts, K. H. (2001). The incident command system: High-reliability organizing for complex and volatile environments. *Academy of Management Journal*, 44, 1281–1299.

Burke, M. J., Sarpy, S. A., Smith-Crowe, K., Chan-Serafin, S., Salvador, R. O., and Islam, G. (2006). Relative effectiveness of worker safety and health training methods. *American Journal of Public Health*, 96, 315–324.

Carroll, J. S. (1998). Organizational learning activities in high-hazard industries: The logics underlying self-analysis. *Journal of Management Studies*, 35, 699–717.

Covello, V. T., and Mumpower, J. (1985). Risk analysis and risk management: An historical perspective. *Risk Analysis*, 5, 103–120.

Deming, W. E. (1986). *Out of the Crisis*. Cambridge, MA: MIT Center for Advanced Engineering Study.

Farjoun, M. (2010). Beyond dualism: Stability and change as duality. *Academy of Management Review*, 35, 202–225.

Feldman, M. S., and Pentland, B. T. (2003). Reconceptualizing organizational routines as a source of flexibility and change. *Administrative Science Quarterly*, 48, 94–118.

Gilbert, C., Amalberti, R., Laroche, H., and Paries, J. (2007). Errors and failures: Towards a new safety paradigm. *Journal of Risk Research*, 10, 959–975.

Goodman, P. S., Ramanujam, R., Carroll, J., Edmondson, A. C., Hofmann, D., and Sutcliffe, K. (2011). Organizational errors: Directions for future research. *Research in Organizational Behavior*, 31, 151–176.

Grote, G. (2007). Understanding and assessing safety culture through the lens of organizational management of uncertainty. *Safety Science*, 45, 637–652.

Grote, G. (2008). Rules management as source for loose coupling in high-risk systems. In: E. Hollnagel, C. Nemeth, and S. Dekker (Eds.), *Remaining Sensitive to the Possibility of Failure* (pp. 91–100). London: Ashgate.

Grote, G. (2009). *Management of Uncertainty—Theory and Application in the Design of Systems and Organizations*. London: Springer.

Grote, G. (2012). Safety management in different high-risk domains—All the same?. *Safety Science*, 50, 1983–1992.

Grote, G. (2015). Promoting safety by increasing uncertainty—Implications for risk management. *Safety Science*, 71, 71–79.

Grote, G., Weichbrodt, J. C., Günter, H., Zala-Mezö, E., and Künzle, B. (2009). Coordination in high-risk organizations: The need for flexible routines. *Cognition, Technology and Work*, 11, 17–27.

Guldenmund, F. (2010). (Mis)understanding safety culture and its relationship to safety management. *Risk Analysis*, 30, 1466–1480.

Hale, A. R., and Borys, D. (2013). Working to rule or working safety? Part 1: A state of the art review. *Safety Science*, 55, 207–221.

Hale, A. R., and Hovden, J. (1998). Management and culture: The third age of safety. A review of approaches to organizational aspects of safety, health and environment. In: A. M. Feyer, and A. Williamson (Eds.), *Occupational Injury: Risk, Prevention and Intervention* (pp. 129–165). London: Taylor & Francis.

Hale, A. R., and Swuste, P. (1998). Safety rules: Procedural freedom or action constraint?. *Safety Science*, 29, 163–177.

Harms-Ringdahl, L. (2004). Relationships between accident investigations, risk analysis, and safety management. *Journal of Hazardous Materials*, 111, 13–19.

Haynes, A. B., Weiser, T. G., Berry, W. R., Lipsitz, S. R., Breizat, A-H. S., Dellinger, E. P., Herbosa, T, Joseph, S., Kibatala, P. L., Lapitan, M. C. M, Merry, A. F, Moorthy, K., Reznick, R. K., Taylor, B., and Gawande, A. A. (2009). A surgical safety checklist to reduce morbidity and mortality in a global population. *New England Journal of Medicine*, 360, 491–499.

Helmreich, R. L., Merritt, A. C., and Wilhelm, J. A. (1999). The evolution of crew resource management training in commercial aviation. *International Journal of Aviation Psychology*, 9, 19–32.

Hollnagel, E., Woods, D. D., and Leveson, N. (2006). *Resilience Engineering: Concepts and Precepts*. Aldershot, UK: Ashgate.

Hopkins, A. (2009). Thinking about process safety indicators. *Safety Science*, 47, 460–465.

Hopkins, A., and Hale, A. (2002). Issues in the regulation of safety: Setting the scene. In: B. Kirwan, A. Hale, and A. Hopkins (Eds.), *Changing Regulation: Controlling Hazards in Society* (pp. 1–12). Oxford: Pergamon.

Hudson, P. (2003). Applying the lessons of high risk industries to health care. *Quality and Safety in Health Care*, 12, i7–i12.

ICAO. (2013). *Safety Management Manual*. Available at: http://www.icao.int

INSAG (International Nuclear Safety Advisory Group). (1999). Management of operational safety in nuclear power plants (INSAG-13). International Atomic Energy Agency, Vienna.

IRGC (International Risk Governance Council). (2005). Risk Governance—Towards an Integrative Approach (White paper No. 1; authored by O. Renn with an appendix by P. Graham). Geneva, IRGC.

ISO 31000 (2009). *Risk Management—Principles and Guidelines*. Geneva: ISO.

Kirwan, B., Hale, A. R., and Hopkins, A. (Eds.). (2002). *Changing Regulation: Controlling Hazards in Society*. Oxford: Pergamon.

Mangelsdorf, D. (1999). Evolution from quality management to an integrative management system based on TQM and its impact on the profession of quality managers in industry. *The TQM Magazine*, 11, 419–424.

Majumdar, S. K., and Marcus, A. A. (2001). Rules versus discretion: The productivity consequences of flexible regulation. *Academy of Management Journal*, 44, 170–179.

May, P. J. (2007). Regulatory regimes and accountability. *Regulation and Governance*, 1, 8–26.

McDonald, N., Corrigan, S., Daly, C., and Cromie, S. (2000). Safety management systems and safety culture in aircraft maintenance organisations. *Safety Science*, 34, 151–176.

Miller, G. A., Galanter, E., and Pribram, K. (1960). *Plans and the Structure of Behavior*. New York: Holt.

Parker, C. (2002). *The Open Corporation: Effective Self-Regulation and Democracy*. Cambridge, UK: Cambridge University Press.

Power, M. (2004). *The Risk Management of Everything: Rethinking the Politics of Uncertainty*. London: Demos.

Reason, J. (1997). *Managing the Risks of Organizational Accidents*. Aldershot, UK: Ashgate.

Reiman, T., and Norros, L. (2002). Regulatory culture: Balancing the different demands of regulatory practice in the nuclear industry. In: B. Kirwan, A. Hale, and A. Hopkins (Eds.), *Changing Regulation: Controlling Hazards in Society* (pp. 175–192). Oxford: Pergamon.

Reiman, T., and Odewald, P. (2007). Assessment of complex sociotechnical systems—Theoretical issues concerning the use of organizational culture and organizational core task concepts. *Safety Science*, 45, 745–768.

Renn, O. (2008). *Risk Governance—Coping with Uncertainty in a Complex World*. London: Earthscan.

Robson, L. S., Clarke, J. A., Cullen, K., Bielecky, A., Severin, C., Bigelow, P. L., Irvin, E., Culyer, A., and Mahood, Q. (2007). The effectiveness of occupational health and safety management system interventions: A systematic review. *Safety Science*, 45, 329–353.

Salas, E., Nichols, D. R., and Driskell, J. E. (2007). Testing three team training strategies in intact teams. A meta-analysis. *Small Group Research*, 38, 471–488.

Salas, E., Wilson, K. A., Burke, C. S., and Wightman, D. C. (2006). Does crew resource management training work? An update, an extension, and some critical needs. *Human Factors*, 48, 392–412.

Schein, E. H. (1992). *Organizational Culture and Leadership* (2nd ed.). San Francisco, CA: Jossey-Bass.

Schein, E. H. (1996). Three cultures of management: The key to organizational learning. *Sloan Management Review*, 38, 9–20.

Thomas, J. W. (2012). *A Systematic Review of the Effectiveness of Safety Management Systems*. Canberra, Australia: Australian Transport Safety Bureau.

Weick, K. E. (1987). Organizational culture as a source of high-reliability. *California Management Review*, 29, 112–127.

Weick, K. E., and R. E. Quinn (1999). Organizational change and development. *Annual Review of Psychology*, 50, 361–386.

Weick, K. E., Sutcliffe, K. M., and Obstfeld, D. (1999). Organizing for high reliability: Processes of collective mindfulness. *Research in Organizational Behavior*, 21, 81–123.

Zeltser, M. V., and Nash, D. B. (2010). Approaching the evidence basis for aviation-derived teamwork training in medicine. *American Journal of Medical Quality*, 25, 13–23.

Zohar, D. (1980). Safety climate in industrial organizations—Theoretical and applied implications. *Journal of Applied Psychology*, 65, 96–102.

FURTHER READING

Amalberti, R. (2013). *Navigating Safety*. Dordrecht, The Netherlands: Springer.

Grote, G., and Carroll, J. (Eds.). (2013). *Safety Management in Context: Cross-Industry Learning for Theory and Practice*. Rüschlikon, Switzerland: Swiss Re Centre for Global Dialogue.

Thomas, J. W. (2012). *A Systematic Review of the Effectiveness of Safety Management Systems*. Canberra, Australia: Australian Transport Safety Bureau.

28

SAFETY CULTURE

Teemu Reiman and Carl Rollenhagen

28.1 INTRODUCTION

A "weak" or deficient safety culture has been identified as a substantive issue or factor in the development of numerous human-created disasters, from BP Texas City refinery explosion and Fukushima Daiichi nuclear disaster to Piper Alpha offshore platform, Challenger and Columbia Space Shuttle explosions, Deepwater Horizon disaster (Waring, 2015). Safety culture as a concept originates from the nuclear industry where it was first used to explain the Chernobyl nuclear accident. In various applications, the concept strives to catch the influence of social and organizational phenomena to safety. It reminds both practitioners and researchers of the fact that most accidents are not solely the result of human error, technical failures, or environmental factors such as weather. Rather, accidents often have their roots in organizational and managerial shortcomings, and could thus be prevented by better safety management systems. The concept illustrates that safety cannot ever be guaranteed by technical means alone, but rather safety depends heavily on management, leadership, and so-called human and organizational factors. It also illustrates the importance of reflecting how safety is perceived and valued in organizations, and what things are considered important in achieving it.

Handbook of Safety Principles, First Edition. Edited by Niklas Möller, Sven Ove Hansson, Jan-Erik Holmberg, and Carl Rollenhagen.

The concept of safety culture is currently used in many fields of safety, including occupational safety, patient safety, and process safety. The concept has been applied to explain accidents and to provide a risk estimate for potential events (cf., Strauch, 2015). The concept is variously used either as a safety management principle (a tool for safety improvement) or as a concept describing a real phenomenon in the organization (something that informs about the level of safety in the organization). Both can be valid and useful applications of the concept, but it is always important to clarify the assumptions underlying a chosen approach to safety culture. Providing such clarifications has proven to be a painstakingly difficult exercise judged from the many interpretations and theoretical orientations found in the literature. As an example of some of the difficulties, consider the quotes below.

Chairman Kurokawa of the Fukushima accident investigation board:

> *"For all the extensive detail it provides, what this report cannot fully convey—especially to a global audience—is the mindset that supported the negligence behind this disaster.*
>
> *What must be admitted—very painfully—is that this was a disaster 'Made in Japan.' Its fundamental causes are to be found in the ingrained conventions of Japanese culture: our reflexive obedience; our reluctance to question authority; our devotion to 'sticking with the program'; our groupism; and our insularity.*
>
> *Had other Japanese been in the shoes of those who bear responsibility for this accident, the result may well have been the same."*
>
> (NAIIC, 2012:9)

> *"BP has not instilled a common, unifying process safety culture among its U.S. refineries. Each refinery has its own separate and distinct process safety culture. While some refineries are far more effective than others in promoting process safety, significant process safety culture issues exist at all five U.S. refineries, not just Texas City. Although the five refineries do not share a unified process safety culture, each exhibits some similar weaknesses. The Panel found instances of a lack of operating discipline, toleration of serious deviations from safe operating practices, and apparent complacency toward serious process safety risks at each refinery."*
>
> (The B.P. U.S. Refineries Independent Safety Review Panel, 2007)

In the quotations above the concept of safety culture is used as a high-level explanation for the accidents of Fukushima in 2011 and Texas City refinery in 2005. The first quote illustrates one of the inherent difficulties associated with the concept of safety culture, namely the problem of demarcating the concept of culture in general from the concept of safety culture as a specific facet of culture. Pending issues include: How is national culture related to organizational culture? How is organizational culture related to various subcultures in organizations and how is all of this related to the specific facet of culture called safety culture? And finally, is it possible to identify such features of safety culture that would be optimal in any macro-context or does the national culture affect what kind of culture is good for safety?

The second quote illustrates the problem of different types of safety, and different types of safety culture. Here some of the debated issues include: Are the properties of safety culture needed to prevent and respond to major accidents different from safety culture needed to prevent occupational accidents? Do organizations have a single safety culture or should we instead inspect different subcultures such as occupational safety culture, process safety culture, etc.? Combining this with the question of the influence of the national culture, a major controversy revolves around the issues of whether there are universal features of good safety culture, and what are the features or elements of (a good or bad) safety culture.

In this chapter, we inspect the different uses of the concept of safety culture and summarize the critique directed toward the term as well as give recommendations for a way forward. We will focus mainly on safety culture rather than on the related concept of safety climate. The two concepts have many similarities but also differences since they emanate from different research traditions. Safety climate is often perceived as a more shallow aspect of safety culture associated with how individuals in an organization directly perceive various states of affairs related to safety. Safety culture, on the other hand, is associated with norms, values, and basic assumptions directly or indirectly related to safety. However, the demarcation between the two concepts and associated measures are often blurred in practice. For example, the use of questionnaires is common in both research traditions.

Many of the technical principles of safety, such as defense-in-depth, are related to safety culture. This is due to the fact that safety culture affects all the independent layers of defense that the defense-in-depth concept deals with. In fact, most of the chapters in this handbook contain aspects that can be related to aspects discussed in the area of safety culture and safety climate.

The disposition of the chapter is as follows. First, we give two examples of events that illustrate various human and organizational features that are associated with safety culture. In Section 28.2, we begin with a short description of the Chernobyl accident in 1986 since this event is claimed to constitute the birth of the safety culture concept. We then trace the roots of the concept of safety culture to the more general concepts of organizational culture and organizational climate. Numerous definitions have been suggested to define safety culture and we exemplify some of these in Section 28.3 followed by a review of theoretical assumptions about the concept in Section 28.4. Section 28.5 gives selected examples of empirical research and Section 28.6 discusses the use and implementation of the safety culture concept. Some of the criticism directed to the safety culture concept is discussed in Section 28.7 followed by discussion and main message in Section 28.8.

Example 1 Saved by the grace of safety culture—Onagawa NPP
The causes of the Fukushima nuclear disaster have been largely attributed to cultural factors (weak safety culture). However, even more interesting is an example of a nuclear power plant that manifested good/strong safety culture during the Great Asian Tsunami, namely, the Onagawa nuclear power plant operated by Tōhoku Electric Power Company.

The plant has three boiling water reactors (BWR Mark I) commissioned in 1984, 1995, and 2002.

Onagawa nuclear power plant was the closest nuclear power plant to the March 2011 earthquake epicenter. The plant experienced "the strongest shaking that any nuclear power plant has ever experienced from an earthquake" (IAEA, 2012), but was able to shut down safely all its three reactors. No radioactive material was released to the environment. The plant units suffered only minor damage, with some internal flooding in unit 2 and turbine damage in all units.

It can be argued that the safe shutdown of Onagawa in many respects can be attributed to the plant's high level of safety culture (Ryu and Meshkati, 2014). Firstly, the plant clearly practiced conservative decision-making and continuous learning: Before beginning construction in 1968, Tohoku Electric reviewed literature and interviewed local people about tsunamis (there was no simulation technology yet available). These initial predictions showed that tsunamis in the region historically had an average height of about 3 meters. Based on that information, the company positioned its plant at 14.8 meters above sea level, which was almost five times the estimated height of an average tsunami.[1] New studies conducted by the power company raised the estimate first to 9.1 meters in 1987 when preparing the license application for unit 2 and then as a result of in separate study in 2002 to 13.6 meters. Several improvements were made based on the new estimates. Further, Tohoku Electric learned from past earthquakes and tsunamis—including one in Chile on February 28, 2010—and continuously improved its countermeasures.

Secondly, Tohoku Electric's emergency response was well organized, collaborative, and controlled. Protocols were in place for emergencies, including tsunami, and operators had been trained to cope with a variety of simulated emergencies, including the loss of off-site power. Tohoku Electric had established emergency response centers both at the Onagawa plant and at company headquarters. They had reinforced the old administration building with braces and were in the process of building a new seismic isolation building to house an emergency response center. Fortunately, the old administration building survived the quake and acted as a response center. There was good support from the headquarters throughout the disaster. Supervisors and chief engineers were dispatched to the main control rooms of the damaged reactors to make decisions, and information was sent in a timely manner to all levels of the response team.

Finally, the successful shutdown of the plant can be attributed to top management commitment to safety and interaction with all stakeholders in matters of nuclear safety. Yanosuke Hirai, vice president of Tohoku Electric from 1960 to 1975—a time period that preceded the 1980 groundbreaking at Onagawa—was adamant about safety protocols and became a member of the Coastal Institution Research Association in 1963 because of his concern about the importance of protecting against natural disasters. With a senior employee in upper management advocating forcefully for safety, a strong safety culture formed within the company. Representatives of Tohoku Electric participated in seminars and panel discussions about earthquake and tsunami disaster prevention held by the Japan Nuclear Energy Safety Organization. The company implemented strict protocols for disaster response, and all workers were familiar with the steps to be taken when a tsunami was approaching.

[1] The 2011 Tsunami was of 14.3 meters height at Onagawa and 13.1 meters at Fukushima.

Example 2 Normalizing deviance—NASA and the *Challenger* space shuttle

NASA Space Transportation System, envisioned in 1969, was supposed to provide the Unites States an easy and convenient access to space. It was marketed as a routine "bus-like" transportation to space with one flight every week. The selling argument for the program at NASA was "safe, cost-effective and routine access to space." The original purpose of the system was twofold: to reduce the cost of spaceflight by replacing the current method of launching "capsules" on expendable rockets with reusable spacecraft; and to support ambitious follow-on programs including permanent orbiting space stations around the Earth and Moon, and a human landing mission to Mars. Space Shuttle Challenger (NASA Orbiter Vehicle Designation: OV-099) was the second in its class. Its sister shuttle, Columbia, the first shuttle to commence operations, made its maiden flight in April 1981. Challenger's maiden flight was on April 4, 1983, and it completed nine missions before its fatal last mission, STS-51-L on January 28, 1986.

Challenger exploded 73 seconds after launch. All of its seven crew members were killed. The technical cause of the accident was as follows: A combustion gas leak through the right solid rocket motor aft field joint initiated at or shortly after ignition eventually weakened and/or penetrated the external tank, initiating vehicle structural breakup and loss of the Space Shuttle Challenger during STS Mission 51-L (Report, 1986). The gas leak was caused by failure in the O-rings of the booster.

The shuttle had several O-rings, made of a rubber compound, which were used to seal the solid rocket booster field joints. The solid rocket boosters are made in sections. There are two types of joints to hold the sections together: the permanent "factory joints" are sealed at the Morton Thiokol factory in Utah; the temporary "field joints" are sealed before each flight—at the Kennedy Space Center in Florida (Feynman, 1988). The O-rings measured 146 inches in diameter and were just 0.280 inch thick. Each one was molded in one piece to span the entire circumference of the booster. Each solid rocket booster had three field joints, and the shuttle had two solid rocket boosters.

The weather on launch day was exceptionally cold (36 F, $\approx 2°C$), 15 degrees lower than that measured for the next coldest previous launch. Ice had formed on the pad during the night. The durability of the O-rings had not been tested at such temperatures and worries about the effect of temperature to the ability of the O-rings to seal effectively were expressed. Thus, before the launch a teleconference was held between NASA and Morton Thiokol, where the dangers of launching at cold temperature were discussed. After the teleconference a decision to launch was made.

The Challenger accident has been investigated by various groups of people. The official investigation by the Presidential Commission on the Space Shuttle Challenger Accident (1986) found numerous rule breakings and deviant behavior at NASA prior to the accident. They also accused NASA of allowing cost and schedule concerns to override safety concerns. Many have blamed the accident on a fundamental design error in the O-rings. The resiliency of the O-rings was directly related to the temperature. The colder the ring, the slower it returns to its original shape after compression.

Vaughan (1996) shows in her analysis of the same accident how most of the actions that employees at NASA conducted were not deviant in terms of the culture at NASA. She also shows how safety remained a priority among the field-level personnel and how the personnel did not see a trade-off between schedule and safety. They perceived the pressure to increase the number of launches and keep the schedule as a matter of workload, not a matter of safety versus schedule (Vaughan, 1996). According to her analysis, the decisions made at NASA from 1977 through 1985 were "normal within the cultural belief systems in which their actions were embedded" (Vaughan, 1996, p. 236). Vaughan (1996, pp. 409–410) summarizes: "The explanation of Challenger launch is a story of how people who worked together developed patterns that blinded them to the consequences of their actions. It is not only about the development of norms but also about the incremental expansion of normative boundaries: how small changes—new behaviours that were slight deviations from the normal course of events—gradually became the norm, providing a basis for accepting additional deviance. No rules were violated; there was no intent to do harm. Yet harm was done." The organization gradually drifted to a state in which it no longer operated safely. Earlier danger signals had become part of "normal" work and they were no longer noted.

28.2 ORIGIN AND HISTORY

Since the concept of safety culture was coined after the Chernobyl accident in 1986 we shall provide some more details about this event followed by a discussion about the concepts of organizational culture and organizational climate: two broad concepts that have influenced theory and practice of safety culture and safety climate.

28.2.1 The Chernobyl Accident

The Chernobyl accident was the worst nuclear power plant accident in history in terms of costs and casualties. It is one of the only two nuclear accidents classified as a level 7 event (the maximum classification) on the International Nuclear Event Scale (INES). The other is the Fukushima Daiichi nuclear disaster in 2011. The accident occurred on April 26, 1986 in the Chernobyl Nuclear Power Plant situated in the village of Pripyat in the former Soviet Union (present day Ukraine).

The accident occurred during a test aimed at evaluating how long the turbines would spin and supply power to the main circulating pumps following a loss of main electrical power supply. The test required shutting down some safety systems (e.g., automatic shutdown mechanisms) and to operate the plant at a lower power than was normal. The test went awry and the operators tried to execute an emergency shutdown of the plant by lowering all the control rods into the reactor. However, this action caused a dramatic power surge due to the inherently unstable nature of the reactor.

IAEA's International Nuclear Safety Advisory Group (INSAG) reviewed the evidence concerning the accident and concluded that the "root cause of the Chernobyl accident … is to be found in the so-called human element" (IAEA, 1986, p. 76), and that "formal procedures must be properly reviewed, approved and supplemented by the creation and maintenance of a 'nuclear safety culture'" (ibid., p. 77). The report offered little definition of what this "nuclear safety culture" could entail, but since many references were made to procedure violations as causal factors in the report, adherence to formal procedures seems to be an important aspect of the report's notion of safety culture. INSAG updated its initial 1986 accident review in 1992 as a result of new information concerning the course of events. For example, it turned out that some of the so-called "procedure violations" were actually approved by the Chief Engineer and was part of the test procedure. Also, some of the hazardous actions made during the accident sequence, and first interpreted as procedure violations by Soviet experts, turned out to have been made without any existing procedural guidance. Thus, the "violations" were probably made without operators' knowledge that they were actually violating something.

The accident turned out to be more systemic in its causes than was envisioned in the first IAEA meeting, and deficiencies in "nuclear safety culture" now seemed an even more reasonable explanation in comparison with when the concept was first introduced. The new information demonstrated lack of feedback of operating experience and problems in communication between designers, engineers, manufacturers, constructors, operators, and regulators. These deficiencies, coupled with a lack of clear lines of responsibility, deference to authority, insufficient procedures, and operators' insufficient understanding of the nuclear hazards and the inherent weaknesses in the RBMK reactors, were found to be critical contributing factors behind the accident. It was also noted that the lessons from the Three Mile Island accident had not been acted upon in the USSR.[2] One can also argue that due to a prevalent belief in the safety of RBMK reactors in the USSR, the emergency preparedness was insufficient. This lack of preparedness for the worst was also illustrated 25 years later at Fukushima accident in Japan (and was also evident to some degree at Three Mile Island, 7 years prior to Chernobyl).

28.2.2 Organizational Culture and Organizational Climate: The Broader Context

In order to further trace the history of safety culture and safety climate we will briefly address the more general terms of organizational culture and organizational climate without the prefix of "safety." Particularly, the concept of organizational climate shall be addressed because there is often confusion about how the concept of climate relates to that of culture. This debate still prevails in some writings about safety culture and safety climate.

Developing a sense of meaning in an organization is supported by many different factors (activities, policies, structures, working conditions, etc.) which then are shared

[2] Many of these contributing factors seem to echo those of the Three Mile Island accident, including the lack of competence of the operators, lack of operating experience feedback concerning similar previous incidents, problems in human–machine interface and inadequate procedures for dealing with emergency situations. Hauntingly, some of these same elements can again be found in the Fukushima accident of 2011; lack of adequate emergency procedures, belief in the inherent safety of nuclear power, inadequate operating experience feedback (cf., the Onagawa example in the main text).

and integrated among the organizational members (Schneider, 1985). Underneath the concept of organizational climate, a main research focus has been to investigate how individuals perceive different meaning shaping factors found in their social work context. The concept of organizational culture, on the other hand, has been applied more to investigate deeper-level basic values and assumptions as part of meaning creation in organizations. Different research strategies follow from this difference in research focus.

Investigations of organizational climate often focus on how individuals draw conclusions based on what they directly perceive in their organizational context. The idea is that such conclusions become shared among the members to constitute a particular organizational climate (Schneider, 1975). This climate then influences individual perceptions of various arrangements in the organization (including those that are associated with safety). The concept of organizational culture goes deeper, below the perceptions that the employees share. It tries to explain the dynamics among individuals and the system as a consequence of shared assumptions, values, and norms that exist and emerge among the members in an organizational setting. The result of this cultural dynamics is partly expressed in symbolic form (Alvesson, 2002) but also as shared cognitive schemas used to interpret the world. It affects how people think, feel, and act (Schein, 1985) in the long term. Climate denotes the surface features of the culture at a given point in time (cf., Flin et al., 2000). Culture is more stable, tacit, and taken for granted than climate.

28.2.3 Safety Climate

The first study of safety climate was made by Zohar (1980) who distinguished a set of factors which he found reflected individual's perception of different safety-related issues (i.e., their relative importance). For example, among the factors identified were the importance of safety-related training, management's attitudes to safety, and the status of safety committees. After Zohar's pioneering work, many psychometric studies have attempted to identify generic safety climate factors by use of questionnaires and factor analysis (Guldenmund, 2007; Zohar, 2010). A frequently found factor in these studies is management commitment (e.g., Cheyne et al., 1998; Cohen, 1977; Dedobbeleer and Beland, 1991; DeJoy et al., 2004; Donald and Canter, 1994; Hofmann and Stetzer, 1996; Ostrom et al., 1993; O'Toole, 2002; Rundmo, 1996; Rundmo and Hale, 2003; Seo et al., 2004; Simonds and Shafari-Sahrai, 1977; Smith et al., 1978; Zohar, 1980; Zohar and Luria, 2005). In a review study by Flin et al. (2000), the most common factors identified were "management/supervision," "safety systems," and "risk." Also, Seo et al. (2004), in their review of previous studies found a core of generic safety climate concepts; management commitment to safety, supervisor safety support, co-worker's safety support, employee participation in safety, and competence. Despite the long research tradition already existing in the area of safety climate, safety culture development in the 1980s and 1990s drew more on the organizational culture literature than the safety climate literature.

28.2.4 Organizational Culture and Safety Culture

Studies of organizational culture (without the prefix "safety") had already begun in the 1970s and they accelerated in the 1980s. The soil was thus well grounded when the concept safety culture appeared in the late 1980s. The roots to the focus on organizational culture can be found in several disciplines from as early on as the 1930s. Among those are symbolic interactionism and social constructivism (Mead, 1934; Berger and Luckman, 1966) and the Tavistock paradigm (Rice, 1958; Miller and Rice, 1967).

The concept of organizational culture has been approached from many perspectives. One such perspective departs from the history of an organization and how this history is reflected, and embedded in assumptions, values, and practices. A second perspective focuses on culture as a variable among others that can be manipulated and changed (i.e., a functionalistic perspective). A third perspective conceptualizes assumptions, beliefs, and values as shared cognitive schemas that are used to interpret the world and act in it. These perspectives are also used in the context of safety culture.

Sociologically oriented researchers also have had an impact on the safety culture and safety climate development. For example, Vaughan's (1996) influential study on the 1986 *Challenger* Space Shuttle accident is cultural in its main approach. Even though she does not use the concept of safety culture, her work provides an interesting view on the effects of culture on safety through an in-depth accident investigation. She recognized several cultural patterns in NASA's and its subcontractors' organizations—cultural patterns that contributed to the accident. For example, the "normalization of deviance" refers to a process where small changes—new behaviors, technical/physical/social anomalies, or other variations that represent slight deviations from the normal course of events—gradually become the norm (Vaughan, 1996). Normalization of deviance produces disregard and misinterpretation—neutralization—of potential danger signals. A signal of potential danger is information that deviates from expectations, contradicting the existing worldview (Vaughan, 1996, p. 243). This is clearly a cultural phenomenon and Vaughan (ibid.) indeed refers to the concept of culture quite often in her analysis. She defines culture as a "set of solutions produced by a group of people to meet specific problems posed by the situations that they face in common" (ibid., p. 64). She further clarifies that the solutions are institutionalized as rules, rituals, and values (ibid., p. 64).

Most cultural approaches today acknowledge the existence of at least three types of subcultures: (a) occupational or professional subcultures based on educational background of the employees, (b) departmental subcultures based on the work unit in question, and (c) age- or tenure-related subcultures (see e.g., Parker, 2000). The existence of subcultures in organizations is something that can create conflicts and problems in communication but also positive differentiation and diversity, as has been highlighted by several researchers (Cooper, 2000; Mearns et al., 1998). Connected to this is the question whether grand organizational cultures really exist. For example, Jones and James (1979) and Rollenhagen et al. (2013) found that different professional groups had developed subcultures related to the type of work they did (professional identity) rather than what organizational unit they belonged to. A subculture can be defined in terms of professional groups that transcend a specific organizational boundary (medical doctors, pilots, etc.) or it can be defined in terms of groups according to an organizational chart, or it can be defined as a specific focus domain among others (safety culture, innovation culture, service culture, etc.) or even in terms of things such as age (youth culture). Thus, any organization can be considered a melting pot of multiple subcultures interacting simultaneously.

Such findings point to the importance of being sensitive to what people actually do in the organization as a determinant of organizational culture in contrast with perceiving an organization as an a priori coherent cultural system (Richter and Koch, 2004). However, some order and coherence is to be expected in organizational settings. Research points out that humans tend to form groups and social identities develop in the context of groups (Haslam et al., 2011). An integral part of a social identity is a social comparison to relevant out-groups ("them"). What is defined as a relevant out-group is then important for the social identity of the in-group. In terms of safety, it would be better to compare one's own group to organizations that have a good safety culture and good past safety record. This should raise a desire to develop one's own organization further. However, there is a natural tendency to compare one's group only to those groups that are perceived inferior. This is good for the in-group self-esteem, but it does not benefit learning. Rather, it assures and reinforces the existing identity.

As illustrated above, the general concepts of organizational culture and organizational climate are in many ways intermingled. This also holds for the concepts of safety climate and safety culture. Both concepts highlight human-oriented factors such as perceptions, attitudes, and values, but they do this somewhat differently. Both culture and climate refer to something that is shared among the organizational members; they both are collective properties of a group or an organization.

28.3 DEFINITIONS AND TERMINOLOGY

The concept of "safety culture" was for the first time defined in IAEA (1986) "Summary Report on the Post-Accident Review Meeting on the Chernobyl Accident" as "that assembly of characteristics and attitudes in organizations and individuals which establishes that, as an overriding priority, nuclear plant safety issues receive the attention warranted by their significance." Several other definitions have followed. The UK Health and Safety Commission (HSC, 1993) defined safety culture as follows: "The product of individual and group values, attitudes, perceptions, competencies, and patterns of behaviour that determine the commitment to, and the style and proficiency of, an organisation's health and safety management." Pidgeon (1998), in turn, defines safety culture as the "set of assumptions, and their associated practices, which permits beliefs about danger and safety to be constructed". Clarke

(1999) sees safety culture as a "subset of organizational culture, where the beliefs and values refer specifically to matters of health and safety". Richter and Koch (2004) in their definition emphasize shared meanings, experiences, and interpretations of work and safety. Hopkins (2005, p. 22) identifies the concept of safety culture with the concepts of mindfulness and risk awareness and argues that these are "largely interchangeable." Based on his extensive review of past research, Guldenmund (ibid., p. 251) adds his own definition of safety culture: "those aspects of the organisational culture which will impact on attitudes and behaviour related to increasing or decreasing risk." Common to most definitions is that they emphasize safety culture as something holistic and shared among the members of the analytic unit in focus.

It is important to emphasize that safety culture, similar to organizational culture, is not intended to be an individual trait, but rather an organizational collective property in the meaning of what individuals share with each other in terms of behavior, attitudes, values, etc. Thus, individuals do not have a safety culture separated from the social context. As a simplification, culture can be referred to as the "personality" of an organization (Hofstede, 1991; Schein, 2010, p. 14), its natural way of being and behaving (interaction, communication, climate, norms, beliefs, values, attitudes, etc.). Schein (2010, p. 14) notes that "just as our personality and character guide and constrain our behavior, so does culture guide and constrain the behavior of members of a group through the shared norms that are held in that group."

This also means that culture is something real. Thus, instead of being a management principle or tool that is applied or a model that is implemented, culture can be perceived as something that exists in the organization, something that the organization has created and continues to create. Culture then affects all the members of the organization, how they see, think, feel, and act in relation to safety.

Today we find large variance in conceptualizations of safety culture, ranging from descriptive studies on the social construction of safety to normative models of ideal safety culture/climate dimensions. By and large, the models of safety culture bear more resemblance to the functionalist theories of organizational culture than to the interpretative ones (Richter and Koch, 2004; cf., Smircich, 1983).

A consensus seems to have been reached that safety culture is a multidimensional construct (Guldenmund, 2000). However, the exact dimensions comprising safety culture have been debated and no agreement has been reached. Meta-analyses of questionnaire studies have identified management commitment to safety as the most common factor (Flin et al., 2000; Flin, 2007). Other widely utilized dimensions of safety culture include communication, competence, resources, quality of rules and instructions, personnel's risk perceptions, organizational learning practices including reporting of near-misses, and the safety management system.

A comprehensive critique of the various conceptualizations of safety culture can be found in Reiman and Rollenhagen (2014). The main argument of the critique is that safety culture discourse has largely failed to integrate itself with the more technically focused safety approaches and other safety considerations in order to capture a holistic view of the sociotechnical system. This critique is elaborated in Section 28.7 of this chapter.

An emerging consensus among many safety culture and safety climate researchers is that ideally it would be better to speak about *cultural influences on safety* rather than to portray safety climate/culture as isolated from its broader organizational cultural context (Grote and Künzler, 2000; Hale, 2000; Haukelid, 2008; Hopkins, 2006). However, as the concept of safety culture is well established in safety management it makes sense to use the concept to highlight the special nature of culture in safety-critical organizations.

28.4 UNDERLYING THEORY AND THEORETICAL ASSUMPTIONS

28.4.1 Some Common Features of Safety Culture Models

Despite disagreements on the exact definition of the safety culture as a concept, we can identify several features of safety culture that are widely considered as important. For example, top management's commitment to safety has been heavily emphasized (Flin, 2003; Guldenmund, 2007; HSE, 1997; IAEA, 1991; Reason, 1997). On the other hand, in organizations where management is not committed to safety, disobedience and whistleblowing by the staff could be a sign of good safety culture at a sub-group level (cf., Sagan, 1993). Thus, a questioning attitude (IAEA, 1991), awareness of safety and hazards (Reiman et al., 2012), openness and trust (Reason, 1997), and vigilance/mindfulness (Hopkins, 2005) are also commonly considered signs of a good safety culture. Finally, it is emphasized that safety needs to be integrated into all activities of the organization and not perceived as a separate task (IAEA, 1991; Grote and Künzler, 2000, p. 134).

A good safety culture does not mean that everyone agrees with everyone else or that different opinions about, for example, safety matters always constitute a risk. In fact, too uniform a culture may become blind to its own weaknesses and seek to find corroboration for the old and familiar opinions (Sagan, 1993; Weick, 1998). For the above-mentioned questioning attitude to work, the organizational climate should be of a kind that allows for open discussions of potential hazards, possible countermeasures, and their effects on production, public image, etc. An effective safety management system is also considered both a sign of good safety culture as well as a necessary precondition for creating a healthy safety culture to the company.

In addition to safety culture studies, certain important insights on the criteria for good safety culture can be gained from the High Reliability Organization (HRO) group, and especially the work of Karl Weick and Kathleen Sutcliffe. They emphasize that organizations that wish to operate reliably have to rely on the expertise of the field, be reluctant to simplify things, put expertise over rank, have a preoccupation with the possibility of failure and thus spent resiliently effort to improve and avoid failures (Weick and Sutcliffe, 2007).

28.4.2 Theoretical Frameworks

A common theoretical framework that researchers and practitioners alike proclaim as an underlying view of safety culture is Edgar Schein's (1985) model of organizational culture.[3] Culture has, according to Schein (ibid.), several levels, with artefacts (visible behavior, technology, work environment, rules and instructions, etc.) presenting the surface level of culture, followed by espoused values (what the people say, slogans, mission statements, justifications of behavior). Schein has defined the deepest layer of organizational culture as a pattern of basic assumptions that a group has invented as it has learned to cope with its problems related to adapting to its environment and integrating the group into a functioning whole. This pattern of basic assumptions has worked well enough to be considered valid, as it is taught to new members of the organization as the correct way to perceive, think, and feel in relation to those problems (Schein, 1985). Such basic assumptions are largely taken for granted but they have an effect on the way people perceive, think, and feel about their organization, including its core task and the various hazards associated with it.

Examples of the issues where the effects of culture manifest include how risks are evaluated and managed, the way the current safety level is interpreted, and what is considered unacceptable behavior. Safety culture also contributes to defining formal and informal norms concerning, for example, safe behavior or proper workplace conduct. Culture "defines" what is normal to a group, and safety culture thus affects what is considered normal work perceived from safety perspective, how work should be carried out, and what the potential warning signals are.[4]

Basic assumptions concern, for example, issues of time perspective, the nature of the company's relationship with its various stakeholders, appropriate ways of responding to critique, criteria for rewarding and punishing people. In safety-critical fields basic assumptions may concern issues such as what constitutes risk, the main hazards the organization is coping with, how one is supposed to speak about risks in the workplace, and what is perceived as a valid warning signal. Further assumptions may concern issues such as the best ways of improving safety, reasons why people make errors, how to deal with the errors once they occur, what is expected of a professional in the organization, and how persons gain respect and/or power in the organization.

The main notion here is that culture is something the company creates for itself, and which, once created, has an effect on the company. This effect is in many ways not perceived by the company itself, since the members of the organization consider all things that happen according to cultural assumptions "business as usual." Culture

[3] Although Schein himself is very critical of the concept of safety culture, see, for example, Reiman and Rollenhagen(2014).

[4] Depending on terminology, culture is sometimes seen as defining certain phenomena or being itself composed of those phenomena. For example, some theories treat the way of working and the organizational practices as being part of the organizational culture, whereas other theories see culture as constraining and affecting those practices.

is repeatedly created and recreated as members behave and communicate in ways that to them seem natural, obvious, and unquestionable and as such contribute to a particular version of risk, danger, and safety (Pidgeon, 2012, pp. 20–21). Culture is a result of shared learning experiences that affects how the group will learn in future (Schein, 2010).

Schein (2010, pp. 16–17) identifies the defined characteristics of culture as "structural stability," "depth," "breadth," and "patterning or integration." Structural stability implies that culture is not only shared but it defines the group's identity, and survives even when some members of the organization depart. Depth refers to the fact that culture denotes the deepest and often unconscious aspects of a group. The essence of culture is invisible and intangible, yet it is tangible in its manifestations. Once culture forms, it covers all of a group's functioning (breadth). Finally, patterning or integration means that culture ties various elements, such as rituals, climate, values, and behavior, together into a coherent and sensible whole (cf., Weick, 1995).

Another way of conceptualizing culture is to view it as a "root metaphor" for the organization itself (Alvesson, 2002; Smirhich, 1983). According to this conceptualization, culture is embedded in the social processes and practices of the organization. It is thus not an element that can be considered, analyzed, and evaluated independently of its context, since culture *is* the context. This is called the interpretive approach to culture.

A middle position, and one that has gained increasing popularity, is to treat safety culture as one element (or variable) within the larger system, but at the same time acknowledging the need to look at other elements and their interaction. In this view, culture is not everything, but it is something that is qualitatively different from other elements of a sociotechnical system. This approach treats safety culture as a contextual phenomenon but also differentiates it from its context, such as the technology, tools, procedures, and the personnel as individuals (Antonsen, 2009a, Guldenmund, 2010; Myers et al., 2014; Nævestad, 2009; Reiman and Rollenhagen, 2014). This view makes it possible to inspect and evaluate safety culture but also treat sociotechnical systems more analytically as comprising qualitatively different elements. This can be considered an interpretive-realist framework of culture.

Many scholars have stressed the importance of treating safety culture as a contextual phenomenon, a property of the organization that is connected to other characteristics in a complex web of interdependent causalities (Tharaldsen and Haukelid, 2009). The deep core of culture, shared values, beliefs, and assumptions not only relate to each other but also influence practices, structures, personnel, and technology which, in turn, may have a causal influence on each other as well as on the shared assumptions.

Guldenmund (2007) argues that within organizations three major forces are operating at the same time on (the behavior of) the people who work there. These generic forces are structure, culture, and processes and they are dynamically interrelated, that is, the particular strength of each force is determined by the other two. Antonsen (2009a) differentiates between structure, culture, and interaction, preserving the concept of culture for values and attitudes. He views culture as "the frames of reference through which information, symbols and behaviour are interpreted and the conventions for behaviour, interaction and communication are generated." Antonsen (2009a) emphasizes that safety culture relates to the informal aspects of an organization, whereas the formal aspects are part of the structure of the organization. Myers et al. (2014) propose a clearer delineation of culture, that is, the socially constructed abstract systems of meaning, norms, beliefs, and values from concrete behaviors, social relations, and other properties of workplaces (e.g., organizational structures) and of society itself.

In a cultural framework, the social world is seen not as objective, tangible, and quantifiable but as constructed by people and reproduced by the networks of symbols and meanings that people share and make shared action possible (Alvesson, 2002, p. 25). This means that even the technological solutions, production technologies, and other man-made structures are given meanings by their designers and users, which affect their subsequent utilization (and subsequent reinterpretation of this meaning).

It further means that concepts such as safety, reliability, sustainability, or cost-effectiveness are not predefined and set in stone. Rather, organizations construct the meaning of these concepts and act in accordance to this constructed meaning. For example, if the organization socially constructs a view that the essence of assuring safety is to prevent individuals—considered as the weakest links in the system—from committing errors, safety management is likely to focus on individuals and include measures such as attitude training, demotion, and blaming. However, we should not fall in the trap of misplaced social constructivism. Hazards are real and not only social constructions (which can also be considered "real," but in a qualitatively different way) although we develop different ways to think, speak, perceive, and act about them.

Figure 28.1 illustrates culture and its relation to structure and behavior. As the figure illustrates, culture is created and maintained by behavior but it also influences behavior. Structures, such as organizational structure, management system,

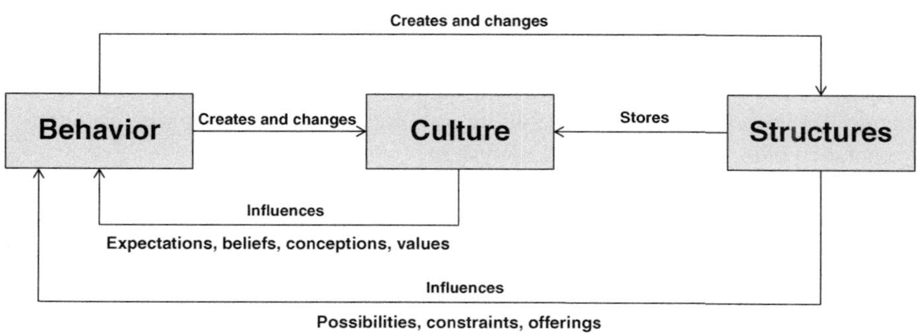

Figure 28.1. The iterative nature of culture.

or production technology, in turn store elements of culture. Structures provide constraints and possibilities for behavior. Culture can be changed by changing both behavior and structures. Behavior is naturally influenced by many other situational and contextual factors. Structure can also be analyzed in more detail, and, for example, organizational processes (Guldenmund, 2007), social relations (Myers et al., 2014), and interaction (Antonsen, 2009a) could be included in the model, but these and other factors (e.g., external environment) have been omitted to highlight the interaction of three central elements: behavior, culture, and structure.

28.5 EMPIRICAL RESEARCH

Safety culture has been a topic of extensive research. Surprisingly few predictive or longitudinal studies exist though (for an overview of predictive studies in health care, see Itoh et al., 2014). Most studies have used past or concurrent safety performance as validation criteria (cross-sectional study design). The reader is referred to reviews by Guldenmund (2000, 2007, 2010), Flin et al. (2000), Sorensen (2002), and Choudhry et al. (2007) for summaries of safety culture and climate studies. Here we shall mention only a few recent studies which we find particularly relevant for the present purposes. We have omitted in this article the large body of empirical research focusing on the content validity and operationalization of the safety culture concept. This branch of safety culture research was briefly addressed in Sections 28.3 and 28.4. Here the focus is on empirical research addressing the predictive validity of safety culture. Studies of safety culture improvement/change are addressed in Section 28.6.

Safety culture studies have often tried to link the safety culture concept to various negative outcomes such as occupational injuries, adverse events, or process disturbances. Using these unwanted outcomes to validate the link between safety culture and safety is problematic in two ways (Reiman and Rollenhagen, 2014): first, if safety culture is only one variable among others in a sociotechnical system, it cannot sensibly be expected to correlate with safety without measuring or controlling the influence of the other variables. Second, using incidents as a measure of safety neglects important details about the nature of safety as a dynamic and emergent property of the functioning of the entire sociotechnical system (Reiman and Pietikäinen, 2012; cf., Itoh et al., 2014). Still, studies using incidents or other negative events as outcome variables can provide valuable information about the relation between the different elements of the sociotechnical system.

Morrow et al. (2014) studied the relationship between safety culture and other indicators of safety in the US nuclear power industry. Their questionnaire study included 2876 responses from 63 nuclear power plant sites. The relationships between safety culture and safety performance were more consistent when safety performance was measured concurrently, as opposed to being measured 1 year after the survey administration. However, some interesting correlations were found. Organizations where employees perceived less of a questioning attitude were more likely to receive higher numbers of allegations from the Nuclear Regulatory Commission in the same year and the following year. Moreover, organizations with lower overall scores on the safety culture survey were more likely to have higher counts of unplanned scrams, and have inspection findings related to inadequacies in problem identification and resolution.

Bergman et al. (2014) studied safety climate as both a leading (climate → incident) and a lagging (incident → climate) indicator of safety-critical incidents. They conducted a survey at a large, multinational chemical manufacturing company, with 7467 responses at 42 worksites in 12 countries linked to over 14,000 incident records during the 2 years prior and 2 years following the survey period. Their main finding was that safety climate predicts incidents of varying levels of severity, but it predicts the most severe incidents over the shortest period of time (cf., Reiman et al., 2013).

Neal and Griffen (2006) identified a relationship between safety climate and subsequent safety-related behaviors and occupational injuries in an Australian hospital. Antonsen (2009b) found in his study of offshore drilling platform before and after an accident happened there that a pre-accident safety culture questionnaire failed to detect safety issues that were identified after the accident. Reiman et al. (2013) found that in a hospital setting, working conditions, management of contractors/external parties (e.g., leased doctors), and management safety leadership correlated with the amount of adverse events, measured independently 16 months after the safety culture survey.

28.6 USE AND IMPLEMENTATION

The concept of safety culture has been used in many different ways in various contexts and for various purposes, as already stated in the introduction. In this section, we illustrate some of the recommended uses of the concept. In Section 28.7 we will point out some misuses of the concept. Before elaborating on these different uses, it should be pointed out that contrary to some of the other safety principles presented in this book, safety culture is a phenomenon and as such it does not need implementation into the organization. Rather, the safety culture that the organization in any case has needs to be developed, maintained, changed, and managed to guarantee optimal safety.

28.6.1 When and Where to Use the Concept?

One reason for using the concept of safety culture is in terms of a leading indicator of the safety level of the organization. That is, it is hoped that by monitoring changes in safety culture, organizational deficiencies can be identified and corrected before they actualize as incidents or accidents. This requires a good model of safety culture as well as its relation to other elements of the sociotechnical system.

Another related use is to use the concept as an analytical tool for safety management. The factors often included as being part of safety culture opens for discussions that otherwise may not have an explicit place in organizations. Safety culture can, despite its often-mentioned definition problems, serve as an umbrella for discussions of safety and can be used to address phenomena that people intuitively may be aware of but which have not been addressed explicitly before.

28.6.2 Safety Culture as an Evaluation Framework

Many international agencies in the nuclear sector such as IAEA (International Atomic Energy Agency), WANO (World Association of Nuclear Operators), and INPO (Institute of Nuclear Power Operations) have produced lists of safety culture characteristics or attributes to be used by the power companies as well as regulators as basis for developing and auditing safety culture. For example, IAEA (2016) defines five characteristics of (good) safety culture: (1) safety is a clearly recognized value, (2) leadership for safety is clear, (3) accountability for safety is clear, (4) safety is integrated into all activities, and (5) safety is learning driven. Similar lists of attributes, or normative criteria, have been developed also in other safety-critical fields.

One of the main reasons for the relevance of the safety culture concept for management of modern organizations can be seen against the observed tendency of organizations to gradually drift into conditions where they have trouble identifying their vulnerabilities and the practices that create or maintain these vulnerabilities. Safety science has illustrated that accidents rarely occur as a result of a single malfunction or error, or an entirely new, one-time threat to safety (Hollnagel, 2004; Rasmussen, 1997; Vicente, 2004). Rather, accidents more often seem to be a consequence of "systematically induced migration in work practices combined with an odd event or coincidence that winds up revealing the degradation in safety that has been steadily increasing all the while" (Vicente, 2004, p. 276). A main point here is that due to, for example, cultural blind spots this degradation can remain undetected and no remedial actions are taken before an accident occurs.

Karl Weick (1998) has emphasized that organizations are in fact defined more by what they ignore than by what they attend to. He points out that a strong culture can also be a safety risk if the culture provides strong social order that encourages, for example, the compounding of small failures (ibid.) or uniformly neglects some critical safety considerations. Safety cultures are thus determined as much, if not more, by what they ignore as by what they pay attention to and what they consider important and meaningful. Unfortunately, often only an accident wakes an organization to realize the risks that it had been ignoring.

Table 28.1 provides examples of a declining, or unhealthy, safety culture (see, e.g., DOE, 2009; Hopkins, 2005; IAEA, 2002; Oedewald et al., 2011; Snook, 2000; Vaughan, 1996; Weick and Sutcliffe, 2007;).

TABLE 28.1. Signs of an Unhealthy or Declining Safety Culture

Safety is given low priority in decision-making concerning, e.g., investments or production issues. Short-term profits dominate the management attention.
Safety is considered to be the safety organization's responsibility or something that is done only for the regulator. Safety function does not have the authority to influence operative or strategic decisions.
Safety is equaled to technical reliability or component quality. Human influence to safety is seen mainly as a source of errors, or it is not considered at all.
Organization sticks to old categories and dismisses new events as "one-off cases" and new evidence as "false" or not convincing enough. New events are interpreted in line with the typical causal attributions used in the past. The same standard solution is applied to every problem.
The organization sees future as relatively predictable repetition of the past. There are few contingency plans.
Many warning signals, alarms, or deviations are ignored since they recur so often; they have become normalized and considered part of everyday life.
The organization glorifies its past and sees itself as already having developed all the risk control measures it needs. Safety is seen as a static state that has already been accomplished.
In the organization there are a lot of "holy cows" and "taboos" that are not open for discussion.
Individuals and departments optimize their own performance without considering how that affects the overall performance of the organization.
The safety management system exists only on paper and does not guide actual practices. Procedures and instructions are not followed in real life.
The organization is in a fire-fighting mentality. All time goes to taking care of issues that are considered acute, leaving no time for developing activities or even normal daily work.
Personnel feel they have too much work and report stress. Sick leaves are higher than average in number and longer in duration than the average.
Organization is condescending to its negative conditions and does not seek to improve them.
There is low tolerance for uncertainty and a strong focus on finding permanent solutions to ambiguous issues. Discussions without immediate corrective actions are discouraged.
There is a fear of punishment if one reports a safety concern, human error, near-miss, or an incident. This leads to fewer reports of near-misses compared to actual incidents.
The organization isolates itself from the environment, does not share information with others, or seek to learn from others. Organization's self-image can be overly positive.
Organization's internal dynamics create vicious circles where small problems or arguments escalate quickly into major problems and conflicts.
The boundary between acceptable and unacceptable behavior is blurry. Personnel and managers may behave erratically and even hostile toward others, and may break rules without the organization considering this unacceptable.

To understand why an organization exhibits signs of a weak safety culture, a systemic theory is needed that can analyze the interplay between various sociotechnical factors (Reiman and Rollenhagen, 2014). This theory should define what can be considered signs of a weak, or weakening, safety culture and what signs imply weaknesses in other elements important to safety.

28.6.3 Developing Safety Culture

Studies of culture change in the safety field are scarce (Hale et al., 2010). It has also been argued that the vagueness of the safety culture concept makes it difficult to translate into change efforts (Nielsen, 2014, p. 7). It is difficult, if not impossible, to influence safety culture directly. Rather, improvement efforts are directed to the manifestations of culture such as organizational structures, practices, or tools with an overall goal of changing the culture via its manifestations. It has also been proposed that safety climate offers "an access" to safety culture, climate being more superficial and more easily influenced than culture (Nielsen, 2014, p. 8).

As in the safety culture literature in general, the role of management is emphasized in safety culture change and improvement (see also Schein, 2010; Zohar, 2002). A related term of "leadership" is sometimes used to denote not only official managers but also other influential people in the organization. Schein (2010, p. 3) calls leadership and culture "two sides of the same coin". Kello (2006) points out that in order to change safety culture, the safety professionals need to redefine their role more broadly as facilitating organizational transformation instead of narrow spot changes. This reflects the growing recognition of the complexity of modern organizations and realization of the limits of traditional top-down management models (Reiman et al., 2015). This view of an organization draws on lessons from complexity science and complex adaptive systems, and instead of structures pays attention to interactions and emergent properties (McMillan, 2008). So far, the applications to safety culture development have been few.

Nielsen (2014, p. 8) argues that "by combining theories on organizational culture, (safety) climate and complex adaptive systems, it is feasible that cultural change can be created by changing the patterns of interactions between organizational members." He shows in his study of a Danish industrial plant how creating more and better safety-related interactions also affected safety climate and culture.

In the studies of safety and safety culture improvement, there has been a lot of discussion about the relation of behaviorally oriented and more structurally oriented development approaches. Often a difference is made between behavioral and cultural improvement programs. Behavioral strategies represent a more direct focus on individual behaviors and human performance whereas cultural strategies focus more on values and attitudes, as well as various structural arrangements in the environment (DeJoy, 2005). Some behavioral-based safety approaches treat safety culture as an individual level behavioral issue and neglect other considerations such as collective norms, competence, or shared conceptions of safety (Reiman and Rollenhagen,

2014). Hopkins (2005) also points out how safety culture has been misused in the industry to denote individual attitudes instead of organization-specific phenomenon. These approaches then dismiss the discussion of differences between safety culture and behavior by essentially treating them as the same thing. Authors such as Tharaldsen and Haukelid (2009) have argued for a "balanced strategy" where both types of interventions are needed, behavioral and cultural.

Safety culture can also be used as a tool for reflection. For example, the influence of shared assumptions to communication can be considered through cultural lenses; many problems in communication stem from differences in cultures, even inside organizational subcultures. These problems may remain unsolved because it is hard to pinpoint the actual problem since the assumptions typically remain implicit for both parties. Just making these differences in perception visible may help in at least understanding why misunderstandings can, and have, taken place. This issue can be extended into coordination and integration of activities. Culture can act as coordinating or disintegrating, depending on its contents and level of sharedness among the members.

28.7 WEAKNESSES AND CRITIQUE

The concept of safety culture can be criticized on many grounds. A main critique targets the fact that despite several decades of research there is no consensus on what the concept actually denotes. We have previously criticized the concept for its lack of integration with classical engineering principles and concepts as well as for failing to facilitate a systems view on safety (Reiman and Rollenhagen, 2014; Rollenhagen, 2010). This is due to several reasons, for example, the neglect of emergent system phenomena, often undefined nature of safety and focus of research on the internal dimensions of safety culture instead of the context where this "safety culture" is supposed to exist (i.e., the other aspects of the sociotechnical system). Next, we will elaborate on a few points that can be considered weaknesses in the various approaches to safety culture.[5]

Safety culture theories treat organizations in simplistic terms. The reality of organizational life is usually very different from that described in formal documents. This is natural in all social contexts and not necessarily a bad thing. Few safety culture approaches share this view however. The underlying assumption rather seems to be that the prescribed logic of the organization has to coincide with the logics of practice in the field. There are several challenges in this view from systems point of view. The search for deviations from the prescribed logic of the organization may in fact

[5] It is important to note that since the approaches to safety culture vary considerably, the critique probably does not fully apply to any single theory or model of safety culture. On the other hand, it can be argued that most approaches suffer from at least some of the mentioned limitations (Reiman and Rollenhagen, 2014).

camouflage the reality since causes are attributed based on observed deviations rather than exposing contextual factors that unfold the reality of organizational activities. Safety culture risks becoming just another prescribed logic that remains disconnected from the daily reality of the organization (Reiman and Rollenhagen, 2014). For example, Schein's model of culture was originally influenced by open systems theory (Katz and Kahn, 1966) as well as the structural functionalism approach by Talcott Parsons (1951).[6] These models have recently been challenged for presenting an oversimplified view of how organizations function (Reiman et al., 2015).

The concept of safety culture has sometimes been treated as too monolithic, as denoting a total agreement on everything. These views have two weaknesses. First, they downplay the role of power in organizations—there are always power plays and people who have more influence than others. Dekker and Nyce (2014) criticize safety science in general for neglecting the role of power (see also Antonsen, 2009a). Power can have both positive and negative influences on safety. Second, strong and cohesive culture does not always mean a safe culture. On the contrary, there is growing evidence that diversity, conceptual slack, and even conflict can in many ways contribute to safety (Grote, 2007; Reiman et al., 2015; Schulman, 1993; Weick et al., 1999). As noted above, all cultures ignore issues they do not consider relevant. Strong cultures strongly agree on those issues, whereas less homogenous culture can incorporate multiple views on what is relevant and what is not.

Little discussion has been devoted to the concept of "safety" in safety culture discussion. It is rather often taken for granted that the concept of safety is self-explanatory which is far from the case. For example, there exists many different types of safety (occupational safety, process safety, etc.) and it is far from self-evident that the measures and the factors or dimensions used to explore the corresponding safety cultures are the same. Development of generic safety culture assessment tools assumed to be relevant for all types of safety and braches should, according to our view, be viewed with a skeptical stance.

The moral/ethical dimension is often implicitly interwoven in safety culture discussions but seldom explicitly so (but see, Reiman and Rollenhagen (2011, 2014), Ersdal and Aven, 2008; Hansson, 2007a,b; Kastenberg, 2014). A stronger focus on ethics in consideration of safety culture could open new areas for inquiry and a closer connection with the research of ethics.

Safety culture theories lack attention to mechanisms and processes that produce systemic meanings, including understandings of risk, safety, authority, and control (Silbey, 2009, p. 360). More interest has been put on defining the factors or criteria of good safety culture than on defining the dynamics of the safety culture itself. This is understandable given the normative and instrumental role of the safety culture

[6] Sometimes this view has led to an approach that conceptualizes safety culture as one variable among others, as one factor to be considered in a similar manner as other factors (technical, social, etc.). This is a limited view if the variables are considered independent of each other, or if these dependencies are not made explicit (Reiman and Rollenhagen, 2014).

concept as a tool for safety evaluations. However, we argue that this emphasis has alienated the concept from its roots in the scientific organizational culture literature, and more importantly, adopted an oversimplified view of cause–effect relations. It also reflects the prevalence of safety culture concept in safety evaluation rather than safety management. The safety culture concept has in some cases become a similar catch-all category as human error used to be (see also Cox and Flin, 1998; Guldenmund, 2010; Pidgeon, 1998;). To blame an organization for having a weak safety culture has become an easy response to problems. This leads us to our next critique.

Safety culture is used in situations and contexts where it is not applicable. For example, individuals do not have safety culture and individual acts may not reflect safety culture at all. Still, people are prone to make quick judgments about safety culture based on very limited evidence. On the other hand, the concept of culture does not easily lend itself to measuring by the traditional psychometric instruments. Guldenmund (2007) points out that researchers need to balance between the empirical properties of the analyzed culture (i.e., the natural unit of analysis, such as a hospital ward) and the psychometric requirements of statistical analysis (i.e., the sample size that is needed for certain analyses, e.g., to confirm the validity of the questionnaire's factor structure). This discussion relates to the debate on whether there are organization-wide cultures or whether the concept of (organizational) culture should in fact be reserved for group-level phenomena. Edgar Schein (2010) has argued that culture is a property of a group of people interacting with each other. Thus, organizations seldom have only one culture. Schein is also very critical of the idea of "safety culture" as a real phenomenon (see Reiman and Rollenhagen, 2014, p. 13). In line with the above critique it can be questioned whether there is a single safety culture (with potential subcultures) in any organization, or is there a "separate" safety culture for each domain of safety; thus we could distinguish, for example, occupational safety culture, process safety culture, environmental safety culture, and product safety culture. These cultures would naturally be partly overlapping and interacting, but could still be considered analytically distinct concepts.

In practice, the concept of safety culture sometimes tends to be used as an "excuse" for not dealing with some more fundamental safety problems involving technological design (Rollenhagen, 2010). Organizations can try to compensate their problems with bad/weak technical solutions by emphasizing only the improvement of the human component (often conceptualized as a "problem of attitude" or "problem of management"). Using the concept of safety culture in this manner can underemphasize the importance of technology and thus be detrimental to safety in the long run.

Everything is treated as a manifestation of culture. Schein (1985) makes a distinction between behavior that is an artefact of culture and behavior that stems from situational or individual variables. In essence, this means that some physical manifestations in a sociotechnical system are cultural, whereas some are not. However, the ontological difference between these two physical properties of the organization is not made clear by Schein. Schein's idea of artefacts can also be interpreted to mean that a physical object can be an artefact of some organizational culture, and not necessarily an artefact of the culture where the object currently resides. An example of this could be certain production technology used by a manufacturing company but designed by an engineering firm and embedding assumptions prevalent in the culture of that engineering firm. The relation between culture and technology is an important one and seldom made explicit in safety culture conceptualizations. The basic "production technology" can be considered as setting the constraints and requirements (Reiman and Oedewald, 2007) for the formation of culture (Schein, 2010). However, once the culture is formed the organization needs to improve, modify, and maintain its technological elements as well as its human elements.

28.8 MAIN MESSAGES AND WHAT THE CONCEPT TELLS ABOUT SAFETY

Safety culture is foremost about people's collective values, beliefs, and assumptions—it is a collective mindset that characterizes a group, not a property of any one individual. When developing or studying safety culture it is necessary to go beyond individual attitudes about safety to the level of shared thinking and the administrative structures and resources that embed ideas of what is means to be safe, and how safe we are now (cf., Pidgeon, 2012. pp. 20–21). A safety culture is built on these mostly implicit assumptions and associated practices that inform beliefs about danger and safety (Pidgeon, 2012. pp. 20–21). And, as Weick (1998) has noted, organizations are defined as much by what they (or their culture) ignores—ignorance that is embedded in assumptions. However, at the same time, it has to be remembered that safety culture is not everything; it does not reduce the importance of other perspectives on safety. Thus, safety culture approach is not intended to replace the other, more technical or administrative, perspectives. Rather, it complements them and it also challenges them to reveal their assumptions about safety. It matters what the people's conceptions about safety are. This is so because how people perceive safety and risks affects what is done and not done, what is talked about, what is communicated and subsequently what is perceived as being real.

Safety management and development of safety does not need to start from a critical event or identification of risks, but from constantly building organizational capacity for increased risk perception and risk management. This starts from a realization that safety is a long slow continuous process that needs encouragement and constant attention. The focus on organizations in often very much on how to manage the identified hazards, but safety culture thinking reminds that there also needs to be attention on the possibility of emerging or unidentified hazards that fall outside the organization's existing view on safety. To identify these emerging risks, the organization needs to have enough requisite variety in imagination and mindfulness to conceive events that have not yet taken place, but could have. Thus, when speaking about a "strong" safety culture it has to be remembered that a way of seeing is always a way of not seeing (something else), and thus "strong" should not refer to unanimously shared views and beliefs. Rather, "strong" should refer to a moral commitment about the importance of safety as a value.

Safety culture affects how the environment is enacted, what kind of causal attributions are made between discrete events, and thus how the meaning and relevance of various external and internal events is constructed. It can be argued that safety culture dictates what constitutes an "event" in the first place, since "business as usual" is not considered "eventful." However, even the business as usual shapes culture in a more subtle way by strengthening the underlying assumptions and the associated norms, practices, and values. Thus, safety culture is simultaneously a result of safety management and a force affecting the way the organization conducts its everyday work, including the management of safety and implementation of the other safety principles described in this handbook.

REFERENCES

Alvesson, M. (2002). *Understanding Organizational Culture*. London: Sage.

Antonsen, S. (2009a). *Safety Culture: Theory, Method and Improvement*. Farnham, UK: Ashgate.

Antonsen, S. (2009b). Safety culture assessment: A mission impossible? *Journal of Contingencies and Crisis Management*, 17, 242–254.

BP U.S. Refineries Independent Safety Review Panel (2007). The Report of the BP U.S. Refineries Independent Safety Review Panel. January 2007.

Berger, P., and Luckmann, T. (1966). *The Social Construction of Reality*. New York: Penguin.

Bergman, M. E., Payne, S. C., Taylor, A. B., and Beus, J. M. (2014). The shelf life of a safety climate assessment: How long until the relationship with safety—critical incidents expires? *Journal of Business and Psychology*, 29, 519–540.

Cheyne, A., Cox, S., Oliver, A., and Tomás, J. M. (1998). Modelling employee attitudes to safety. *Work and Stress*, 12, 255–271.

Choudhry, R. M., Fang, D., and Mohamed, S. (2007). The nature of safety culture: A survey of the state-of-the-art. *Safety Science*, 45, 993–1012.

Clarke, S. (1999). Perceptions of organizational safety: Implications for the development of safety culture. *Journal of Organizational Behavior*, 20, 185–198.

Cohen, A. (1977). Factors in successful occupational safety programs. *Journal of Safety Research*, 9, 168–178.

Cooper, M. D. (2000). Towards a model of safety culture. *Safety Science*, 36, 111–136.

Cox, S., and Flin, R. (1998). Safety culture: Philosopher's stone or man of straw? *Work and Stress*, 12, 189–201.

Dedobbeleer, N., and Beland, F. (1991). A safety climate measure for construction sites. *Journal of Safety Research*, 22, 97–103.

DeJoy, D. M. (2005). Behavior change versus culture change: Divergent approaches to managing workplace safety. *Safety Science*, 43, 105–129.

DeJoy, D. M., Schaffer, B. S., Wilson, M. G., Vandenberg, R. J., and Butts, M. M. (2004). Creating safer workplaces: Assessing the determinants and role of safety climate. *Journal of Safety Research*, 35, 81–90.

Dekker, S. W. A., and Nyce, J. M. (2014). There is safety in power, or power in safety. *Safety Science*, 67, 44–49.

DOE (2009). *DOE Standard. Human Performance Improvement Handbook. Volume 1: Concepts and Principles* (DOE-HDBK-1028-2009). Washington, DC: U.S. Department of Energy.

Donald, I., and Canter, D. (1994). Employee attitudes and safety in the chemical industry. *Journal of Loss Prevention in the Process Industries*, 7, 203–208.

Ersdal, G., and Aven, T. (2008). Risk informed decision-making and its ethical basis. *Reliability Engineering and System Safety*, 93, 197–205.

Feynman, R. P. (1988). "What Do You Care What Other People Think?": Further Adventures of a Curious Character. W.W. Norton.

Flin, R. (2003). "Danger—Men at work": Management influence on safety. *Human Factors and Ergonomics in Manufacturing*, 13, 261–268.

Flin, R. (2007). Measuring safety culture in healthcare: A case for accurate diagnosis. *Safety Science*, 45, 653–667.

Flin, R., Mearns, K., O'Connor, P., and Bryden, R. (2000). Measuring safety climate: Identifying the common features. *Safety Science*, 34, 177–192.

Grote, G. (2007). Understanding and assessing safety culture through the lens of organizational management of uncertainty. *Safety Science*, 45, 637–652.

Grote, G., and Künzler, C. (2000). Diagnosis of safety culture in safety management audits. *Safety Science*, 34, 131–150.

Guldenmund, F. W. (2000). The nature of safety culture: A review of theory and research. *Safety Science*, 34, 215–257.

Guldenmund, F. (2007). The use of questionnaires in safety culture research—An evaluation. *Safety Science*, 45, 723–743.

Guldenmund, F. W. (2010). (Mis)understanding safety culture and its relationship to safety management. *Risk Analysis*, 30, 1466–1480.

Hale, A. R. (2000) Culture's confusions. *Safety Science*, 34, 1–14.

Hale, A. R., Guldenmund, F. W., van Loenhout, P. L. C. H., and Oh, J. I. H. (2010). Evaluating safety management and culture interventions to improve safety: Effective intervention strategies. *Safety Science*, 48, 1026–1035.

Hansson, S. O. (2007a). Risk. Stanford Encyclopedia of Philosophy. Available at: http://plato.stanford.edu/entries/risk/

Hansson, S. O. (2007b). Ethics and radiation protection. *Journal of Radiological Protection*, 27, 147–156.

Haslam, S. A., Reicher, S. D., and Platow, M. J. (2011). *The New Psychology of Leadership. Identity, Influence and Power.* East Sussex, UK: Psychology Press.

Haukelid, K. (2008) Theories of (safety) culture revisited—An anthropological approach. *Safety Science*, 46, 413–426.

Hofmann, D. A., and Stetzer, A. (1996). A cross-level investigation of factors influencing unsafe behaviours and accidents personnel. *Psychology*, 49, 307–339.

Hofstede, G. (1991), *Cultures in Organizations: Software of the Mind*. London: McGraw-Hill.

Hollnagel, E. (2004). *Barriers and Accident Prevention*. Aldershot, UK: Ashgate.

Hopkins, A. (2005). *Safety, Culture and Risk. The Organisational Causes of Disasters*. Sydney: CCH.

Hopkins, A. (2006) Studying organisational cultures and their effects on safety. *Safety Science*, 44, 875–889.

HSC (Health and Safety Commission). (1993). Third report: Organising for safety. ACSNI Study Group on Human Factors, HMSO, London.

HSE. (1997). *Successful Health and Safety Management. Health and Safety Executive*. London: HMSO.

IAEA. (1986). Summary Report on the Post-accident Review Meeting on the Chernobyl Accident. A Report by the International Nuclear Safety Advisory Group. INSAG Series No. 1.

IAEA. (1991). Safety Series No. 75-INSAG-4. Safety culture. International Atomic Energy Agency, Vienna.

IAEA. (2002). Safety culture in nuclear installations. Guidance for use in the enhancement of safety culture. IAEA-TECDOC-1329. International Atomic Energy Agency, Vienna.

International Atomic Energy Agency (IAEA) (2012). IAEA Mission to Onagawa Nuclear Power Station to Examine the Performance of Systems, Structures and Components Following the Great East Japanese Earthquake and Tsunami. IAEA Mission Report (Onawaga and Tokyo, July 30-August 11, 2012), Department of Nuclear Safety and Security, Department of Nuclear Energy, Vienna: IAEA.

IAEA (2016). OSART. Independent Safety Culture Assessment (ISCA) Guidelines. IAEA Services Series 32. International Atomic Energy Agency, Vienna.

Itoh, K., Andersen, H. B., and Mikkelsen, K. L. (2014). Safety culture dimensions, patient safety outcomes and their correlations. In: P. Waterson (Ed.), *Patient Safety Culture. Theory, Methods and Application*. Farnham, UK: Ashgate.

Jones, A. P., and James, L. R. (1979). Psychological climate: Dimensions and relationships of individual and aggregated work environment perceptions. *Organizational Behaviour and Human Performance*, 23, 201–250.

Kastenberg, W. E. (2014). Ethics, risk and safety culture: Reflection on Fukushima and beyond. *Journal of Risk Research*, 18(3), 304–316.

Katz, D., and Kahn, R. L. (1966). The social psychology of organizations. New York: John Wiley.

Kello, J. (2006). Changing the safety culture. Safety professional as change agent. *International Journal of Knowledge, Culture, and Change Management*, 6, 151–156.

Mead, G. H. (1934). *Mind, Self, and Society*. Chicago, IL: University of Chicago Press.

Mearns, K., Flin, R., Gordon, R., Fleming, M. (1998). Measuring safety climate on offshore installations. *Work and Stress*, 12, 238–254.

McMillan, E. (2008). *Complexity, Management and the Dynamics of Change*. London: Routledge.

Miller, E., Rice, A. K. (1967). *Systems of Organization: Task and Sentient Systems and Their Boundary Control*. London: Tavistock.

Morrow, S. L., Koves, G. K., and Barnes, V. E. (2014). Exploring the relationship between safety culture and safety performance in U.S. nuclear power operations. *Safety Science*, 69, 37–47.

Myers, D. J., Nyce, J. M., and Dekker, S. W. A. (2014). Setting culture apart: Distinguishing culture from behavior and social structure in safety and injury research. *Accident Analysis and Prevention*, 68, 25–29.

Nævestad, T.-O. (2009). Mapping research on culture and safety in high-risk organizations: Arguments for a sociotechnical understanding of safety culture. *Journal of Contingencies and Crisis Management*, 7, 126–136.

Neal, A., and Griffin, M. A. (2006). A study of the lagged relationships among safety climate, safety motivation, safety behavior, and accidents at the individual and group levels. *Journal of Applied Psychology*, 91, 946–953.

The National Diet of Japan Fukushima Nuclear Accident Independent Investigation Commission (NAIIC) (2012). Executive Summary of The Official Report of Fukushima Nuclear Accident Independent Investigation Commission, July 5, 2012. Available at: http://warp.da.ndl.go.jp/info:ndljp/pid/3856371/naiic.go.jp/en/report/index.html&gt

Nielsen, K. J. (2014). Improving safety culture through the health and safety organization: A case study. *Journal of Safety Research*, 48, 7–17.

Oedewald, P., Pietikäinen, E., and Reiman, T. (2011). A guidebook for evaluating organizations in the nuclear industry—an example of safety culture evaluation. Research Report 2011:20, Swedish Radiation Safety Authority.

Ostrom, L., Wilhelmsen, C., and Kaplan, B. (1993). Assessing safety culture. *Nuclear Safety*, 34, 163–172.

O'Toole, M. (2002). The relationship between employees' perceptions of safety and organizational culture. *Journal of Safety Research*, 33, (2), 231–243.

Parker, M. (2000). Organizational culture and identity. London: Sage.

Parsons, T. (1951). The social system. London: Routledge & Kegan Paul.

Pidgeon, N. (1998). Safety culture: Key theoretical issues. *Work and Stress*, 12, 202–216.

Pidgeon, N. (2012). Complex organizational failures: Culture, high reliability and lessons from Fukushima. *The Bridge*, 42, 17–22.

Presidential Commission on the Space Shuttle Challenger Accident (1986). "Report to the President By the Presidential Commission on the Space Shuttle Challenger Accident (Rogers Commission Report)". Challenger STS-51L: Reports. 2.

Rasmussen, J. (1997). Risk management in a dynamic society: A modelling problem. *Safety Science*, 27, 183–213.

Reason, J. (1997). *Managing the Risks of Organizational Accidents*. Aldershot, UK: Ashgate.

Reiman, T., and Oedewald, P. (2007). Assessment of complex sociotechnical systems—Theoretical issues concerning the use of organizational culture and organizational core task concepts. *Safety Science*, 45, 745–768.

Reiman, T., and Pietikäinen, E. (2012). Leading indicators of system safety—Monitoring and driving the organizational safety potential. *Safety Science*, 50, 1993–2000.

Reiman, T., and Rollenhagen, C., (2011). Human and organizational biases affecting the management of safety. *Reliability Engineering and System Safety*, 96, 1263–1274.

Reiman, T., and Rollenhagen, C. (2014). Does the concept of safety culture help or hinder systems thinking in safety? *Accident Analysis and Prevention*, 68, 5–15.

Reiman, T., Pietikäinen, E., Oedewald, P., and Gotcheva, N. (2012). System modeling with the DISC framework: Evidence from safety-critical domains. *Work*, 41, 3018–3025.

Reiman, T., Silla, I., and Pietikäinen, E. (2013). The validity of the Nordic patient safety culture questionnaire TUKU. *International Journal of Risk and Safety in Medicine*, 25, 169–184.

Reiman, T., Rollenhagen, C., Pietikäinen, E. and Heikkilä, J. (2015). Principles of adaptive management in complex safety critical organizations. *Safety Science*, 71, 80–92.

Rice, A. K. (1958). *Productivity and Social Organisation: The Ahmedabad Experiment*. London: Tavistock Publications.

Richter, A., and Koch, C. (2004). Integration, differentiation and ambiguity in safety cultures. *Safety Science*, 42, 703–722.

Rollenhagen, C. (2010). Can focus on safety culture become an excuse for not rethinking design of technology? *Safety Science*, 48, 268–278.

Rollenhagen, C., Westerlund, J., and Näswall, K. (2013). Professional subcultures in nuclear power plants. *Safety Science*, 59, 78–85.

Rundmo, T. (1996). Associations between risk perception and safety. *Safety Science*, 24, 197–209.

Rundmo, T., and Hale, A. (2003). Managers' attitudes towards safety and accident prevention. *Safety Science*, 41, 557–574.

Ryu, A., and Meshkati, N. (2014). Why you haven't heard about Onagawa Nuclear Power Station after the earthquake and tsunami of March 11, 2011. Vitebi School of Engineering, University of Southern California (USC).

Sagan, S. D. (1993). The limits of safety. Organizations, accidents, and nuclear weapons. New Jersey: Princeton University Press.

Schein, E. H. (1985). *Organizational Culture and Leadership*. San Francisco, CA: Jossey-Bass.

Schein, E. H. (2010). *Organizational Culture and Leadership*. (4th ed.). San Francisco, CA: Jossey-Bass.

Schneider, B. (1975). Organizational climates: An essay. *Personnel Psychology*, 28, 447–479.

Schneider, B. (1985). Organizational behaviour. *Annual Review of Psychology*, 36, 573–611.

Schulman, P. R. (1993). The negotiated order of organizational reliability. *Administration and Society*, 25, 353–372.

Seo, D. C., Torabi, M. R., Blair, E. H., and Ellis, N. T. (2004). A cross-validation of safety climate scale using confirmatory factor analytic approach. *Journal of Safety Research*, 35, 427–445.

Silbey, S. S. (2009). Taming Prometheus: Talk About Safety and Culture. *Annual Review of Sociology*, 35, 341–369.

Simonds, R. H., and Shafari-Sahrai, Y. (1977). Factors apparently affecting injury frequency in eleven matched pairs of companies. *Journal of Safety Research*, 9, 120–127.

Smircich, L. (1983). Concepts of culture and organizational analysis. *Administrative Science Quarterly*, 28, 339–358.

Smith, M. J., Cohen, A., Cohen, H. H., and Cleveland, R. S. (1978). Characteristics of successful safety programs. *Journal of Safety Research*, 10, 5–15.

Snook, S. A. (2000). *Friendly Fire. The Accidental Shootdown of U.S. Black Hawks over Northern Iraq*. New Jersey: Princeton University Press.

Sorensen, J. N. (2002). Safety culture: A survey of the state-of-the-art. *Reliability Engineering and System Safety*, 76, 189–204.

Strauch, B. (2015). Can we examine safety culture in accident investigations, or should we? *Safety Science*, 77, 102–111.

Tharaldsen, J. E., and Haukelid, K. (2009). Cultural and behavioral perspectives on safety—Towards a balanced approach. *Journal of Risk Research*, 12(3–4), 375–388.

Vaughan, D. (1996). *The Challenger Launch Decision*. Chicago, IL: University of Chicago Press.

Vicente, K. (2004). *The Human Factor. Revolutionizing the way People Live with Technology*. New York: Routledge.

Waring, A. (2015). Managerial and non-technical factors in the development of human-created disasters: A review and research agenda. *Safety Science*, 79, 254–267.

Weick, K. E. (1995). Sensemaking in organizations. Thousand Oaks: Sage.

Weick, K. E. (1998). Foresights of failure: An appreciation of Barry Turner. *Journal of Contingencies and Crisis Management*, 6, 72–75.

Weick, K. E., and Sutcliffe, K. M. (2007). *Managing the Unexpected. Resilient Performance in an Age of Uncertainty*. (2nd ed.). San Francisco, CA: Jossey-Bass.

Weick, K. E., Sutcliffe, K. M., and Obstfeld, D. (1999). Organizing for high reliability: Processes of collective mindfulness. *Research in Organizational. Behavior*, 21, 81–123.

Zohar, D. (1980). Safety climate in industrial organizations: Theoretical and applied implications. *Journal of Applied Psychology*, 65, 96–102.

Zohar, D. (2002). Modifying supervisory practices to improve subunit safety: A leadership-based intervention model. *Journal of Applied Psychology*, 87, 156–163.

Zohar, D. (2010). Thirty years of safety climate research: Reflections and future directions. *Accident Analysis and Prevention*, 42, 1517–1522.

Zohar, D., and Luria, G. (2005). A multilevel model of safety climate: Cross level relationships between organization and group-level climates. *Journal of Applied Psychology*, 90(4), 616–628.

29

PRINCIPLES OF BEHAVIOR-BASED SAFETY

Steve Roberts and E. Scott Geller

29.1 INTRODUCTION

You have removed the hazards, provided better equipment, and enforced safety-related procedures. Now, how can you improve safety further? In the past, organizations have focused on improving safety by addressing the work environment, removing or mitigating hazards, providing better tools and equipment, and enforcing safety-related rules and regulations. These are all approaches that, understandably, have worked well at improving safety. But many organizations have reached an injury-rate plateau, finding that relying primarily on these approaches without taking a more comprehensive view of safety produces only marginal gains.

Minimizing or controlling environmental hazards is certainly a necessary precondition for an ideal safety culture, but it is not enough. We need to consider how safety is supported through the interaction of equipment, facilities, procedures, and people, and focus on the factors that influence occurrences of at-risk behavior. Behavior-based safety (BBS) addresses these interactions.

Handbook of Safety Principles, First Edition. Edited by Niklas Möller, Sven Ove Hansson, Jan-Erik Holmberg, and Carl Rollenhagen.

29.2 ORIGIN AND HISTORY OF BBS

The BBS approach is founded on behavioral science as conceptualized and researched by B.F. Skinner (1938, 1953, 1974). Experimental behavior analysis, and later applied behavior analysis, emerged from Skinner's research and teaching, and laid the groundwork for numerous therapies and interventions designed to improve the quality of life among individuals, groups, and entire communities (Goldstein and Krasner, 1987; Greene et al., 1987). Whether working one-on-one in a clinical setting or with work teams throughout an organization, the intervention procedures always target specific behaviors relevant to producing constructive change.

In other words, BBS focuses on what people do, analyzes why they do it, and then applies a research-supported intervention strategy to improve what people do. Thus, whatever the intervention approach, the focus is on behavior. It is behavior-based training, behavior-based leadership development, behavior-based goal-setting, behavior-based feedback, behavior-based incident analysis, behavior-based incentives and rewards, and so on.

For more than three decades, BBS has been flourishing in industrial settings across the United States, and subsequently throughout the world. Several books detail the principles and procedures of BBS (e.g., Geller, 1996, 1998, 2001a; Geller and Williams, 2001; Krause, 1995; Krause et al., 1996; McSween, 1995; Petersen, 1989; Sulzer-Azaroff, 1998; Williams, 2010). Moreover, a number of systematic reviews of the literature provide solid evidence for the success of this approach at preventing injury (e.g., Al-Hemoud and Al-Asfoor, 2006; Choudhry, 2014, Chen and Tian, H., 2012; Foster et al., 2008; Grindle et al., 2000; Ismail et al., 2012; Lingard and Rowlinson, 1997; McAfee and Winn, 1989; Parand and Foster, 2006; Petersen, 1989; Sulzer-Azaroff and Austin, 2000).

One of the most common misperceptions of BBS is that it disregards critical components of safety such as environmental conditions or leadership behavior by focusing exclusively on the behaviors of employees. However, identifying critical behaviors is only the starting point. We often begin BBS workshops by explaining that the work environment and physical hazards are important, but we should begin with a focus on behavior.

Let us say we have a piece of equipment with the potential to cause severe personal injury on one side of the facility. On the other side of the facility, there is also a piece of equipment that could cause serious injury—just less severe. Question: Given we have a limited amount of time, money, and resources, which piece of equipment should we focus on first to best improve safety? Should we focus on the equipment that could lead to the most severe injury first?

Let us look at the same situation from a behavioral perspective. Consider people only work with and around the riskiest equipment an average of 1 hour every 6 months. However, people work with and around the moderately risky equipment 24 hours per day, 7 days per week. With this perspective, although both conditions should eventually be properly controlled, we should focus on the moderately risky equipment first. Thus, our primary concern should not simply be with the latent injury potential inherent within the equipment, but rather with the probability that this injury potential will actually cause harm. This requires an examination of not simply the physical hazards, but also on how the interactions of people's behavior with those hazards influence the potential for injury.

At a very basic level, BBS includes three primary components. First, BBS involves more people identifying critical behaviors that could lead to injury. Next, BBS includes more people analyzing factors that facilitate the occasion of at-risk behaviors and inhibit the occurrence of safe behaviors. Finally, BBS involves more people providing supportive feedback to increase the frequency of safe behavior, corrective feedback to decrease the occurrence of at-risk behavior, the identification and removal of barriers to safe behavior, and the development of organizational systems to motivate and support safe behavior.

Although BBS starts by identifying critical behaviors that could influence occupational safety, BBS considers a broad spectrum of potential contributing factors to workplace injuries. However, organizations often take a narrow a view of injury prevention. An ideal safety culture develops continuous improvement activities around the six critical components depicted in Figure 29.1 *leadership, physical environment/conditions, systems, behaviors, employee engagement, and internal person factors*. Taking a BBS approach to each of these components leads to more comprehensive and effective solutions to building an ideal safety culture and reducing injury rates.

Figure 29.1. Components of an ideal safety culture.

29.3 LEADERSHIP

Responsible leaders foster a work environment of actively caring for people (AC4P). [1] When leaders communicate effectively they do not want unnecessary risks to be taken because they care about people and want to prevent harm to all employees, customers, and the larger community, they facilitate the nurturing of a comprehensive safety culture as opposed to focusing primarily on reducing injury-rate numbers. In fact, focusing primarily on injury rate can inadvertently lead to underreporting of incidents and at- risk behaviors.

Although senior leaders are often held accountable for outcome numbers such as injury rate, these numbers are not diagnostic regarding injury prevention. Plus, attempts to manage safety by using these same outcome numbers to motivate others are not effective. These numbers are typically based on the relatively rare occurrence of an injury, and such failure-focused reactivity provides no direction for beneficial change.

The most effective leaders go beyond a focus on outcome numbers and hold people accountable for accomplishing proactive process activities that contribute to eventual group and organizational success in productivity, quality, and/or safety. When people see a clear link between their behaviors and organizational performance, they feel intrinsically rewarded for their efforts and are more likely to develop a sense of personal responsibility for continued contributions toward improvement. Thus, by focusing on the process, leaders help people see the influence they have to contribute to organizational success.

The topics leaders choose to discuss in meetings or include in formal communications set the stage for what is viewed as important to the organization. More importantly, what gets discussed, recognized, and rewarded *on the job* should be consistent with the stated organizational values. In other words, people at all levels of the organization need to understand safety is not just a topic they use to kick off meetings. Rather, the criteria for organizational success include the integration of safety outcomes with relevant process activities as a key component of organizational performance.

The role of integrating safety-related activities as part of the job often falls on the supervisor. However, too often supervisors do not give significant attention to safety-related activities nor guide employees into making safety a key part of their jobs. This is not typically because they do not care about safety. In fact, if you ask most senior leaders or supervisors whether they care about safety, you would most certainly receive a resounding, "*Yes!*" But if you asked a group of employees, "Does your supervisor care about safety?," the sentiment may not be a strong "Yes." The actions and words a leader believes communicate a commitment to safety may not be interpreted as such by the observers and listeners.

[1] www.ac4p.org

The most likely reason supervisors do not perform behaviors most critical to showing support for safety: The organization has not identified or adequately defined the supervisory activities that demonstrate genuine concern for safety. Plus, supervisors rarely receive feedback as to whether they are performing these activities. Nor do they know how their activities are interpreted by line workers regarding a commitment to safety.

Furthermore, leaders or supervisors may inadvertently contradict true support for safety through their interpersonal communication. For example, it is too common for leaders to overlook and fail to correct at-risk work practices they observe. Plus, they often fail to praise or acknowledge the occurrence of safe behavior, even when they wished most of the workforce was performing such behavior. At the same time, supervisors frequently acknowledge increased productivity when this was achieved by taking short cuts and putting people at risk for injury. Leaders may also model at-risk behaviors themselves, believing they do not have the same degree of exposure or they have more protective knowledge, ability, or experience.

Each of these situations can inadvertently send the wrong unspoken message regarding safety, and cause people to believe leaders are actually encouraging at-risk behaviors in order to get the job done on time. Therefore, leaders need to consider carefully the variety of ways their own behavior can influence others' perceptions of their commitment to safety or lack thereof.

To show real support for safety, leaders need to do more than say, "I care about your safety." They need to "actively care" (Geller, 1991). In other words, what you say is not enough; what you *do* is most important. But what behaviors best demonstrate actively caring for safety? A number of assessments and cultural development projects have revealed what employees want to see from their leaders in order to show safety is important to them personally and for the organization. Here is a brief summary of what thousands of employees say it takes:

Employees want their leaders to: (1) *show up* by visiting site/work locations in person and talk to the people on the job, (2) *get their hands dirty* by getting out of the meeting rooms and into the production/operations areas to see first-hand the conditions, equipment, and procedures employees must deal with, and (3) *bring the checkbook*. This does not mean simply throw money at the problem, but rather demonstrate a willingness to provide the appropriate resources for tools/equipment/personnel/rosters/scheduling to support safety, considering specific recommendations from the employees who work in the given conditions.

To identify a comprehensive list of behaviors that determine whether the safety intentions of leadership are cascading through the culture, a more formal assessment is advised. A more formal assessment should include an identification of critical leader behaviors related to occupational safety. Leaders should evaluate themselves on the same behaviors assessed by their employees. Then the personnel above, below, and parallel to this target leader in the organizational structure should provide anonymous evaluations of that leader's behavior.

Target leaders review these feedback evaluations to gain insight into how others perceive their safety-related behaviors in comparison to their self-perceptions. They can see whether their support is being interpreted as intended, and what specific safety-supportive behaviors are expected of them and may not have been observed. Workshops or educational sessions can then be developed to support common leadership behaviors shown to be lacking among the leaders of an organization.

We once heard the senior executive of a large organization say publicly during a staff meeting, "I want to set the right example for our employees. However, up here in my office, the typical employee has little exposure to what I do on a daily basis. Therefore it is difficult for me to set a safe example."

We challenged this leader's assumptions by explaining that the employees are paying more attention than one may think. Perhaps the day-to-day behaviors of corporate leaders are not seen. However, the decisions made by leaders are closely watched by employees and can send a very clear message about safety support or the lack thereof. For example, we use an SPS Leadership Development Exercise to make this point. A number of different scenarios have been used with this exercise to best fit each organization. Two representative examples are given in Figure 29.2.

Read each scenario and discuss how leader decisions are interpreted by the workforce

Scenario 1: *One of your employees refuses to clean the cutting blades in the upper production area. The blades are 40 feet in the air, awkward to access, and with limited tie off points. The safest way to clean the blades is to lower them so they can be accessed from the ground. However, to do this, production lines need to be shut down and equipment has to be moved, resulting in significant loss of production and time. Therefore, the blades are always cleaned in the upper position. As employees are allowed to refuse hazardous tasks, you typically need to ask three to four employees before someone will agree to do this job. After two more employees refuse the job, one of the older employees comes to you and says he is willing to do it.*

➢ Q: What would you do?
➢ Q: How would your decision be interpreted by the general workforce?

Scenario 2: *The cooling tower basins need to be "de-mucked." Because of concerns with Legionnaires Disease, company procedures require company employees to have extensive knowledge and protection (e.g., specialized training, PPE, etc.). Therefore, to bypass some of these issues, it is easier to simply assign this task to a contractor (as contractors are not required to have the same level of training/equipment). Therefore, it is now standard procedure to assign contractors this task. The schedule now requires you to send two employees to de-muck the cooling towers.*

➢ *Q: What would you do?*
➢ *Q: How would your decision be interpreted by the general workforce?*

Figure 29.2. Safety Performance Solutions, Inc. leadership development exercise.

29.4 PHYSICAL ENVIRONMENT/CONDITIONS

Historically, organizations have focused a great deal of effort to improve safety by addressing the work environment. Eliminating or reducing hazards, providing safer tools and equipment, and developing and enforcing adequate procedures have worked well at preventing work injuries. But many organizations have reached a plateau whereby relying primarily on these approaches without taking a more comprehensive view of safety has produced only marginal gains.

Obviously, a work environment that minimizes, contains, or controls serious hazards should be a necessary precondition for an ideal safety culture. However, we need to consider how safety is supported or inhibited through the interaction of equipment, facilities, procedures, and people. A number of environmental conditions increase the likelihood of a workplace injury. These factors are listed in Table 29.1 as derived from a number of sources, including the Tripod Delta technique described by Reason (1990)and the consulting experiences of SPS, Inc. (SPS, 2015).

A proactive approach to occupational safety addresses these factors systematically before an injury occurs, instead of waiting until someone gets hurt before analyzing what went wrong. By taking a proactive approach to the physical environment, new and improved rules/procedures, equipment fixes, and hazard controls can be developed before they lead to problems.

Participation in this proactive approach is often more difficult to initiate or maintain because the consequences of preventive activities are rarely soon or certain. We will never know how many injuries are prevented by our proactive behaviors. Nevertheless, such behavior is required to continuously move toward an injury-free workplace.

Employees at all levels of the organization need to continually search for, discuss, and help address the factors listed in Table 29.1. For safety to progress beyond mere hazard recognition and rule enforcement, people need to show interdependent compassion and courage to speak up and identify work-related environmental and situational concerns. Employees also need to see that voicing such concerns leads to their serious consideration and to relevant beneficial change.

29.5 SYSTEMS

Organizations rely on a number of safety management systems to manage risk and thereby decrease the chance of incidents and injuries. These systems generally include *safety rules and procedures, safety training, hazard identification and correction, discipline, incident reporting and analysis, safety communications, safety suggestions, and rewards/recognition*.

Each of these safety management systems contributes significantly to improving workplace safety and influencing the organization's safety culture. At best, when the system is poorly designed or operating ineffectively, safety-related benefits are

TABLE 29.1. Factors Contributing to the Probability of a Workplace Injury

Physical Environment. Are there uncontrolled hazards such as energy sources? Is working at unsecured heights required? Is there uncontrolled exposure to harmful substances? Is there extreme heat, cold, or noise? Is there adequate lighting? Is there sufficient workspace (e.g., not cramped) and proper housekeeping?

Hardware. Are tools, equipment, materials available? Is non-standard or at-risk use of equipment required to accomplish tasks?

Human–Machine Interface. Are the controls and displays adequate?

Equipment Design. Does the design of equipment provide external cues for proper use? Are the ranges of safe actions clear, and is feedback provided to the user regarding safe use?

Task Design. Is the workload too high or too low? Is the task performed at inopportune times (end of shift, off shift)? Are operators exposed to distractions? Is the task beyond the operators expected capabilities?

Supervision. Are there adequate pre-job briefings, checks, and scheduling. Is there excessive production pressure? Is progress checked and feedback provided?

Selection of Worker(s). Are workers qualified? Are they fatigued? Do work teams include people with an adequate combination of the needed skills?

New Technology. Does new technology (e.g., automation, new equipment) eliminate some problems while increasing others?

Maintenance Management. Are maintenance activities properly planned? Is maintenance carried out in a timely fashion?

Procedures. Are the procedures accurate, relevant, and practical?

Goal Compatibility. Conflict between an individual's work and personal responsibilities occurs when informal norms conflict with the organization's goals, and when different goals of the organization are inconsistent. Does goal incompatibility cause confusion?

Communication. Are sufficient channels of communication available? Is appropriate information transmitted through the communication channels? Is the information timely and understandable? Is there coordination among interdependent groups?

Training. Have task and needs analyses been performed? Has adequate education training been provided on all relevant knowledge/skills? Has sufficient practice with feedback been provided? Was testing for competence performed? Was transfer of training evaluated and periodic refresher training provided?

Feedback. Does the task/job provide intrinsic (i.e., natural) feedback or are systems in place to provide external feedback?

Employee Engagement. Do employees have choice, flexibility, discretion, and feel personal ownership?

Fault-Tolerant Systems. Are errors detectable and recoverable? Are there adequate warnings, protection, containment, escape, and rescue strategies?

depreciated. At worst, a poorly designed, badly implemented, or ill-functioning system can actually have a destructive influence on the organization's overall safety culture. To further compound the situation, these systems are interactive and often overlap. For example, *hazard identification and correction* requires an atmosphere that fosters *employee participation*, sufficient *education/training* so employees can recognize and correct hazards, ample *communication* of the hazard, and its sufficient resolution.

Poor features of one system can have negative influences on other systems, making the problem areas more difficult to isolate and correct. Therefore, an ideal safety culture not only has well-functioning safety management systems with regard to performing its primary function, but also works interdependently with other systems to advance the safety culture. Too often, however, this is not the case. In many organizations, *rewards* for safety focus on outcomes (i.e., injury rates) rather than behaviors and processes. If employee incentive/reward programs and/or leaders' performance evaluations are based primarily on injury rates, it is unreasonable to expect those employees to embrace a system that encourages frank and open discussions, reporting of injuries, and or peer-to-peer observation of safety-related work behaviors. Yet such discussions and observations are critical to assessing factors related to injuries and their prevention.

If an assessment of existing safety management systems shows deficiencies, two options should be considered. First, examine the actual practices of the organization, department, or team to assess strengths and weaknesses. For example, *safety training* concerns may be caused by a variety of issues, each with different solutions. Safety training may be too short, too complicated, poorly conducted, or too general for application on the job. Training may be given by employees who lack credibility, or may be conducted on required overtime either for the trainees or their counterparts back on the job, leading to decreased attention or negative feelings regarding the training.

The training itself may be top-notch but treated by the employees' supervisor as a nuisance or as secondary to "getting the work done." Each of these issues and more can give training a bad reputation and, more importantly, cause it to be ineffective at improving or maintaining employee safety and health. Note also that authentic training implies that participants perform relevant behaviors and receive behavior-based feedback, perhaps through simulations or role-playing exercises. Without a behavior-feedback component, a "training" session should be considered education or awareness.

The second option is to analyze how *perceptions* of the system are being managed (or not managed). For example, a *safety suggestion process* might be viewed as beneficial only by those whose suggestions have been implemented or have received feedback. Suggestions may be actively solicited, objectively evaluated by a cross-functional team of employees, amply funded, and quickly acted on *but* poorly communicated to the rest of the workforce. Employees would evaluate such a system negatively, because they are unaware of its benefits. This issue can be readily addressed with proper site-wide communication.

Another safety-relevant system which is often very telling of an organization's culture is *incident reporting and analysis*. The frequency of first aid cases and "close-call" reporting is higher in organizations where employees share trust and a problem-solving perspective. If incident and close-call reporting are suppressed, fact-finding analyses may be less than thorough, communication of findings may be spotty, penalty-focused discipline may be feared, and safety may be viewed as entirely "fear of failure," with only reactive solutions. Initiating and maintaining a BBS process in this type of culture takes more time, effort, and patience, and could be relatively ineffective.

However, organizational leadership should not wait until an injury or other incident spotlights a defective system. Instead of waiting for incidents to occur, and then reactively assessing the systems that may have influenced the event, organizations should critically analyze each system proactively to ensure each carries out its primary function in a way that builds an actively caring culture (Geller, 2014, 2015, 2016).

To assess safety management systems, we recommend organizations select key groups (e.g., senior leaders, supervisors, and hourly safety committee members) and assess the maturity and effectiveness of each organizational system. If various groups are selected for the assessment, a gap analysis can be performed to identify consistent perceptions as well as differences across groups. Relevant data can then be used to create a new management system or refine existing systems to improve their overall effectiveness, including their influence on the organization's safety culture.

Figure 29.3 depicts sample items from the SPS management systems maturity path many organizations have used to assess their incident reporting and analysis. A

MATURITY PATH FOR… INCIDENT REPORTING & ANALYSIS		
Beginning (1)		**Leading (4)**
1. All but the most serious incidents go unreported	1 2 3 4	All incidents (e.g., near misses, property damage, injuries) are reported in a timely fashion.
2. No training is provided for incident reporting and analysis.	1 2 3 4	All employees receive training in the philosophy and overall process of investigating incidents. Investigation team leaders receive detailed training in investigation skills such as interviewing and root cause analysis.
3. The focus of the investigation is on documenting the event.	1 2 3 4	The focus of the investigation is on finding system-level root causes.
4. Analyses often result in identifying "who's to blame." Corrective measures such as discipline or "counseling" are common.	1 2 3 4	Incident analyses focus on determining system-level root causes and minimizing or eliminating them. Individuals are not assumed to be at fault. Appropriate behavior-based corrective actions are introduced where warranted.

Figure 29.3. Sample items from the SPS incident reporting and analysis maturity path.

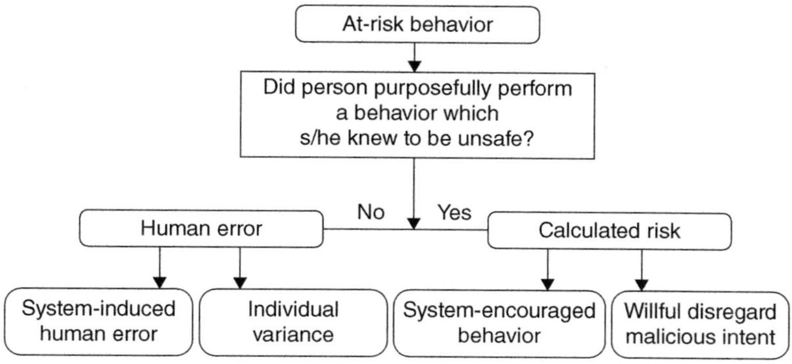

Figure 29.4. Flow chart of issues to consider after identification of a critical at-risk behavior.

maturity path analysis presents a beginning or "immature" behavioral anchor as well as a leading or "mature" behavioral anchor, and asks key groups to rate the maturity level of their culture according to its proximity to each behavioral anchor.

Furthermore, to consider system factors adequately, a proper analysis of the occurrence of at-risk behaviors is necessary. For example, after a critical at-risk behavior is identified, consider the underlying reasons for the at-risk behavior by asking the right questions. A lack of proper questioning often leads to an incomplete analysis, inadequate consideration of system factors, and thus misdirected corrective action. Figure 29.4 depicts a flow chart to guide questions to be asked and issues to consider after a critical at-risk behavior or behavioral trend is identified.

In determining the most appropriate strategy for addressing the occurrence of a risky behavior, it is necessary to consider first whether employees were aware they were performing an at-risk behavior. Of course, if an injury results, an analysis would determine the behavior was at-risk after the fact. But there are times when employees do not realize a behavior they are performing is at risk.

However, sometimes people perform behaviors they know are risky. But even when this is the case, the calculated risk may not have included a malicious disregard for safety. The purposeful at-risk behavior could have been system induced or encouraged. In other words, when someone mindfully performs an at-risk behavior, we see two possible influences: the behavior was encouraged or rewarded by the system (e.g., a reasonable person could believe the behavior was benefiting the company through something like increased productivity and was not explicitly discouraged by management), or the individual was acting outside the system through willful negligence or malicious intent. Malicious intent is very rare (or if not, you have serious organizational or leadership issues beyond safety).

Most workers who intentionally perform at-risk behavior are trying to be a good employee, get a job done more quickly and be more productive, solve a problem in the way they consider to be best, believe there is no other way to perform the task effectively, or they are otherwise encouraged by the system or work environment. Therefore, it is critical that organizational leaders set clear expectations regarding safety-related behaviors. Although creativity and problem solving are important, solving problems with unnecessary risk taking must be specifically discouraged.

Before labeling at at-risk behavior as a willful violation, consider the formal and informal reward/recognition systems that may have encouraged the behavior, the type and quality of the education/training the employee had received, the nature of the tools and equipment available to perform the job, the type of corrective feedback given to others for similar at-risk behaviors, and the extent a "just get 'er done" mindset is prevalent in the culture.

If the employee did not realize the behavior was at-risk, we have an instance of *human error*. The behavior may have been performed intentionally, but the employee did not realize it was risky. Two possible behavioral determinants within this category: (1) The behavior was somehow enabled or facilitated by the system (e.g., formal non-standard interface designs, inadequate training or tools, etc.) making it easy for people to make mistakes; or (2) The behavior was a product of human variance (e.g., a cognitive failure, or a temporary lapse in focus or awareness).

Corrective action should address the potential causal factors identified. When a system-level factor is identified, the corrective action should focus on fixing the system, rather than focusing on the employee. A classic example, explained to us by an older operator at a nuclear power plant, involved the layout of his control panel. At this facility, the control panel had two prominent red buttons situated beside each other. Pressing one button would flood the reactor. The other button would drain the reactor. Neither of these buttons should be pushed except in emergency situations.

Because of the layout, it should have been predictable that someone would eventually push the wrong button. The obvious solution to this situation is to move one of the buttons or at least change the color of one button. It is noteworthy, one self-described nuclear safety expert told us a layout such as this would never occur in a nuclear plant, yet others have told us the example is accurate. Regardless of its validity, the example serves to illustrate an important point: When system factors lead to error, the appropriate solution is to change the system.

As for human variance, the cognitive failures of others are most difficult to address. We know people are going to have them; it is not a matter of "if" but "when." Personal, situational, and dispositional factors contribute to human error due to a cognitive failure. For example, illnesses, divorce, problems with kids, financial burdens, or drug problems can make it difficult to focus vigilantly and continuously on the job and avoid error. For this reason, employees and leaders need to understand the capabilities and availability of employee assistance programs.

The potential for human error indicates the need for a broad and effective "safety net"—fault tolerant systems and recoverability options that lessen the negative consequences of a "brain cramp" when they inevitably occur. An example of a fault-tolerant system with a recovery option many readers have experienced is an ATM machine. If you had intended to withdraw $20, but by error entered $200, the feedback from the ATM screen would immediately reflect this error by providing the amount selected and prompting you to select "yes" if the amount is correct. The recoverability option is reflected in the option to select "no" and try again for the $20.

In the workplace, employees can observe each other to identify errors and other at-risk behaviors. Then, behavior-focused feedback provides an opportunity to recover or improve. This highlights the need for employees to maintain awareness of those around them and show the courage to step in and actively care when corrective action is called for. In other words, the ideal safety net is a brother's/sister's keeper's culture in which employees continually look out for the safety and health of each other with an interdependent mindset.

29.6 BEHAVIORS

BBS obviously focuses on behavior. However, the methods, perspectives, and techniques for this focus are critical in determining a successful versus unsuccessful BBS process. This section details the most important principles for understanding and improving behavior.

Look for External Factors to Understand and Improve Behavior

We all do what we do because of both external/environmental and internal/person factors. However, given the difficulty in objectively defining internal states or traits, it is more cost effective to identify environmental conditions that influence behavior and then address those factors when behavior change is desired. Examining external factors to explain and improve behavior is a primary focus of organizational behavior management (e.g., Austin, 2000; Austin et al., 1999; Bailey and Austin, 1996; Gilbert, 1978). In occupational safety, this approach has been termed behavioral safety analysis (Geller, 2000, 2001b).

When BBS became a popular topic at company safety meetings and professional development conferences in the 1990s, Hans (1996) reported this quote from the Director of the National Safety Council's consulting division: "The all-consuming focus on employee behaviors can mask management inadequacies that otherwise might come to light" (pp. 45–46). But in fact, when BBS is implemented correctly, the opposite is true.

A systematic behavioral analysis of at-risk work practices can pinpoint many determinants of such behavior, including inadequate management systems, supervisor behaviors that promote or inadvertently encourage risky on-the-job behaviors, or management behaviors and decisions that are inconsistent with safety messages from organizational leaders. Without the upstream and objective problem-solving perspective fostered by the BBS principles, these inadequacies may never be identified (or they might be revealed only after a "close call" or injury).

Direct Behaviors with Activators and Motivate with Consequences

Behavior is influenced by two distinct factors: *activators* and *consequences*. Activators precede behavior and serve to guide, prompt, or direct a behavior. Activators often tell us what, when, or how we should be doing something. While driving to work, roadway signs instruct us to comply with the speed limit, come to complete stops, and yield to oncoming traffic.

However, other activators can influence drivers to take shortcuts: "I'm 20 minutes late for an important meeting and if I don't hurry, I'll be caught in the rush-hour traffic generated by all the other morning commuters." Other drivers are also speeding. Faced with these activators, what will the driver do? It is a matter of consequences. What does the driver expect to gain or avoid by speeding on this morning's commute.

Let us examine the potential consequences of speeding. Receiving a speeding ticket from the local police officer would be a significant negative consequence. Compare that negative consequence with the positive consequence of saving time. Which of these two consequences is more powerful? For most drivers, the "positive" consequence would likely be most influential. While receiving a ticket is a very aversive consequence, the probability a police officer would detect this risky driving from a number of other speeding vehicles is low.

The at-risk driver says to himself, "My experience tells me I can save a few minutes by increasing my speed. I have driven this route daily for the last 10 years and I know if I drive 15 miles an hour over the posted limit, I can save 15 minutes. My experience tells me the chances of getting a ticket are quite low compared to the likelihood I will save time."

In this example, the *probability* of the expected consequence greatly influenced the driver's behavior. In addition to probability, people also consider the *timing* and *significance* of the consequence. Consequences that occur sooner rather than later are more influential, and consequences that are personally significant are more motivating than those that are insignificant. In the speeding example, the driver expects to save time immediately, and those 15 minutes are significant. If he got caught, he would receive the ticket immediately, and it would be significant. However, because he does not expect that consequence, the timing or the significance of the ticket are irrelevant. To make matters worse, many other drivers are engaging in the same at-risk behavior, encouraging this driver to continue his speeding.

Employees are faced daily with similar decisions to perform safe or at-risk behaviors. A behavior-based process can provide the analysis tools to determine the contributing factors for certain at-risk behavior, which in turn informs the nature of interventions needed to improve the safety of the situation. A number of intervention options are possible.

Apply the Scientific Method to Improve Behavior

The occurrence of specific behaviors can be observed and measured objectively before and after the implementation of an intervention process. This application of the scientific method provides feedback that enables improvement. The acronym "DO IT" illustrates this process, and is useful for remembering and teaching this principle of BBS.

"D" for Define. The process begins by defining specific behaviors to target. These are safe behaviors that need to occur more often or at-risk behaviors that need to occur less often. Avoiding at-risk behavior often requires certain safe behavior, and therefore a safe target might be behavior to substitute for a particular at-risk behavior. On the other hand, a safe target behavior can be defined independently of an associated at-risk behavior. The definition of a safe target might be as basic as using certain personal protective equipment (PPE) or "operating mobile equipment at a safe speed." Or, the safe target could be a process requiring a particular sequence of safe behaviors, as when lifting a heavy load or locking out an energy source.

Deriving a precise definition of a DO IT target is facilitated with the development of a critical behavior checklist (CBC) that can be used to evaluate whether a certain target behavior or process is being performed safely. Developing behavioral definitions for a CBC provides a valuable learning experience. When people get involved in deriving a CBC, they own a training process that can improve human dynamics on both the outside (behaviors) and the inside (feelings and attitudes) of people.

From the relevant operating procedures, work teams define the specific behaviors on their job that are safe and at-risk. They derive their own customized CBC, with the critical behaviors listed and a column to check for safe observations and at-risk observations, as well as a column for recording comments or factors that may have influenced the behavior(s) observed. A worksheet for developing a CBC is depicted in Figure 29.5.

"O" for Observe. When people observe each other for certain safe and at-risk behaviors, they realize everyone occasionally performs at-risk behavior, sometimes without even realizing it. The observation stage is not a faultfinding procedure; it is a fact-finding learning process to facilitate the discovery of behaviors and conditions that need to be changed or continued in order to prevent injuries. Thus, no behavioral observation is made without awareness and explicit permission from the person observed. The observer should be open to learning as much (if not more) from the post-observation feedback conversation as s/he expects to teach from completing the CBC.

"I" for Intervene. During this stage, interventions are designed and implemented in an attempt to increase the occurrence of safe behavior and/or decrease the frequency of at-risk behavior. A BBS intervention implies changing external conditions of the system in order to make safe behavior more likely than at-risk behavior. When designing interventions, it is important to realize the most motivating consequences

Observer______________________ Date (M/Y):______________________

____________________ ____________________ ____________________

Behavioral Category	Safe	At-Risk	Comments
1) Personal Protective Equipment: a. Head b. Eye/Face c. Hearing d. Hand e. Foot f. Protective Clothing g.			
2) Tool and Equipment Use: a. Correct Tool for the Task b. Proper Use of Tool c. Tool in Good Condition d. Lock Out / Tag Out e.			
3) Body Positioning/Protecting: a. Lifting, Bending b. Reaching, Twisting c. Pushing, Pulling d. Neutral, Stable Posture e. Line of fire, Pinch points f.			
4) Housekeeping: a. Slip/trip Hazards b. Orderly Storage c. Clear aisles/exits d.			
5) Pace of Work: a.			
6) Communications: a.			
7) Visual Focus a.			
8) a.			

Figure 29.5. Sample critical checklist development worksheet

are soon, certain, and sizable; and positive consequences are preferable to negative consequences.

The process of observing and recording the frequency of safe and at-risk behavior on a CBC provides an opportunity to give individuals and groups valuable behavior-based feedback. When the results of a behavioral observation are shown to individuals or groups, they receive the kind of information that enables practice for performance improvement.

Considerable research has shown that providing workers with feedback regarding their ongoing behavior is a very cost-effective intervention approach. See, for example, the seminal analysis of the Hawthorne Effect by Parsons (1974), comprehensive reviews by Alvero et al. (2001) and Balcazar et al. (1986), as well as a historical review and meta-analysis of feedback interventions by Kluger and DeNisi (1996). Furthermore, occupational safety has significantly improved following the feedback display of workers' percentages of safe versus at-risk behavior (e.g., Austin et al., 1996; Sulzer-Azaroff and de Santamaria, 1980; Williams and Geller, 2000; Zohar et al., 1980).

Feedback should also move beyond an interpersonal, workgroup, or site level. Once significant at-risk behaviors or hazards are identified or once solutions to these problems are achieved, safety management systems (e.g., communications) should ensure the organization as a whole receives feedback on these issues, allowing others to avoid these same hazards and enabling solutions to be generalized across the organization.

"T" for Test. The test phase of DO IT provides work teams with the information they need to refine or replace a behavior-change intervention, and thereby improve the process. If observations indicate significant improvement in the target behavior has not occurred, the work team analyzes and discusses the situation. The workers might refine the intervention or choose another intervention approach. On the other hand, if a target behavior reaches the desired frequency level, the participants turn their attention to one or more other target behaviors.

Beyond Completing Checklists

As the "DO IT" process illustrates, BBS includes a variety of steps. However, leaders too often overemphasize the importance of simply turning in behavioral observation checklists at the exclusion of other key steps. Although calculating the number of observations conducted can provide a useful metric that is easy to generate, over focusing on this metric can be a major pitfall to a successful BBS implementation. The case study below provides a useful illustration of this point.

Case Study: BBS Is More Than Completing Checklists

A corporate vice president responsible for three industrial sites was leading a divisional meeting with the plant managers and key staff members from each site. He was reviewing the number of behavioral observations from each site over the last 3 months. The employees at Site 1 submitted approximately 1000 behavior checklists. The employees at Site 2 submitted approximately 400 behavior checklists. The employees at Site 3 submitted only 90 behavior checklists.

The VP congratulated the manager from Site 1 and announced, "Great job! That's exactly what I'm looking for." He said Site 2 had done fairly well but needed to "step it up a bit." Then, he told the manager from Site 3, "You really needed to put more effort into the process. In fact, I want you to have an action plan on my desk within a week detailing how your site can be more like Site 1."

However, during the next part of the meeting, each site's representatives made short presentations about the successes and challenges of their BBS processes. The presentation from Site 1 focused on how few behavior checklists they were initially receiving, but after they initiated an incentive program, checklist submissions increased. But there was no mention of what they actually did during the observations or with the checklists. Site 2 had a similar presentation.

In contrast, the presentation from Site 3, where only 90 behavior checklists had been submitted, focused on how their employees were talking to each other about critical safety issues in ways they had never done before, how the BBS team reviewed each behavior checklist, searched for behavioral trends, and selected the most critical at-risk behavior to address. They then formed a cross-functional team to analyze the factors contributing to that at-risk behavior. The team selected several interventions they thought would help reduce this at-risk behavior and increase occurrences of the safer alternative. After several weeks, the frequency of the safe behavior had increased, and their success was shared with all personnel during an all-hands meeting.

After the presentations were complete, it was obvious to all that Site 3 was actually the exemplar regarding BBS performance. It is not about how many cards are completed, but about the discussions they facilitate, the safety-relevant data they provide, and how the data are used to drive safety-related improvements in the work process and/or management system.

The Intervention Hierarchy

Focusing on behavior change is often an effective way to prevent personal injury. However, it should not always be the first choice. Sometimes behavior change can be difficult because such change is contrary to basic human nature. At-risk behavior is often more comfortable, more convenient, faster, or easier than the safer alternative. Plus, past and present reward structures, past learning and habitual routines, and cultural influences can promote the maintenance of at-risk behavior. Also, barriers to safe behavior such as improper tool/equipment availability or inadequate training can make desirable behavior change difficult.

Therefore, before attempting to change an at-risk behavior to prevent personal injury, other considerations should typically come first. That is, before targeting behavior change, other interventions should be considered. Consider the following intervention hierarchy (ASSE, 2003), which is commonly used:

1st—Eliminate the hazard (substitution of materials, automation)
2nd—Remove people from the hazard
3rd—Install engineering controls (guarding, interlocks)
4th—Provide warnings (signs, alarms)
5th—Implement behavior change interventions to avoid the hazard, administrative controls, and training (following procedures, job rotation, lockout/tagout training, equipment inspections)
6th—Provide and require PPE (eye protection, respiratory protection, fall protection)

This hierarchy clearly regards behavior change and PPE use as the least effective. However, it is not always feasible to: (1) eliminate all hazards, (2) remove people from all risky environments, or (3) provide engineering controls to create an acceptable risk level. Therefore, even when workplaces have been "designed" to reduce hazards, personal injuries, and fatalities can still occur.

Complex systems require a great deal of human contribution to maintain productivity, quality, and safety. Human error is the inevitable by-product of our necessary involvement in complex systems. To eliminate human error completely would require an elimination of the best source of creativity, flexibility, and problem-solving ability. People are not perfect and will occasionally make mistakes, despite their best intentions and working in the best of surroundings.

No matter how safe the design of a workplace, how thorough the training, or how consistently safety rules are enforced, at-risk behavior will occur. Therefore, as long as we have people involved in a workplace, behavior-based observation and feedback will be an important intervention for reducing personal injury.

In addition, in an ideal safety culture, BBS means not only eliminating or reducing hazards and at-risk behavior, but also ensuring the methods used to bring about that change leave people feeling better about themselves, their co-workers, their situation, and the organization as a whole. This requires the development of intervention strategies that promote choice and personal control, encourage employee involvement and ownership, and focus on behaviors employees can perform proactively to achieve a safe workplace in lieu of putting most attention on what behaviors to avoid in order to prevent injuries.

29.7 EMPLOYEE INVOLVEMENT AND OWNERSHIP

Authoritarian directives may communicate action plans efficiently, but they may also stifle self-motivation or self-persuasion (Aronson, 1999). Behaviors performed to comply with a prescribed standard, policy, or mandate are other-directed (Watson and

Tharp, 1997). Such behaviors are accomplished to satisfy someone else, and they are likely to cease when compliance cannot be monitored.

When the development of an action plan involves the people expected to carry out that plan, ownership for both the process and the outcome is likely to develop. In other words, when leaders give a reasonable rationale for a desired outcome, allow for input and discussion, and then offer opportunities for others to customize methods for achieving that outcome, they facilitate self-directed or self-motivated behavior. Such behavior is likely to continue in the absence of an other-directed accountability system (Geller, 2001c.).

Input from Relevant Others

Recently, the first author observed a CEO speaking to the presidents and VPs of each of division and subsidiary organization of his company. When referring to employee involvement, he announced to the group … *"I know you all, I've worked with each of you, and to be honest, you're just not that smart! If we are to become world class with regard to safety, we need help and input from everyone in the organization, especially those with the time and experience doing the hazardous work…"*

Although the insult was said in jest, the point was made and seemingly accepted by all in the room. When employees are involved in activities, they are more motivated to help make them succeed; people care about the things they own. Also, involving employees in the development of safety-related activities or systems will increase the likelihood they will be seen as relevant/important. The more relevance people see in their safety activities, the more likely they will participate in performing them.

Also, creativity and flexibility are often needed for many tasks. Mindlessly following rules and routines in the same setting contribute to fatigue, burnout, and an increase in error probability (Langer, 1989).Therefore, facilitating ownership, autonomy, and empowerment helps increase mindful attention toward safety.

It is also critical that efforts to gain employee input include helping people feel their input is sincerely valued, needed, and seriously considered. If you ask for people's input, use it. To ask for input and not to use it creates frustration, resentment, and less willingness to provide input in the future.

How to Motivate Ownership and Involvement

A variety of techniques can encourage more employee involvement in safety-related activities. First, provide opportunities for people at all levels of the organization to offer contributions. Enable those involved to see how their efforts are making a difference, and show genuine appreciation for these efforts. Listed below are additional methods for promoting involvement in safety-related activities. Each has been successfully applied at a variety of organizations.

- Communicate regularly with employees about the organization's safety-related processes, both in group sessions and in one-on-one conversations.
- Personally request employee participation.
- Pair experienced employees with those less experienced on a safety-related task.
- Use various communication channels (e.g., safety training, newsletters, and bulletin boards) to regularly update all employees on the progress of a particular improvement process.
- Solicit input from employees through one-on-one communication, safety suggestion boxes, and safety meetings.
- Make safety relevant for both work and home life, and encourage family involvement whenever possible.
- Send memos to supervisors that encourage them to discuss key safety topics with their employees.
- Post reminders throughout the facility that encourage employees to participate in safety-related projects.
- Involve employees in designing safety signs, posters, newsletters, and emails to promote safety awareness and involvement.
- Affix highly visible tags or stickers with safety team logos to equipment or areas that have been fixed or improved as a result of a particular safety process.
- Display the names, photographs, and contact information of safety-team members who can provide information regarding certain safety issues.
- Provide relevant safety education and training to all managers, supervisors, employees, new hires, and contractors (when appropriate).
- Provide relevant and timely refresher education and training to all managers, supervisors, and employees that emphasizes any safety-related refinements and the beneficial outcomes expected.
- Present regularly the results of safety-improvement processes to employees, supervisors and managers.
- Share the latest data that demonstrate both site-wide and area-specific levels of employee involvement in safety-related activities and resultant improvements.
- Display the latest safety-relevant process data (quantitative and qualitative) in safety meetings, newsletters, bulletin boards, and other communication outlets.
- Advertise safety-process successes through various channels such as safety meetings, bulletin boards, newsletters, and group emails.
- Provide one-on-one supportive feedback and positive recognition for employees who are actively engaged in safety processes.
- Send personal thank-you cards to employees who frequently provide high-quality participation in safety processes.

- Provide group celebrations (e.g., steak dinners) for safety-process achievements.
- Provide surprise “now-that” rewards (e.g., caps, shirts with a relevant safety message or logo) to employees who demonstrate notable dedication toward safety-process activities.
- Hold supervisors accountable to support designated safety processes.
- Praise supervisors who effectively encourage and support a specific safety process.
- Provide adequate time, financial resources, and moral support for the process committees/teams.
- Hold regular meetings with relevant leadership and process committees/teams to discuss successes, barriers, and overall progress of specific safety-improvement programs.
- Ensure the safety-process committees/teams have adequate representation from various shifts, areas, and organizational levels.
- Encourage the safety-process committees/teams to recruit new members in order to maintain diversity, fresh outlooks, and new ideas.

Facilitating Self-Motivation and Empowerment

The key to developing involvement in and ownership of a safety-improvement process is to provide opportunities for personal choice regarding relevant goal-setting and achievement. Therefore, if an employee is not taking appropriate ownership and personal responsibility, it may be useful to assess how that employee feels about the safety process and related target behaviors they are expected to perform. For example, it may be useful to ask if the employee can answer affirmatively to three questions derived from social learning theory (Bandura, 1997).

First question: "Can you do it? Do you have the resources, the time, the knowledge, and the skills to handle the assignment?” But knowing you have the relevant skills and resources to perform a particular task is not enough to feel empowered and self-motivated. Thus, the second question: “Will it work? Do you believe that performing the behavior or completing the assignment will lead to the desired outcome or contribute to a valued mission of the organization, work team, or the individual?” In the domain of safety, this translates to believing a particular injury-prevention process (e.g., a BBS coaching process) will contribute to achieving the vision of a workplace with zero harm to employees.

While training may be necessary for a “yes” answer to the first question, education might be needed to persuade a person that “it will work.” People might believe they can accomplish a particular process or task, but not believe such an accomplishment will make a beneficial difference in an ultimate desirable outcome. The education might include an explanation of an evidence-based principle, and the presentation of convincing data.

The third question is a motivational question: "Is it worth it?" The performance of relevant behavior is motivated by anticipating a positive consequence to achieve or a negative consequence to avoid. Are the consequences sufficient to motivate the effort needed to accomplish the task?

Consider an emergency stop button on an assembly line. If we ask the line-workers whether they believe they are physically able to hit the E-stop, we would almost certainly get a "yes" answer. Although certain actions require a great deal of training, such is not the case for the E-stop, and this condition could be readily met. Likewise, it is reasonable to expect that employees believe hitting the E-stop would actually stop the line and control/eliminate a hazard. We would not need to lay out the electrical schematics to show how the button can disengage power to the line.

However, in this example, a "yes" to the motivational question is not obvious and depends somewhat on the anticipated consequences. What happened to the last several people who hit the E-stop? Were they thanked? Were they chastised for "slowing down production" or wasting product? Were they put on a less desirable line or job? The answer to these questions from past experiences will greatly influence employee perceptions when determining the cost/benefit analysis related to this behavior—and will therefore greatly influence its occurrence.

29.8 PERSON STATES

Intervention procedures influence how we feel, and how we feel can be pleasant or unpleasant, desirable or undesirable. In other words, internal feelings or attitudes are influenced indirectly by the type of behavior-focused intervention procedure implemented, including the delivery method. Relationships between techniques and feelings or person-states require careful consideration by the developers and managers of a BBS process.

The rationale for using more positive than negative consequences to motivate behavior is based on the differential feeling states provoked by positive versus negative consequences. Similarly, the way an intervention process is implemented can increase or decrease feelings of empowerment, build or destroy trust, or cultivate or inhibit a sense of teamwork or belonging (Geller, 2001c, 2002, 2014). Therefore, it can be useful to assess feeling states or perceptions that occur concomitantly with an intervention process. This can be accomplished informally by merely imagining oneself going through a particular set of intervention procedures and asking, "How would I feel?" More formal assessment methods could include structured interviews and group discussions or a perception survey (O'Brien, 2000; Petersen, 2001).

Three types of questions appear on the SPS Safety Culture Survey relevant to assessing different person factors: (1) Do participants agree with the concept? This is accomplished by asking whether people feel they *should* perform a certain safety-related behavior (e.g., "Should employees caution their co-workers when they are observed working at-risk"). (2) Are participants *willing* to perform certain

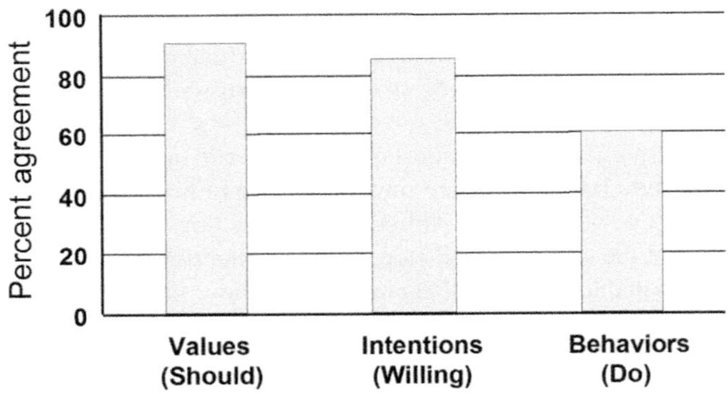

Figure 29.6. Percent agreement as a function whether one "should," "is willing," and "does" actively care for others.

behaviors? (e.g., "I am willing to caution my co-workers about working at-risk"). (3) Do participants *actually perform* the behavior? (e.g., "When I see a co-worker working at-risk, I caution him/her").

As shown in Figure 29.6, the results of approximately 300,000 respondents show most people respond favorably to the first two questions. Most have the necessary values and intentions. But far fewer respondents agree with the final question, indicating there are personal and organizational barriers to actually performing these critical safety-related behaviors.

Analogous to Gilbert's PIP (performance improvement potential) described in his classic text, *Engineering Worthy Performance* (Gilbert, 1978), the difference between the "should," "willing," and "do" can be used to estimate the potential for improvement, as well as informing ways to design the most appropriate intervention strategy. When employees believe they should or are willing to perform these behaviors more than they currently do, a potential for relatively quick improvement can be expected following basic education and training. But if employees say they do not feel they should, they are not willing, and they actually do not perform such behaviors, some one-on-one coaching is called for.

One possible explanation of the function of person states is to consider them establishing operations (i.e., setting events). A setting event is an environmental event, operation, or stimulus condition that affects an organism by altering (1) the reinforcing effectiveness of other events, and (2) the frequency of occurrence of that part of the organism's repertoire relevant to those events as consequences (Michael, 1993). In other words, setting events are conditions that make a positive consequence more or less rewarding.

For example, if a person has not eaten for several hours, a setting event has been established that increases the reinforcing value of food. Along these lines, having a high sense of belonging could set the stage for pro-social or actively caring behavior or make helping behavior more rewarding. Furthermore, if you like certain people (e.g., members of your in-group), helping these individuals is likely to be more internally rewarding than if you do not like them or have neutral feelings toward them.

Likewise, feelings of self-efficacy ("I can do it"), personal control ("I'm in control"), and optimism ("I expect the best") could make job enrichment (i.e., expanded duties and the offer of new challenges) more rewarding. For example, a key element of a BBS process is for co-workers to observe each other, identify both safe and at-risk behaviors, and give appropriate supportive and corrective feedback. Some people are confident in their ability to identify both safe and at-risk behaviors as they observe a co-worker, whereas others feel they do not know another person's job well enough or they are not effective communicators. Therefore, for individuals low in self-efficacy, personal control, and optimism, participation in a BBS observation/feedback process is unlikely without an extrinsic accountability or incentive/reward system.

One of our favorite quotes from W. Edwards Deming (1991): *"It is more important to have an imprecise measure of the right thing than a perfect measure of the wrong thing."* This quote is relevant here because although it is difficult if not impossible to get a true measure of the level of the person factors contributing to an ideal safety culture, these factors are part of the "right thing" we need to consider when developing safety-improvement activities.

29.9 THE BENEFITS OF BEHAVIOR-BASED SAFETY

We referred to BBS throughout the discussion of the six critical components of an ideal safety culture. Let us review benefits one can expect to gain from a successful BBS process. The most obvious benefit is the reduction of personal injuries due to an increase in occurrences in safe behavior and a reduction in frequencies of at-risk behavior. However, five other benefits are likely to result from people contributing interdependently to an effective BBS process.

Benefit 1. The Evaluation Focuses on the Right Numbers

How is safety performance measured at most facilities? It is measured by final outcomes—injuries that occur during a certain time period. Companies keep score by trying to improve their injury rates. Do workers walk around the job thinking about lowering the company's injury rates? Can they relate to a statistical number? Of course not. It is too abstract and remote. Plus, it is not really under the workers' immediate perceived control.

Top management needs to keep worrying about the outcome numbers, but not the people doing the jobs. Workers need to focus on the process—the day-to-day operations. That is what they can control, and that is the focus of BBS. When employees concentrate on what they can actually **do** for safety, they will reach the outcome everyone wants—fewer injuries.

Benefit 2: Positive Attitudes are Nurtured

Have you ever noticed how safety-related conversations often resemble an adult–child confrontation? One person holds another accountable for at-risk behavior, as in: "I saw what you did." Then safety becomes a"gotcha" game, supported by corporate rules and OSHA regulations. This heavy-handed approach only diminishes a person's feelings of empowerment, importance, belonging, and interpersonal trust.

BBS focuses on the use of rewards, positive feedback, interpersonal recognition, improvement of management systems, and the removal of barriers to motivate and support safe behavior. This encourages people to get engaged in a safety improvement process because they want to, not because they feel threatened and think they have to.

Benefit 3: Personal Responsibility for Safety Increases

People often think "accountability" and "responsibility" are the same thing. They are not. When you are held accountable, you are asked to reach a certain objective or goal, often within a designated time period. But that does not mean you feel any personal responsibility or self-motivation to meet the deadline. You might feel responsible enough to complete the assignment, but that is all. You do only what is required and no more.

When people have tools they can use on a daily basis to prevent injuries and they have support to use these tools, they have real personal control over safety. If they are held accountable for process numbers they can control, and they believe their efforts will prevent workplace injuries, they will feel responsible and do more than what is required. They will feel empowered and self-motivated to get involved in an improvement process.

Benefit 4: Interpersonal Coaching and Teamwork is Enhanced

Imagine a workplace where everyone coaches each other about the safest way to perform a job. When workers depend on others in this way to improve safety, they understand teamwork. They appreciate how everyone's safe and at-risk behavior influences the safety of everyone else. With this interdependent mindset, they are willing to use behavior-based coaching to actively care for their coworkers.

Benefit 5: Systems Thinking is Promoted

It is easy to get bogged down with handling immediate short-term demands—like production deadlines— and lose sight of the bigger picture. Systems thinkers take a broad and long-term perspective. They look beyond immediate payoffs, like the ease, speed, or comfort they get by taking a risky shortcut. They consider the possibility of a bigger payoff in the distant future. They realize their safe behavior teaches others by example and protects them from injury.

Systems thinkers understand the link between behavior and attitude. A small change in behavior can result in a beneficial change in attitude, followed by more behavior change and then more attitude change—eventually resulting in total commitment to support and maintain the safety-related behaviors required to achieve an ideal safety culture. So BBS sets the stage for systems thinking and interdependent teamwork, while determining safe-behaviors to support and at-risk behaviors to correct in themselves and their co-workers.

29.10 WEAKNESSES, LIMITATIONS, AND CRITICISMS

The following quotations from *Professional Safety* illustrate a number of criticisms of BBS. We disagree with these criticisms, as they seem to reflect a basic misinterpretation of the tools and methods of BBS, presume BBS is narrow and only about the behavior of line workers or wage employees, or they assume poor applications of the process reflect the underlying principles. However, these criticisms are still worth noting to help ensure your BBS process does not fall victim to a narrow or poor application of the principles.

- "Behavior-based safety is not an upstream venture. It is as far downstream as one can go" (Manuele, 1998, p. 34).
- "...as firms purchase packages...behavior-based safety moves toward becoming nothing more than the next program of the month" (Petersen, 1998, p. 34).
- "...too many behavior-based approaches seem to encourage management to abandon its responsibility" (Manuele, 1998, p. 35).
- "...the focus is improving worker behavior rather than an organization's culture or the work system" (Manuele, 1998, p. 33).
- "...'behavior-based' concepts...in many settings, have little practical use in productivity, quality, and other areas" (Petersen, 1998, p. 35).
- "...not enough attention is given to the reasons for the at-risk behavior...Behavior-based safety is but a small element of a successful initiative" (Zeller, 1998, p. 41).

More critical of BBS, however, have been certain labor unions, in particular the United Automobile Workers and the Oil, Chemical, and Atomic Workers International Union. Consider, for example, the following quotations from union documents. Although they are inaccurate, they further reflect the misunderstanding of the behavioral approach to occupational safety and health.

From the UAW Health and Safety Department (1999):

- "Behavior-based safety is one of the oldest and most outdated approaches to health and safety" (p. 1).
- "Behavior-based safety programs will drive problems underground, inject fear into the workplace, and discourage workers from reporting injuries and illnesses" (p. 7).
- "Behavior-based safety programs sentence workers to a work life of exposure to serious health and safety hazards" (p. 7).

From the Assistant Director, UAW Health and Safety Department (Howe, 1998):

- "Behavior-based safety programs are just a retread of old outdated ideas and strategies that have never been proven effective" (p. 5).
- "Behavior-based safety advocates…blaming workers for almost all health and safety problems" (p. 9).
- "Behavior-based programs emphasize working downstream…(and) are completely reactive recommending the use of the most outdated, ineffective methods of control" (p. 10).
- "Many behavior-based safety proponents say that regulations and standards are not needed" (p. 12).

From a representative of the Oil, Chemical, and Atomic Workers International Union (Hoyle, 1998):

- "Behavior modification programs are the least effective way to prevent accidents because they focus on the narrow issue of correcting worker mistakes" (p. 4).
- "Changing worker behavior is one of the least effective methods for accident prevention" (p. 7).
- "In all behavior modification programs the central thing that is looked for is the use of personal protective equipment" (p. 10).
- "The research shows that positive reinforcement is as bad as negative reinforcement. It is just a different side of the same coin" (p. 13).
- "Behavior modification safety programs are a perfect fit for management's avoidance of responsibility for health, safety, and accidents" (p. 15).

Moreover, the chairman of the Labor Division, U. S. Occupational Safety and Health Administration (Foster, 2000) urged the National Safety Council to develop an official policy statement for "Behavior-Based Safety Training" with the following introduction:

The National Safety Council urges caution in the implementation and administration of Behavior Modification Programs, Behavior-Based Safety Training (BBS), incentive and/or performance-based safety, peer-group safety or other similar behavior modification programs, hereinafter referred to as BBS or Behavior-Based Safety Training programs. If such programs are improperly implemented and administered, they may place blame on employees, pit worker against worker for reward competition, cause underreporting of accidents and the ability to learn from accidents and close calls. Key components of a good safety program must include adequately identifying hazards, promoting training for hazard recognition, promoting employee buy in for safe practices and hazard reduction.

These criticisms of BBS are based on either a misunderstanding of the BBS principles or the presumption that BBS leads to blaming employees or does not adequately consider the full range of contributing factors to at-risk behavior. As we have discussed in this chapter, a variety of contributing factors should be considered when determining the reasons for safe and at-risk behaviors.

These factors should include the interaction of physical environment/conditions, leadership, organizational systems, and the behaviors of all people in the organization, employee engagement, and person-states (e.g., knowledge, skills, abilities, beliefs, expectancies, and personal dispositions). The vision of zero injuries is more likely to be achieved when safety is treated as multidimensional, and when BBS principles are integrated into each dimension.

Thus, BBS addresses much more than behavior. It is about developing ownership, empowerment, personal control, systems thinking, interdependency, and a sense of belonging. It is about "actively caring."

However, a legitimate criticism of BBS may be a direct by-product of its strengths. Because autonomy, choice, and personal control are central to BBS, and most successful BBS processes attempt to customize the process to best match the existing organizational culture, there have undoubtedly been poor interpretations of the BBS principles, poor implementations, and inappropriate choices made by those acting autonomously to lead BBS efforts.

Whenever autonomy and choice are encouraged, people will sometimes choose inappropriately. However, this is not an argument to limit choice and control of the process by those leading BBS efforts. Instead, it highlights the need to ensure the appropriate principles are taught when first implementing BBS and the process includes a well-trained Steering Team to help keep the process on track. Regular improvement cycles are also recommended to ensure the process does not drift too far from the principles and does not begin to exhibit some of the flaws described above by BBS critics.

29.11 RELATIONSHIP WITH OTHER PRINCIPLES

To increase motivation to stay involved in a BBS process, it is critical to use employee input when determining behaviors and hazards to target for improvement as well as their ideas on how to best intervene. In more formal processes, observation data are also collected and used to identify behavioral trends, select target behaviors, and help guide interventions. However, more complex and thorough methods may also be used to help identify critical behaviors and hazards such as a Quantitative Risk Analysis (Chapter 22) or a Qualitative Risk Analysis (Chapter 27b).

At-risk behaviors are performed for many reasons, including (1) barriers that make the safe alternative more difficult or even impossible, (2) poorly designed systems, tools, and equipment, (3) activators in the environment directing a person to perform at-risk, or (4) consequences that punish safe or reward at-risk behavior. Further, some at-risk behaviors are performed in error. If tools and equipment are not well matched to the capabilities of the people performing the work, it may lead to at-risk behavior. This concept is detailed in Chapter 7 (Principles of Human Factors Engineering).

Many activators and consequences of at-risk behavior are created by or a function of organizational systems. Although organizational systems were discussed extensively in this chapter as possible contributing factors to safe and at-risk behavior, Chapter 30 (Safety Management Principles) provides additional background, detail, and history of safety management systems.

Regarding human error as a possible cause for at-risk behavior, Human Reliability Analysis (Chapter 23) provides more detail for understanding and methods for reducing human error. Another link between BBS and Human Reliability Analysis includes a similar history of misunderstanding and mistrust.

BBS includes involving employees in identifying and analyzing the contributing factors to at-risk behavior; it is not about blaming employees. However, as the criticisms listed in the previous section highlighted, BBS has a history wherein critics have misunderstood the reasons for focusing on at-risk behavior or assumed poor applications of BBS reflect underlying flaws. Human Reliability Analysis has a similar history, where its methods and goals have been misunderstood as blaming employees for their errors. In fact, as discussed in Chapter 23, one of the reasons the term Human Reliability Analysis is used instead of Human Error Analysis, as was used more commonly in the past, is to reduce a "blame the worker" connotation and ensure the focus on contextual factors in understanding human error.

The BBS principles of employee involvement, continuous improvement, and considering contextual or situational factors when attempting to understand safe and at-risk behaviors are more likely to be applied effectively in a more open and trusting safety culture. Furthermore, employees seeing their concerns taken seriously, viewing their BBS efforts valued by their coworkers and leaders, and knowing their ideas for improvements are acted upon and appreciated can also help a less mature safety culture evolve to a more advanced state. A more detailed history and discussion of Safety Culture is provided in Chapter 32.

Interpersonal feedback is an integral component of BBS. More formal applications of BBS also include collecting and sharing feedback with work teams, sites, and the organization as a whole. Chapter 6 (Experience Feedback) provides additional background and methods for this invaluable communication tool from a management systems perspective.

REFERENCES

Al-Hemoud, A. M., and Al-Asfoor, M. M. (2006). A behavior based safety approach at a Kuwait research institution. *Journal of Safety Research*, 37, 201–206.

Alvero, A. M., Bucklin, B. R., and Austin, J. (2001). An objective review of the effectiveness and characteristics of performance feedback in organizational settings (1985–1998). *Journal of Organizational Behavior Management*, 21(1), 3–29.

Aronson, E. (1999). The power of self-persuasion. *American Psychologist*, 54, 875–884.

ASSE. (2003). Control of Hazardous Energy-Lockout/Tagout and Alternative Methods. ANSI/ASSE Z244.1-2003. ASSE, Des Plaines, IL.

Austin, J. (2000). Performance analysis and performance diagnostics. In: J. Austin, and J. E. Carr (Eds.), *Handbook of Applied Behavior Analysis* (pp. 321–349). Reno, NV: Context Press.

Austin, J., Carr, J. E., and Agnew, J. (1999). The need for assessing maintaining variables in OBM. *Journal of Organizational Behavior Management*, 19(2), 59–87.

Austin, J., Kessler, M. L., Riccobono, J. E., and Bailey, J. S. (1996). Using feedback and reinforcement to improve the performance and safety of a roofing crew. *Journal of Organizational Behavior Management*, 16(2), 49–75.

Bailey, J. S., and Austin, J. (1996). Evaluating and improving productivity in the workplace. In: B. Thyer, and M. Mattaini (Eds.), *Behavior Analysis and Social Work* (pp. 179–200). Washington, DC: American Psychological Association.

Balcazar, F., Hopkins, B. L., and Suarez, I. (1986). A critical, objective review of performance feedback. *Journal of Organizational Behavior Management*, 7(3/4), 65–89.

Bandura, A. (1997). *Self-Efficacy: The Exercise of Control*. New York: W. H. Freeman and Company.

Chen, D., and Tian, H. (2012). Behavior based safety for accidents prevention and positive study in china construction project. *Procedia Engineering*, 43, 528–534.

Choudhry, R. M. (2014). Behavior-based safety on construction sites: A case study. *Accident Analysis and Prevention*, 70, 14–23.

Deming, W. E. (1991). Quality, productivity, and competitive position. Four-day workshop presented in Cincinnati, Ohio by Quality Enhancement Seminars, Inc, Los Angeles, CA.

Foster, P.J, Parand, A., and Bennett, J. (2008). Improving the safety performance of the UK quarrying industry through a behavioural-based safety intervention. *The Southern African Institute of Mining and Metallurgy*: 108(11), 683–690.

Foster, R. D. (2000, May). *National Safety Council Position/Policy Request Form*. Dallas, TX: US/DOL OSHA.

Geller, E. S. (1991). If only more would actively care. *Journal of Applied Behavior Analysis*, 24, 607–612.

Geller, E. S. (1996). *The Psychology of Safety: How to Improve Behaviors and Attitudes on the Job*. Radnor, PA: The Chilton Book Company.

Geller, E. S. (1998). *Beyond Safety Accountability: How to Increase Personal Responsibility*. Neenah, WI: J. J. Keller and Associates, Inc.

Geller, E. S. (2000). Ten leadership qualities for a total safety culture: Safety management is not enough. *Professional Safety*, 45(5), 38–41.

Geller, E. S. (2001a). *The Psychology of Safety Handbook*. Boca Raton, FL: CRC Press.

Geller, E. S. (2001b). Behavior-based safety in industry: Realizing the large-scale potential of psychology to promote human welfare. *Applied and Preventive Psychology*, 10, 87–105.

Geller, E. S. (2001c.). From conversation to commitment. *Occupational Health and Safety*,70(1), 58–63.

Geller, E. S. (2002). *The Participation Factor: How to Increase Involvement in Occupational Safety*. Des Plaines, IL: American Society of Safety Engineers.

Geller, E. S. (2014). The psychology of AC4P. *Actively Caring for People: Cultivating a Culture of Compassion* (4th ed., pp. 35–62). Newport, VA: Make-A-Difference, LLC.

Geller, E. S. (2015). Seven life lessons from humanistic behaviorism: How to bring the best out of yourself and others. *Journal of Organizational Behavior Management*, 35(1–2), 151–170.

Geller, E. S., and Williams, J. H. (Eds.). (2001). *Keys to Behavior-Based Safety from Safety Performance Solutions*. Rockville, MD: Government Institutes.

Gilbert, T. F. (1978). *Human Competence—Engineering Worthy Performance*. New York: McGraw-Hill.

Goldstein, A. P., and Krasner, L. (1987). *Modern Applied Psychology*. New York: Pergamon Press.

Greene, B. F., Winett, R. A., Van Houten, R., Geller, E. S., and Iwata, B. A. (Eds.). (1987). *Behavior Analysis in the Community: Readings from the Journal of Applied Behavior Analysis*. Lawrence, KS: University of Kansas.

Grindle, A. C., Dickinson, A. M., and Boettcher, W. (2000). Behavioral safety research in manufacturing settings: A review of the literature. *Journal of Organizational Behavior Management*, 20(1), 29–68.

Howe, J. (1998, January). A union critique of behavioral safety. Paper presented at the ASSE Behavioral Safety Symposium, Orlando, FL.

Hoyle, B. (1998). *Fixing the Workplace, Not the Worker: A Workers' Guide to Accident Prevention*. Lakewood, CO: Oil, Chemical and Atomic Workers International Union.

Hans, M. (1996, June). Does behavior-based safety work? *Safety and Health*, 44–49.

Ismail, F., Hashim, A. E., Zuriea, W., Ismail, W., Kamarudin, H., and Baharom, Z.A. (2012). Behaviour-based approach for quality and safety environment improvement: Malaysian experience in the oil and gas industry. *Procedia—Social and Behavioral Sciences*, 35, 586–594.

Kluger, A. N., and DeNisi, A. (1996). The effects of feedback interventions on performance: A historical review, a meta-analysis, and a preliminary feedback intervention theory. *Psychological Bulletin*, 119(2), 254.

Krause, T. R. (1995). *Employee-Driven Systems for Safe Behavior: Integrating Behavioral and Statistical Methodologies*. New York: Van Nostrand Reinhold.

Krause, T. R., Hidley, J. H., and Hodson, S. J. (1996). *The Behavior-Based Safety Process: Managing Improvement for an Injury-Free Culture* (2nd ed.). New York: Van Nostrand Reinhold.

Langer, E. J. (1989). *Mindfulness*. Reading, MA, Perseus Books.

Lingard, H., and Rowlinson, S. (1997). Behavior-based in Hong-Kong's safety management construction industry. *Journal of Safety Research*, 2(4), 243–256.

Manuele, F. A. (1998). Perspectives on behavioral safety: Observations of ASSE's behavior safety symposium. *Professional Safety*, 43(8), 32–37.

McAfee, R. B., and Winn, A. R. (1989). The use of incentives/feedback to enhance workplace safety: A critique of the literature. *Journal of Safety Research*, 20(1), 7–19.

McSween, T. E. (1995). *The Values-Based Safety Process: Improving your Safety Culture with a Behavioral Approach*. New York: Van Nostrand Reinhold.

Michael, J. (1993). Establishing operations. *The Behavior Analyst*, 16, 191–206.

O'Brien, D. P. (2000). *Business Measurements for Safety Performance*. New York: Lewis Publishers.

Parand, A., and Foster, P.J. (2006) Behavioural-based safety in the minerals industry: A research based methodology carried out in the UK quarrying sector. In: Queensland Resources Council Mining Health and Safety Conference, Townsville, Queensland.

Parsons, H. M. (1974). What happened at Hawthorne? *Science*, 183, 922–932.

Petersen, D. (1989). *Safe Behavior Reinforcement*. Goshen, NY: Aloray, Inc.

Petersen, D. (1998). Behavior-based safety: Magic pill, fad, or what? *Professional Safety*, 43(8), 34.

Petersen, D. (2001). *Authentic Involvement*. Itasca, IL: National Safety Council.

Reason, J. T. (1990). *Human Error*. Cambridge, England: Cambridge University Press.

Skinner, B. F. (1938). *The Behavior of Organisms: An Experimental Analysis*. Acton, MA: Copley Publishing Group.

Skinner, B. F. (1953). *Science and Human Behavior*. New York: Macmillan.

Skinner, B. F. (1974). *About Behaviorism*. New York: Alfred A. Knopf.

SPS. (2015). Safety Performance Solutions. Available at: www.safetyperformance.com

Sulzer-Azaroff, B. (1998). *Who Killed my Daddy? A Behavioral Safety Fable*. Cambridge, MA: Cambridge Center for Behavioral Studies.

Sulzer-Azaroff, B., and Austin, J. (2000). Does BBS work? Behavior-based safety and injury reduction: A survey of the evidence. *Professional Safety*, 45(7), 19–24.

Sulzer-Azaroff, B., and de Santamaria, M. C. (1980). Industrial safety hazard reduction through performance feedback. *Journal of Applied Behavior Analysis*, 13, 287–295.

UAW Health and Safety Department. (1999). Warning: Behavior-based safety can be hazardous to your health and safety program. Available at: http://www.uaw.org/publications/h&s/0998/hs_03.html

Watson, D.L., and Tharp, R. G. (1997). *Self-Directed Behavior: Self-Modification for Personal Adjustment* (7th ed.). Pacific Grove, CA: Brooks/Cole Publishing Company.

Williams, J. H. (2010). *Keeping People Safe: The Human Dynamics of Injury Prevention*. Lanham, MD: The Scarecrow Press, Inc.

Williams, J. H., and Geller, E. S. (2000). Behavior-based intervention for occupational safety: Critical impact of social comparison feedback. *Journal of Safety Research*, 31(3), 135–142.

Zeller, D. (1998). Still Searching. *Professional Safety*, 43(11), 41.

Zohar, D., Cohen, A., and Azar, N. (1980). Promoting increased use of ear protectors in noise through information feedback. *Human Factors*, 22, 69–79.

FURTHER READING

Biglan, A. (2015). *The Nurture Effect: How the Science of Human Behavior Can Improve Our Lives and Our World*. Oakland, CA: New Harbinger Publications.

Clarke, S., Probst, T. M., Guldenmund, F., and Passmore, J. (Eds.). (2016). *The Wiley Blackwell Handbook of the Psychology of Occupational Safety and Workplace Health*. Chichester, West Sussex, UK: John Wiley & Sons.

Geller, E. S. (Ed.). (2016). *Applied Psychology: Actively Caring for People*. New York, NY: Cambridge University Press.

Holdsambeck, R. D., and Pennypacker, H. S. (2016). *Behavioral Science: Tales of Inspiration, Discovery, and Service*. Beverly, MA: The Cambridge Center for Behavioral Studies.

30

PRINCIPLES OF EMERGENCY PLANS AND CRISIS MANAGEMENT

Ann Enander

30.1 INTRODUCTION

An *emergency plan* specifies procedures for dealing with sudden or unexpected events. This kind of plan may also be referred to as an *emergency response plan*, an *emergency operations plan* or, used in a broader sense, a *contingency plan*. The plan itself is generally a written document which has been developed as the result of a planning process. The term *crisis management* refers broadly to the application of strategies and procedures to deal with sudden or unexpected negative events once they have occurred. Thus, the emergency plan is oriented toward preparing for a possible future event, while crisis management deals with the actual event or situation as it develops.

At a very basic level, individuals or households may develop an emergency plan relating to such events as outbreak of fire in the home, power-cuts, or flooding. For example, in Canada citizens are actively encouraged to make plans and collect an emergency toolkit, preparing their household to survive without outside assistance for a minimum of 72 hours. Similarly, people living in areas subject to outbreaks of bushfire in Australia are expected to have, and to be able to activate, a bushfire

Handbook of Safety Principles, First Edition. Edited by Niklas Möller, Sven Ove Hansson, Jan-Erik Holmberg, and Carl Rollenhagen.

survival plan relating to their own home and safety (Wilkinson and Eriksen, 2015). For an organization or community, however, emergency planning becomes a rather more complex process. This chapter examines some of the key aspects and challenges in planning collectively for emergencies and crises which can affect a workplace, organization, or an entire community.

When an event has occurred, we seldom use the term crisis management at an individual or household level. Dealing with the consequences of a fire in the home, loss of electricity or flood damage is generally described in terms of individual or household coping. The term crisis management has broader connotations involving the protection of threatened values and recovery of functions critical for an organization or community. One part of an emergency plan can be to prepare for and designate the various responsibilities of managing a crisis event. A key problem of emergency planning is, however, the difficulty in foreseeing what kind of event may occur and what the consequences may be. Thus, the link between what we plan for and what we are actually faced with in a critical situation to a great extent defines the value of emergency plans.

This first section of the chapter provides an overview of key components of emergency planning and relates these to the reality of actual crisis management. In the following sections, the origin and history of emergency planning is traced; key definitions, terminology, and classifications are presented; and underlying theory and theoretical assumptions are outlined. Implementation of emergency planning and crisis management, and empirical research on use and efficiency are then discussed, highlighting the lessons learned from case studies of serious events and crises. Finally, some weaknesses and limitations in the field are pointed out, closing with some brief remarks on relations to other safety principles.

30.1.1 Components in an Emergency Plan

An emergency plan will be designed and developed according to the organization involved, the range of identified responsibilities, and the perceived risk panorama. Plans may be specific to a certain type of threat, for example, planning for terror threats or pandemics. Plans may be specifically designed for a certain geographical site or activity, for example, an industrial plant, a ferry, a school, or a major sports event. Plans can also be more limited in scope, but detailed to deal with a particular aspect of emergency response, for example, psychosocial support or decontamination of an area. Thus, plans may differ considerably. Nevertheless, the basic function of the plan is to support those responsible for dealing with an emergency by providing relevant information (such as contact details, site plans), instructions as to established procedures, and guidance as a basis for decision-making. Plans need to provide necessary information, but also be as simple, practical, and realistic as possible (Choularton, 2007). Some generic features of most emergency plans (Alexander, 2002; Perry and Lindell, 2007) can be summarized to include:

- Assessment of risks and potential threats
- Identification of vulnerabilities
- Identification of protective equipment and measures
- Definition of warning and alarm systems, routines
- Preparation for evacuation and/or protection in place
- Clarification of operative responsibilities (e.g., search and rescue)
- Clarification of overall leadership/chain of command
- Preparation for information and communication (internal and external)
- Preparation for care of victims
- Procedures for documentation and follow-up of measures taken

Focusing on community planning, Alexander (2002) organizes the basic components into five categories: context (legislation, participating organizations); scenarios (e.g., hazards, vulnerabilities); emergency needs (e.g., search and rescue, shelter, damage prevention); available resources (in terms of structures, items, and competencies); and resource utilization (e.g., application, dissemination, and testing of the plan). He also provides a useful generic structure of a typical emergency plan which can be adapted to apply for different organizations and contexts (Alexander, 2002, p. 98). An example applying a similar structure, but adapted to a specific context is found in the procedures for dealing with emergencies in the chemical process industry (Mannan, 2012). A wide range of adapted templates for developing emergency plans for different contexts can be found at official and stakeholder websites such as that of the UK government (community emergency plan), FEMA, OSHA, business continuity organizations, and others.

30.1.2 Emergency Planning as a Process

Although the emergency plan is generally the more tangible product of emergency planning, it is repeatedly emphasized that it is the process, rather than the plan itself, that is of significant value (Alexander, 2002; Perry and Lindell, 2003; Drennan et al., 2015). Plans need to be continuously updated and revised to keep pace with a shifting risk panorama and organizational changes. Plans must be exercised and tested under as realistic conditions as possible. Key personnel must be familiar not only with the written plan, but also with the reasoning and considerations underpinning the plan. Thus involvement of key personnel in the development, updating, and testing of emergency plans is generally held to be crucial.

Emphasizing the process aspect of emergency planning and drawing on much of the previous work in the area, Perry and Lindell (2003) suggest 10 basic guidelines for emergency preparedness planning. According to their suggestions, the emergency planning process should:

- be based on accurate knowledge of the threat and of likely human responses
- encourage appropriate actions by emergency managers
- acknowledge the dynamic nature of disaster environments and that it is impossible to cover every contingency that might arise in the future
- address inter-organizational coordination
- integrate plans for each individual hazard scenario into a comprehensive approach for multi-hazard management
- have a training component
- provide for testing the proposed response operations
- be a continuing process
- incorporate awareness that emergency planning is most often conducted in the face of conflict and resistance
- recognize that planning and management are different functions and that the real test of the plan comes when it is implemented during an emergency

Following these guidelines in actual practice and implementing a planning process involve a number of challenges. Thus, there are a number of "tensions between the ideals of crisis preparedness and the realities of crisis" (McConnell and Drennan, 2006). One such fundamental tension relates to the high potential impact of a crisis versus the low priority given to crisis management in the everyday running of an organization. The low probability of crisis events, the perceived lack of sufficient resources, and a tendency toward preoccupation with the aftermath of previous threats are all factors contributing to low prioritization of planning for future crises (Boin, t'Hart et al., 2005). Although organizational resources are important supporting factors, the personal mindset and values of individual decision-makers have also been demonstrated to be significant in promoting involvement in emergency planning and crisis issues (Enander et al., 2015).

30.1.3 Crisis Management in Theory and Practice

At a very basic level, crisis management can be viewed as the actual implementation of the emergency plan. The match between the plan and the crisis event is, however, seldom perfect, in many cases very far from perfect. Thus, crisis management is something more than simply implementing the plan. One reason for the mismatch may be that the threat appears in a different form to that expected. For example, in 1986, the emergency planning for radioactive fallout in Sweden was designed around the

Swedish nuclear power plants, which constituted the then expected source of possible radiation. The cloud from Chernobyl passing over the country and releasing radioactive particles with falling rain in quite other areas was a scenario no one at that time appears to have seriously envisaged. Another reason for mismatch may be that the resources identified in the emergency plan are for some reason not available. When in 1998, a freak snowstorm in the space of a few hours practically buried the city of Gävle on the Swedish east coast, the municipal emergency plan was activated. However, it soon became clear that many of the persons with key functions identified in the plan lived outside the city and had no possibility to make their way in through the snowdrifts in order to assume their responsibilities.

Alarm systems can malfunction, warnings may not be heeded, unforeseen vulnerabilities may emerge. Thus, crisis management needs to be thought through and planned, but will almost always also require an element of flexibility and improvisation. Arguing the need for both *discipline* and *agility*, Harrald (2006) suggests an organizational typology based on these two attributes. Organizations low on discipline but high on agility will tend to be reactive, applying ad hoc solutions as events unfold. Organizations low on agility but high on discipline will tend toward the bureaucratic and sticking to protocol, regardless of the situational demands. The ideal organization combining discipline with agility will manage to combine structure with creativity (Pearson and Sommer, 2011), acting in a balanced but adaptive manner. Capturing the intrinsic tension between, on the one hand, planning and on the other hand, meeting the unexpected, Kreps (1991) noted:

> "Without preparedness, emergency management loses clarity and efficiency in meeting essential disaster-related demands. Without improvisation, emergency management loses flexibility in the face of changing conditions."
>
> (Kreps, 1991, p. 34).

30.1.4 Crisis Leadership

Balancing the need for structure with the need to adapt and improvise is a leadership task. Boin et al. (2005) identify five key challenges that are fundamental for leadership in crisis situations and that crisis managers need to master. The first concerns making sense of the crisis as it unfolds, recognizing the threat and the response needs. Decision-making is a further challenge, particularly with regard to the conditions of uncertainty and threatened values involved. Meaning-making captures a third challenge of communicating with others and putting the situation into a context to support understanding and coping. The two remaining challenges refer to ending the crisis and dealing with issues of accountability and blame, then to drawing conclusions and learning from the experience.

The challenges of crisis leadership make considerable demands on the cognitive and emotional competence of crisis managers. Kapucu et al. (2008) identify no less than 12 critical competencies—including creativity, flexibility, teambuilding, strategic planning, ability to motivate—required of leaders in extreme circumstances. The list can appear daunting and serves as an illustration of why crisis leadership is sometimes described as "mission impossible" (Boin and t'Hart, 2003; McConnell and Drennan, 2006). Examining an actual incident in offshore oil drilling, Crichton et al. (2005) emphasize four main categories of competencies: ability to analyze the situation, decision-making, teamwork, and leadership. Interestingly, the authors note that these cognitive and social competencies are seldom the focus in training and exercises, which instead tend to emphasize procedures and more technical aspects. Other analyses of the demands on crisis leadership have highlighted the need for both cognitive abilities (Comfort, 2007) and emotional awareness, empathy, and sensitivity (Eyre, 2006; Enander et al., 2010; Kanov et al., 2004; Madera and Smith, 2009).

30.2 ORIGIN AND HISTORY

Planning for emergencies—more accurately perhaps the *lack* of such planning—has been a topic throughout history. When Noah built his ark, he was planning for an emergency event, in his case one of which he had prior warning from a source he deemed highly reliable. Beliefs about the root causes of accidents and disasters (see, e.g., Green, 1997) and the values underlying interpretations of warning signs have fundamentally influenced how the possibilities for preparing have been assessed by different societies. Tracing beliefs about the causes of accidents and disasters and the implications for planning and management, Quarantelli (2000) describes a progression from beliefs centered around Acts of God through Acts of Nature to Acts of Men and Women, and today increasingly attributions to Acts of Society. Earlier views of emergencies and disasters as Acts of God or Nature, combined with an often fatalistic mindset, tended to render the idea of elaborate planning efforts rather meaningless. During the Enlightenment period, such views were increasingly brought into question. Writing to one another after the devastating Lisbon earthquake of 1755, Voltaire and Rousseau exchanged views on the meaning of the event. In this exchange, Rousseau provides an early example of a social scientific view of disaster (Dynes, 1999), pointing out that it was man, and not nature, that constructed "twenty thousand houses of six to seven stories" in a closely packed city. The emphasis today on the role of human intervention and social systems as causes of negative events brings the corresponding responsibilities for prevention and preparedness clearly into the public societal arena (Kapucu and Van Wart, 2006; Tierney, 2014).

In many countries, modern emergency planning at the community level can trace its roots in civil protection and defense, with strong influences from measures taken for protection of the civilian population during World War II and later the Cold War (Kapucu, 2012; McEntire, 2012b; Stone Wyman, 2012). Remnants of the thinking related to the nuclear threat influenced much of early community emergency planning, for example, in terms of building shelters, creating supply depots, and informing citizens on appropriate reactions to warning sirens. During the 1970s, a shift and broadening in the view of the threat panorama led to an increasing emphasis on generic planning and an all-hazards approach. During this period, the critical thinking of leading disaster sociologists such as Henry Quarantelli and Russell Dynes regarding the principles of planning and managing disasters was influential in challenging previous military principles of command and control in emergencies (Dynes, 1994). Their work opened up for ideas about the need for flexibility and adaptive solutions and brought emergency planning into a broader context.

The more recent development of the emergency and crisis field can be closely linked to the impact of some key events during the late twentieth century. For example, in the crisis management literature, the Three Mile Island nuclear accident in 1979 is often identified as a landmark event. In many countries, the consequences of the Chernobyl accident of 1986 revealed major gaps and flaws in community preparedness. After 9/11 (the World Trade Center attack in New York in September 2001), the perceived need to prepare for terror threats on a major scale has influenced emergency planning at many levels. These developments have also raised issues of democratic principles and ethics, such as the conflict of societal security versus personal integrity (Alexander, 2002b; Svedin, 2011). From an industrial perspective, major accidents such as the toxic disaster in 1984 at Bhopal in India brought into sharp focus the responsibilities of preparing not only for consequences within the industrial plant site, but also for those affecting surrounding communities. Events such as the In Amenas terror attack against the joint BP and Statoil gas facility in Algeria 2013 correspondingly demonstrated the need to take a broad range of possible external threats and societal conflicts into account in planning for emergencies on site (Statoil, 2013).

30.3 DEFINITIONS AND TERMINOLOGY

There is a notable lack of consensus concerning how to define many of the key concepts in the emergency and crisis field. An *emergency* can generally be defined as an exceptional event that exceeds the capacity of normal resources and organization to cope with it (Alexander, 2002). More disputed are the concepts of *crisis*, *disaster*, and *catastrophe*. While an event categorized as a disaster always involves destruction or casualties in some form, it has proved difficult to define boundaries regarding the degree of loss or suffering justifying use of the concept (Quarantelli, 1998). Similarly, catastrophe is often conceived of as a very major disaster, but again the distinguishing boundary remains difficult to pin down (Alexander, 2002). A crisis may be closely linked to a disaster event, but may arise from other causes such as loss of reputation for a business or public administration, or from an extreme threat to an organization or society. Identifying key components of a crisis, Boin et al. (2005) suggest that such events arise suddenly and unexpectedly, include a threat to fundamental values (such as lives, environment, or basic trust in societal institutions), and require resources over and above the normal. To add to the complexity in the application of nomenclature, Van Wart and Kapucu (2011) point out that use of the terms emergency management and crisis management differs between the private and public sectors. They suggest that crisis management tends to be used broadly in the private sector to encompass all types of untoward events. In the public sector, on the other hand, there is a clearer distinction between emergency management as pertaining to anticipated and planned for incidents and disasters, and crisis management, which is reserved for situations which overwhelm the organization and call for extraordinary resources and efforts.

A key principle in emergency planning is the concept of preparedness and the conceptualization of it being possible to prepare for unknown and unexpected events. Defining preparedness has also proved to be challenging (Jackson, 2008). Critical questions concern, first, what indicators are relevant in order to capture the existence of a state of preparedness, and second, how these can be measured. Applying quantitative indicators such as the number of personnel on duty, vehicles available for evacuation transportation, etc. has obvious weaknesses. If these resources cannot be promptly activated in an emergency situation then such plans have little real preparedness value, as termed by Clarke (1999) they are mere "fantasy plans," possibly also creating a false sense of security. Discussing this problem, Jackson (2008) argues for the need for qualitative measures and assessments of reliability as an integral part of emergency preparedness. Counting the resources available is only one step, assessing the degree to which they can be relied on to be in function in an emergency situation is the crucial test of preparedness.

Other important, but also contested, concepts relating to emergency plans and crisis management are *vulnerability*, *capability*, and *resilience*. The effects of a hazard, and thus the magnitude of the emergency or disaster it causes, depend to a great extent on the vulnerability of the exposed people and environment. An earthquake in an uninhabited desert area is an event, but not a crisis or disaster. Thus, there is agreement on the importance of understanding vulnerability, but many different ways of defining the concept. Summarizing some of these differences, McEntire (2012a) points to two common components in the use of the vulnerability concept: factors that determine a disposition, susceptibility or proneness to be affected or hurt (*liabilities*), and factors that capture the ability to withstand, resist, or cope with stress and damage (*capacity* or *capability*). The concept of capability is in turn also defined in different ways. Reviewing the use of the concept, Lindbom et al. (2015) position their definition of capability within a broader framework and point to the importance of relating capability to the particular agent and task involved in the specific context.

The concept of resilience, as applied in emergency and crisis contexts, is generally conceived of as a societal factor related to vulnerability, primarily focusing on the capacity or capability aspect. Societal resilience is often defined rather simply in terms of a capacity to recover quickly from a damaging blow (Boin and McConnell, 2007). In a thoughtful discussion of various definitions of resilience in relation to societal crises, Drennan et al. (2015) point out the different implications that implicitly accompany each specific definition. For example, defining resilience in terms of "bouncing back" emphasizes reconstruction efforts to ensure recovery back to "normal" after an event. Resilience viewed in terms of adaptation moves the focus to continuous adaption to environmental changes, placing more emphasis on planning and preparedness phases. Resilience is increasingly being used in conjunction with community activities, emphasizing the importance of community networks and social capital as the foundation of "community resilience" (Kapucu and Özerdem, 2013). More recent definitions also emphasize continued functioning as a significant aspect of resilience (see also Chapter 3). Thus, for example, Tierney (2014) defines resilience as:

> "...the ability of social entities (for example, individuals, households, firms, communities, economies) to absorb the impacts of external and internal system shocks without losing the ability to function, and failing that, to cope, adapt, and recover from those shocks."
>
> (Tierney, 2014, p. 8).

30.3.1 Classifications and Typologies

In addition to attempts at defining these key concepts, a further important and debated issue has concerned development of disaster or crisis typologies. A classical distinction here is that between natural events and those that are caused by human action. Although seemingly clear, in practice, this difference is not so easy to uphold. With an increasingly anthropocentric view of the world, the role of human actions is deeply intertwined in the perceived causes and consequences of events which might at first appear "natural." Thus the effects of an earthquake depend on the degree of warning that people may have received, and on urban planning and building techniques. Natural disasters such as extreme flooding or drought are increasingly being attributed to effects of climate change and thus indirectly also to the inadequacy of human countermeasures. Despite the difficulties in applying typological concepts systematically, these attempts do reflect an important issue in relation to management of crises, namely that perceptions of the cause of an event have profound implications for the impact on those affected. The intention behind an event is significant in this regard, thus it may be relevant to make a distinction between events perceived as having causes which are natural, and those that are attributed to human error, to human negligence, or at worst to human malevolence (Weisaeth et al., 2002). Actions of direct malevolence or aggression, such as terror acts, are suggested to arouse the strongest reactions and societal repercussions and thus to make particular demands on a balanced crisis management.

The changing panorama of hazards and potential crises has stimulated further attempts to develop useful ways to classify crises to cover also dynamically changing events. One such suggestion is based on the dimensions of *predictability* and *possibilities for influence* before and during the crisis (Gundel, 2005).

30.4 UNDERLYING THEORY AND THEORETICAL ASSUMPTIONS

Theoretically, the field of emergency and crisis management draws upon a broad range of theories and frameworks from different disciplines. For example, McEntire (2004) illustrates how the concept of vulnerability alone can be viewed from at least 15 or so different disciplinary perspectives including geography, meteorology, engineering, anthropology, economics, sociology, and psychology. In a similar vein, Drabek (2004) points out that there is no one integrated theory of emergency management, but that the field draws on a broad repertoire of theoretical and conceptual inputs which he categorizes broadly as *normative theories*, *broad perspectives*, and *micro theories*. Normative theories are numerous in the literature, focusing on the challenges of crisis management and presenting prescriptive lessons for action. Examples of broader perspectives are particularly found in the area of social science, focusing on explaining and predicting human behavior. Micro theories have a more limited application, for example, relating to specific areas such as communication of risk or responses to warnings. Drabek also includes a category of *embryonic theories*, pointing to developments in the emergency management field which, although promising, remain sketchy and underdeveloped.

It has been suggested that one weakness in the theoretical development of the crisis management field has been an overly strong focus on the actual crisis as a unique event, neglecting to explore the broader temporal, organizational, or societal context (Roux-Dufort and Lalonde, 2013). Some attempts have, however, been made to place the field of crisis management in a more general theoretical context, for example, in relation to mainstream organizational theory (Lalonde, 2007; Roux-Dufort, 2007) or as a special case of change management (Van Wart and Kapucu, 2011).

30.4.1 The Emergency Response Cycle

One commonly accepted framework in the emergency/crisis field identifies four phases in the disaster/crisis cycle: mitigation, preparedness, response, and recovery. These phases are linked to different strategies and activities, although the boundaries between them may be indistinct and they are to a considerable extent interdependent. Mitigation refers to efforts to eliminate the causes of an emergency or crisis and is thus oriented toward prevention. Mitigation activities may be directed toward reducing the likelihood of an event or to limiting the negative consequences should it occur. Building dams and levees to limit flooding or storing dangerous goods under rigorous protective measures to prevent leakage or explosion are examples of mitigation activities. Preparedness is oriented toward building up measures to protect threatened values such as lives, health, property, or the environment in the event of an emergency occurring. Having warning systems in place, training personnel, and building up stored resources such as protective equipment and reserves are examples of preparedness activities. Both mitigation and preparedness are linked to preplanning prior to an event actually occurring.

The third phase, response, is linked to the materialization of a threat and involves activities to deal with the ensuing impact and consequences. Response activities are initiated when the threat is detected, for example, rising water levels threatening a community or an uncontrolled industrial process in imminent danger of explosion. Dealing with the source of the threat to limit the damage and protecting the values that are threatened are the two key categories of response activities. In the case of a flood scenario, this might involve minimizing the flood damage by transporting sandbags, building up barriers, evacuating people, and salvaging property. Defining when the response phase transfers into the final recovery phase is not always completely straightforward. Generally, this transition is linked to the stabilization of the immediate threat situation, for example, when the risk of flooding has been abated or the industrial process is again under control. Recovery activities then deal with restoration of functions affected by the event. These functions may be material, in the form of damaged infrastructure and environment, but are also often social and psychological. It can be difficult to assess when the recovery phase can be considered to be completed, since the long-term impact of emergencies and crises may cast a long shadow on the organizations and communities affected (Enander, 2006; Enander et al., 2009). It should be said, however, that the post-crisis impact is not always entirely negative, since a crisis can also present an opportunity for change and development (Boin et al., 2008; Brockner and James, 2008; Lerbinger 2011).

30.5 USE AND IMPLEMENTATION

Facing a complex risk panorama, increased media exposure, and pressure of public expectations, few organizations or communities are unaware of the need to plan for and be able to deal with emergencies and crises. The road from awareness to implementation is, however, paved with a number of hindrances, including low prioritization of rare and unusual events, lack of coordination between and within organizations, and a focus on symbolic rather than active planning measures (McConnell and Drennan, 2006). Thus although examples of emergency plans abound, their value is difficult to assess. Examining over 30 community emergency plans and finding a striking variety and lack of homogeneity among these documents, Alexander (2005) argues for the need to develop standards for creating, evaluating, and approving emergency plans. Basing his proposals on guidelines and criteria such as those of Perry and Lindell (2003, see Section 1), Alexander suggests 18 principles as criteria for emergency planning standards. While many of these principles can be viewed as basic common sense, the call for criteria and standards to evaluate emergency plans is relevant in an area where the term "emergency plan" can cover a multitude of weaknesses.

Experience of an actual crisis provides an opportunity to learn from the event and implement these lessons in planning for future emergencies. While this may seem self-evident, there are a number of barriers to learning from crises at both the individual and organizational level (Boin et al., 2005; Deverell, 2009; Smith and Elliott,

2008). An interesting example where an organization commissioned an investigation into a crisis event with the express purpose of improving future emergency preparedness is that of the Statoil company and the attack by terrorists on the In Amenas gas facility[1] (Statoil, 2013). Among the recommendations concerning emergency planning which are made public in the investigation report, the commission highlights the overall need to distinguish between safety and security risks, to include low probability risks in planning scenarios if the potential impact of these risks is great, and to develop awareness of local context in terms of culture and customs. The commission also points to the need to establish standards for emergency planning and to review and assure existing joint venture emergency response plans.

30.6 EMPIRICAL RESEARCH ON USE AND EFFICIENCY

Emergency planning and crisis management as topics of research have a relatively recent history. Much of the discussion regarding the use and efficiency of emergency planning is based on anecdotal evidence rather than empirically grounded research. Clearly, there are considerable methodological problems in studying and evaluating the planning for and realization of management of extreme and rare events. Thus, the research field has evolved predominantly through case studies of particular disasters and crises, focusing on drawing lessons regarding the factors related to success or failure in the management of the situation.

One significant issue has concerned the possibility to draw general conclusions from an array of completely different events and circumstances. Advocating more work in this vein in order to draw cross-sectoral lessons, Crichton et al. (2009) review seven incidents (from the United Kingdom and international) to identify recurring themes which can be applied more generally to explore the resilience of emergency plans. Key themes include emphasizing the process of preparing for emergencies rather than the outcome in the form of a plan; recognizing tendencies to underestimation of the risk scenarios; integrating a strong safety culture into emergency response systems; developing and clarifying command and control; communicating with the public; attending to welfare long term; training responders in non-technical skills; and assuring the capability and availability of resources. Although drawn from widely differing sectors, threats, and circumstances, the authors suggest that these lessons can be broadly applied to examine emergency planning and increase organizational resilience. Similarly, Howitt and Leonard (2009) draw some general conclusions from a collection of in-depth case studies from a number of different events including hurricane Katrina, SARS in Toronto, the 1992 Los Angeles riots, and the Anthrax crisis in

[1] In January 2013, the In Amenas gas facility in Algeria, jointly owned by BP, Statoil and the Algerian company Sonatrach, was attacked by 32 heavily armed terrorists. Of the 800 people on the site, many were taken hostage in a siege that lasted four days. 40 innocent people from 10 countries lost their lives in the attack.

the US postal service. In their analysis, the authors emphasize the different response modes relating to, on the one hand, routine emergencies (familiar and relatively frequent) and on the other hand, crises (novel serious events).

Training and exercises form an important part of the emergency planning process. Although the significance of exercises for managing future crisis situations is acknowledged, it has been suggested that exercises tend to produce weak learning results that are not applicable on a variety of crisis situations (Borodzicz and van Haperen, 2002; Borell and Eriksson, 2013). This criticism refers to the fact that exercises may produce "symbolic readiness," a notion of operational preparedness, leading exercise participants to think they are fully prepared for a real contingency (McConnell and Drennan, 2006). In contrast, however, there are examples of empirical case studies that have shown that exercises can produce learning among the participants. Thus, exercises have been demonstrated to create conditions for individual and organizational learning (Nilsson, 2009) and furthermore can increase the participants' perceived knowledge about incident management systems and the perceived level of confidence regarding ability to work as a team (Perry, 2004). Participants in the planning process need also to maintain awareness of societal context and reactions outside the immediate organization in the event of a crisis, thus finding ways to capture citizen participation and the role of the citizen perspective have come to be emphasized also in exercises (Scott et al., 2015).

30.7 WEAKNESSES, LIMITATIONS, AND CRITICISM

The basic purpose of emergency planning is to create a state of preparedness within an organization or community. The process of planning is assumed to lead to a higher level of preparation to meet a realized threat. Preparation is in turn assumed to result in improved performance in actually dealing with the threat when it materializes. Although intuitively feasible, this line of reasoning has been questioned on a number of counts. One issue concerns whether planning in fact does lead to better preparedness. Providing some perspective to the perceived panacea of contingency planning, Eriksson and McConnell (2011) point out that such planning does not guarantee a successful response once the crisis occurs, nor does lack of planning automatically lead to failure. While acknowledging the importance of planning for emergencies, this perspective also calls for realistic expectations as to what planning can achieve.

Emergency planning is almost inevitably influenced by the experiences and expectations of those involved in the process, thus raising the risk of "planning for the last crisis" (McConnell and Drennan, 2006; Brändström, 2004). In the wake of a disaster event, there is a strong and immediate demand for measures to prevent that particular event from ever happening again. This can lead to measures being taken which are geared toward the last event, but which may create new weaknesses relating to other kinds of events. An interesting example is the issue of flight safety and access to the cockpit during flight. After 9/11 2001, the focus on decreasing the risk of hijackings led to a number of measures to increase the security of the cockpit and to limit unauthorized access from within the aircraft. When in 2015, the captain of a Germanwings aircraft was unable to re-enter the cockpit to prevent his co-pilot from crashing the plane into the alpine mountainside, these security measures immediately came under debate. The focus was now turned to the risks of a (potentially unstable or suicidal) pilot being left alone in the cockpit. A task force set up by the European Aviation Safety Agency recommended that the principle of "two persons in the cockpit at all times" should be maintained, but also emphasized the need to improve medical and psychological monitoring of pilots (European Aviation Safety Association, 2015). In their deliberations, the task force considers the possible risks of increasing the pool of crew members with access to the cockpit, pointing out the need for training and possibly additional security screening as extra precautionary measures. Interestingly, one of the arguments highlighted by the task force in formulating their recommendation is the reactions of the general public, citing "public confidence in and acceptance of the measure" (EASA, 2015, p. 7) as a significant aspect.

30.7.1 Myths and Misconceptions

A more subtle weakness in both emergency planning and actual crisis management may be erroneous expectations regarding reactions and needs among the public. A long tradition of social science research has demonstrated persistent myths, misconceptions, and stereotypical images in reports of citizen behaviors in disasters (Quarantelli and Dynes, 1972). For instance, crude stereotypes of irrational citizens involved in rampant looting and violence permeated American media during Hurricane Katrina, images which were much exaggerated and oversimplified (Tierney, 2006). In contrast, a strong body of empirical evidence convincingly demonstrates that individuals do not generally panic, loot, or act disruptively in disaster situations, but rather tend to act in a rational manner in relation to how they perceive the situation (Helsloot and Ruitenberg, 2004). However, in spite of this contradicting evidence, it nevertheless appears that misconceptions about human behavior tend to be tenacious among crisis managers (Fischer, 1998; Alexander, 2007; Wester, 2011). One contributing factor may be the pervasive, but subtle framing of citizen reactions to crisis in media reporting from emergencies and disasters, conveying implicit messages concerning vulnerability as well as expected and appropriate/inappropriate reactions (Nilsson et al., 2016).

Clearly, if emergency planning is based on incorrect assumptions, management responses will tend to be dysfunctional rather than adaptive, and use of resources will be less efficient (Alexander, 2007). Misleading images and unrealistically negative beliefs concerning the public may hamper the planning of efficient emergency response (Ockerby, 2001), resulting in misdirected resources and failure to adequately disseminate appropriate crisis information (Tierney et al., 2001). The assumption of public panic may, for instance, lead to reluctance in giving timely warnings or information about an approaching threat to citizens at risk (Fischer, 1998). This could, in turn, lead to a further complication in emergency response since people tend to be less inclined to follow instructions if the information is unclear or incomplete (Perry and Lindell, 2003).

30.7.2 Success or Failure

In the aftermath of a crisis, the management of the event inevitably comes under scrutiny and is often evaluated in terms of success or failure, both in more official inquiries and by the media. For example, the national and federal management of Hurricane Katrina has been popularly termed "the grand failure" (Farazmand, 2007), echoing the title of the Select Bipartisan Committee report on the disaster ("A failure of initiative," U.S. House of Representatives, 2006). A similar labeling has come to be associated with the inadequate actions of the Swedish Government after the 2004 tsunami, in which over 500 Swedish citizens lost their lives (Swedish Tsunami Commission 2005:104, Strömbäck and Nord, 2006). Although these kinds of judgments often quickly become established truths, McConnell (2011) points to a problematic lack of academic work to analyze and discuss the relevant criteria for success or failure of crisis management. McConnell suggests three key dimensions pertaining to the success or failure of crisis management: the *processes*, the *decisions*, and the *politics*. In this suggested framework, processes refer to the plans and procedures that were or were not followed, decisions refer to both operational and political-strategic levels of decision-making, and the dimension of politics captures issues of acceptance, support, and blame. This latter dimension illustrates that there can be situations where processes and decisions are in line with planning and accepted "best practice," but may in the public debate be questioned and fail to gain acceptance due to later developments, conflicting views, or lack of trust.

Over time, the evaluations of success or failure may also be revised. The actions of Swedish authorities in arranging for mass vaccination of the population when the A(H1N1) influenza developed into a global pandemic in 2009 was a schoolbook example of the implementation of a well-established contingency plan which had been in place for a number of years. The vaccination campaign was initially by most accounts judged a success, since over 60% of the Swedish population became vaccinated (Börjesson and Enander, 2014). However, the later identification of a number of cases of narcolepsy among young people linked to the vaccination brought this view into serious question, at least in the public debate (Scott and Enander, 2017).

30.8 RELATIONS TO OTHER SAFETY PRINCIPLES

Given the broad scope and multidisciplinary nature of emergency planning and crisis management, there are numerous links to other principles of safety in different areas.

Emergency planning and crisis management deal specifically with awareness of, preparation for, and management of sudden uncertain and demanding conditions. The underlying ideas are in many ways similar to principles of reliability in organizations dealing with risky activities on a daily basis. A more recent strand of research specifically advocates a new approach to crisis management, linking more closely to organizational theory and development (Lalonde, 2007) and emphasizing structured and continuous learning processes (Robert and Lajtha, 2002), some of which closely resemble the principles characterizing such high reliability organizations (HROs). In the HRO literature, the focus is on explaining why some organizations conducting activities associated with a high degree of risk still are able to maintain a strong safety record. In the emergency planning and crisis management literature, the focus is on understanding why some organizations are better prepared and equipped to deal with sudden critical events when these do occur. Key features of systems which can successfully manage the unexpected (Le Coze, Chapter 32), which include confronting cultural constructs and establishing favorable power configurations, also closely resemble some central themes relating to successful sense-making and decision-making in crisis management (Boin et al., 2005).

The prerequisites for establishing and maintaining a high level of preparedness in an organization can be clearly linked to the concept of safety culture. Indeed, many of the key features considered important in achieving a strong safety culture, such as management commitment, awareness of safety and hazards, vigilance/mindfulness (see Chapter 28) are equally significant for emergency planning and crisis preparedness.

Communication is a key aspect of managing crises, thus applying the principles of risk and crisis communication can be seen as an integral part of emergency planning and crisis management. A recurring criticism in case studies of crisis management relates to the inadequate preparation for and implementation of communication, in particular with the public. A particular criticism has also been directed toward crisis communication from public organizations when it is characterized by a so-called "crisis management logic." This type of crisis communication takes its inspiration from the corporate world and from situations where the focus is primarily on restoring image and reputation. Particularly in crisis situations, information that comes delivered in the form of a "company message" runs the serious risk of undermining citizens' trust in public organizations (Olsson, 2014). The increasing awareness of the need for citizen input and a "bottom-up" approach in preparing and training for emergencies (Scott et al., 2015) echo current developments in risk communication theory and practice (see Chapter 11). Although a top-down approach for communication may be relevant under extreme and urgent circumstances, it has become increasingly clear that effective emergency planning and crisis management needs to be built on principles of cooperation, vigilance, and a strong awareness of vulnerabilities, reactions, and needs at all levels of an organization or community.

REFERENCES

Alexander, D. (2002). *Principles of Emergency Planning and Management.* Edinburgh: Dunedin Academic Press.

Alexander, D. (2002b). From civil defence to civil protection—And back again. *Disaster Prevention and Management*, 11(3), 209–213.

Alexander, D. (2005). Towards the development of a standard in emergency planning. *Disaster Prevention and Management*, 14(2), 158–175.

Alexander, D. E. (2007). Misconception as a barrier to teaching about disasters. *Prehospital and Disaster Medicine*, 22(2), 95–103.

Boin, A., and t'Hart, P. (2003). Public leadership in times of crisis: Mission impossible? *Public Administration Review*, 63(5), 544–553.

Boin, A., t'Hart, P., Stern, E., and Sundelius, B. (2005). *The Politics of Crisis Management*. Cambridge: Cambridge University Press.

Boin, A., and McConnell, A. (2007). Preparing for critical infrastructure breakdowns: The limits of crisis management and the need for resilience. *Journal of Contingencies and Crisis Management*, 15(1), 50–59.

Boin, A., McConnell, A., and t'Hart, P. (2008). *Governing After Crisis: The Politics of Investigation, Accountability and Learning*. Cambridge: Cambridge University Press.

Borell, J., and Eriksson, K. (2013). Learning effectiveness of discussion-based crisis management exercises. *International Journal of Disaster Risk Reduction*, 5, 28–37.

Borodzicz, E. P., and van Haperen, K. (2002). Individual and group learning in crisis simulations. *Journal of Contingencies and Crisis Management*, 10(3), 139–147.

Brockner, J., and James, E. H. (2008). Toward an understanding of when executives see crisis as opportunity. *Journal of Applied Behavioral Science*, 44, 94–115.

Brändström, A., Bynander, F., and t'Hart, P. (2004). Governing by looking back: Historical analogies and crisis management. *Public Administration*, 81(1), 191–210.

Börjesson, M., and Enander, A. (2014). Perceptions and sociodemographic factors influencing vaccination uptake and precautionary behaviours in response to the A/H1N1 influenza in Sweden. *Scandinavian Journal of Public Health*, 42, 215–222.

Choularton, R. (2007). Contingency planning and humanitarian action: A review of practice. Paper no. 59, Humanitarian Practice Network. Overseas Development Institute, London.

Clarke, L. (1999). *Mission Improbable: Using Fantasy Documents to Tame Disasters*. Chicago, IL: University of Chicago Press.

Comfort, L. (2007, December). Crisis management in hindsight: Cognition, communication, coordination and control. *Public Administration Review*, 67(s1), 189–197.

Crichton, M. T., Lauche, K., and Flin, R. (2005). Incident command skills in the management of an oil industry drilling incident: A case study. *Journal of Contingencies and Crisis Management*, 13(3), 116–128.

Crichton, M. T., Ramsay, C. G., and Kelly, T. (2009). Enhancing organizational resilience through emergency planning: Learnings from cross-sectoral lessons. *Journal of Contingencies and Crisis Management*, 17(1), 24–37.

Deverell, E. (2009). Crises as learning triggers: Exploring a conceptual framework of crisis-induced learning. *Journal of Contingencies and Crisis Management*, 17(3), 179–188.

Drabek, T. E. (2004). Theories relevant to emergency management versus a theory of emergency management. Paper presented at the Annual Emergency Management Higher Education Conference, National Emergency Training Center, Emmitsburg, MD.

Drennan, L. T., McConnell, A., and Stark, A. (Eds.). (2015). *Risk and Crisis Management in the Public Sector* (2nd ed.). New York: Routledge.

Dynes, R. R. (1994). Community emergency planning: False assumptions and inappropriate analogies. *International Journal of Mass Emergencies and Disasters*, 12(2), 141–158.

Dynes, R. (1999). The dialogue between Voltaire and Rousseau on the Lisbon earthquake: The emergence of a social science view. Preliminary paper 293, Disaster Research Centre. University of Delaware.

Enander, A. (2006). Recalling Chernobyl: Reflections among Swedish farmers. *International Journal of Mass Emergencies and Disasters*, 24(2), 251–269.

Enander, A., Hede, S., and Lajksjö, Ö. (2009). One crisis after another: Municipal experiences of storm in the shadow of the tsunami. *Disaster Prevention and Management*, 18(2), 137–149.

Enander, A., Hede, S., and Lajksjö, Ö. (2015). Why worry? Motivation for crisis preparedness work among municipal leaders in Sweden. *Journal of Contingencies and Crisis Management*, 23(1), 1–10.

Enander, A., Lajksjö, Ö., and Tedfeldt, E.-L. (2010). A tear in the social fabric: Local communities dealing with socially generated crises. *Journal of Contingencies and Crisis Management*, 18(1), 39–48.

Eriksson, K., and McConnell, A. (2011). Contingency planning for crisis management: Recipe for success or political fantasy? *Policy and Society*, 30, 89–99.

European Aviation Safety Association. (2015). Task force on measures following the accident of Germanwings flight 9525: Final report. Available at: http://ec.europa.eu/transport/modes/air/news/doc/2015-07-17-germanwings-report/germanwings-task-force-final-report.pdf

Eyre, A. (2006). Remembering: Community commemoration after disaster. In: H. Rodriguez, E. L. Quarantelli, and R. R. Dynes (Eds.), *Handbook of Disaster Research* (Chapter 26, pp. 441–455). New York: Springer.

Farazmand, A. (2007). Learning from the Katrina crisis: A global and international perspective with implications for future crisis management. *Public Administration Review*, 67, 149–159.

Fischer, H. W., III. (1998). *Response to Disaster: Fact Versus Fiction and Its Perpetuation* (2nd ed.). Lanham, MD: University Press of America.

Green, J. (1997). *Risk and Misfortune: The Social Construction of Accidents*. London: University College London.

Gundel, S. (2005). Towards a new typology of crises. *Journal of Contingencies and Crisis Management*, 13(3), 106–115.

Harrald, J. R. (2006). Agility and discipline: Critical success factors for disaster response. *Annals of the American Academy of Political and Social Science*, 604, 256–272.

Helsloot, I., and Ruitenberg, A. (2004). Citizen responses to disasters: A survey of literature and some practical implications. *Journal of Contingencies and Crisis Management*, 12(3), 98–111.

Howitt, A. M., and Leonard, H. B. (Eds.). (2009). *Managing Crises: Responses to Large-Scale Emergencies*. Washington, DC: Sage.

Jackson, B. (2008). The problem of measuring emergency preparedness: The need for assessing "response reliability" as part of homeland security planning. Occasional Paper, Rand Corporation, Santa Monica, CA.

Kanov, J. M., Maitlis, S., Worline, M. C., Dutton, J. E., Frost, P. J., and Lilius, J. M. (2004). Compassion in organizational life. *American Behavioral Scientist*, 47, 808–827.

Kapucu, N. (2012). Emergency and crisis management in the United Kingdom: Disasters experienced, lessons learned and recommendations for the future. In: D. McEntire (Ed.), *Comparative Emergency Management: Understanding Disaster Policies, Organizations, and Initiatives from Around the World.* Available at: https://www.training.fema.gov/hiedu/aemrc/booksdownload/compemmgmtbookproject/

Kapucu, N., and Van Wart, M. (2006). The evolving role of the public sector in managing catastrophic disasters: Lessons learned. *Administration and Society*, 38, 279–308.

Kapucu, N., and Van Wart, M. (2008). Making matters worse. An anatomy of leadership failures in managing catastrophic events. *Administration and Society*, 40(7), 711–740.

Kapucu, N., and Özerdem, A. (2013). *Managing Emergencies and Crises.* Burlington, MA: Jones & Bartlett.

Kreps, G. A. (1991). Organizing for emergency management. In: T. Drabek, E. Thomas, and G. J. Hoetmer (Eds.), *Emergency Management: Principles and Practice for Local Government.* Washington, DC: International City/County Management Association.

Lalonde, C. (2007). The potential contribution of the field of organizational development to crisis management. *Journal of Contingencies and Crisis Management*, 15(2), 95–104.

Lerbinger, O. (2011). *The Crisis Manager: Facing Disasters, Conflicts and Failures* (2nd ed.). Routledge.

Lindbom, H., Tehler, H., Eriksson, K., and Aven, T. (2015). The capability concept—On how to define and describe capability in relation to risk, vulnerability and resilience. *Reliability Engineering and System Safety*, 135, 45–54.

Madera, J. M., and Smith, D. B. (2009). The effects of leader negative emotions on evaluations of leadership in a crisis situation: The role of anger and sadness. *The Leadership Quarterly*, 20, 103–114.

Mannan, S. (Ed.) (2012). Emergency planning. *Lees' Loss Prevention in the Process Industries* (4th ed., Vol. 2, Chapter 24). Oxford: Elsevier.

McConnell, A. (2011). Success? Failure? Something in-between? A framework for evaluating crisis management. *Policy and Society*, 30, 63–76.

McConnell, A., and Drennan, L. (2006). Mission impossible? Planning and preparing for crisis. *Journal of Contingencies and Crisis Management*, 14(2), 59–70.

McEntire, D. A. (2004, June 8). The status of emergency management theory: issues, barriers, and recommendations for improved scholarship. Paper presented at the FEMA Higher Education Conference, Emmitsburg, MD.

McEntire, D. (2012a). Understanding and reducing vulnerability: From the approach of liabilities and capabilities. *Disaster Prevention and Management*, 12(2), 206–225.

McEntire, D. (2012b). Emergency management in the United States: Disasters experienced, lessons learned and recommendations for the future. In: D. McEntire (Ed.), *Comparative Emergency Management: Understanding Disaster Policies, Organizations, and Initiatives from around the World*. Available at: https://www.training.fema.gov/hiedu/aemrc/booksdownload/compemmgmtbookproject/

Nilsson, J. (2009). Using tabletop exercises to learn about crisis management: Empirical evidence. *International Journal of Emergency Management*, 6(2), 136–151.

Nilsson, S., Alvinius, A., and Enander, A. (2016). Frames of public reactions in crisis. *Journal of Contingencies and Crisis Management*, 24(1), 14–26.

Ockerby, P. (2001). Evacuation of a passenger ship—Is panic a major factor? *Australian Journal of Emergency Management*, 16(1), 8–14.

Olsson, E. K. (2014). Crisis communication in public organizations: Dimensions of crisis communication revisited. *Journal of Contingencies and Crisis Management*, 22(2), 113–125.

Pearson, C. M., and Sommer, S. A. (2011). Infusing creativity into crisis management: An essential approach today. *Organizational Dynamics*, 40, 27–33.

Perry, R. W. (2004). Disaster exercise outcomes for professional emergency personnel and citizen volunteers. *Journal of Contingencies and Crisis Management*, 12(2), 64–75.

Perry, R. W., and Lindell, M. K. (2003). Preparedness for emergency response: Guidelines for the emergency planning process. *Disasters*, 27(4), 336–350.

Perry, R. W., and Lindell, M. K. (2007). *Emergency Planning*. John Wiley & Sons.

Quarantelli, E. L. (Ed.) (1998). *What is a Disaster? Perspectives on the Question*. London: Routledge.

Quarantelli, E. L. (2000). Disaster planning, emergency management and civil protection: The historical development of organized efforts to plan for and to respond to disasters. Preliminary Paper #301, Disaster Research Centre, University of Delaware.

Quarantelli, E. L., and Dynes, R. R. (1972). When disaster strikes: It isn't much like you've heard and read about. *Psychology Today*, 5(9), 66–71.

Robert, B., and Lajtha, C. (2002). A new approach to crisis management. *Journal of Contingencies and Crisis Management*, 10(4), 181–191.

Roux-Dufort, C. (2007). Is crisis management (only) a management of exceptions? *Journal of Contingencies and Crisis Management*, 15(2), 105–114.

Roux-Dufort, C., and Lalonde, C. (2013). Editorial: Exploring the theoretical foundations of crisis management. *Journal of Contingencies and Crisis Management*, 21(1), 1–3.

Scott, D., Brandow, C., Hobbins, J., Nilsson, S., and Enander, A. (2015). Capturing the citizen perspective in crisis management exercises: Possibilities and challenges. *International Journal of Emergency Services*, 4(1), 86–102.

Scott, D, and Enander, A. (2017). Post-pandemic nightmare: A framing analysis of authorities and narcolepsy victims in Swedish press. *Journal of Contingencies and Crisis Management*, 25(2), 91–102.

Smith, D., and Elliott, D. (2008). Exploring the barriers to learning from crisis. *Management Learning*, 38(5), 519–538.

Statoil. (2013). The In Amenas attack. Report of the investigation into the terrorist attack on In Amenas. Prepared for Statoil ASA's board of directors.

Swedish Tsunami Commission. (2005). Sweden and the Tsunami: Examination and proposals. Swedish Government Official Report, Number SOU 2005:104.

Strömbäck, J., and Nord, L. W. (2006). Mismanagement, mistrust and missed opportunities: A study of the 2004 tsunami and Swedish political communication. *Media, Culture and Society*. 28(5), 789–800.

Svedin, L. (Ed.) (2011). *Ethics and Crisis Management*. Charlotte, NC: Information Age Publishing.

Tierney, K. (2014). *The Social Roots of Risk*. Stanford, CA: Stanford University Press.

Tierney, K., Bevc, C., and Kuligowski, E. (2006). Metaphors matter: Disaster myths, media frames and their consequences in hurricane Katrina. *The Annals of the American Academy of Political and Social Science*, 604, 57–80.

Tierney, K. J., Lindell, M. K., and Perry, R.W. (2001). *Facing the Unexpected: Disaster Preparedness and Response in the United States. Washington, DC*: Joseph Henry Press.

U.S. House of Representatives. (2006). A failure of initiative. Final report of the Select Bipartisan Committee to investigate the preparation for and response to Hurricane Katrina. U.S. Government Printing Office, Washington, DC.

Van Wart, M., and Kapucu, N. (2011). Crisis management competencies. *Public Management Review*, 13(4), 489–511.

Weisaeth, L., Knudsen, Ø., and Tønnessen, A. (2002). Technological disasters, crisis management and leadership. *Journal of Hazardous Materials*, 93(1), 33–45.

Wester, M. (2011). Fight, flight or freeze: Assumed reactions of the public during a crisis. *Journal of Contingencies and Crisis Management*, 19(4), 207–214.

Wilkinson, C., and Eriksen, C. (2015). Fire, water and everyday life: Bushfire and household defence in a changing climate. *Fire Safety Journal*, 78, 102–110.

Wyman, J. S. (2012). Emergency management in Scandinavia: Lessons learned at home and abroad. In: D. McEntire (Ed.), *Comparative Emergency Management: Understanding Disaster Policies, Organizations, and Initiatives from around the World*. Available at: https://www.training.fema.gov/hiedu/aemrc/booksdownload/compemmgmtbookproject/

FURTHER READING

Alexander, D. (2002). *Principles of Emergency Planning and Management*. Edinburgh: Dunedin Academic Press.

Boin, A., t'Hart, P., Stern E., and Sundelius B. (2005). *The Politics of Crisis Management*. Cambridge: Cambridge University Press.

Lerbinger, O. (2011). *The Crisis Manager: Facing Disasters, Conflicts and Failures* (2nd ed.). Routledge.

Perry, R. W., and Lindell, M. K. (2007). *Emergency Planning*. John Wiley & Sons.

31

SAFETY STANDARDS: CHRONIC CHALLENGES AND EMERGING PRINCIPLES

Ibrahim Habli

31.1 INTRODUCTION

Standards are highly influential in the safety domain. They aim to capture and share best or accepted practice in safety engineering and management as determined by experts in the domain. They define safety goals, objectives, or requirements for development, operation, maintenance, and decommissioning. They also describe the quality and quantity of the evidence needed for compliance. Many standards even specify the means by which the evidence should be generated by prescribing specific methods and techniques.

Standards impact, and often constrain, both industrial practice and academic research. When a company has to demonstrate the safety of their systems or operations, a central argument typically is provided based on satisfying the requirements of national or international standards. This is often used as a means for demonstrating due diligence based on compliance with best industrial practice as embodied in the relevant standards (Habli et al., 2010).

The impact of standards on academic research is also significant. New requirements or techniques, which are specified by the standards, often become the subject of vibrant research themes. For example, safety and assurance cases have emerged

Handbook of Safety Principles, First Edition. Edited by Niklas Möller, Sven Ove Hansson, Jan-Erik Holmberg, and Carl Rollenhagen.

as a key theme in safety science (Kelly, 1999), and is researched and debated extensively through dedicated research projects and international forums (Denney et al., 2013). This is, in part, due to the dramatic increase in the number of safety standards and industrial guidelines that require the submission of explicit safety and assurance cases (Rushby, 2015).

However, standards vary dramatically (McDermid and Pumfrey, 2001). Variations can take different forms, in terms of *scope*, *depth*, and *theory*. The *scope* of some standards is specific to a particular domain, problem, or type of technology. For example, the assurance document DO 178C was developed in order to provide guidance on the development of airborne software (RTCA, 2011). The scope of other standards is more generic and tends to be technology neutral. For example, the assurance standard ARP 4754A considers the development of complex aircraft systems in isolation of any particular technology (i.e., software or hardware; SAE, 2010).

In terms of *depth*, standards in the published literature are often split into the two categories: prescriptive and goal- or performance-based (Hawkins et al., 2013). Some standards are very detailed and specific in defining the basis for compliance. They prescribe the requirements, the means for compliance, and the necessary evidence. For example, the functional safety standard IEC 61508 provides extensive details about what is expected in order to achieve compliance, for example, in terms of the techniques and criteria that should be used (IEC, 2010). Other standards define high-level goals and leave it to the engineers to decide on, and justify the use of suitable means for compliance. The UK Defence Standard 00-56 is an example of goal-based standard (MoD, 2007). The requirements for compliance are limited to high-level principles and a set of mandatory safety deliverables, for example, a safety case report (see Chapter 18 for a more detailed discussion of safety cases).

Variations in *theory* are perhaps not surprising. The safety discipline does not have a unifying theory of assurance and certification, and as such, standards tend to subscribe to different principles and schools of thought. Take for instance software safety (McDermid, 2001; Habli et al., 2010a). It is a highly debated area about which safety and software standards differ in how they answer basic questions such as (Rushby, 2011): Is the treatment of safety meaningful at the software level given that the software cannot directly, that is, by itself, lead to harm? Is it counter-effective to treat the software contribution to safety and hazards as a correctness and assurance problem and assume that safety is exclusively addressed at the system level?

The lack of consensus between standards can be seen as both a hurdle and a necessity. It is a hurdle for developers of technologies that are intended for applications in different domains. These developers often complain about the cost involved in demonstrating compliance with different standards, which use different terminology and specify different requirements. Transferring compliance credit from one standard to another is never straightforward. This is particularly the case when the process involves the use of safety integrity levels (SILs) or development assurance levels (DALs) that tend to include different quantitative measures (e.g., failure rates) and qualitative indicators (e.g., development process rigour) (Redmill, 2000). In such a case, the highest SIL in one standard does not equate with its counterpart in another standard, for example, a SIL 4 system developed according to IEC 61508 is not equivalent to a DAL A system developed according to ARP 4751A.

However, the lack of consensus can also be seen as a necessity. Different nations have to comply with different legal principles (see Chapter 26). For example, legal systems vary in their treatment of cost–benefit analysis in risk acceptance decisions, as something that has to be used and justified explicitly for pragmatic reasons, or avoided for moral reasons. Further, different industries have their own business models (bespoke vs. volume-produced systems) and engineering terminology and it is inevitable that the safety practices vary in order to integrate the safety considerations into the organizational and engineering processes. We are always reminded that safety is a context-specific property. Safety standards are not an exception. They vary in order to address the specific conventions of the domain and environment in which they are used.

31.2 DEFINITIONS AND TERMINOLOGY

The Oxford English Dictionary defines a standard as a "*required or agreed level of quality or attainment*" (OED, 2016). In the context of this chapter, the quality of concern is safety. Attainment is typically described in terms of factors such as the levels of risk, integrity, and confidence that should be achieved.

In this chapter, we do not make a significant distinction between the terms "standards," "guidelines," "guidance," or "recommended practice." Instead, we focus on the unifying concepts underlying these terms, that is, objectives, evidence, and means for compliance (where applicable). For simplicity, we will also avoid the discussion as to whether a standard is mandatory or optional. For example, many standards are developed by international bodies and are adopted by some nations, regulatory bodies, or domains but not others.

31.3 ORGANIZATION OF SAFETY STANDARDS

Although the authors of standards try to avoid the prescription of a specific engineering process (e.g., agile or waterfall processes), many standards provide an overall lifecycle model in order to clarify the scope of the safety considerations (e.g., types of system failures and environmental conditions that are excluded), phases (e.g., development, maintenance, and decommissioning), and dependencies (e.g., between the different parts of the document or technologies). In addition, standards typically include a mandatory part covering the necessary requirements for compliance. This is often supplemented by an informative part that provides guidance on means for interpreting and achieving the mandatory requirements.

31.3.1 Safety Lifecycle Models

The functional safety standard IEC 61505 is one of the most widely used standards in the safety domain. The standard covers all safety lifecycle activities for "*systems comprised of electrical and/or electronic and/or programmable electronic (E/E/PE) elements that are used to perform safety functions*" (IEC, 2010).

A noticeable feature of the standard is the wide coverage of activities from concept, design, and implementation to operation, maintenance, and decommissioning. The whole approach is intended to be risk-based, driven by the results of the hazard and risk analysis and the overall safety requirements.

One of the key objectives of the IEC 61505 standard is to provide a generic basis for the development of domain-specific standards. IEC 61508 derivatives include standards in rail (CENELEC, 2011), health care (Jordan, 2006), machinery (IEC, 2005), nuclear (IEC, 2001), and automotive (ISO, 2011). The perceived advantage of this approach is that it improves consistency, for example, in terms of principles and terminology, between the different domains, leading to both safety and economic benefits.

A recent derivative of IEC 61508 is the automotive functional safety standard ISO 26262, which targets electrical and/or electronic (E/E) systems embedded in road vehicles. Similar to IEC 61505, ISO 26262 provides a full coverage of the safety lifecycle activities.

The ISO 26262 standard is centered on three interlinked V-lifecycle models corresponding to development activities at the system, hardware, and software levels. Being an adaptation of IEC 61508, the approach is risk-based and is driven by the results of the hazard analysis and risk assessment, leading to various safety goals and requirements that are defined at different hierarchies (Birch et al., 2013):

- *Safety Goals (Hierarchy 1)*—the vehicle in its environment;
- *Functional Safety Requirements (Hierarchy 2)*—the vehicle and its systems;
- *Technical Safety Requirements (Hierarchy 3)*—the E/E system; and
- *Hardware and Software Requirements (Hierarchy 4)*—component and part level.

For each hierarchy, ISO 26262 specifies items of evidence, in the form of *work products*, for satisfying these requirements. For example, in order to show that the technical safety requirements have been correctly implemented at the hardware–software level, evidence should be generated by means such as requirements-based testing, fault injection testing, or back-to-back testing.

One important aspect that safety standards emphasize is the need to demonstrate an explicit and traceable integration between the development processes and the safety activities. The lifecycles in both IEC 61508 and ISO 26262 start with a definition and an analysis of the system and environment (i.e., concept and scope definition in IEC 61505 and item definition in ISO 26262). This aims to provide an understanding of the system boundary, functions, structures and material and the dependencies between the system and its environment. An inadequate understanding of the system and environment could easily undermine confidence in the subsequent safety activities, particularly the results of hazard analysis and risk assessment.

The traceable relationship between the development and safety processes is not limited to early lifecycle activities. The expectation is for the system design and safety analyses to coevolve and drive the activities of one another. This is best exemplified in the Guidelines for Development of Civil Aircraft and Systems ARP 4754A (SAE, 2010). The ARP4754A safety assessment model emphasizes the bidirectional relationship between the main development activities (function definition, system architecture, and implementation) and the primary safety analyses, namely: functional hazard assessment (FHA), preliminary aircraft safety assessment (PASA), preliminary system safety assessment (PSSA), system safety assessment (SSA), and aircraft safety assessment (ASA). Independence between functions, systems, or items is often used as a key risk reduction strategy. As such, common cause analysis (CCA) is treated as an ongoing exercise that is used to detect and examine unintended interferences between systems that are required to be independent.

Finally, the interrelationship between the system and the safety processes is not limited to design-time activities. Safety analysis is an ongoing process. Feedback from operation forms an indispensable source of data for updating and maintaining the validity of the safety analysis results and the safety case of live systems and services (Denney et al., 2015). The relationship between safety at the design stage and safety at the operational stage is illustrated in the Federal Aviation Administration's (FAA) framework for Safety Management Systems (SMS) (AC 120-92A) (FAA, 2010), depicted in Figure 31.1. The framework makes a distinction between two interrelated processes: safety risk management (SRM) and safety assurance (SA). SRM covers the primary activities, performed during design, for hazard identification and risk assessment, including risk control. SA complements SRM by providing a process for monitoring risk levels and the effectiveness of the risk control strategies and implementations. The feedback between SRM and SA is not only used for providing ongoing safety assurance, but also as a means for understanding design and operational weaknesses and improving the effectiveness of the safety controls, aiming to help institutionalize a learning and evolving safety culture.

31.4 DOMAIN SPECIFIC PRINCIPLES

The standards discussed above tend to agree on the following overarching principle of safety assurance (Habli et al., 2010):

> *The safety management, development and operational processes should be hazard-directed and risk-based, in which rigour should be proportionate to the potential risk posed by the system.*

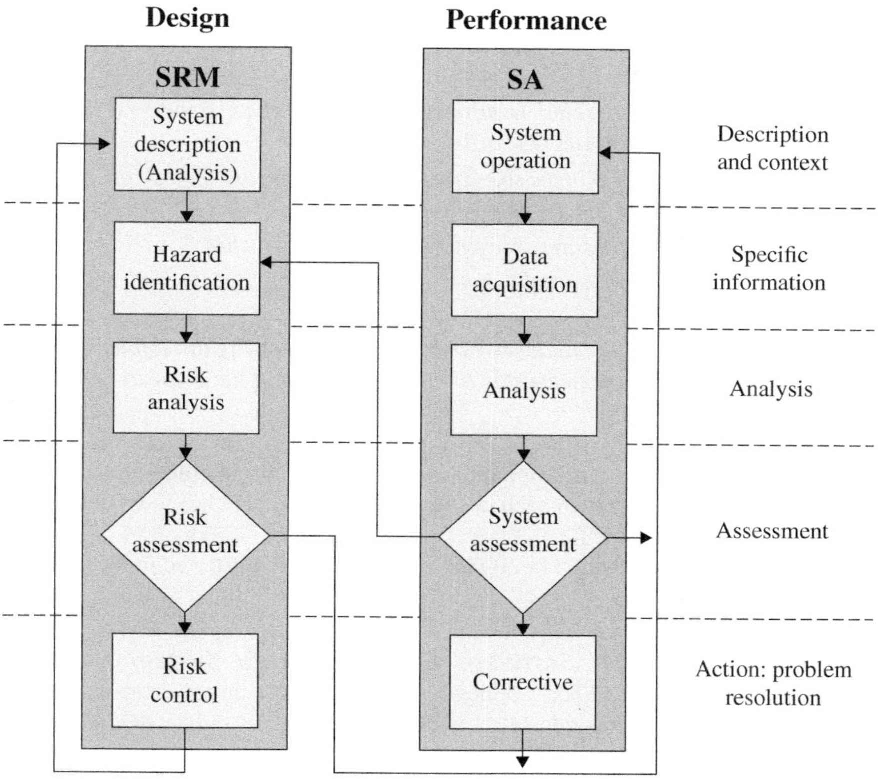

Figure 31.1. FAA SMS framework (AC 120-92A). *Source*: FAA 2010. Public domain.

Developing universal principles for safety standards is beyond the scope of this chapter. Instead, we provide examples of safety and assurance principles that relate to safety certification and approval practices and standards in two specific domains: software systems and complex automotive applications. For these principles, we also discuss how they can be used to drive the common structure and contents of the safety case for the intended domain.

31.4.1 Software Safety Assurance Principles

Software safety is perhaps one of the areas in safety science where standards vary greatly in how they approach and assure the contribution that the software makes to safety hazards (McDermid and Pumfrey, 2001). In order to help maintain a systematic and consistent understanding of the fundamental issues in software safety, Hawkins et al. (2013) defined a set of common principles that are generated from software assurance standards and industry best practice. Of course, developers will still have to comply with relevant standards. However, the principles offer a clear basis for cross-sector software certification. The principles are as follows (Hawkins et al., 2013):

- "*Principle 1: Software safety requirements shall be defined to address the software contribution to system hazards.*"
- "*Principle 2: The intent of the software safety requirements shall be maintained throughout requirements decomposition.*"
- "*Principle 3: Software safety requirements shall be satisfied.*"
- "*Principle 4: Hazardous behaviour of the software shall be identified and mitigated.*"
- "*Principle 4 + 1: The confidence established in addressing the software safety principles shall be commensurate to the contribution of the software to system risk.*"

The last principle cuts across the implementation of the other principles, dictating the degree of confidence with which each of the first four principles has to be achieved, hence the "4 + 1" and not "5"!

Below, each principle is first briefly discussed and then illustrated in the context of relevant safety standards.

31.4.1.1 Principle 1: Software Contribution to System Hazards Software in many safety-critical applications can make a direct contribution to the occurrence of safety hazards (e.g., commanding braking when not required; McDermid, 2012). As part of the safety assessment process, it is essential for the software contributions to be identified and analyzed, that is, describing the specific software failure modes that can lead to hazards. Otherwise, we are in danger of merely defining generic correctness requirements (e.g., requirements about freedom of deadlocks or stack overflows), which, although they are necessary, can fail to address the specific hazardous software failure modes that affect the safety of the system.

Safety standards cover Principle 1 by requiring the demonstration of a clear and traceable link between the hazard analysis at the system level and the safety requirements at the software level. This is clearly shown in the way in which safety requirements are allocated to software components in DO 178C, IEC 61508, and ISO 26262. For example, DO 178C requires that "*high-level requirements that address system requirements allocated to software to preclude system hazards should be defined*" (RTCA, 2011). For DO 178C, the system hazards are typically identified in the companion standard ARP 4754A.

31.4.1.2 Principle 2: Maintaining Safety Intent The development of software is typically carried out as a series of refinements, from software requirements to architecture and design, followed by coding, compilation, and the generation of the executable object code (i.e., the artifact that will ultimately run on the hardware platform). Throughout this process and at each refinement stage, the software engineers have to re-express the safety requirements in order to suit the relevant artifact, for example, requirements specified in natural language at the software/system level and then re-expressed in a mathematical notation at the coding level. During the refinement process, it is essential for the software requirements to maintain their original safety intent, that is, the meaning and significance from the safety perspective. In other words, it is important to preserve why and how the requirements relate to potential safety hazards.

Many standards deal with Principle 2 through enforcing strict objectives concerning traceability. For instance, ISO 26262 requires "*the fulfilment of the software safety requirements as allocated to the software units (in accordance with 7.4.9) through traceability*" (ISO, 2011). However, Principle 2 goes beyond mere structural traceability. It highlights the significance of upholding the meaning of, and the rationale behind, the requirements throughout the software refinement process (e.g., the correct interpretation of the system state "Aircraft on Ground" at the software level through a set of software variables).

31.4.1.3 Principle 3: Software Safety Requirements Satisfaction A combination of testing, analysis, and review results is often necessary in order to provide the verification evidence that shows that the software system has satisfied its allocated or derived safety requirements. Standards, particularly prescriptive ones, are typically detailed in specifying the techniques and the coverage criteria for satisfying the software safety requirements.

Testing in particular, that is, executing the software system and observing its behavior, tends to offer the most accepted form of evidence, at least by the regulators, since it relates directly to the actual behavior of the software system. For example, DO 178C explicitly defines the test coverage criteria, for example, in the form statement coverage, decision coverage, or modified condition/decision coverage, that are expected to be demonstrated during the verification process (RTCA, 2011). However, there is an increased interest, supported by various success stories, in the use of mathematical verification techniques for the generation of the software evidence (Bennion and Habli, 2014).

31.4.1.4 Principle 4: Hazardous Behavior of the Software The search for potential hazardous software behavior should not be limited to system-level safety analysis, but should continue during the software engineering process. That is, software is no exception compared to other technologies (e.g., structural, electrical, and mechanical components). Systematic errors and unintended behaviors of the software can result from decisions made at any stage in software development and therefore such errors and behaviors have to be identified and analyzed for their potential effects on the safe behavior of the system. For example, a decision can be made at the design level to increase software monitoring and as such increase the frequency of data sampling at certain periods. This could result in performance delays at the system level, which might not have been considered as a particular safety issue in the original safety analysis.

Safety standards vary in how they approach safety analysis at the software level. On the one hand, some imply that safety analysis is only meaningful at the system level, as performed by the system domain experts, where hazards and risks are better understood (e.g., DO 178C). Standards such as SW 01 (CAA, 2014) and the UK Defence Standard 00-55 (MoD, 2014), on the other hand, highlight the need for explicit safety analysis not only at the system level, but also at the software level as a way for dealing with emergent software behavior and errors.

31.4.1.5 Principle 4 + 1: Proportionality The first four safety assurance principles can be demonstrated with different degrees of confidence. It is well understood that absolute confidence in the safe behavior of the software is unattainable. As such, the degree of confidence in the software behavior should vary with the contribution that the software can make to system risk. That is, the quantity and rigor of the software evidence should be commensurate to the system risk, for example, higher risks imply more rigorous processes and diverse items of evidence.

The notion of proportionality is often captured in software standards through integrity or assurance levels (e.g., SILs in IEC 61508). In these instances, the standards define the measures and techniques that are necessary or are recommended to demonstrate a particular degree of confidence, for example, statement coverage for the test evidence for Level C in DO 178C. This can imply that the relationship between confidence at the software level and risk at the system level has been predefined by the standard. This is of course a highly debatable issue and tends to be avoided by goal-based standards. These standards leave it to the developers to justify, by means of explicit safety or assurance arguments, why they believe that the confidence established at the software level is commensurate to the risk at the system level, for example, Defence Standard 00-55.

31.4.1.6 The 4 + 1 Principles and Safety Cases The 4 + 1 principles can be treated as the basis for the generation of the software safety or assurance cases. The guidance on this is provided through a number of reusable argument patterns that are documented in detail in (Hawkins and Kelly, 2013). These patterns are represented using the goal structuring notation (GSN) (GSN Working Group, 2011).

Figure 31.2 shows one of these patterns, in which the claims *sw contribution, SSRidentify*, *SSRnSat*, and *hazCont* deal with Principles 1, 2, 3, and 4 respectively (Hawkins et al., 2013a). Principle 4 + 1 is covered by a separate confidence argument for justifying why the level of confidence achieved at the software level is sufficient, that is, commensurate to system risk. The loop in the argument pattern (*n*++) indicates that the claims concerning the consideration of the principles should be supported at the different tiers of development (e.g., specification, design, and coding).

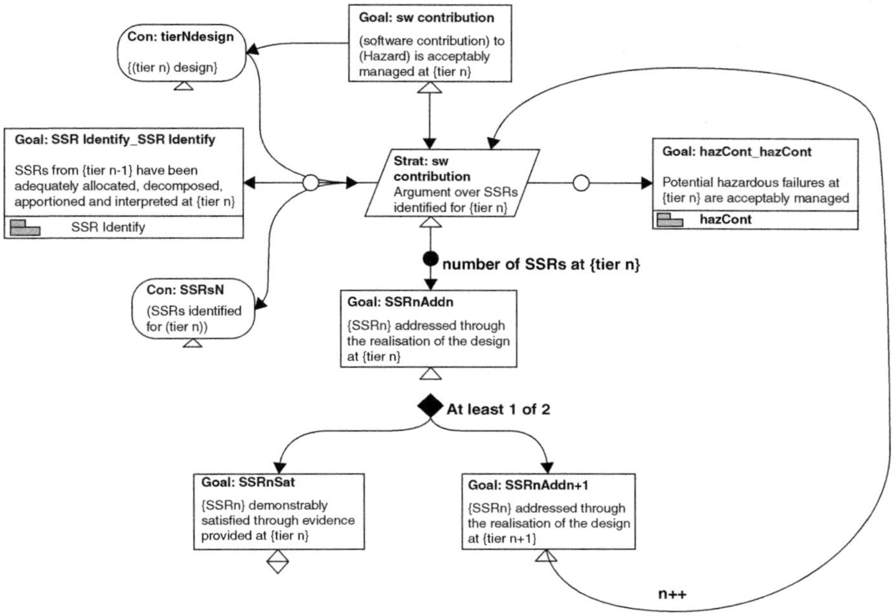

Figure 31.2. Software safety argument pattern.

31.4.2 Automotive Functional Safety Principles

The automotive functional safety standard ISO 26262 was issued in 2011 in order to provide guidance, in the form of requirements and processes, for avoiding unreasonable residual risk caused by the malfunctioning behavior of E/E systems (ISO, 2011). Compliance with the standard requires the submission of a safety case for any safety-related systems. However, ISO 26262 does not provide practical guidance on the development and review of the safety argument, nor does it describe how the safety argument should be evaluated in the functional safety assessment process (Birch et al., 2013). The use of safety cases is relatively new for the assessment of road vehicles, unlike in the nuclear and defense domains, for example, and therefore this has been seen as a challenge by the automotive industry.

To address this challenge, a working group within the Motor Industry Software Reliability Association (MISRA) is in the process of producing guidelines for the assurance of E/E systems by means of explicit safety cases (Birch et al., 2013, 2014). These guidelines are driven by four overarching principles for defining the safety claims that are necessary for any well-reasoned safety argument.

The types of these claims are captured in a *Layered Argument Model* that defines the following primary factors that have to be supported in the safety case (Birch et al., 2014):

- *Rationale*—claims concerning the adequacy of the safety requirements and the basis on which they were defined, that is, analysis and risk classification of hazardous events;
- *Satisfaction*—claims concerning the achievement of the safety requirements;
- *Means*—claims concerning the processes, methods, people, or tools used to carry out or support the definition and satisfaction of the safety requirements; and
- *Environment*—claims concerning the organizational context in which the safety-related activities have been performed.

Similar to the 4 + 1 principles for software safety assurance, a well-reasoned safety argument should address the claims that cover all of the above at each level of abstraction, for example, from system definition and hazard analysis and risk assessment to hardware and software safety requirements, their implementation, and production. Importantly, it is insufficient, from an assurance point of view, for a system to meet its safety requirements while the rationale behind these requirements is not captured and explicitly justified.

31.5 DEVELOPMENT OF STANDARDS

An important question to pose at this stage is the following: where do standards come from? Standards, particularly those providing technical guidelines, are developed in order to capture the consensus of a community concerning best practice and compliance within a particular domain. Membership of standardization committees varies. For example, it can be open, by invitation, or through nomination by a national body. Committees tend to be dominated by experts from industry and the major design, operational, and regulatory organizations. These organizations are keen to contribute to the body of work and influence the nature of the requirements and the means for compliance, including the degree of flexibility in the choice of the safety techniques and evidence. Strategically, organizations have an interest in reducing the impact of changes in standards on their current business models (e.g., to avoid retraining, reassessment, and restructuring costs).

Knight has highlighted a number of problems with existing models of safety standardization (Knight, 2014). In particular, safety is a specialized discipline and technical experts cannot afford to engage in a process that can take up to seven years. Experts in academia, who can play a key independent role, typically lack the funding necessary for engaging in such an expensive and lengthy exercise. Critical views of outsiders are commonly dismissed as belonging to those who do not have an adequate understanding and appreciation of the practice and the organizational and business constraints within the domain, that is, instead of dealing directly with the safety and technical dimension of these critical views.

Concerns also exist about the lack of empirical evidence in support of the effectiveness of the majority of safety standards. This is problematic particularly when compliance with standards is cited as primary evidence in safety cases. Generally, there is a lack of experiments and case studies for either supporting the introduction of new certification objectives or guidelines or demonstrating the effectiveness of long-established standards that have been in use within a particular domain. In order to start addressing this deficit, a set of experiments has been proposed in Graydon and Holloway (2015).

31.6 RATIONALE IN STANDARDS

Safety specialists, particularly those involved in standardization and regulatory activities, make a distinction between the "*letter*" and the "*spirit*" of the standards or highlight the importance of "*reading between the lines*." For the majority of the standards, the rationale behind the objectives and the means for compliance remain implicit. Uncovering this rationale is a challenge even for those involved in the standardization exercise, particularly when faced with legacy objectives, that is, statements or criteria whose rationale is unclear and therefore hard to maintain or remove. Many objectives can be the result of compromises. This often leads to a lack of agreement on the significance and relevance of particular parts of the safety standards.

The lack of an explicit documentation of the rationale is a key limiting factor for organizations wishing to deploy new means for compliance, which might not have been covered by the relevant standard. A common example here was the use of testing as the primary verification technique in DO 178B, which made it difficult for developers to use alternative analytical approaches, for example, Formal Methods (RTCA 1992; Galloway et al., 2005). In order to justify the adequacy of alternative means for compliance, engineers needed to have an understanding of the intent of the certification objectives and the recommended means for compliance in order to show that the alternative means for compliance were at least as effective as those published in the standard. This of course can complicate the certification or approval process and deter companies from the use of potentially more effective, though unconventional, techniques (Habli and Kelly, 2009).

The lack of a published rationale also often results in making safety standards open to different interpretations (Dodd and Habli, 2012; Graydon et al., 2012). It is likely for developers and analysts to over- or under-engineer a system in their effort to fulfill the objectives of the standards. To this end, they may be overspending to achieve certification credit or under-spending and risking an audit failure. A company can seek advice from the certification authorities, but, to maintain their independence, the authorities are restricted in the advice they can offer. Many companies can make recurring mistakes within the domain due to a poor understanding of the certification objectives that the authorities aim to assess. This can be due to lack of experience (perhaps due to turnover of staff) or lack of engagement with the standardization community.

To help overcome this challenge, efforts are currently made by both researchers and practitioners to uncover and publish the implicit assurance case underlying standards. For detailed examples, interested readers are encouraged to refer to the work of Holloway on the assurance case for DO 178C (Holloway, 2015) and the MISRA safety case group on the underlying assurance principles of ISO 26262 (Birch et al., 2013).

31.7 CHAPTER SUMMARY

Despite the lack of consensus about safety objectives and means for compliance, safety standards continue to provide valuable guidance and advice on the development and operation of safety-critical systems. This chapter has reviewed key conceptual and practical challenges that the safety-critical industry faces and discussed overarching principles for software assurance and automotive safety. The increased emphasis on greater transparency is a promising sign, particularly in highlighting the need for more inclusive standardization committees, more rigorous evidence about the effectiveness of safety standards, and an explicit documentation of the rationale behind the assurance objectives and means for compliance.

REFERENCES

Bennion, M., and Habli, I. (2014). A candid industrial evaluation of formal software verification using model checking. In: Proceedings of the 36th International Conference on Software Engineering (pp. 175–184), Hyderabad, India, May 31–June 7, 2014.

Birch, J., Rivett, R., Habli, I., Bradshaw, B., Botham, J., Higham, D., and Palin, R. (2013). Safety cases and their role in ISO 26262 functional safety assessment. In: *Computer Safety, Reliability, and Security* (pp. 154–165). Berlin/Heidelberg: Springer.

Birch, J., Rivett, R., Habli, I., Bradshaw, B., Botham, J., Higham, D., and Palin, R. (2014). A layered model for structuring automotive safety arguments. In: Tenth European Dependable Computing Conference (EDCC), Newcastle, UK, May 13–16, 2014.

CAA. (2014). CAP 670: Air Traffic Services Safety Requirements. UK Civil Aviation Authority.

CENELEC. (2011). EN50128: Railway applications-communication, signaling and processing systems-software for railway control and protection systems.

Denney, E., Pai, G., and Habli, I. (2015). Dynamic safety cases for through-life safety assurance. In: 37th International Conference on Software Engineering (ICSE 2015), Florence, Italy, May 16–24, 2015.

Denney, E., Pai, G., Habli, I., Kelly, T., and Knight, J. (2013). 1st International Workshop on Assurance Cases for Software-Intensive Systems (ASSURE, 2013). In: 35th International Conference on Software Engineering (ICSE) (pp. 1505–1506), IEEE, San Francisco, CA, May 18–26, 2013.

Dodd, I., and Habli, I. (2012). Safety certification of airborne software: An empirical study. *Reliability Engineering & System Safety*, 98(1), 7–23.

FAA. (2010). Framework for safety management system (SMS). AC 120-92A. Federal Aviation Administration.

Galloway, A., Paige, R. F., Tudor, N. J., Toyn, I., and McDermid, J. (2005). Proof vs. testing in the context of safety standards. In: Digital Avionics Systems Conference 2005, Washington, DC, October 30–November 3, 2005.

Graydon, P., Habli, I., Hawkins, R., Kelly, T., and Knight, J. (2012). Arguing conformance. *Software, IEEE*, 29(3), 50–57.

Graydon, P. J., and Holloway, C. M. (2015). Planning the unplanned experiment: Towards assessing the efficacy of standards for safety-critical software. In: Proceedings of the 33rd International System Safety Conference, San Diego, CA.

GSN Working Group. (2011). GSN Community Standard. Version 1.

Habli, I., Hawkins, R., and Kelly, T. (2010a). Software safety: Relating software assurance and software integrity. *International Journal of Critical Computer-Based Systems*, 1(4), 364–383.

Habli, I., Ibarra, I., Rivett, R. S., and Kelly, T. (2010). Model-based assurance for justifying automotive functional safety. In: 2010 SAE World Congress, Detroit, MI.

Habli, I., and Kelly, T. (2009). A generic goal-based certification argument for the justification of formal analysis. *Electronic Notes in Theoretical Computer Science*, 238(4), 27–39.

Hawkins, R., Habli, I., and Kelly, T. (2013). The principles of software safety assurance. In: 31st International System Safety Conference, Boston, MA.

Hawkins, R., Habli, I., and Kelly, T. (2013a). Principled construction of software safety cases. In: SAFECOMP 2013-Workshop SASSUR (Next Generation of System Assurance Approaches for Safety-Critical Systems) of the 32nd International Conference on Computer Safety, Reliability and Security.

Hawkins, R., Habli, I., Kelly, T., and McDermid, J. (2013). Assurance cases and prescriptive software safety certification: A comparative study. *Safety Science*, 59, 55–71.

Hawkins, R., and Kelly, T. (2013). A software safety argument pattern catalogue. YCS-2013-482. Technical Report, Department of Computer Science, University of York.

Holloway, C. M. (2015). Explicate '78: Uncovering the implicit assurance case in DO–178C. Safety-Critical Systems Club (SCSC) Annual Symposium.

IEC. (2001). 61513: 2001. Nuclear power plants-instrumentation and control for systems important to safety—General requirements for systems. International Electrotechnical Commission.

IEC. (2005). IEC 62061. Safety of machinery—Functional safety of safety-related electrical, electronic and programmable electronic control systems. International Electrotechnical Commission.

IEC. (2010). IEC 61508. Functional safety of electrical/electronic/programmable electronic safety-related systems. International Electrotechnical Commission.

ISO. (2011). ISO 26262. Road vehicles—Functional safety. International Organization for Standardization.

Jordan, P. (2006). Standard IEC 62304. Medical device software—Software lifecycle processes. In: The Institution of Engineering and Technology Seminar on Software for Medical Devices 2006 (pp. 41–47).

Kelly, T. P. (1999). *Arguing Safety—A Systematic Approach to Managing Safety Cases*. University of York.

Knight, J. (2014). Safety standards—A new approach. In: Safety-Critical Systems Club (SCSC) Annual Symposium.

McDermid, J. A. (2001). Software safety: Where's the evidence? In: Sixth Australian Workshop on Safety Critical Systems and Software (pp. 1–6), Australian Computer Society, Inc., Brisbane, Australia.

McDermid, A. (2012). Safety critical software. In: *Encyclopedia of Aerospace Engineering*. John Wiley & Sons.

McDermid, J. A., and Pumfrey, D. J. (2001). Software safety: Why is there no consensus? In: ISSC 2001, System Safety Society, Huntsville, AL.

MoD, Defence Standard. (2007). 00-56 Safety management requirements for defence systems. UK Ministry of Defence.

MoD. (2014). Requirements for Safety of Programmable Elements (PE) in Defence Systems Part 1: Requirements and Guidance, Interim Defence Standard 00-55. UK Ministry of Defence.

OED, http://www.oed.com, Oxford English Dictionary, (accessed on January 31, 2016).

Redmill, F. (2000). Understanding the use, misuse and abuse of safety integrity levels. In: Proceedings of the Eighth Safety-critical Systems Symposium (pp. 8–10).

RTCA. (1992). DO178B—Software considerations in airborne systems and equipment certification. Radio and Technical Commission for Aeronautics.

RTCA. (2011). DO178C: Software considerations in airborne systems and equipment certification. Radio and Technical Commission for Aeronautics.

Rushby, J. (2011). New challenges in certification for aircraft software. In: Proceedings of the Ninth ACM International Conference on Embedded software (pp. 211–218), ACM.

Rushby, J. (2015). The interpretation and evaluation of assurance cases. Technical Report, SRI-CSL-15-01.

SAE. (2010). ARP4754A: Guidelines for Development of Civil Aircraft and Systems.

FURTHER READING

Birch, J., Rivett, R., Habli, I., Bradshaw, B., Botham, J., Higham, D., and Palin, R. (2013). Safety cases and their role in ISO 26262 functional safety assessment. In: *Computer Safety, Reliability, and Security* (pp. 154–165). Berlin/Heidelberg: Springer.

Hawkins, R., Habli, I., and Kelly, T. (2013). The principles of software safety assurance. In: 31st International System Safety Conference, Boston, MA.

Knight, J. (2014). Safety standards—A new approach. In: Safety-Critical Systems Club (SCSC) Annual Symposium.

McDermid, A. (2012). Safety critical software. In: *Encyclopedia of Aerospace Engineering*. John Wiley & Sons.

32

MANAGING THE UNEXPECTED

Jean-Christophe Le Coze

32.1 INTRODUCTION

Somehow, many, if not all, of the chapters of this handbook bring some element of answer to how the unexpected is (or should be) managed in high-risk systems and the different ways to conceptualize this through *safety principles*. I introduce this idea in this section before opening to the problem of definition of the unexpected and the contribution of social sciences in this area. Risk assessment (Chapters 20–22), defense-in-depth principles and safety barriers (Chapters 4 and 5), safety management systems (Chapter 27) including learning (or "experience feedback," Chapter 7), maintenance (Chapters 18 and 24) also indicators (Chapter 8), resilience engineering (Chapter 3), human factors (Chapters 9, 13, 14, and 25) or safety culture (Chapter 28) are some of these key examples which define the management of the unexpected, and, of course, crisis management (Chapter 30) as well as communication about risks (Chapter 11).

They all have their specific angle. *The engineering view* details how risk assessments are (or should be) performed to produce quantitative estimates of the unexpected. First, it brings a qualitative analysis of what could possibly happen; second,

Handbook of Safety Principles, First Edition. Edited by Niklas Möller, Sven Ove Hansson, Jan-Erik Holmberg, and Carl Rollenhagen.

it tries to work out the likelihood of these events when taking into account the barriers designed to prevent or mitigate the consequences. Socio-economical studies can help justify choices. *The human factors view* (or cognitive engineering) tackles the problem of designing work situations while composing with strength and weaknesses of humans in specific material, informational, and social contexts. Recommendations for display design and functionalities, recommendations for procedures, and also team coordination and cooperation are produced to buttress safe performance (in some companies, behavior-based safety approaches are included to this human factor perspective).

The managerial view describes expectations of systems in terms of what processes (should) deliver to support the management of safety in companies. These include expected links between risk assessment, learning from experience, management of change, etc., activities which constitute the backbone of any systematic approach to safety, and described in international standards. Management keeps track of these processes through indicators which reflect the state of the system, conveyed by channels of communication which produce flow of information throughout the organization, relying on formal moments of the organizational life (e.g., learning from events, audits, inspection, meetings). When complying with these processes (or activities) which describe normatively what is expected from a system's point of view, an organization is seen as behaving rationally. Hale's graphical representation (Hale, 2003; Figure 32.1) captures well the logic behind this managerial view.

To this approach is very often now added a cultural one, which breathes life into the abstract, almost "engineering like" approach to management processes as described above. Safety culture programs are concerned with behaviors which should ensure that people act the way that is expected from them. Based on a hierarchical or nested relationship between beliefs, values, norms, and behavior (as, for instance, illustrated by the onion, Figure 32.2), the idea is to find possibilities to work on these relationships to reach expected results. Beliefs determine values and values determine norms which determine behavior. Of course, there are many debates against a simplistic perspective of the phenomenon of culture, especially in relation to the notion safety culture itself, and the possibility to influence it directly through incentives.

Finally, actors of high-risk organizations are also strongly concerned about the interactions with outsiders (e.g., regulators, civil society), deploying therefore strategies to address this topic which are not indicated in this figure (Figure 32.2), such as risk communication. This is the *governance view*, looking at the management of high-risk systems from the point of view of the interactions between the regulator, the media, civil society, justice, and the industry.

One is entitled to wonder after this short presentation "What more is there to say in the last chapter of the handbook that has not yet been said about managing the unexpected?" When putting together the chapters, one finds indeed a type of answer

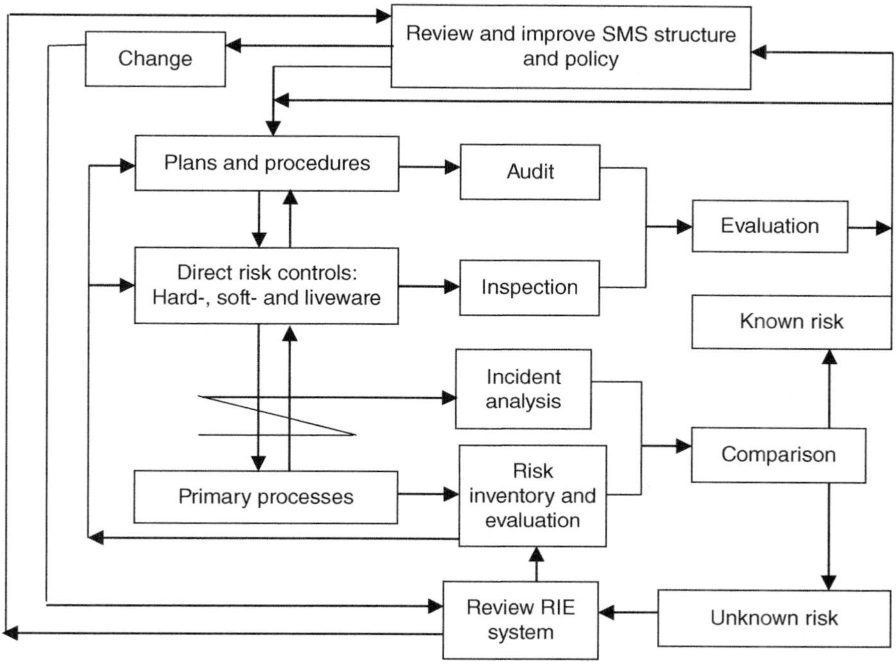

Figure 32.1. Example of safety management system model. Adapted from Hale, 2003.

to this question. Engineering, human factors, managerial, and governance principles have to be applied to ensure safe performance, to manage the unexpected. This is indeed one possible strategy.[1]

But there have been, in the last 30–40 years, the production of a rich literature analyzing disasters and daily operations of high-risk systems. This literature is based on empirical studies combined with theoretical developments, equipped with the methodological and conceptual tools of the social sciences. These works offer insights into the problem of the definition of the unexpected, its conceptualization, and examples of case studies and theoretical proposition to characterize how organizations manage the unexpected. It is this second strategy that is followed in this chapter. Let us start with three disasters, three examples to discuss the problem of the definition of the unexpected.

[1] Note that it was how Management Oversight and Risk Tree (MORT), in the 1970s (Johnson, 1973), was designed: putting together good practices into a coherent framework (see, Le Coze, 2008 for a short presentation of this normative rationale in comparison to other ones).

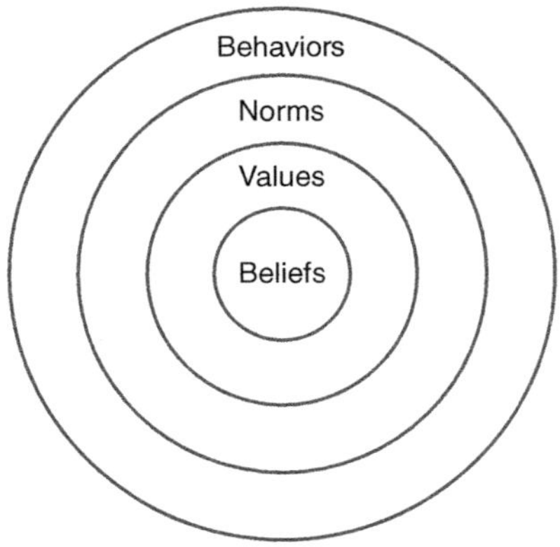

Figure 32.2. Onion representation of (safety) culture.

32.2 DEFINING THE UNEXPECTED

32.2.1 The Unexpected, What Are We Dealing With? Three Examples

Example 1:
In March 2011, a tsunami in Japan hit a wall that was designed to protect the nuclear power plant located behind it. But the wall was too low. Sea water flooded the premises where backup batteries were located, and a loss of electric power occurred as a consequence. Without power, the process operators, engineers, managers, regulators, media, officials, and civil society acted in the face of the behavior of a complex system that they had never experienced under these specific circumstances. Meltdown of the core of some of the reactors of the nuclear power plant ensued. Operators risked their lives in emergency actions, thousands of people were evacuated in the area, and the environment contaminated to an extent that is for now still left uncertain.

Example 2:
In April 2010, highly flammable pressurized gases from deep geological layers shot up to an offshore platform (Deepwater Horizon, DWH), forming a cloud which ignited and exploded. The measures designed to prevent such an event, including a blowout preventer (BOP), did not function as intended. The purpose of such a device is to cut the line connecting the offshore platform to the well in order to stop the gas from reaching the platform in case of a lack of containment of the cement designed to seal the well. Several people were killed, the platform sunk, and a pollution followed. Operators, engineers, managers, authorities, regulators, shareholders, media, and civil society faced the biggest oil pollution of US history.

Example 3:
In June 2009, sensors indicating speed in the Air France aircraft flying from Rio to Paris froze and sent false indications to pilots about plane's speed. The aircraft was,

at the time, going through a known storm area with freezing cold temperatures, lightning, and strong winds. The shift to manual handling of the flight by the crew in these circumstances was subsequently impaired by interface design issues (e.g., alarms) conveying contradictory information to pilots about aircraft's highly dynamic status while crew struggled with the understanding of the situation. The crash left no survivors and became the deadliest crash in the company's history.

32.2.2 Were These Disasters Unexpected, Surprising?

Were these disasters unexpected or surprising? We know that planes can crash, nuclear reactors can melt down, and offshore platforms can blow. If one follows the philosopher Paul Virilio, "Inventing the sailing or steam boat, it is inventing shipwreck. Inventing the train, it is inventing the railway accident of derailing (…) Objects heavier than air taking off, an aircraft but also an airship, it is inventing the crash, the air disaster. When it comes to Challenger, its explosion the same year as the Chernobyl disaster, it is the accident of a new machine, the equivalent of the first shipwreck of the first ship …" (Virilio, 2005). In that sense, these events are not totally unexpected or surprising. Their potentialities are fairly known, they come with technology.

This conundrum was aptly pointed at by Turner, one of the very first pioneers to the topic: "It becomes clear that the answer to the question 'who is surprised when the large-scale accident does occur?' is an important one for the present analysis" (Turner and Pidgeon, 1997, 1972). Large-scale man-made disasters remain relatively rare and our societies rely, on a daily basis, on the safe operations of an array of high-risk systems dealing successfully with a range of complexities, dealing successfully with the unexpected.

In this respect, Anthony Giddens, a British sociologist, characterizes one aspect of our late modernity by the trust one needs to put, in order to wake up every morning, in many organizations and institutions (Giddens, 1990). We eat food, we fly, we go to hospitals, we take the train, we drink water, we use bank services, we drive cars on diverse infrastructures (e.g., motorway, bridges, tunnels), etc. expecting these artifacts and services to meet safety requirements which depends on the expertise of interacting individuals within a diversity of organizations and institutions.

In the three types of events used here as examples in this introduction, many preventive and coping measures indeed existed for these safety-critical systems. They were built-in capacities for a range of potential situations resulting from the implementation of the safety principles briefly introduced above and developed in this handbook. A certain idea of the unexpected had indeed shaped a particular design of nuclear power plant (e.g., walls, backup batteries); offshore platforms (e.g., blow out preventer, cement), and aircraft (e.g., sensors, automation). However, on the day of the events, the specific circumstances exceeded or defeated these socially and technologically designed capabilities which had proven so far to be adequate.

One finds many reasons for this. First, there are limitations in our *engineering knowledge* of the socio-technological artifacts that we create, especially when they reach a certain degree of complexity or innovation. As a result, the practice of risk assessment contains limits in itself. Second, the complexities of human cognition still defy our best *human factors models*. As a consequence, predicting human adaptive behaviors in evolving contexts of an increasingly technological environment remains highly problematic and might be for a long time (probably forever). The prediction of the aggregation of these adaptive behaviors in large-scale networks is even more daunting for sociologists (or economists).

Third, *managerial decisions* always involve ambiguous, uncertain, and imperfect situations and resources, something well documented in the organizational literature. Therefore, it questions the rational and mechanistic view of organizations that safety management relies on, and challenges the idea that any principle could be applied perfectly, idealistically: organizations are messier than their official presentations. Fourth, interactions between civil society, the media, regulators, justice, and industry can subvert well-established *governance processes*. Governance principles remain guidance to steer a rather complex reality which depends on the configuration of powers between states, companies, and citizens which creates in turn more or less favorable conditions for safety.

I will come back later more precisely on some of these reasons and how they have been conceptualized by various authors in the social sciences through empirical studies. Why the implementation of safety principles such as the ones introduced in this handbook is subverted by human and social realities will be presented, and the converse also: what human and social properties facilitate them.

32.2.3 The Unexpected, a Highly Relative Category

One problem is, as a consequence, that the notion of the "unexpected" is highly relative, depending on the appreciations of observers. And this issue is one core or crucial tension in the field of industrial safety. *How unexpected or surprising are incidents, accidents, or disasters, and for who?* We are in fact, pretty good these days at imagining the worst case scenarios. A lot of efforts are dedicated to them. See, for instance, the range of existential risks (Bostrom, 2008), in these times of globalization and ecological awareness, that are today considered (Box 32.1). But the materialization of these kinds of high profile events is always the result of very specific circumstances that no one would be able to precisely predict in very fine details deterministically. This level of precision, of exactly where, when, how, and under which specific circumstances unwanted events will happen in complex systems; only the retrospective posture, that is, history, can offer.

It is in this sense that the unexpected has to be understood in this chapter: the future remains uncertain despite often knowing about what could potentially happen (Figure 32.3). A wave of the size of the tsunami in Japan had never hit a nuclear power plant in the past (Example 1); deep water drilling in the Gulf

Box 32.1 Range of Existential Risks

Natural catastrophes

- Collapsing star (10)
- Comets/asteroids: deadly impact (9)
- Supervolcano eruption (8)
- Hostile extraterrestrial (7)
- Natural pandemic (6)

Anthropic catastrophes (more likely in the next 100 years)

- Physics experiment (5)
- Climate catastrophe (4)
- Doomsday war (3)
- Machine superintelligence (2)
- Synthetic biology - Bioerror (1)

Ranked in terms of likelihood in Sjöström, 2010.

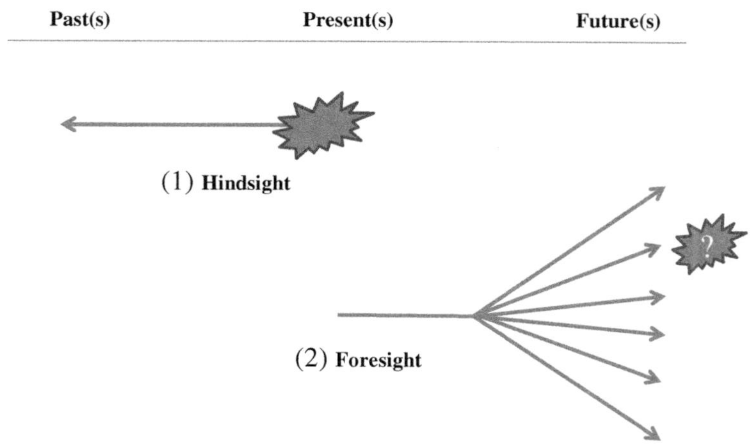

Figure 32.3. Hindsight and foresight.

of Mexico had so far proven to be safe before DWH while being an incredible achievement of managerial, scientific, and engineering expertise in hazardous endeavors (Example 2)[2]; and Air France had strong safety records over several

[2] Let us recall (it is a striking anecdote) that Obama, US president at the time, gave a speech few days before the disaster to acknowledge how safe this industry was, while giving a "political go" to expanding the offshore drilling activity in the Gulf of Mexico. This decision to grant new licenses was, of course, immediately put on hold after the event.

decades of flying complex aircrafts around the world in many different (sometimes hostile) environments (Example 3). For many people, insiders and outsiders (but certainly not all), the level of resources, expertise, and management devoted to the safe operation of these systems were adequate, until the events showed otherwise.

This is precisely why there is space for a lot of different appreciations in daily operations. As the future remains uncertain, so do any claim that would pretend to produce exact predictions. Managing the unexpected in high-risk systems is confronted to this problem of regular debates, voices, concerns, and controversies inside and outside organizations about how safe they are, for example, how tall enough a wall is (Example 1), how robust a blowout preventer is (Example 2), how well-designed a cockpit display is (Example 3). And these systems include, without any attempt to be exhaustive: nuclear power, chemical plants, aircraft, airways, dams, mining, space missions, marine transport, nuclear weapons, toxic wastes, military weapons, medical equipment, financial markets, wildfires, oil spills, food contamination (Perrow, 1999, p. 354). They have potential for wide scale impact on populations and/or the environment.

32.3 THIRTY YEARS OF RESEARCH ON THE UNEXPECTED

32.3.1 Conceptualizing the Unexpected: Four Different Threads

In this section, I come back more accurately on how has "the unexpected" been conceptualized in the field of industrial safety, beyond the description and combination of safety principles. This will serve as a basis to approach the problem of managing the unexpected. A close look at the literature indeed reveals that this notion has been sensitized in different ways by authors in the past 30–40 years. In this section, I offer a historical appreciation of these sensitivities, following three steps (and extending a first version of this classification, Le Coze, 2015a). First, I identify pioneers who developed specific lenses to offer, from a system and social point of view, an interpretation of disasters. There are, in my opinion, three of them: Barry Turner, Charles Perrow, and Jens Rasmussen. Together, they delineate four threads (Perrow accounts for two). Second, I extract what I believe to be the intellectual background of these three pioneers' contributions. By broadening the horizon with references to key thinkers in the history of the social sciences (e.g., philosophy, history, or sociology), authors such as Jacques Ellul, Karl Marx, Thomas Kuhn, or Ross Ashby, I create links to other authors in the field of safety who developed similar ideas, sometimes independently, but drawing on these thinkers. These major intellectual figures provide the basis for naming these four threads: Ellulian, Marxian, Kuhnian, and Ashbyan. Here is a box presenting the four threads (Box 32.2, Figure 32.4) before turning to explanations.

Box 32.2 Four Threads about the "Unexpected" in High-Risk Systems

- **Perrow (1) and its technological thesis**
  - An "Ellulian" thread
  - Technology out of control
- **Perrow (2) and his critical stance**
  - A "Marxian" thread
  - Failure of executives (and regulators)
- **Turner and the "failure of foresight"**
  - A "Kuhnian" thread
  - Fallible constructs
- **Rasmussen and complexity**
  - An "Ashbyan" thread
  - Self organized and emergent properties

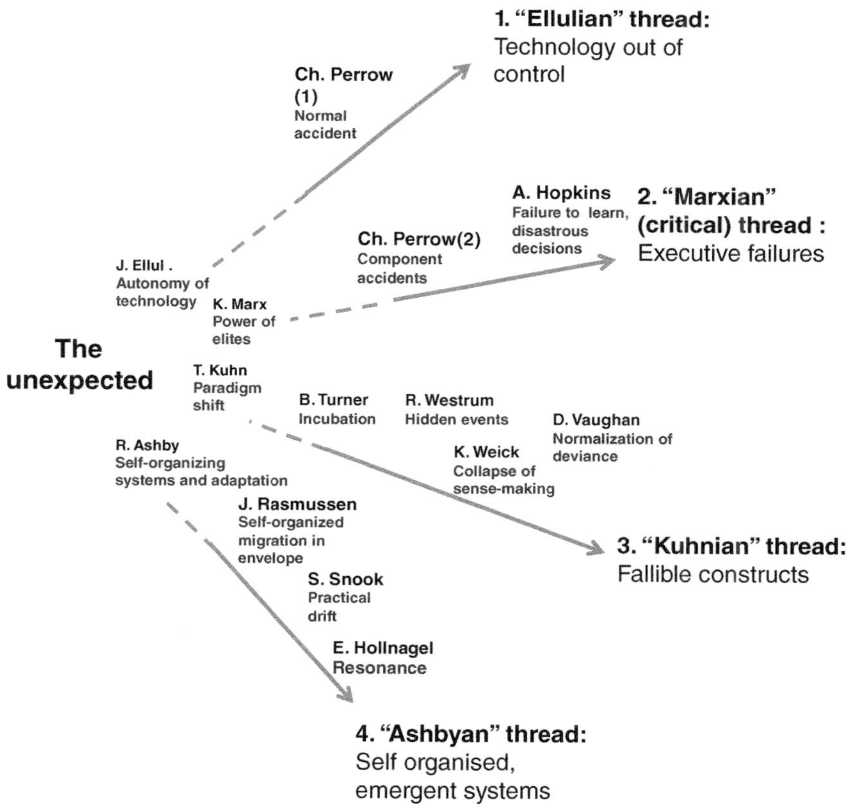

Figure 32.4. Conceptualizing the unexpected: four threads, initiators, and followers.

32.3.2 Charles Perrow and Normal Accident

32.3.2.1 An "Ellulian" Thread Chronologically, it is Turner who published his book first in 1978, but it is Perrow, an American sociologist of organizations, in 1984 who created a momentum with his thesis associating disasters with the unexpected (Perrow, 1984). I therefore start with his work. His thesis is that in tightly coupled and complex systems, accidents are, from time to time, inevitable. One conclusion that follows is that tightly coupled and complex systems with catastrophic potential should be abandoned, or made less coupled and less interactive. Where does this idea come from? It comes from Perrow's experience of taking part in a panel of experts looking into the Three Miles Island accident (TMI) (Perrow, 1982).

Impressed by the technological complexity of the nuclear power plant (revealed by how operators, designers, and engineers seemed to be baffled by the unexpected interactions leading to the TMI accident), Perrow characterized this event as a "Normal Accident." Unexpected sequences of interactions within complex and tightly coupled systems lead to, from time to time, accidents which could not be predicted, despite best efforts from everyone. They are "normal." This is one possibility of characterization of the unexpected. It is based on a strong deterministic assumption, with technology at the heart of it.

In the words of Perrow "Besides being a book about organizations (but painlessly, without the jargon and the sacred texts), this is a book about technology. (…) As the saying goes, man's reach has always exceeded his grasp. (…) Systems are too complex, and our reach has surpassed our ability to grasp. There is a bit of this philosophy in this book" (Perrow, 1999, 11, 339). This corresponds to a certain idea of technological determinism as found in the writings of philosophers during the 1970s, for example, Winner (1977), who was inspired by authors who introduced the debate on the autonomy of technology, such as Jacques Ellul (1954, 1977).

Ellul considered that technology had become, since the industrial revolution, a major influence on modern societies. For this author, technology had the ability to shape these modern societies as an independent influence, colonizing practices with a certain degree of autonomy, out of human control. Technology could be seen, along these lines, as considerably shaping societies, in a subversive way. But he was not the only writer concerned by the spread of technological systems in the 1960s, as Thomas Hughes, an American historian of technology, describes in the United States, "numerous other academics and professional writers similarly considered omnipresent technological systems out of control" (Hughes, 2005, p. 89).

Although this view applied to safety has been subsequently largely questioned by social scientists (e.g., Pinch, 1991; Vaughan, 1996; Bourrier, 1999; Hopkins, 1999a), one can see how it makes sense to characterize Perrow's emphasis on technology out of control as an "Ellulian" perspective on normal accident, even if he never refers to this author. Accidents happen because technology escapes human control. This is a quite popular version of the unexpected, and the rise of new technological systems based on artificial intelligence certainly triggers new concern about our ability to master our creation to an extent probably unseen in human history up to now (e.g., Bostrom, 2014; Hawking, 2014). In this context, the "Ellulian" thread might take on more relevance in the future, this is a first thread conceptualizing the unexpected.

32.3.2.2 A "Marxian" Thread However, Perrow had a second agenda in normal accident, hidden from sight at the time of its publication because of the novelty and radicalism of his deterministic argument which triggered debates (Le Coze, 2015b). As he lamented in the second edition of his book in the afterword "I would say that most of the work in the risk area is systematically detoxing the power aspects of my work" (Perrow, 1999, p. 379). There was indeed another critical view of disasters and of the unexpected in normal accident, something which makes sense in the context of Perrow's wider contribution to organizational sociology (e.g., Perrow, 1970, 1986a, 1986b, 2002).

In this second message of normal accident, "component accidents," instead of normal (or systemic) ones, do not result from unexpected interactions but lack of commitment to safety related to the issue of elite using their power to other purpose than maintaining safe operations. This view is entirely consistent with his material, wider, macro-social, critical, and power view of organizations and society that he has been developing throughout several decades of research.

> "Following the publication of the book in 1984, we had a series of notable accidents: Bhopal, Chernobyl, and in the United States, the Challenger and the Exxon Valdez oil spill. I was told I had bragging rights, but this was not true. All of them were just waiting to happen: the result of managerial negligence and, in addition, in the case of the Challenger, of extreme production pressures. Normal accidents are fairly uncommon and, I add, cannot be completely prevented. These accidents were all preventable. The vast majority of accidents are preventable."
>
> (Perrow, 1999a, p. 23).

A good example is his interpretation of the Challenger accident. His comments are of a very different nature than a focus on the technology. It targets instead the "master" of NASA and its proximity to political powers. "NASA's decline seems to have begun in 1971 with the administration of James Fletcher. A multimillionaire, he sits on the board of directors of many corporations that hold large contracts with NASA, and he has served in executive positions with some. He is from Utah, the home state of shuttle booster Senator Jake Garn, and the location of the important Morton Thiokol plant" (Perrow, 1986b, p. 354).

To support this, he cites a journalist of *The New York Times*, Stuart Diamond (1986a,b), who published a series of two articles revealing many problems with the money spent as well as a number of audit reports indicating latent issues in the organization of the agency. But there is more about Challenger, if one, this time, introduces the issue of "organizational goal": "to be safe again, the space program will have to return to its original mission of scientific exploration, and separate itself from both commercial and military pressures" (Perrow, 1986a, p. 354). This is not quite a technological argument.

The power of the elite, the political environment, and the goals of NASA explain the accident, not technology. This interpretation, Perrow says in the 1990s, is "different from those of Milliken (1988) and Vaughan (1990)" (Perrow, 1994, p. 19). The difference is that his view is much more critical and targets explicitly failures of executives, confronting top management with the impact of their decisions on high-risk systems, something that is an enduring trait of Perrow's sociological stance, and to be distinguished from the coupling/interactivity frame of the first reading of normal accident. It is this second reading of normal accident which frames Perrow's reaction in the wake of Fukushima Daïchi disaster.

"Nothing is perfect, no matter how hard people try to make things work, and in the industrial arena, there will always be failures of design, components, or procedures. There will always be operator errors and unexpected environmental conditions. Because of the inevitability of these failures and because there are often economic incentives for business not to try very hard to play it safe, government regulates risky systems in an attempts to make them less so" (Perrow, 2011, p. 44). Although main sociological influences of Perrow can be found both in Weber and Marx, this critical view of disasters is more appropriately defined as "Marxian." Its critical tone is what makes it more distinguishable from the next threads to be introduced below (and the one above). Marx had of course a critical view of capitalism, and Perrow extends this perspective by including modern for-profit organizations in the contemporary picture as they represent the new social forms of power, serving a capitalism, which did not exist in Marx's times (Perrow, 2002).

This is what Hopkins concludes about an accident that he interpreted. "Despite the technological complexities of the Longford site, the accident was not inevitable. The principles listed above are hardly novel—they emerge time and again in disasters studies. As the commission said, measures to prevent the accident were plainly practicable" (Hopkins, 2000). Hopkins has developed this type of analysis for a diversity of case studies, arguing how preventable these disasters were in principle from a managerial point of view. They were far from being inevitable (Hopkins, 1999b, 2000, 2008, 2012). This view is shared more generally by other writers who assert that control of operations can be achieved, but is not because of lack of appropriate managerial commitment. This is a quite popular narrative, of the accident "waiting to happen" or of the "reckless companies" that journalists frequently contribute to purport in the aftermath of disasters (e.g., Lustgarten, 2012). This thread is therefore quite a popular one.

32.3.3 Barry Turner and Man-Made Disaster: A "Kuhnian" Thread

As already hinted, Perrow released his now classic book 6 years after the publication of an equally important one by a British sociologist in 1978, Turner. If Perrow was a critical sociologist with strong roots in Marxism, Turner was not, and his reading of disasters was different and innovative for at least two reasons. First, he looked into this phenomenon through a managerial and sociological angle. No one else before had theorized accident from this perspective, which distinguished himself from human factors and engineering authors. Second, contrary to Perrow, he applied a cultural interpretation to accidents. And there again, he relied to do so on a personal account of culture in organizations that no one else had advocated before so explicitly in organization theory (Turner, 1971). His idea was to conceptualize the possibility for disasters by stressing the links between technology, organizational, cognitive, and cultural dimensions. To do so, he found, among others, inspiration in the work of the American historian of science, Thomas Kuhn (1962).

Kuhn produced one of the most influential thesis in the field of philosophy and history of science of the twentieth century. In a nutshell, he described patterns of paradigm shifts when scientific fields are revolutionized by new theories. The most common example is the move from heliocentric to geocentric understanding of earth–sun dynamic. For similar observations of sun's movements, two different interpretations were possible about the earth, until one finally prevailed. The process described by Kuhn is pictured as slow, consisting in scientists dealing with anomalies seen at first as external to the validity of the theory but then gradually challenging it. Scientists acknowledge that anomalies will not go away. Only a shift in theory or paradigm can accommodate these anomalies coherently.

In philosophy, the implications of his thesis are profound but are not the focus of this chapter. The main idea to retain is that we, individually and collectively, make sense of reality through constructs which are fallible as time passes. Data are not picked up independently of established worldviews which organize meaning. This idea helped Turner to frame the principle of an "incubation period." During an incubation period, signals of a pending disaster accumulate but are not seen as strong enough to challenge current assumptions of organization members.

The link to Kuhn by Turner was therefore explicit. "The term anomaly suggests links with Kuhn discussions of the way in which scientists come to accept new theories or paradigms about the world, and this association is a conscious one, for the problem faced by both administrators and scientists, in building up and maintaining a coherent picture of the world within which they are trying to operate, display a number of similarities in the context of this discussion (…) both in science and in administration, anomalies are noted as curiosities, as events which are not readily assimilable, but also as events which there is no immediate and pressing urgency about elucidating. Anomalies may be labelled and shelved, to wait until a superior explanation comes along to include them, and sometimes their existence 'flags up', to an inquiring mind, the possibility that there may be data here which could serve as a starting point in a search for a new level of information" (Turner, 1978, p. 127). For this reason, I describe this as a Kuhnian thread.

Identifying this thread can accommodate a diversity of writers who favor this type of interpretation of the unexpected. For instance, and independently, Ron Westrum, an

American sociologist, developed a closely related concept although he never conceptualized the use of anomalies into an incubation model applied to disasters as Turner did (Westrum, 1988, 1992). His interest as a sociologist for themes such as meteorites or the child battered syndrome exemplified what he conceptualized as "hidden events" (Westrum, 1978, 1982, 1986). "Thus the present stage of knowledge and even the professional intuition of scientists and practitioners make these events into anomalies; they are implausible because they do not articulate with current frame of knowledge" (Westrum, 1982, p. 384). It laid the ground for a fruitful translation later to the topic of accident and safety (Westrum, 1988, 1992, 1997).

Other authors who nicely fit in this thread are, first, Karl Weick and his sense-making approach to socio-cognition applied to reliability, safety, and disasters (Weick, 1987, 1990, 1993) and second, Diane Vaughan who also relied on both Kuhn and Turner for her groundbreaking study of the Challenger accident (Vaughan, 1996). This study extended Perrow's argument of the unexpected. "The Challenger disaster can justifiably be classed as a normal accident: an organisational-technical system failure that was the inevitable product of the two complex systems. But this case extends Perrow's notion of system to include aspects of both environment and organisation that affect the risk assessment process and decision making" (Vaughan, 1996, pp. 239, 415). A third author can be mentioned with John Downer, a British sociologist, because of his explicit constructivist or epistemic extension to Perrow's argument. Following earlier sociologists of science and technology such as Pinch (1991) or Wynne (1988), Downer sees some accidents as unexpected not because of complex technologies but because of our intrinsic limited knowledge of technological behaviors of innovative artifacts (Downer, 2011).

32.3.4 Jens Rasmussen and Complexity: An Ashbyan Thread

Contrary to Turner and Perrow, Jens Rasmussen is a cognitive engineer but he, in my view, explicitly framed a fourth thread at the social level of analysis (Le Coze, 2015a). Relying on ideas derived from his cybernetic background, and combining it with a specific interpretation of the defense-in-depth engineering principle, Rasmussen formulated a fruitful conceptualization of the unexpected in the context of high-risk systems. Providing powerful visuals associated with his analytical insights made his contributions highly heuristic to many researchers and practitioners.

One important move toward this thread is from his micro view of errors to a macro (socio-technical) perspective on accidents, where the notion of "degree of freedom" of individuals in accomplishing their tasks played a central role. "In real-life situations, a large degree of freedom is left to the human even though the overall goal is stated unambiguously" (Rasmussen, 1980). This sentence reflects the findings of his empirical study of electronic troubleshooting (Rasmussen and Jensen, 1974) and has implications for a macro approach of safety and accident. The notion of self-organization, from the development of the cybernetics movement in the 1960s, made up the next building block, which stood with the degree of freedom principle.

His manner of associating degree of freedom and self-organization is articulated: "it follows directly from this discussion that the structuring of work processes through on the job training by an individual will be a self-organising, evolutionary process, simply because an optimising search is the only way in which the large number of degrees of freedom in a complex situation can be resolved" (Rasmussen, 1990).

The interesting twist is when he then associates these self-organizing properties and degree of freedom with the principle of defense-in-depth. In the period of reflection following the technical disasters of the 1980s (Bhopal, Chernobyl, Piper Alpha, Challenger, etc.), it is apparent that the defense-in-depth concept was an important one, as James Reason, a British psychologist, for instance, had ruminated over the subject of catastrophe beyond human error by providing what is probably the most renown and advocated model in the field for accident investigation (which became known as the "Swiss Cheese," Figure 32.5).

> "Reason introduces the discussion of the key problems of industrial safety by an emphasis on the sensitivity of the 'defence-in-depth' design philosophy to combinations of human failures' (…) 'Two important conclusions emerge from Reason's review: first, disasters are rarely caused by any one factor, either mechanical or human; second, most of the significant root causes are present within a system long before an accident sequence is identified. In short, violations very likely turn into 'resident pathogens' to use Reason's very illustrative medical metaphor."
>
> (Rasmussen, 1989).

However, instead of sticking with the medical metaphor, Rasmussen combined self-organization and defense-in-depth together to characterize what he called the "fallacy of defense-in-depth." "One basic problem is that in such a system having functionally redundant protective defenses, a local violation of one of the defenses has no immediate, visible effect and then may not be observed in action. In this

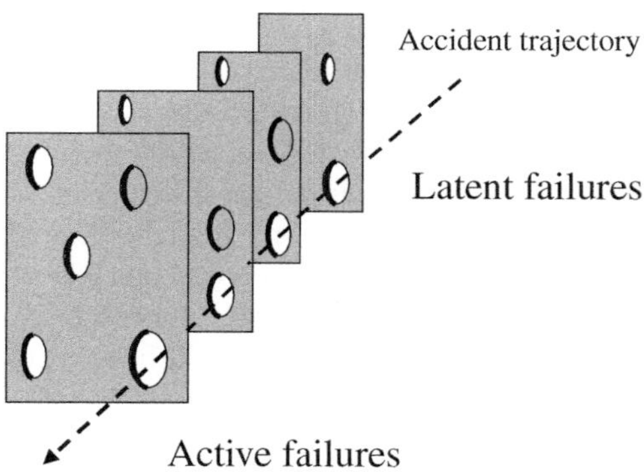

Figure 32.5. Defense-in-depth model. Adapted from Reason, 1997.

situation, the boundary of safe behaviour of one particular actor depends on the possible violation of defenses by other actors" (Rasmussen, 1997b).

From there, two consequences follow, one is that it is possible to see accidents as unexpected (or "normal" in the sense of Perrow but for other reasons), with the idea that "catastrophic system breakdown is a normal feature of systems which have self-organising features and at the same time, depend on protection against rare combinations of conditions which are individually effected by adaptation" (Rasmussen, 1990b). Another is that this adaptation, based on the self-organization and degree of freedom of local agents at the level of operations, can indeed be transferred to different actors (e.g., managers) within high-risk systems. "Analogy can be drawn between the adaptive mechanisms involved in the skill attainment of individual (…) and the role of management decisions—which may be errors in a safety point of view—in the adaptation to efficiency measures of an organisation. Errors in management and planning are intimately related to organisational attempts to adapt to the requirement of a competitive environment."

This helps define a model of migration toward the boundary of acceptable performance served by an analogy from the domain of physical sciences. "Activity will be characterised by local, situation induced variations within work space calling to mind the 'Brownian movements' of the molecules of gas. Such variability will give the actors themselves ample opportunity to identify an 'effort gradient' while management is likely to build up a 'cost gradient'" (Rasmussen, 1993). This is an analogy and a translation, first, from a physical phenomenon to techno-social one, second, from a vision of operators as self-organized agents with degrees of freedom to a (macro) vision of safety as a global migration of a system (Figure 32.6).

When conceptualizing the unexpected way, Rasmussen drew heavily on Ashby and his cybernetics, where feedforward thinking goes hand in hand with requisite variety (Ashby, 1956). Requisite variety concerns the issue of regulating self-organized complex systems and is combined with the black box concept, for which the internal mechanism cannot be fully described and predicted. We thus see the influence of Ross Ashby's cybernetics: "Systems with a high degree of autonomous internal functioning, with self organising and adaptive features, may change their internal functional organisation frequently in order to meet the requirements of the environment and to suit their internal goals or performance criteria. Even though such systems are basically causal and controlled by law of nature, their complexity makes it impractical, if not impossible, to explain or predict their performance by functional analysis during real life decision making" (Rasmussen, 1985).

Today, the principle of self-organization is of great contemporary significance (in many fields including physics, biology, and social sciences and expanding through the science—and philosophy—of complexity) and that he recognized it early in his career as being extremely valuable for the field is quite prescient. It was his knowledge of Ashby's work, the first to pin down the properties of self organization, which allowed him to pioneer this interpretation, hence the Ashbyan title of the fourth thread of the unexpected. And many authors found inspirations in the metaphors of

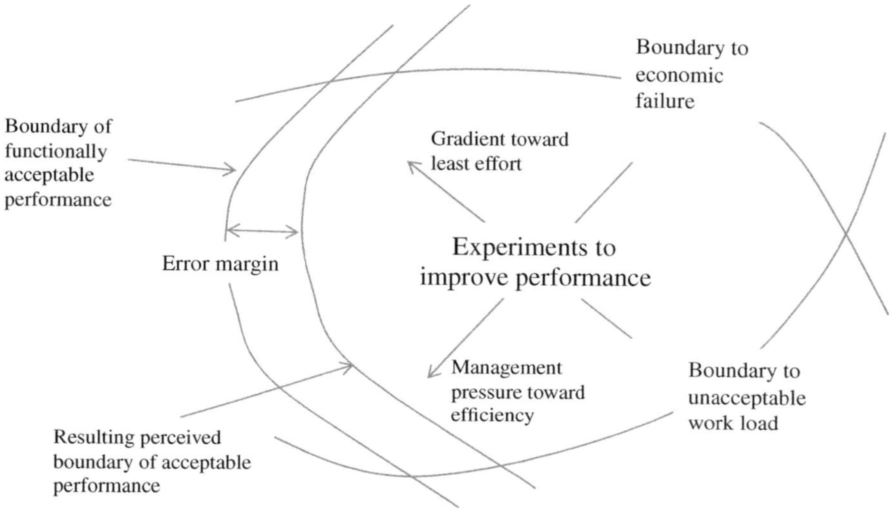

Figure 32.6. Migration model. Adapted from Rasmussen, 1997.

envelope, boundaries, migration, self-organized properties, and adaption, and followed his lead. I am inclined, for instance, to classify an author such as Scott Snook (2000), a behavioral scientist, into this fourth line of thought. This author refers directly to the original writings of Rasmussen on self-organization applied to accident and safety.

One cannot help but think of the "defense-in-depth fallacy" when reading Snook's presentation of "practical drift" (Snook, 2000). Snook stood back to produce a big picture of "friendly fire," which he investigated in order to revise the official report of this accident. "Practical drift" consists of localized drift in an individual's practices. Whereas the organization is at first designed to ensure the consistency of the whole by proceduralizing the behavior of the parts, in fact, these parts (individuals) slowly drift from their expected course to create a path for an accident. This is no different from Rasmussen interpretation of the Clapham Junction accident, applying his new vision of safety and accident and its "defense-in-depth fallacy" (see box 32.3).

Box 32.3 Rasmussen's Clapham Junction Interpretation
"The Clapham Junction railway accident presents a clear example of how a safe work procedure for signal system modifications, including multiple precautions against human errors, gradually degenerates due to adaptation at all levels of the organisation to locally more efficient work practice (…) safety checks following modifications of signal system wiring were planned to be

independently performed by three different persons, the technician, his supervisor, and the system engineer. Work force constraints and tight work schedules, however, led to a more 'efficient' division of work. The supervisor took part in the actual, physical work and the independent check by him as well as by the engineer was abandoned. In addition, the technician integrated the check, i.e. a 'wire count' into the modification task although it was intended to be his final separate check. In short, adaptation to a more effective division of work under time pressure causes the redundancy required for high reliability to deteriorate."

Source: Rasmussen, 1994.

Snook pushed further this idea of the normality of accidents in his book (an interpretation endorsed by Perrow, 1999b) through the catch phrase that it was "a normal accident in a highly reliable organisation" (Snook, 2000), but adding that "it was normal because it occurred as the result of normal people behaving in normal ways in normal organisations" (Snook, 2000). System thinking and complexity references pervade Snook's explanations. There are also direct references to Rasmussen when it comes to issues of stop rules and causality in investigations (a point discussed in the next section). But Erik Hollnagel distinguishes himself (Hollnagel, 2004). Whereas Snook relies on an empirical case study, in which the concept of "practical drift" only comes after a grounded analysis based on qualitative data, Hollnagel is deliberately both more conceptually and practically oriented.

Building on ideas similar to those of Rasmussen and Snook, he moves to another level of conceptualization by introducing the principle of "resonance." It is an analogy with a physical sciences phenomenon that, as its counterpart the "Brownian" analogy from Rasmussen, seeks to support the identification of unwanted patterns creating paths to disasters. The concept of emergence (vs. resulting) is then added to advocate a nonreductionist approach to the problem of safety. Although bringing a next level of conceptualization with practical ambitions (i.e., functional resonance accident model), Hollnagel's ideas have strong affinities with Rasmussen's. In my view, Snook and Hollnagel show the relevance of exploring safety and accident through the application of self-organization, emergence, and complexity. They explore and extend the Ashbyan thread opened by Rasmussen. Dekker (2011) and Woods (2015) are other examples.

32.3.5 Four Threads, Four Sensitivities, But Not Exclusive: A Synthesis

The four threads can be grouped together in the following figure (Figure 32.4).

This classification does not mean that the threads are exclusive although they do stress specific interpretations that author might favor in the aftermath of events. For instance, the Marxian thread, with Perrow and its critical tone can reject interpretations of the unexpected that would rely too much on the constructivist or cultural sensitivity of the Kuhnian thread. When considering the Challenger accident, Perrow is explicit in his rejection of the cultural interpretation. "They were told to take off their engineering hats and put on their managerial hats and make a managerial and not engineering decision. To me, that is a case of sheer power, it is not the routinisation of deviance" (Perrow, 1999, p. 369). As a consequence, his emphasis over power in organizations and society leads to the trenchant assertion that "we miss a great deal when we substitute culture for power" (Perrow, 1999, p. 369).

This is just an illustration of how the threads influence understandings of the unexpected. The interest of identifying them is therefore to stress the complexity of this problem. The diversity of conceptualizations and interpretations reflects the wealth of views, but also backgrounds including education, experiences, and preconceptions about an array of topics such as organizations, society, technology, human nature, determinism, etc. This issue cannot be developed in this chapter but one aspect that should be kept in mind about the unexpected is that the interpretation of such events should always, as much as possible, rely on such a conscious or reflexive approach. How do I interpret events, from which of the four thread? Why?

Moreover, we understand now why the descriptions in this handbook of engineering, human factors, managerial, and governance (safety) principles that should be applied to ensure safe operations have to be understood in reality, and not only in theory. Principles are normative ideals that empirical descriptions and conceptualizations prove to be, sometimes, rather complex to implement in practice or with limitations. There are indeed cognitive, social, and political realities, illustrated by the four threads of this chapter, behind any effort to perform safely according to normative principles (even if there is no other option than to describe them to serve as yardsticks).

And, if the threads point to different sensitivities, they are not exclusive and have a potential to be combined to express the complex nature of the unexpected in high-risk systems (for instance, although Hopkins has been associated in this chapter with the Marxian thread for his rejection of the inevitability of accidents, he also strongly relies on Turner's idea of the unexpected through the culture lenses, e.g., Hopkins, 1999b). This is what the following synthesis attempts to grasp (Le Coze, 2015b).

1. High-risk systems face strong internal and external constraints and pressures (e.g., technological, social, political, and economic), and, as result, operators, engineers, managers, executives, and regulators interact and deal with a wide range of uncertainties, leading to innumerate, local, and simultaneous decisions:
   - Technological uncertainties for engineers in design and operational phases
   - Components, automations, and working conditions uncertainties for operators and middle managers

- Organizational, market, and environment uncertainties for top managers and executives.

2. Because of the complexity of these interactions within these socio-technological systems (and networks), no one can predict with certainty their behavior; no one can pretend to produce the big picture that would help to entirely describe these interactions in order to obtain full control of them; and the imperfections of organizations have become accepted (and expected) by many.
3. Consequently, the boundary between operating safely and operating unsafely is probably more ambiguous that many (from private companies to regulators) are willing to publicly admit in a world that promotes technological, managerial, and financial innovations translated in (oftentimes) ambitious company strategies.
4. Whether one favors a critical, power view of organizations or prefers a seemingly more neutral approach emphasising the weight of uncertainties, cultural, and institutional influences, one's understanding of managerial, executive, and regulatory implications and their degree of failure (and of responsibility) will differ.

This synthesis to characterize the unexpected combines several notions together by introducing uncertainties which defies our knowledge and our ability to treat anomalies to make sense of situations (Kuhnian thread, point 1 above), by stressing the notion adaptation under an array of different constraints (Ashbyan thread, points 1 and 2 above) while including the reality of powerful actors orienting sometimes ambitious strategies for high-risk systems (Marxian thread, points 1 and 3 above) and the large-scale complexity of these systems (Ellulian thread, point 1).

32.4 MANAGING THE UNEXPECTED

Authors who theorized the unexpected in high-risk systems were also very often interested in conceptualizing the opposite, namely the cognitive and social dynamics which allow actors in these organizations to manage the unexpected. A very good example is Karl Weick, an American social-psychologist, who has always paid as much interest to failures (Weick, 1990, 1993) than to successes (Weick, 1987, 1989), and consequently building on these studies (and many others) to create, in collaboration, the model of collective mindfulness (Weick et al., 1999; Weick and Sutcliffe, 2003).

This model theorized five processes of high reliability[3] which were empirically studied in the 1980s in several safety critical organizations (nuclear powerplant,

[3] These processes are preoccupation with failures, reluctance to simplify, commitment to resilience, sensitivity to operations and deference to expertise (previously "unspecified structures").

aviation traffic control, aircraft carriers) by a team of social scientists in the 1980s (Roberts, 1989, 1990; Rochlin, 1989; LaPorte and Consolini, 1991). The model can then in turn also serve to interpret accidents (e.g., Weick, 2005). So researchers have produced concepts that could explain why, despite the presence of intrinsic limitations due to cognition, organization, or complexity, some systems successfully managed the unexpected.

They came up, for some independently, with several compatible (sometimes overlapping) topics that are introduced and discussed in relation to the threads in this section. I group them into three categories. First, building favorable power configurations (limiting the possibility of executive (and regulatory) failure) as a response to the Marxian thread; second, confronting our fallible (cultural) constructs against the Kuhnian thread; and third, keeping sight of the relation between parts and whole as an answer to the Ashbyan interpretation of the unexpected.[4]

32.4.1 Building Favorable Power Configurations (vs. Marxian Thread)

To limit the possibility of a failure of executives or/and regulation (Marxian thread), Perrow followed a macro systemic view of high-risk systems stressing the degree to which the networks of actors and organizations, namely the interactions between civil society (and associations), unions, legal and state entities as well as private companies (including insurance), contributed to maintaining safe practices, to managing the unexpected. He contrasts for that matter, "error avoiding" and the "error inducing" systems and locates aviation (Box 32.4) and maritime sectors (Box 32.5) as two ends of this spectrum. In effect, he reaches the conclusion that "there is an enormous incentive to make commercial aviation safe (...) no other high-risk system is so well positioned to effectively pursue safety as a goal" (Perrow, 1999, p. 127), whereas in marine transport "only a wholesale reconfiguration could make the parts fit together in an error-neutral or error-avoiding manner" (Perrow, 1999, p. 172).

Box 32.4 Aviation According to Perrow's Macro View of High-Risk Industry

Victims are neither hidden, random, nor delayed, and can include influential members of the industry and Congress. A remarkable *voluntary reporting system exists* (Air Safety Reporting System). There is a *strong union* at work to protest unsafe conditions—Alpa, the airline pilots association. It even conducts its own studies and makes its own safety recommendations. *Lawsuits* follow immediately if the investigation of the NTSB [National Transport Safety Board]

[4] The Ellulian thread has a specific status comparatively to the three others. The solution against the unexpected as seen through the lenses of a technological determinism is obviously to limit technological developments, or to make them both less coupled and simpler (according to Perrow's criteria, Perrow, 1996).

hints at vendor or airline culpability. *The FAA* is charged with both safety and facilitating air travel and air transport, and spends significant amounts of tax dollars pursuing safety studies and regulations. *An independent board, the NTSB*, conducts investigations and prods the FAA to set new safety requirements. Airline travel drops after large accidents. *Airline companies* suffer if one of their models appears to have more than its share of accidents. *Experience is extensive* and the repetitive cycle of take-offs, cruising, and landing promotes rapid training, precise experience with failures, and trial and errors for new deigns and conditions.

Source: Perrow, 1999, pp 172–173.

Box 32.5 Marine According to Perrow's Macro View of High-Risk Industry

"*The identifiable victims* are primarily low status, unorganized, or poorly organized seamen. *The third party victims of pollution and toxic spills* are anonymous, random, and the effects delayed. *Elites* do not sail on Liberian tankers. *The marine courts* exist to establish legal liability and settle material claims, not to investigate the cause of accidents and compensate seamen. *Shippers* do not avoid risky 'bottoms' but pick the cheapest and most convenient, and cannot choose to stop shipping for a time because the last cargo was lost. *The federal* presence is minor and appears inept in the US. It sets standards for those ships that want to use our ports, but the United States ranks fourteenth among nations in ship safety, so the standards cannot be very high. And finally, the only *international association* concerned with safety is advisory and concerned primarily with nationalist economic goals. Regulations succumb to economic and nationalistic pressures, and is highly ineffective. *Unions* are by and large weak. *The insurance company* is a passive contributor, passing on the costs to the final consumer."

Source: Perrow, 1999.

As this analytical framework shows, managing the unexpected is an activity that needs to be conceptualized as resulting not only from internal, but also from external interactions of high-risk systems within their environments. Regulating systems through state inspections to ensure that known approved design or management safety practices (translated in regulations) are implemented is crucial, and depends on power relationship between states and industries (Perrow, 2015). Indeed, indications of safety concerns and issues have to be taken seriously into consideration despite their potential implications in terms of costs, time, resources, etc. within organizations.

Certainly, properties such as *deference to expertize* as put forward by Weick and Sutcliffe (2003) (and part of the model of *collective mindfulness*) play a strong role to ensure that messages are heard by managers and executives. This property of deference indicates the possibility for a diversity of actors to express their views and for their expertise recognized within organizational structures and information flows.

The balance of power between different categories of employees such as managers, scientists, engineers, foremen, or fieldworkers in different activities including maintenance, production, quality, safety will allow messages to circulate horizontally or vertically. Warnings against impending issues, such as noncompliance to important safety standards are raised by specific actors, in specific organizational contexts, which are more or less favorable given power configurations.

But Perrow's reading of the problem of safety makes also evident that the environments of high-risk systems are at least as important as the internal functioning of these systems. With aviation and marine transport (Boxes 32.3 and 32.4), we have an empirical illustration of what is today described as governance mechanisms. The role of the network of outsiders, whether *regulators*, *watchdog groups* (e.g., NGO, independent media, researchers), or *professional associations* (e.g., engineers, scientists, etc.), has to be seen as constitutive and not external to the ability for high-risk systems to manage the unexpected, something also argued by other authors, such as Evan and Manion (2002).

32.4.2 Confronting Our Fallible (Cultural) Constructs (vs. Kuhnian Thread)

Perrow's approach offers invaluable insights into the power configurations that determine the degree to which one can expect the unexpected to be managed in high-risk systems, but this approach partly avoids the central problem to know how a diversity of actors establishes specific understanding of situations, and how they deal with uncertainty. One needs to turn to other authors when it comes to this problem, and to consider their answers to the next thread. To cope with the fact that we individually and collectively produce fallible constructs (Kuhnian thread), Turner and his colleagues (Pidgeon et al., 1986; Pidgeon and Turner, 1986) saw in *learning* a theoretical construct that could help grasp the ability of actors in organizations to imagine problems and defeat the processes of incubation periods (Turner and Pidgeon, 1997; Pidgeon and O Leary, 2000).

Learning is therefore critical in high-risk systems (Hopkins, 2009; Le Coze, 2013a). Learning is an essential feature broadly shared by the living and central to the transformation of cultures. By constantly revising the schemes of actions in relation to experience, learning processes allow adaptations to ever changing environments. Engineers, managers, and operators all develop learning abilities to cope with evolving environments whether these concern technological, organization, or market behaviors. And, these learning processes need the support of active cognitive, individual, social, and collective dynamics which stimulates thinking.

Translated into the field of safety, this idea is found in Westrum's work with the notion of *requisite imagination* (Westrum, 1992; Adamski and Westrum, 2003), or also in the Weick and Sutcliffe (2003) concept of *collective mindfulness* which combines several processes, and in particular for learning, *preoccupation with failure* and *reluctance to simplify* (Weick and Sutcliffe, 2003). All these authors insist on the importance of *socio-cognitive processes which support the imaginative and creative anticipation* that operators, engineers, and managers deploy to make sense of cues, signals that could reveal potential problems.

Carl McRae, a British psychologist, through a qualitative approach based on observations and interviews, indicates that there are at least four cognitive strategies that employees of an airline company develop to imagine and learn about the possibility of failure. *Making patterns* consists in identifying events that once combined define a trend to be investigated and potentially corrected. *Drawing connections* consists in linking specific events with similar past major air accidents.

Recognizing novelty implies a subtle attention to problems not experienced before, indicating a lack of awareness about specific types of operational failure. *Sensing discrepancy* describes the analogical process of comparing events reported with expectations about normal operations, about the way "things should work." How intuition, experience, and imagination shape this process is stressed by the author "as previously indicated, warnings are not so much there for the taking but must be actively constructed and uncovered through vigilant interpretive work" (Macrae, 2009, p. 115). This study exemplifies the socio-cognitive complexities behind the process of learning.

32.4.3 Keeping Sight of the Relation Between Parts and Whole (vs. Ashbyan Thread)

Faced with the issue of self-organized and adaptive behaviors of individuals in high-risk systems (Ashbyan thread), Roberts and Weick refer to concepts such as *having the bubble* or *heedful interaction* (Roberts, 1990; Robert, Weick, 1993). These concepts concern the ability to keep a broad vision of the potential negative outcomes of interactions between various actors that adapts locally in different places and time and who do not see the broad implications of their adaptations[5]. When turning to the model of *collective mindfulness* of Weick and Sutcliffe (2003), this property is formulated under the item of *sensitivity to operations*, which is close to a notion such as *situation awareness* (found predominantly in the aviation domain).

In this respect, Westrum refers to a *human envelope of thought* that is created by so called *maestros* "probably the single most important function of the maestro is 'having the bubble' in the larger sense. In large projects, some of the components teams may have an undue concentration on their own tasks. Without someone who has a larger vision, this concentration can lead to sub-optimisation or even to serious imbalances that imperil safety" (Westrum, 1997, p. 247). When multiplying actors

[5] Recall that these unwanted patterns have been defined by Rasmussen as the "defense-in-depth fallacy," by Snook as a "practical drift" then by Hollnagel as a "resonance." Tools, methodologies and concepts are developed for instance by Hollnagel in the Resilience Engineering tradition to tackle this problem specifically (see Chapter 2).

and organisations, the problem becomes obviously more challenging, and the number of maestros increases.

By maintaining an updated vision of what is going on between different activities, key actors in organisations allow smooth handling of situations. By directing flow of information, by prioritising actions, by anticipating perturbations, by allocating resources in real time, by keeping an eye on the coherence of the whole, these individuals play a crucial role. They are located at supervisory and managerial levels and also have a detailed understanding of what strategies are developed by operators, what shortcuts are taken to perform under work constraints, and are able to put them in the context of safety more broadly. They manage the parts in relation to the whole.

32.4.4 Limitations and Opening

The three generic categories (1) building favorable power configurations, (2) confronting our fallible (cultural) constructs, and (3) keeping sight of the relation between parts and whole depend of course very much on the specific technological, social and historical contexts of high-risk systems.

Whether one studies safety critical activities such as space flights, firefighting or surgery, the way the categories can be observed in reality will vary accordingly. The technology used to flight spaceships, to practice surgery or to extinguish fires differs because developed to tackle various uncertain and complex type of phenomena and environment: space, body or flames.

The scientists, engineers and managers populating NASA differ from the nurses, anaesthetist and surgeons found in hospitals as much as the firefighters and officers in a fire department. Socialised differently through education, interacting within specific work cultures and (hierarchical) power status, performing under different public and political scrutiny, etc these actors manage the unexpected in a variety of social contexts.

Finally, the historical moments when these activities take place also matter. Working at NASA, in a hospital or a fire department today or in the 80s are two different stories. The world we live in now is not the world of the past (Le Coze, 2013b). The conditions under which the management of the unexpected is taking place has to include many of the transformations of the last 30 years which affected organisations under processes of globalisation (Le Coze, 2016a).

Without being exhaustive, a certain number of features can be listed which indicates that history matters in this respect: standardisation, open markets and financialisation, IT technology, networked organisations including subcontracting.

32.5 RELATION TO OTHER PRINCIPLES: FURTHER READING

This chapter is in relation with many chapters of this handbook, as introduced above because managing the unexpected is about, in theory, applying reasonable safety principles. But this chapter has also been about developing the insights from the human and social sciences. So, although many chapters could be mentioned, those addressing issues related to "safety management systems" (Gudela Grote, Chapter 27), "safety culture" (Carl Rollenhagen and Teemu Reiman, Chapter 28) and "resilience engineering" (Erik Hollnagel, Chapter 3) are the closest to managing the unexpected as conceptualised here. For further refinement on the complexities involved in grasping the unexpected as experienced in socio-technological systems and networks, see Le Coze (2016b).

32.6 CONCLUSION

Planes can crash, nuclear reactors can melt, boats can sink but the future remains uncertain despite often knowing about what could potentially happen. This chapter has introduced and explored this specific problem of the unexpected and its relativistic aspect, which entails its appreciation by observers. In industrial safety, the social sciences, with pioneers such as Charles Perrow, Barry Turner and Jens Rasmussen, have conceptualised the unexpected while reflecting the relativity of this notion. Ellulian, Marxian, Kuhnian and Ashbyan threads represent in this respect several possibilities of interpretation of disasters in relation to the notion of the unexpected. Technological determinism implies the possibility for machines to escape human control (Ellulian thread). Failure of executives entails the lack of appropriate safety management practices, and approaches this issue with a critical angle (Marxian tread). Fallible cultural constructs indicates that we establish individual and collective worldviews which require to be challenged through learning and imaginative processes (Kuhnian thread). Finally, self organised properties of socio-technological systems and networks produce emergent properties that defy our ability to anticipate them (Ashbyan thread). In this chapter, managing the unexpected was described through cognitive and social properties of high risk systems which help counteract these sources of the unexpected. Building favorable power configurations (1), confronting our fallible construct (2), and keeping sight of the relation between parts and whole (3) were described as three features which are conducive to strong safety achievements, to managing the unexpected.

REFERENCES

Adamski, A. J., and Westrum, R. (2003). Requisite imagination: The fine art of anticipating what might go wrong. In: E. Hollnagel (Ed.), *Handbook of Cognitive Task Design*. Mahwah, NJ: Lawrence Erlbaum Associates.

Ashby, R. (1956). *Introduction to Cybernetics*. London: Chapman & Hall.

Bostrom, N. (2014). *Superintelligence: Paths, Dangers, Strategies*. Oxford: Oxford University Press.

Bostrom, N., and Ćirković, M. (Eds.). (2008). *Global Catastrophic Risks*. Oxford: Oxford University Press.

Bourrier, M. (1999). *Le nucléaire à l'épreuve de l'organisation* [*Nuclear Industry from an Organisational Point of View*]. Presses Universitaires de France.

Dekker, S. (2011). *Drift into Failure: From Hunting Broken Components to Understanding Complex Systems*. Ashgate.

Diamond, S. (1986a). NASA cut or delayed safety spending. *The New York Times*, April 24, 1986.

Diamond, S. (1986b). NASA wasted billions, federal audits disclose. *The New York Times*, April 23, 1986.

Downer, J. (2011). "737-cabriolet": The limit of knowledge and the sociology of inevitable failure. *American Journal of Sociology*, 117(3), 725–762.

Ellul, J. (1954). *La technique ou l'enjeu du siècle*. Paris: Armand Colin.

Ellul, J. (1977). *Le système technicien*. Paris: Calmann-Lévy.

Evan, W. M., and Manion, M. (2002). Minding the machines. In: *Preventing Technological Disasters*. Upper Saddle River, NJ: Prentice Hall.

Giddens, A. (1990). *The Consequences of Modernity*. Stanford, CA: Stanford University Press.

Hale, A. R. (2003). Safety management in production. *Human Factors and Ergonomics in Manufacturing & Service Industries*, 13(3), 185–201.

Hawking, S. (2014). *BBC News*, December 2, 2014.

Hollnagel, E. (2004). *Barriers and Prevention*. Aldershot, UK: Ashgate.

Hopkins, A. (1999a). The limits of normal accident theory. *Safety Science*, 32, 93–102.

Hopkins, A. (1999b). *Managing Major Hazards: The Lessons of the Moura Mines Disaster*. St Leonards, Australia: Allen & Unwin.

Hopkins, A. (2000). *The Longford Explosion*. Sidney: CCH.

Hopkins, A. (2008). *Failure to Learn: The BP Texas City Refinery Disaster*. Sydney: CCH.

Hopkins, A. (2009). *Learning from High Reliability Organizations*. Sidney: CCH.

Hopkins, A. (2012). *Disastrous Decisions: The Human and Organisational Causes of the Gulf of Mexico Blowout*. Sydney: CCH.

Hughes, T. (2005). *Human-Built World: How to Think About Technology and Culture*. Chicago, IL: University of Chicago Press.

Khun, T. (1962). *The Structure of Scientific Revolutions*. Chicago, IL: University of Chicago Press.

La Porte, T. R., and Consolini, P. (1991). Working in practice but not in theory: Theoretical challenges of high reliability organizations. *Journal of Public Administration Research and Theory*, 1(1), 19–47.

Le Coze, J. C. (2008). Organisations and disasters: From lessons learnt to theorizing. *Safety Science*, 46, 132–149.

Le Coze, J. C. (2013a). What have we learned about learning from accidents. *Safety Science*, 51(1), 441–453.

Le Coze, J. C. (2013b). New models for new times. An anti-dualist move. *Safety Science*, 59, 200–218.

Le Coze, J. C. (2015a). Reflecting on Jens Rasmussen's legacy. A strong program for a hard problem. *Safety Science*, 71, 123–141.

Le Coze, J. C. (2015b). 1984–2014. Normal accidents. Was Charles Perrow right for the wrong reasons? *Journal of Contingencies and Crisis Management*, 23(4), 275–286.

Le Coze, J. C. (2016a). Risk management: Sociotechnological risks and disasters. In: A. Burgess, A. Alemanno, and J. O. Zinn (Eds.), *Handbook of Risk Studies*. London: Taylor & Francis.

Le Coze, J. C. (2016b). Trente ans d'accidents. In: *Le nouveau visage des risques sociotechnologiques* [*Thirty Years of Accidents: The New Face of Sociotechnological Risks*]. Toulouse: Octarès.

Lustgarten, A. (2012). *Run to Failure: BP and the Making of the Deepwater Horizon Disaster*. New York: W. W. Norton & Company.

Macrae, C. (2009). From risk to resilience: Assessing flight safety incidents in airlines. In: A. Hopkins (Ed.), *Learning from High Reliability Organisations*. Sidney: CCH.

Perrow, C. (1970). *Organizational Analysis: A Sociological View*. Tavistock Publications.

Perrow, C. (1984). *Normal Accidents: Living with High Risk Technologies*. New York: Basic Books.

Perrow, C. (1986a). *Complex Organizations: A Critical Essay* (3rd ed.). McGraw-Hill.

Perrow, C. (1986b). The habit of courting disaster. *The Nation*, October 1986.

Perrow, C. (1994) Accidents in high-risk systems. *Technology Studies*, 1, 1–20.

Perrow, C. (1999). *Normal Accidents: Living with High Risk Technologies* (2nd ed.). Princeton, NJ: Princeton University Press.

Perrow, C. (2002). *Organising America: Wealth, Power and the Origins of Corporate Capitalism*. Princeton, NJ: Princeton University Press.

Perrow, C. (2011). Fukushima and the inevitability of accidents. *Bulletin of the Atomic Scientists*, 67(6), 44–52.

Perrow, C. (2015). Cracks in the "regulatory state." *Social Currents*, 2, 203–212.

Pidgeon, N. F., Blockley, D. I., and Turner, B. A. (1986). Design practice and snow loading: Lessons from a roof collapse. *The Structural Engineer*, 64(A), 67–71.

Pidgeon, N., and O'Leary, M. (2000). Man-made disasters: Why technology and organizations (sometimes) fail. *Safety Science*, 34(1–3), 15–30.

Pidgeon, N. F., and Turner, B. A. (1986). Human error and socio-technical system failure in structural engineering. In: A. Nowak (Ed.), *Modelling Human Error in Structural Design and Construction* (pp. 193–203). New York: American Society of Civil Engineers.

Pinch, T. (1991). How do we treat technical uncertainty in technical failure? The case of the Space Shuttle Challenger. In: T. La Porte (Ed.), *Social Responses to Large Technical Systems. NATO ASI Series (Series D: Behavioural and Social Sciences)* (Vol. 58, pp. 143–158). Dordrecht, The Netherlands: Springer.

Rasmussen, I. (1980). What can be learned from human error reports. In: K. Duncan, M. Gruneberg, and D. Wallis (Eds.), *Changes in Working Life*. New York. John Wiley & Sons.

Rasmussen, J. (1985). Trend in human reliability analysis. *Ergonomics*, 28(8), 1185–1195.

Rasmussen, J. (1990). The role of error in organizing behavior. *Ergonomics*, 33, 1185–1199.

Rasmussen, J. (1993). Market economy, management culture and accident causation: New research issues? In: Proceedings Second International Conference on Safety Science. Meeting Budapest Organizer Ltd., Budapest, Hungary.

Rasmussen, J. (1997). Risk management in a dynamic society: A modelling problem. *Safety Science*, 27(2/3), 183–213.

Rasmussen, J., and Batstone, R. (1989). Why do complex organizational systems fail? Environment Working Paper No. 20, The World Bank Policy Planning and Research Staff.

Rasmussen, J., and Jensen, A. (1974). Mental procedures in real life tasks: A case study of electronic trouble shooting. *Ergonomics*, 17(3), 293–307.

Reason, J. (1997). *Managing the Risk of Organisational Accidents*. London: Ashgate.

Roberts, K. H. (1989). New challenges in organisational research: High reliability organizations. *Industrial Crisis Quarterly*, 3, 111–125.

Roberts, K. H. (1990). Some characteristics of one type of high reliability in organisation. *Organization Science*, 1(2), 160–176.

Roberts, K. H., and Weick, K. E. (1993). Collective mind in organizations: Heedful interrelating on flight decks. *Administrative Science Quarterly*, 38(3), 357–381.

Rochlin, G. I. (1989). Informal organizational networking as a crisis-avoidance strategy: US naval flight operations as a case study. *Industrial Crisis Quarterly*, 3, 159–176.

Sjöström, M. (2010). 10 Ways to End the World. National Geographic, UK.

Snook, S. A. (2000). Friendly *Fire: The Accidental Shootdown of US Black Hawks over Northern Iraq*. Princeton University Press.

Starbuck, W. H., and Milliken, F. (1988). Challenger: Changing the odds until something breaks. *Journal of Management Studies*, 25, 319–340.

Turner, B. A. (1971). *Exploring the Industrial Subcultures*. London: Macmillan Press.

Turner, B. A., and Pidgeon, N. (1997). *Man-Made Disaster: The Failure of Foresight* (2nd ed.). London: Butterworth-Heinmann.

Vaughan, D. (1990). Autonomy, interdependence, and social control: NASA and the Space Shuttle Challenger. *Administrative Science Quarterly*, 35, 225–257.

Vaughan, D. (1996). *The Challenger Launch Decision: Risky Technology, Culture and Deviance at NASA*. Chicago, IL: University of Chicago Press.

Virilio, P. (2005). L'accident originel. Paris, Editions Galilée.

Weick, K. (1987). Organizational culture as a source of high reliability. *California Management Review*, 29(2), 112–127.

Weick, K. E. (1989). Mental models of high reliability systems. *Industrial Crisis Quarterly*, 3, 127–142.

Weick, K. E. (1990). The vulnerable system: An analysis of the Tenerife air disaster. *Journal of Management*, 16(3), 571–593.

Weick, K. E. (1993). The collapse of sensemaking in organisation. *Administrative Science Quarterly*, 38, 628–652.

Weick, K. E. (2005). Making sense of blurred images: Mindful organizing in mission STS107. In: W. Starbuck and M. Farjoun (Eds.), *Organization at the Limit*. Oxford: Blackwell.

Weick, K., and Sutcliffe, K. (2003). *Managing the Unexpected*. San Francisco, CA: Jossey-Bass.

Weick, K., Sutcliffe, K. M., and Obstfeld, D. (1999). Organising for high reliability: Processes of collective mindfullness. *Research in Organisational Behavior*, 21, 81–123.

Westrum, R. (1978). Science and social intelligence about anomalies: The case of meteorites. *Social Studies of Science*, 8, 461–493.

Westrum, R. (1982). Social intelligence about hidden events. *Knowledge*, 3(3), 381–400.

Westrum, R. (1986). Management strategies and information failure. In: NATO Advanced Research Workshop on Failure Analysis of Information Systems, Bad Winsheim, Germany, August 1986.

Westrum, R. (1988). Organisational and inter-organisational thought. In: World Bank Workshop on Safety Control and Risk Management, Washington, DC, October 1988.

Westrum, R. (1992). Cultures with requisite imagination. In: J. Wise, D. Hopkin, and P. Stager (Eds.), *Verification and Validation of Complex Systems: Human Factors Issues* (pp. 401–416). New York: Springer-Verlag.

Westrum, R. (1997). Social factors in safety-critical systems. In: R. Redmill and J. Rajan (Eds.), *Human Factors in Safety Critical Systems*. Oxford: Butterworth-Heinemann.

Winner, L. (1977). *Autonomous Technology: Technics-out-of-Control as a Theme in Political Thought*. Cambridge, MA: MIT Press.

Wynne, B. (1988). Unruly technology: Practical rules, impractical discourses and public understanding. *Social Studies of Science*, 18, 147–167.

INDEX

Handbook of Safety Principles, First Edition. Edited by Niklas Möller, Sven Ove Hansson, Jan-Erik Holmberg, and Carl Rollenhagen.

Printed and bound by CPI Group (UK) Ltd, Croydon, CR0 4YY

17/09/2026

14964287-0001